GREEK ALPH[illegible]

Α	α	**Alpha**	[illegible]	[illegible]	[illegible]
Β	β	**Beta**	Ξ	ξ	**Xi**
Γ	γ	**Gamma**	Ο	o	**Omicron**
Δ	δ	**Delta**	Π	π	**Pi**
Ε	ϵ	**Epsilon**	Ρ	ρ	**Rho**
Ζ	ζ	**Zeta**	Σ	σ	**Sigma**
Η	η	**Eta**	Τ	τ	**Tau**
Θ	θ	**Theta**	Υ	υ	**Upsilon**
Ι	ι	**Iota**	Φ	ϕ	**Phi**
Κ	κ	**Kappa**	Χ	χ	**Chi**
Λ	λ	**Lambda**	Ψ	ψ	**Psi**
Μ	μ	**Mu**	Ω	ω	**Omega**

HANDBOOK OF MATHEMATICAL TABLES AND FORMULAS

HANDBOOK OF MATHEMATICAL TABLES AND FORMULAS

FIFTH EDITION

RICHARD STEVENS BURINGTON, Ph.D.

McGRAW-HILL BOOK COMPANY
New York St. Louis San Francisco Düsseldorf Johannesburg
Kuala Lumpur London Mexico Montreal New Delhi
Panama Rio de Janeiro Singapore Sydney Toronto

HANDBOOK OF
MATHEMATICAL TABLES
AND FORMULAS

678910 MUBP 89876543210

This book was set in Bodoni Book by York Graphic Services, Inc.
The editors were Jack L. Farnsworth and Michael Gardner;
the designer was Nicholas Krenitsky;
and the production supervisor was John A. Sabella.
The drawings were done by Bertrick Associate Artists, Inc.
The printer was The Murray Printing Company;
the binder, The Book Press, Inc.

Library of Congress Cataloging in Publication Data

Burington, Richard Stevens, 1901–
Handbook of mathematical tables and formulas.

Bibliography: p.
1. Mathematics—Tables, etc. I. Title.
II. Title: Mathematical tables and formulas.
QA47.B8 1972 510'.21'2 78-39634
ISBN 0-07-009015-7

CONTENTS

PREFACE

The excellent reception and wide use of the earlier editions of this book coupled with the valuable suggestions from its many users for additions have resulted in the publisher's requesting a revision and enlargement of the fourth edition.

As with the earlier editions, this book has been constructed to meet the needs of students and workers in mathematics, engineering, physics, chemistry, science, and other fields in which mathematical reasoning, processes, and computations are required.

The content and manner in which mathematical subjects are taught and used are constantly undergoing changes. This is true at all levels of instruction and use, from the elementary and secondary schools to colleges, institutes, and universities, as well as the many industries and laboratories where mathematics is so widely used. The applications of mathematics to many diverse fields continue to grow at an accelerated pace.

The large-scale use of various types of computing devices has greatly widened the use of certain kinds of mathematical methods and has had a marked impact on the teaching and utilization of mathematics. However, there are many aspects of the tradition, language, use, and teaching of mathematics which remain much the same as they have been for many years. The degree of acceptance and use of contemporary changes vary widely. The existence of many high-speed computers has not reduced the need for compilations of a mathematical character, or for résumés of mathematical methods, but rather it has increased the need for a variety of types of compilations, of which this book covers certain widely used areas.

These and other matters of concern have been carefully weighed by the author in preparing this edition. The present edition should meet the needs of both the contemporary and the more traditional users of mathematics.

The general spirit and arrangement of the fourth edition has been retained. The elementary character of the book has been preserved. The size remains modest.

The first part of the book includes a summary of the more important formulas and theorems of algebra, trigonometry, analytical geometry, calculus, and vector analysis. The sections on sets, logic, relations, functions, algebraic structures, Boolean algebra, matrices, number systems, and elementary statistics have been retained. Comprehensive tables of derivatives and integrals are included. New sections on annuities, linear algebra, linear vector spaces, numerical analysis, solutions of linear and nonlinear systems of equations, approximate differentiation and integration, interpolation and approximation, calculus of finite differences, partial differential equations, Legendre and Bessel functions, Fourier analysis, Laplace transforms, and complex variable theory have been added. Tables of finite Fourier cosine and sine functions, and Laplace transforms have been included.

The second part of the book contains logarithmic and trigonometric tables to five places; tables of natural logarithms and exponential and hyperbolic functions; tables of squares, cubes, square roots, and cube roots; reciprocals and other numerical quantities; normal probability distribution functions; χ^2 distribution. The annuity tables have been enlarged to cover higher interest rates. A table of the summed Poisson distribution function has been added.

This edition is designed to serve as a companion to the *Handbook of Probability and Statistics with Tables,* second edition, by Richard S. Burington and Donald C. May, McGraw-Hill Book Company.

The entire book has been gone over, and numerous changes and additions have been made within the text as well as in the tables themselves. Much effort has been made to arrange the tables in such a manner that the user can and will interpret them properly, and with ease. Every effort has been taken to ensure accuracy.

The author wishes to acknowledge the valuable suggestions made by many users of the book, and to express his appreciation to his daughter Artha Jean Snyder and son-in-law Grant Snyder for their help and critical suggestions.

The author is greatly indebted to his wife, Jennet Mae Burington, for her continued interest and major assistance in the preparation of the manuscript and in the reading of the proof.

RICHARD STEVENS BURINGTON

How to Use the Table Locator

Each locator block at the side of this page corresponds in position to an index block on the title page of each group of tables. These readily appear along the side of the book when the pages are flexed as shown.

To locate any of these groups, flex the pages and turn to the page whose index block shows opposite the locator block on this page.

LOGARITHMS

TRIGONOMETRIC FUNCTIONS

EXPONENTIAL AND HYPERBOLIC FUNCTIONS

POWERS, ROOTS, RECIPROCALS, AREAS, AND CIRCUMFERENCES

STATISTICS AND PROBABILITY

INTEREST AND ACTUARIAL EXPERIENCE

CONSTANTS, EQUIVALENTS, AND CONVERSION FACTORS

FOUR-PLACE TABLES

THIS BOOK CONTAINS

Part One: *Summaries of formulas and theorems of*

ALGEBRA
ANNUITIES AND INTEREST
ELEMENTARY GEOMETRY
TRIGONOMETRY
ANALYTIC GEOMETRY
DIFFERENTIAL CALCULUS
TABLE OF DERIVATIVES
TABLE OF SERIES
INTEGRAL CALCULUS
TABLE OF INTEGRALS
VECTOR ANALYSIS
SETS, RELATIONS, AND FUNCTIONS
LOGIC
CALCULUS OF PROPOSITIONS
CALCULUS OF RELATIONS
ALGEBRAIC STRUCTURES
MATHEMATICAL SYSTEMS
GROUPS
POSTULATES FOR COMMON STRUCTURES
BOOLEAN ALGEBRAS
NUMBER SYSTEMS
MATRICES
LINEAR ALGEBRA
LINEAR VECTOR SPACES
NUMERICAL ANALYSIS
SOLUTION OF LINEAR SYSTEMS OF EQUATIONS
SOLUTION OF NONLINEAR EQUATIONS
INTERPOLATION AND APPROXIMATION
NUMERICAL DIFFERENTIATION
NUMERICAL INTEGRATION
FINITE DIFFERENCE CALCULUS
DIFFERENTIAL EQUATIONS
LEGENDRE POLYNOMIALS
BESSEL FUNCTIONS
PARTIAL DIFFERENTIAL EQUATIONS
FOURIER SERIES AND TRANSFORMS
TABLE OF FINITE SINE AND COSINE TRANSFORMS
LAPLACE TRANSFORMS
TABLE OF LAPLACE TRANSFORMS
COMPLEX VARIABLE THEORY
STATISTICS

Part Two: *Tables of*

LOGARITHMS
TRIGONOMETRIC FUNCTIONS
EXPONENTIAL AND HYPERBOLIC FUNCTIONS
POWERS, ROOTS, RECIPROCALS, AREAS, AND CIRCUMFERENCES
STATISTICS AND PROBABILITY
INTEREST AND ACTUARIAL EXPERIENCE
CONSTANTS, EQUIVALENTS, AND CONVERSION FACTORS
FOUR-PLACE TABLES

PART ONE

Formulas, definitions, and theorems from elementary mathematics

1. ALGEBRA

1. FUNDAMENTAL LAWS

(a) Commutative law: $a + b = b + a \qquad ab = ba$

(b) Associative law: $a + (b + c) = (a + b) + c,$

$$a(bc) = (ab)c$$

(c) Distributive law: $a(b + c) = ab + ac$

2. LAWS OF EXPONENTS

$$a^x a^y = a^{x+y} \qquad (ab)^x = a^x b^x \qquad (a^x)^y = a^{xy}$$

$$a^0 = 1 \quad \text{if} \quad a \neq 0 \qquad a^{-x} = \frac{1}{a^x} \qquad \frac{a^x}{a^y} = a^{x-y}$$

$$a^{x/y} = \sqrt[y]{a^x} \qquad a^{1/y} = \sqrt[y]{a}$$

3. OPERATIONS WITH ZERO

$$a - a = 0 \qquad a \times 0 = 0 \times a = 0$$

If $a \neq 0$, $\dfrac{0}{a} = 0$, $0^a = 0$, $a^0 = 1$ (Division by zero undefined)

4. COMPLEX NUMBERS (a number of the form $a + bi$ where a and b are real)

$$i = \sqrt{-1},\ i^2 = -1,\ i^3 = -i,\ i^4 = 1,\ i^5 = i, \text{ etc.}$$

$$a + bi = c + di \text{ if and only if } a = c,\ b = d$$

$$(a + bi) + (c + di) = (a + c) + (b + d)i$$

$$(a + bi)(c + di) = (ac - bd) + (ad + bc)i$$

$$\frac{a + bi}{c + di} = \frac{(a + bi)(c - di)}{(c + di)(c - di)} = \frac{ac + bd}{c^2 + d^2} + \frac{bc - ad}{c^2 + d^2} i$$

5. LAWS OF LOGARITHMS (see explanation of Table 1).

If M, N, b are positive numbers and $b \neq 1$:

$$\log_b MN = \log_b M + \log_b N \qquad \log_b \frac{M}{N} = \log_b M - \log_b N$$

$$\log_b M^p = p \log_b M \qquad \log_b \sqrt[q]{M} = \frac{1}{q} \log_b M$$

$$\log_b \frac{1}{M} = -\log_b M \qquad \log_b b = 1 \qquad \log_b 1 = 0$$

Change of base of logarithms ($c \neq 1$):

$$\log_b M = \log_c M \ \log_b c = \frac{\log_c M}{\log_c b}$$

6. BINOMIAL THEOREM (n a positive integer)

$$(a + b)^n = a^n + na^{n-1}b + \frac{n(n-1)}{2!}a^{n-2}b^2 + \frac{n(n-1)(n-2)}{3!}a^{n-3}b^3 + \cdots + nab^{n-1} + b^n$$

where $\qquad n! = \underline{|n} = 1 \cdot 2 \cdot 3 \cdots (n-1)n$

7. EXPANSIONS AND FACTORS

$$(a \pm b)^2 = a^2 \pm 2ab + b^2$$

$$(a \pm b)^3 = a^3 \pm 3a^2b + 3ab^2 \pm b^3$$

$$(a + b + c)^2 = a^2 + b^2 + c^2 + 2ab + 2ac + 2bc$$

$$a^2 - b^2 = (a - b)(a + b)$$

$$a^3 - b^3 = (a - b)(a^2 + ab + b^2)$$

$$a^3 + b^3 = (a + b)(a^2 - ab + b^2)$$

$$a^n - b^n = (a - b)(a^{n-1} + a^{n-2}b + \cdots + b^{n-1})$$

$$a^n - b^n = (a + b)(a^{n-1} - a^{n-2}b + \cdots - b^{n-1})$$

for n an even integer

$$a^n + b^n = (a + b)(a^{n-1} - a^{n-2}b + \cdots + b^{n-1})$$

for n an odd integer

$$a^4 + a^2b^2 + b^4 = (a^2 + ab + b^2)(a^2 - ab + b^2)$$

8. RATIO AND PROPORTION

If $\quad a:b = c:d \quad$ or $\quad \dfrac{a}{b} = \dfrac{c}{d}$, then $\quad ad = bc, \ \dfrac{a}{c} = \dfrac{b}{d}$

If
$$\frac{a}{b} = \frac{c}{d} = \frac{e}{f} = \cdots = k,$$

then
$$k = \frac{a + c + e + \cdots}{b + d + f + \cdots} = \frac{pa + qc + re + \cdots}{pb + qd + rf + \cdots}$$

9. CONSTANT FACTOR OF PROPORTIONALITY (OR VARIATION), k

If y varies directly as x, or y is proportional to x,

$$y = kx$$

If y varies inversely as x, or y is inversely proportional to x,

$$y = \frac{k}{x}$$

If y varies jointly as x and z,

$$y = kxz$$

If y varies directly as x and inversely as z,

$$y = \frac{kx}{z}$$

10. ARITHMETIC PROGRESSION

$$a \qquad a + d \qquad a + 2d \qquad a + 3d \qquad \ldots$$

If a is the first term, d the common difference, n the number of terms, l the last term, and s the sum of n terms,

$$l = a + (n - 1)d \qquad s = \frac{n}{2}(a + l)$$

The *arithmetic mean* of a and b is $(a + b)/2$.

11. GEOMETRIC PROGRESSION

$$a, ar, ar^2, ar^3 \ldots$$

If a is the first term, r the common ratio, n the number of terms, l the last term, and S_n the sum of n terms,

$$l = ar^{n-1} \qquad S_n = a\frac{r^n - 1}{r - 1} = \frac{rl - a}{r - 1}$$

If $r^2 < 1$, S_n approaches the limit S_∞ as n increases without limit,

$$S_\infty = \frac{a}{1 - r}$$

The *geometric mean* of a and b is $\sqrt{ab}$.

12. HARMONIC PROGRESSION A sequence of numbers whose reciprocals form an arithmetic progression is called an *harmonic progression.* Thus

$$\frac{1}{a} \qquad \frac{1}{a + d} \qquad \frac{1}{a + 2d} \qquad \cdots$$

The *harmonic mean* of a and b is $2ab/(a + b)$.

13. PERMUTATION Each different arrangement of all or a part of a set of things is called a *permutation.* The number of permutations of n different things taken r at a time is

$$P(n,r) = {}_nP_r = n(n - 1)(n - 2) \cdots (n - r + 1) = \frac{n!}{(n - r)!}$$

where $$n! = n(n - 1)(n - 2) \cdots (1)$$

14. COMBINATIONS Each of the groups or relations which can be made by taking part or all of a set of things, without regard to the arrangement of the things in a group, is called a *combination.* The number of combinations of n different things taken r at a time is

$$C(n,r) = {}_nC_r = \binom{n}{r} = \frac{{}_nP_r}{r!}$$

$$= \frac{n(n - 1) \cdots (n - r + 1)}{r(r - 1) \cdots 1} = \frac{n!}{r!(n - r)!}$$

15. PROBABILITY If an event may occur in p ways and may fail in q ways, all ways being equally likely, the *probability* of its occurrence is $p/(p + q)$ and that of its failure to occur is $q/(p + q)$. (For further details see Richard S. Burington and Donald C. May, *Handbook of Probability and Statistics with Tables,* 2d ed., McGraw-Hill Book Company, New York, 1970.) (See Secs. 418 to 424.)

16. REMAINDER THEOREM (see Sec. 30) If the polynomial $f(x)$ is divided by $x - a$, the remainder is $f(a)$. Hence, if a is a root of the equation $f(x) = 0$, then $f(x)$ is divisible by $x - a$.

17. DETERMINANTS* The determinant D of order n,

$$D = \begin{vmatrix} a_{11} & a_{12} & \cdots & a_{1n} \\ a_{21} & a_{22} & \cdots & a_{2n} \\ \cdot & \cdot & \cdots & \cdot \\ a_{n1} & a_{n2} & \cdots & a_{nn} \end{vmatrix}$$

is defined to be the sum

$$\Sigma\ (\pm)\ a_{1i}a_{2j}a_{3k} \cdots a_{nl}$$

of $n!$ terms, the sign in a given term being taken plus or minus according as the number of inversions (of the numbers 1, 2, 3, . . . , n) in the corresponding sequence $i, j, k, \ldots, l$ is even or odd.

The *cofactor* A_{ij} of the element a_{ij} is defined to be the product of $(-1)^{i+j}$ by the determinant obtained from D by deleting the ith row and the jth column.

The following theorems are true:

(*a*) If the corresponding rows and columns of D are interchanged, D is unchanged.

(*b*) If any two rows (or columns) of D are interchanged, D is changed to $-D$.

(*c*) If any two rows (or columns) are alike, then $D = 0$.

(*d*) If each element of a row (or column) of D is multiplied by m, the new determinant is equal to mD.

(*e*) If to each element of a row (or column) is added m times the corresponding element in another row (or column), D is unchanged.

(*f*) $D = a_{1j}A_{1j} + a_{2j}A_{2j} + \cdots + a_{nj}A_{nj},\ j = 1, 2, \ldots, n$

(*g*) $0 = a_{1k}A_{1j} + a_{2k}A_{2j} + \cdots + a_{nk}A_{nj}$, if $j \neq k$

(*h*) The solution of the system of equations

$$a_{i1}x_1 + a_{i2}x_2 + \cdots + a_{in}x_n = c_i \qquad i = 1, 2, \ldots, n$$

is unique if $D \neq 0$. The solution is given by the equations

$$Dx_1 = C_1 \qquad Dx_2 = C_2, \ldots \qquad Dx_n = C_n$$

where C_k is what D becomes when the elements of its kth column are replaced by $c_1, c_2, \ldots, c_n$, respectively.

* See Chap. 13, Matrices, and Secs. 275 to 282 for further theorems.

Example 1.

$$\begin{vmatrix} a_{11} & a_{12} & a_{13} \\ a_{21} & a_{22} & a_{23} \\ a_{31} & a_{32} & a_{33} \end{vmatrix} = a_{11}\begin{vmatrix} a_{22} & a_{23} \\ a_{32} & a_{33} \end{vmatrix} - a_{12}\begin{vmatrix} a_{21} & a_{23} \\ a_{31} & a_{33} \end{vmatrix} + a_{13}\begin{vmatrix} a_{21} & a_{22} \\ a_{31} & a_{32} \end{vmatrix}$$

$$= a_{11}(a_{22}a_{33} - a_{32}a_{23}) - a_{12}(a_{21}a_{33} - a_{31}a_{23}) + a_{13}(a_{21}a_{32} - a_{31}a_{22})$$

Example 2. Find the values of x_1, x_2, x_3 which satisfy the system

$$\begin{aligned} 2x_1 + x_2 - 2x_3 &= -6 \\ x_1 + x_2 + x_3 &= 2 \\ -x_1 - 2x_2 + 3x_3 &= 12 \end{aligned}$$

By 17(***h***), we find

$$x_1 = \frac{\begin{vmatrix} -6 & 1 & -2 \\ 2 & 1 & 1 \\ 12 & -2 & 3 \end{vmatrix}}{\begin{vmatrix} 2 & 1 & -2 \\ 1 & 1 & 1 \\ -1 & -2 & 3 \end{vmatrix}} = \frac{8}{8} = 1; \quad x_2 = \frac{\begin{vmatrix} 2 & -6 & -2 \\ 1 & 2 & 1 \\ -1 & 12 & 3 \end{vmatrix}}{\begin{vmatrix} 2 & 1 & -2 \\ 1 & 1 & 1 \\ -1 & -2 & 3 \end{vmatrix}} = \frac{-16}{8} = -2;$$

$$x_3 = \frac{\begin{vmatrix} 2 & 1 & -6 \\ 1 & 1 & 2 \\ -1 & -2 & 12 \end{vmatrix}}{\begin{vmatrix} 2 & 1 & -2 \\ 1 & 1 & 1 \\ -1 & -2 & 3 \end{vmatrix}} = 3$$

Interest, Annuities, Sinking Funds

In this section, n is the number of years and r the rate of interest expressed as a decimal.

18. AMOUNT A principal P placed at a rate of interest r for n years accumulates to an amount A_n as follows:

At simple interest: $A_n = P(1 + nr)$

At interest compounded annually:* $A_n = P(1 + r)^n$

At interest compounded q times a year: $A_n = P\left(1 + \frac{r}{q}\right)^{nq}$

* See Table 21.

19. NOMINAL AND EFFECTIVE RATES The rate of interest quoted in describing a given variety of compound interest is called the *nominal rate.* The rate per year at which interest is earned during each year is called the *effective rate.* The effective rate i corresponding to the nominal rate r, compounded q times a year, is:

$$i = \left(1 + \frac{r}{q}\right)^q - 1$$

20. PRESENT OR DISCOUNTED VALUE OF A FUTURE AMOUNT The present quantity P which in n years will accumulate to the amount A_n at the rate of interest r is:

At simple interest: $$P = \frac{A_n}{1 + nr}$$

At interest compounded annually:* $$P = \frac{A_n}{(1 + r)^n}$$

At interest compounded q times a year: $$P = \frac{A_n}{(1 + r/q)^{nq}}$$

P is called the *present value* of A_n due in n years at rate r.

The *true discount* is $D = A_n - P$.

21. SIMPLE ANNUITIES A fixed sum of money paid at regular intervals is called an *annuity*. Let

$$s_{\overline{n}|} \text{ at } r = \frac{(1 + r)^n - 1}{r} \qquad a_{\overline{n}|} \text{ at } r = \frac{1 - (1 + r)^{-n}}{r}$$

21a. AMOUNT OF AN ANNUITY† If an annuity P is deposited at the end of each successive year (beginning one year hence), and the interest at rate r, compounded annually, is paid on the accumulated deposit at the end of each year, the total amount N accumulated at the end of n years is

$$N = P(s_{\overline{n}|} \text{ at } r)$$

N is called the *amount* of an annuity P.

* See Table 22. † See Table 23.

21*b*. PRESENT VALUE OF AN ANNUITY* The total present amount P which will supply an annuity N at the end of each year for n years, beginning one year hence (assuming that in successive years the amount not yet paid out earns interest at rate r, compounded annually), is

$$P = \frac{N(s_{\overline{n}|} \text{ at } r)}{(1+r)^n} = N(a_{\overline{n}|} \text{ at } r)$$

P is called the *present value* of an annuity.

21*c*. AMOUNT OF A SINKING FUND If a fixed investment N is made at the end of each successive year (beginning at the end of the first year), and interest paid at rate r, compounded annually, is paid on the accumulated amount of the investment at the end of each year, the total amount S accumulated at the end of n years is

$$S = N(s_{\overline{n}|} \text{ at } r)$$

S is called the *amount of the sinking fund.*

21*d*. FIXED INVESTMENT, OR ANNUAL INSTALLMENT† The amount N that must be placed at the end of each year (beginning one year hence), with compound interest paid at rate r on the accumulated deposit, in order to accumulate a sinking fund S in n years is

$$N = S(s_{\overline{n}|}^{-1} \text{ at } r)$$

N is called a *fixed investment* or *annual installment.*

22. GENERAL ANNUITY FORMULAS Let

W = periodic payment of the annuity
r = interest rate per *conversion period* (the time between successive conversions of interest into principal)
n = term of the annuity expressed in conversion periods
p = interest period divided by payment interval
= number of payment intervals in each conversion period
$R = pW$ = *rent per conversion period*

In simple annuity problems $p = 1$.

* See Table 24. † See Table 25.

23. If interest is compounded *once* in each payment interval at the rate c per interval, then compound interest at rate c gives the same value as compound interest at rate r with its corresponding conversion period provided c is selected so that

$$(1 + c)^p = 1 + r \qquad \text{or} \qquad c = (1 + r)^{1/p} - 1$$

24. AMOUNT OF AN ANNUITY The first payment of an annuity whose rent is 1 per interest (conversion) period occurs at the end of $1/p$ interest periods. Each periodic payment is $1/p$. The npth payment is made at the end of n conversion periods. The *amount* of this annuity (whose rent per conversion period is 1) is

$$s_{\overline{n}|}^{(p)} \text{ at } r = \frac{1}{p} \sum_{i=1}^{i=np} (1 + r)^{(np-i)/p} = \frac{(1 + r)^n - 1}{j_p \text{ at } r} = \frac{r(s_{\overline{n}|} \text{ at } r)}{j_p \text{ at } r}$$

where j_p at $r = p[(1 + r)^{1/p} - 1]$.

The *present value* of this annuity is

$$a_{\overline{n}|}^{(p)} \text{ at } r = (1 + r)^{-n}(s_{\overline{n}|}^{(p)} \text{ at } r) = \frac{1 - (1 + r)^{-n}}{j_p \text{ at } r} = \frac{r(a_{\overline{n}|} \text{ at } r)}{j_p \text{ at } r}$$

If the rent per conversion period is R, the *amount* S and the *present value* A of the annuity are

$$S = R(s_{\overline{n}|}^{(p)} \text{ at } r) \qquad A = R(a_{\overline{n}|}^{(p)} \text{ at } r)$$

25. SINKING FUND A *sinking fund* is a fund formed to pay an obligation falling due at some future date. Unless otherwise stated it is assumed that a sinking fund is created by investing equal periodic payments. In this case, the amount in the sinking fund at any time is the amount of the annuity formed by the payments. The formulas developed for annuities may be used in sinking-fund problems.

26. DETERMINATION OF PERIODIC PAYMENT *Example.* A man borrows an amount A. He agrees to cancel his obligation by paying equal installments at the end of each given fixed payment interval, for a total period of n payment intervals. The first payment is due one payment interval after the loan is made. Interest is at rate r

per payment (the conversion) interval. The amount R of his payment is $R = A(a_{\overline{n|}}^{-1}$ at $r)$.

Thus, from Table 25, if $A = \$1{,}000$, $r = 0.02$ per quarter, and $n = 40$, the quarterly installments are each $R = Aa_{\overline{40|}}^{-1} = \$1{,}000$ $(0.03656) = \$36.56$.

*Algebraic Equations**

27. QUADRATIC EQUATIONS If

$$ax^2 + bx + c = 0 \qquad a \neq 0$$

then

$$x = \frac{-b \pm \sqrt{b^2 - 4ac}}{2a}$$

If a, b, c are real and

if $b^2 - 4ac > 0$	the roots are real and unequal
if $b^2 - 4ac = 0$	the roots are real and equal
if $b^2 - 4ac < 0$	the roots are imaginary

28. CUBIC EQUATIONS The cubic equation

$$y^3 + py^2 + qy + r = 0$$

may be reduced by the substitution

$$y = x - \frac{p}{3}$$

to the normal form

$$x^3 + ax + b = 0$$

where $a = \frac{1}{3}(3q - p^2) \qquad b = \frac{1}{27}(2p^3 - 9pq + 27r)$

which has the solutions x_1, x_2, x_3,

$$x_1 = A + B,\ x_2,\ x_3 = -\frac{1}{2}(A + B) \pm \frac{i\sqrt{3}}{2}(A - B)$$

* For linear equations, see Sec. 17.

where

$$i^2 = -1 \qquad A = \sqrt[3]{-\frac{b}{2} + \sqrt{\frac{b^2}{4} + \frac{a^3}{27}}}$$

$$B = \sqrt[3]{-\frac{b}{2} - \sqrt{\frac{b^2}{4} + \frac{a^3}{27}}}$$

If p, q, r are real (and hence if a and b are real) and

if $\frac{b^2}{4} + \frac{a^3}{27} > 0$ there are one real root and two conjugate imaginary roots

if $\frac{b^2}{4} + \frac{a^3}{27} = 0$ there are three real roots of which at least two are equal

if $\frac{b^2}{4} + \frac{a^3}{27} < 0$ there are three real and unequal roots

If

$$\frac{b^2}{4} + \frac{a^3}{27} < 0$$

the above formulas are impractical. The real roots are

$$x_k = 2\sqrt{-\frac{a}{3}}\cos\left(\frac{\phi}{3} + 120^\circ\, k\right) \qquad k = 0, 1, 2$$

where

$$\cos\phi = \mp\sqrt{\frac{b^2/4}{-a^3/27}}$$

and where the upper sign is to be used if b is positive and the lower sign if b is negative.

If $\frac{b^2}{4} + \frac{a^3}{27} > 0$ and $a > 0$

the real root is

$$x = 2\sqrt{\frac{a}{3}}\cot 2\phi$$

where ϕ and ψ are to be computed from

$$\cot 2\psi = \mp\sqrt{\frac{b^2/4}{a^3/27}} \qquad \tan\phi = \sqrt[3]{\tan\psi}$$

and where the upper sign is to be used if b is positive and the lower sign if b is negative.

If

$$\frac{b^2}{4} + \frac{a^3}{27} = 0$$

the roots are

$$x = \mp 2\sqrt{-\frac{a}{3}}, \qquad \pm\sqrt{-\frac{a}{3}}, \qquad \pm\sqrt{-\frac{a}{3}}$$

where the upper sign is to be used if b is positive and the lower sign if b is negative.

29. BIQUADRATIC (QUARTIC) EQUATION The quartic equation

$$y^4 + py^3 + qy^2 + ry + s = 0$$

may be reduced to the form

$$x^4 + ax^2 + bx + c = 0$$

by the substitution

$$y = x - \frac{p}{4}$$

Let l, m, and n denote the roots of the resolvent cubic

$$t^3 + \frac{a}{2}t^2 + \frac{a^2 - 4c}{16}t - \frac{b^2}{64} = 0$$

The required roots of the reduced quartic are

$$x_1 = +\sqrt{l} + \sqrt{m} + \sqrt{n} \qquad x_2 = +\sqrt{l} - \sqrt{m} - \sqrt{n}$$
$$x_3 = -\sqrt{l} + \sqrt{m} - \sqrt{n} \qquad x_4 = -\sqrt{l} - \sqrt{m} + \sqrt{n}$$

where the selection of the square root to be attached to each of the quantities $\sqrt{l}$, $\sqrt{m}$, $\sqrt{n}$ must give $\sqrt{l}\sqrt{m}\sqrt{n} = -b/8$.

30. GENERAL EQUATIONS OF THE *n*th DEGREE*

$$P \equiv a_0x^n + a_1x^{n-1} + a_2x^{n-2} + \cdots + a_{n-1}x + a_n = 0$$

If $n > 4$, there is no formula which gives the roots of this general equation. The following methods may be used to advantage:

(a) *Roots by factors.* By trial, find a number r such that $x = r$ satisfies the equation, that is, such that

$$a_0r^n + a_1r^{n-1} + a_2r^{n-2} + \cdots + a_{n-1}r + a_n = 0 \qquad a_0 \neq 0$$

Then $x - r$ is a factor of the left-hand member P of the equation. Divide P by $x - r$, leaving an equation of degree one less than that of the original equation. Next, proceed in the same manner with the reduced equation. (All integer roots of $P = 0$ are divisors of a_n/a_0.)

(b) *Roots by approximation.* Suppose the coefficients a_i in P are real. Let a and b be real numbers. If for $x = a$ and $x = b$ the left member P of the equation has opposite signs, then a root lies between a and b. By repeated application of this principle, real roots to any desired degree of accuracy may be obtained.

(c) *Roots by graphing.* If a graph of P is plotted as a function of x, the real roots are the values of x where the graph crosses the x axis. By increasing the scale of the portion of the graph near an estimated root, the root may be obtained to any desired degree of accuracy.

(d) *Descartes' rule.* The number of positive real roots of an equation with real coefficients either is equal to the number of its variations of sign or is less than that number by a positive even integer. A root of multiplicity m is here counted as m roots.

31. THE EQUATION $x^n = a$

The n roots of this equation are

$$x = \begin{cases} \sqrt[n]{a}\left(\cos \dfrac{2k\pi}{n} + \sqrt{-1}\sin \dfrac{2k\pi}{n}\right) & \text{if } a > 0 \\ \sqrt[n]{-a}\left(\cos \dfrac{(2k+1)\pi}{n} + \sqrt{-1}\sin \dfrac{(2k+1)\pi}{n}\right) & \text{if } a < 0 \end{cases}$$

where k takes successively the values $0, 1, 2, 3, \ldots, n - 1$.

* See Secs. 294 and 305 to 308.

31*a*. **STIRLING'S FORMULA** For large values of n,

$$\sqrt{2n\pi}\left(\frac{n}{e}\right)^n < n! < \sqrt{2n\pi}\left(\frac{n}{e}\right)^n\left(1 + \frac{1}{12n - 1}\right)$$

where $\pi = 3.14159 \cdots$ and $e = 2.71828 \cdots$.

$$\log n! \cong (n + \tfrac{1}{2}) \log n - n \log e + \log \sqrt{2\pi}$$

2. ELEMENTARY GEOMETRY

Let a, b, c, d, and s denote lengths, A denote areas, and V denote volumes.

32. TRIANGLE (see Sec. 65) $A = bh/2$, where b denotes the base and h the altitude.

33. RECTANGLE $A = ab$, where a and b denote the lengths of the sides.

34. PARALLELOGRAM (OPPOSITE SIDES PARALLEL) $A = ah = ab \sin\theta$, where a and b denote the sides, h the altitude, and θ the angle between the sides.

35. TRAPEZOID (FOUR SIDES, TWO PARALLEL) $A = \frac{1}{2}h(a + b)$, where a and b are the sides and h the altitude.

36. REGULAR POLYGON OF n SIDES (Fig. 1, see Sec. 37)

$$A = \frac{1}{4} na^2 \cot \frac{180°}{n}$$ where a is length of side

$$R = \frac{a}{2} \csc \frac{180°}{n}$$ where R is radius of circumscribed circle

$$r = \frac{a}{2} \cot \frac{180°}{n}$$ where r is radius of inscribed circle

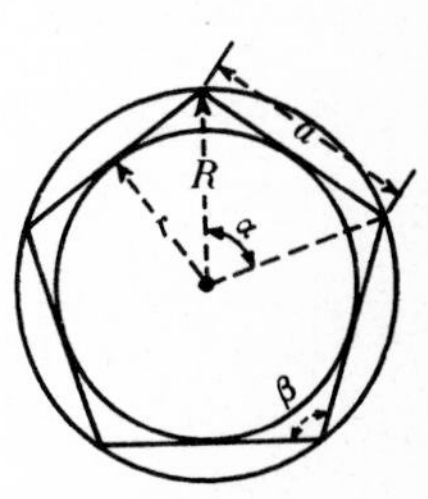

Fig. 1

$$\alpha = \frac{360°}{n} = \frac{2\pi}{n}$$ radians

$$\beta = \frac{n-2}{n} \times 180° = \frac{n-2}{n}\pi$$ radians where α and β are the angles indicated in Fig. 1

$$a = 2r \tan \frac{\alpha}{2} = 2R \sin \frac{\alpha}{2}$$

37. CIRCLE (Fig. 2)

Let C = circumference
R = radius
D = diameter
A = area

S = length of arc subtended by θ
l = chord subtended by arc S
h = rise
θ = central angle, radians

$$C = 2\pi R = \pi D \qquad \pi = 3.14159 \cdots$$

$$S = R\theta = \tfrac{1}{2}D\theta = D \cos^{-1} \frac{d}{R}$$

$$l = 2\sqrt{R^2 - d^2} = 2R \sin \frac{\theta}{2} = 2d \tan \frac{\theta}{2}$$

$$d = \tfrac{1}{2}\sqrt{4R^2 - l^2} = R \cos \frac{\theta}{2} = \tfrac{1}{2}l \cot \frac{\theta}{2}$$

$$h = R - d$$

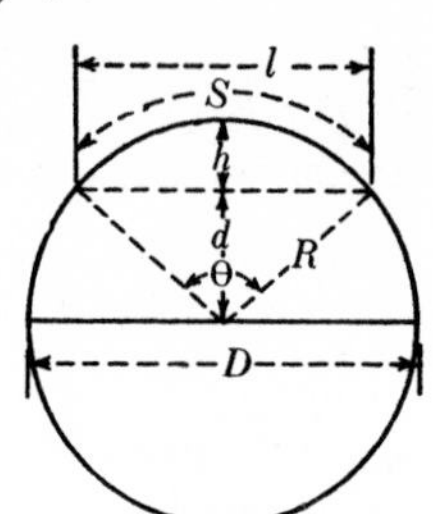

Fig. 2

$$\theta = \frac{S}{R} = \frac{2S}{D} = 2 \cos^{-1} \frac{d}{R} = 2 \tan^{-1} \frac{l}{2d} = 2 \sin^{-1} \frac{l}{D}$$

$$A \text{ (circle)} = \pi R^2 = \tfrac{1}{4}\pi D^2$$

$$A \text{ (sector)} = \tfrac{1}{2}Rs = \tfrac{1}{2}R^2\theta$$

$$A \text{ (segment)} = A \text{ (sector)} - A \text{ (triangle)} = \tfrac{1}{2}R^2(\theta - \sin\theta)$$

$$= R^2 \cos^{-1} \frac{R - h}{R} - (R - h)\sqrt{2Rh - h^2}$$

Perimeter of an n-side regular polygon inscribed in a circle

$$= 2nR \sin \frac{\pi}{n}$$

Area of inscribed polygon $= \tfrac{1}{2}nR^2 \sin \dfrac{2\pi}{n}$

Perimeter of an n-side regular polygon circumscribed about a circle

$$= 2nR \tan \frac{\pi}{n}$$

Area of circumscribed polygon $= nR^2 \tan \dfrac{\pi}{n}$

Radius of circle inscribed in a triangle of sides a, b, and c is

$$r = \sqrt{\frac{(s - a)(s - b)(s - c)}{s}} \qquad s = \tfrac{1}{2}(a + b + c)$$

Radius of circle circumscribed about a triangle is

$$R = \frac{abc}{4\sqrt{s(s-a)(s-b)(s-c)}}$$

38. ELLIPSE (see Sec. 84) Perimeter $= 4aE(k)$, where $k^2 = 1 - (b^2/a^2)$ and $E(k)$ is the complete elliptic integral E given in Table 34. $A = \pi ab$, where a and b are lengths of semimajor and semiminor axes, respectively.

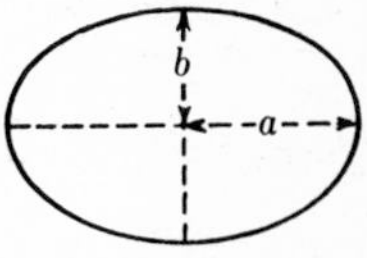

Fig. 3

39. PARABOLA (see Sec. 83)

$$A = \frac{2ld}{3}$$

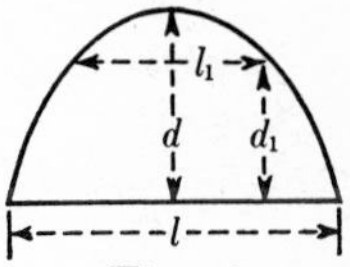

Fig. 4

$$\text{Height of } d_1 = \frac{d}{l^2}(l^2 - l_1^2)$$

$$\text{Width of } l_1 = l\sqrt{\frac{d - d_1}{d}}$$

$$\text{Length of arc} = l\left[1 + \frac{2}{3}\left(\frac{2d}{l}\right)^2 - \frac{2}{5}\left(\frac{2d}{l}\right)^4 + \cdots\right]$$

40. CATENARY, CYCLOID, ETC. (see Secs. 91 to 101)

41. AREA BY APPROXIMATION If $y_0, y_1, y_2, \ldots, y_n$ are the lengths of a series of equally spaced parallel chords and if h is their distance apart, the area enclosed by boundary is given approximately by any one of the following formulas:

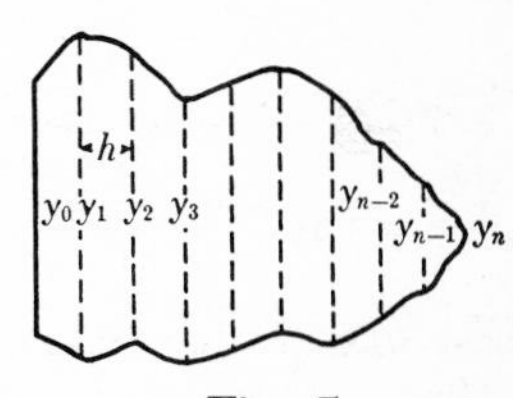

Fig. 5

$$A_T = h[\tfrac{1}{2}(y_0 + y_n) + y_1 + y_2 + \cdots + y_{n-1}]$$

(trapezoidal rule)

$$A_D = h[0.4(y_0 + y_n) + 1.1(y_1 + y_{n-1}) + y_2 + y_3 + \cdots + y_{n-2}]$$

(Durand's rule)

$$A_S = \tfrac{1}{3}h[(y_0 + y_n) + 4(y_1 + y_3 + \cdots + y_{n-1}) + 2(y_2 + y_4 + \cdots + y_{n-2})]$$

(n even, Simpson's rule)

In general, A_S gives the more accurate approximation.
The greater the value of n, the greater the accuracy of approximation.

42. **CUBE** $V = a^3$; $d = a\sqrt{3}$; total surface area $= 6a^2$, where a is length of side and d is length of diagonal.

43. **RECTANGULAR PARALLELOPIPED**

$$V = abc \qquad d = \sqrt{a^2 + b^2 + c^2}$$

Total surface area $= 2(ab + bc + ca)$, where a, b, and c are lengths of sides and d is length of diagonal.

44. **PRISM OR CYLINDER** $V =$ (area of base) $\times$ (altitude)
Lateral area $=$ (perimeter of right section) $\times$ (lateral edge)

45. **PYRAMID OR CONE** $V = \frac{1}{3}$(area of base) $\times$ (altitude)
Lateral area of regular pyramid
$= \frac{1}{2}$(perimeter of base) $\times$ (slant height)

46. **FRUSTUM OF PYRAMID OR CONE** $V = \frac{1}{3}(A_1 + A_2 + \sqrt{A_1A_2})h$, where h is the altitude and A_1 and A_2 are the areas of the bases.
Lateral area of a regular figure
$= \frac{1}{2}$(sum of perimeters of base) $\times$ (slant height)

46*a*. **PRISMOID** $V = (h/6)(A_1 + A_2 + 4A_3)$, where $h =$ altitude, A_1 and A_2 are the areas of the bases, and A_3 is the area of the midsection parallel to bases.

47. **AREA OF SURFACE AND VOLUME OF REGULAR POLYHEDRA OF EDGE l**

Name	*Type of surface*	*Area of surface*	*Volume*
Tetrahedron	4 equilateral triangles	$1.73205l^2$	$0.11785l^3$
Hexahedron (cube)	6 squares	$6.00000l^2$	$1.00000l^3$
Octahedron	8 equilateral triangles	$3.46410l^2$	$0.47140l^3$
Dodecahedron	12 pentagons	$20.64578l^2$	$7.66312l^3$
Icosahedron	20 equilateral triangles	$8.66025l^2$	$2.18169l^3$

48. SPHERE

$A \text{ (sphere)} = 4\pi R^2 = \pi D^2$

$A \text{ (zone)} = 2\pi R h_1 = \pi D h_1$

$V \text{ (sphere)} = \frac{4}{3}\pi R^3 = \frac{1}{6}\pi D^3$

$V \text{ (spherical sector)} = \frac{2}{3}\pi R^2 h_1$

$= \frac{1}{6}\pi D^2 h_1$

V (spherical segment of one base)

$= \frac{1}{6}\pi h_3(3r_3^2 + h_3^2)$

V (spherical segment of two bases)

$= \frac{1}{6}\pi h_2(3r_3^2 + 3r_2^2 + h_2^2)$

$A \text{ (lune)} = 2R^2\theta$, where θ is angle, radians of lune

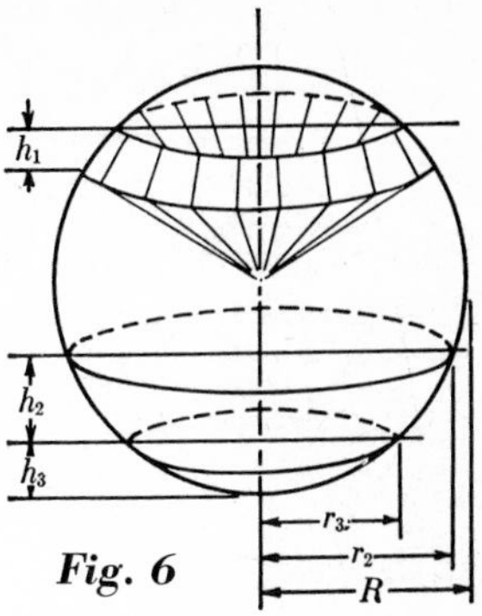

Fig. 6

49. SOLID ANGLE ψ

The solid angle ψ at any point P subtended by a surface S is equal to the area A of the portion of the surface of a sphere of unit radius, center at P, which is cut out by a conical surface, with vertex at P, passing through the perimeter of S.

The unit solid angle ψ is called the *steradian*.

The total solid angle about a point is 4π *steradians*.

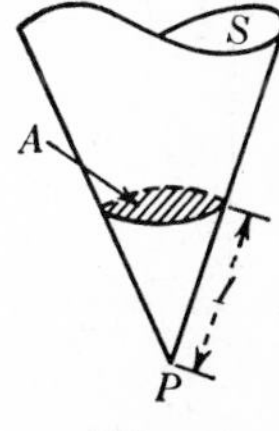

Fig. 7

50. ELLIPSOID

$V = \frac{4}{3}\pi abc$, where a, b, and c are the lengths of the semiaxes.

51. TORUS

$$V = 2\pi^2 R r^2$$

$$\text{Area of surface} = S = 4\pi^2 R r$$

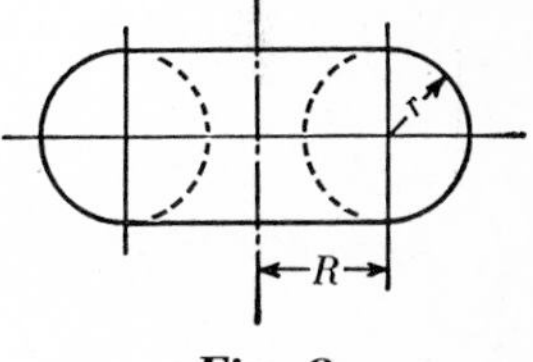

Fig. 8

51a. THEOREMS OF PAPPUS

(a) If a plane area A is rotated about a line l in the plane of A and not cutting A, the volume of the solid generated is equal to the product of A and the distance traveled by the center of gravity of A.

(b) If a plane curve C is rotated about a line l in the plane of C and not cutting C, the area of the surface generated is equal to the product of the length of C and the distance traveled by the center of gravity of C.

3. TRIGONOMETRY

52. ANGLE If two lines intersect, one line may be rotated about their point of intersection through the *angle* which they form until it coincides with the other line.

The angle is said to be *positive* if the rotation is counterclockwise, and *negative* if clockwise.

A complete revolution of a line is a rotation through an angle of 360°. Thus

A *degree* is $\frac{1}{360}$ of the plane angle about a point.

A *radian* is the angle subtended at the center of a circle by an arc whose length is equal to that of the radius.

$$180° = \pi \text{ radians}; \; 1° = \frac{\pi}{180} \text{ radians}; \; 1 \text{ radian} = \frac{180}{\pi} \text{ degrees}$$

53. TRIGONOMETRIC FUNCTIONS OF AN ANGLE α Let α be any angle whose initial side lies on the positive x axis and whose vertex is at the origin and (x, y) be any point on the terminal side of the angle. (x is positive if measured along OX to the right from the y axis and negative if measured along OX' to the left from the y axis. Likewise, y is positive if measured parallel to OY and negative if measured parallel to OY'.) Let r be the positive distance from the origin to the point. The trigonometric functions of an angle are defined as follows:

$$\text{sine } \alpha = \sin \alpha = \frac{y}{r}$$

$$\text{cosine } \alpha = \cos \alpha = \frac{x}{r}$$

$$\text{tangent } \alpha = \tan \alpha = \frac{y}{x}$$

$$\text{cotangent } \alpha = \cot \alpha = \text{ctn } \alpha = \frac{x}{y}$$

$$\text{secant } \alpha = \sec \alpha = \frac{r}{x}$$

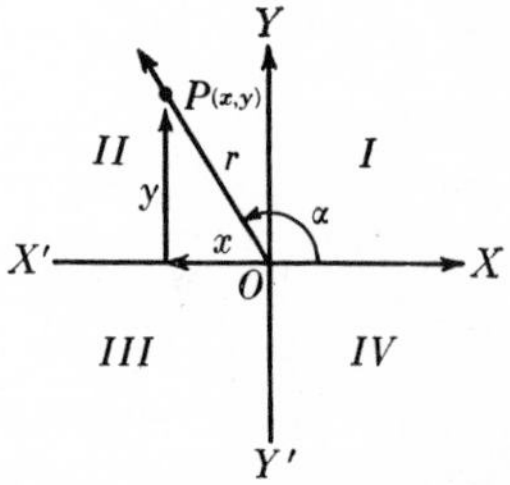

Fig. 9

$$\text{cosecant } \alpha = \csc \alpha = \frac{r}{y}$$

$$\text{exsecant } \alpha = \text{exsec } \alpha = \sec \alpha - 1$$

$$\text{versine } \alpha = \text{vers } \alpha = 1 - \cos \alpha$$

$$\text{coversine } \alpha = \text{covers } \alpha = 1 - \sin a$$

$$\text{haversine } \alpha = \text{hav } \alpha = \tfrac{1}{2} \text{ vers } \alpha$$

For graphs of trigonometric functions see Secs. 86 and 87.

54. SIGNS OF THE FUNCTIONS

Quadrant	sin	cos	tan	cot	sec	csc
I	+	+	+	+	+	+
II	+	−	−	−	−	+
III	−	−	+	+	−	−
IV	−	+	−	−	+	−

55. FUNCTIONS OF 0°, 30°, 45°, 60°, 90°, 180°, 270°, 360°

	0°	*30°*	*45°*	*60°*	*90°*	*180°*	*270°*	*360°*
sin	0	$\frac{1}{2}$	$\frac{\sqrt{2}}{2}$	$\frac{\sqrt{3}}{2}$	1	0	−1	0
cos	1	$\frac{\sqrt{3}}{2}$	$\frac{\sqrt{2}}{2}$	$\frac{1}{2}$	0	−1	0	1
tan	0	$\frac{\sqrt{3}}{3}$	1	$\sqrt{3}$		0		0
cot		$\sqrt{3}$	1	$\frac{\sqrt{3}}{3}$	0		0	
sec	1	$\frac{2\sqrt{3}}{3}$	$\sqrt{2}$	2		−1		1
csc		2	$\sqrt{2}$	$\frac{2\sqrt{3}}{3}$	1		−1	

56. VARIATIONS OF THE FUNCTIONS

Quadrant	sin	cos	tan	cot	sec	csc
I	$0 \to +1$	$+1 \to 0$	$0 \to +\infty$	$+\infty \to 0$	$+1 \to +\infty$	$+\infty \to +1$
II	$+1 \to 0$	$0 \to -1$	$-\infty \to 0$	$0 \to -\infty$	$-\infty \to -1$	$+1 \to +\infty$
III	$0 \to -1$	$-1 \to 0$	$0 \to +\infty$	$+\infty \to 0$	$-1 \to -\infty$	$-\infty \to -1$
IV	$-1 \to 0$	$0 \to +1$	$-\infty \to 0$	$0 \to -\infty$	$+\infty \to +1$	$-1 \to -\infty$

57. FUNCTIONS OF ANGLES IN ANY QUADRANT IN TERMS OF ANGLES IN FIRST QUADRANT

	$-\alpha$	$90° \pm \alpha$	$180° \pm \alpha$	$270° \pm \alpha$	$n(360)° \pm \alpha$
sin	$-\sin \alpha$	$+\cos \alpha$	$\mp\sin \alpha$	$-\cos \alpha$	$\pm\sin \alpha$
cos	$+\cos \alpha$	$\mp\sin \alpha$	$-\cos \alpha$	$\pm\sin \alpha$	$+\cos \alpha$
tan	$-\tan \alpha$	$\mp\cot \alpha$	$\pm\tan \alpha$	$\mp\cot \alpha$	$\pm\tan \alpha$
cot	$-\cot \alpha$	$\mp\tan \alpha$	$\pm\cot \alpha$	$\mp\tan \alpha$	$\pm\cot \alpha$
sec	$+\sec \alpha$	$\mp\csc \alpha$	$-\sec \alpha$	$\pm\csc \alpha$	$+\sec \alpha$
csc	$-\csc \alpha$	$+\sec \alpha$	$\mp\csc \alpha$	$-\sec \alpha$	$\pm\csc \alpha$

n denotes any integer.

For example:

$$\tan(270° + \alpha) = -\cot \alpha$$

58. FUNDAMENTAL IDENTITIES

$$\sin \alpha = \frac{1}{\csc \alpha}; \ \cos \alpha = \frac{1}{\sec \alpha}; \ \tan \alpha = \frac{1}{\cot \alpha} = \frac{\sin \alpha}{\cos \alpha}$$

$$\csc \alpha = \frac{1}{\sin \alpha}; \ \sec \alpha = \frac{1}{\cos \alpha}; \ \cot \alpha = \frac{1}{\tan \alpha} = \frac{\cos \alpha}{\sin \alpha}$$

$$\sin^2 \alpha + \cos^2 \alpha = 1; \ 1 + \tan^2 \alpha = \sec^2 \alpha; \ 1 + \cot^2 \alpha = \csc^2 \alpha$$

$$\sin 2\alpha = 2 \sin \alpha \cos \alpha$$

$$\cos 2\alpha = 2\cos^2\alpha - 1 = 1 - 2\sin^2\alpha = \cos^2\alpha - \sin^2\alpha$$

$$\tan 2\alpha = \frac{2\tan\alpha}{1 - \tan^2\alpha}$$

$$\sin 3\alpha = 3\sin\alpha - 4\sin^3\alpha$$

$$\cos 3\alpha = 4\cos^3\alpha - 3\cos\alpha$$

$$\sin n\alpha = 2\sin(n-1)\alpha\cos\alpha - \sin(n-2)\alpha$$

$$\cos n\alpha = 2\cos(n-1)\alpha\cos\alpha - \cos(n-2)\alpha$$

$$\sin(\alpha \pm \beta) = \sin\alpha\cos\beta \pm \cos\alpha\sin\beta$$

$$\cos(\alpha \pm \beta) = \cos\alpha\cos\beta \mp \sin\alpha\sin\beta$$

$$\tan(\alpha \pm \beta) = \frac{\tan\alpha \pm \tan\beta}{1 \mp \tan\alpha\tan\beta}$$

$$\sin\alpha + \sin\beta = 2\sin\tfrac{1}{2}(\alpha+\beta)\cos\tfrac{1}{2}(\alpha-\beta)$$

$$\sin\alpha - \sin\beta = 2\cos\tfrac{1}{2}(\alpha+\beta)\sin\tfrac{1}{2}(\alpha-\beta)$$

$$\cos\alpha + \cos\beta = 2\cos\tfrac{1}{2}(\alpha+\beta)\cos\tfrac{1}{2}(\alpha-\beta)$$

$$\cos\alpha - \cos\beta = -2\sin\tfrac{1}{2}(\alpha+\beta)\sin\tfrac{1}{2}(\alpha-\beta)$$

$$\sin\frac{\alpha}{2} = \pm\sqrt{\frac{1-\cos\alpha}{2}}$$

positive if $\alpha/2$ in I or II quadrants, negative otherwise

$$\cos\frac{\alpha}{2} = \pm\sqrt{\frac{1+\cos\alpha}{2}}$$

positive if $\alpha/2$ in I or IV, negative otherwise

$$\tan\frac{\alpha}{2} = \frac{1-\cos\alpha}{\sin\alpha} = \frac{\sin\alpha}{1+\cos\alpha} = \pm\sqrt{\frac{1-\cos\alpha}{1+\cos\alpha}}$$

positive if $\alpha/2$ in I or III, negative otherwise

$$\sin^2\alpha = \tfrac{1}{2}(1 - \cos 2\alpha);\quad \cos^2\alpha = \tfrac{1}{2}(1 + \cos 2\alpha)$$

$$\sin^3\alpha = \tfrac{1}{4}(3\sin\alpha - \sin 3\alpha);\quad \cos^3\alpha = \tfrac{1}{4}(\cos 3\alpha + 3\cos\alpha)$$

$$\sin\alpha\sin\beta = \tfrac{1}{2}\cos(\alpha-\beta) - \tfrac{1}{2}\cos(\alpha+\beta)$$

$$\cos\alpha\cos\beta = \tfrac{1}{2}\cos(\alpha-\beta) + \tfrac{1}{2}\cos(\alpha+\beta)$$

$$\sin\alpha\cos\beta = \tfrac{1}{2}\sin(\alpha+\beta) + \tfrac{1}{2}\sin(\alpha-\beta)$$

59. EQUIVALENT EXPRESSIONS FOR sin α, cos α, tan α, etc.

Function	$\sin\alpha$	$\cos\alpha$	$\tan\alpha$	$\cot\alpha$	$\sec\alpha$	$\csc\alpha$
$\sin\alpha$	$\sin\alpha$	$\pm\sqrt{1-\cos^2\alpha}$	$\dfrac{\tan\alpha}{\pm\sqrt{1+\tan^2\alpha}}$	$\dfrac{1}{\pm\sqrt{1+\cot^2\alpha}}$	$\dfrac{\pm\sqrt{\sec^2\alpha-1}}{\sec\alpha}$	$\dfrac{1}{\csc\alpha}$
$\cos\alpha$	$\pm\sqrt{1-\sin^2\alpha}$	$\cos\alpha$	$\dfrac{1}{\pm\sqrt{1+\tan^2\alpha}}$	$\dfrac{\cot\alpha}{\pm\sqrt{1+\cot^2\alpha}}$	$\dfrac{1}{\sec\alpha}$	$\dfrac{\pm\sqrt{\csc^2\alpha-1}}{\csc\alpha}$
$\tan\alpha$	$\dfrac{\sin\alpha}{\pm\sqrt{1-\sin^2\alpha}}$	$\dfrac{\pm\sqrt{1-\cos^2\alpha}}{\cos\alpha}$	$\tan\alpha$	$\dfrac{1}{\cot\alpha}$	$\pm\sqrt{\sec^2\alpha-1}$	$\dfrac{1}{\pm\sqrt{\csc^2\alpha-1}}$
$\cot\alpha$	$\dfrac{\pm\sqrt{1-\sin^2\alpha}}{\sin\alpha}$	$\dfrac{\cos\alpha}{\pm\sqrt{1-\cos^2\alpha}}$	$\dfrac{1}{\tan\alpha}$	$\cot\alpha$	$\dfrac{1}{\pm\sqrt{\sec^2\alpha-1}}$	$\pm\sqrt{\csc^2\alpha-1}$
$\sec\alpha$	$\dfrac{1}{\pm\sqrt{1-\sin^2\alpha}}$	$\dfrac{1}{\cos\alpha}$	$\pm\sqrt{1+\tan^2\alpha}$	$\dfrac{\pm\sqrt{1+\cot^2\alpha}}{\cot\alpha}$	$\sec\alpha$	$\dfrac{\csc\alpha}{\pm\sqrt{\csc^2\alpha-1}}$
$\csc\alpha$	$\dfrac{1}{\sin\alpha}$	$\dfrac{1}{\pm\sqrt{1-\cos^2\alpha}}$	$\dfrac{\pm\sqrt{1+\tan^2\alpha}}{\tan\alpha}$	$\pm\sqrt{1+\cot^2\alpha}$	$\dfrac{\sec\alpha}{\pm\sqrt{\sec^2\alpha-1}}$	$\csc\alpha$

Note: The quadrant in which α terminates determines the sign to be used. See table of signs in Sec. 54.

60. INVERSE TRIGONOMETRIC FUNCTIONS The complete solution of $x = \sin y$ is (in radians)

$$y = (-1)^n \operatorname{Sin}^{-1} x + n\pi \qquad -\frac{\pi}{2} \leqq \operatorname{Sin}^{-1} x \leqq \frac{\pi}{2}$$

where $\operatorname{Sin}^{-1} x$ is the *principal value* of the angle whose sine is x, n an integer. Similarly, if $x = \tan y$,

$$y = \operatorname{Tan}^{-1} x + n\pi \qquad -\frac{\pi}{2} < \operatorname{Tan}^{-1} x < \frac{\pi}{2}$$

The symbol $\sin^{-1} x$ (or arc sin x) is used to denote any angle y whose sine is x; and $\operatorname{Sin}^{-1} x$ (or $\sin^{-1} x$) is commonly used to designate the *principal value.* Similarly for other inverse relations, as defined below:

Relation	*Principal value*	
$\sin^{-1} x$	$-\frac{\pi}{2} \leqq \operatorname{Sin}^{-1} x \leqq \frac{\pi}{2}$	
$\tan^{-1} x$	$-\frac{\pi}{2} < \operatorname{Tan}^{-1} x < \frac{\pi}{2}$	
$\sec^{-1} x$	$0 \leqq \operatorname{Sec}^{-1} x < \frac{\pi}{2}$	if $x > 0$
	$-\pi \leqq \operatorname{Sec}^{-1} x < -\frac{\pi}{2}$	if $x < 0$
$\cosh^{-1} x$	$0 \leqq \operatorname{Cosh}^{-1} x$	
$\cos^{-1} x$	$0 \leqq \operatorname{Cos}^{-1} x \leqq \pi$	
$\cot^{-1} x$	$0 < \operatorname{Cot}^{-1} x < \pi$	
$\csc^{-1} x$	$0 < \operatorname{Csc}^{-1} x \leqq \frac{\pi}{2}$	if $x > 0$
	$-\pi < \operatorname{Csc}^{-1} x \leqq -\frac{\pi}{2}$	if $x < 0$
$\operatorname{sech}^{-1} x$	$0 \leqq \operatorname{Sech}^{-1} x$	

61. CERTAIN RELATIONS For principal values as in Sec. 60; $a > 0$:

$$\sin^{-1} a = \cos^{-1}\sqrt{1-a^2} = \tan^{-1}\frac{a}{\sqrt{1-a^2}} = \cot^{-1}\frac{\sqrt{1-a^2}}{a}$$

$$= \sec^{-1}\frac{1}{\sqrt{1-a^2}} = \csc^{-1}\frac{1}{a}$$

$$\cos^{-1} a = \sin^{-1}\sqrt{1-a^2} = \tan^{-1}\frac{\sqrt{1-a^2}}{a} = \cot^{-1}\frac{a}{\sqrt{1-a^2}}$$

$$= \sec^{-1}\frac{1}{a} = \csc^{-1}\frac{1}{\sqrt{1-a^2}}$$

$$\tan^{-1} a = \sin^{-1}\frac{a}{\sqrt{1+a^2}} = \cos^{-1}\frac{1}{\sqrt{1+a^2}} = \cot^{-1}\frac{1}{a}$$

$$= \sec^{-1}\sqrt{1+a^2} = \csc^{-1}\frac{\sqrt{1+a^2}}{a}$$

$$\cot^{-1} a = \tan^{-1}\frac{1}{a};\ \sec^{-1} a = \cos^{-1}\frac{1}{a};\ \csc^{-1} a = \sin^{-1}\frac{1}{a}$$

62. SOLUTION OF TRIGONOMETRIC EQUATIONS To solve a trigonometric equation, reduce the given equation by means of the relations expressed in Secs. 58 to 60 to an equation containing only a single function of a single angle. Solve the resulting equation by algebraic methods (Sec. 30) for the remaining function, and from this find the values of the angle, by Sec. 57 and Table 5. All these values should then be tested in the original equation, discarding those which do not satisfy it.

63. RELATIONS BETWEEN SIDES AND ANGLES OF PLANE TRIANGLES Let a, b, c denote the sides and α, β, γ the corresponding opposite angles with

$2s = a + b + c$

A = area

h_b = altitude on side b

r = radius of inscribed circle

R = radius of circumscribed circle

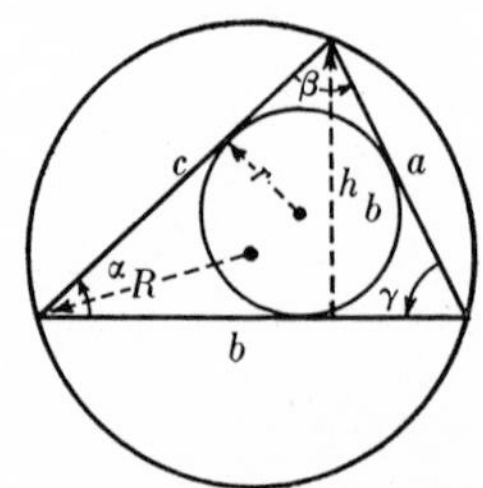

Fig. 10

$$\alpha + \beta + \gamma = 180°$$

$$\frac{a}{\sin \alpha} = \frac{b}{\sin \beta} = \frac{c}{\sin \gamma} \qquad \text{(law of sines)}$$

$$\frac{a+b}{a-b} = \frac{\tan \frac{1}{2}(\alpha + \beta)}{\tan \frac{1}{2}(\alpha - \beta)} \qquad \text{(law of tangents)*}$$

$$a^2 = b^2 + c^2 - 2bc \cos \alpha \qquad \text{(law of cosines)*}$$

$$a = b \cos \gamma + c \cos \beta^*$$

$$\cos \alpha = \frac{b^2 + c^2 - a^2}{2bc}^* \qquad \sin \alpha = \frac{2}{bc}\sqrt{s(s-a)(s-b)(s-c)}^*$$

$$\sin \frac{\alpha}{2} = \sqrt{\frac{(s-b)(s-c)}{bc}}^* \qquad \cos \frac{\alpha}{2} = \sqrt{\frac{s(s-a)}{bc}}^*$$

$$\tan \frac{\alpha}{2} = \sqrt{\frac{(s-b)(s-c)}{s(s-a)}} = \frac{r}{s-a}^*$$

where $r = \sqrt{\dfrac{(s-a)(s-b)(s-c)}{s}}$

$$A = \tfrac{1}{2}bh_b^* = \tfrac{1}{2}ab \sin \gamma^* = \frac{a^2 \sin \beta \sin \gamma^*}{2 \sin \alpha}$$

$$= \sqrt{s(s-a)(s-b)(s-c)} = rs$$

$$R = \frac{a^*}{2 \sin \alpha} = \frac{abc}{4A} \qquad h_b = c \sin \alpha^* = a \sin \gamma^* = \frac{2rs}{b}$$

64. SOLUTION OF A RIGHT TRIANGLE Given one side and any acute angle α, or any two sides, to find the remaining parts.

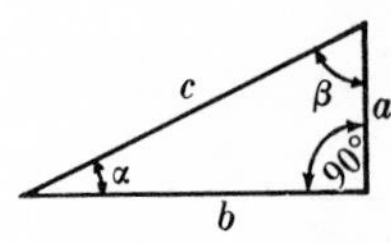

Fig. 11

$$a = \sqrt{(c+b)(c-b)} = c \sin \alpha = b \tan \alpha$$

$$b = \sqrt{(c+a)(c-a)} = c \cos \alpha = \frac{a}{\tan \alpha}$$

$$\sin \alpha = \frac{a}{c} \qquad \cos \alpha = \frac{b}{c} \qquad \tan \alpha = \frac{a}{b} \qquad \beta = 90° - \alpha$$

* Two more formulas may be obtained by replacing a by b, b by c, c by a, α by β, β by γ, γ by α.

$$c = \frac{a}{\sin \alpha} = \frac{b}{\cos \alpha} = \sqrt{a^2 + b^2}$$

$$A = \tfrac{1}{2}ab = \frac{a^2}{2 \tan \alpha} = \frac{b^2 \tan \alpha}{2} = \frac{c^2 \sin 2\alpha}{4}$$

65. SOLUTION OF OBLIQUE TRIANGLES (Fig. 12) The formulas of the preceding Sec. 63 will suffice to solve any oblique triangle. Use the trigonometric tables for numerical work. We give one method. Solutions should be checked by some other method.

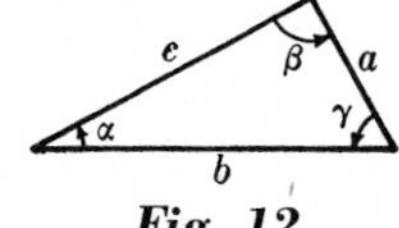

Fig. 12

(a) Given any two sides b and c and included angle α,

$$\tfrac{1}{2}(\beta + \gamma) = 90° - \tfrac{1}{2}\alpha \qquad \tan \tfrac{1}{2}(\beta - \gamma) = \frac{b - c}{b + c} \tan \tfrac{1}{2}(\beta + \gamma)$$

$$\beta = \tfrac{1}{2}(\beta + \gamma) + \tfrac{1}{2}(\beta - \gamma) \qquad \gamma = \tfrac{1}{2}(\beta + \gamma) - \tfrac{1}{2}(\beta - \gamma)$$

$$a = \frac{b \sin \alpha}{\sin \beta}$$

(b) Given any two angles α and β and any side c,

$$\gamma = 180° - (\alpha + \beta) \qquad a = \frac{c \sin \alpha}{\sin \gamma} \qquad b = \frac{c \sin \beta}{\sin \gamma}$$

(c) Given any two sides a and c and an angle opposite one of these, say, α,

$$\sin \gamma = \frac{c \sin \alpha}{a} \qquad \beta = 180° - (\alpha + \gamma) \qquad b = \frac{a \sin \beta}{\sin \alpha}$$

This case may have two sets of solutions, for γ may have two values, $\gamma_1 < 90°$ and $\gamma_2 = 180° - \gamma_1 > 90°$. If $\alpha + \gamma_2 > 180°$, use only γ_1.

(d) Given the three sides a, b, and c,

$$s = \tfrac{1}{2}(a + b + c) \qquad r = \sqrt{\frac{(s - a)(s - b)(s - c)}{s}}$$

$$\tan \tfrac{1}{2}\alpha = \frac{r}{s - a} \qquad \tan \tfrac{1}{2}\beta = \frac{r}{s - b} \qquad \tan \tfrac{1}{2}\gamma = \frac{r}{s - c}$$

Spherical Trigonometry

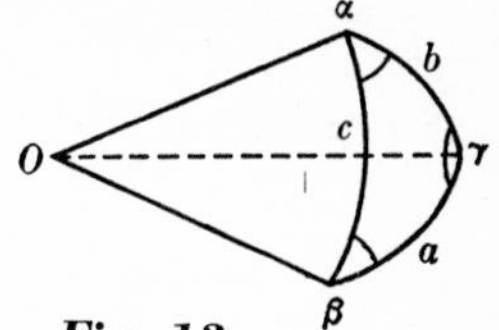

Fig. 13

66. THE RIGHT SPHERICAL TRIANGLE Let O be the center of a sphere and a, b, c the sides of a right triangle with opposite angles α, β, and $\gamma = 90°$, respectively, the sides being measured by the angle subtended at the center of the sphere.

$$\sin a = \sin \alpha \sin c \qquad \sin b = \sin \beta \sin c$$
$$\sin a = \tan b \cot \beta \qquad \sin b = \tan a \cot \alpha$$
$$\cos \alpha = \cos a \sin \beta \qquad \cos \beta = \cos b \sin \alpha$$
$$\cos \alpha = \tan b \cot c \qquad \cos \beta = \tan a \cot c$$
$$\cos c = \cot \alpha \cot \beta \qquad \cos c = \cos a \cos b$$

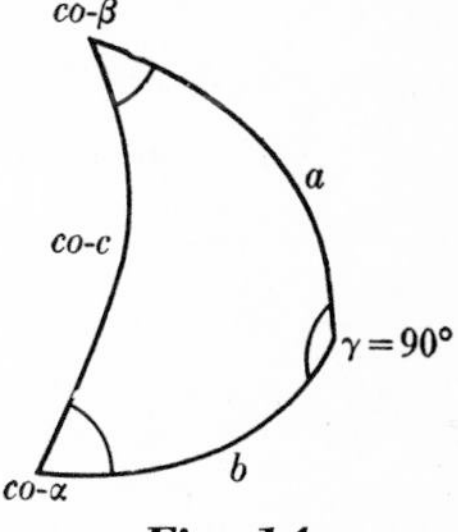

Fig. 14

67. NAPIER'S RULES OF CIRCULAR PARTS (see Fig. 14) Let the five quantities a, b, co-α (complement of α), co-c, co-β be arranged in order as indicated in the figure. If we denote any one of these quantities a *middle* part, then two of the other parts are *adjacent* to it, and the other two parts are *opposite* to it. The above formulas may be remembered by means of the following rules:

(a) *The sine of a middle part is equal to the product of the tangents of the adjacent parts.*

(b) *The sine of a middle part is equal to the product of the cosines of the opposite parts.*

68. THE OBLIQUE SPHERICAL TRIANGLE Let a, b, c denote the sides and α, β, γ the corresponding opposite angles of the spherical triangle, Δ its area, E its spherical excess, R the radius of the sphere upon which the triangle lies, and α', β', γ', a, b, c the corresponding parts of the polar triangle.

$$0° < a + b + c < 360° \qquad 180° < \alpha + \beta + \gamma < 540°$$
$$\alpha = 180° - a' \qquad \beta = 180° - b' \qquad \gamma = 180° - c'$$
$$a = 180° - \alpha' \qquad b = 180° - \beta' \qquad c = 180° - \gamma'$$

$$\frac{\sin \alpha}{\sin a} = \frac{\sin \beta}{\sin b} = \frac{\sin \gamma}{\sin c} \qquad \text{(law of sines)}$$

$$\cos a = \cos b \cos c + \sin b \sin c \cos \alpha$$

$$\cos \alpha = -\cos \beta \cos \gamma + \sin \beta \sin \gamma \cos a \qquad \text{(law of cosines)}$$

$$\tan \frac{\alpha}{2} = \sqrt{\frac{\sin (s - b) \sin (s - c)}{\sin s \sin (s - a)}}, \text{ where } s = \tfrac{1}{2}(a + b + c)$$

$$\tan \frac{a}{2} = \sqrt{\frac{-\cos \sigma \cos (\sigma - \alpha)}{\cos (\sigma - \beta) \cos (\sigma - \gamma)}}, \text{ where } \sigma = \tfrac{1}{2}(\alpha + \beta + \gamma)$$

$$\frac{\sin \frac{1}{2}(\alpha + \beta)}{\sin \frac{1}{2}(\alpha - \beta)} = \frac{\tan \frac{1}{2}c}{\tan \frac{1}{2}(a - b)} \qquad \frac{\cos \frac{1}{2}(\alpha + \beta)}{\cos \frac{1}{2}(\alpha - \beta)} = \frac{\tan \frac{1}{2}c}{\tan \frac{1}{2}(a + b)}$$

$$\frac{\sin \frac{1}{2}(a + b)}{\sin \frac{1}{2}(a - b)} = \frac{\cot \frac{1}{2}\gamma}{\tan \frac{1}{2}(\alpha - \beta)} \qquad \frac{\cos \frac{1}{2}(a + b)}{\cos \frac{1}{2}(a - b)} = \frac{\cot \frac{1}{2}\gamma}{\tan \frac{1}{2}(\alpha + \beta)}$$

$$\sin \tfrac{1}{2}(\alpha + \beta) \cos \tfrac{1}{2}c = \cos \tfrac{1}{2}(a - b) \cos \tfrac{1}{2}\gamma$$

$$\cos \tfrac{1}{2}(\alpha + \beta) \cos \tfrac{1}{2}c = \cos \tfrac{1}{2}(a + b) \sin \tfrac{1}{2}\gamma$$

$$\sin \tfrac{1}{2}(\alpha - \beta) \sin \tfrac{1}{2}c = \sin \tfrac{1}{2}(a - b) \cos \tfrac{1}{2}\gamma$$

$$\cos \tfrac{1}{2}(\alpha - \beta) \sin \tfrac{1}{2}c = \sin \tfrac{1}{2}(a + b) \sin \tfrac{1}{2}\gamma$$

$$\tan \frac{E}{4} = \sqrt{\tan \tfrac{1}{2}s \tan \tfrac{1}{2}(s - a) \tan \tfrac{1}{2}(s - b) \tan \tfrac{1}{2}(s - c)}$$

$$\Delta = \frac{\pi R^2 E}{180} \qquad \alpha + \beta + \gamma - 180° = E$$

*Hyperbolic Functions**

69. DEFINITIONS (for definition of e see Sec. 108)

$$\text{Hyperbolic sine of } x = \sinh x = \tfrac{1}{2}(e^x - e^{-x})$$

$$\text{Hyperbolic cosine of } x = \cosh x = \tfrac{1}{2}(e^x + e^{-x})$$

$$\text{Hyperbolic tangent of } x = \tanh x = \frac{e^x - e^{-x}}{e^x + e^{-x}} = \frac{\sinh x}{\cosh x}$$

$$\operatorname{csch} x = \frac{1}{\sinh x} \qquad \operatorname{sech} x = \frac{1}{\cosh x} \qquad \coth x = \frac{1}{\tanh x}$$

* For derivatives and integrals of hyperbolic functions see Secs. 108 and 121.

70. INVERSE OR ANTIHYPERBOLIC FUNCTIONS (see Sec. 60)

If $x = \sinh y$, then y is the *inverse hyperbolic sine* of x, written $y = \sinh^{-1} x$ or arc sinh x.

$$\sinh^{-1} x = \log_e (x + \sqrt{x^2 + 1})^*$$

$$\cosh^{-1} x = \log_e (x + \sqrt{x^2 - 1})^\dagger$$

$$\tanh^{-1} x = \tfrac{1}{2} \log_e \left(\frac{1 + x}{1 - x}\right) \qquad \coth^{-1} x = \tfrac{1}{2} \log_e \left(\frac{x + 1}{x - 1}\right)$$

$$\operatorname{sech}^{-1} x = \log_e \left(\frac{1 + \sqrt{1 - x^2}}{x}\right)^\dagger$$

$$\operatorname{csch}^{-1} x = \log_e \left(\frac{1 + \sqrt{1 + x^2}}{x}\right)^\ddagger$$

71. FUNDAMENTAL IDENTITIES

$$\sinh(-x) = -\sinh x \qquad \cosh^2 x - \sinh^2 x = 1$$

$$\cosh(-x) = \cosh x \qquad \operatorname{sech}^2 x + \tanh^2 x = 1$$

$$\tanh(-x) = -\tanh x \qquad \operatorname{csch}^2 x - \coth^2 x = -1$$

$$\sinh(x \pm y) = \sinh x \cosh y \pm \cosh x \sinh y$$

$$\cosh(x \pm y) = \cosh x \cosh y \pm \sinh x \sinh y$$

$$\tanh(x \pm y) = \frac{\tanh x \pm \tanh y}{1 \pm \tanh x \tanh y}$$

$$\sinh 2x = 2 \sinh x \cosh x \qquad \cosh 2x = \cosh^2 x + \sinh^2 x$$

$$2 \sinh^2 \frac{x}{2} = \cosh x - 1 \qquad 2 \cosh^2 \frac{x}{2} = \cosh x + 1$$

72. CONNECTION WITH CIRCULAR FUNCTIONS

$$\sin x = \frac{e^{ix} - e^{-ix}}{2i} \qquad \cos x = \frac{e^{ix} + e^{-ix}}{2} \qquad i^2 = -1$$

$$\sin x = -i \sinh ix \qquad \cos x = \cosh ix \qquad \tan x = -i \tanh ix$$

* If $x \geqq 0$. † For principal values. ‡ If $x > 0$.

$$\sin ix = i \sinh x \qquad \cos ix = \cosh x \qquad \tan ix = i \tanh x$$
$$\sinh ix = i \sin x \qquad \cosh ix = \cos x \qquad \tanh ix = i \tan x$$
$$\sinh (x \pm iy) = \sinh x \cos y \pm i \cosh x \sin y$$
$$\cosh (x \pm iy) = \cosh x \cos y \pm i \sinh x \sin y$$
$$\sinh (x + 2i\pi) = \sinh x \qquad \cosh (x + 2i\pi) = \cosh x$$
$$\sinh (x + i\pi) = -\sinh x \qquad \cosh (x + i\pi) = -\cosh x$$
$$\sinh (x + \tfrac{1}{2}i\pi) = i \cosh x \qquad \cosh (x + \tfrac{1}{2}i\pi) = i \sinh x$$
$$e^{\theta} = \cosh \theta + \sinh \theta \qquad e^{-\theta} = \cosh \theta - \sinh \theta$$
$$e^{i\theta} = \cos \theta + i \sin \theta \qquad e^{-i\theta} = \cos \theta - i \sin \theta$$

$$x + iy = re^{i\theta}$$

where $r = +\sqrt{x^2 + y^2}$, $\theta = \text{arc cos}\ \dfrac{x}{r} = \text{arc sin}\ \dfrac{y}{r}$

$$\log_e (x \pm iy) = \tfrac{1}{2} \log_e (x^2 + y^2) \pm i \text{ arc tan } \frac{y}{x}$$
$$(\cos \theta + i \sin \theta)^n = \cos n\theta + i \sin n\theta$$

(De Moivre's theorem)

$$\left(\cos \frac{2k\pi}{n} + i \sin \frac{2k\pi}{n}\right)^n = 1 \qquad k = 0, 1, \ldots, n - 1$$

$$\sinh x + \sinh y = 2 \sinh \frac{x + y}{2} \cosh \frac{x - y}{2}$$

$$\sinh x - \sinh y = 2 \cosh \frac{x + y}{2} \sinh \frac{x - y}{2}$$

$$\cosh x + \cosh y = 2 \cosh \frac{x + y}{2} \cosh \frac{x - y}{2}$$

$$\cosh x - \cosh y = 2 \sinh \frac{x + y}{2} \sinh \frac{x - y}{2}$$

4. ANALYTIC GEOMETRY

73. RECTANGULAR COORDINATES Let $X'X$ (x axis) and $Y'Y$ (y axis) be two perpendicular lines meeting in the point O called the origin. The point $P(x, y)$ in the plane of the x and y axes is fixed by the distances x (abscissa) and y (ordinate) from $Y'Y$ and $X'X$, respectively, to P. x is positive to the right and negative to the left of the y axis, and y is positive above and negative below the x axis.

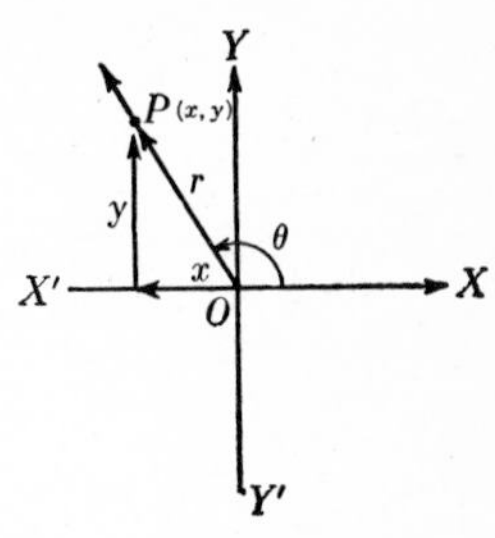

Fig. 15

74. POLAR COORDINATES Let OX (initial line) be a fixed line in the plane and O (pole or origin) a point on this line. The position of any point $P(r,\theta)$ in the plane is determined by the distance r (radius vector) from O to the point together with the angle θ (vectorial angle) measured from OX to OP. θ is positive if measured counterclockwise, negative if measured clockwise; r is positive if measured along the terminal side of θ and negative if measured along the terminal side of θ produced through the pole.

75. RELATIONS BETWEEN RECTANGULAR AND POLAR COORDINATES (see Sec. 113)

$$\begin{cases} x = r\cos\theta \\ y = r\sin\theta \end{cases} \qquad \begin{cases} r = \sqrt{x^2 + y^2} \\ \theta = \tan^{-1}\dfrac{y}{x} \end{cases} \qquad \begin{cases} \sin\theta = \dfrac{y}{\sqrt{x^2 + y^2}} \\ \cos\theta = \dfrac{x}{\sqrt{x^2 + y^2}} \end{cases}$$

76. POINTS AND SLOPES Let $P_1(x_1, y_1)$ and $P_2(x_2, y_2)$ be any two points and α_1 be the angle measured counterclockwise from OX to P_1P_2:

Distance between P_1 and $P_2 = P_1P_2$

$$= d = \sqrt{(x_2 - x_1)^2 + (y_2 - y_1)^2}$$

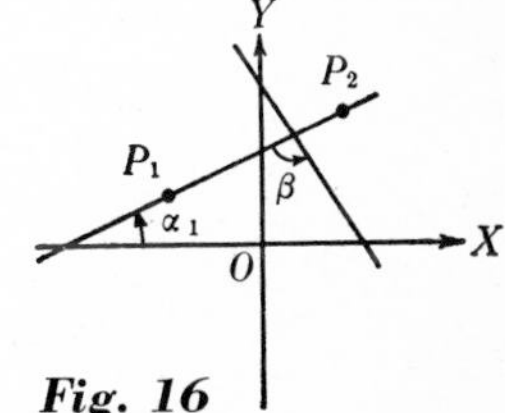

Fig. 16

$$\text{Slope of } P_1P_2 = \tan \alpha_1 = m = \frac{y_2 - y_1}{x_2 - x_1}$$

Point dividing P_1P_2 in ratio $m_1\!:\!m_2$

$$= \left(\frac{m_1x_2 + m_2x_1}{m_1 + m_2}, \frac{m_1y_2 + m_2y_1}{m_1 + m_2} \right)$$

$$\text{Midpoint of } P_1P_2 = \left(\frac{x_1 + x_2}{2}, \frac{y_1 + y_2}{2} \right)$$

The angle β between lines of slopes m_1 and m_2, respectively, is given by

$$\tan \beta = \frac{m_2 - m_1}{1 + m_1m_2}$$

Two lines of slopes m_1 and m_2 are perpendicular if $m_2 = -(1/m_1)$ and parallel if $m_1 = m_2$.

77. **AREA OF TRIANGLE** If the vertices are the points (x_1, y_1), (x_2, y_2), (x_3, y_3), then the area is equal to the numerical value of

$$\frac{1}{2}\begin{vmatrix} x_1 & y_1 & 1 \\ x_2 & y_2 & 1 \\ x_3 & y_3 & 1 \end{vmatrix} = \tfrac{1}{2}(x_1y_2 - x_1y_3 + x_2y_3 - x_2y_1 + x_3y_1 - x_3y_2)$$

78. **LOCUS AND EQUATION** The set of all points which satisfy a given condition is called the *locus* of that condition. An *equation* is called the equation of the locus if it is satisfied by the coordinates of every point on the locus and by no other points. There are three common representations of the locus by means of equations:

(a) *Rectangular equation,* which involves the rectangular coordinates (x, y)

(b) *Polar equation,* which involves the polar coordinates (r,θ)

(c) *Parametric equations,* which express x and y or r and θ in terms of a third independent variable called a parameter

79. **TRANSFORMATION OF COORDINATES** To transform an equation or a curve from one system of coordinates in x, y to another such system in x', y', substitute for each variable its value in terms of variables of the new system.

(*a*) *Rectangular system: old axes parallel to new axes*
The coordinates of new origin in terms of old system are (h,k).

$$\begin{cases} x = x' + h \\ y = y' + k \end{cases}$$

(*b*) *Rectangular system: old origin coincident with new origin and the x' axis making an angle θ with the x axis*

$$\begin{cases} x = x' \cos \theta - y' \sin \theta \\ y = x' \sin \theta + y' \cos \theta \end{cases}$$

(*c*) *Rectangular system: old axes not parallel with new. New origin at (h,k) in old system*

$$\begin{cases} x = x' \cos \theta - y' \sin \theta + h \\ y = x' \sin \theta + y' \cos \theta + k \end{cases}$$

80. STRAIGHT LINE The equations of the straight line may assume the following forms:

(*a*) $y = mx + b$ (m = slope, b = intercept on y axis)

(*b*) $y - y_1 = m(x - x_1)$ [m = slope, line passes through point (x_1, y_1)]

(*c*) $\dfrac{y - y_1}{x - x_1} = \dfrac{y_2 - y_1}{x_2 - x_1}$ [line passes through points (x_1, y_1) and (x_2, y_2)]

(*d*) $\dfrac{x}{a} + \dfrac{y}{b} = 1$ (a and b are the intercepts on x and y axes, respectively)

(*e*) $x \cos \alpha + y \sin \alpha = p$ (*normal form*, p is distance from origin to the line, α is angle which normal to the line makes with x axis)

(*f*) $Ax + By + C = 0$ [*general form*, slope $= -(A/B)$]

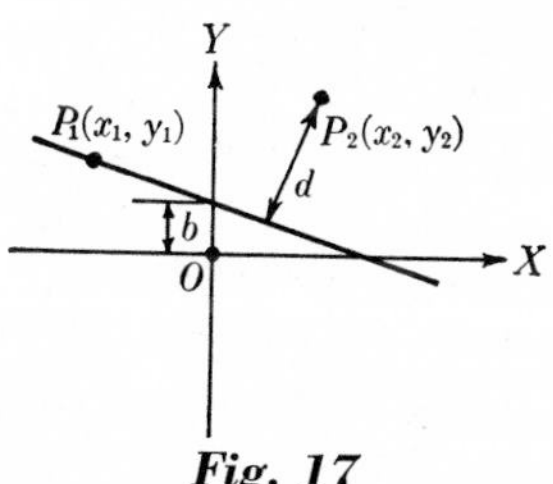

Fig. 17

To reduce $Ax + By + C = 0$ to normal form **(e)**, divide by $\pm\sqrt{A^2 + B^2}$, where the sign of the radical is taken opposite to that of C when $C \neq 0$.

The distance from the line $Ax + By + C = 0$ to the point $P_2(x_2, y_2)$ is

$$d = \frac{Ax_2 + By_2 + C}{\pm\sqrt{A^2 + B^2}}$$

81. CIRCLE (see Sec. 37) General equation of circle with radius R and center at (h,k) is

$$(x - h)^2 + (y - k)^2 = R^2$$

82. CONIC The locus of a point P which moves so that its distance from a fixed point F (focus) bears a constant ratio e (eccentricity) to its distance from a fixed straight line (directrix) is a conic.

If d is the distance from focus to directrix and F is at the origin,

$$x^2 + y^2 = e^2(d + x)^2$$

$$r = \frac{de}{1 - e\cos\theta}$$

If $e = 1$, the conic is a *parabola;* if $e > 1$, a *hyperbola;* and if $e < 1$, an *ellipse.*

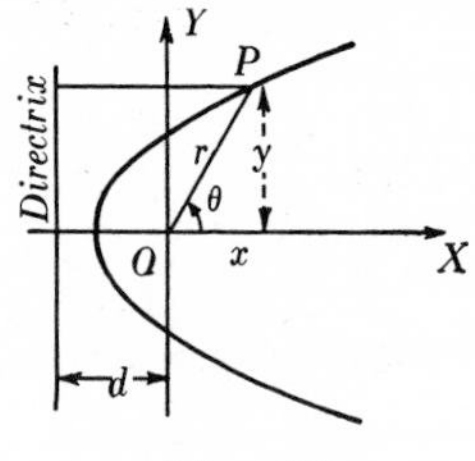

Fig. 18

83. PARABOLA (see Sec. 39) $e = 1$

(a) $(y - k)^2 = 4a(x - h)$. Vertex at (h,k), axis $\parallel OX$ (Fig. 19). Figure 19 is drawn for the case when a is positive.

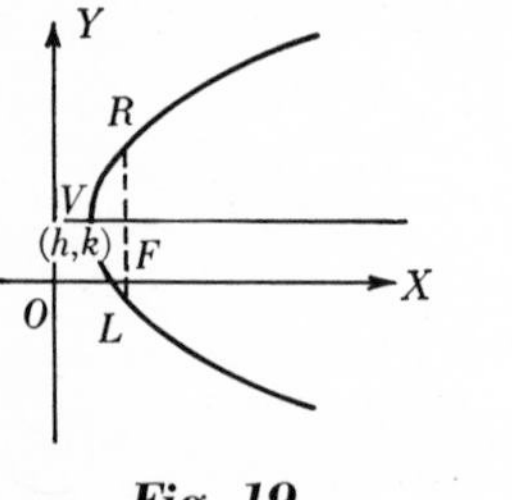

Fig. 19

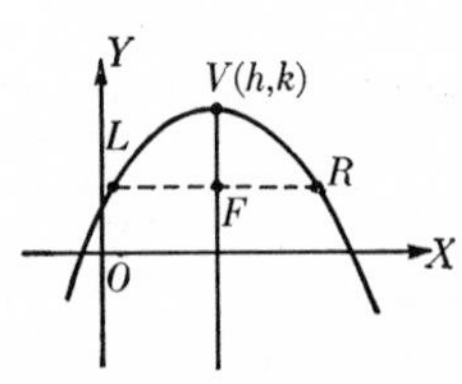

Fig. 20

(*b*) $(x - h)^2 = 4a(y - k)$. Vertex at (h,k), axis $\parallel OY$ (Fig. 20). Figure 20 is drawn for the case when a is negative.

$$\text{Distance from vertex to focus} = VF = a$$
$$\text{Distance from vertex to directrix} = a$$
$$\text{Latus rectum} = LR = 4a$$

84. ELLIPSE (see Sec. 38) $e < 1$

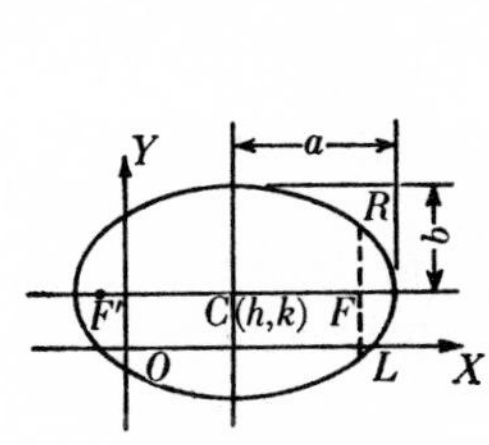

Fig. 21

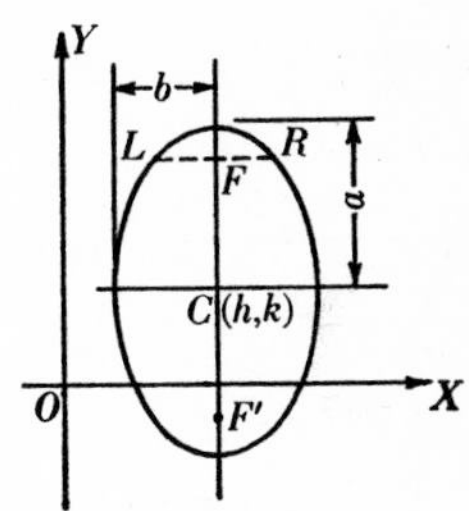

Fig. 22

(a) $\dfrac{(x-h)^2}{a^2} + \dfrac{(y-k)^2}{b^2} = 1.$ Center at (h,k), major axis $\parallel OX$ (Fig. 21).

(b) $\dfrac{(y-k)^2}{a^2} + \dfrac{(x-h)^2}{b^2} = 1.$ Center at (h,k), major axis $\parallel OY$ (Fig. 22).

Major axis $= 2a$ Minor axis $= 2b$

Distance from center to either focus $= \sqrt{a^2 - b^2}$

Distance from center to either directrix $= a/e$

Eccentricity $= e = \sqrt{a^2 - b^2}/a$

Latus rectum $= 2b^2/a$

Sum of distances from any point P on ellipse to foci, $PF' + PF = 2a$

85. HYPERBOLA $e > 1$

(a) $\dfrac{(x-h)^2}{a^2} - \dfrac{(y-k)^2}{b^2} = 1.$

Center at (h,k), transverse axis $\parallel OX$.

Slopes of asymptotes $= \pm b/a$ (Fig. 23).

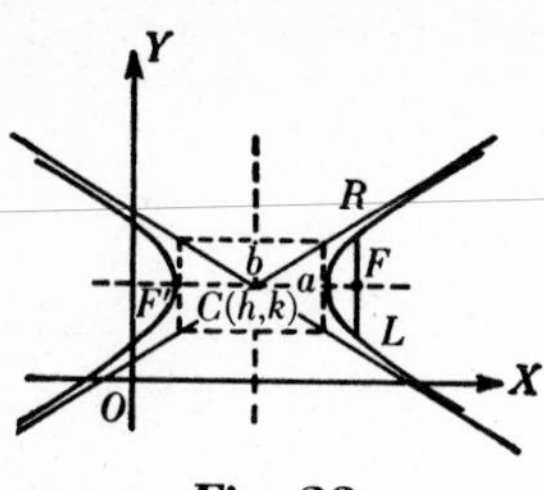

Fig. 23

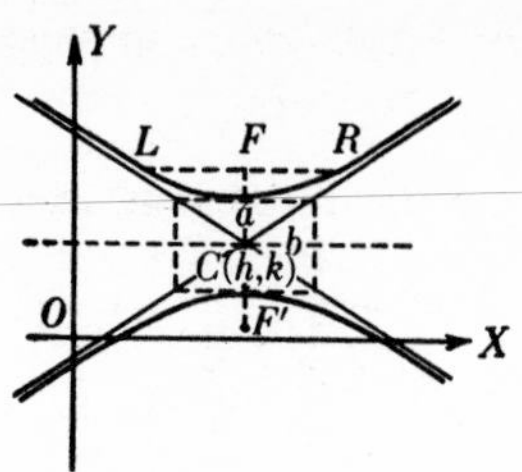

Fig. 24

(b) $$\frac{(y-k)^2}{a^2} - \frac{(x-h)^2}{b^2} = 1.$$

Center at (h,k), transverse axis $\parallel OY$.
Slopes of asymptotes $= \pm a/b$ (Fig. 24).

Transverse axis $= 2a$ Conjugate axis $= 2b$

Distance from center to either focus $= \sqrt{a^2 + b^2}$

Distance from center to either directrix $= a/e$

Difference of distances of any point on hyperbola from the foci $= 2a$

Eccentricity $= e = \dfrac{\sqrt{a^2 + b^2}}{a}$

Latus rectum $= 2b^2/a$

86. SINE CURVE

$$y = a \sin (bx + c)$$

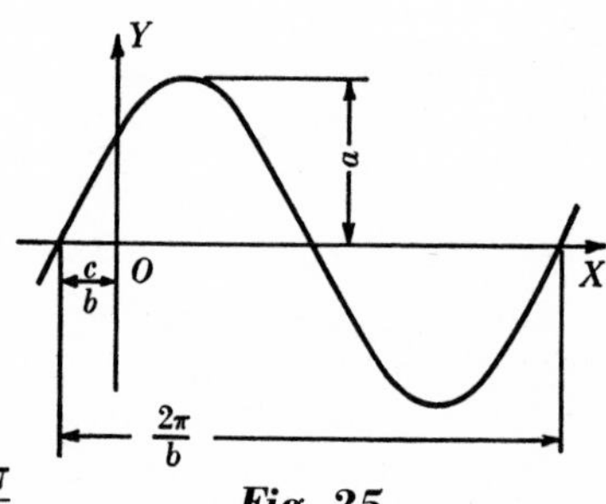

Fig. 25

$$y = a \cos (bx + c')$$

$= a \sin (bx + c)$, where $c = c' + \dfrac{\pi}{2}$

$$y = p \sin bx + q \cos bx = a \sin (bx + c)$$

where $c = \tan^{-1} \dfrac{q}{p}$, $a = \sqrt{p^2 + q^2}$

$a =$ amplitude $=$ maximum height of wave

$\dfrac{2\pi}{b} =$ wavelength $=$ distance from any point on wave to the corresponding point on the next wave

$x = -\frac{c}{b}$ = phase, indicates a point on x axis from which the positive half of the wave starts

87. TRIGONOMETRIC CURVES

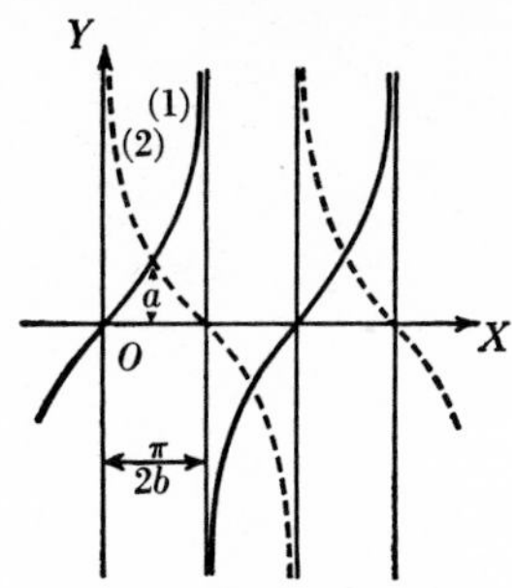

Fig. 26

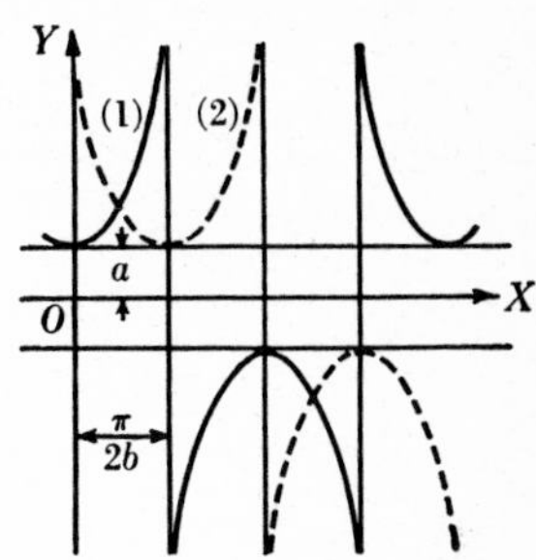

Fig. 27

In Fig. 26,

(1) $y = a \tan bx$ or $x = \frac{1}{b} \tan^{-1} \frac{y}{a}$

(2) $y = a \cot bx$ or $x = \frac{1}{b} \cot^{-1} \frac{y}{a}$

In Fig. 27,

(1) $y = a \sec bx$ or $x = \frac{1}{b} \sec^{-1} \frac{y}{a}$

(2) $y = a \csc bx$ or $x = \frac{1}{b} \csc^{-1} \frac{y}{a}$

88. LOGARITHMIC AND EXPONENTIAL CURVES

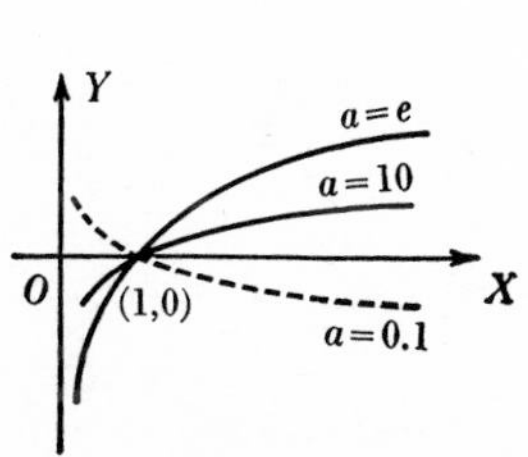

Fig. 28. ***Logarithmic curve.*** $y = \log_a x$ ***or*** $x = a^y$

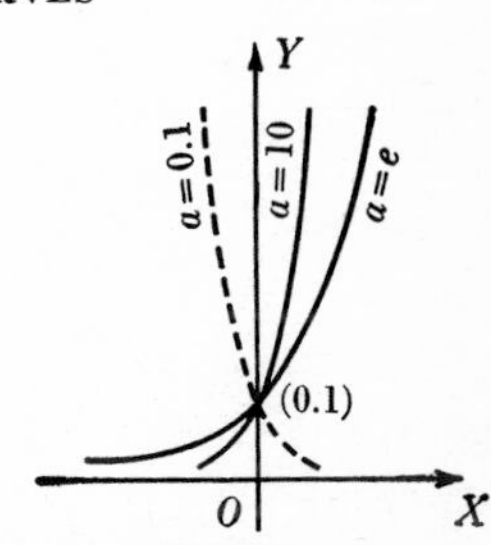

Fig. 29. ***Exponential curve.*** $y = a^x$ ***or*** $x = \log_a y$

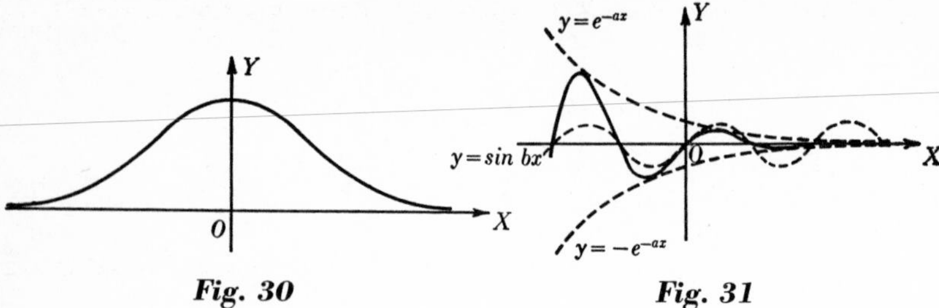

Fig. 30 *Fig. 31*

89. **PROBABILITY CURVE** (Fig. 30) $y = e^{-x^2}$

90. **OSCILLATORY WAVE OF DECREASING AMPLITUDE** (Fig. 31)

$$y = e^{-ax} \sin bx$$

91. **CATENARY** (Fig. 32)

$$y = \frac{a}{2}\left(e^{x/a} + e^{-(x/a)}\right)$$

A curve made by a cord of uniform weight suspended freely between two points. (See Sec. 69.)

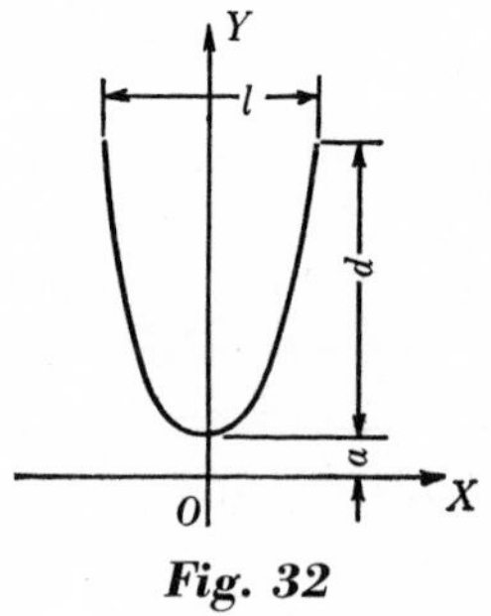

Fig. 32

$$\text{Length of arc} = s = l\left[1 + \frac{2}{3}\left(\frac{2d}{l}\right)^2\right]$$

approximately, if d is small in comparison with l.

92. **CYCLOID** (Fig. 33)

$$\begin{cases} x = a(\phi - \sin \phi) \\ y = a(1 - \cos \phi) \end{cases}$$

A curve described by a point on the circumference of circle which rolls along a fixed straight line.

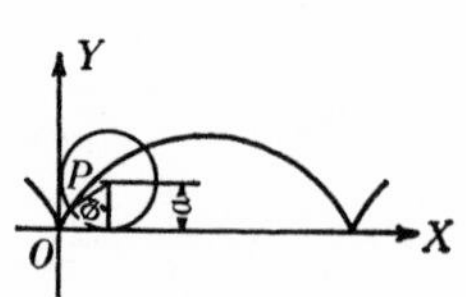

Fig. 33

$$\text{Area one arch} = 3\pi a^2$$

$$\text{Length of arc of one arch} = 8a$$

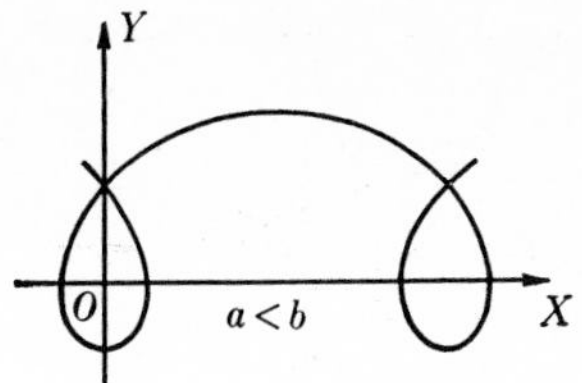

Fig. 34. Prolate cycloid.

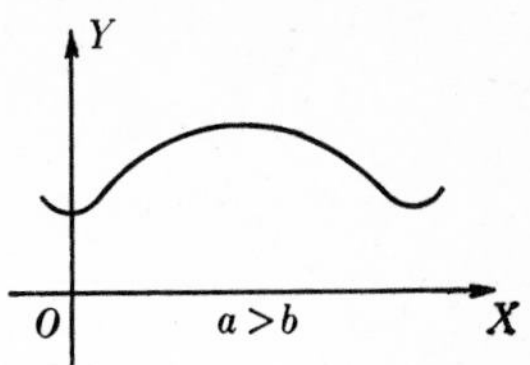

Fig. 35. Curtate cycloid.

93. PROLATE AND CURTATE CYCLOID

$$\begin{cases} x = a\phi - b \sin \phi \\ y = a - b \cos \phi \end{cases}$$

A curve described by a point on a circle at a distance b from the center of the circle of radius a which rolls along a fixed straight line.

94. EPICYCLOID (Fig. 36)

$$\begin{cases} x = (a + b) \cos \phi - a \cos \left(\dfrac{a + b}{a} \phi\right) \\ y = (a + b) \sin \phi - a \sin \left(\dfrac{a + b}{a} \phi\right) \end{cases}$$

A curve described by a point on the circumference of a circle which rolls along the outside of a fixed circle.

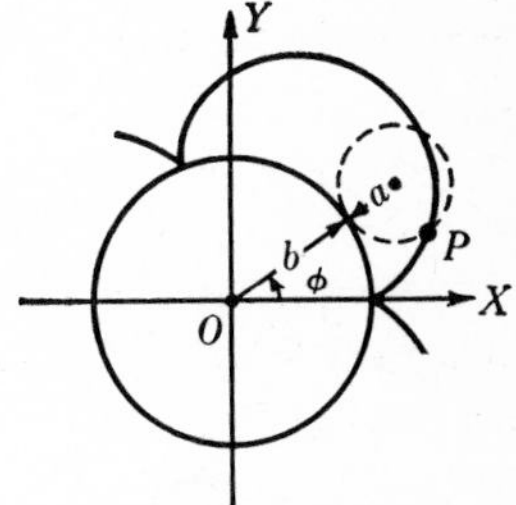

Fig. 36

95. CARDIOID (Fig. 37)

$$r = a(1 + \cos \theta)$$

An epicycloid in which both circles have the same radii is called a cardioid.

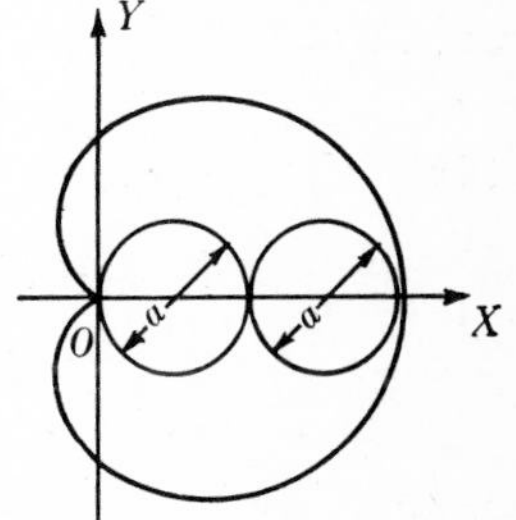

Fig. 37

96. HYPOCYCLOID

$$\begin{cases} x = (a - b) \cos \phi + b \cos \left(\dfrac{a - b}{b} \phi\right) \\ y = (a - b) \sin \phi - b \sin \left(\dfrac{a - b}{b} \phi\right) \end{cases}$$

A curve described by a point on the circumference of a circle which rolls along the inside of a fixed circle.

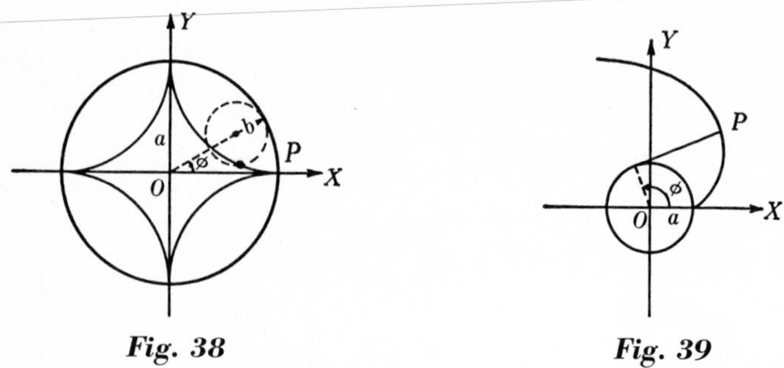

Fig. 38

Fig. 39

97. **HYPOCYCLOID OF FOUR CUSPS** (Fig. 38)

$$x^{2/3} + y^{2/3} = a^{2/3}$$

$$x = a \cos^3 \phi \qquad y = a \sin^3 \phi$$

The radius of fixed circle is four times the radius of rolling circle.

98. **INVOLUTE OF THE CIRCLE** (Fig. 39)

$$\begin{cases} x = a \cos \phi + a \phi \sin \phi \\ y = a \sin \phi - a \phi \cos \phi \end{cases}$$

A curve generated by the end of a string which is kept taut while being unwound from a circle.

99. **LEMNISCATE** (Fig. 40)

$$r^2 = 2a^2 \cos 2\theta$$

100. ***n*-LEAVED ROSE** (Fig. 41)

(1) $$r = a \sin n\theta$$

(2) $$r = a \cos n\theta$$

If n is an odd integer, there are n leaves; if n is even, $2n$ leaves, of which the figure shows one.

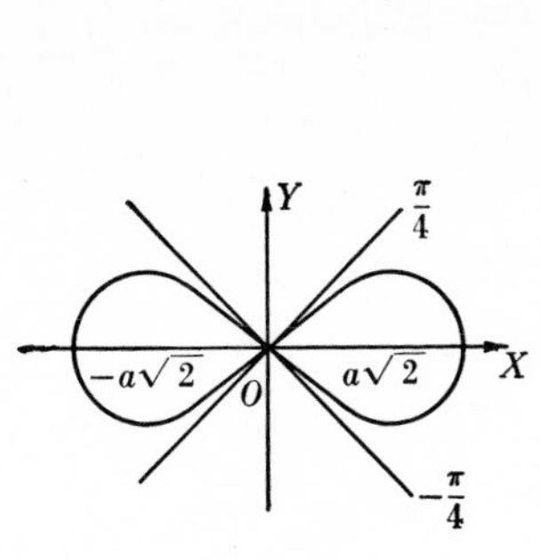

Fig. 40

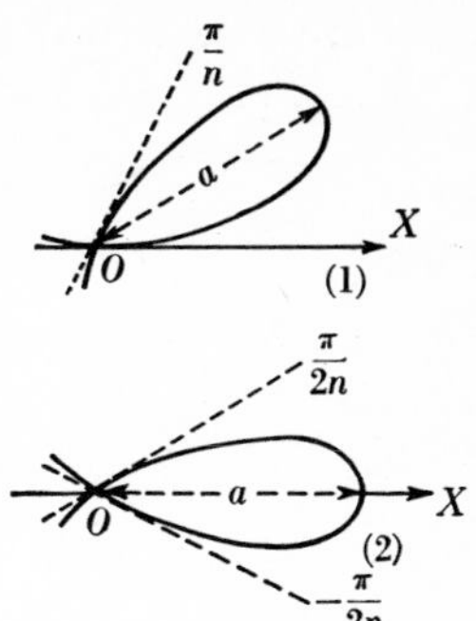

Fig. 41

101. SPIRALS

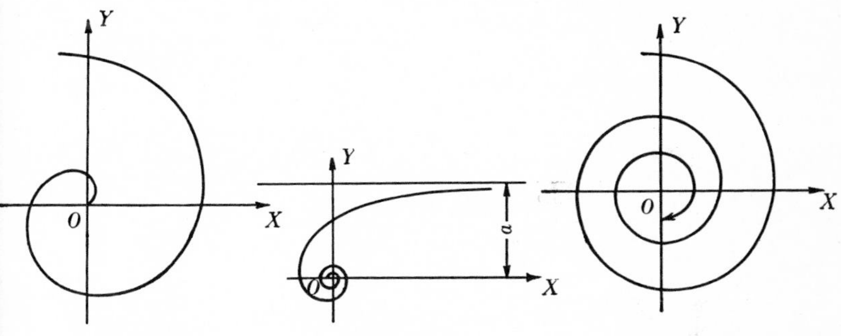

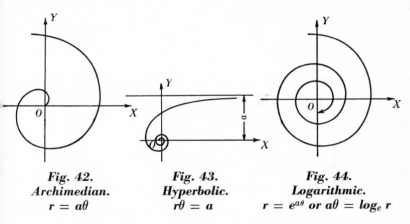

Fig. 42. Archimedian. $r = a\theta$

Fig. 43. Hyperbolic. $r\theta = a$

Fig. 44. Logarithmic. $r = e^{a\theta}$ or $a\theta = \log_e r$

Solid Analytic Geometry

102. COORDINATES (Fig. 45)

(a) *Rectangular system.* The position of a point $P(x, y, z)$ in space is fixed by its three distances x, y, and z from three coordinate planes XOY, XOZ, ZOY, which are mutually perpendicular and meet in a point O (origin).

(b) *Cylindrical system.* The position of any point $P(r, \theta, z)$ is fixed by (r,θ), the polar coordinates of the projection of P in the XOY plane, and by z, its distance from the XOY plane.

(c) *Spherical* (*or polar or geographical*) *system.* The position of any point $P(\rho,\theta,\phi)$ is fixed by the distance $\rho = \overline{OP}$, the angle $\theta = \angle XOM$, and the angle $\phi = \angle ZOP$. θ is called the *longitude* and ϕ the *colatitude.*

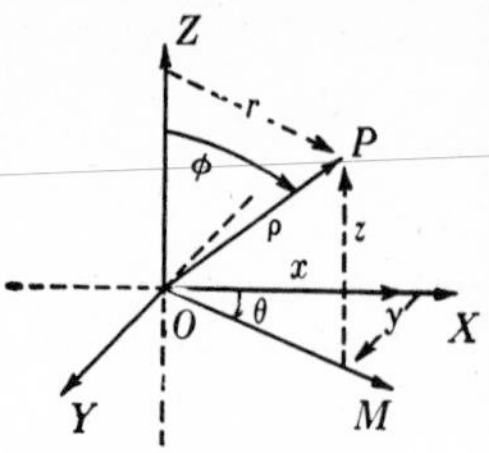

Fig. 45

The following relations exist between the three coordinate systems:

$$\begin{cases} x = \rho \sin \phi \cos \theta \\ y = \rho \sin \phi \sin \theta \\ z = \rho \cos \phi \end{cases} \qquad \begin{cases} r = \rho \sin \phi \\ z = \rho \cos \phi \end{cases} \qquad \begin{cases} \rho = \sqrt{r^2 + z^2} \\ \cos \phi = \dfrac{z}{\sqrt{r^2 + z^2}} \end{cases}$$

103. POINTS, LINES, AND PLANES Distance between two points $P_1(x_1,y_1,z_1)$ and $P_2(x_2,y_2,z_2)$ is

$$d = \sqrt{(x_2 - x_1)^2 + (y_2 - y_1)^2 + (z_2 - z_1)^2}$$

Direction cosines of a line are the cosines of the angles α, β, γ which the line or any parallel line makes with the coordinate axes.

The direction cosines of the line segment $P_1(x_1,y_1,z_1)$ to $P_2(x_2,y_2,z_2)$ are

$$\cos \alpha = \frac{x_2 - x_1}{d}, \quad \cos \beta = \frac{y_2 - y_1}{d}, \quad \cos \gamma = \frac{z_2 - z_1}{d}$$

If $\cos \alpha : \cos \beta : \cos \gamma = a:b:c$, then

$$\cos \alpha = \frac{a}{\sqrt{a^2 + b^2 + c^2}}, \quad \cos \beta = \frac{b}{\sqrt{a^2 + b^2 + c^2}},$$

$$\cos \gamma = \frac{c}{\sqrt{a^2 + b^2 + c^2}}$$

$$\cos^2 \alpha + \cos^2 \beta + \cos^2 \gamma = 1$$

Angle θ between two lines, whose direction angles are α_1, β_1, γ_1 and α_2, β_2, γ_2, is given by

$$\cos \theta = \cos \alpha_1 \cos \alpha_2 + \cos \beta_1 \cos \beta_2 + \cos \gamma_1 \cos \gamma_2$$

Equation of a plane is

$$Ax + By + Cz + D = 0$$

where A, B, C are proportional to the direction cosines of a normal (a line perpendicular to the plane) to the plane.

Angle between two planes is the angle between their normals.

Equation of a straight line through the point $P_1(x_1, y_1, z_1)$ is

$$\frac{x - x_1}{a} = \frac{y - y_1}{b} = \frac{z - z_1}{c}$$

where a, b, c are proportional to the direction cosines of the line. a, b, c are called *direction numbers* of the line.

104. FIGURES IN THREE DIMENSIONS

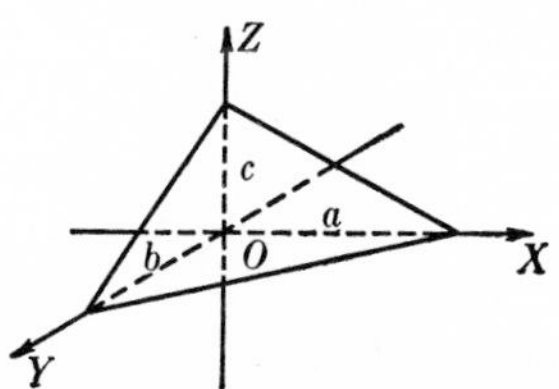

Fig. 46. **Plane.***

$$\frac{x}{a} + \frac{y}{b} + \frac{z}{c} = 1$$

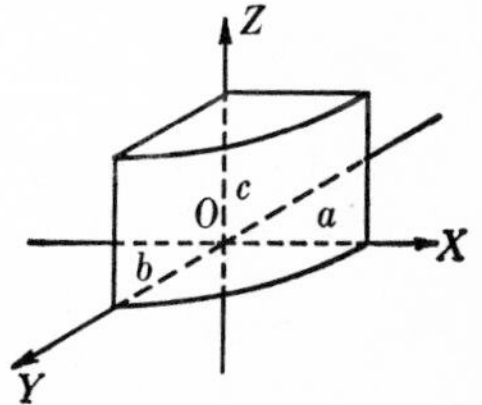

Fig. 47. **Elliptic cylinder.***

$$\frac{x^2}{a^2} + \frac{y^2}{b^2} = 1$$

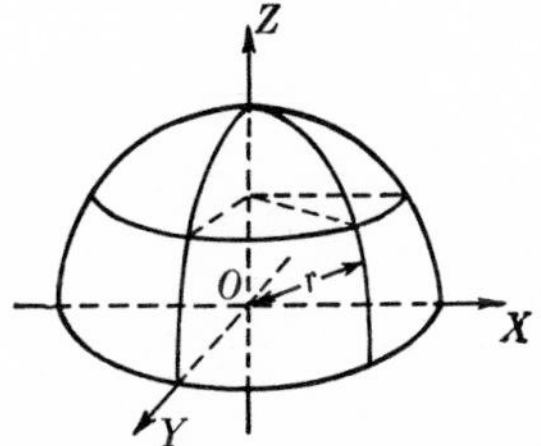

Fig. 48. **Sphere.***

$$x^2 + y^2 + z^2 = r^2$$

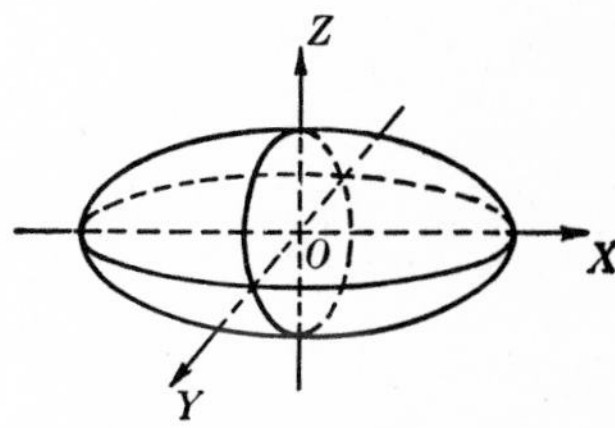

Fig. 49. **Ellipsoid.**

$$\frac{x^2}{a^2} + \frac{y^2}{b^2} + \frac{z^2}{c^2} = 1$$

* Only a portion of the figure is shown.

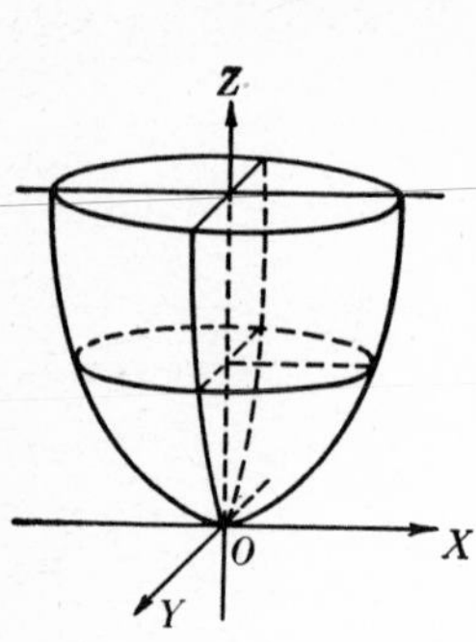

Fig. 50.
Elliptic paraboloid.

$$\frac{x^2}{a^2} + \frac{y^2}{b^2} = cz$$

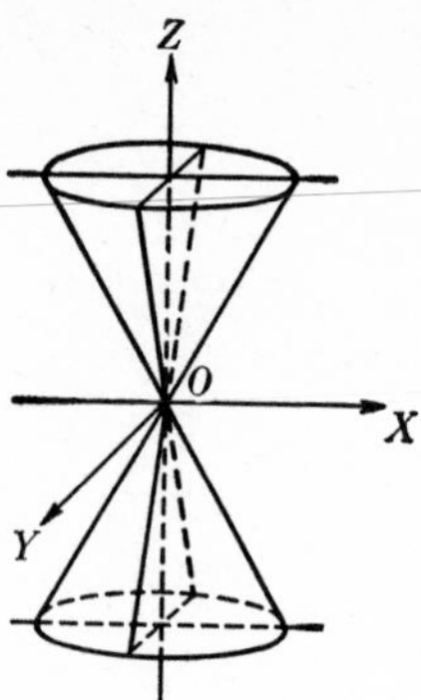

Fig. 51.
Portion of cone.

$$\frac{x^2}{a^2} + \frac{y^2}{b^2} - \frac{z^2}{c^2} = 0$$

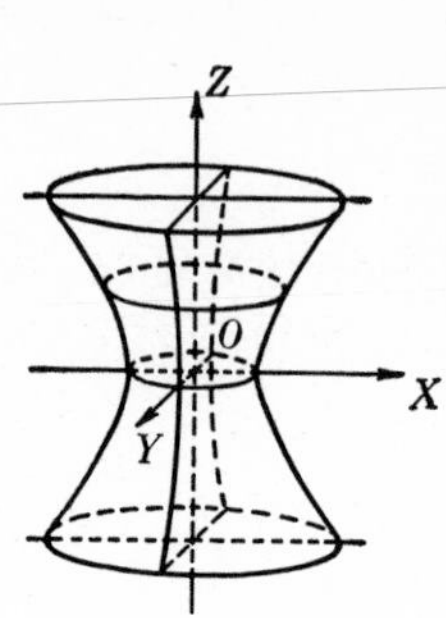

Fig. 52.
Hyperboloid of one sheet.

$$\frac{x^2}{a^2} + \frac{y^2}{b^2} - \frac{z^2}{c^2} = 1$$

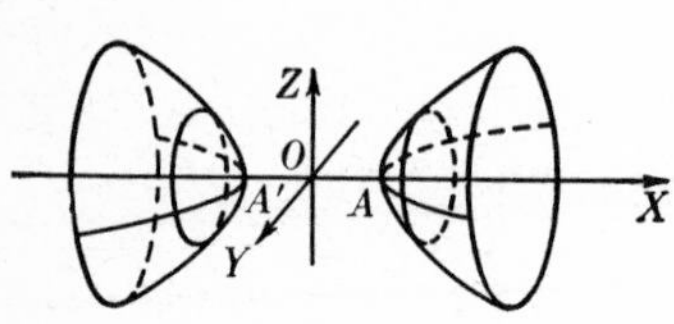

Fig. 53.
Hyperboloid of two sheets.

$$\frac{x^2}{a^2} - \frac{y^2}{b^2} - \frac{z^2}{c^2} = 1$$

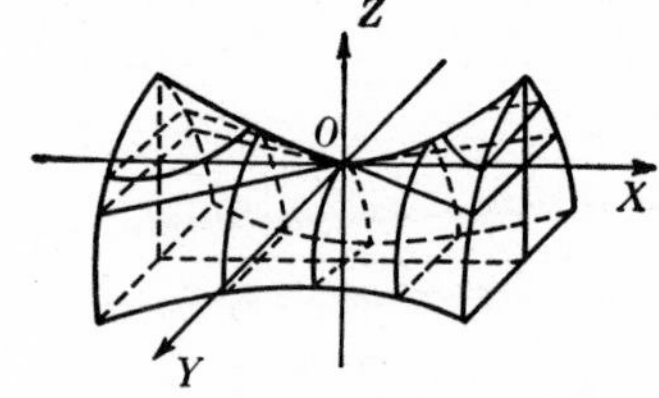

Fig. 54.
Hyperbolic paraboloid.

$$\frac{x^2}{a^2} - \frac{y^2}{b^2} = cz$$

5. DIFFERENTIAL CALCULUS

105. DEFINITIONS The symbols $f(x)$, $F(x)$, $g(x)$, $\phi(x)$, . . . are used to denote the values of the (real) functions f, F, g, ϕ, . . . , respectively, at x. $f(a)$ and $f(a + h)$ represent the values of $f(x)$ at $x = a$ and $x = a + h$, respectively.

For definitions of function, limit, and continuous function, see Chap. 9.

106. DERIVATIVE Let $f(x)$ be the value of f at x. Consider the relation $f'(a)$ defined by

$$f'(a) = \lim_{h \to 0} \frac{f(a + h) - f(a)}{h}$$

The value of $f'(a)$, if it exists, may depend on how $h \to 0$. If $f'(a)$ exists as a unique finite limit, independent of how $h \to 0$, then $f(x)$ is said to be *differentiable* at $x = a$, and $f'(a)$ is called the *derivative of f with respect to x at x = a*. The phrase *the derivative of f* is also used to mean the function f' whose value at x is the number $f'(x)$.

The derivative of $y = f(x)$ at x is written in various ways, e.g.,

$$\frac{dy}{dx} = \lim_{\Delta x \to 0} \frac{\Delta y}{\Delta x} = \lim_{\Delta x \to 0} \frac{f(x + \Delta x) - f(x)}{\Delta x} = f'(x) = D_x y = y'$$

where $\Delta x = h$ is called an *increment* in x and $\Delta y = k$ an *increment* in y.

Higher derivatives are defined as follows:

$$\frac{d^2y}{dx^2} = \frac{d}{dx}\left(\frac{dy}{dx}\right) = \frac{d}{dx} f'(x) = f''(x) \qquad \text{(2d derivative)}$$

$$\frac{d^3y}{dx^3} = \frac{d}{dx}\left(\frac{d^2y}{dx^2}\right) = \frac{d}{dx} f''(x) = f'''(x) \qquad \text{(3d derivative)}$$

$$\frac{d^ny}{dx^n} = \frac{d}{dx}\left(\frac{d^{n-1}y}{dx^{n-1}}\right) = \frac{d}{dx} f^{(n-1)}(x) = f^{(n)}(x) \qquad \text{(nth derivative)}$$

The symbol $f^{(n)}(a)$ represents the value of $f^{(n)}(x)$ when $x = a$.

107. CERTAIN RELATIONS AMONG DERIVATIVES

If $x = f(y)$, then

$$\frac{dy}{dx} = \frac{1}{dx/dy}$$

If $y = f(u)$ and $u = F(x)$, then

$$\frac{dy}{dx} = \frac{dy}{du}\frac{du}{dx}$$

If $x = f(\alpha)$ and $y = \phi(\alpha)$, then

$$\frac{dy}{dx} = \frac{\phi'(\alpha)}{f'(\alpha)} \qquad \frac{d^2y}{dx^2} = \frac{f'(\alpha)\,\phi''(\alpha) - \phi'(\alpha)\,f''(\alpha)}{[f'(\alpha)]^3}$$

108. TABLE OF DERIVATIVES

In this table, u and v represent functions of x; a, n, e represent constants ($e = 2.7183\cdots$); and all angles are measured in radians.

$$\frac{d}{dx}x = 1 \qquad \frac{d}{dx}a = 0$$

$$\frac{d}{dx}(u \pm v \pm \cdots) = \frac{du}{dx} \pm \frac{dv}{dx} \pm \cdots$$

$$\frac{d}{dx}(au) = a\frac{du}{dx} \qquad \frac{d}{dx}(uv) = u\frac{dv}{dx} + v\frac{du}{dx}$$

$$\frac{d}{dx}\left(\frac{u}{v}\right) = \frac{v\dfrac{du}{dx} - u\dfrac{dv}{dx}}{v^2} \qquad \frac{d}{dx}\sin u = \cos u\frac{du}{dx}$$

$$\frac{d}{dx}u^n = nu^{n-1}\frac{du}{dx} \qquad \frac{d}{dx}\cos u = -\sin u\frac{du}{dx}$$

$$\frac{d}{dx}\log_a u = \frac{\log_a e}{u}\frac{du}{dx} \qquad \frac{d}{dx}\tan u = \sec^2 u\frac{du}{dx}$$

$$\frac{d}{dx}\log_e u = \frac{1}{u}\frac{du}{dx} \qquad \frac{d}{dx}\cot u = -\csc^2 u\frac{du}{dx}$$

$$\frac{d}{dx}a^u = a^u\log_e a\frac{du}{dx} \qquad \frac{d}{dx}\sec u = \sec u\tan u\frac{du}{dx}$$

$$\frac{d}{dx} e^u = e^u \frac{du}{dx} \qquad \frac{d}{dx} \csc u = -\csc u \cot u \frac{du}{dx}$$

$$\frac{d}{dx} u^v = vu^{v-1} \frac{du}{dx} + u^v \log_e u \frac{dv}{dx} \qquad \frac{d}{dx} \operatorname{vers} u = \sin u \frac{du}{dx}$$

$$\lim_{x \to 0} \frac{\sin x}{x} = 1, \ \lim_{x \to 0} (1 + x)^{1/x} = e = 2.71828 \cdots$$
$$= 1 + 1 + \frac{1}{2!} + \frac{1}{3!} + \cdots$$

$$\frac{d}{dx} \sin^{-1} u = \frac{1}{\sqrt{1 - u^2}} \frac{du}{dx} \qquad \frac{-\pi}{2} \leqq \sin^{-1} u \leqq \frac{\pi}{2}$$

$$\frac{d}{dx} \cos^{-1} u = -\frac{1}{\sqrt{1 - u^2}} \frac{du}{dx} \qquad 0 \leqq \cos^{-1} u \leqq \pi$$

$$\frac{d}{dx} \tan^{-1} u = \frac{1}{1 + u^2} \frac{du}{dx} \qquad \frac{d}{dx} \cot^{-1} u = -\frac{1}{1 + u^2} \frac{du}{dx}$$

$$\frac{d}{dx} \sec^{-1} u = \frac{1}{u\sqrt{u^2 - 1}} \frac{du}{dx}, \ -\pi \leqq \sec^{-1} u < -\frac{\pi}{2},$$
$$0 \leqq \sec^{-1} u < \frac{\pi}{2}$$

$$\frac{d}{dx} \csc^{-1} u = -\frac{1}{u\sqrt{u^2 - 1}} \frac{du}{dx}, \ -\pi < \csc^{-1} u \leqq -\frac{\pi}{2},$$
$$0 < \csc^{-1} u \leqq \frac{\pi}{2}$$

$$\frac{d}{dx} \operatorname{vers}^{-1} u = \frac{1}{\sqrt{2u - u^2}} \frac{du}{dx} \qquad 0 \leqq \operatorname{vers}^{-1} u \leqq \pi$$

$$\frac{d}{dx} \sinh u = \cosh u \frac{du}{dx} \qquad \frac{d}{dx} \cosh u = \sinh u \frac{du}{dx}$$

$$\frac{d}{dx} \tanh u = \operatorname{sech}^2 u \frac{du}{dx} \qquad \frac{d}{dx} \coth u = -\operatorname{csch}^2 u \frac{du}{dx}$$

$$\frac{d}{dx} \operatorname{sech} u = -\operatorname{sech} u \tanh u \frac{du}{dx} \qquad \frac{d}{dx} \operatorname{csch} u$$
$$= -\operatorname{csch} u \coth u \frac{du}{dx}$$

$$\frac{d}{dx}\sinh^{-1}u = \frac{1}{\sqrt{u^2+1}}\frac{du}{dx} \qquad \frac{d}{dx}\cosh^{-1}u = \frac{1}{\sqrt{u^2-1}}\frac{du}{dx}{}^{*}, \quad u > 1$$

$$\frac{d}{dx}\tanh^{-1}u = \frac{1}{1-u^2}\frac{du}{dx} \qquad \frac{d}{dx}\coth^{-1}u = -\frac{1}{u^2-1}\frac{du}{dx}$$

$$\frac{d}{dx}\operatorname{sech}^{-1}u = -\frac{1}{u\sqrt{1-u^2}}\frac{du}{dx}{}^{\dagger} \qquad \frac{d}{dx}\operatorname{csch}^{-1}u = -\frac{1}{u\sqrt{u^2+1}}\frac{du}{dx}, \quad u > 0$$

109. SLOPE OF A CURVE. EQUATION OF TANGENT AND NORMAL (RECTANGULAR COORDINATES) The slope of the curve $y = f(x)$ at the point $P(x, y)$ is defined as the slope of the tangent line to the curve at P.

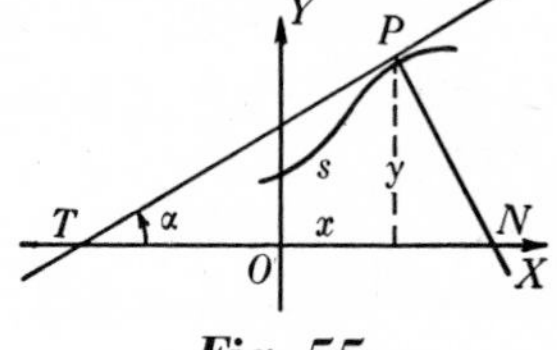

Fig. 55

$$\text{Slope} = m = \tan\alpha = \frac{dy}{dx} = f'(x)$$

$$\text{Slope at } x = x_1 \text{ is } m_1 = f'(x_1)$$

Equation of *tangent line* to curve at $P_1(x_1, y_1)$ is

$$y - y_1 = m_1(x - x_1)$$

Equation of *normal line* to curve at $P_1(x_1, y_1)$ is

$$y - y_1 = -\frac{1}{m_1}(x - x_1)$$

Angle θ of intersection of two curves whose slopes are m_1 and m_2 at the common point is given by

$$\tan\theta = \frac{m_2 - m_1}{1 + m_1 m_2}$$

The sign of $\tan\theta$ determines whether the acute or obtuse angle is meant.

* $\cosh^{-1}u > 0$. † $\operatorname{sech}^{-1}u > 0$.

110. DIFFERENTIAL Let $y = f(x)$. $f'(x_0)$ is the derivative of f with respect to x at x_0. The *differential of x*, written dx, is the increment of x, that is, $dx = \Delta x$; the *differential of f* at x_0, written df, is the product $df = f'(x_0)\,\Delta x$. The differential of f at x is written in various ways, e.g.,

$$dy = df = f'(x)\,dx = \frac{df(x)}{dx}\,dx = \frac{dy}{dx}\,dx \qquad \text{where } \frac{dy}{dx} = f'(x)$$

If $x = f(t)$, $y = \phi(t)$, then $dx = f'(t)\,dt$ and $dy = \phi'(t)\,dt$.

Every derivative formula has a corresponding differential formula. For example, from the table in Sec. 108,

$$d(\sin u) = \cos u\,du \qquad d(uv) = u\,dv + v\,du$$

111. MAXIMUM AND MINIMUM VALUES OF A FUNCTION A *maximum* (*minimum*) value of a function $f(x)$ in the interval (a,b) is a value of the function which is greater (less) than the values of the function in the immediate vicinity.

The values of x which give a maximum or minimum value to $y = f(x)$ are found by solving the equations $f'(x) = 0$ or ∞. If a is a root of $f'(x) = 0$ and if $f''(a) < 0$, $f(a)$ is a maximum; if $f''(a) > 0$, $f(a)$ is a minimum. If $f''(a) = 0$, $f'''(a) \neq 0$, $f(a)$ is neither a maximum nor a minimum, but if $f''(a) = f'''(a) = 0$, $f(a)$ is maximum or minimum according as $f^4(a) \lesseqgtr 0$.*

In general, if the first derivative which does not vanish for $x = a$ is of an odd order, $f(a)$ is neither a maximum nor a minimum; but if it is of an even order, the $2n$th, say, then $f(a)$ is a maximum or minimum according as $f^{(2n)}(a) \lesseqgtr 0$.*

To find the largest or smallest values of a function in an interval (a,b), find $f(a)$, $f(b)$ and compare with the maximum and minimum values as found in the interval.

112. POINTS OF INFLECTION OF A CURVE.* The curve is said to have a *point of inflection* at $x = a$ if $f''(a) = 0$ and $f''(x) < 0$ on one side of $x = a$ and $f''(x) > 0$ on the other side of $x = a$. Wherever $f''(x) < 0$, the curve is *concave downward*, and wherever $f''(x) > 0$, the curve is *concave upward*.

* Here it is assumed that $f'(x), f''(x), \ldots$ exist.

113. **DERIVATIVE OF ARC LENGTH. RADIUS OF CURVATURE** (see Secs. 75, 107, 126) Let s be the length of arc measured along the curve $y = f(x)$ [or in polar coordinates $r = \phi(\theta)$] from some fixed point to any point $P(x, y)$ and α be the angle of inclination of the tangent line at P with OX. Then

$$\frac{dx}{ds} = \cos\alpha = \frac{1}{\sqrt{1 + \left(\dfrac{dy}{dx}\right)^2}}, \quad \frac{dy}{ds} = \sin\alpha = \frac{1}{\sqrt{1 + \left(\dfrac{dx}{dy}\right)^2}}$$

$$ds = \sqrt{dx^2 + dy^2} = \sqrt{1 + \left(\frac{dy}{dx}\right)^2}\,dx = \sqrt{1 + \left(\frac{dx}{dy}\right)^2}\,dy$$

$$ds = \sqrt{dr^2 + r^2 d\theta^2} = \sqrt{r^2 + \left(\frac{dr}{d\theta}\right)^2}\,d\theta = \sqrt{1 + r^2\left(\frac{d\theta}{dr}\right)^2}\,dr$$

If $\qquad x = r\cos\theta \qquad$ and $\qquad y = r\sin\theta$

$$dx = \cos\theta\,dr - r\sin\theta\,d\theta, \quad dy = \sin\theta\,dr + r\cos\theta\,d\theta$$

The *radius of curvature* R at any point $P(x, y)$ of the curve $y = f(x)$ is

$$R = \frac{ds}{d\alpha} = \frac{\left[1 + \left(\dfrac{dy}{dx}\right)^2\right]^{3/2}}{\dfrac{d^2y}{dx^2}} = \frac{\{1 + [f'(x)]^2\}^{3/2}}{f''(x)}$$

$$= \frac{\left[r^2 + \left(\dfrac{dr}{d\theta}\right)^2\right]^{3/2}}{r^2 + 2\left(\dfrac{dr}{d\theta}\right)^2 - r\dfrac{d^2r}{d\theta^2}}$$

The *curvature* (K) at (x, y) is $K = 1/R$.

The *center of curvature* corresponding to the point (x_1, y_1) on $y = f(x)$ is (h, k) where

$$h = x_1 - \frac{f'(x_1)\{1 + [f'(x_1)]^2\}}{f''(x_1)}$$

$$k = y_1 + \frac{1 + [f'(x_1)]^2}{f''(x_1)}$$

114. THEOREM OF THE MEAN. ROLLE'S THEOREM If $y = f(x)$ and its derivative $f'(x)$ are continuous on the interval (a,b), there exists a value of x somewhere between a and b such that

$$f(b) = f(a) + (b - a)f'(x_1) \qquad a < x_1 < b$$

Rolle's theorem is a special case of the theorem of the mean with $f(a) = f(b) = 0$; that is, there exists at least one value of x between a and b for which $f'(x_1) = 0$.

115. EVALUATION OF INDETERMINATE FORMS If $f(x)$ and $F(x)$ are two continuous functions of x having continuous derivatives, $f'(x)$ and $F'(x)$, then

(a) If $\lim_{x\to a} f(x) = 0$, $\lim_{x\to a} F(x) = 0$, and $\lim_{x\to a} F'(x) \neq 0$ [or if $\lim_{x\to a} f(x) = \lim_{x\to a} F(x) = \infty$], then

$$\lim_{x\to a} \frac{f(x)}{F(x)} = \lim_{x\to a} \frac{f'(x)}{F'(x)}$$

(b) If $\lim_{x\to a} f(x) = 0$ and $\lim_{x\to a} F(x) = \infty$ [that is, $F(x)$ becomes infinite as x approaches a as a limit], then $\lim_{x\to a} [f(x)F(x)]$ may often be determined by writing

$$f(x)F(x) = \frac{f(x)}{1/F(x)}$$

thus expressing the functions in the form of **(a)**.

(c) If $\lim_{x\to a} f(x) = \infty$ and $\lim_{x\to a} F(x) = \infty$, then $\lim_{x\to a} [f(x) - F(x)]$ may often be determined by writing

$$f(x) - F(x) = \frac{[1/F(x)] - [1/f(x)]}{1/[f(x)F(x)]}$$

and using **(a)**.

(d) The $\lim_{x\to a} [f(x)^{F(x)}]$ may frequently be evaluated upon writing

$$f(x)^{F(x)} = e^{F(x)\log_e f(x)}$$

When one factor of the last exponent approaches zero and the other becomes infinite, the exponent is of the type considered in **(b)**.

Thus we are led to the indeterminate forms which are symbolized by $0/0$, ∞/∞, 0^0, 1^∞, ∞^0.

116. TAYLOR'S AND MACLAURIN'S THEOREMS Any function (continuous and having derivatives) may, in general, be expanded into a *Taylor's series:*

$$f(x) = f(a) + f'(a)\frac{x-a}{1!} + f''(a)\frac{(x-a)^2}{2!} + f'''(a)\frac{(x-a)^3}{3!} + \cdots + f^{(n-1)}(a)\frac{(x-a)^{n-1}}{(n-1)!} + R_n$$

where a is any quantity for which $f(a), f'(a), f''(a), \ldots$ are finite.

If the series is to be used for approximating $f(x)$ for some value of x, then a should be picked so that the difference $(x - a)$ is numerically very small, and thus only a few terms of the series need be used. The remainder, after n terms, is $R_n = f^{(n)}(x_1)(x - a)^n/n!$, where x_1 lies between a and x. R_n gives the limits of error in using n terms of the series for the approximation of the function.

$$n! = \underline{\lfloor n} = 1 \cdot 2 \cdot 3 \cdot 4 \cdots n$$

If $a = 0$, the above series is called *Maclaurin's series.*

$$f(x) = f(0) + f'(0)\frac{x}{1!} + f''(0)\frac{x^2}{2!} + f'''(0)\frac{x^3}{3!} + \cdots + f^{(n-1)}(0)\frac{x^{n-1}}{(n-1)!} + R_n$$

117. SERIES The following series may be obtained through the expansion of the functions by Taylor's or Maclaurin's theorem. The expressions following a series indicate the region of convergence of the series, that is, the values of x for which R_n approaches zero as n becomes infinite, so that an approximation of the function may be obtained by using a number of terms of the series. If the region of convergence is not indicated, the series converges for all finite values of x. ($n! = 1 \cdot 2 \cdot 3 \cdots n$). $\log_e u \equiv \log u$. (See Sec. 186.)

(a) *Binomial series*

$$(a + x)^n = a^n + na^{n-1}x + \frac{n(n-1)}{2!}a^{n-2}x^2 + \frac{n(n-1)(n-2)}{3!}a^{n-3}x^3 + \cdots \qquad x^2 < a^2$$

If n is a positive integer, the series consists of $(n + 1)$ terms; otherwise, the number of terms is infinite.

$$(a - bx)^{-1} = \frac{1}{a}\left(1 + \frac{bx}{a} + \frac{b^2x^2}{a^2} + \frac{b^3x^3}{a^3} \cdots\right) \qquad b^2x^2 < a^2$$

(b) *Exponential, logarithmic, and trigonometric series**

$$e = 1 + \frac{1}{1!} + \frac{1}{2!} + \frac{1}{3!} + \frac{1}{4!} + \cdots$$

$$e^x = 1 + x + \frac{x^2}{2!} + \frac{x^3}{3!} + \frac{x^4}{4!} + \cdots$$

$$a^x = 1 + x \log a + \frac{(x \log a)^2}{2!} + \frac{(x \log a)^3}{3!} + \cdots \qquad a > 0$$

$$e^{-x^2} = 1 - x^2 + \frac{x^4}{2!} - \frac{x^6}{3!} + \frac{x^8}{4!} - \cdots$$

$$\log x = (x - 1) - \tfrac{1}{2}(x - 1)^2 + \tfrac{1}{3}(x - 1)^3 - \cdots \qquad 0 < x \leqq 2$$

$$\log x = \frac{x - 1}{x} + \frac{1}{2}\left(\frac{x - 1}{x}\right)^2 + \frac{1}{3}\left(\frac{x - 1}{x}\right)^3 + \cdots \qquad x > \frac{1}{2}$$

$$\log x = 2\left[\frac{x - 1}{x + 1} + \frac{1}{3}\left(\frac{x - 1}{x + 1}\right)^3 + \frac{1}{5}\left(\frac{x - 1}{x + 1}\right)^5 + \cdots\right] \qquad x > 0$$

$$\log (1 + x) = x - \frac{x^2}{2} + \frac{x^3}{3} - \frac{x^4}{4} + \cdots \qquad -1 < x \leqq 1$$

$$\log (a + x) = \log a + 2\left[\frac{x}{2a + x} + \frac{1}{3}\left(\frac{x}{2a + x}\right)^3 + \frac{1}{5}\left(\frac{x}{2a + x}\right)^5 + \cdots\right] \qquad a > 0, -a < x < +\infty$$

$$\log \left(\frac{1 + x}{1 - x}\right) = 2\left(x + \frac{x^3}{3} + \frac{x^5}{5} + \frac{x^7}{7} + \cdots\right) \qquad x^2 < 1$$

$$\log \left(\frac{x + 1}{x - 1}\right) = 2\left[\frac{1}{x} + \frac{1}{3}\left(\frac{1}{x}\right)^3 + \frac{1}{5}\left(\frac{1}{x}\right)^5 + \frac{1}{7}\left(\frac{1}{x}\right)^7 + \cdots\right] \qquad x^2 > 1$$

* $\log u \equiv \log_e u$. For principal values of inverse functions, see Sec. 60.

$$\log\left(\frac{x+1}{x}\right) = 2\left[\frac{1}{2x+1} + \frac{1}{3(2x+1)^3} + \frac{1}{5(2x+1)^5} + \cdots\right] \qquad (2x+1)^2 > 1$$

$$\log\left(x + \sqrt{1+x^2}\right) = x - \frac{1}{2}\frac{x^3}{3} + \frac{1\cdot 3}{2\cdot 4}\frac{x^5}{5} - \frac{1\cdot 3\cdot 5}{2\cdot 4\cdot 6}\frac{x^7}{7} + \cdots \qquad x^2 < 1$$

$$\sin x = x - \frac{x^3}{3!} + \frac{x^5}{5!} - \frac{x^7}{7!} + \cdots$$

$$\cos x = 1 - \frac{x^2}{2!} + \frac{x^4}{4!} - \frac{x^6}{6!} + \cdots$$

$$\tan x = x + \frac{x^3}{3} + \frac{2x^5}{15} + \frac{17x^7}{315} + \frac{62x^9}{2835} + \cdots \qquad x^2 < \frac{\pi^2}{4}$$

$$\sin^{-1} x = x + \frac{x^3}{6} + \frac{1}{2}\frac{3}{4}\frac{x^5}{5} + \frac{1}{2}\frac{3}{4}\frac{5}{6}\frac{x^7}{7} + \cdots \qquad x^2 < 1$$

$$\tan^{-1} x = \begin{cases} x - \frac{1}{3}x^3 + \frac{1}{5}x^5 - \frac{1}{7}x^7 + \cdots & x^2 < 1 \\ \frac{\pi}{2} - \frac{1}{x} + \frac{1}{3x^3} - \frac{1}{5x^5} + \cdots & x > 1 \end{cases}$$

$$\log|\sin x| = \log|x| - \frac{x^2}{6} - \frac{x^4}{180} - \frac{x^6}{2835} - \cdots \qquad x^2 < \pi^2$$

$$\log \cos x = -\frac{x^2}{2} - \frac{x^4}{12} - \frac{x^6}{45} - \frac{17x^8}{2520} - \cdots \qquad x^2 < \frac{\pi^2}{4}$$

$$\log|\tan x| = \log|x| + \frac{x^2}{3} + \frac{7x^4}{90} + \frac{62x^6}{2835} + \cdots \qquad x^2 < \frac{\pi^2}{4}$$

$$e^{\sin x} = 1 + x + \frac{x^2}{2!} - \frac{3x^4}{4!} - \frac{8x^5}{5!} - \frac{3x^6}{6!} + \cdots$$

$$e^{\cos x} = e\left(1 - \frac{x^2}{2!} + \frac{4x^4}{4!} - \frac{31x^6}{6!} + \cdots\right)$$

$$e^{\tan x} = 1 + x + \frac{x^2}{2!} + \frac{3x^3}{3!} + \frac{9x^4}{4!} + \frac{37x^5}{5!} + \cdots \qquad x^2 < \frac{\pi^2}{4}$$

$$\sinh x = x + \frac{x^3}{3!} + \frac{x^5}{5!} + \frac{x^7}{7!} + \cdots$$

$$\cosh x = 1 + \frac{x^2}{2!} + \frac{x^4}{4!} + \frac{x^6}{6!} + \cdots$$

$$\tanh x = x - \frac{x^3}{3} + \frac{2x^5}{15} - \frac{17x^7}{315} + \cdots \qquad x^2 < \frac{\pi^2}{4}$$

$$\sinh^{-1} x = x - \frac{1}{2}\frac{x^3}{3} + \frac{1 \cdot 3}{2 \cdot 4}\frac{x^5}{5} - \frac{1 \cdot 3 \cdot 5}{2 \cdot 4 \cdot 6}\frac{x^7}{7} + \cdots \qquad x^2 < 1$$

$$\sinh^{-1} x = \log 2x + \frac{1}{2}\frac{1}{2x^2} - \frac{1 \cdot 3}{2 \cdot 4}\frac{1}{4x^4} + \frac{1 \cdot 3 \cdot 5}{2 \cdot 4 \cdot 6}\frac{1}{6x^6} \cdots \qquad x > 1$$

$$\cosh^{-1} x = \log 2x - \frac{1}{2}\frac{1}{2x^2} - \frac{1 \cdot 3}{2 \cdot 4}\frac{1}{4x^4} - \frac{1 \cdot 3 \cdot 5}{2 \cdot 4 \cdot 6}\frac{1}{6x^6} - \cdots \qquad x > 1$$

$$\tanh^{-1} x = x + \frac{x^3}{3} + \frac{x^5}{5} + \frac{x^7}{7} + \cdots \qquad x^2 < 1$$

118. PARTIAL DERIVATIVES. DIFFERENTIALS If $z = f(x, y)$ is a function of two variables, then the derivative of z with respect to x, treating y as a constant, is called the *first partial derivative of z with respect to x* and is denoted by $\partial z/\partial x$. Similarly, the derivative of z with respect to y, treating x as a constant, is called the *first partial derivative of z with respect to y* and is denoted by $\partial z/\partial y$.

Similarly, if $z = f(x, y, u, \ldots)$, then the first derivative of z with respect to x, treating y, u, . . . as constants, is called the first partial derivative of z with respect to x and is denoted by $\partial z/\partial x$. Likewise, the second partial derivatives are defined as indicated below:

$$\frac{\partial^2 z}{\partial x^2} = \frac{\partial}{\partial x}\frac{\partial z}{\partial x}; \quad \frac{\partial^2 z}{\partial y^2} = \frac{\partial}{\partial y}\frac{\partial z}{\partial y}; \quad \frac{\partial^2 z}{\partial x\,\partial y} = \frac{\partial}{\partial x}\frac{\partial z}{\partial y} = \frac{\partial}{\partial y}\frac{\partial z}{\partial x} = \frac{\partial^2 z}{\partial y\,\partial x}$$

If $z = f(x, y, \ldots, u)$ and $x, y, \ldots, u$ are functions of a single variable t, then

$$\frac{dz}{dt} = \frac{\partial z}{\partial x}\frac{dx}{dt} + \frac{\partial z}{\partial y}\frac{dy}{dt} + \cdots + \frac{\partial z}{\partial u}\frac{du}{dt}$$

$$dz = \frac{\partial z}{\partial x}dx + \frac{\partial z}{\partial y}dy + \cdots + \frac{\partial z}{\partial u}du$$

If $F(x, y, z, \ldots, u) = 0$, then $\dfrac{\partial F}{\partial x}dx + \dfrac{\partial F}{\partial y}dy + \cdots + \dfrac{\partial F}{\partial u}du = 0.$

119. SURFACES. SPACE CURVES (see Analytic Geometry, Secs. 97 and 98) The *tangent plane* to the surface $F(x, y, z) = 0$ at the point (x_1, y_1, z_1) on the surface is

$$(x - x_1)\left(\frac{\partial F}{\partial x}\right)_1 + (y - y_1)\left(\frac{\partial F}{\partial y}\right)_1 + (z - z_1)\left(\frac{\partial F}{\partial z}\right)_1 = 0$$

where $(\partial F/\partial x)_1$ is the value of $\partial F/\partial x$ at (x_1, y_1, z_1), etc.

The equations of the *normal* to the surface at (x_1, y_1, z_1) are

$$\frac{x - x_1}{(\partial F/\partial x)_1} = \frac{y - y_1}{(\partial F/\partial y)_1} = \frac{z - z_1}{(\partial F/\partial z)_1}$$

The *direction cosines* of the normal to the surface at the point (x_1, y_1, z_1) are proportional to

$$\left(\frac{\partial F}{\partial x}\right)_1, \quad \left(\frac{\partial F}{\partial y}\right)_1, \quad \left(\frac{\partial F}{\partial z}\right)_1$$

Given the *space* curve $x = x(t)$, $y = y(t)$, $z = z(t)$. The direction cosines of the tangent line to the curve at any point are proportional to

$$\frac{dx}{dt}, \frac{dy}{dt}, \frac{dz}{dt} \qquad \text{or to} \qquad dx, dy, dz$$

The equations of the *tangent line* to the curve at (x_1, y_1, z_1) on the curve are

$$\frac{x - x_1}{(dx/dt)_1} = \frac{y - y_1}{(dy/dt)_1} = \frac{z - z_1}{(dz/dt)_1}$$

where $(dx/dt)_1$ is the value of dx/dt at (x_1, y_1, z_1), etc.

6. INTEGRAL CALCULUS

120. DEFINITION OF INDEFINITE INTEGRAL $F(x)$ is said to be an *indefinite integral* of $f(x)$ if the derivative of $F(x)$ is $f(x)$ or if the differential of $F(x)$ is $f(x)\,dx$; symbolically:

$$F(x) = \int f(x)\,dx \quad \text{if} \quad \frac{dF(x)}{dx} = f(x) \quad \text{or} \quad dF(x) = f(x)\,dx$$

In general, $\int f(x)\,dx = F(x) + C$, where C is an arbitrary constant.

121. FUNDAMENTAL THEOREMS ON INTEGRALS. SHORT TABLE OF INTEGRALS* (u and v denote functions of x; a, b, and C denote constants.) Interpret symbols for inverse relations (Sec. 60) to be principal values.

1. $\int df(x) = f(x) + C$

2. $d\int f(x)\,dx = f(x)\,dx$

3. $\int 0\,dx = C$

4. $\int a f(x)\,dx = a\int f(x)\,dx$

5. $\int (u \pm v)\,dx = \int u\,dx \pm \int v\,dx$

6. $\int u\,dv = uv - \int v\,du$

7. $\int \frac{u\,dv}{dx}\,dx = uv - \int v\frac{du}{dx}\,dx$

8. $\int f(y)\,dx = \int \frac{f(y)\,dy}{dy/dx}$

* See Sec. 148. Where an integration gives a term of the form $A \log N$, where A is a constant and $N < 0$, use $A \log |N|$; if $N > 0$, use $A \log N$.

9. $\int u^n\,du = \dfrac{u^{n+1}}{n+1} + C \qquad n \neq -1$

10. $\int \dfrac{du}{u} = \log_e u + C,\ u > 0 \qquad \text{or, } \log_e |u| + C \qquad u \neq 0$

11. $\int e^u\,du = e^u + C$

12. $\int b^u\,du = \dfrac{b^u}{\log_e b} + C \qquad b > 0,\ b \neq 1$

13. $\int \sin u\,du = -\cos u + C$

14. $\int \cos u\,du = \sin u + C$

15. $\int \tan u\,du = \log_e \sec u + C = -\log_e \cos u + C$

16. $\int \cot u\,du = \log_e \sin u + C = -\log_e \csc u + C$

17. $\int \sec u\,du = \log_e (\sec u + \tan u) + C$

$$= \log_e \tan\left(\frac{u}{2} + \frac{\pi}{4}\right) + C$$

18. $\int \csc u\,du = \log_e (\csc u - \cot u) + C = \log_e \tan \dfrac{u}{2} + C$

19. $\int \sin^2 u\,du = \dfrac{1}{2}u - \dfrac{1}{2}\sin u \cos u + C$

20. $\int \cos^2 u\,du = \dfrac{1}{2}u + \dfrac{1}{2}\sin u \cos u + C$

21. $\int \sec^2 u\,du = \tan u + C$

22. $\int \csc^2 u\,du = -\cot u + C$

23. $\int \tan^2 u\,du = \tan u - u + C$

24. $\int \cot^2 u\,du = -\cot u - u + C$

25.* $\displaystyle\int \frac{du}{u^2 + a^2} = \frac{1}{a}\tan^{-1}\frac{u}{a} + C$

26. $\displaystyle\int \frac{du}{u^2 - a^2} = \begin{cases} \dfrac{1}{2a}\log_e \dfrac{u-a}{u+a} + C = -\dfrac{1}{a}\coth^{-1}\dfrac{u}{a} + C & u^2 > a^2 \\ \dfrac{1}{2a}\log_e \dfrac{a-u}{a+u} + C = -\dfrac{1}{a}\tanh^{-1}\dfrac{u}{a} + C & u^2 < a^2 \end{cases}$

27. $\displaystyle\int \frac{du}{\sqrt{a^2 - u^2}} = \begin{cases} \sin^{-1}\dfrac{u}{a} + C & a > 0 \\ \sin^{-1}\dfrac{u}{|a|} + C & |a| \neq 0 \end{cases}$

28. $\displaystyle\int \frac{du}{\sqrt{u^2 \pm a^2}} = \log_e (u + \sqrt{u^2 \pm a^2})\dagger + C$

29.* $\displaystyle\int \frac{du}{\sqrt{2au - u^2}} = \cos^{-1}\frac{a-u}{a} + C$

30. $\displaystyle\int \frac{du}{u\sqrt{u^2 - a^2}} = \frac{1}{a}\sec^{-1}\frac{u}{a} + C = \frac{1}{a}\cos^{-1}\frac{a}{u} + C$

31. $\displaystyle\int \frac{du}{u\sqrt{a^2 \pm u^2}} = -\frac{1}{a}\log_e \frac{a + \sqrt{a^2 \pm u^2}\dagger}{u} + C$

32. $\displaystyle\int \sqrt{a^2 - u^2}\,du = \frac{1}{2}\left(u\sqrt{a^2 - u^2} + a^2\sin^{-1}\frac{u}{a}\right) + C$

33. $\displaystyle\int \sqrt{u^2 \pm a^2}\,du = \frac{1}{2}[u\sqrt{u^2 \pm a^2}$
$\qquad\pm a^2 \log_e (u + \sqrt{u^2 \pm a^2})]\dagger + C$

34. $\displaystyle\int \sinh u\,du = \cosh u + C$

* In formulas 25 to 32 and footnote†, use $a > 0$.

† $\log_e \dfrac{u + \sqrt{u^2 + a^2}}{a} = \sinh^{-1}\dfrac{u}{a}$ $\qquad \log_e \dfrac{a + \sqrt{a^2 - u^2}}{u} = \operatorname{sech}^{-1}\dfrac{u}{a}$

$\log_e \dfrac{u + \sqrt{u^2 - a^2}}{a} = \cosh^{-1}\dfrac{u}{a}$ $\qquad \log_e \dfrac{a + \sqrt{a^2 + u^2}}{u} = \operatorname{csch}^{-1}\dfrac{u}{a}$

35. $\int \cosh u \, du = \sinh u + C$

36. $\int \tanh u \, du = \log_e (\cosh u) + C$

37. $\int \coth u \, du = \log_e (\sinh u) + C$

38. $\int \operatorname{sech} u \, du = \sin^{-1} (\tanh u) + C$

39. $\int \operatorname{csch} u \, du = \log_e \left(\tanh \frac{u}{2} \right) + C$

40. $\int \operatorname{sech} u \tanh u \, du = -\operatorname{sech} u + C$

41. $\int \operatorname{csch} u \coth u \, du = -\operatorname{csch} u + C$

Definite Integrals

122. DEFINITION OF DEFINITE INTEGRAL Let $f(x)$ be continuous for the interval from $x = a$ to $x = b$ inclusive. Divide this interval into n equal parts by the points $a, x_1, x_2, \ldots, x_{n-1}, b$ such that $\Delta x = (b - a)/n$. The definite integral of $f(x)$ with respect to x between the limits $x = a$ to $x = b$ is

$$\int_a^b f(x)\, dx = \lim_{n \to \infty} [f(a)\, \Delta x + f(x_1)\, \Delta x + f(x_2)\, \Delta x + \cdots + f(x_{n-1})$$

$$= \left[\int f(x)\, dx \right]_a^b = \left[F(x) \right]_a^b = F(b) - F(a)$$

where $F(x)$ is a function whose derivative with respect to x is $f(x)$.

123. APPROXIMATE VALUES OF DEFINITE INTEGRAL Approximate values of the above definite integral are given by the rules of Sec. 41, where $y_0, y_1, y_2, \ldots, y_{n-1}, y_n$ are the values of $f(x)$ for $x = a, x_1, x_2, \ldots, x_{n-1}, b$, respectively, and $h = (b - a)/n$.

124. SOME FUNDAMENTAL THEOREMS

$$\int_a^b [f_1(x) + f_2(x) + \cdots + f_n(x)]\, dx = \int_a^b f_1(x)\, dx + \int_a^b f_2(x)\, dx + \cdots + \int_a^b f_n(x)\, dx$$

$$\int_a^b kf(x)\, dx = k\int_a^b f(x)\, dx \qquad \text{if } k \text{ is a constant}$$

$$\int_a^b f(x)\, dx = -\int_b^a f(x)\, dx$$

$$\int_a^b f(x)\, dx = \int_a^c f(x)\, dx + \int_c^b f(x)\, dx$$

$$\int_a^b f(x)\, dx = (b - a)\, f(x_1) \qquad \text{where } x_1 \text{ lies between } a \text{ and } b$$

$$\int_a^{\infty} f(x)\, dx = \lim_{t\to\infty} \int_a^t f(x)\, dx$$

$$\int_{-\infty}^b f(x)\, dx = \lim_{t\to\infty} \int_{-t}^b f(x)\, dx$$

$$\int_{-\infty}^{+\infty} f(x)\, dx = \int_{-\infty}^c f(x)\, dx + \int_c^{\infty} f(x)\, dx$$

If $f(x)$ has a singular point* at $x = b$, $b \neq a$,

$$\int_a^b f(x)\, dx = \lim_{e\to 0} \int_a^{b-e} f(x)\, dx$$

The mean value of the function $f(x)$ on the interval (a,b) is

$$\frac{1}{b-a}\int_a^b f(x)\, dx$$

Some Applications of the Definite Integral

125. PLANE AREA The area bounded by $y = f(x)$, $y = 0$, $x = a$, $x = b$, where y has the same sign for all values of x between a and b, is

$$A = \int_a^b f(x)\, dx \qquad dA = f(x)\, dx$$

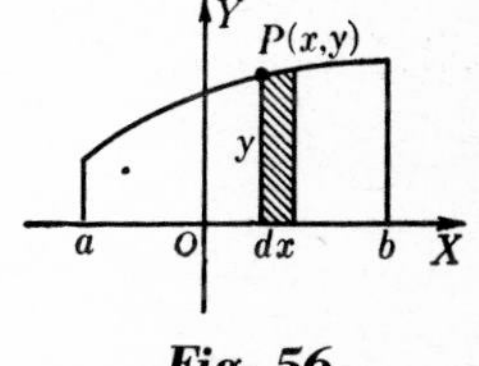

Fig. 56

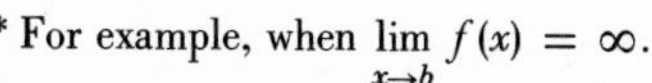

* For example, when $\lim_{x\to b} f(x) = \infty$.

The area bounded by the curve $r = f(\theta)$ and the two radii $\theta = \alpha$ and $\theta = \beta$ is

$$A = \frac{1}{2}\int_{\alpha}^{\beta} [f(\theta)]^2 \, d\theta \qquad dA = \frac{1}{2} r^2 \, d\theta$$

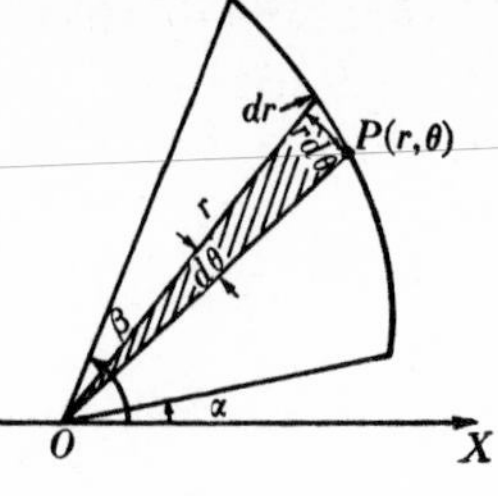

Fig. 57

126. LENGTH OF ARC (see Sec. 113) The length s of arc of a plane curve $f(x,y) = 0$ from the point (a,c) to the point (b,d) is

$$s = \int_a^b \sqrt{1 + \left(\frac{dy}{dx}\right)^2} \, dx = \int_c^d \sqrt{1 + \left(\frac{dx}{dy}\right)^2} \, dy$$

If the equation of the curve is $x = f(t)$, $y = f(t)$, the length of arc from $t = a$ to $t = b$ is

$$s = \int_a^b \sqrt{\left(\frac{dx}{dt}\right)^2 + \left(\frac{dy}{dt}\right)^2} \, dt$$

If the equation of the curve is $r = f(\theta)$, then

$$s = \int_{\theta_1}^{\theta_2} \sqrt{r^2 + \left(\frac{dr}{d\theta}\right)^2} \, d\theta = \int_{r_1}^{r_2} \sqrt{r^2\left(\frac{d\theta}{dr}\right)^2 + 1} \, dr$$

127. VOLUME BY PARALLEL SECTIONS If the plane perpendicular to the x axis at $(x,0,0)$ cuts from a given solid a section whose area is $A(x)$, then the volume of that part of the solid between $x = a$ and $x = b$ is

$$\int_a^b A(x) \, dx$$

128. VOLUME OF REVOLUTION The volume of a solid of revolution generated by revolving that portion of the curve $y = f(x)$ between $x = a$ and $x = b$

(a) about the x axis is $\pi \int_a^b y^2 \, dx$

(b) about the y axis is $\pi \int_c^d x^2 \, dy$, where c and d are the values of y corresponding to the values a and b of x

129. **AREA OF SURFACE OF REVOLUTION** The area of the surface of a solid of revolution generated by revolving the curve $y = f(x)$ between $x = a$ and $x = b$

(a) about the x axis is $2\pi \int_a^b y \sqrt{1 + \left(\frac{dy}{dx}\right)^2}\, dx$

(b) about the y axis is $2\pi \int_c^d x \sqrt{1 + \left(\frac{dx}{dy}\right)^2}\, dy$

130. **PLANE AREAS BY DOUBLE INTEGRATION**

(a) Rectangular coordinates

$$A = \int_a^b \qquad dy\, dx \qquad \text{or} \qquad \int_c^d \int_{\xi(y)}^{\Psi(y)} dx\, dy$$

(b) Polar coordinates

$$A = \int_{\theta_1}^{\theta_2} \int_{f_1(\theta)}^{f_2(\theta)} r\, dr\, d\theta \qquad \text{or} \qquad \int_{r_1}^{r_2} \int_{\phi_1(r)}^{\phi_2(r)} r\, d\theta\, dr$$

131. **VOLUMES BY DOUBLE INTEGRATION** If $z = f(x, y)$,

$$V = \int_a^b \int_{\phi(x)}^{\Psi(x)} f(x, y)\, dy\, dx \qquad \text{or} \qquad \int_c^d \int_{\alpha(y)}^{\beta(y)} f(x, y)\, dx\, dy$$

132. **VOLUMES BY TRIPLE INTEGRATION** (see Sec. 102)

(a) Rectangular coordinates

$$V = \iiint dx\, dy\, dz$$

(b) Cylindrical coordinates

$$V = \iiint r\, dr\, d\theta\, dz$$

(c) Spherical coordinates

$$V = \iiint \rho^2 \sin\phi\, d\theta\, d\phi\, d\rho$$

where the limits of integration must be supplied. Other formulas may be obtained by changing the order of integration.

Area of surface $z = f(x, y)$

$$A = \iint \sqrt{\left(\frac{\partial z}{\partial x}\right)^2 + \left(\frac{\partial z}{\partial y}\right)^2 + 1}\, dy\, dx$$

where the limits of integration must be supplied.

133. MASS* The mass of a body of density δ is

$$m = \int dm \qquad dm = \delta\, dA, \quad \text{or} \quad \delta\, ds, \quad \text{or} \quad \delta\, dV, \quad \text{or} \quad \delta\, dS$$

where dA, ds, dV, dS are, respectively, the elements of area, length, volume, and surface of Secs. 125 to 132.

134. DENSITY* If δ is a variable (or constant) density (mass per unit of element) and $\bar{\delta}$ is the mean density of a solid of volume V, then

$$\bar{\delta} = \frac{\int \delta\, dV}{\int dV}$$

135. MOMENT* The moments M_{yz}, M_{xz}, M_{xy} of a mass m with respect to the coordinate planes (as indicated by the subscripts) are

$$M_{yz} = \int x\, dm, \; M_{xz} = \int y\, dm, \; M_{xy} = \int z\, dm$$

136. CENTROID OF MASS OR CENTER OF GRAVITY* The coordinates $(\bar{x}, \bar{y}, \bar{z})$ of the centroid of a mass m are

$$\bar{x} = \frac{\int x\, dm}{\int dm}, \; \bar{y} = \frac{\int y\, dm}{\int dm}, \; \bar{z} = \frac{\int z\, dm}{\int dm}$$

Note: In the above equations x, y, z are the coordinates of the center of gravity of the element dm.

* The limits of integration are to be supplied.

137. CENTROID OF SEVERAL MASSES The x coordinate ($\bar{x}$) of the centroid of several masses $m_1, m_2, \ldots, m_n$, having $\bar{x}_1, \bar{x}_2, \ldots, \bar{x}_n$, respectively, as the x coordinates of their centroids, is

$$\bar{x} = \frac{m_1\bar{x}_1 + m_2\bar{x}_2 + \cdots + m_n\bar{x}_n}{m_1 + m_2 + \cdots + m_n}$$

Similar formulas hold for the other coordinates $\bar{y}$, $\bar{z}$.

138. MOMENT OF INERTIA (SECOND MOMENT)* The moments of inertia (I):

(*a*) for a plane curve about the x axis, y axis, and origin, respectively, are

$$I_x = \int y^2\,ds, \quad I_y = \int x^2\,ds, \quad I_0 = \int (x^2 + y^2)\,ds$$

(*b*) for a plane area about the x axis, y axis, and origin, respectively, are

$$I_x = \int y^2\,dA, \quad I_y = \int x^2\,dA, \quad I_0 = \int (x^2 + y^2)\,dA$$

(*c*) for a solid of mass m about the yz, xz, xy planes, x axis, etc., respectively, are

$$I_{yz} = \int x^2\,dm, \; I_{xz} = \int y^2\,dm, \; I_{xy} = \int z^2\,dm, \; I_x = I_{xz} + I_{xy}, \text{ etc.}$$

138*a*. THEOREM OF PARALLEL AXES Let L be any line in space and L_g a line parallel to L, passing through the centroid of the body of mass m. If d is the distance between the lines L and L_g, then

$$I_L = I_{Lg} + d^2m$$

where I_L and I_{Lg} are the moments of inertia of the body about the lines L and L_g, respectively.

139. RADIUS OF GYRATION If I is the moment of inertia of a mass m and K is the radius of gyration, $I = mK^2$. A similar relation holds for areas, lengths, volumes, etc.

* The limits of integration are to be supplied.

If masses (or areas, etc.) $m_1, m_2, \ldots, m_n$ have, respectively, the radii of gyration $k_1, k_2, \ldots, k_n$, with respect to a line or plane, then with respect to this line or plane, the several masses taken together have the radius of gyration K, where

$$K^2 = \frac{m_1k_1^2 + m_2k_2^2 + \cdots + m_nk_n^2}{m_1 + m_2 + \cdots + m_n}$$

140. WORK The work W done in moving a particle from $s = a$ to $s = b$ by a force whose component expressed as a function of s in the direction of motion is F_s is

$$W = \int_{s=a}^{s=b} F_s\,ds \qquad dW = F_s\,ds$$

141. PRESSURE The pressure (p) against an area vertical to the surface of a liquid and between the depths a and b is

$$P = \int_{y=a}^{y=b} wly\,dy \qquad dp = wly\,dy$$

where w is the weight of liquid per unit volume, y is the depth beneath the surface of liquid of a horizontal element of area, and l is the length of the horizontal element of area expressed in terms of y.

142. CENTER OF PRESSURE The depth $\bar{y}$ of the center of pressure against an area vertical to the surface of the liquid and between the depths a and b is

$$\bar{y} = \frac{\int_{y=a}^{y=b} y\,dp}{\int_{y=a}^{y=b} dp} \qquad \text{(for } dp \text{ see Sec. 141)}$$

7. TABLE OF INTEGRALS

Certain Elementary Processes

143. To integrate $\int R(x)\,dx$, where $R(x)$ is rational function of x Write $R(x)$ in the form of a fraction whose terms are polynomials. If the fraction is improper (i.e., the degree of the denominator is less than or equal to the degree of the numerator), divide the numerator by the denominator and thus write $R(x)$ as the sum of a quotient $Q(x)$ and a proper fraction $P(x)$. (In a proper fraction the degree of the numerator is less than the degree of the denominator.) The polynomial $Q(x)$ is readily integrated. To integrate $P(x)$, separate it into a sum of partial fractions (as indicated below) and integrate each term of the sum separately.

To separate the proper fraction $P(x)$ into partial fractions, write $P(x) = f(x)/\psi(x)$, where $f(x)$ and $\psi(x)$ are polynomials,

$$\psi(x) = (x - a)^p(x - b)^q(x - c)^r \cdots$$

and the constants a, b, c are all different. By algebra, there exist constants $A_1, A_2, \ldots, B_1, B_2, \ldots$ such that

$$\frac{f(x)}{\psi(x)} = \frac{A_1}{x - a} + \frac{A_2}{(x - a)^2} + \cdots + \frac{A_p}{(x - a)^p} + \frac{B_1}{x - b} + \frac{B_2}{(x - b)^2} + \cdots + \frac{B_q}{(x - b)^q} + \cdots$$

The separate terms of this sum may be integrated by the formulas

$$\int \frac{dx}{(x - \alpha)^t} = \frac{-1}{(t - 1)(x - \alpha)^{t-1}}, \; t > 1, \quad \int \frac{dx}{x - \alpha} = \log (x - \alpha)$$

If $f(x)$ and $\psi(x)$ have real coefficients and $\psi(x) = 0$ has imaginary roots, the above method leads to imaginary quantities. To avoid this, separate $P(x)$ in a different way. As before, corresponding to each p-fold real root a of $\psi(x)$, use the sum

$$\frac{A_1}{x - a} + \frac{A_2}{(x - a)^2} + \cdots + \frac{A_p}{(x - a)^p}$$

To each λ-fold real quadratic factor $x^2 + \alpha x + \beta$ of $\psi(x)$ which does not factor into real linear factors, use, instead of the two sets of terms occurring in the first expansion of $R(x)$ and dependent on the conjugate complex roots of $x^2 + \alpha x + \beta = 0$, sums of the forms

$$\frac{D_1 x + E_1}{x^2 + \alpha x + \beta} + \frac{D_2 x + E_2}{(x^2 + \alpha x + \beta)^2} + \cdots + \frac{D_\lambda x + E_\lambda}{(x^2 + \alpha x + \beta)^\lambda}$$

where the quantities $D_1, D_2, \ldots, E_1, E_2, \ldots$ are real constants.

These new sums may be separately integrated by means of integral formulas 149 to 156, etc.

144. TO INTEGRATE AN IRRATIONAL ALGEBRAIC FUNCTION If no convenient method of integration is apparent, the integration may frequently be performed by means of a change of variable. For example, if R is a rational function of two arguments and n is an integer, then to integrate

(a) $\int R[x, (ax + b)^{1/n}]\, dx$, let $(ax + b) = y^n$, whence the integral reduces to $\int P(y)\, dy$, where $P(y)$ is a rational function of y and may be integrated as in Sec. 143;

(b) $\int R[x, (x^2 + bx + c)^{1/2}]\, dx$, let $(x^2 + bx + c)^{1/2} = z - x$, reduce R to a rational function of z, and proceed as in Sec. 143;

(c) $\int R(\sin x, \cos x)\, dx$, let $\tan \dfrac{x}{2} = t$, whence

$$\sin x = \frac{2t}{1 + t^2} \qquad \cos x = \frac{1 - t^2}{1 + t^2} \qquad dx = \frac{2\, dt}{1 + t^2}$$

reduce R to a rational function of t and proceed as in Sec. 143.

145. TO INTEGRATE EXPRESSIONS CONTAINING

$$\sqrt{a^2 - x^2} \qquad \sqrt{x^2 \pm a^2}$$

Expressions containing these radicals can frequently be integrated after making the following transformations:

(a) if $\sqrt{a^2 - x^2}$ occurs, let $x = a \sin t$
(b) if $\sqrt{x^2 - a^2}$ occurs, let $x = a \sec t$
(c) if $\sqrt{x^2 + a^2}$ occurs, let $x = a \tan t$

146. TRIGONOMETRIC EXPRESSIONS The identities of Sec. 58 frequently facilitate the integration of such functions.

147. INTEGRATION BY PARTS The formula $\int u\,dv = uv - \int v\,du$ is often effective in reducing an integral to simpler integrals.

148. TABLE OF INDEFINITE INTEGRALS (see Sec. 121) In this table the constant C of integration is omitted, but C *should be added to the result of every integration.* The letter x represents any variable; u, any function of x; a, b, . . . arbitrary constants; and $\log_e u \equiv \log u$, unless otherwise stated. All angles are in radians. In general, interpret symbols for inverse relations (Sec. 60) as principal values (for example, $\sin^{-1} x$ means $\mathrm{Sin}^{-1} x$); interpret $\sqrt{\ }$ as the positive value (for example, $t = \sqrt{x^2 - a^2}$ means the positive value of t). Where an integration gives a constant A times the logarithm of a negative number N, say, $A \log N$, use $A \log |N|$. [For example, if $X = ax + b$, $\int (1/X)\,dx = (1/a) \log_e X + C$, which is used if $X > 0$. If $X < 0$, use $\int (1/X)\,dx = (1/a) \log_e |X| + C$, since $\log(-1) \equiv (2k+1)\pi i$, k an integer, may be considered a part of the constant of integration.]

Expressions Containing $(ax + b)$

42. $\displaystyle\int (ax+b)^n\,dx = \frac{1}{a(n+1)}(ax+b)^{n+1} \qquad n \neq -1$

43. $\displaystyle\int \frac{dx}{ax+b} = \frac{1}{a} \log_e (ax+b)$

44. $\displaystyle\int \frac{dx}{(ax+b)^2} = -\frac{1}{a(ax+b)}$

45. $\displaystyle\int \frac{dx}{(ax+b)^3} = -\frac{1}{2a(ax+b)^2}$

46. $$\int x(ax+b)^n\,dx = \frac{1}{a^2(n+2)}(ax+b)^{n+2} - \frac{b}{a^2(n+1)}(ax+b)^{n+1} \qquad n \neq -1, -2$$

47. $$\int \frac{x\,dx}{ax+b} = \frac{x}{a} - \frac{b}{a^2}\log(ax+b)$$

48. $$\int \frac{x\,dx}{(ax+b)^2} = \frac{b}{a^2(ax+b)} + \frac{1}{a^2}\log(ax+b)$$

49. $$\int \frac{x\,dx}{(ax+b)^3} = \frac{b}{2a^2(ax+b)^2} - \frac{1}{a^2(ax+b)}$$

50. $$\int x^2(ax+b)^n\,dx$$
$$= \frac{1}{a^3}\left[\frac{(ax+b)^{n+3}}{n+3} - 2b\frac{(ax+b)^{n+2}}{n+2} + b^2\frac{(ax+b)^{n+1}}{n+1}\right] \qquad n \neq -1, -2, -3$$

51. $$\int \frac{x^2\,dx}{ax+b}$$
$$= \frac{1}{a^3}\left[\frac{1}{2}(ax+b)^2 - 2b(ax+b) + b^2\log(ax+b)\right]$$

52. $$\int \frac{x^2\,dx}{(ax+b)^2} = \frac{1}{a^3}\left[ax+b-2b\log(ax+b) - \frac{b^2}{ax+b}\right]$$

53. $$\int \frac{x^2\,dx}{(ax+b)^3} = \frac{1}{a^3}\left[\log(ax+b) + \frac{2b}{ax+b} - \frac{b^2}{2(ax+b)^2}\right]$$

54. $$\int x^m(ax+b)^n\,dx$$
$$= \frac{1}{a(m+n+1)}\left[x^m(ax+b)^{n+1} - mb\int x^{m-1}(ax+b)^n\,dx\right]$$
$$= \frac{1}{m+n+1}\left[x^{m+1}(ax+b)^n + nb\int x^m(ax+b)^{n-1}\,dx\right] \qquad m>0,\ m+n+1\neq 0$$

55. $\int \frac{dx}{x(ax + b)} = \frac{1}{b} \log \frac{x}{ax + b}$

56. $\int \frac{dx}{x^2(ax + b)} = -\frac{1}{bx} + \frac{a}{b^2} \log \frac{ax + b}{x}$

57. $\int \frac{dx}{x^3(ax + b)} = \frac{2ax - b}{2b^2x^2} + \frac{a^2}{b^3} \log \frac{x}{ax + b}$

58. $\int \frac{dx}{x(ax + b)^2} = \frac{1}{b(ax + b)} - \frac{1}{b^2} \log \frac{ax + b}{x}$

59. $\int \frac{dx}{x(ax + b)^3} = \frac{1}{b^3}\left[\frac{1}{2}\left(\frac{ax + 2b}{ax + b}\right)^2 + \log \frac{x}{ax + b}\right]$

60. $\int \frac{dx}{x^2(ax + b)^2} = -\frac{b + 2ax}{b^2x(ax + b)} + \frac{2a}{b^3} \log \frac{ax + b}{x}$

61. $\int \sqrt{ax + b}\, dx = \frac{2}{3a}\sqrt{(ax + b)^3}$

62. $\int x\sqrt{ax + b}\, dx = \frac{2(3ax - 2b)}{15a^2}\sqrt{(ax + b)^3}$

63. $\int x^2\sqrt{ax + b}\, dx = \frac{2(15a^2x^2 - 12abx + 8b^2)\sqrt{(ax + b)^3}}{105a^3}$

64. $\int x^3\sqrt{ax + b}\, dx$

$$= \frac{2(35a^3x^3 - 30a^2bx^2 + 24ab^2x - 16b^3)\sqrt{(ax + b)^3}}{315a^4}$$

65. $\int x^n\sqrt{ax + b}\, dx = \frac{2}{a^{n+1}}\int u^2(u^2 - b)^n\, du \qquad u = \sqrt{ax + b}$

66. $\int \frac{\sqrt{ax + b}}{x}\, dx = 2\sqrt{ax + b} + b\int \frac{dx}{x\sqrt{ax + b}}$

67. $\int \frac{dx}{\sqrt{ax + b}} = \frac{2\sqrt{ax + b}}{a}$

68. $$\int \frac{x\,dx}{\sqrt{ax+b}} = \frac{2(ax-2b)}{3a^2}\sqrt{ax+b}$$

69. $$\int \frac{x^2\,dx}{\sqrt{ax+b}} = \frac{2(3a^2x^2-4abx+8b^2)}{15a^3}\sqrt{ax+b}$$

70. $$\int \frac{x^3\,dx}{\sqrt{ax+b}}$$
$$= \frac{2(5a^3x^3-6a^2bx^2+8ab^2x-16b^3)}{35a^4}\sqrt{ax+b}$$

71. $$\int \frac{x^n\,dx}{\sqrt{ax+b}} = \frac{2}{a^{n+1}}\int (u^2-b)^n\,du \qquad u=\sqrt{ax+b}$$

72. $$\int \frac{dx}{x\sqrt{ax+b}} = \frac{1}{\sqrt{b}}\log\frac{\sqrt{ax+b}-\sqrt{b}}{\sqrt{ax+b}+\sqrt{b}} \qquad b>0$$

73. $$\int \frac{dx}{x\sqrt{ax+b}} = \begin{cases} \dfrac{2}{\sqrt{-b}}\tan^{-1}\sqrt{\dfrac{ax+b}{-b}} & b<0 \\[2ex] \dfrac{-2}{\sqrt{b}}\tanh^{-1}\sqrt{\dfrac{ax+b}{b}} & b>0 \end{cases}$$

74. $$\int \frac{dx}{x^2\sqrt{ax+b}} = -\frac{\sqrt{ax+b}}{bx} - \frac{a}{2b}\int\frac{dx}{x\sqrt{ax+b}}$$

75. $$\int \frac{dx}{x^3\sqrt{ax+b}}$$
$$= -\frac{\sqrt{ax+b}}{2bx^2} + \frac{3a\sqrt{ax+b}}{4b^2x} + \frac{3a^2}{8b^2}\int\frac{dx}{x\sqrt{ax+b}}$$

76. $$\int \frac{dx}{x^n(ax+b)^m} = -\frac{1}{b^{m+n-1}}\int\frac{(u-a)^{m+n-2}\,du}{u^m}$$
$$u=\frac{ax+b}{x}$$

77. $$\int (ax+b)^{\pm(n/2)}\,dx = \frac{2(ax+b)^{(2\pm n)/2}}{a(2\pm n)}$$

78. $\int x(ax+b)^{\pm(n/2)}\,dx$

$$= \frac{2}{a^2}\left[\frac{(ax+b)^{(4\pm n)/2}}{4\pm n} - \frac{b(ax+b)^{(2\pm n)/2}}{2\pm n}\right]$$

79. $\int \frac{dx}{x(ax+b)^{n/2}} = \frac{1}{b}\int \frac{dx}{x(ax+b)^{(n-2)/2}} - \frac{a}{b}\int \frac{dx}{(ax+b)^{n/2}}$

80. $\int \frac{x^m\,dx}{\sqrt{ax+b}} = \frac{2x^m\sqrt{ax+b}}{(2m+1)a} - \frac{2mb}{(2m+1)a}\int \frac{x^{m-1}\,dx}{\sqrt{ax+b}}$

81. $\int \frac{dx}{x^n\sqrt{ax+b}}$

$$= \frac{-\sqrt{ax+b}}{(n-1)bx^{n-1}} - \frac{(2n-3)a}{(2n-2)b}\int \frac{dx}{x^{n-1}\sqrt{ax+b}}$$

82. $\int \frac{(ax+b)^{n/2}}{x}\,dx = a\int (ax+b)^{(n-2)/2}\,dx + b\int \frac{(ax+b)^{(n-2)/2}}{x}\,dx$

83. $\int \frac{dx}{(ax+b)(cx+d)} = \frac{1}{bc-ad}\log\frac{cx+d}{ax+b} \qquad bc-ad\neq 0$

84. $\int \frac{dx}{(ax+b)^2(cx+d)}$

$$= \frac{1}{bc-ad}\left(\frac{1}{ax+b} + \frac{c}{bc-ad}\log\frac{cx+d}{ax+b}\right) \qquad bc-ad\neq 0$$

85. $\int (ax+b)^n(cx+d)^m\,dx = \frac{1}{(m+n+1)a}\Big[(ax+b)^{n+1}(cx+d)^m$

$$- m(bc-ad)\int (ax+b)^n(cx+d)^{m-1}\,dx\Big]$$

86. $\int \frac{dx}{(ax+b)^n(cx+d)^m} = \frac{-1}{(m-1)(bc-ad)}\Big[\frac{1}{(ax+b)^{n-1}(cx+d)^{m-1}}$

$$+ a(m+n-2)\int \frac{dx}{(ax+b)^n(cx+d)^{m-1}}\Big], \quad m>1,\ n>0,\ bc-ad\neq 0$$

87. $$\int \frac{(ax+b)^n}{(cx+d)^m}\,dx = -\frac{1}{(m-1)(bc-ad)}\left[\frac{(ax+b)^{n+1}}{(cx+d)^{m-1}} + (m-n-2)a\int\frac{(ax+b)^n\,dx}{(cx+d)^{m-1}}\right] = \frac{-1}{(m-n-1)c}\left[\frac{(ax+b)^n}{(cx+d)^{m-1}} + n(bc-ad)\int\frac{(ax+b)^{n-1}}{(cx+d)^m}\,dx\right]$$

88. $$\int \frac{x\,dx}{(ax+b)(cx+d)} = \frac{1}{bc-ad}\left[\frac{b}{a}\log(ax+b) - \frac{d}{c}\log(cx+d)\right] \qquad bc-ad \neq 0$$

89. $$\int \frac{x\,dx}{(ax+b)^2(cx+d)} = \frac{1}{bc-ad}\left[-\frac{b}{a(ax+b)} - \frac{d}{bc-ad}\log\frac{cx+d}{ax+b}\right] \qquad bc-ad \neq 0$$

90. $$\int \frac{cx+d}{\sqrt{ax+b}}\,dx = \frac{2}{3a^2}(3ad-2bc+acx)\sqrt{ax+b}$$

91. $$\int \frac{\sqrt{ax+b}}{cx+d}\,dx = \frac{2\sqrt{ax+b}}{c} - \frac{2}{c}\sqrt{\frac{ad-bc}{c}}\tan^{-1}\sqrt{\frac{c(ax+b)}{ad-bc}} \qquad c>0,\ ad>bc$$

92. $$\int \frac{\sqrt{ax+b}}{cx+d}\,dx = \frac{2\sqrt{ax+b}}{c} + \frac{1}{c}\sqrt{\frac{bc-ad}{c}}\log\frac{\sqrt{c(ax+b)}-\sqrt{bc-ad}}{\sqrt{c(ax+b)}+\sqrt{bc-ad}} \qquad c>0,\ bc>ad$$

93. $$\int \frac{dx}{(cx+d)\sqrt{ax+b}} = \frac{2}{\sqrt{c}\sqrt{ad-bc}}\tan^{-1}\sqrt{\frac{c(ax+b)}{ad-bc}} \qquad c>0,\ ad>bc$$

94. $\int \frac{dx}{(cx + d)\sqrt{ax + b}}$

$$= \frac{1}{\sqrt{c}\sqrt{bc - ad}} \log \frac{\sqrt{c(ax + b)} - \sqrt{bc - ad}}{\sqrt{c(ax + b)} + \sqrt{(bc - ad}}, c > 0, bc > ad$$

Expressions Containing $ax^2 + c$, $ax^n + c$, $x^2 \pm p^2$, and $p^2 - x^2$

95. $\int \frac{dx}{p^2 + x^2} = \frac{1}{p} \tan^{-1} \frac{x}{p} \quad \text{or} \quad -\frac{1}{p} \cot^{-1} \frac{x}{p}$

96. $\int \frac{dx}{p^2 - x^2} = \frac{1}{2p} \log \frac{p + x}{p - x} \quad \text{or} \quad \frac{1}{p} \tanh^{-1} \frac{x}{p}$

97. $\int \frac{dx}{ax^2 + c} = \frac{1}{\sqrt{ac}} \tan^{-1} \left(x \sqrt{\frac{a}{c}} \right) \qquad a, c > 0$

98. $$\int \frac{dx}{ax^2 + c} = \begin{cases} \dfrac{1}{2\sqrt{-ac}} \log \dfrac{x\sqrt{a} - \sqrt{-c}}{x\sqrt{a} + \sqrt{-c}} & a > 0, c < 0 \\ \dfrac{1}{2\sqrt{-ac}} \log \dfrac{\sqrt{c} + x\sqrt{-a}}{\sqrt{c} - x\sqrt{-a}} & a < 0, c > 0 \end{cases}$$

99. $$\int \frac{dx}{(ax^2 + c)^n} = \frac{1}{2(n - 1)c} \frac{x}{(ax^2 + c)^{n-1}} + \frac{2n - 3}{2(n - 1)c} \int \frac{dx}{(ax^2 + c)^{n-1}} \qquad n > 1$$

100. $\int x(ax^2 + c)^n \, dx = \frac{1}{2a} \frac{(ax^2 + c)^{n+1}}{n + 1} \qquad n \neq -1$

101. $\int \frac{x}{ax^2 + c} \, dx = \frac{1}{2a} \log (ax^2 + c)$

102. $\int \frac{dx}{x(ax^2 + c)} = \frac{1}{2c} \log \frac{x^2}{ax^2 + c}$

103. $\int \frac{dx}{x^2(ax^2 + c)} = -\frac{1}{cx} - \frac{a}{c} \int \frac{dx}{ax^2 + c}$

104. $\int \frac{x^2 \, dx}{ax^2 + c} = \frac{x}{a} - \frac{c}{a} \int \frac{dx}{ax^2 + c}$

105. $$\int \frac{x^n\,dx}{ax^2+c} = \frac{x^{n-1}}{a(n-1)} - \frac{c}{a}\int \frac{x^{n-2}\,dx}{ax^2+c} \qquad n \neq 1$$

106. $$\int \frac{x^2\,dx}{(ax^2+c)^n} = -\frac{1}{2(n-1)a}\frac{x}{(ax^2+c)^{n-1}} + \frac{1}{2(n-1)a}\int \frac{dx}{(ax^2+c)^{n-1}}$$

107. $$\int \frac{dx}{x^2(ax^2+c)^n} = \frac{1}{c}\int \frac{dx}{x^2(ax^2+c)^{n-1}} - \frac{a}{c}\int \frac{dx}{(ax^2+c)^n}$$

108. $$\int \sqrt{x^2 \pm p^2}\,dx = \frac{1}{2}[x\sqrt{x^2 \pm p^2} \pm p^2 \log (x + \sqrt{x^2 \pm p^2})]$$

109. $$\int \sqrt{p^2 - x^2}\,dx = \frac{1}{2}\left(x\sqrt{p^2 - x^2} + p^2 \sin^{-1}\frac{x}{p}\right)$$

110. $$\int \frac{dx}{\sqrt{x^2 \pm p^2}} = \log (x + \sqrt{x^2 \pm p^2})$$

111. $$\int \frac{dx}{\sqrt{p^2 - x^2}} = \sin^{-1}\frac{x}{p} \qquad \text{or} \qquad -\cos^{-1}\frac{x}{p}$$

112. $$\int \sqrt{ax^2+c}\,dx = \frac{x}{2}\sqrt{ax^2+c} + \frac{c}{2\sqrt{a}}\log (x\sqrt{a} + \sqrt{ax^2+c}) \qquad a > 0$$

113. $$\int \sqrt{ax^2+c}\,dx = \frac{x}{2}\sqrt{ax^2+c} + \frac{c}{2\sqrt{-a}}\sin^{-1}\left(x\sqrt{\frac{-a}{c}}\right) \qquad a < 0$$

114. $$\int \frac{dx}{\sqrt{ax^2+c}} = \frac{1}{\sqrt{a}}\log (x\sqrt{a} + \sqrt{ax^2+c}) \qquad a > 0$$

115. $$\int \frac{dx}{\sqrt{ax^2+c}} = \frac{1}{\sqrt{-a}}\sin^{-1}\left(x\sqrt{\frac{-a}{c}}\right) \qquad a < 0$$

116. $$\int x\sqrt{ax^2+c}\,dx = \frac{1}{3a}(ax^2+c)^{3/2}$$

117. $\int x^2\sqrt{ax^2+c}\,dx = \frac{x}{4a}\sqrt{(ax^2+c)^3} - \frac{cx}{8a}\sqrt{ax^2+c} - \frac{c^2}{8\sqrt{a^3}}\log\left(x\sqrt{a}+\sqrt{ax^2+c}\right)$ $\quad a > 0$

118. $\int x^2\sqrt{ax^2+c}\,dx = \frac{x}{4a}\sqrt{(ax^2+c)^3} - \frac{cx}{8a}\sqrt{ax^2+c} - \frac{c^2}{8a\sqrt{-a}}\sin^{-1}\left(x\sqrt{\frac{-a}{c}}\right)$ $\quad a < 0$

119. $\int \frac{x\,dx}{\sqrt{ax^2+c}} = \frac{1}{a}\sqrt{ax^2+c}$

120. $\int \frac{x^2\,dx}{\sqrt{ax^2+c}} = \frac{x}{a}\sqrt{ax^2+c} - \frac{1}{a}\int\sqrt{ax^2+c}\,dx$

121. $\int \frac{\sqrt{ax^2+c}}{x}\,dx = \sqrt{ax^2+c} + \sqrt{c}\log\frac{\sqrt{ax^2+c}-\sqrt{c}}{x}$ $\quad c > 0$

122. $\int \frac{\sqrt{ax^2+c}}{x}\,dx = \sqrt{ax^2+c} - \sqrt{-c}\tan^{-1}\frac{\sqrt{ax^2+c}}{\sqrt{-c}}$ $\quad c < 0$

123. $\int \frac{dx}{x\sqrt{p^2 \pm x^2}} = -\frac{1}{p}\log\frac{p+\sqrt{p^2\pm x^2}}{x}$

124. $\int \frac{dx}{x\sqrt{x^2-p^2}} = \frac{1}{p}\cos^{-1}\frac{p}{x}$ or $-\frac{1}{p}\sin^{-1}\frac{p}{x}$

125. $\int \frac{dx}{x\sqrt{ax^2+c}} = \frac{1}{\sqrt{c}}\log\frac{\sqrt{ax^2+c}-\sqrt{c}}{x}$ $\quad c > 0$

126. $\int \frac{dx}{x\sqrt{ax^2+c}} = \frac{1}{\sqrt{-c}}\sec^{-1}\left(x\sqrt{-\frac{a}{c}}\right)$ $\quad c < 0$

127. $\int \frac{dx}{x^2\sqrt{ax^2+c}} = -\frac{\sqrt{ax^2+c}}{cx}$

128. $$\int \frac{x^n\,dx}{\sqrt{ax^2+c}} = \frac{x^{n-1}\sqrt{ax^2+c}}{na} - \frac{(n-1)c}{na}\int \frac{x^{n-2}\,dx}{\sqrt{ax^2+c}} \qquad n>0$$

129. $$\int x^n\sqrt{ax^2+c}\,dx = \frac{x^{n-1}(ax^2+c)^{3/2}}{(n+2)a} - \frac{(n-1)c}{(n+2)a}\int x^{n-2}\sqrt{ax^2+c}\,dx \qquad n>0$$

130. $$\int \frac{\sqrt{ax^2+c}}{x^n}\,dx = -\frac{(ax^2+c)^{3/2}}{c(n-1)x^{n-1}} - \frac{(n-4)a}{(n-1)c}\int \frac{\sqrt{ax^2+c}}{x^{n-2}}\,dx \qquad n>1$$

131. $$\int \frac{dx}{x^n\sqrt{ax^2+c}} = -\frac{\sqrt{ax^2+c}}{c(n-1)x^{n-1}} - \frac{(n-2)a}{(n-1)c}\int \frac{dx}{x^{n-2}\sqrt{ax^2+c}} \qquad n>1$$

132. $$\int (ax^2+c)^{3/2}\,dx = \frac{x}{8}(2ax^2+5c)\sqrt{ax^2+c} + \frac{3c^2}{8\sqrt{a}}\log\left(x\sqrt{a}+\sqrt{ax^2+c}\right) \qquad a>0$$

133. $$\int (ax^2+c)^{3/2}\,dx = \frac{x}{8}(2ax^2+5c)\sqrt{ax^2+c} + \frac{3c^2}{8\sqrt{-a}}\sin^{-1}\left(x\sqrt{\frac{-a}{c}}\right) \qquad a<0$$

134. $$\int \frac{dx}{(ax^2+c)^{3/2}} = \frac{x}{c\sqrt{ax^2+c}}$$

135. $$\int x(ax^2+c)^{3/2}\,dx = \frac{1}{5a}(ax^2+c)^{5/2}$$

136. $$\int x^2(ax^2+c)^{3/2}\,dx = \frac{x^3}{6}(ax^2+c)^{3/2} + \frac{c}{2}\int x^2\sqrt{ax^2+c}\,dx$$

137. $\int x^n(ax^2+c)^{3/2}\,dx$

$$= \frac{x^{n+1}(ax^2+c)^{3/2}}{n+4} + \frac{3c}{n+4}\int x^n\sqrt{ax^2+c}\;dx$$

138. $\displaystyle\int \frac{x\,dx}{(ax^2+c)^{3/2}} = -\frac{1}{a\sqrt{ax^2+c}}$

139. $\displaystyle\int \frac{x^2\,dx}{(ax^2+c)^{3/2}} = -\frac{x}{a\sqrt{ax^2+c}}$

$$+\frac{1}{a\sqrt{a}}\log\left(x\sqrt{a}+\sqrt{ax^2+c}\right) \qquad a>0$$

140. $\displaystyle\int \frac{x^2\,dx}{(ax^2+c)^{3/2}} = -\frac{x}{a\sqrt{ax^2+c}}$

$$+\frac{1}{a\sqrt{-a}}\sin^{-1}\left(x\sqrt{\frac{-a}{c}}\right) \qquad a<0$$

141. $\displaystyle\int \frac{x^3\,dx}{(ax^2+c)^{3/2}} = -\frac{x^2}{a\sqrt{ax^2+c}} + \frac{2}{a^2}\sqrt{ax^2+c}$

142. $\displaystyle\int \frac{dx}{x(ax^n+c)} = \frac{1}{cn}\log\frac{x^n}{ax^n+c}$

143. $\displaystyle\int \frac{dx}{(ax^n+c)^m} = \frac{1}{c}\int\frac{dx}{(ax^n+c)^{m-1}} - \frac{a}{c}\int\frac{x^n\,dx}{(ax^n+c)^m}$

144. $\displaystyle\int \frac{dx}{x\sqrt{ax^n+c}} = \frac{1}{n\sqrt{c}}\log\frac{\sqrt{ax^n+c}-\sqrt{c}}{\sqrt{ax^n+c}+\sqrt{c}} \qquad c>0$

145. $\displaystyle\int \frac{dx}{x\sqrt{ax^n+c}} = \frac{2}{n\sqrt{-c}}\sec^{-1}\sqrt{\frac{-ax^n}{c}} \qquad c<0$

146. $\int x^{m-1}(ax^n+c)^p\,dx$

$$= \frac{1}{m+np}\left[x^m(ax^n+c)^p + npc\int x^{m-1}(ax^n+c)^{p-1}\,dx\right]$$

$$= \frac{1}{cn(p+1)}\left[-x^m(ax^n+c)^{p+1} + (m+np+n)\int x^{m-1}(ax^n+c)^{p+1}\,dx\right]$$

$$= \frac{1}{a(m+np)}\left[x^{m-n}(ax^n+c)^{p+1} - (m-n)c\int x^{m-n-1}(ax^n+c)^p\,dx\right]$$

$$= \frac{1}{mc}\left[x^m(ax^n+c)^{p+1} - (m+np+n)a\int x^{m+n-1}(ax^n+c)^p\,dx\right]$$

147. $\int \frac{x^m\,dx}{(ax^n+c)^p} = \frac{1}{a}\int \frac{x^{m-n}\,dx}{(ax^n+c)^{p-1}} - \frac{c}{a}\int \frac{x^{m-n}\,dx}{(ax^n+c)^p}$

148. $\int \frac{dx}{x^m(ax^n+c)^p} = \frac{1}{c}\int \frac{dx}{x^m(ax^n+c)^{p-1}} - \frac{a}{c}\int \frac{dx}{x^{m-n}(ax^n+c)^p}$

Expressions Containing $(ax^2 + bx + c)$

149. $\int \frac{dx}{ax^2+bx+c}$

$$= \frac{1}{\sqrt{b^2-4ac}}\log\frac{2ax+b-\sqrt{b^2-4ac}}{2ax+b+\sqrt{b^2-4ac}} \qquad b^2 > 4ac$$

150. $\int \frac{dx}{ax^2+bx+c}$

$$= \frac{2}{\sqrt{4ac-b^2}}\tan^{-1}\frac{2ax+b}{\sqrt{4ac-b^2}} \qquad b^2 < 4ac$$

151. $\int \frac{dx}{ax^2+bx+c} = -\frac{2}{2ax+b} \qquad b^2 = 4ac$

152. $\int \frac{dx}{(ax^2+bx+c)^{n+1}} = \frac{2ax+b}{n(4ac-b^2)(ax^2+bx+c)^n}$

$$+ \frac{2(2n-1)a}{n(4ac-b^2)}\int \frac{dx}{(ax^2+bx+c)^n}$$

153. $\int \frac{x\,dx}{ax^2+bx+c} = \frac{1}{2a}\log(ax^2+bx+c) - \frac{b}{2a}\int \frac{dx}{ax^2+bx+c}$

154. $\int \frac{x^2\,dx}{ax^2+bx+c} = \frac{x}{a} - \frac{b}{2a^2}\log(ax^2+bx+c)$

$$+ \frac{b^2-2ac}{2a^2}\int \frac{dx}{ax^2+bx+c}$$

155. $$\int \frac{x^n\,dx}{ax^2+bx+c} = \frac{x^{n-1}}{(n-1)a} - \frac{c}{a}\int \frac{x^{n-2}\,dx}{ax^2+bx+c} - \frac{b}{a}\int \frac{x^{n-1}\,dx}{ax^2+bx+c}$$

156. $$\int \frac{x\,dx}{(ax^2+bx+c)^{n+1}} = \frac{-(2c+bx)}{n(4ac-b^2)(ax^2+bx+c)^n} - \frac{b(2n-1)}{n(4ac-b^2)}\int \frac{dx}{(ax^2+bx+c)^n}$$

157. $$\int \frac{x^m\,dx}{(ax^2+bx+c)^{n+1}} = -\frac{x^{m-1}}{a(2n-m+1)(ax^2+bx+c)^n} - \frac{n-m+1}{2n-m+1}\,\frac{b}{a}\int \frac{x^{m-1}\,dx}{(ax^2+bx+c)^{n+1}} + \frac{m-1}{2n-m+1}\,\frac{c}{a}\int \frac{x^{m-2}\,dx}{(ax^2+bx+c)^{n+1}}$$

158. $$\int \frac{dx}{x(ax^2+bx+c)} = \frac{1}{2c}\log\left(\frac{x^2}{ax^2+bx+c}\right) - \frac{b}{2c}\int \frac{dx}{ax^2+bx+c}$$

159. $$\int \frac{dx}{x^2(ax^2+bx+c)} = \frac{b}{2c^2}\log\left(\frac{ax^2+bx+c}{x^2}\right) - \frac{1}{cx} + \left(\frac{b^2}{2c^2} - \frac{a}{c}\right)\int \frac{dx}{ax^2+bx+c}$$

160. $$\int \frac{dx}{x^m(ax^2+bx+c)^{n+1}} = -\frac{1}{(m-1)cx^{m-1}(ax^2+bx+c)^n} - \frac{n+m-1}{m-1}\,\frac{b}{c}\int \frac{dx}{x^{m-1}(ax^2+bx+c)^{n+1}} - \frac{2n+m-1}{m-1}\,\frac{a}{c}\int \frac{dx}{x^{m-2}(ax^2+bx+c)^{n+1}}$$

161. $$\int \frac{dx}{x(ax^2+bx+c)^n} = \frac{1}{2c(n-1)(ax^2+bx+c)^{n-1}} - \frac{b}{2c}\int \frac{dx}{(ax^2+bx+c)^n} + \frac{1}{c}\int \frac{dx}{x(ax^2+bx+c)^{n-1}}$$

162. $$\int \frac{dx}{\sqrt{ax^2 + bx + c}}$$
$$= \frac{1}{\sqrt{a}} \log (2ax + b + 2\sqrt{a}\sqrt{ax^2 + bx + c}) \qquad a > 0$$

163. $$\int \frac{dx}{\sqrt{ax^2 + bx + c}} = \frac{1}{\sqrt{-a}} \sin^{-1} \frac{-2ax - b}{\sqrt{b^2 - 4ac}} \qquad a < 0$$

164. $$\int \frac{x\,dx}{\sqrt{ax^2 + bx + c}} = \frac{\sqrt{ax^2 + bx + c}}{a} - \frac{b}{2a} \int \frac{dx}{\sqrt{ax^2 + bx + c}}$$

165. $$\int \frac{x^n\,dx}{\sqrt{ax^2 + bx + c}} = \frac{x^{n-1}}{an} \sqrt{ax^2 + bx + c}$$
$$- \frac{b(2n - 1)}{2an} \int \frac{x^{n-1}\,dx}{\sqrt{ax^2 + bx + c}} - \frac{c(n - 1)}{an} \int \frac{x^{n-2}\,dx}{\sqrt{ax^2 + bx + c}}$$

166. $$\int \sqrt{ax^2 + bx + c}\,dx = \frac{2ax + b}{4a} \sqrt{ax^2 + bx + c} + \frac{4ac - b^2}{8a} \int \frac{dx}{\sqrt{ax^2 + bx + c}}$$

167. $$\int x\sqrt{ax^2 + bx + c}\,dx = \frac{(ax^2 + bx + c)^{3/2}}{3a} - \frac{b}{2a} \int \sqrt{ax^2 + bx + c}\,dx$$

168. $$\int x^2\sqrt{ax^2 + bx + c}\,dx = \left(x - \frac{5b}{6a}\right) \frac{(ax^2 + bx + c)^{3/2}}{4a} + \frac{5b^2 - 4ac}{16a^2} \int \sqrt{ax^2 + bx + c}\,dx$$

169. $$\int \frac{dx}{x\sqrt{ax^2 + bx + c}}$$
$$= -\frac{1}{\sqrt{c}} \log \left(\frac{\sqrt{ax^2 + bx + c} + \sqrt{c}}{x} + \frac{b}{2\sqrt{c}} \right) \qquad c > 0$$

170. $\int \frac{dx}{x\sqrt{ax^2 + bx + c}} = \frac{1}{\sqrt{-c}} \sin^{-1} \frac{bx + 2c}{x\sqrt{b^2 - 4ac}} \qquad c < 0$

171. $\int \frac{dx}{x\sqrt{ax^2 + bx}} = -\frac{2}{bx}\sqrt{ax^2 + bx} \qquad c = 0$

172. $\int \frac{dx}{x^n\sqrt{ax^2 + bx + c}} = -\frac{\sqrt{ax^2 + bx + c}}{c(n-1)x^{n-1}}$

$$+ \frac{b(3-2n)}{2c(n-1)} \int \frac{dx}{x^{n-1}\sqrt{ax^2 + bx + c}}$$

$$+ \frac{a(2-n)}{c(n-1)} \int \frac{dx}{x^{n-2}\sqrt{ax^2 + bx + c}}$$

173. $\int \frac{dx}{(ax^2 + bx + c)^{3/2}} = -\frac{2(2ax + b)}{(b^2 - 4ac)\sqrt{ax^2 + bx + c}}$

$b^2 \neq 4ac$

174. $\int \frac{dx}{(ax^2 + bx + c)^{3/2}} = -\frac{1}{2\sqrt{a^3}(x + b/2a)^2} \qquad b^2 = 4ac$

Miscellaneous Algebraic Expressions

175. $\int \sqrt{2px - x^2}\, dx = \frac{1}{2}\left[(x-p)\sqrt{2px - x^2} + p^2 \sin^{-1} \frac{x - p}{p}\right]$

176. $\int \frac{dx}{\sqrt{2px - x^2}} = \cos^{-1} \frac{p - x}{p}$

177. $\int \frac{dx}{\sqrt{ax + b}\sqrt{cx + d}} = \frac{2}{\sqrt{-ac}} \tan^{-1} \sqrt{\frac{-c(ax + b)}{a(cx + d)}}$

or $\frac{2}{\sqrt{ac}} \tanh^{-1} \sqrt{\frac{c(ax + b)}{a(cx + d)}}$

178. $\int \sqrt{ax + b}\sqrt{cx + d}\, dx$

$$= \frac{(2acx + bc + ad)\sqrt{ax + b}\sqrt{cx + d}}{4ac}$$

$$- \frac{(ad - bc)^2}{8ac} \int \frac{dx}{\sqrt{ax + b}\sqrt{cx + d}}$$

179. $$\int \sqrt{\frac{cx+d}{ax+b}}\,dx = \frac{\sqrt{ax+b}\sqrt{cx+d}}{a} + \frac{ad-bc}{2a}\int \frac{dx}{\sqrt{ax+b}\sqrt{cx+d}}$$

180. $$\int \sqrt{\frac{x+b}{x+d}}\,dx = \sqrt{x+d}\sqrt{x+b} + (b-d)\log\left(\sqrt{x+d}+\sqrt{x+b}\right)$$

181. $$\int \sqrt{\frac{1+x}{1-x}}\,dx = \sin^{-1} x - \sqrt{1-x^2}$$

182. $$\int \sqrt{\frac{p-x}{q+x}}\,dx = \sqrt{p-x}\sqrt{q+x} + (p+q)\sin^{-1}\sqrt{\frac{x+q}{p+q}}$$

183. $$\int \sqrt{\frac{p+x}{q-x}}\,dx = -\sqrt{p+x}\sqrt{q-x} - (p+q)\sin^{-1}\sqrt{\frac{q-x}{p+q}}$$

184. $$\int \frac{dx}{\sqrt{x-p}\sqrt{q-x}} = 2\sin^{-1}\sqrt{\frac{x-p}{q-p}}$$

Expressions Containing sin *ax*

185. $$\int \sin u\,du = -\cos u$$ where u is any function of x

186. $$\int \sin ax\,dx = -\frac{1}{a}\cos ax$$

187. $$\int \sin^2 ax\,dx = \frac{x}{2} - \frac{\sin 2ax}{4a}$$

188. $$\int \sin^3 ax\,dx = -\frac{1}{a}\cos ax + \frac{1}{3a}\cos^3 ax$$

189. $$\int \sin^4 ax\,dx = \frac{3x}{8} - \frac{3\sin 2ax}{16a} - \frac{\sin^3 ax\cos ax}{4a}$$

190. $\int \sin^n ax\,dx = -\frac{\sin^{n-1} ax \cos ax}{na} + \frac{n-1}{n}\int \sin^{n-2} ax\,dx$

n positive integer

191. $\int \frac{dx}{\sin ax} = \frac{1}{a}\log\tan\frac{ax}{2} = \frac{1}{a}\log(\csc ax - \cot ax)$

192. $\int \frac{dx}{\sin^2 ax} = \int \csc^2 ax\,dx = -\frac{1}{a}\cot ax$

193. $\int \frac{dx}{\sin^n ax} = -\frac{1}{a(n-1)}\frac{\cos ax}{\sin^{n-1} ax} + \frac{n-2}{n-1}\int \frac{dx}{\sin^{n-2} ax}$

n integer > 1

194. $\int \frac{dx}{1 \pm \sin ax} = \mp\frac{1}{a}\tan\left(\frac{\pi}{4} \mp \frac{ax}{2}\right)$

195. $\int \frac{dx}{b + c\sin ax}$

$= \frac{-2}{a\sqrt{b^2 - c^2}}\tan^{-1}\left[\sqrt{\frac{b-c}{b+c}}\tan\left(\frac{\pi}{4} - \frac{ax}{2}\right)\right], \quad b^2 > c^2$

196. $\int \frac{dx}{b + c\sin ax}$

$= \frac{-1}{a\sqrt{c^2 - b^2}}\log\frac{c + b\sin ax + \sqrt{c^2 - b^2}\cos ax}{b + c\sin ax}, \quad c^2 > b^2$

197. $\int \sin ax \sin bx\,dx = \frac{\sin(a-b)x}{2(a-b)} - \frac{\sin(a+b)x}{2(a+b)}, \quad a^2 \neq b^2$

198. $\int \sqrt{1 + \sin x}\,dx = \pm 2\left(\sin\frac{x}{2} - \cos\frac{x}{2}\right)$; use + sign

when $(8k-1)\frac{\pi}{2} < x \leqq (8k+3)\frac{\pi}{2}$, otherwise −, k an integer

199. $\int \sqrt{1 - \sin x}\,dx = \pm 2\left(\sin\frac{x}{2} + \cos\frac{x}{2}\right)$; use + sign

when $(8k-3)\frac{\pi}{2} < x \leqq (8k+1)\frac{\pi}{2}$, otherwise −, k an integer

Expressions Involving cos *ax*

200. $\int \cos u\, du = \sin u$ where u is any function of x

201. $\int \cos ax\, dx = \frac{1}{a}\sin ax$

202. $\int \cos^2 ax\, dx = \frac{x}{2} + \frac{\sin 2ax}{4a}$

203. $\int \cos^3 ax\, dx = \frac{1}{a}\sin ax - \frac{1}{3a}\sin^3 ax$

204. $\int \cos^4 ax\, dx = \frac{3x}{8} + \frac{3\sin 2ax}{16a} + \frac{\cos^3 ax \sin ax}{4a}$

205. $\int \cos^n ax\, dx = \frac{\cos^{n-1} ax \sin ax}{na} + \frac{n-1}{n}\int \cos^{n-2} ax\, dx$

206. $\int \frac{dx}{\cos ax} = \frac{1}{a}\log\tan\left(\frac{ax}{2} + \frac{\pi}{4}\right) = \frac{1}{a}\log(\tan ax + \sec ax)$

207. $\int \frac{dx}{\cos^2 ax} = \frac{1}{a}\tan ax$

208. $\int \frac{dx}{\cos^n ax} = \frac{1}{a(n-1)}\frac{\sin ax}{\cos^{n-1} ax} + \frac{n-2}{n-1}\int \frac{dx}{\cos^{n-2} ax}$

n integer > 1

209. $\int \frac{dx}{1 + \cos ax} = \frac{1}{a}\tan\frac{ax}{2}$

210. $\int \frac{dx}{1 - \cos ax} = -\frac{1}{a}\cot\frac{ax}{2}$

211. $\int \sqrt{1 + \cos x}\, dx = \pm\sqrt{2}\int \cos\frac{x}{2}\, dx = \pm 2\sqrt{2}\sin\frac{x}{2}$

Use + when $(4k - 1)\pi < x \leqq (4k + 1)\pi$, otherwise −, k an integer

212. $\int \sqrt{1 - \cos x}\, dx = \pm\sqrt{2}\int \sin\frac{x}{2}\, dx = \mp 2\sqrt{2}\cos\frac{x}{2}$

Use top signs when $4k\pi < x \leqq (4k + 2)\pi$, otherwise bottom signs

213. $$\int \frac{dx}{b + c \cos ax} = \frac{1}{a\sqrt{b^2 - c^2}} \tan^{-1} \frac{\sqrt{b^2 - c^2} \sin ax}{c + b \cos ax} \qquad b^2 > c^2$$

214. $$\int \frac{dx}{b + c \cos ax} = \frac{1}{a\sqrt{c^2 - b^2}} \tanh^{-1} \frac{\sqrt{c^2 - b^2} \sin ax}{c + b \cos ax} \qquad c^2 > b^2$$

215. $$\int \cos ax \cos bx \, dx = \frac{\sin (a - b)x}{2(a - b)} + \frac{\sin (a + b)x}{2(a + b)} \qquad a^2 \neq b^2$$

Expressions Containing sin *ax and* cos *ax*

216. $$\int \sin ax \cos bx \, dx = -\frac{1}{2}\left[\frac{\cos (a - b)x}{a - b} + \frac{\cos (a + b)x}{a + b}\right] \qquad a^2 \neq b^2$$

217. $$\int \sin^n ax \cos ax \, dx = \frac{1}{a(n + 1)} \sin^{n+1} ax \qquad n \neq -1$$

218. $$\int \cos^n ax \sin ax \, dx = -\frac{1}{a(n + 1)} \cos^{n+1} ax \qquad n \neq -1$$

219. $$\int \frac{\sin ax}{\cos ax} dx = -\frac{1}{a} \log \cos ax$$

220. $$\int \frac{\cos ax}{\sin ax} dx = \frac{1}{a} \log \sin ax$$

221. $$\int (b + c \sin ax)^n \cos ax \, dx = \frac{1}{ac(n + 1)} (b + c \sin ax)^{n+1} \qquad n \neq -1$$

222. $\int (b + c \cos ax)^n \sin ax\,dx$

$$= -\frac{1}{ac(n+1)}(b + c \cos ax)^{n+1} \qquad n \neq -1$$

223. $\int \frac{\cos ax\,dx}{b + c \sin ax} = \frac{1}{ac} \log (b + c \sin ax)$

224. $\int \frac{\sin ax}{b + c \cos ax}\,dx = -\frac{1}{ac} \log (b + c \cos ax)$

225. $\int \frac{dx}{b \sin ax + c \cos ax}$

$$= \frac{1}{a\sqrt{b^2 + c^2}}\left[\log \tan \frac{1}{2}\left(ax + \tan^{-1} \frac{c}{b}\right)\right]$$

226. $\int \frac{dx}{b + c \cos ax + d \sin ax} = \frac{-1}{a\sqrt{b^2 - c^2 - d^2}} \sin^{-1} U$

$$U \equiv \left[\frac{c^2 + d^2 + b(c \cos ax + d \sin ax)}{\sqrt{c^2 + d^2}(b + c \cos ax + d \sin ax)}\right]$$

$$\text{or} = \frac{1}{a\sqrt{c^2 + d^2 - b^2}} \log V$$

$$V \equiv \left[\frac{c^2 + d^2 + b(c \cos ax + d \sin ax) + \sqrt{c^2 + d^2 - b^2}(c \sin ax - d \cos ax)}{\sqrt{c^2 + d^2}(b + c \cos ax + d \sin ax)}\right]$$

$$b^2 \neq c^2 + d^2, \; -\pi < ax < \pi$$

227. $\int \frac{dx}{b + c \cos ax + d \sin ax}$

$$= \frac{1}{ab}\left[\frac{b - (c + d) \cos ax + (c - d) \sin ax}{b + (c - d) \cos ax + (c + d) \sin ax}\right], \; b^2 = c^2 + d^2$$

228. $\int \frac{\sin^2 ax\,dx}{b + c \cos^2 ax} = \frac{1}{ac}\sqrt{\frac{b + c}{b}} \tan^{-1}\left(\sqrt{\frac{b}{b + c}} \tan ax\right) - \frac{x}{c}$

229. $\int \frac{\sin ax \cos ax\,dx}{b \cos^2 ax + c \sin^2 ax} = \frac{1}{2a(c - b)} \log (b \cos^2 ax + c \sin^2 ax)$

230. $\int \frac{dx}{b^2 \cos^2 ax - c^2 \sin^2 ax} = \frac{1}{2abc} \log \frac{b \cos ax + c \sin ax}{b \cos ax - c \sin ax}$

231. $\int \frac{dx}{b^2 \cos^2 ax + c^2 \sin^2 ax} = \frac{1}{abc} \tan^{-1} \frac{c \tan ax}{b}$

232. $\int \sin^2 ax \cos^2 ax \, dx = \frac{x}{8} - \frac{\sin 4\,ax}{32a}$

233. $\int \frac{dx}{\sin ax \cos ax} = \frac{1}{a} \log \tan ax$

234. $\int \frac{dx}{\sin^2 ax \cos^2 ax} = \frac{1}{a}(\tan ax - \cot ax)$

235. $\int \frac{\sin^2 ax}{\cos ax}\, dx = \frac{1}{a}\left[-\sin ax + \log \tan\left(\frac{ax}{2} + \frac{\pi}{4}\right)\right]$

236. $\int \frac{\cos^2 ax}{\sin ax}\, dx = \frac{1}{a}\left(\cos ax + \log \tan \frac{ax}{2}\right)$

237. $\int \sin^m ax \cos^n ax \, dx = -\frac{\sin^{m-1} ax \cos^{n+1} ax}{a(m+n)}$

$$+ \frac{m-1}{m+n} \int \sin^{m-2} ax \cos^n ax \, dx \qquad m, n > 0$$

238. $\int \sin^m ax \cos^n ax \, dx = \frac{\sin^{m+1} ax \cos^{n-1} ax}{a(m+n)}$

$$+ \frac{n-1}{m+n} \int \sin^m ax \cos^{n-2} ax \, dx \qquad m, n > 0$$

239. $\int \frac{\sin^m ax}{\cos^n ax}\, dx = \frac{\sin^{m+1} ax}{a(n-1)\cos^{n-1} ax}$

$$- \frac{m-n+2}{n-1} \int \frac{\sin^m ax}{\cos^{n-2} ax}\, dx \qquad m, n > 0, n \neq 1$$

240. $\int \frac{\cos^n ax}{\sin^m ax}\, dx = \frac{-\cos^{n+1} ax}{a(m-1)\sin^{m-1} ax}$

$$+ \frac{m-n-2}{m-1} \int \frac{\cos^n ax}{\sin^{m-2} ax}\, dx \qquad m, n > 0, m \neq 1$$

241. $$\int \frac{dx}{\sin^m ax \cos^n ax} = \frac{1}{a(n-1)} \frac{1}{\sin^{m-1} ax \cos^{n-1} ax} + \frac{m+n-2}{n-1} \int \frac{dx}{\sin^m ax \cos^{n-2} ax}$$

242. $$\int \frac{dx}{\sin^m ax \cos^n ax} = -\frac{1}{a(m-1)} \frac{1}{\sin^{m-1} ax \cos^{n-1} ax} + \frac{m+n-2}{m-1} \int \frac{dx}{\sin^{m-2} ax \cos^n ax}$$

243. $$\int \frac{\sin^{2n} ax}{\cos ax} dx = \int \frac{(1-\cos^2 ax)^n}{\cos ax} dx$$

(Expand, divide, and use 205.)

244. $$\int \frac{\cos^{2n} ax}{\sin ax} dx = \int \frac{(1-\sin^2 ax)^n}{\sin ax} dx$$

(Expand, divide, and use 190.)

245. $$\int \frac{\sin^{2n+1} ax}{\cos ax} dx = \int \frac{(1-\cos^2 ax)^n}{\cos ax} \sin ax\, dx$$

(Expand, divide, and use 218.)

246. $$\int \frac{\cos^{2n+1} ax}{\sin ax} dx = \int \frac{(1-\sin^2 ax)^n}{\sin ax} \cos ax\, dx$$

(Expand, divide, and use 217.)

Expressions Containing tan ax *or* cot ax (tan $ax = 1/\cot ax$)

247. $$\int \tan u\, du = -\log \cos u \qquad \text{or} \qquad \log \sec u$$

where u is any function of x

248. $$\int \tan ax\, dx = -\frac{1}{a} \log \cos ax$$

249. $$\int \tan^2 ax\, dx = \frac{1}{a} \tan ax - x$$

250. $$\int \tan^3 ax\, dx = \frac{1}{2a} \tan^2 ax + \frac{1}{a} \log \cos ax$$

251. $\int \tan^n ax\,dx = \frac{1}{a(n-1)} \tan^{n-1} ax - \int \tan^{n-2} ax\,dx$

n integer > 1

252. $\int \cot u\,du = \log \sin u$ or $-\log \csc u$

253. $\int \cot^2 ax\,dx = \int \frac{dx}{\tan^2 ax} = -\frac{1}{a} \cot ax - x$

254. $\int \cot^3 ax\,dx = -\frac{1}{2a} \cot^2 ax - \frac{1}{a} \log \sin ax$

255. $\int \cot^n ax\,dx = \int \frac{dx}{\tan^n ax}$

$= -\frac{1}{a(n-1)} \cot^{n-1} ax - \int \cot^{n-2} ax\,dx$ n integer > 1

256. $\int \frac{dx}{b + c \tan ax} = \int \frac{\cot ax\,dx}{b \cot ax + c}$

$= \frac{1}{b^2 + c^2} \left[bx + \frac{c}{a} \log (b \cos ax + c \sin ax) \right]$

257. $\int \frac{dx}{b + c \cot ax} = \int \frac{\tan ax\,dx}{b \tan ax + c}$

$= \frac{1}{b^2 + c^2} \left[bx - \frac{c}{a} \log (c \cos ax + b \sin ax) \right]$

258. $\int \frac{dx}{\sqrt{b + c \tan^2 ax}} = \frac{1}{a\sqrt{b - c}} \sin^{-1} \left(\sqrt{\frac{b - c}{b}} \sin ax \right)$

b positive, $b^2 > c^2$

Expressions Containing sec $ax = 1/\cos ax$ *or* csc $ax = 1/\sin ax$

259. $\int \sec u\,du = \log (\sec u + \tan u) = \log \tan \left(\frac{u}{2} + \frac{\pi}{4} \right)$

where u is any function of x

260. $\int \sec ax\,dx = \frac{1}{a} \log \tan \left(\frac{ax}{2} + \frac{\pi}{4} \right)$

261. $\int \sec^2 ax\, dx = \frac{1}{a} \tan ax$

262. $\int \sec^3 ax\, dx = \frac{1}{2a}\left[\tan ax \sec ax + \log \tan\left(\frac{ax}{2} + \frac{\pi}{4}\right)\right]$

263. $\int \sec^n ax\, dx = \frac{1}{a(n-1)} \frac{\sin ax}{\cos^{n-1} ax}$

$$+ \frac{n-2}{n-1} \int \sec^{n-2} ax\, dx \qquad n \text{ integer} > 1$$

264. $\int \csc u\, du = \log(\csc u - \cot u) = \log \tan \frac{u}{2}$

265. $\int \csc ax\, dx = \frac{1}{a} \log \tan \frac{ax}{2}$

266. $\int \csc^2 ax\, dx = -\frac{1}{a} \cot ax$

267. $\int \csc^3 ax\, dx = \frac{1}{2a}\left(-\cot ax \csc ax + \log \tan \frac{ax}{2}\right)$

268. $\int \csc^n ax\, dx = -\frac{1}{a(n-1)} \frac{\cos ax}{\sin^{n-1} ax}$

$$+ \frac{n-2}{n-1} \int \csc^{n-2} ax\, dx \qquad n \text{ integer} > 1$$

Expressions Containing tan ax *and* sec ax *or* cot ax *and* csc ax

269. $\int \tan u \sec u\, du = \sec u$ where u is any function of x

270. $\int \tan ax \sec ax\, dx = \frac{1}{a} \sec ax$

271. $\int \tan^n ax \sec^2 ax\, dx = \frac{1}{a(n+1)} \tan^{n+1} ax \qquad n \neq -1$

272. $\int \tan ax \sec^n ax\, dx = \frac{1}{an} \sec^n ax \qquad n \neq 0$

273. $\int \cot u \csc u \, du = -\csc u$

274. $\int \cot ax \csc ax \, dx = -\frac{1}{a} \csc ax$

275. $\int \cot^n ax \csc^2 ax \, dx = -\frac{1}{a(n+1)} \cot^{n+1} ax \qquad n \neq -1$

276. $\int \cot ax \csc^n ax \, dx = -\frac{1}{an} \csc^n ax \qquad n \neq 0$

277. $\int \frac{\csc^2 ax \, dx}{\cot ax} = -\frac{1}{a} \log \cot ax$

Expressions Containing Algebraic and Trigonometric Functions

278. $\int x \sin ax \, dx = \frac{1}{a^2} \sin ax - \frac{1}{a} x \cos ax$

279. $\int x^2 \sin ax \, dx = \frac{2x}{a^2} \sin ax + \frac{2}{a^3} \cos ax - \frac{x^2}{a} \cos ax$

280. $\int x^3 \sin ax \, dx$

$$= \frac{3x^2}{a^2} \sin ax - \frac{6}{a^4} \sin ax - \frac{x^3}{a} \cos ax + \frac{6x}{a^3} \cos ax$$

281. $\int x \sin^2 ax \, dx = \frac{x^2}{4} - \frac{x \sin 2ax}{4a} - \frac{\cos 2ax}{8a^2}$

282. $\int x^2 \sin^2 ax \, dx = \frac{x^3}{6} - \left(\frac{x^2}{4a} - \frac{1}{8a^3}\right) \sin 2ax - \frac{x \cos 2ax}{4a^2}$

283. $\int x^3 \sin^2 ax \, dx = \frac{x^4}{8} - \left(\frac{x^3}{4a} - \frac{3x}{8a^3}\right) \sin 2ax$

$$- \left(\frac{3x^2}{8a^2} - \frac{3}{16a^4}\right) \cos 2ax$$

284. $\int x \sin^3 ax \, dx = \frac{x \cos 3ax}{12a} - \frac{\sin 3ax}{36a^2} - \frac{3x \cos ax}{4a} + \frac{3 \sin ax}{4a^2}$

285. $\int x^n \sin ax \, dx = -\frac{1}{a} x^n \cos ax + \frac{n}{a} \int x^{n-1} \cos ax \, dx$

286. $$\int \frac{\sin ax\,dx}{x} = ax - \frac{(ax)^3}{3\cdot 3!} + \frac{(ax)^5}{5\cdot 5!} - \cdots$$

287. $$\int \frac{\sin ax\,dx}{x^m} = \frac{-1}{m-1}\frac{\sin ax}{x^{m-1}} + \frac{a}{m-1}\int \frac{\cos ax\,dx}{x^{m-1}}$$

288. $$\int x\cos ax\,dx = \frac{1}{a^2}\cos ax + \frac{1}{a}x\sin ax$$

289. $$\int x^2\cos ax\,dx = \frac{2x}{a^2}\cos ax - \frac{2}{a^3}\sin ax + \frac{x^2}{a}\sin ax$$

290. $$\int x^3\cos ax\,dx = \frac{(3a^2x^2-6)\cos ax}{a^4} + \frac{(a^2x^2-6x)\sin ax}{a^3}$$

291. $$\int x\cos^2 ax\,dx = \frac{x^2}{4} + \frac{x\sin 2ax}{4a} + \frac{\cos 2ax}{8a^2}$$

292. $$\int x^2\cos^2 ax\,dx = \frac{x^3}{6} + \left(\frac{x^2}{4a} - \frac{1}{8a^3}\right)\sin 2ax + \frac{x\cos 2ax}{4a^2}$$

293. $$\int x^3\cos^2 ax\,dx = \frac{x^4}{8} + \left(\frac{x^3}{4a} - \frac{3x}{8a^3}\right)\sin 2ax + \left(\frac{3x^2}{8a^2} - \frac{3}{16a^4}\right)\cos 2ax$$

294. $$\int x\cos^3 ax\,dx = \frac{x\sin 3ax}{12a} + \frac{\cos 3ax}{36a^2} + \frac{3x\sin ax}{4a} + \frac{3\cos ax}{4a^2}$$

295. $$\int x^n\cos ax\,dx = \frac{1}{a}x^n\sin ax - \frac{n}{a}\int x^{n-1}\sin ax\,dx, \quad n \text{ positive}$$

296. $$\int \frac{\cos ax}{x}\,dx = \log ax - \frac{(ax)^2}{2\cdot 2!} + \frac{(ax)^4}{4\cdot 4!} - \cdots$$

297. $$\int \frac{\cos ax}{x^m}\,dx = -\frac{1}{m-1}\frac{\cos ax}{x^{m-1}} - \frac{a}{m-1}\int \frac{\sin ax}{x^{m-1}}\,dx$$

Expressions Containing Exponential and Logarithmic Functions

298. $$\int e^u\,du = e^u \qquad \text{where } u \text{ is any function of } x$$

299. $\int b^u\,du = \frac{b^u}{\log b}$

300. $\int e^{ax}\,dx = \frac{1}{a}e^{ax} \qquad \int b^{ax}\,dx = \frac{b^{ax}}{a\log b}$

301. $\int xe^{ax}\,dx = \frac{e^{ax}}{a^2}(ax-1), \quad \int xb^{ax}\,dx = \frac{xb^{ax}}{a\log b} - \frac{b^{ax}}{a^2(\log b)^2}$

302. $\int x^2e^{ax}\,dx = \frac{e^{ax}}{a^3}(a^2x^2 - 2ax + 2)$

303. $\int x^ne^{ax}\,dx = \frac{1}{a}x^ne^{ax} - \frac{n}{a}\int x^{n-1}e^{ax}\,dx$ $\qquad n$ positive

304. $\int x^ne^{ax}\,dx = \frac{e^{ax}}{a^{n+1}}\Big[(ax)^n - n(ax)^{n-1} + n(n-1)(ax)^{n-2} - \cdots + (-1)^n n!\Big]$ $\qquad n$ positive integer

305. $\int x^ne^{-ax}\,dx = -\frac{e^{-ax}}{a^{n+1}}\Big[(ax)^n + n(ax)^{n-1} + n(n-1)(ax)^{n-2} + \cdots + n!\Big]$ $\qquad n$ positive integer

306. $\int x^nb^{ax}\,dx = \frac{x^nb^{ax}}{a\log b} - \frac{n}{a\log b}\int x^{n-1}b^{ax}\,dx$ $\qquad n$ positive

307. $\int \frac{e^{ax}}{x}\,dx = \log x + ax + \frac{(ax)^2}{2\cdot 2!} + \frac{(ax)^3}{3\cdot 3!} + \cdots$

308. $\int \frac{e^{ax}}{x^n}\,dx = \frac{1}{n-1}\left(-\frac{e^{ax}}{x^{n-1}} + a\int \frac{e^{ax}}{x^{n-1}}\,dx\right)$ $\qquad n$ integ. > 1

309. $\int \frac{dx}{b+ce^{ax}} = \frac{1}{ab}\Big[ax - \log(b+ce^{ax})\Big]$

310. $\int \frac{e^{ax}\,dx}{b+ce^{ax}} = \frac{1}{ac}\log(b+ce^{ax})$

311. $\int \frac{dx}{be^{ax}+ce^{-ax}} = \frac{1}{a\sqrt{bc}}\tan^{-1}\left(e^{ax}\sqrt{\frac{b}{c}}\right)$ $\qquad b, c$ positive

312. $\int e^{ax}\sin bx\,dx = \frac{e^{ax}}{a^2+b^2}(a\sin bx - b\cos bx)$

313. $\int e^{ax} \sin bx \sin cx \, dx$

$$= \frac{e^{ax}\{(b - c) \sin [(b - c)x] + a \cos [(b - c)x]\}}{2[a^2 + (b - c)^2]} - \frac{e^{ax}\{(b + c) \sin [(b + c)x] + a \cos [(b + c)x]\}}{2[a^2 + (b + c)^2]}$$

314. $\int e^{ax} \cos bx \, dx = \frac{e^{ax}}{a^2 + b^2}(a \cos bx + b \sin bx)$

315. $\int e^{ax} \cos bx \cos cx \, dx$

$$= \frac{e^{ax}\{(b - c) \sin [(b - c)x] + a \cos [(b - c)x]\}}{2[a^2 + (b - c)^2]} + \frac{e^{ax}\{(b + c) \sin [(b + c)x] + a \cos [(b + c)x]\}}{2[a^2 + (b + c)^2]}$$

316. $\int e^{ax} \sin bx \cos cx \, dx$

$$= \frac{e^{ax}\{a \sin [(b - c)x] - (b - c) \cos [(b - c)x]\}}{2[a^2 + (b - c)^2]} + \frac{e^{ax}\{a \sin [(b + c)x] - (b + c) \cos [(b + c)x]\}}{2[a^2 + (b + c)^2]}$$

317. $\int e^{ax} \sin bx \sin (bx + c) \, dx$

$$= \frac{e^{ax} \cos c}{2a} - \frac{e^{ax}[a \cos (2bx + c) + 2b \sin (2bx + c)]}{2(a^2 + 4b^2)}$$

318. $\int e^{ax} \cos bx \cos (bx + c) \, dx$

$$= \frac{e^{ax} \cos c}{2a} + \frac{e^{ax}[a \cos (2bx + c) + 2b \sin (2bx + c)]}{2(a^2 + 4b^2)}$$

319. $\int e^{ax} \sin bx \cos (bx + c) \, dx$

$$= -\frac{e^{ax} \sin c}{2a} + \frac{e^{ax}[a \sin (2bx + c) - 2b \cos (2bx + c)]}{2(a^2 + 4b^2)}$$

320. $$\int e^{ax} \cos bx \sin (bx + c)\, dx = \frac{e^{ax} \sin c}{2a} + \frac{e^{ax}[a \sin (2bx + c) - 2b \cos (2bx + c)]}{2(a^2 + 4b^2)}$$

321. $$\int xe^{ax} \sin bx\, dx = \frac{xe^{ax}}{a^2 + b^2}(a \sin bx - b \cos bx) - \frac{e^{ax}}{(a^2 + b^2)^2}[(a^2 - b^2) \sin bx - 2ab \cos bx]$$

322. $$\int xe^{ax} \cos bx\, dx = \frac{xe^{ax}}{a^2 + b^2}(a \cos bx + b \sin bx) - \frac{e^{ax}}{(a^2 + b^2)^2}[(a^2 - b^2) \cos bx + 2ab \sin bx]$$

323. $$\int e^{ax} \cos^n bx\, dx = \frac{(e^{ax} \cos^{n-1} bx)(a \cos bx + nb \sin bx)}{a^2 + n^2b^2} + \frac{n(n-1)b^2}{a^2 + n^2b^2} \int e^{ax} \cos^{n-2} bx\, dx$$

324. $$\int e^{ax} \sin^n bx\, dx = \frac{(e^{ax} \sin^{n-1} bx)(a \sin bx - nb \cos bx)}{a^2 + n^2b^2} + \frac{n(n-1)b^2}{a^2 + n^2b^2} \int e^{ax} \sin^{n-2} bx\, dx$$

325. $$\int \log ax\, dx = x \log ax - x$$

326. $$\int x \log ax\, dx = \frac{x^2}{2} \log ax - \frac{x^2}{4}$$

327. $$\int x^2 \log ax\, dx = \frac{x^3}{3} \log ax - \frac{x^3}{9}$$

328. $$\int (\log ax)^2\, dx = x(\log ax)^2 - 2x \log ax + 2x$$

329. $$\int (\log ax)^n\, dx = x(\log ax)^n - n\int (\log ax)^{n-1}\, dx \qquad n \text{ pos.}$$

330. $\int x^n \log ax\, dx = x^{n+1}\left[\frac{\log ax}{n+1} - \frac{1}{(n+1)^2}\right] \qquad n \neq -1$

331. $\int x^n (\log ax)^m\, dx = \frac{x^{n+1}}{n+1}(\log ax)^m - \frac{m}{n+1}\int x^n (\log ax)^{m-1}\, dx$

332. $\int \frac{(\log ax)^n}{x}\, dx = \frac{(\log ax)^{n+1}}{n+1} \qquad n \neq -1$

333. $\int \frac{dx}{x \log ax} = \log(\log ax)$

334. $\int \frac{dx}{x(\log ax)^n} = -\frac{1}{(n-1)(\log ax)^{n-1}}$

335. $\int \frac{x^n\, dx}{(\log ax)^m} = \frac{-x^{n+1}}{(m-1)(\log ax)^{m-1}} + \frac{n+1}{m-1}\int \frac{x^n\, dx}{(\log ax)^{m-1}} \qquad m \neq 1$

336. $\int \frac{x^n\, dx}{\log ax} = \frac{1}{a^{n+1}}\int \frac{e^y\, dy}{y} \qquad y = (n+1)\log ax$

337. $\int \frac{x^n\, dx}{\log ax} = \frac{1}{a^{n+1}}\left[\log|\log ax| + (n+1)\log ax + \frac{(n+1)^2(\log ax)^2}{2 \cdot 2!} + \frac{(n+1)^3(\log ax)^3}{3 \cdot 3!} + \cdots\right]$

338. $\int \frac{dx}{\log ax} = \frac{1}{a}\left[\log|\log ax| + \log ax + \frac{(\log ax)^2}{2 \cdot 2!} + \frac{(\log ax)^3}{3 \cdot 3!} + \cdots\right]$

339. $\int \sin(\log ax)\, dx = \frac{x}{2}\left[\sin(\log ax) - \cos(\log ax)\right]$

340. $\int \cos(\log ax)\, dx = \frac{x}{2}\left[\sin(\log ax) + \cos(\log ax)\right]$

341. $\int e^{ax} \log bx\, dx = \frac{1}{a} e^{ax} \log bx - \frac{1}{a}\int \frac{e^{ax}}{x}\, dx$

Expressions Containing Inverse Trigonometric Functions

342. $\int \sin^{-1} ax\,dx = x \sin^{-1} ax + \frac{1}{a}\sqrt{1 - a^2x^2}$

343. $\int (\sin^{-1} ax)^2\,dx = x\,(\sin^{-1} ax)^2 - 2x + \frac{2}{a}\sqrt{1 - a^2x^2}\,\sin^{-1} ax$

344. $\int x \sin^{-1} ax\,dx = \frac{x^2}{2}\sin^{-1} ax - \frac{1}{4a^2}\sin^{-1} ax + \frac{x}{4a}\sqrt{1 - a^2x^2}$

345. $\int x^n \sin^{-1} ax\,dx = \frac{x^{n+1}}{n+1}\sin^{-1} ax - \frac{a}{n+1}\int \frac{x^{n+1}\,dx}{\sqrt{1 - a^2x^2}} \qquad n \neq -1$

346. $\int \frac{\sin^{-1} ax\,dx}{x} = ax + \frac{1}{2\cdot 3\cdot 3}(ax)^3 + \frac{1\cdot 3}{2\cdot 4\cdot 5\cdot 5}(ax)^5 + \frac{1\cdot 3\cdot 5}{2\cdot 4\cdot 6\cdot 7\cdot 7}(ax)^7 + \cdots \qquad a^2x^2 < 1$

347. $\int \frac{\sin^{-1} ax\,dx}{x^2} = -\frac{1}{x}\sin^{-1} ax - a \log\left|\frac{1 + \sqrt{1 - a^2x^2}}{ax}\right|$

348. $\int \cos^{-1} ax\,dx = x \cos^{-1} ax - \frac{1}{a}\sqrt{1 - a^2x^2}$

349. $\int (\cos^{-1} ax)^2\,dx = x(\cos^{-1} ax)^2 - 2x - \frac{2}{a}\sqrt{1 - a^2x^2}\,\cos^{-1} ax$

350. $\int x \cos^{-1} ax\,dx = \frac{x^2}{2}\cos^{-1} ax - \frac{1}{4a^2}\cos^{-1} ax - \frac{x}{4a}\sqrt{1 - a^2x^2}$

351. $\int x^n \cos^{-1} ax\,dx = \frac{x^{n+1}}{n+1}\cos^{-1} ax + \frac{a}{n+1}\int \frac{x^{n+1}\,dx}{\sqrt{1 - a^2x^2}} \qquad n \neq -1$

352. $$\int \frac{\cos^{-1} ax\, dx}{x} = \frac{\pi}{2} \log |ax| - ax - \frac{1}{2 \cdot 3 \cdot 3}(ax)^3 - \frac{1 \cdot 3}{2 \cdot 4 \cdot 5 \cdot 5}(ax)^5 - \frac{1 \cdot 3 \cdot 5}{2 \cdot 4 \cdot 6 \cdot 7 \cdot 7}(ax)^7 - \cdots$$
$a^2x^2 < 1$

353. $$\int \frac{\cos^{-1} ax\, dx}{x^2} = -\frac{1}{x} \cos^{-1} ax + a \log \left| \frac{1 + \sqrt{1 - a^2x^2}}{ax} \right|$$

354. $$\int \tan^{-1} ax\, dx = x \tan^{-1} ax - \frac{1}{2a} \log (1 + a^2x^2)$$

355. $$\int x^n \tan^{-1} ax\, dx = \frac{x^{n+1}}{n+1} \tan^{-1} ax - \frac{a}{n+1} \int \frac{x^{n+1}\, dx}{1 + a^2x^2}$$
$n \neq -1$

356. $$\int \frac{\tan^{-1} ax\, dx}{x^2} = -\frac{1}{x} \tan^{-1} ax - \frac{a}{2} \log \frac{1 + a^2x^2}{a^2x^2}$$

357. $$\int \cot^{-1} ax\, dx = x \cot^{-1} ax + \frac{1}{2a} \log (1 + a^2x^2)$$

358. $$\int x^n \cot^{-1} ax\, dx = \frac{x^{n+1}}{n+1} \cot^{-1} ax + \frac{a}{n+1} \int \frac{x^{n+1}\, dx}{1 + a^2x^2}$$
$n \neq -1$

359. $$\int \frac{\cot^{-1} ax\, dx}{x^2} = -\frac{1}{x} \cot^{-1} ax + \frac{a}{2} \log \frac{1 + a^2x^2}{a^2x^2}$$

360. $$\int \sec^{-1} ax\, dx = x \sec^{-1} ax - \frac{1}{a} \log (ax + \sqrt{a^2x^2 - 1})$$

361. $$\int x^n \sec^{-1} ax\, dx = \frac{x^{n+1}}{n+1} \sec^{-1} ax \pm \frac{1}{n+1} \int \frac{x^n\, dx}{\sqrt{a^2x^2 - 1}}$$
$n \neq -1$

Use + sign when $\frac{\pi}{2} < \sec^{-1} ax < \pi$;

− sign when $0 < \sec^{-1} ax < \frac{\pi}{2}$.

362. $$\int \csc^{-1} ax\, dx = x \csc^{-1} ax + \frac{1}{a} \log (ax + \sqrt{a^2x^2 - 1})$$

363. $\int x^n \csc^{-1} ax\, dx = \frac{x^{n+1}}{n+1} \csc^{-1} ax \pm \frac{1}{n+1} \int \frac{x^n\, dx}{\sqrt{a^2x^2 - 1}}$ $\qquad n \neq -1$

Use + sign when $0 < \csc^{-1} ax < \frac{\pi}{2}$;

− sign when $-\frac{\pi}{2} < \csc^{-1} ax < 0$.

Remarks on Use of Integral Tables

Many of the integrals apply to both real and complex variables. Often the relations to be integrated may be multiple-valued and have several branches, discontinuities, multiple inverses, etc.; and questions arise concerning the proper signs to use. A graph of the relations concerned helps. As a general rule integration should not be carried out across a point of discontinuity or from a point on one branch of a relation to a point on another branch. [For example, $\int (1/x)\, dx$ should not be integrated from a negative value of x to a positive x, since the interval of integration includes $x = 0$, where $1/x$ is undefined.]

Definite Integrals

364. $$\int_0^\infty \frac{a\, dx}{a^2 + x^2} = \begin{cases} \frac{\pi}{2} & a > 0 \\ 0 & a = 0 \\ \frac{-\pi}{2} & a < 0 \end{cases}$$

365. $\int_0^\infty x^{n-1} e^{-x}\, dx = \int_0^1 \left(\log_e \frac{1}{x}\right)^{n-1} dx = \Gamma(n)$

$\Gamma(n+1) = \begin{cases} n\,\Gamma(n), \text{ if } n > 0 \\ n!, \text{ if } n \text{ is an integer} \end{cases}$ $\qquad \Gamma(2) = \Gamma(1) = 1$

$\qquad \Gamma(\tfrac{1}{2}) = \sqrt{\pi}$

$\Gamma(n) = \Pi(n-1)$ $\qquad Z(y) = D_y[\log_e \Gamma(y)]$

$Z(1) = -0.5772157 \cdots$ $\qquad$ (see integral 418)

366. $\int_0^\infty z^n x^{n-1} e^{-zx}\, dx = \Gamma(n)$ $\qquad z > 0$

367. $\int_0^1 x^{m-1}(1-x)^{n-1}\,dx = \int_0^\infty \frac{x^{m-1}\,dx}{(1+x)^{m+n}} = \frac{\Gamma(m)\Gamma(n)}{\Gamma(m+n)}$

368. $\int_0^\infty \frac{x^{n-1}}{1+x}\,dx = \frac{\pi}{\sin n\pi}$ $\qquad 0 < n < 1$

369. $$\int_0^{\pi/2} \sin^n x\,dx = \begin{cases} \int_0^{\pi/2} \cos^n x\,dx & \\ \frac{1}{2}\sqrt{\pi}\,\frac{\Gamma\left(\frac{n}{2}+\frac{1}{2}\right)}{\Gamma\left(\frac{n}{2}+1\right)} & \text{if } n > -1 \\ \frac{1\cdot3\cdot5\cdots(n-1)}{2\cdot4\cdot6\cdots(n)}\,\frac{\pi}{2} & \text{if } n \text{ is an even integer} \\ \frac{2\cdot4\cdot6\cdots(n-1)}{1\cdot3\cdot5\cdot7\cdots n} & \text{if } n \text{ is an odd integer} \end{cases}$$

370. $\int_0^\infty \frac{\sin^2 x}{x^2}\,dx = \frac{\pi}{2}$

371. $\int_0^\infty \frac{\sin ax}{x}\,dx = \frac{\pi}{2}$ $\qquad$ if $a > 0$

372. $$\int_0^\infty \frac{\sin x \cos ax}{x}\,dx = \begin{cases} 0 & \text{if } a < -1 \text{ or } a > 1 \\ \frac{\pi}{4} & \text{if } a = -1 \text{ or } a = 1 \\ \frac{\pi}{2} & \text{if } -1 < a < 1 \end{cases}$$

373. $\int_0^\pi \sin^2 ax\,dx = \int_0^\pi \cos^2 ax\,dx = \frac{\pi}{2}$

374. $\int_0^{\pi/a} \sin ax \cos ax\,dx = \int_0^\pi \sin ax \cos ax\,dx = 0$

375. $\int_0^\pi \sin ax \sin bx\,dx = \int_0^\pi \cos ax \cos bx\,dx = 0$ $\qquad a \neq b$

376. $\displaystyle\int_0^{\pi} \sin ax \cos bx \, dx = \begin{cases} \dfrac{2a}{a^2 - b^2} & \text{if } a - b \text{ is odd} \\ 0 & \text{if } a - b \text{ is even} \end{cases}$

377. $\displaystyle\int_0^{\infty} \frac{\sin ax \sin bx}{x^2} \, dx = \frac{1}{2}\pi a$ $\qquad a < b$

378. $\displaystyle\int_0^{\infty} \cos (x^2) \, dx = \int_0^{\infty} \sin (x^2) \, dx = \frac{1}{2}\sqrt{\frac{\pi}{2}}$

379. $\displaystyle\int_0^{\infty} e^{-a^2x^2} \, dx = \frac{\sqrt{\pi}}{2a} = \frac{1}{2a}\Gamma\left(\frac{1}{2}\right)$ $\qquad a > 0$

380. $\displaystyle\int_0^{\infty} x^n e^{-ax} \, dx = \frac{\Gamma(n + 1)}{a^{n+1}}$

$\displaystyle = \frac{n!}{a^{n+1}}$ $\qquad$ if n is a positive integer, $a > 0$

381. $\displaystyle\int_0^{\infty} x^{2n} e^{-ax^2} \, dx = \frac{1 \cdot 3 \cdot 5 \cdots (2n - 1)}{2^{n+1}a^n}\sqrt{\frac{\pi}{a}}$

382. $\displaystyle\int_0^{\infty} \sqrt{x}\, e^{-ax} \, dx = \frac{1}{2a}\sqrt{\frac{\pi}{a}}$

383. $\displaystyle\int_0^{\infty} \frac{e^{-ax}}{\sqrt{x}} \, dx = \sqrt{\frac{\pi}{a}}$

384. $\displaystyle\int_0^{\infty} e^{(-x^2 - a^2/x^2)} \, dx = \frac{1}{2}e^{-2a}\sqrt{\pi}$ $\qquad a > 0$

385. $\displaystyle\int_0^{\infty} e^{-ax} \cos bx \, dx = \frac{a}{a^2 + b^2}$ $\qquad a > 0$

386. $\displaystyle\int_0^{\infty} e^{-ax} \sin bx \, dx = \frac{b}{a^2 + b^2}$ $\qquad a > 0$

387. $\displaystyle\int_0^{\infty} \frac{e^{-ax} \sin x}{x} \, dx = \cot^{-1} a$ $\qquad a > 0$

388. $\displaystyle\int_0^{\infty} e^{-a^2x^2} \cos bx \, dx = \frac{\sqrt{\pi}e^{-b^2/4a^2}}{2a}$ $\qquad a > 0$

389. $\displaystyle\int_0^{1} (\log x)^n \, dx = (-1)^n n!$ $\qquad$ n positive integer

390. $\int_0^1 \frac{\log x}{1-x}\,dx = -\frac{\pi^2}{6}$

391. $\int_0^1 \frac{\log x}{1+x}\,dx = -\frac{\pi^2}{12}$

392. $\int_0^1 \frac{\log x}{1-x^2}\,dx = -\frac{\pi^2}{8}$

393. $\int_0^1 \frac{\log x}{\sqrt{1-x^2}}\,dx = -\frac{\pi}{2}\log 2$

394. $\int_0^1 \log\left(\frac{1+x}{1-x}\right)\frac{dx}{x} = \frac{\pi^2}{4}$

395. $\int_0^\infty \log\left(\frac{e^x+1}{e^x-1}\right)dx = \frac{\pi^2}{4}$

396. $\int_0^1 \frac{dx}{\sqrt{\log(1/x)}} = \sqrt{\pi}$

397. $\int_0^1 \log|\log x|\,dx$

$$= \int_0^\infty e^{-x}\log x\,dx = -\gamma \equiv -0.5772157\cdots$$

398. $\int_0^{\pi/2} \log\sin x\,dx = \int_0^{\pi/2}\log\cos x\,dx = -\frac{\pi}{2}\log_e 2$

399. $\int_0^\pi x\log\sin x\,dx = -\frac{\pi^2}{2}\log_e 2$

400. $\int_0^1 \left(\log\frac{1}{x}\right)^{1/2} dx = \frac{\sqrt{\pi}}{2}$

401. $\int_0^1 \left(\log\frac{1}{x}\right)^{-1/2} dx = \sqrt{\pi}$

402. $\int_0^1 x^m\left(\log\frac{1}{x}\right)^n dx = \frac{\Gamma(n+1)}{(m+1)^{n+1}}, \quad m+1>0,\ n+1>0$

403. $\int_0^\pi \log(a \pm b\cos x)\,dx = \pi\log\left(\frac{a+\sqrt{a^2-b^2}}{2}\right), \quad a \geqq b$

404. $\int_0^{\pi} \frac{\log(1 + \sin a \cos x)}{\cos x}\, dx = \pi a$

405. $\int_0^1 \frac{x^b - x^a}{\log x}\, dx = \log \frac{1 + b}{1 + a}$

406. $\int_0^{\pi} \frac{dx}{a + b \cos x} = \frac{\pi}{\sqrt{a^2 - b^2}}$ $\qquad a > b > 0$

407. $\int_0^{\pi/2} \frac{dx}{a + b \cos x} = \frac{\cos^{-1} \frac{b}{a}}{\sqrt{a^2 - b^2}}$ $\qquad a > b$

408. $$\int_0^{\infty} \frac{\cos ax\, dx}{1 + x^2} = \begin{cases} \frac{\pi}{2} e^{-a} & a > 0 \\ \frac{\pi}{2} e^{a} & a < 0 \end{cases}$$

409. $\int_0^{\infty} \frac{\cos x\, dx}{\sqrt{x}} = \int_0^{\infty} \frac{\sin x\, dx}{\sqrt{x}} = \sqrt{\frac{\pi}{2}}$

410. $\int_0^{\infty} \frac{e^{-ax} - e^{-bx}}{x}\, dx = \log \frac{b}{a}$

411. $\int_0^{\infty} \frac{\tan^{-1} ax - \tan^{-1} bx}{x}\, dx = \frac{\pi}{2} \log \frac{a}{b}$

412. $\int_0^{\infty} \frac{\cos ax - \cos bx}{x}\, dx = \log \frac{b}{a}$

413. $\int_0^{\pi/2} \frac{dx}{a^2 \cos^2 x + b^2 \sin^2 x} = \frac{\pi}{2ab}$

414. $\int_0^{\pi/2} \frac{dx}{(a^2 \cos^2 x + b^2 \sin^2 x)^2} = \frac{\pi(a^2 + b^2)}{4a^3b^3}$

415. $$\int_0^{\pi} \frac{(a - b \cos x)\, dx}{a^2 - 2ab \cos x + b^2} = \begin{cases} 0 & a^2 < b^2 \\ \frac{\pi}{a} & a^2 > b^2 \\ \frac{\pi}{2a} & a = b \end{cases}$$

416. $\int_0^1 \frac{1 + x^2}{1 + x^4}\,dx = \frac{\pi}{4}\sqrt{2}$

417. $\int_0^1 \frac{\log(1 + x)}{x}\,dx = \frac{1}{1^2} - \frac{1}{2^2} + \frac{1}{3^2} - \frac{1}{4^2} + \cdots = \frac{\pi^2}{12}$

418. $\int_{+\infty}^1 \frac{e^{-xu}}{u}\,du = \gamma + \log x - x$

$$+ \frac{x^2}{2 \cdot 2!} - \frac{x^3}{3 \cdot 3!} + \frac{x^4}{4 \cdot 4!} - \cdots$$

where $\gamma = \lim_{t \to \infty} \left(1 + \frac{1}{2} + \frac{1}{3} + \cdots + \frac{1}{t} - \log t\right)$

$$= 0.5772157 \cdots \qquad 0 < x < \infty$$

419. $\int_{+\infty}^1 \frac{\cos xu}{u}\,du = \gamma + \log x - \frac{x^2}{2 \cdot 2!} + \frac{x^4}{4 \cdot 4!} - \frac{x^6}{6 \cdot 6!} + \cdots$

$$0 < x < \infty$$

420. $\int_0^1 \frac{e^{xu} - e^{-xu}}{u}\,du = 2\left(x + \frac{x^3}{3 \cdot 3!} + \frac{x^5}{5 \cdot 5!} + \cdots\right)$

$$0 < x < \infty$$

421. $\int_0^1 \frac{1 - e^{-xu}}{u}\,du = x - \frac{x^2}{2 \cdot 2!} + \frac{x^3}{3 \cdot 3!} - \frac{x^4}{4 \cdot 4!} + \cdots$

$$0 < x < \infty$$

422. $\int_0^{\pi/2} \frac{dx}{\sqrt{1 - K^2 \sin^2 x}} = \frac{\pi}{2}\left[1 + \left(\frac{1}{2}\right)^2 K^2\right.$

$$\left. + \left(\frac{1 \cdot 3}{2 \cdot 4}\right)^2 K^4 + \left(\frac{1 \cdot 3 \cdot 5}{2 \cdot 4 \cdot 6}\right)^2 K^6 + \cdots\right] \qquad K^2 < 1$$

423. $\int_0^{\pi/2} \sqrt{1 - K^2 \sin^2 x}\,dx = \frac{\pi}{2}\left[1 - \left(\frac{1}{2}\right)^2 K^2\right.$

$$\left. - \left(\frac{1 \cdot 3}{2 \cdot 4}\right)^2 \frac{K^4}{3} - \left(\frac{1 \cdot 3 \cdot 5}{2 \cdot 4 \cdot 6}\right)^2 \frac{K^6}{5} - \cdots\right] \qquad K^2 < 1$$

424. $$f(x) = \frac{1}{2}a_0 + a_1 \cos\frac{\pi x}{c} + a_2 \cos\frac{2\pi x}{c} + \cdots$$

$$+ b_1 \sin\frac{\pi x}{c} + b_2 \sin\frac{2\pi x}{c} + \cdots \qquad -c < x < +c$$

where $a_m = \dfrac{1}{c}\displaystyle\int_{-c}^{+c} f(x) \cos\frac{m\pi x}{c}\,dx$

$b_m = \dfrac{1}{c}\displaystyle\int_{-c}^{+c} f(x) \sin\frac{m\pi x}{c}\,dx$ (Fourier series)

425.* $$\int_0^\infty e^{-ax} \cosh bx\,dx = \frac{a}{a^2 - b^2} \qquad 0 \leqq |b| < a$$

426. $$\int_0^\infty e^{-ax} \sinh bx\,dx = \frac{b}{a^2 - b^2} \qquad 0 \leqq |b| < a$$

427. $$\int_0^\infty xe^{-ax} \sin bx\,dx = \frac{2ab}{(a^2 + b^2)^2} \qquad a > 0$$

428. $$\int_0^\infty xe^{-ax} \cos bx\,dx = \frac{a^2 - b^2}{(a^2 + b^2)^2} \qquad a > 0$$

429. $$\int_0^\infty x^2e^{-ax} \sin bx\,dx = \frac{2b(3a^2 - b^2)}{(a^2 + b^2)^3} \qquad a > 0$$

430. $$\int_0^\infty x^2e^{-ax} \cos bx\,dx = \frac{2a(a^2 - 3b^2)}{(a^2 + b^2)^3} \qquad a > 0$$

431. $$\int_0^\infty x^3e^{-ax} \sin bx\,dx = \frac{24ab(a^2 - b^2)}{(a^2 + b^2)^4} \qquad a > 0$$

432. $$\int_0^\infty x^3e^{-ax} \cos bx\,dx = \frac{6(a^4 - 6a^2b^2 + b^4)}{(a^2 + b^2)^4} \qquad a > 0$$

* The (*one-sided*) *Laplace transform* of $f(t)$, t real, is (if it exists)

$$F(s) \equiv \mathfrak{L}[f(t)] = \int_0^{+\infty} f(t)e^{-st}\,dt$$

where $s = \sigma + i\omega$ is a complex variable. The *inverse Laplace transform* $\mathfrak{L}^{-1}[F(s)]$ of a (suitable) function $F(s)$ is (if it exists) a function $f(t)$ such that $F(s) \equiv \mathfrak{L}[f(t)]$. See chap. 20.

433. $\int_0^\infty x^n e^{-ax} \sin bx\, dx = \frac{n![(a - ib)^{n+1} - (a + ib)^{n+1}]i}{2(a^2 + b^2)^{n+1}}$ $\qquad a > 0$

434. $\int_0^\infty x^n e^{-ax} \cos bx\, dx = \frac{n![(a - ib)^{n+1} + (a + ib)^{n+1}]}{2(a^2 + b^2)^{n+1}}$ $\qquad a > 0$

435. $\int_0^\infty e^{-x} \log x\, dx = -\gamma = -0.5772157\cdots$ (see 418)

436. $\int_0^\infty \left(\frac{1}{1 - e^{-x}} - \frac{1}{x}\right) e^{-x}\, dx = \gamma = 0.5772157\cdots$

437. $\int_0^\infty \frac{1}{x}\left(\frac{1}{1 + x} - e^{-x}\right) dx = \gamma = 0.5772157\cdots$

438. $\int_0^1 \frac{1 - e^{-x} - e^{-1/x}}{x}\, dx = \gamma = 0.5772157\cdots$

δ-function. Let a be a constant. The δ-function is a "function" $\delta(x - a)$ such that

439. $\int_{-\infty}^\infty \delta(t - a)\, dt = 1$ and $\delta(x - a) = 0$ $\qquad x \neq a$

Let $g(t)$ be a real function of t.

440. $\int_{-\infty}^\infty g(t)\, \delta(t - a)\, dt = g(a)$

The "derivative" of the δ-function with respect to x is defined as an entity $\delta'(x - a)$ such that

441. $\int_{-\infty}^\infty g(t)\, \delta'(t - a)\, dt = -g'(a)$

442. $\int_{-\infty}^\infty g(t)\, \delta^k(t - a)\, dt = (-1)^k g^{(k)}(a)$ $\qquad k = 1, 2, \cdots$

Derivative of a Definite Integral. Let $a(\alpha)$ and $b(\alpha)$ be continuous and differentiable functions of α in $c \leqq \alpha \leqq d$, and let $f(x,\alpha)$ be a continuous function of the independent variables x and α in $a(\alpha) \leqq x \leqq b(\alpha)$, $c \leqq \alpha \leqq d$.

Let $$\phi(\alpha) = \int_{x=a(\alpha)}^{x=b(\alpha)} f(x,\alpha)\, dx$$

Then

443. **(a)** $$\frac{d\phi}{d\alpha} = -f(a,\alpha)\frac{da}{d\alpha} + \int_{x=a}^{x=b} \frac{\partial f(x,\alpha)}{\partial \alpha}\, dx + f(b,\alpha)\frac{db}{d\alpha}$$

If b and a are each independent of α, **(a)** becomes

(b) $$\frac{d\phi}{d\alpha} = \int_a^b \frac{\partial f(x,\alpha)}{\partial \alpha}\, dx$$

If $f(t)$ is continuous on $a \leqq t \leqq b$ and $a < x < b$, then

$$\frac{d}{dx}\int_a^x f(t)\, dt = f(x)$$

Theorem of the Mean for Integrals. If $f(x)$ and $g(x)$ are continuous and integrable on $a \leqq x \leqq b$, and if $g(x)$ retains the same sign on $a \leqq x \leqq b$, then there exists a number ξ, $a < \xi < b$, such that

444. **(a)** $$\int_a^b f(x)\, g(x)\, dx = f(\xi)\int_a^b g(x)\, dx$$

If $g(x) \equiv 1$, **(a)** is the ordinary theorem of the mean.

(b) $$\int_a^b f(x)\, dx = (b - a) f(\xi) \qquad a < \xi < b$$

Green's Theorem. If P, Q, $\partial P/\partial y$, $\partial Q/\partial x$ are continuous within and on the boundary C of a region A,

445. $$\iint\limits_A \left(\frac{\partial P}{\partial y} - \frac{\partial Q}{\partial x}\right) dx\, dy = -\int_C (P\, dx + Q\, dy)$$

where the (*line*) integral is taken in the positive direction along C. A point (x,y) traces out C in a *positive* direction if it traces counterclockwise.

8. VECTOR ANALYSIS

149. DEFINITIONS. ANALYTIC REPRESENTATION A *vector* $\boldsymbol{V}$ is a quantity which is completely specified by a magnitude and a direction. A vector may be represented geometrically by a directed line segment $\boldsymbol{V} = \overrightarrow{OA}$. A *scalar* S is a quantity which is completely specified by a magnitude.

Let $\boldsymbol{i}, \boldsymbol{j}, \boldsymbol{k}$ represent three vectors of unit magnitude along the three mutually perpendicular lines OX, OY, OZ, respectively. Let $\boldsymbol{V}$ be a vector in space, and a, b, c the magnitudes of the projections of $\boldsymbol{V}$ along the three lines OX, OY, OZ, respectively. Then $\boldsymbol{V}$ may be represented by $\boldsymbol{V} = a\boldsymbol{i} + b\boldsymbol{j} + c\boldsymbol{k}$. The magnitude of $\boldsymbol{V}$ is

$$|\boldsymbol{V}| = +\sqrt{a^2 + b^2 + c^2}$$

and the direction cosines of $\boldsymbol{V}$ are such that $\cos\alpha:\cos\beta:\cos\gamma = a:b:c$.

150. VECTOR SUM $\boldsymbol{V}$ OF n VECTORS Let $\boldsymbol{V}_1, \boldsymbol{V}_2, \ldots, \boldsymbol{V}_n$ be n vectors given by $\boldsymbol{V}_1 = a_1\boldsymbol{i} + b_1\boldsymbol{j} + c_1\boldsymbol{k}$, etc. Then the *sum* is

$$\boldsymbol{V} = \boldsymbol{V}_1 + \boldsymbol{V}_2 + \cdots + \boldsymbol{V}_n = (a_1 + a_2 + \cdots + a_n)\boldsymbol{i} + (b_1 + b_2 + \cdots + b_n)\boldsymbol{j} + (c_1 + c_2 + \cdots + c_n)\boldsymbol{k}$$

151. PRODUCT OF A SCALAR S AND A VECTOR $\boldsymbol{V}$

$$S\boldsymbol{V} = (Sa)\boldsymbol{i} + (Sb)\boldsymbol{j} + (Sc)\boldsymbol{k}$$

$$(S_1 + S_2)\boldsymbol{V} = S_1\boldsymbol{V} + S_2\boldsymbol{V} \qquad (\boldsymbol{V}_1 + \boldsymbol{V}_2)S = \boldsymbol{V}_1 S + \boldsymbol{V}_2 S$$

152. SCALAR PRODUCT OF TWO VECTORS, $\boldsymbol{V}_1 \cdot \boldsymbol{V}_2$

$$\boldsymbol{V}_1 \cdot \boldsymbol{V}_2 = |\boldsymbol{V}_1||\boldsymbol{V}_2| \cos\phi \qquad \text{where } \phi \text{ is the angle from } \boldsymbol{V}_1 \text{ to } \boldsymbol{V}_2$$

$$\boldsymbol{V}_1 \cdot \boldsymbol{V}_2 = a_1a_2 + b_1b_2 + c_1c_2 = \boldsymbol{V}_2 \cdot \boldsymbol{V}_1 \qquad \boldsymbol{V}_1 \cdot \boldsymbol{V}_1 = |\boldsymbol{V}_1|^2$$

$$(\boldsymbol{V}_1 + \boldsymbol{V}_2) \cdot \boldsymbol{V}_3 = \boldsymbol{V}_1 \cdot \boldsymbol{V}_3 + \boldsymbol{V}_2 \cdot \boldsymbol{V}_3$$

$$\boldsymbol{V}_1 \cdot (\boldsymbol{V}_2 + \boldsymbol{V}_3) = \boldsymbol{V}_1 \cdot \boldsymbol{V}_2 + \boldsymbol{V}_1 \cdot \boldsymbol{V}_3$$

$$\boldsymbol{i}\cdot\boldsymbol{i} = \boldsymbol{j}\cdot\boldsymbol{j} = \boldsymbol{k}\cdot\boldsymbol{k} = 1 \qquad \boldsymbol{i}\cdot\boldsymbol{j} = \boldsymbol{j}\cdot\boldsymbol{k} = \boldsymbol{k}\cdot\boldsymbol{i} = 0$$

153. VECTOR PRODUCT OF TWO VECTORS, $\boldsymbol{V}_1 \times \boldsymbol{V}_2$

$\boldsymbol{V}_1 \times \boldsymbol{V}_2 = |\boldsymbol{V}_1||\boldsymbol{V}_2| \sin\phi\, \mathbf{1}$, where ϕ is the angle from $\boldsymbol{V}_1$ to $\boldsymbol{V}_2$ and $\mathbf{1}$ is a unit vector perpendicular to the plane of $\boldsymbol{V}_1$ and $\boldsymbol{V}_2$ and so directed that a right-handed screw driven in the direction of $\mathbf{1}$ would carry $\boldsymbol{V}_1$ into $\boldsymbol{V}_2$.

$$\boldsymbol{V}_1 \times \boldsymbol{V}_2 = -\boldsymbol{V}_2 \times \boldsymbol{V}_1$$

$$= (b_1c_2 - b_2c_1)\boldsymbol{i} + (c_1a_2 - c_2a_1)\boldsymbol{j} + (a_1b_2 - a_2b_1)\boldsymbol{k}$$

$$(\boldsymbol{V}_1 + \boldsymbol{V}_2) \times \boldsymbol{V}_3 = \boldsymbol{V}_1 \times \boldsymbol{V}_3 + \boldsymbol{V}_2 \times \boldsymbol{V}_3$$

$$\boldsymbol{V}_1 \times (\boldsymbol{V}_2 + \boldsymbol{V}_3) = \boldsymbol{V}_1 \times \boldsymbol{V}_2 + \boldsymbol{V}_1 \times \boldsymbol{V}_3$$

$$\boldsymbol{V}_1 \times (\boldsymbol{V}_2 \times \boldsymbol{V}_3) = \boldsymbol{V}_2(\boldsymbol{V}_1 \cdot \boldsymbol{V}_3) - \boldsymbol{V}_3(\boldsymbol{V}_1 \cdot \boldsymbol{V}_2)$$

$$\boldsymbol{i} \times \boldsymbol{i} = \boldsymbol{j} \times \boldsymbol{j} = \boldsymbol{k} \times \boldsymbol{k} = 0, \quad \boldsymbol{i} \times \boldsymbol{j} = \boldsymbol{k}, \quad \boldsymbol{j} \times \boldsymbol{k} = \boldsymbol{i}, \quad \boldsymbol{k} \times \boldsymbol{i} = \boldsymbol{j}$$

$$\boldsymbol{V}_1 \cdot (\boldsymbol{V}_2 \times \boldsymbol{V}_3) = (\boldsymbol{V}_1 \times \boldsymbol{V}_2) \cdot \boldsymbol{V}_3 = \boldsymbol{V}_2 \cdot (\boldsymbol{V}_3 \times \boldsymbol{V}_1) = [\boldsymbol{V}_1\boldsymbol{V}_2\boldsymbol{V}_3]$$

$$= \begin{vmatrix} a_1 & a_2 & a_3 \\ b_1 & b_2 & b_3 \\ c_1 & c_2 & c_3 \end{vmatrix}$$

154. DIFFERENTIATION OF VECTORS

$\boldsymbol{V} = a\boldsymbol{i} + b\boldsymbol{j} + c\boldsymbol{k}$. If $\boldsymbol{V}_1$, $\boldsymbol{V}_2, \ldots$ are functions of a scalar variable t, then

$$\frac{d}{dt}(\boldsymbol{V}_1 + \boldsymbol{V}_2 + \cdots) = \frac{d\boldsymbol{V}_1}{dt} + \frac{d\boldsymbol{V}_2}{dt} + \cdots$$

where $\dfrac{d\boldsymbol{V}_1}{dt} = \dfrac{da_1}{dt}\boldsymbol{i} + \dfrac{db_1}{dt}\boldsymbol{j} + \dfrac{dc_1}{dt}\boldsymbol{k}$, etc.

$$\frac{d}{dt}(\boldsymbol{V}_1 \cdot \boldsymbol{V}_2) = \frac{d\boldsymbol{V}_1}{dt} \cdot \boldsymbol{V}_2 + \boldsymbol{V}_1 \cdot \frac{d\boldsymbol{V}_2}{dt}$$

$$\frac{d}{dt}(\boldsymbol{V}_1 \times \boldsymbol{V}_2) = \frac{d\boldsymbol{V}_1}{dt} \times \boldsymbol{V}_2 + \boldsymbol{V}_1 \times \frac{d\boldsymbol{V}_2}{dt}$$

$\boldsymbol{V} \cdot \dfrac{d\boldsymbol{V}}{dt} = |\boldsymbol{V}|\dfrac{d|\boldsymbol{V}|}{dt}$. If $|\boldsymbol{V}|$ is constant, $\boldsymbol{V} \cdot \dfrac{d\boldsymbol{V}}{dt} = 0$.

$$\text{grad } S \equiv \nabla S \equiv \frac{\partial S}{\partial x}\boldsymbol{i} + \frac{\partial S}{\partial y}\boldsymbol{j} + \frac{\partial S}{\partial z}\boldsymbol{k} \qquad \text{where } S \text{ is a scalar}$$

$$\text{div } \boldsymbol{V} \equiv \nabla \cdot \boldsymbol{V} \equiv \frac{\partial a}{\partial x} + \frac{\partial b}{\partial y} + \frac{\partial c}{\partial z} \qquad \text{(divergence of } \boldsymbol{V}\text{)}$$

$$\text{curl } \boldsymbol{V} \equiv \text{rot } \boldsymbol{V} \equiv \begin{vmatrix} \boldsymbol{i} & \boldsymbol{j} & \boldsymbol{k} \\ \dfrac{\partial}{\partial x} & \dfrac{\partial}{\partial y} & \dfrac{\partial}{\partial z} \\ a & b & c \end{vmatrix} \equiv \nabla \times \boldsymbol{V}$$

$$\text{div grad } S \equiv \nabla^2 S \equiv \frac{\partial^2 S}{\partial x^2} + \frac{\partial^2 S}{\partial y^2} + \frac{\partial^2 S}{\partial z^2}$$

$$\nabla^2 \boldsymbol{V} \equiv \boldsymbol{i}\nabla^2 a + \boldsymbol{j}\nabla^2 b + \boldsymbol{k}\nabla^2 c$$

$$\text{curl grad } S = 0 \qquad \text{div curl } \boldsymbol{V} = 0$$

$$\text{curl curl } \boldsymbol{V} = \text{grad div } \boldsymbol{V} - \nabla^2 \boldsymbol{V}$$

155. STOKES' THEOREM Let $\mathcal{S}$ be a *convex surface*, that is, one which is cut in at most two points by a line parallel to any of the coordinate axes. Let C be a simple closed curve on $\mathcal{S}$ bounding a region K, and $\boldsymbol{F}$ a vector. Then

$$\int_C \boldsymbol{F} \cdot d\boldsymbol{r} = \iint_K \text{curl } \boldsymbol{F} \cdot d\boldsymbol{S}$$

where the integrations are carried out over region K and C, the direction of the line integral being related to that of $d\boldsymbol{S}$ by a right-hand corkscrew rule. Here $d\boldsymbol{r} = dx\,\boldsymbol{i} + dy\,\boldsymbol{j} + dz\,\boldsymbol{k}$, $d\boldsymbol{S} = dS_1\boldsymbol{i} + dS_2\boldsymbol{j} + dS_3\boldsymbol{k}$, dS_1, dS_2, dS_3 being the projections of $d\boldsymbol{S}$ on the yz, zx, xy planes, respectively. If $l = \cos\alpha$, $m = \cos\beta$, $n = \cos\gamma$ are the direction cosines of the outward normal $\boldsymbol{n}$ at any point of $\mathcal{S}$ with respect to the x, y, z axes, $d\boldsymbol{S} = l dS\,\boldsymbol{i} + m dS\,\boldsymbol{j} + n dS\,\boldsymbol{k}$. Here dS is the magnitude of $d\boldsymbol{S}$.

156. GREEN'S THEOREM Let $\boldsymbol{F}$ be a vector and V a volume bounded by a convex surface $\mathcal{S}$. Then

$$\iiint_V \text{div } \boldsymbol{F}\, dV = \iint_{\mathcal{S}} \boldsymbol{F} \cdot d\boldsymbol{S}$$

where the integrations are to be carried out over V and $\mathcal{S}$. Here dV is the element of volume and $d\boldsymbol{S}$ is defined as in Sec. 155.

9. SETS, RELATIONS, AND FUNCTIONS

157. SETS A *set* (*class, aggregate, ensemble*) S is a well-defined collection of objects, or symbols, called *elements* or *members* of the set. No restriction is placed on the nature of the objects.

Set A is a *subset* of set B, written $A \subset B$ or $B \supset A$, if and only if each element of A is also an element of B. A is a *proper subset* of B if at least one member of B is not a member of A. $A \not\subset B$ means that at least one element of A is not included in B. Some writers use $A \subseteq B$ to indicate that A is a subset of B and reserve $A \subset B$ to indicate that A is a proper subset of B.

Set A is said to *equal* set B, written $A = B$, if every element of A is an element of B and every element of B is an element of A. Two equal sets are said to be *identical.* If A and B are not equal, we write $A \neq B$.

A *universal set* I (or *universe* or *space*) is the term given to an overall set which includes all the sets of concern in a given study.

An *empty* (*null*) set, written $\emptyset$, is a set that has no elements. It is usually assumed that for every set A, $\emptyset \subset A$.

The *complement* A' of a set A with respect to a given universal set I is the set of elements in I that are not in A. [Some writers use $\bar{A}$ or $\mathcal{C}(A)$ or $I - A$ to denote the complement A'.]

Notation. The elements of a set are usually denoted by small letters. Capital letters are used as the names of sets. The elements of a set are often listed by enclosing them in braces. The notation $x \in S$ means that x is a member of set S; $x \notin S$ means that x is not a member of S.

To indicate a set of objects x having the property P, the symbol $\{x \mid x \text{ has the property } P\}$ is used. $\{x \mid \quad\}$ is called a *set builder.* The bar "$\mid$" is read "such that."

Example. $S = \{1,2,4,6\}$ is a set S consisting of the elements 1, 2, 4, 6.

Example. $\{(x,y) \mid (x - y) > 2\}$ denotes the set of number pairs (x,y) such that $(x - y)$ exceeds 2.

158. UNION AND INTERSECTION The *union* (*join*) of sets A and B, written $A \cup B$ or "A cup B," is the set of all elements that are in A or in B or in both A and B.

The *intersection* (*meet*) of the sets A and B, written $A \cap B$, is the set of all elements which are in both A and B.

The sets A and B are said to be *disjoint* or *mutually exclusive* if they have no elements in common, that is, if their intersection is the empty set. A and B *overlap* if they have some elements in common.

159. VENN DIAGRAM One scheme for representing a subset V of ordered pairs (x, y) consists in using a rectangle to represent the universal set, and sets of points inside the rectangle to represent V. For example, Figs. 58 and 59 illustrate the commutative law; Fig. 60, the complement.

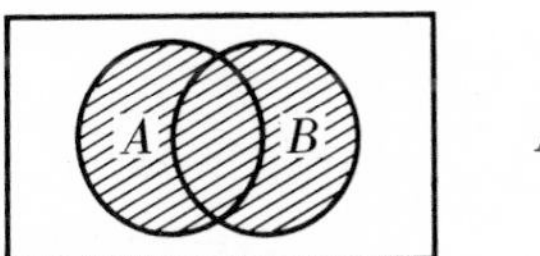

$A \cup B = B \cup A$

Fig. 58. *Crosshatched section illustrates* $A \cup B = B \cup A$.

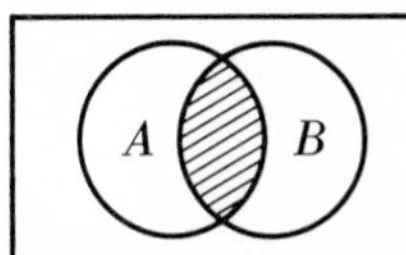

$A \cap B = B \cap A$

Fig. 59. *Crosshatched section illustrates* $A \cap B = B \cap A$.

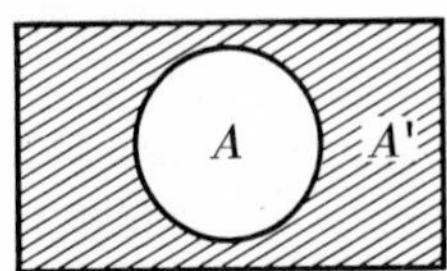

A'

Fig. 60. *Crosshatched section illustrates* A'.

160. OPERATIONS ON SETS Consider the ordered n-tuple $X = \{x_1, x_2, \ldots, x_n\}$ of n elements, where the elements x_i are members of set S_i. An *operation* on X is a *rule* which assigns to each X a uniquely defined element in some set S. If $n = 1$, the operation is *unary;* if $n = 2$, the operation is *binary;* if $n = m$, the operation is *m-ary*.

161. **ALGEBRA OF SETS** If A, B, C are any three subsets of a universe I, then, for the operations $\cup$ and $\cap$, the following relationships (laws) hold:

1. Closure laws
   (a) There exists a unique set $A \cup B$
   (b) There exists a unique set $A \cap B$
2. Commutative laws
   (a) $A \cup B = B \cup A$ **(b)** $A \cap B = B \cap A$
3. Associative laws
   (a) $(A \cup B) \cup C = A \cup (B \cup C)$
   (b) $(A \cap B) \cap C = A \cap (B \cap C)$
4. Distributive laws
   (a) $A \cup (B \cap C) = (A \cup B) \cap (A \cup C)$
   (b) $A \cap (B \cup C) = (A \cap B) \cup (A \cap C)$
5. Idempotent laws
   (a) $A \cup A = A$ **(b)** $A \cap A = A$
6. Properties of universe I. Identity laws
   (a) $A \cup I = I$ **(b)** $A \cap I = A$
7. Properties of null set $\emptyset$. Identity laws
   (a) $A \cup \emptyset = A$ **(b)** $A \cap \emptyset = \emptyset$
8. Properties of inclusion $\subset$
   (a) $A \subset I$ **(b)** $\emptyset \subset A$
   (c) If $A \subset B$, then $A \cup B = B$
   (d) If $B \subset A$, then $A \cap B = B$
   (e) $A \subset (A \cup B)$ **(f)** $(A \cap B) \subset A$
9. Properties of complement
   (a) For every set A there exists a unique set A' with respect to I.
   (b) $A \cup A' = I$
   (c) $A \cap A' = \emptyset$
   (d) $(A \cup B)' = A' \cap B'$ De Morgan's law (dualization)
   (e) $(A \cap B)' = A' \cup B'$ De Morgan's law (dualization)
   (f) $\emptyset' = I$
   (g) $I' = \emptyset$
   (h) $(A')' = A$ (involution)
10. Duality principle

    If in any "correct" formula involving the symbols $\cup$, $\cap$, $\emptyset$, I, $\subset$, and $\supset$ we interchange $\cup$ and $\cap$, $\emptyset$ and I, and $\subset$ and $\supset$ throughout the formula, the resulting formula is correct.

11. Absorption laws
 (*a*) $A \cup (A \cap B) = A$ (*b*) $A \cap (A \cup B) = A$
12. Consistency property
 The conditions $A \subset B$, $A \cap B = A$, and $A \cup B = B$ are mutually equivalent.

162. **ONE-TO-ONE CORRESPONDENCE BETWEEN SETS** Two sets A and B are said to be in *one-to-one* (*reciprocal*) *correspondence* when a pairing of the elements of A with the elements of B is such that each element of A corresponds to exactly one element of B and each element of B corresponds to exactly one element of A. ("Exactly" means "one and only one.")

163. **EQUIVALENT SETS** Two sets A and B are said to be *equivalent,* written $A \approx B$, if and only if they can be placed in one-to-one correspondence.

164. **CARDINAL NUMBERS** Two sets are said to have the same *cardinal number* if and only if they are equivalent.

A set A is said to have the cardinal number n if and only if there exists a one-to-one correspondence between the elements of A and the natural numbers 1, 2, 3, . . . , n.

165. **FINITE AND INFINITE SETS** A nonempty set is *finite* if and only if its cardinal number is one of the cardinal numbers 1, 2, 3, ... An *infinite set* is a nonempty set which is *not finite.*

166. **$\aleph_0$ AND $\aleph$** The cardinal number of the set of all natural numbers 1, 2, 3, . . . is denoted by $\aleph_0$. The cardinal number of the set R of all real numbers is denoted by $\aleph$. $\aleph_0$ and $\aleph$ are examples of *transfinite numbers.*

167. **DENUMERABLE SETS** If a set has the cardinal number $\aleph_0$, it is said to be *denumerable* (*enumerable*).

168. **COUNTABLE SETS** A set A is *countable* if and only if there exists a one-to-one correspondence between the elements of A and the elements of a set of natural numbers.

If a set is either finite or denumerable, it is *countable.* The elements of a countable set can be listed in a sequence.

Example. Every finite set A is countable, the number of elements in A being identical with the cardinal number of A.

Example. The set of all rational numbers is countable (and denumerable) with cardinal number $\aleph_0$.

Note. The meanings of *denumerable* and *countable* vary with authors.

Theorem. A set A which has the same cardinal number as one of its proper subsets is an infinite set. A set which does not have the same cardinal number as any of its proper subsets is a finite set.

169. NONCOUNTABLE SETS A set which is not countable is said to be *noncountable* (*nondenumerable*).

Example. The set of all real numbers between 0 and 1 is noncountable with cardinal number $\aleph$.

170. MAPPING OF SET A INTO SET B Let A and B be two sets not necessarily disjoint. A collection of pairs (a,b), where $a \in A$ and $b \in B$, such that each a of A occurs in one and only one pair (a,b) and such that any b of B may occur in any number of pairs is called a *mapping* of A *into* B. Such a mapping is known as a *many-one* (or *many-to-one*) *correspondence* between the elements of A and some or all of the elements of B. A mapping of a set A into a set B is a matching procedure that assigns to each element $a \in A$ a unique element $b \in B$. This matching is written $a \to b$, b being called the *image* of a.

If every element of B is used as an image, then the mapping is said to be "A *onto* B," as well as "A *into* B."

171. PRODUCT SET Let $A_1, A_2, \ldots, A_n$ be n sets. Let $a_i \in A_i$, for $i = 1, 2, \ldots, n$. The collection of all ordered n-ples $(a_1, a_2, \ldots, a_n)$ is called the *product set* $A_1 \times A_2 \times \cdots \times A_n$.

The set of all ordered pairs (x, y), where $x \in X$ and $y \in Y$, is called the *cartesian product* of X and Y, is denoted by $X \times Y$, and is read "X cross Y."

$U \times U$. If $x \in U$ and $y \in U$, the set of all ordered pairs (x, y) that can be found using the elements of U is a *cartesian product set* on U and is denoted by $U \times U$.

Equal Ordered Pairs. The ordered pair (a,b) *equals* the ordered pair (x,y); that is, $(a,b) = (x,y)$, if and only if $a = x$ and $b = y$.

172. VARIABLE A symbol in a sentence, phrase, or expression is a *variable* (or *placeholder*) if and only if it is replaceable by the name of an element (number) of a set called the *replacement set* (or *domain* or *universal set*) specifically defined for the variable. If the set contains only a single element, the symbol used to represent this element is called a *constant.* Variables are used as placeholders for numbers and other mathematical quantities in algebraic sentences, such as equations and inequalities.

173. RELATION A *relation* R in a set U is a subset of ordered pairs (x,y) from $U \times U$. The *domain* of the relation R is the subset of U for which x is a placeholder; the *range* of the relation is the subset of U for which y is a placeholder.

174. GRAPH Let x and y be real numbers. The *graph* of a relation whose ordered pairs are (x,y) is the set of points in the XY plane whose coordinates are the given pairs.

175. LOCUS A *locus* is the set of those points, and no other points, that satisfy a given condition. A locus is a graph of a relation.

176. SOLUTION SET Suppose x belongs to the universal set I. Consider the set $\{x \mid S\}$, where S is a statement involving x, or a condition placed on x. Any member of I for which the statement S is true is said to *satisfy* S, or to be a *solution set* of S. S is called a *defining* condition, or *set selector.*

Example. If x belongs to the set of real numbers, the solution set of $\{x \mid 2x = 3\}$ is $\{3/2\}$.

Example. If $I = \{1,3,5\}$ and x belongs to I, the solution set of $\{x \mid x > 2\}$ is $\{3,5\}$.

177. FUNCTION A *function* f in a set U is a subset of ordered pairs (x,y) from $U \times U$ having exactly one y for each x. In other words, a *function* f in U is a relation R in U such that for every x in the domain of R there corresponds exactly one y in the range of R. (This definition of function corresponds to the notion of a single-valued function given by many writers.)

The notation $f: (x, y)$ is sometimes used to mean "the function f whose ordered pairs are (x, y)."

The notation $f(x)$, read "f at x," is often used to denote the second element y of the ordered pair (x, y) whose first element is x. $f(x)$ is called "the value of the function f at x." It is common to write $y = f(x)$, meaning y is the value of f at x.

Example. If $U = \{0,1\}$, the set of ordered pairs $U \times U$ is the *lattice* points (0,0), (0,1), (1,0), (1,1). If the relation R_1 in U is given by $R_1 = \{(x, y) \mid (x^2 + y^2) < 1\}$, then R_1 selects from $U \times U$ the subset (0,0). If $R_2 = \{(x, y) \mid (x^2 + y^2) \leqq 1\}$, then R_2 selects the subset (0,0), (0,1), (1,0). The domain of the relation R_1 is $\{0\}$, the range of R_1 is $\{0\}$; the domain of R_2 is $\{0,1\}$, the range of R_2 is $\{0,1\}$. R_1 is a function; R_2 is *not* a function.

Example. The relation $\{(x, y) \mid y = x^2 + 1\}$ in the set U of real numbers defines a function F. $F(1) = 2$ is the value of $F(x) = x^2 + 1$ at $x = 1$. The domain of the function is U, the range of F is the set of all real numbers equal to or greater than 1.

178. INVERSE RELATION The set R^{-1} obtained by interchanging the coordinates of all the ordered pairs of a given relation R is called the *inverse relation of* R.

Example. The inverse R^{-1} of the relation $R = \{(1,2), (1,3), (2,4)\}$ is $R^{-1} = \{(2,1), (3,1), (4,2)\}$. R is not a function; R^{-1} is a function.

179. INVERSE FUNCTION The inverse of a function is obtained by interchanging the coordinates of each of its ordered pairs; the inverse is a relation, but may or may not be a function.

A function F is a set of ordered pairs (x, y) no two of which have the same first component x with different second components. If a function F is such that no two of its ordered pairs have the same second component with different first components, then the set of ordered pairs obtained from F by interchanging in each ordered pair the first and second elements is called the *inverse function* and is denoted by F^{-1}.

The range of F is the domain of F^{-1}, and the domain of F is the range of F^{-1}. We write $F: (a,b)$ and $F^{-1}: (b,a)$.

Example. If the function F has elements (ordered pairs) of the form (a_1,b_1), (a_2,b_2), . . . , the inverse function F^{-1}, if it exists, has elements (ordered pairs) of the form (b_1,a_1), (b_2,a_2), ...

Example. Let x and y be real numbers. The inverse to the relation $R = \{(x,y) \mid y = 2x\}$ is $R^{-1} = \{(y,x) \mid x = y/2\}$. R^{-1} and R are each functions. The inverse function is then sometimes written $R^{-1} = \{(x,y) \mid y = x/2\}$.

Example. Let x and y be real numbers. The inverse relation to the relation $R = \{(x,y) \mid y = 3x^2\}$ is $R^{-1} = \{(y,x) \mid x = \pm\sqrt{y/3}\}$. R is a function; R^{-1} is not a function. The inverse relation may be written $R^{-1} = \{(x,y) \mid y = \pm\sqrt{x/3}\}$. If the domain of R is restricted so that $x \geqq 0$, then $R^{-1} = \{(x,y) \mid y = \sqrt{x/3}\}$ is a function.

Example. The inverse relation to $R = \{(x,y) \mid y - 2 = |x| - x\}$ is $R^{-1} = \{(x,y) \mid x - 2 = |y| - y\}$. R^{-1} is not a function.

Example. Let x and y be real numbers. The inverse R^{-1} of the relation $R = \{(x,y) \mid y = \sin x\}$ is $R^{-1} = \{(x,y) \mid x = \sin y\}$. Here y is a number whose sine is x. R^{-1} is not a function.

If R is restricted to values of x such that $-\pi/2 \leqq x \leqq \pi/2$, then R^{-1} is a function. In this case we may write (see Sec. 60)

$$R^{-1} = \{(x,y) \mid y = \operatorname{Sin}^{-1} x\} \qquad \frac{-\pi}{2} \leqq y \leqq \frac{\pi}{2}$$

180. COMPOSITE OF TWO FUNCTIONS Let $f\colon (x,y)$ and $g\colon (y,z)$ be two given functions. Construct an ordered pair (x,z) as follows: Select an x; f assigns to x a value y. If y is in the domain of g, g assigns to y a value z. This gives an ordered pair (x,z). The set of all such pairs (x,z) which can be constructed by this scheme describes a function known as the *composite function* $g \circ f$. It is written $g \circ f\colon (x,z)$.

Example. Let the values of the functions $f\colon (x,y)$ and $g\colon (y,z)$ be $y = f(x) = x^2$ and $z = g(y) = 3y - 2$, where x, y, z are real numbers. The composite function $g \circ f\colon (x,z)$ is given by $z = g(y) = 3x^2 - 2$; that is,

$$g \circ f\colon (x,z) = g \circ f\colon (x, 3x^2 - 2)$$

The composite function $f \circ g\colon (x,z) = f \circ g\colon [x, (3x - 2)^2]$, since the values of $g\colon (x,y)$ and $f\colon (y,z)$ are

$$y = g(x) = 3x - 2 \qquad \text{and} \qquad z = f(y) = y^2 = (3x - 2)^2$$

181. ALGEBRA OF FUNCTIONS Let $f\colon(x,y)$ and $g\colon(x,z)$ be two functions whose domains are d_f and d_g, respectively. Let d_f and d_g consist of sets of real or complex numbers for which the elementary operations of $+$, $-$, $\times$, $\div$ are defined. The *sum* $f+g$, the *difference* $f-g$, the *product* fg, and the *quotient* f/g are defined as follows:

$$f\colon(x,y) + g\colon(x,z) = (f+g)\colon(x, y+z)$$
$$f\colon(x,y) - g\colon(x,z) = (f-g)\colon(x, y-z)$$
$$f\colon(x,y) \times g\colon(x,z) = (fg)\colon(x, yz)$$
$$\frac{f\colon(x,y)}{g\colon(x,z)} = \left(\frac{f}{g}\right)\colon\left(x, \frac{y}{z}\right)$$

Let $y = f(x)$ and $z = g(x)$ be the functional values of f and g, respectively. Then the functional values of the sum, difference, product, and quotient are, respectively,

$$(f+g)(x) = f(x) + g(x)$$
$$(f-g)(x) = f(x) - g(x)$$
$$(fg)(x) = f(x) \times g(x)$$
$$\left(\frac{f}{g}\right)(x) = \frac{f(x)}{g(x)}$$

Domain. The domain of $f+g$, $f-g$, and fg is the set of all elements x common to the domains of f and g, that is, the intersection of the sets d_f and d_g. This is the domain of f/g, except that those xs for which $g(x) = 0$ must be excluded since division by zero is not possible.

Identity Function. The *identity function* is the function i whose elements are the ordered pairs (x,x). i is sometimes written $i = \{(x,i(x))|i(x) = x\}$.

182. COMPOSITE OF FUNCTION AND ITS INVERSE Let $F\colon(x,y)$ and $F^{-1}\colon(y,x)$ be a function and its inverse, respectively. The composite function $F^{-1} \circ F$ sends (under F) each x into some y, and then (under F^{-1}) sends y back into x. That is, $F^{-1} \circ F = i$. Likewise $F \circ F^{-1} = i$.

183. ADDITIVE SET FUNCTION Let $A, B, C, \ldots$ be subsets of a finite set I. Suppose f is a relation that assigns to any subset S of I a unique real number denoted by $f(S)$ such that if A and B are disjoint subsets of I, $f(A \cup B) = f(A) + f(B)$, then f is said to be an *additive set function.*

Theorem. If f is an additive set function and $\emptyset$ is a null set, then $f(\emptyset) = 0$; that is, the value of f at $\emptyset$ is zero.

Theorem. If A and B are any two subsets of I, not necessarily disjoint, and if f is an additive set function, then

$$f(A \cup B) = f(A) + f(B) - f(A \cap B)$$

Weight Function. An additive set function w, all of whose values are positive or zero, is called a *weight function*; i.e., for all subsets S of I, $w(S) \geqq 0$.

Probability Function. If P is a weight function, I the universal set, and $P(I) = 1$, then P is called a *probability function.*

Theorem. If f is an additive set function, $f(A) + f(A') = f(I)$, where A is a subset of the universal set I and A' the complement of A with respect to I.

Theorem. If P is a probability function, $P(A) + P(A') = 1$.

184. LIMITS Let x take on the sequence of real values $x_1, x_2, \ldots, x_n, \ldots$. The variable x is said to approach the number a as a limit, written $x \to a$, if for every positive number ϵ there exists an integer N such that for every $n > N$, $|x_n - a| < \epsilon$.

Let $f(x)$ be the value of the function f at x. $f(x)$ is said to approach L as a limit as $x \to a$, written $\lim_{x \to a} f(x) = L$, if for every $\epsilon > 0$ there exists a positive number δ such that for all x satisfying $0 < |x - a| < \delta$, $|f(x) - L| < \epsilon$.

185. CONTINUOUS FUNCTION The function f is *continuous* at $x = a$ if $f(a)$ is defined, $\lim_{x \to a} f(x) = L$ exists, and $f(a) = L$.

A function $f(x)$ is *uniformly continuous* on (a,b) if for every $\epsilon > 0$ there exists a number $\delta > 0$ such that $|f(x_1) - f(x_2)| < \epsilon$ for every pair of numbers x_1, x_2 on $[a,b]$ closed for which $|x_1 - x_2| < \delta$.

A *closed interval* $[a,b]$ is a set of real numbers x such that $a \leqq x \leqq b$.

186. SEQUENCES AND SERIES If to every integer n there corresponds a number s_n, then $s_1, s_2, \ldots, s_n, \ldots$ form a *sequence* $\{s_n\}$.

A sequence $\{s_n\}$ *converges* to the number S, (i.e., is *convergent* with *limit* S), if for every preassigned $\epsilon > 0$ there exists a number N such that for every $n > N$, $|s_n - S| < \epsilon$. A sequence which is not convergent is *divergent.*

If for any $G > 0$ an N exists such that for all $n > N$, $s_n > G$, then $\{s_n\}$ *definitely diverges positively,* i.e., *increases without bound.* This is written $s_n \to +\infty$ as $n \to +\infty$.

If for any $-G < 0$, $s_n < -G$ for all $n > N$, then $\{s_n\}$ *definitely diverges negatively,* i.e., *decreases without bound.* This is written $s_n \to -\infty$ as $n \to +\infty$.

A divergent sequence which is not definitely divergent is *indefinitely divergent* (i.e., it may *oscillate*).

The *infinite series* $\sum\limits_{i=1}^{\infty} a_i \equiv a_1 + a_2 + \cdots + a_n + \cdots$ is a symbol for the sequence $\{s_n\}$, where $s_n = \sum\limits_{i=1}^{n} a_i$. The series is convergent, definitely divergent, or indefinitely divergent according as $\{s_n\}$ exhibits the behavior indicated by those names. If $\{s_n\}$ converges to S, then S is called the *sum* of the series $\sum\limits_{i=1}^{\infty} a_i$.

The series $\sum\limits_{i=1}^{\infty} a_i$ is *absolutely convergent* if $\sum\limits_{i=1}^{\infty} |a_i|$ is convergent.

187. UNIFORM CONVERGENCE Let $s_n \equiv s_n(x)$ be functions of a real variable x defined on I. Suppose $\{s_n\}$ is convergent to $S(x)$ for every value of x in I. $\{s_n\}$ is *uniformly convergent* in a subinterval J of I if for every $\epsilon > 0$ there exists a single number N, independent of x in J, such that $|s_n(x) - S(x)| < \epsilon$ for all $n > N$ and for all x in J.

Consider a series $\sum_{i=1}^{\infty} a_i(x)$ of functions $a_i(x)$ of x, convergent in I. Let $s_n(x) = \sum_{i=1}^{n} a_i(x)$. The series $\sum_{i=1}^{\infty} a_i(x)$ is *uniformly convergent* in subinterval J of I if $\{s_n(x)\}$ is uniformly convergent in J.

Theorem. A necessary and sufficient condition for the convergence of $\{s_n\}$ is that for any $\epsilon > 0$ there exists a number N such that for every $n > N$ and for every $k \geqq 1$, $|s_{n+k} - s_n| < \epsilon$, that is, that $|a_{n+1} + \cdots + a_{n+k}| < \epsilon$.

188. **ASYMPTOTIC EQUALITY** If $a_n \equiv a(n)$ and $b_n \equiv b(n)$ are two functions of integer n and if $\lim_{n\to\infty} a_n/b_n = 1$, a_n is said to be *asymptotically equal* to b_n. In this case, $a_n \cong b_n$ is written.

10. LOGIC

189. SYMBOLIC LOGIC A mathematical theory results from the interplay of a set of postulates and a logic. The theory starts with a set of postulates as the basis. The logic constitutes the rules by which such a basis may be developed into a body of theorems. The mathematical theory, or system, consists of the collection of statements composed of the postulates and the theorems.

A *symbolic logic* (*mathematical logic*, *logistics*) is a *skeleton of language* (i.e., a system of symbols and rules for their use). Symbolic logic is not a theory (i.e., a system of assertions about objects). In symbolic logic, relations among propositions, sets, and so on, are represented by symbols and formulas whose interpretations are free from the ambiguities of ordinary language.

The signs of a symbolic logic remain uninterpreted as long as we are concerned only with a *pure logic* (i.e., the building of the language of a symbolic logic). In *applied logic* we are concerned with the application and interpretation of a symbolic logic with respect to a given theory.

190. BASIC SYMBOLS; PROPOSITIONS AND PROPOSITIONAL FORMS We are concerned with the elementary logical properties of statements. Ordinary declaratory statements (assertions, propositions) are denoted by the letters p, q, r, ... Statements may involve constants and variables. A *constant* has a fixed specific meaning. A *variable* serves to refer to unspecified objects, properties, etc., and has associated with it a valid domain of values. A statement which contains variables is said to be *open*; otherwise, it is *closed.* A *proposition* is any statement concerning which it is meaningful to say that its content is "true" or "false". (What is meant by "true" and by "false" will not be discussed here.)

A statement which becomes a proposition when specific values (i.e., constants) are submitted for all variables is called a *propositional form* (*propositional function*, *sentential function*, *sentential formula*). The term *constant* is also called a *propositional constant* or *sentential constant.* The term *propositional* or *sentential variable* is used for a variable whose domain is a set of whole sentences.

Assigned to propositions are *truth-values,* the truth-value of a proposition being *truth* if it is true and *falsehood* if it is false. A statement about which it is meaningless, or ridiculous, etc., to say that its content is true or false is specifically excluded in our definition of a proposition. Statements requiring other types of symbolic logic, for example, n-way logics, and situations where many gradations of possibilities may occur, as contrasted with the exclusive possibility of true or false, yes or no, etc., are specifically excluded here.

When we assert a proposition p, without qualification as to its truth or falsity, we mean to imply that p is true. The *assertion* of the truth of a proposition p is written $\vdash . p$.

Statements can be combined in various ways to form new statements. Among the grammatical connectives used to combine phrases, statements, etc., are such common words as "not", "and", "or", "is", "every", "some", "if then", and "if and only if". These words are used in many ways, and their meanings in a given context are often vague and ambiguous. In symbolic logic, the concern is to establish the precise meaning of such terms and to lay down the laws in which these terms are involved. The logical meanings of "and", "or", etc., differ somewhat from the uses of these connectives in ordinary language.

Calculus of Propositions

191. The object of the *calculus of propositions* (*propositional* or *sentential calculus*) is the development of logical inference. As with statements in general, propositions p, q, r, . . . may be combined in many ways to form new propositions called *composite propositions.* We introduce the following definitions:

1. *Conjunction.* A proposition, represented by $p \wedge q$, which is true when and only when both proposition p and proposition q are true is called a *conjunction* (*logical product*). $p \wedge q$ is read "p and q".

2. *Disjunction.* A proposition, represented by $p \vee q$, which is true when and only when at least one of the two propositions p and q is true is called an (*inclusive*) *disjunction* (*logical sum*). The symbol $p \vee q$ is read "p or q".

3. *Implication.* A proposition, represented by $p \to q$, which is false when and only when p is true and q is false is called a (*material*) *implication* (or *conditional sentence*). $p \to q$ is read "if p, then q".

4. *Equivalence.* A proposition, represented by $p \leftrightarrow q$, which is true when and only when p and q are both true or both false is called an *equivalence.* $p \leftrightarrow q$ is read "p if and only if q".

5. *Negation.* The denial or contradiction of a proposition p is a proposition represented by the symbol p'. p' is read "*not* p". The proposition p' is true when p is false, and p' is false when p is true. A proposition of the form p' is called a *negation.*

6. *Exclusive Disjunction.* A proposition, represented by $p \veebar q$, which is true when and only when proposition p or proposition q is true, but not both, is called an *exclusive disjunction.*

192. TRUTH TABLES The symbols $\wedge$, $\vee$, $\rightarrow$, and $\leftrightarrow$ denote binary operations performed on two propositions. The symbol $'$ denotes a unary operation performed on propositions.

A compact way of summarizing the definitions of conjunction, etc., is by use of *truth tables.* In such tables the symbol T denotes that the corresponding proposition is true, F that it is false. To form such a table, all possible combinations of T and F for the component propositions involved must be listed and then the truth or falsity of each combination is entered into the table. Table 10.1 gives the truth tables for the operations of conjunction, disjunction, implication, equivalence, negation, and exclusive disjunction. For example, the table states that $p \wedge q$ is true if p and q are each true; that $p \wedge q$ is false in every other case.

TABLE 10.1. TRUTH TABLE FOR OPERATIONS 1 TO 6

p	q	$p \wedge q$	$p \vee q$	$p \rightarrow q$	$p \leftrightarrow q$	p'	$p \veebar q$
T	T	T	T	T	T	F	F
T	F	F	T	F	F	F	T
F	T	F	T	T	F	T	T
F	F	F	F	T	T	T	F

193. RELATIONS TO ORDINARY LANGUAGE OF DEFINITIONS 1 THROUGH 6 "*and*". In logic *and* is used to connect any two propositions whatever without regard to their possible bearing on one another. In ordinary language *and* is commonly used to conjoin propositions that are relevant to each other, as, for example, in a connected discourse.

"*or*". In logic *or* is used in the inclusive sense of the Latin word *vel* (which means "p or q or both p and q"), and the possible bearing of one proposition on another is disregarded. In ordinary language *or* is used in the exclusive sense of the Latin word *aut* (which means "p or q but not both p and q") and in the inclusive sense of *vel*.

"*if . . . , then . . .*". In logic "if . . . , then . . ." is considered meaningful even if no connection exists between the *antecedent* (the subordinate clause to which "if" is prefixed) and the *consequent* (the principal clause introduced by "then"). Contemporary logic uses implications in the material meaning defined in 3; the concern is with the truth-values of propositions, not with the meanings of the propositions; whether there is some necessary connection between p and q is not of importance.

In ordinary language "if . . . , then . . ." is often used only when there is a connection between the forms and contents of the antecedent and consequent, a relation between a premise and a conclusion, or a sort of cause and effect relation. (In mathematics, theorems tend to have the form of an implication; the antecedent may be called a *hypothesis*, the consequent a *conclusion*.) Often in ordinary language if the antecedent is assumed to be true, the consequent must be assumed to be true too; implications are commonly asserted without exact knowledge of whether or not the antecedent and consequent are true.

Implication is commonly used in ordinary language in situations where the presence of a formal connection between antecedent and consequent is a necessary condition for the meaningfulness and truth of the implication. Such an implication is a *formal implication*, or an *implication in formal meaning*. A formal implication which is based on a relation of premise and conclusion, or of cause and effect, really depends upon the structure of the component propositions rather than upon their truth or falsity, and has been termed a *strict implication*.

"*if and only if*". In logic the assertion $p \leftrightarrow q$ means that p and q have the same truth-values; that is, $p \leftrightarrow q$ is true when and only when p and q are both true or both false; $p \leftrightarrow q$ is not intended to imply that p and q have the same significance or meaning. In ordinary language $p \leftrightarrow q$, that is, "if and only if", is often used with a looser meaning; e.g., it may imply that the antecedent and consequent have the same significance or meaning.

"*not*". The negation of a sentence is formed with the help of "not". Two sentences, the first being the negation of the other, are called *contradictory sentences.*

194. **INVERSE SENTENCE** If in the sentence "if p, then q" the antecedent p is replaced by the negation p' and the consequent q is replaced by the negation q', then the resulting sentence "if p', then q'" is called the *inverse sentence* to "if p, then q".

195. **COUNTERPOSITIVE SENTENCE** If in the inverse sentence "if p', then q'" the antecedent and consequent are interchanged, the resulting sentence "if q', then p'" is called the *counterpositive sentence* to "if p, then q".

196. **CONVERSE SENTENCE** If in the sentence "if p, then q" the antecedent and consequent are interchanged, the resulting sentence "if q, then p" is called the *converse sentence* to "if p, then q".

197. **TRUTH-FUNCTIONS** Many propositional forms can be constructed from simple propositional forms with the help of variables, constants, $\wedge$, $\vee$, $\rightarrow$, etc. Every propositional form is a *truth-function.* The truth or falsehood of any whole sentence obtained from a given propositional form depends exclusively on the truth or falsehood of the sentences substituted for the propositional variables p, q, r, etc.

198. **TRUTH-VALUES OF COMPOSITE PROPOSITIONS** Propositions may be combined or modified into more complicated *composite* propositions. The truth-value of a composite proposition can be determined for any combination of truth-values of the fundamental propositions $p, q, \ldots$ by means of the truth tables for the operations $\wedge$, $\vee$, $\rightarrow$, $\leftrightarrow$, $'$ involved.

199. **TAUTOLOGIES** A composite proposition involving the propositions $p, q, \ldots$ which has the truth-value T no matter what the truth values of $p, q, \ldots$ is called a *tautology* and is said to be "logically true". A sentence is called a *tautology*, a *contingency*, or a *contradiction* (*absurdity*, "logically false") as its distribution of truth-values shows, respectively, only T, at least one T, or only F.

TABLE 10.2. SOME LAWS OF LOGIC (TAUTOLOGIES)

For any propositions p, q, r, irrespective of their truth-values, each of the following propositions is a tautology, i.e., has the truth-value T:

1. Law of excluded middle	$p \vee p'$
2. Law of contradiction	$(p \wedge p')'$
3. Law of syllogism	$[(p \to q) \wedge (q \to r)] \to (p \to r)$
4. Law of double negation	$p \leftrightarrow (p')'$
5. Law of contraposition	$[p \to q] \leftrightarrow [q' \to p']$

200. LOGICALLY EQUIVALENT PROPOSITIONS Let m and n be two composite (or fundamental) propositions. Form the truth tables for m and n in terms of $p, q, \ldots$. If the truth tables for m and n agree with each other in point of truth and falsity, the propositions are said to be *logically equivalent.* In such a case the proposition $m \leftrightarrow n$ is a tautology, or law of logic.

Example. The propositions $p \to q$ and $q' \to p'$ are logically equivalent; and $[p \to q] \leftrightarrow [q' \to p']$ is a tautology.

Example. The proposition $(p')'$ is logically equivalent to p; and $(p')' \leftrightarrow p$ is a tautology.

TABLE 10.3. PAIRS OF LOGICALLY EQUIVALENT PROPOSITIONS

For any propositions p and q, the propositions m and n listed on a given row are logically equivalent; that is, $m \leftrightarrow n$.

m $\longleftrightarrow$	n
$p \wedge q$	$(p' \vee q')'$
$p \to q$	$p' \vee q$
$p \leftrightarrow q$	$[(p' \vee q)' \vee (q' \vee p)']'$
$p \vee q$	$(p' \wedge q')'$
$p \to q$	$(p \wedge q')'$
$p \leftrightarrow q$	$(p \wedge q')' \wedge (q \wedge p')'$
$p \wedge q$	$(p \to q')'$
$p \vee q$	$p' \to q$
$p \leftrightarrow q$	$[(p \to q) \to (q \to p)']'$

201. QUANTIFIERS In addition to "and", "or", etc., there are phrases such as "for all x . . ." and "there exists an x such that . . ."

involving the *quantifiers* "for all" and "there exists" which must be considered in any development of logic.

The following notations for quantifiers are in common use:

$\forall$ "for every element (of the domain)" or "for any . . ." or "for all . . ." or "for each . . ."

$\exists$ "there is at least one element (of the domain) such that" or "there exists an element (of the domain)" or "there are . . ." or "for some . . ."

The symbol $\forall$ is called a *universal quantifier.* The symbol $\exists$ is called the *existential quantifier.*

Example. The phrase "for any x, p is true" may be written $\forall\, x$, p is true.

Example. The phrase "there exists an element x such that p is true" may be written $\exists\, x$, p is true.

202. **UNIVERSAL AND EXISTENTIAL SENTENCES** A sentence of the form "for all things $x, y, \ldots, p$ is true" is a *universal sentence.*

A sentence of the form "there exists things $x, y, \ldots$ such that p is true" is an *existential sentence.*

203. **LIMITATIONS OF CALCULUS OF PROPOSITIONS** The calculus of propositions is adequate for the rendering of logical connections in which the propositions appear as unanalyzed wholes. However, this calculus is not adequate for more general logical purposes, since there are logical inferences which depend on the inner content of the propositions themselves as well as upon the propositions as wholes. To take care of this type of logical structure the calculus of sets (or classes) is needed.

204. **OTHER LOGICS** Statements for which it is not meaningful to say that their content is true or false lead to various types of logic (e.g., many-valued logics). In such cases the truth tables for a proposition may have entries which are *undecidable.*

Calculus of Relations

205. **RELATION** A *relation* is a form of connectivity, or association, between two or more things. A relation connecting two elements is a *dyadic* (*binary*) *relation*; connecting three elements, a *triadic relation*; . . . , connecting many elements, a *polyadic relation.*

The phrase "the thing x has the dyadic (binary) relation R to the thing y" may be written $x\ R\ y$. x is called a *predecessor* and y a *successor* with respect to the relation R. The set of all predecessors with respect to R is the *domain*, and the set of all successors is the *range* of R.

If x is not related to y by the relation R, we write

$$x \not R\, y \qquad \text{or} \qquad \sim(x\ R\ y)$$

For mathematical purposes a binary relation R is defined by a set of ordered pairs (see Sec. 160).

206. INCLUSION If, whenever R holds between two things, S also holds between them, the relation R is *included* in the relation S. This is written $R \subset S$.

Example. The relation R of "being smaller" is included in the relation S of "diversity"; that is, $R \subset S$.

207. IDENTICAL RELATIONS If the relations R and S both hold between the same things, that is, if $R \subset S$ and $S \subset R$, then the *relations are identical*, and this is written $R = S$.

208. UNIVERSAL AND NULL RELATIONS A relation which holds between any two individuals is a *universal relation.* A relation which holds between no individuals is a *null relation.*

209. UNION OF RELATIONS The relation $R \cup S$ is said to hold between two things if and only if at least one of the relations R and S holds between the things. $R \cup S$ is the *union* (or *sum*) of two relations R and S.

210. INTERSECTION OF RELATIONS The relation $R \cap S$ is said to hold between two things if and only if each of the relations R and S holds between the things. $R \cap S$ is the *intersection* (or *product*) of two relations R and S.

211. NEGATION OF RELATIONS A relation R' which holds between two things if and only if the relation R does not hold between them is the *negation* R' (or the *complement*) of relation R.

212. IDENTITY AND DIVERSITY If two individuals x and y are identical, the relation (*identity*) is written $x\,I\,y$. If x and y are different, the relation (*diversity*) is written $x\,D\,y$. The symbols $=$ and $\neq$ are

TABLE 10.4. PROPERTIES OF RELATIONS

Let x, y, z be any elements of set K. With respect to K:

If	*Then relation R is*
for each x, $x\,R\,x$	Reflexive
for each x, $x\,\not R\,x$	Irreflexive
for some but not all elements x, $x\,R\,x$; and for some but not all x, $x\,\not R\,x$	Nonreflexive*
for each x and y, whenever $x\,R\,y$, then $y\,R\,x$	Symmetric
for each x and y, whenever $x\,R\,y$, then $y\,\not R\,x$	Asymmetric
for each x and y, $x\,R\,y$ and $y\,R\,x$, then $x = y$	Antisymmetric
for some but not all elements x and y, $x\,R\,y$ implies $y\,R\,x$; and for some but not all x and y, $x\,R\,y$ implies $y\,\not R\,x$	Nonsymmetric*
for each x, y, and z, whenever $x\,R\,y$ and $y\,R\,z$, then $x\,R\,z$	Transitive
for each x, y, and z, whenever $x\,R\,y$ and $y\,R\,z$, then $x\,\not R\,z$	Intransitive
for some but not all elements x, y, and z, $x\,R\,y$ and $y\,R\,z$ imply $x\,R\,z$; and for some but not all x, y, and z, $x\,R\,y$ and $y\,R\,z$ imply $x\,\not R\,z$	Nontransitive*
for each x and y, $x \neq y$, then either $x\,R\,y$ or $y\,R\,x$	Connected or determinate
for each x and y, then either $x\,R\,y$ or $y\,R\,x$	Strongly connected*

* Some writers define these terms in slightly different ways.

used in the calculus of relations to denote the identity and diversity between relations.

213. CONVERSE The relation $\check{R}$, called the *converse* of R, is said to hold between things x and y if and only if R holds between y and x.

214. PROPERTIES OF RELATIONS Table 10.4 lists properties of relations in common use. For example, a relation R is reflexive with respect to set K if each element x of K has the relation R to itself, that is, $x\ R\ x$.

215. ORDERING RELATIONS With respect to the dyadic relation R a set K is said to be *ordered* in the sense shown in Table 10.6 if in K the relation R has the properties indicated. For example, the set K is *quasi-ordered* with respect to R if R is reflexive and transitive in K.

216. INDEPENDENCE OF CERTAIN PROPERTIES OF RELATIONS The reflexive, symmetric, and transitive sets of properties are independent of each other. There are 27 triples of these properties that dyadic relations may possess.

TABLE 10.5. EXAMPLES OF RELATIONS

Relation	*Properties*
"is father of"	Asymmetric, intransitive
"is a brother of"	Nonsymmetric
"admires"	Nonreflexive, nonsymmetric, nontransitive
"is married to"	Symmetric
"implies"	Nonsymmetric
"is son of"	Irreflexive, asymmetric
"is greater than" ($>$)	Irreflexive, asymmetric, transitive
"is smaller than" ($<$)	Irreflexive, asymmetric, transitive
"is different from" ($\neq$)	Irreflexive, symmetric
"is included in"	Reflexive, transitive
"is older than"	Irreflexive, asymmetric, transitive, not necessarily connected
"is equivalent to"	Reflexive, symmetric, transitive
"is between"	A triadic relation

TABLE 10.6. TYPES OF ORDERING

Type of ordering	*Properties of relation R*
Quasi-ordering	Reflexive, transitive
Partial ordering	Reflexive, antisymmetric, transitive
Simple ordering	Antisymmetric, transitive, strongly connected
Strict partial ordering	Asymmetric, transitive
Strict simple ordering	Asymmetric, transitive, connected

TABLE 10.7. EXAMPLES OF ORDERING

Type of ordering	*Relation*
Quasi-ordering	"being at least as tall as" for a set of persons
Partial ordering	"inclusion $\subseteq$" for set of all sets
Simple ordering	$\leqq$ for set of all real numbers
Strict partial ordering	"being taller than" or "of being a husband" for set of human beings
Strict simple ordering	$<$ for set of all real numbers

217. EQUIVALENCE RELATIONS A dyadic relation R is an *equivalence relation* for a set K of elements if and only if R is reflexive, symmetric, and transitive.

Examples. The relations "is similar to", "is identical with", "is the same age as", "equals", "equality", and "congruent to" are equivalence relations.

An equivalence relation $\approx$ in a set K separates the elements of K into subsets. All elements equivalent to some particular element x of K belong to one subset; all elements equivalent to some other element y of K, not already in the first subset, belong to a second subset, and so on. Such a separation of K into subsets is called the (*class* or *subset*) *decomposition* of K corresponding to the equivalence relation $\approx$.

11. ALGEBRAIC STRUCTURES

Groups

218. **TYPICAL MATHEMATICAL SYSTEM** A typical *mathematical system* M consists of *elements* $a, b, c, \ldots$, an *equals relation*, and certain *operations*. The equals relation separates the elements into classes. Two elements a and b are *equal* if and only if they belong to the same class.

An equals relation $a = b$ is characterized by the following properties: (*a*) *Determinative.* Either $a = b$ or $a \neq b$. (*b*) *Reflexive.* $a = a$. (*c*) *Symmetric.* If $a = b$, then $b = a$. (*d*) *Transitive.* If $a = b$ and $b = c$, then $a = c$. (See Sec. 214.)

219. **BINARY OPERATION** Consider a binary operation $\circ$ which relates to every ordered pair of elements a and b of M a uniquely defined element c of some set N, written $a \circ b = c$. If $N \subseteq M$, then M is *closed* under the operation $\circ$. (In defining operation some writers require that set N be included in set M.) The operation $\circ$ is said to be *properly* or *well defined* relative to the "equals" relation if, when either a or b or both are replaced by elements equal to them respectively, c is replaceable by an element equal to it. A variety of symbols are used for an operation $\circ$, such as $\odot$, $\oplus$, $\otimes$, $+$, $-$, $*$, ...

220. **GROUPS** Consider a mathematical system G containing the elements $a, b, c, \ldots$ and a binary operation $\circ$, which is used to pair two elements of G, $a \circ b$, in the order indicated. The operation is assumed to be well defined relative to the equals relation. The system G is called a *group* if the following four postulates hold. For each a, b, c of G:

1. *Closure.* $a \circ b$ is a unique element c in G.
2. *Associative law.* $(a \circ b) \circ c = a \circ (b \circ c)$.
3. *Identity element.* There exists in G an element i, called the *identity* (or *unit*), such that for every a in G, $a \circ i = a$.
4. *Inverse element.* For each a in G there exists in G an element a^{-1}, called the *inverse* of a, such that $a \circ a^{-1} = i$.

If in addition a group satisfies the postulate

5. *Commutativity.* $a \circ b = b \circ a$,

it is called a *commutative* (or *abelian*) group.

The relation $a \circ b$ is called the "product of a and b."

221. A group consisting of but a finite number of elements is called a *finite group.* The *order* of a finite group is the number of elements which it contains.

A *subgroup* of G is a set of elements of G which by themselves constitute a group. A subgroup of G different from G and from the identity is called a *proper subgroup* of G. A subgroup of G which is not proper is *improper.*

If G_1 and G_2 are subgroups of G, the elements common to G_1 and G_2 form a group called the *intersection* of G_1 and G_2, written $G_1 \wedge G_2$.

The symbol for $a \circ a \circ a \circ \cdots \circ a$ to n factors is a^n.

The least positive integer r such that $a^r = i$ is called the *period* of a.

A set of elements of a group G are said to be *independent* if no element of the set can be expressed as a product of the remaining elements of the set, allowing any number of repetitions.

A set of independent elements of G which have the property that every element of G can be represented in terms of them as a finite product is called a set of *independent generators* of G.

A *cyclic group* is a group generated by a single element.

Examples. The following sets form a group under the operation indicated:

(*a*) The set of all integers, under addition; the unit element is 0, the inverse of a is $-a$.

(*b*) The set of all positive rational numbers, under multiplication; the unit element is 1.

(*c*) The elements of a ring, under addition (see Sec. 226).

(*d*) The nonzero elements of a field, under multiplication (see Sec. 226).

(*e*) The set of three letters a, b, c, under permutations.

(*f*) The set of all Euclidean rotations of the plane about the origin,

$$R\colon\ x' = x\cos\theta + y\sin\theta \qquad y' = -x\sin\theta + y\cos\theta$$

the operation being composition of transformations. (The result of performing first rotation R_1 and then rotation R_2 is denoted by $R_2 \circ R_1$.)

222. THEOREMS ON GROUPS For each $a, b, c, \ldots$ in group G:

(a) $a \circ i = i \circ a = a$.

(b) $a \circ a^{-1} = a^{-1} \circ a = i$.

(c) $(a \circ b)^{-1} = b^{-1} \circ a^{-1}$.

(d) $(a^{-1})^{-1} = a$.

(e) The identity i is unique.

(f) The inverse of a is unique.

(g) If $a \circ b = a \circ c$, then $b = c$.

(h) If $b \circ a = c \circ a$, then $b = c$.

(i) The equation $a \circ x = b$ has a unique solution $x = a^{-1} \circ b$ in G.

(j) The equation $y \circ a = b$ has a unique solution $y = b \circ a^{-1}$ in G.

(k) If $a = b$, then for any c in G, $a \circ c = b \circ c$.

(l) If $a = b$ and $c = d$, then $a \circ c = b \circ d$.

(m) If a is an element of a finite group, there exists a positive integer r such that $a^r = i$ and $a^{-1} = a^{r-1}$.

(n) The order of a subgroup of a group G is a divisor of the order of G.

(o) Every finite group possesses a set of independent generators.

Algebra of Positive Integers

223. ADDITION Let a and b be any two positive integers. To each ordered pair a and b there is assigned a unique positive integer e, called the *sum* of a and b, denoted by $e = a + b$. The operation $+$ of *addition* is a binary operation on the set of positive integers and possesses properties 1, 2, and 3 of Table 11.1. Here $=$ means identity.

MULTIPLICATION To each ordered pair a and b of positive integers there is assigned a unique positive integer f, called the *product* of a and b, denoted by $f = a \times b$. The operation $\times$ of multiplication is a binary operation on the set of positive integers and possesses properties 4, 5, and 6 of Table 11.1.

DISTRIBUTIVE LAW The operations of addition and multiplication possess property 7 of Table 11.1.

ALGEBRA OF POSITIVE INTEGERS The statements made in Table 11.1 always hold in the algebra of positive integers.

224. OTHER ALGEBRAS The statements of Table 11.1 hold in other situations where the elements $a, b, c, \ldots$ belong to a set S other than the positive integers and the operations $+$ and $\times$ are suitably defined.

Examples. If S is the set of all real numbers with $+$ and $\times$ denoting the usual addition and multiplication operations, then the statements of Table 11.1 always hold in the algebra of real numbers. When S is the set of all real polynomials in the real variable x, the statements of Table 11.1 always hold in the algebra of real polynomials.

TABLE 11.1. BASIC PROPERTIES OF BINARY OPERATIONS OF ADDITION AND MULTIPLICATION OF POSITIVE INTEGERS

Let a, b, and c denote arbitrary positive integers and $=$ be the symbol for identity.

1. *Closure law for addition.* $a + b$ is a positive integer.
2. *Commutative law for addition.* $a + b = b + a$.
3. *Associative law for addition.* $(a + b) + c = a + (b + c)$.
4. *Closure law for multiplication.* $a \times b$ is a positive integer.
5. *Commutative law for multiplication.* $a \times b = b \times a$.
6. *Associative law for multiplication.* $(a \times b) \times c = a \times (b \times c)$.
7. *Distributive law for multiplication and addition.* $a \times (b + c) = (a \times b) + (a \times c)$.

Postulates for Common Types of Algebraic Structures

225. There are a number of algebraic structures in common use which have associated with them one or more of the postulates listed in Table 11.2.

TABLE 11.2. LIST OF POSTULATES, ONE OR MORE OF WHICH ARE ASSOCIATED WITH ALGEBRAIC STRUCTURES IN COMMON USE

Let S be a set of elements $a, b, c, \ldots$. Let $\oplus$ and $\otimes$ denote two binary operations on the set S. Let the "equality" symbol $=$ be used in the sense of identity; that is, $a = b$ means that a and b are the same element and $a \neq b$ means that a and b are not the same element.

(The operations $\oplus$ and $\otimes$ are called "addition" and "multiplication," respectively, but this does not mean that $\oplus$ and $\otimes$ always stand for ordinary addition and ordinary multiplication.)

For each $a, b, c, \ldots$ of S:

A_1. *Closure.* $a \oplus b$ is a unique element in S.

A_2. *Associativity.* $(a \oplus b) \oplus c = a \oplus (b \oplus c)$.

A_3. *Identity element.* There exists in S an element z, called the *identity*, or *unit*, such that for any element in S, $a \oplus z = a$. (z is called "zero," or 0.)

A_4. *Inverse element under* $\oplus$. For each element a in S there exists an element $\bar{a}$ in S, called the *inverse* of a with respect to operation $\oplus$, such that $a \oplus \bar{a} = z$. [$\bar{a}$ is often written $(-a)$, so that $a \oplus (-a) = 0$.]

A_5. *Commutativity.* $a \oplus b = b \oplus a$.

M_1. *Closure.* $a \otimes b$ is a unique element in S.

M_2. *Associativity.* $(a \otimes b) \otimes c = a \otimes (b \otimes c)$.

M_3. *Identity element.* There exists in S an element u, called the *identity* or *unit* for the operation $\otimes$, such that for every element a in S, $a \otimes u = a$. (u is called "unity.")

M_4. *Inverse element under* $\otimes$. For each element a of S, except z, there exists an element a^{-1} in S, called the *inverse* of a with respect to operation $\otimes$, such that $a \otimes a^{-1} = u$.

M_5. *Commutativity.* $a \otimes b = b \otimes a$.

M_6. *No proper divisors.* The set S contains no proper divisors of zero.

D_1. *Left-hand distributive law of* $\otimes$ *over* $\oplus$.
$a \otimes (b \oplus c) = (a \otimes b) \oplus (a \otimes c)$.

D_2. *Right-hand distributive law of* $\otimes$ *over* $\oplus$.
$(b \oplus c) \otimes a = (b \otimes a) \oplus (c \otimes a)$.

C_1. *Left-hand cancellation law for the operation* $\otimes$. If $c \neq z$ and $c \otimes a = c \otimes b$, then $a = b$.

C_2. *Right-hand cancellation law for the operation* $\otimes$. If $c \neq z$ and $a \otimes c = b \otimes c$, then $a = b$.

Laws of Order

O_1. There exists a subset P of the set S such that if $a \neq z$, then one and only one of a and $\bar{a}$ is in P.

O_2. If a and b are in P, then $a \oplus b$ is in P.

O_3. If a and b are in P, then $a \otimes b$ is in P.

Definitions of Symbols $\ominus$ *and* $\oslash$. The elements of set P are called the *positive* elements of S; all other elements of S, except the *zero* z, are called the *negative* elements of S. If a and b are in S and if $a \oplus \bar{b}$ is positive, then we write $a \ominus b$ and $b \oslash a$.

Divisors of Zero. If $a \otimes b = z$ and $b \neq z$, a is called a *divisor of zero.* A number $a \neq z$ which has this property is called a *proper divisor of zero.* Nonzero elements a and b having product $a \otimes b = z$ are called "divisors of zero."

226. POSTULATE SETS FOR SEVERAL TYPES OF ALGEBRAIC STRUCTURES Different selections from the postulates listed in Table 11.2 have been used to define a variety of algebraic structures. Several systems are defined below. The sets of postulates listed are in some cases redundant (i.e., fewer postulates could be used).

Ordinary Elementary Alegbra. The 18 postulates of Table 11.2 constitute a postulate set for ordinary elementary algebra.

Basic Properties of the Positive Integers. The basic properties of the positive integers are given by postulates A_1, A_2, A_5; M_1, M_2, M_3, M_5; D_1, D_2.

Group. $M_1, \ldots, M_4$ constitute a postulate set for a group. If M_5 holds, the group is an *abelian* or *commutative group.*

Field. $A_1, \ldots, A_5$; $M_1, \ldots, M_5$; D_1, D_2 constitute a postulate set for a *field.*

Ordered Field. $A_1, \ldots, A_5$; $M_1, \ldots, M_5$; D_1, D_2; O_1, O_2, O_3 constitute a postulate set for an *ordered field.*

Noncommutative Field. $A_1, \ldots, A_5$; M_1, M_2, M_3, M_4; D_1, D_2 constitute a postulate set for a *noncommutative field.*

Ring. $A_1, \ldots, A_5$; M_1, M_2; D_1, D_2 constitute a postulate set for a *ring.*

Commutative Ring. A ring satisfying postulate M_5 is a *commutative ring.*

Ring with Unit (Identity) Element. A ring satisfying postulate M_3 is a *ring with unit element.*

Integral Domain. $A_1, \ldots, A_5$; M_1, M_2, M_3, M_5; C_1, C_2; D_1, D_2 constitute a postulate set for an *integral domain,* or *domain of integrity.*

227. ALTERNATE DEFINITIONS A more informative way of defining the structures listed above is as follows:

A *field* is a set S of at least two elements for which two binary operations $\oplus$ and $\otimes$ are defined such that (1) the elements of S constitute an abelian group under addition $\oplus$; (2) the nonzero elements of S constitute an abelian group under multiplication $\otimes$; and (3) multiplication is distributive over addition. An *ordered field* is a field for which postulates O_1, O_2, O_3 hold.

A *ring* is a set R of at least two elements for which two binary operations $\oplus$ and $\otimes$ are defined such that (1) the elements of R constitute an abelian group under addition $\oplus$, the identity element being denoted by z, and the inverse of a by $-a$; (2) multiplication is distributive with respect to addition; (3) multiplication is associative.

An *integral domain* D is a commutative ring with unit element which is not necessarily a field because postulate M_4 is replaced by postulate M_6. $A_1, \ldots, A_5$; M_1, M_2, M_3, M_5, M_6; D_1, D_2 constitute a postulate set for an integral domain. Instead of M_6, the postulates C_1 and C_2 can be used.

A *noncommutative* field is a ring with unit element "unity" satisfying M_4.

Remark. Slight variations of the above definitions are given by different writers.

Examples: With respect to addition and multiplication, each of the indicated sets forms the following:

Field: (*a*) all rational numbers; (*b*) all real numbers; (*c*) all numbers $a + b\sqrt{2}$, where a and b are rational.

Ring: (*a*) the positive and negative integers and zero; (*b*) the integers modulo m; (*c*) the even integers.

Integral domain: (*a*) the polynomials whose coefficients lie in an integral domain; (*b*) the set of all numbers $a + b\sqrt{2}$, where a and b are integers.

228. THEOREMS ON FIELDS Let $a, b, c, \ldots$ be elements of a field F, with identity element ("zero") z and "unity" element u; let $-a$ denote the addition inverse $\overline{a}$ of a and a^{-1} denote the multiplicative inverse of a:

(*a*) $a \otimes b = z$ if and only if $a = z$ or $b = z$, or both $a = z$ and $b = z$.

(*b*) $a \oplus x = b$ has a unique solution in F.

(*c*) $a \otimes x = b$ has a unique solution in F if $a \neq z$.

(d) If $a = b$ and $c = d$, then $a \oplus c = b \oplus d$.
(e) If $a = b$ and $c = d$, then $a \otimes c = b \otimes d$.
(f) $[-(-a)] = a$.
(g) $[a^{-1}]^{-1} = a$, provided $a \neq z$.
(h) $a \otimes z = z$ for all a in F.
(i) $a \otimes (-b) = -(a \otimes b)$.
(j) $(-a) \otimes (-b) = a \otimes b$, or $\bar{a} \otimes \bar{b} = a \otimes b$.

229. THEOREMS ON INTEGRAL DOMAINS Let $a, b, c, \ldots$ be elements of an integral domain D:
(a) The zero element z is unique.
(b) The unit element u is unique.
(c) If $a \oplus b = a \oplus c$, then $b = c$.
(d) The additive inverse $\bar{a}$ of a is unique.
(e) $a \oplus x = b$ has exactly one solution x.
(f) For all a in D, $a \otimes z = z \otimes a = z$.
(g) For all a and b in D, $\bar{a} \otimes \bar{b} = a \otimes b$.
(h) $a \otimes b = z$ if and only if $a = z$ or $b = z$ or both $a = z$ and $b = z$.

230. THEOREMS ON RINGS Many of the theorems on integral domains, which do not depend on the cancellation law or the commutative law, apply to rings.

231. ISOMORPHIC MODELS OF POSTULATE SYSTEMS Two models (interpretations) M_1 and M_2 of a postulate (axiom) system are said to be *isomorphic with respect to the postulate system* Σ if there exists a one-to-one correspondence between the elements of M_1 and M_2 which preserves the postulate statements of Σ (i.e., preserves the relations and the operations of Σ).

232. HOMOMORPHIC CORRESPONDENCE Let A and B denote two mathematical systems. Let $\bigcirc_A$ be a well-defined operation which relates to every ordered pair of elements a_i, a_j of A a unique element a_{ij} of A, and denote this by $a_i \bigcirc_A a_j = a_{ij}$. Similarly, let $\bigcirc_B$ be a well-defined operation relating every ordered pair of elements b_i, b_j of B with a unique element b_{ij} of B, written $b_i \bigcirc_B b_j = b_{ij}$. Suppose there is a correspondence, denoted by $a_i \to b_i$, by means of which every element a_i of A determines a unique element b_i of B, such that every element b_i of B is the correspondent of at least one a_i of A. If the correspondence $a_i \to b_i$, $a_j \to b_j$ implies the correspondence $a_{ij} \to b_{ij}$, that is, the correspondence $a_i \bigcirc_A a_j \to b_i \bigcirc_B b_j$, then the correspondence is called a *homomorphism* (or a *mapping*) *of A onto B*. We write $A \sim B$ to mean that a homomorphism of A onto B exists.

In case each element b_i of B is the correspondent of one and only one element a_i of A, the correspondence is called a *biunique correspondence*, or a *one-to-one reciprocal correspondence*. This is written $a_i \leftrightarrow b_i$. If the correspondence $a_i \leftrightarrow b_i$, $a_j \leftrightarrow b_j$ implies $a_i \bigcirc_A a_j = a_{ij} \leftrightarrow b_i \bigcirc_B b_j = b_{ij}$, the homomorphism is an *isomorphism*. In this case we write $A \cong B$.

An *endomorphism* is a homomorphism of a system onto itself, or onto a part of itself.

An *automorphism* is an isomorphism of a system onto (or with) itself.

Boolean Algebra

233. A POSTULATE SET FOR BOOLEAN ALGEBRA A Boolean algebra is a set B of elements $a, b, c, \ldots$ with two binary operations $\cup$ and $\cap$, satisfying the following postulates. (Some writers use the symbols $\oplus$ and $\otimes$, others $+$ and $\times$, to denote the operations $\cup$ and $\cap$.)

For all elements $a, b, \ldots$ of B:

1. $a \cup b$ and $a \cap b$ are each unique elements in B. (*closure*)
2. $a \cup b = b \cup a$ and $a \cap b = b \cap a$. (*commutative*)
3. There exists an element (*zero*) z in B and an element u (*unity*) in B such that $a \cup z = a$, $a \cap u = a$, $z \neq u$.
4. $a \cup (b \cap c) = (a \cup b) \cap (a \cup c)$ and
   $a \cap (b \cup c) = (a \cap b) \cup (a \cap c)$. (*distributive*)
5. For each a there exists an element a' of B such that $a \cup a' = u$ and $a \cap a' = z$. (a' is *complement* of a.)

The elements $a, b, c, \ldots$ and operations $\cup$ and $\cap$ may be considered undefined. $a \cup b$ is called the (logical) *sum* and $a \cap b$ the (logical) *product* of a and b. For any two elements a and b, $a \cup b$ and $a \cap b$ are each unique elements of B.

234. The subsets of a given set B under the operations of union and intersection satisfy the postulates for a Boolean algebra. In this application, $a, b, c, \ldots$ are the subsets, z is represented by the null set $\emptyset$

and u by the universe I, and $=$ means "is" or "equals." The algebra of sets is one representation of a Boolean algebra.

235. THEOREMS Each formula listed in Sec. 161, with $A, B, C, \ldots,$ $\emptyset$, and I replaced by $a, b, c, \ldots, z$, and u, respectively, is a theorem derivable from the postulates for a Boolean algebra. Thus:

Duality Principle. If in any theorem of Boolean algebra the symbols $\cup$ and $\cap$, the elements z and u, and $\subset$ and $\supset$ are interchanged throughout, the resulting statement is valid.

236. FURTHER PROPERTIES

$(a \cup b) \cup c = a \cup (b \cup c)$ (associative law)
$(a \cap b) \cap c = a \cap (b \cap c)$ (associative law)
$a \cup a = a,\ a \cap a = a$ (idempotent property)
$a \cup u = u,\ a \cap z = z$
$a \cap u = a,\ a \cup z = a$
$a \cap (a \cup b) = a$ (absorption law)
$a \cup (a \cap b) = a$ (absorption law)
a' is unique
$(a \cup b)' = a' \cap b'$ (dualization)
$(a \cap b)' = a' \cup b'$ (dualization)
$(a')' = a,\ u' = z,\ z' = u$
$a \cup (a' \cap b) = a \cup b$
$(a \cap b) \cup (a \cap c) \cup (b \cap c') = (a \cap c) \cup (b \cap c')$

237. DEFINITION OF INCLUSION $\subset$ Let a and b be elements of B. $a \subset b$ if and only if $a \cup b = b$.

Theorems. Let $a, b, c, \ldots$ be elements of B:

(*a*) $a \subset a$ for all a. (reflexive)
(*b*) If $a \subset b$ and $b \subset a$, then $a = b$. (antisymmetric)
(*c*) If $a \subset b$ and $b \subset c$, then $a \subset c$. (transitive)
(*d*) The elements of a Boolean algebra are partially ordered with respect to relation $\subset$.
(*e*) For each a of B, $z \subset a \subset u$. (universal bound)
(*f*) If $a \subset x$ and $b \subset x$ where x is in B, then $(a \cup b) \subset x$.
(*g*) $a \cup b = b$ if and only if $a \cap b = a$. (consistency)
(*h*) $a \subset b$ if and only if $a \cap b = a$. (consistency)

238. BOOLEAN FUNCTIONS Let $x_1, x_2, \ldots, x_n$ be n Boolean variables, each of which can equal any element of a Boolean algebra B. A *Boolean function* (or *Boolean polynomial*), written $Y = F(x_1, x_2, \ldots, x_n)$, is an expression built up from $x_1, x_2, \ldots, x_n$ through the operations of $\cup$, $\cap$ and complementation $'$.

A *minimal polynomial* in n variables $x_1, \ldots, x_n$ is a logical product of n letters in which the ith letter is either x_i or the complement x_i'.

Theorem. Every Boolean function can be expressed uniquely as a logical sum of minimal polynomials or is identically equal to 0.

239. MEASURE ALGEBRA A Boolean algebra B with elements A_1, $A_2, A_3, \ldots$ is *completely additive* if and only if every sum $A_1 \cup A_2 \cup \cdots$ is uniquely defined in B. A completely additive Boolean algebra B is a *measure algebra* if and only if there exists a real-valued *measure function* M defined for all elements A of B, such that

$$M(A) \geqq 0 \qquad M(0) = 0$$
$$M(A_1 \cup A_2 \cup \cdots) = M(A_1) \cup M(A_2) \cup \cdots$$

for every countable set of disjoint elements $A_1, A_2, \ldots$. Here $M(A)$ denotes the *value* of the measure M associated with the element A.

240. ALGEBRA OF EVENTS ASSOCIATED WITH A GIVEN EXPERIMENT Let S denote a set (space) of sample descriptions of all possible *outcomes* (states, events) of a given experiment. An *event* E is any subset of S. To say that the event E *is realized,* or *has occurred,* is to say that the outcome of a specific experiment under consideration has a description that is a member of subset E.

Suppose the events $E_1, E_2, \ldots$ are such as to permit the following definitions:

1. The *union* (*logical sum,* or *join*) of a set of events $E_1, E_2, \ldots$, written $E_1 \cup E_2 \cup E_3 \cup \cdots$, is the outcome of realizing *at least one* of the events $E_1, E_2, \ldots$.
2. The *intersection* (*logical product,* or *meet*) of two events E_1 and E_2, written $E_1 \cap E_2$, is the joint outcome of realizing *both* E_1 and E_2.
3. The (*logical*) *complement* of an event E is the outcome E' of not realizing E. $E' \cup E = S$. $E \cap E' = \emptyset$.
4. The *certain event* I is the set S (i.e., at least one of the outcomes of S is realized in a specific experiment).

5. The *impossible event* $\emptyset$ is an event that contains no descriptions in S and therefore cannot occur in the experiment. $S' = \emptyset$.

241. DEFINITIONS Events E_1 and E_2 are said to be *mutually exclusive,* or *disjoint,* if and only if $E_1 \cap E_2$ is $\emptyset$, that is, if and only if the joint event of simultaneously realizing *both* E_1 and E_2 is impossible.

If E_1 and E_2 are events of the set S, we say $E_1 \subset E_2$ if and only if $E_1 \cup E_2 = E_2$. ($E_1 \subset E_2$ is read "E_1 is included in E_2.") The symbol $=$ here means "is". In other words, the event E_1 is included in the event E_2 if and only if the event of realizing at least one of the two events E_1 and E_2 is the realization of the event E_2.

Theorem. $E_1 \subset E_2$ if and only if $E_1 \cap E_2 = E_1$. In other words, the event E_1 is included in the event E_2 if and only if the event of realizing both events E_1 and E_2 is the realization of the event E_1.

Theorem. For any event E, $\emptyset \subset E \subset I$.

242. DEFINITION If a set B, consisting of set S and the impossible event $\emptyset$, has the structure of a completely additive Boolean algebra and has all the properties defined in Secs. 233 and 239, then B is called an *algebra of events* associated with the given experiment.

The algebra of events associated with the given experiment under the hypothesis that event E_1 occurs is the set B_1 of joint events $E \cap E_1$. In this case the certain event in B_1 is $E_1 \cap E_1 = E_1$.

243. EVENT ALGEBRAS AND SYMBOLIC LOGIC In two-valued (Aristotelian) logic an algebra $\mathfrak{B}$ of events (logical propositions, assertions) is related to a simpler Boolean algebra $\mathfrak{T}$ of truth-values. A truth-value $T[E]$ is taken to be equal to 1 if an event E is realized (i.e., if E is true) and equal to 0 if the event E is not realized (i.e., if E is false). Let I denote a certain event and $\emptyset$ denote the impossible event. The relation between the algebras $\mathfrak{B}$ and $\mathfrak{T}$ is given by the homomorphism

$$(1) \qquad T[I] = 1 \qquad\qquad T[\emptyset] = 0$$
$$T[E_1 \cup E_2] = T[E_1] \oplus T[E_2] \qquad T[E_1 \cap E_2] = T[E_1] \otimes T[E_2]$$

The truth tables for the events $E_1 \cup E_2$ and $E_1 \cap E_2$ and the truth-values T, with associated arithmetic, are shown in Tables 11.3 and 11.4.

TABLE 11.3. TRUTH TABLE

E_1	E_2	$E_1 \cup E_2$	$E_1 \cap E_2$
T	T	T	T
T	F	T	F
F	T	T	F
F	F	F	F

TABLE 11.4. ARITHMETIC OF TRUTH-VALUES

$T[E_1] \oplus T[E_2] = T[E_1 \cup E_2]$	$T[E_1] \otimes T[E_2] = T[E_1 \cap E_2]$
$1 \oplus 1 = 1$	$1 \otimes 1 = 1$
$1 \oplus 0 = 1$	$1 \otimes 0 = 0$
$0 \oplus 1 = 1$	$0 \otimes 1 = 0$
$0 \oplus 0 = 0$	$0 \otimes 0 = 0$

The truth-value $T[E]$ of any proposition E logically related to a set of events $E_1, E_2, \ldots$, and expressible as a (Boolean) function $E = F(E_1, E_2, \ldots)$ of events $E_1, E_2, \ldots$, may be written $T[E] = T[G\{T[E_1], T[E_2], \ldots\}]$, where G is the homomorphic image of F obtained by replacing $\cup$ and $\cap$ by $\oplus$ and $\otimes$, respectively. In the arithmetic of truth-values, $0' = 1, 1' = 0$. In view of the assumptions made with regard to the algebras used here, any proposition E (i.e., an event) is either true or false.

244. APPLICATION TO SWITCHING CIRCUITS The two-valued Boolean algebras mentioned above have been used extensively in the theory of electric switching. Consider a simple electric circuit X joining terminal point α to terminal point β (Fig. 61). The circuit X is either *open* or *closed*. In the event E that circuit X is open, the truth value $T[E]$ of the event E is taken to be 0; and in the event E that the circuit X is closed, the truth-value $T[E]$ is taken to be 1. A similar

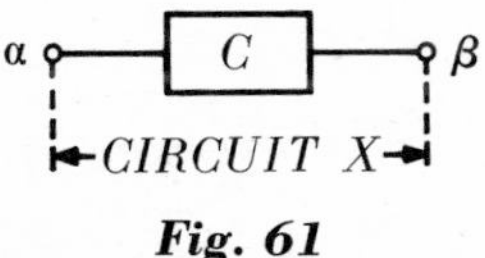

Fig. 61

statement holds if the word "circuit" is replaced by the word "switch." The box C in Fig. 61 may be considered a network of circuits with a terminal at each end.

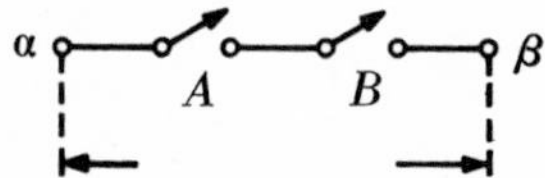

Fig. 62. *Series circuit. Truth-value for circuit joining* α *to* β *is the truth-value of* $A \cap B$*, that is,* $T[A \cap B] = T[A] \otimes T[B]$*. When* $T[A \cap B] = 0$*, the circuit is open; when* $T[A \cap B] = 1$*, the circuit is closed.*

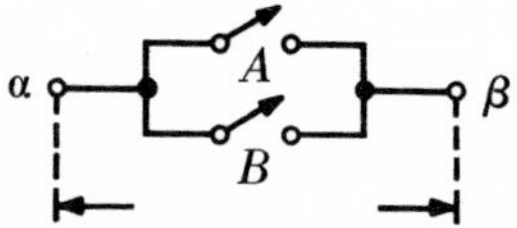

Fig. 63. *Parallel circuit. Truth-value for circuit joining* α *to* β *is the truth-value of* $A \cup B$*, that is,* $T[A \cup B] = T[A] \oplus T[B]$*. When* $T[A \cup B] = 0$*, the circuit is open; when* $T[A \cup B] = 1$*, the circuit is closed.*

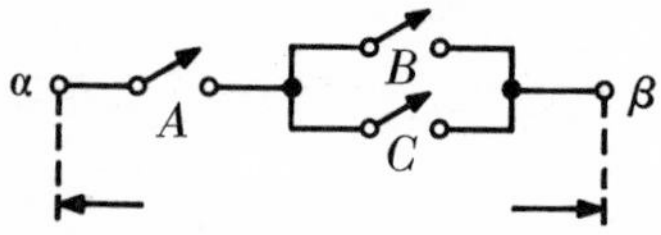

Boolean Polynomial: $A \cap (B \cup C)$

Fig. 64. *Series-parallel circuit. When* $T[A \cap (B \cup C)] = 0$*, the circuit joining* α *and* β *is open; when* $T[A \cap (B \cup C)] = 1$*, the circuit is closed.*

245. SERIES CIRCUIT Consider the series circuit in Fig. 62. In the event switch A is open, the corresponding truth-value is $T[A] = 0$; if A is closed, $T[A] = 1$. If B is open, $T[B] = 0$; if B is closed, $T[B] = 1$. In the event both A and B are closed, the series circuit joining α to β of Fig. 62 is closed; if either A or B or both A and B are open, then the circuit joining α and β is open. In other words the truth-value for the series circuit joining α and β is $T[A \cap B] =$

$T[A] \otimes T[B]$. When $T[A \cap B] = 1$, the series circuit is closed; when $T[A \cap B] = 0$, the series circuit is open. (Compare with Table 11.4.)

246. PARALLEL CIRCUIT The parallel circuit of Fig. 63 may be treated similarly.

247. OTHER TYPES OF CIRCUITS In electrical problems the actual circuit diagrams are used to help set up the corresponding Boolean functions (polynomials) whose truth-values reflect when a circuit is open or closed in terms of the state of the open or closed conditions of the component circuit or switches. Figure 64 is such an example.

248. There are several elementary theorems involving number systems which are fundamental in the field of digital computers.

Algorithms. An *algorithm* is a step-by-step iteration process for solving a given problem. Some algorithms involve trial or tentative steps (e.g., division), others are straightforward (e.g., multiplication). Before a digital computer can be programmed to solve problems of a given type, a complete algorithm must be devised.

249. EUCLIDEAN ALGORITHM Let m and n be two positive integers, $m > n$. By Euclid's theorem, integers q_i and n_i exist, $q_i \geqq 0$, such that

$$\begin{aligned} m &= q_0 n + n_1 & 0 &< n_1 < n \\ n &= q_1 n_1 + n_2 & 0 &< n_2 < n_1 \\ &\cdots & &\cdots \\ n_{k-2} &= q_{k-1} n_{k-1} + n_k & 0 &< n_k < n_{k-1} \\ n_{k-1} &= q_k n_k \end{aligned} \tag{1}$$

If m is a multiple of n, write $n = n_0$ when $k = 0$ in (1).

Theorem. The number n_k to which the above process leads is the *greatest common divisor* of m and n.

250. SCALES OF NOTATION If m and r are positive integers and $r > 1$, then m can be represented in terms of r uniquely in the form

$$m = p_h r^h + p_{h-1} r^{h-1} + \cdots + p_1 r + p_0 \tag{2}$$

where $\quad p_h \neq 0 \qquad 0 \leqq p_i < r \qquad i = 0, 1, \ldots, h$

Representation (2) may be written

$$m = (p_h p_{h-1} \cdots p_1 p_0)_r \tag{3}$$

Here r, called the *base* or *radix,* is the number of different digits used. In any positional number system the digits 0, 1, 2, . . . , $(r - 1)$ have an order. *Advancing the digit* 0 means replacing it by the digit 1;

advancing the digit 1 means replacing it by the digit 2 (if $r > 2$), . . . ; advancing the digit $(r - 1)$ means replacing it by the digit 0. If any digit advances to 0, the digit at its left advances.

251. If N is any real positive number and r is a positive integer, $r > 1$, then N can be represented uniquely as

$$(4) \quad N = p_h r^h + p_{h-1} r^{h-1} + \cdots + p_1 r + p_0 + p_{-1} r^{-1} + \cdots + p_{-m} r^{-m} + \cdots$$

and (4) may be written

$$(5) \qquad N = (p_h p_{h-1} \cdots p_1 p_0 \cdot p_{-1} p_{-2} \cdots p_{-m} \cdots)_r$$

where $p_h \neq 0 \qquad 0 \leqq p_i < r$

$$i = h, h - 1, \ldots, 1, 0, -1, -2, \ldots, -m, \ldots$$

If N is a rational fraction, that is, the ratio of two integers, (4) either terminates or recurs; if N is irrational, (4) is nonrecurring and nonterminating.

The principal systems in use are those having radix 2, 3, 8, 10, 12, 16, or 20.

Table 12.1 gives examples of several common number systems. The decimal system with radix 10 is the one most commonly used. From the elementary notions of addition and multiplication and rules for counting, tables for addition and multiplication can be constructed. Thus, in the ternary system $2 + 1 = 10$, $2 + 2 = 11$, $2 \times 1 = 2$, $2 \times 2 = 11$. In the octal system, $6 + 7 = 15$, $7 \times 2 = 16$. Samples of addition and multiplication tables are given in Table 12.2.

252. CONVERSION RULES To convert a representation of an *integer m*:

1. In a system with radix r to the equivalent integer in decimal system, use Eq. (2).

2. In a decimal system to equivalent representation in a system with radix r, divide m repeatedly by r and take the remainders in *reverse order* (i.e., use a division algorithm).

To convert a representation of a real *positive number:*

3. From a system with radix r to equivalent representation in decimal system, use Eq. (4).

TABLE 12.1. EXAMPLES OF NUMBER SYSTEMS*

	Binary	*Ternary*	*Octal*	*Decimal*	*Duodecimal*
Radix	2	3	8	10	12
Digits used	0, 1	0, 1, 2	0, 1, 2, 3, 4, 5, 6, 7	0, 1, 2, 3, 4, 5, 6, 7, 8, 9	0, 1, . . . , 8, 9, t, e
	00000	0000	0000	0000	0000
	00001	0001	0001	0001	0001
	00010	0002	0002	0002	0002
	00011	0010	0003	0003	0003
	00100	0011	0004	0004	0004
	00101	0012	0005	0005	0005
	00110	0020	0006	0006	0006
	00111	0021	0007	0007	0007
	01000	0022	0010	0008	0008
	01001	0100	0011	0009	0009
	01010	0101	0012	0010	000t
	01011	0102	0013	0011	000e
	01100	0110	0014	0012	0010
	01101	0111	0015	0013	0011
	01110	0112	0016	0014	0012

* The quantities in the same horizontal row have the same cardinal value.

4. From a decimal system to equivalent representation in system of radix r, convert its integral and fractional parts separately. Use Rule 2 to convert integral part. To convert (decimal) fractional part, multiply repeatedly by r and take out the resulting integral parts in *forward order* (i.e., use multiplication algorithm).

Examples. ***(a)*** Convert $(1012)_7$ to decimal system: $(1012)_7 = 1(7)^3 + 0(7)^2 + 1(7)^1 + 2(7)^0 = (352)_{10}$.

(b) Convert $(18)_{10}$ to binary system: $18 = 2(9) + 0,\ \ 9 = 2(4) + 1,\ \ 4 = 2(2) + 0,\ \ 2 = 2(1) + 0,$ $1 = 2(0) + 1$. Hence $(18)_{10} = (10010)_2$.

(c) Convert $(21.201)_5$ to decimal system: $(21.201)_5 = 2(5)^1 + 1(5)^0 + 2(5)^{-1} + 0(5)^{-2} + 1(5)^{-3} = (11.408)_{10}$.

(d) Convert $(0.375)_{10}$ to binary system: $2(0.375) = 0 + 0.75,$

TABLE 12.2. SAMPLES OF ADDITION AND MULTIPLICATION TABLES

Binary Addition

+	0	1
0	0	1
1	1	10

Ternary Addition

+	0	1	2
0	0	1	2
1	1	2	10
2	2	10	11

Binary Multiplication

×	0	1	10
0	0	0	0
1	0	1	10
10	0	10	100

Ternary Multiplication

×	0	1	2	10
0	0	0	0	0
1	0	1	2	10
2	0	2	11	20
10	0	10	20	100

$2(0.75) = 1 + 0.5$, $2(0.5) = 1 + 0.0$, $2(0.0) = 0 + 0.0$. Hence $(0.375)_{10} = (.0110)_2$.

(e) Convert $(18.375)_{10}$ to binary system: By Examples **(*b*)** and **(*d*)**, $(18)_{10} = (10010)_2$ and $(0.375)_{10} = (0.0110)_2$. Hence $(18.375)_{10} = (10010)_2 + (0.0110)_2 = (10010.0110)_2$.

Remark. In converting from a binary to a decimal representation it is often simpler to convert from binary to octal to decimal.

Exactly eight numbers can be represented with three binary digits. Triplets of binary digits may be used as marks to represent octal digits. Conversion from binary to octal notation simply implies changing the notation used for the marks. The change is made triplet by triplet, beginning at the right. Thus

$$(10{,}001{,}001)_2 = (211)_8 \qquad (101{,}100{,}111)_2 = (547)_8$$

253. SUBTRACTION BY COMPLEMENTS Some computers subtract by adding the complement of the subtrahend to the minuend and then making suitable adjustments.

Let k be any n-order decimal integer. The *tens complement* of k is $10^n - k$. The *nines complement* of k is $(10^n - k) - 1$.

Let b be any n-order binary integer. The *twos complement* of b is $2^n - b$. The *ones complement* of b is $(2^n - b) - 1$.

TABLE 12.3. EXAMPLES ILLUSTRATING COMPLEMENTS

Decimal number k	Tens complement*	Nines complement†
021	979	978
0021	9979	9978
05555	94445	94444

Binary number b	Twos complement‡	Ones complement§
1	1	0
100	100	011
101	011	010
0101	1011	1010
10110	01010	01001

* *Algorithm:* Subtract each digit in k from 9; then add 1 to the resulting number.
† *Algorithm:* Subtract each digit in k from 9.
‡ *Algorithm:* Change each digit in b; then add 1 to the resulting number.
§ *Algorithm:* Change each digit in b.

254. RULE FOR SUBTRACTION BY TENS (TWOS) COMPLEMENT To the minuend add the tens (twos) complement of the subtrahend. If there is an extra column with digit 1, replace 1 by +; otherwise tens-(twos-)complement the result and prefix −.

255. RULE FOR SUBTRACTION BY NINES (ONES) COMPLEMENT To the minuend add the nines (ones) complement of the subtrahend. If there is an extra column with digit 1, then "end-around-carry"; otherwise, nines- (ones-)complement the result and prefix −. To "end-around-carry" means to replace the extra column with digit 1 by + and carry the 1 to the units column, where it is added to the number.

Examples. **(a)** Subtract $(076)_{10}$ from $(324)_{10}$. The nines complement of 076 is 923. Add. $923 + 324 = 1247$. Replace extra column 1 by +, getting 247. Carry 1 to units column and add, giving the answer 248.

(b) Subtract $(324)_{10}$ from $(076)_{10}$. The nines complement of 324 is 675. Add. $675 + 076 = 751$. There is no extra column 1. The nines complement of 751 is 248. Prefix −, giving the answer −248.

(c) Subtract $(01110001)_2$ from $(11001101)_2$. The ones complement of 01110001 is 10001110. Add. $10001110 + 11001101 = 101011011$. There is an extra column 1. Hence the answer is $+01011011 + 1 = (01011100)_2$.

(d) Subtract $(11001101)_2$ from $(01110001)_2$. The ones complement of 11001101 is 00110010. Add. $00110010 + 01110001 = 10100011$. There is no extra column 1. Hence the answer is $-(01011100)_2$.

256. FIXED- AND FLOATING-POINT ARITHMETIC The decimal (n-ary) point divides the integral and fractional parts of a decimal (n-ary) number. In fixed-point notation, all zeros between the point and the first significant digit must be written down. In floating-point notation, only the significant digits and a power of the base of the number system are indicated. Thus:

Notation for	*Fixed-point*		*Floating-point*
Decimal	3,078,400.0	0.30784×10^7	30,784 + 7
	0.0034123	0.34123×10^{-2}	34,123 − 2
Binary	10101000.0	0.10101×2^8	10,101 + 8
	0.00010101	0.10101×2^{-3}	10,101 − 3

257. CONGRUENCES If a and b are any two integers (positive, zero, or negative) whose difference is divisible by m, a and b are said to be *congruent modulo m*, written

$$a \equiv b \bmod m$$

Each of the numbers a and b is said to be a *residue* of the other. m is called the *modulus*.

Theorems. If $a \equiv b \bmod m$, $d \equiv b \bmod m$, $\alpha \equiv \beta \bmod m$, c and n are any two integers, $n > 0$, and $f(x)$ is any polynomial in x with integer coefficients, then

$$a \equiv d \bmod m \qquad a \pm \alpha \equiv (b \pm \beta) \bmod m$$

$$ca \equiv cb \bmod m \qquad a\alpha \equiv b\beta \bmod m$$

$$a^n \equiv b^n \bmod m \qquad f(a) \equiv f(b) \bmod m$$

13. MATRICES

258. A *matrix* is an array of $m \times n$ numbers (real or complex)

$$\begin{pmatrix} a_{11} & a_{12} & \cdots & a_{1n} \\ a_{21} & a_{22} & \cdots & a_{2n} \\ \cdots & \cdots & \cdots & \cdots \\ a_{m1} & a_{m2} & \cdots & a_{mn} \end{pmatrix}$$

arranged in m rows and n columns which obey the rules of addition and multiplication given in Sec. 259. The matrix may be denoted by $[a_{ij}]$ or (a_{ij}) or $\|a_{ij}\|$ or A, where a_{ij} denotes the element in the ith row and jth column. The *order* of the matrix is the number pair (m,n), written $m \times n$. The matrix is *square* of order n if $m = n$.

A *column matrix,* or *column vector,* is an $m \times 1$ matrix, written $X = \{x_i\}$ or $x_i\uparrow$.

A *row matrix,* or *row vector,* is a $1 \times n$ matrix, written $U = (u_i)$ or $[u_i]$ or $\vec{u_i}$ or $U = (u_1, u_2, \ldots, u_n)$.

259. OPERATIONS Let $A = [a_{ij}]$ and $B = [b_{ij}]$ be two matrices of the same order $m \times n$.

1. *Equals.* A and B are *equal* if and only if the elements in like positions in the arrays are equal (that is, if $a_{ij} = b_{ij}$ for $i = 1, \ldots, m$ and $j = 1, \ldots, n$).

2. *Addition.* The *sum* of A and B is the matrix $C = A + B$, where $C \equiv [c_{ij}]$, with $c_{ij} = a_{ij} + b_{ij}$.

3. *Multiplication.* If A is of order $m \times n$ and B is of order $n \times p$, the *product* AB is the matrix $D = AB$, where

$$D \equiv [d_{ij}] \qquad \text{with } d_{ij} = \sum_{k=1}^{n} a_{ik}\, b_{kj}$$

4. *Product of Scalar and Matrix.* The product of a (real or complex) number λ, called a *scalar,* and a matrix A is the matrix in which each element is λ times the corresponding element of A; that is,

$$\lambda A \equiv \lambda[a_{ij}] = [(\lambda a_{ij})] = A\lambda$$

260. A *null* (*zero*) *matrix* 0 is a matrix in which every element is zero.

An *identity* (*unit*) *matrix* I is a square matrix $[a_{ij}]$ in which $a_{ij} = \delta_{ij}$, where $\delta_{ij} = 1$ if $i = j$, $\delta_{ij} = 0$ if $i \neq j$. (δ_{ij} is the *Kronecker delta.*)

A *diagonal matrix* is a matrix all of whose elements off the *main diagonal* (the set of positions a_{ii}) are zero.

A *scalar matrix* is a diagonal matrix whose main diagonal elements are all equal. If λ is a scalar, λI is a scalar matrix.

261. DETERMINANT Associated with a square matrix A is a number $d(A)$, called the *determinant* of A, defined by

$$d(A) = \Sigma(-1)^h a_{1i_1} a_{2i_2} \cdots a_{ni_n}$$

where the summation is taken over all permutations $i_1, i_2, \ldots, i_n$ of the integers $1, 2, \ldots, n$, and h is the number of successive interchanges of two is which would be necessary to permute the order $i_1, i_2, \ldots, i_n$ into the order $1, 2, \ldots, n$. See Sec. 17 for theorems on determinants.

262. RANK OF A MATRIX Let $A^{r_1 \cdots r_k}_{s_1 \cdots s_k}$ be the square *minor* matrix (array) formed from the square matrix A by selecting from A the elements in rows $r_1 \cdots r_k$ and columns $s_1 \cdots s_k$, in those orders. If the same columns as rows are selected, the minor matrix is called a *principal minor matrix.*

The order ρ of a minor square array of A of maximum order whose determinant is not zero is the *rank* of A. A matrix whose order exceeds its rank is *singular.* If $d(A) \neq 0$, the rank of A is equal to its order.

Theorem. The rank of AB cannot exceed the rank of either A or B. If B and C are nonsingular, then A, AB, CA have the same rank.

263. LAPLACE'S EXPANSION Let any distinct rows $r_1, r_2, \ldots, r_k$ be selected from square matrix A. Then

$$d(A) = \Sigma(-1)^h \, d\left(A^{r_1 \cdots r_k}_{i_1 \cdots i_k}\right) d\left(A^{r_{k+1} \cdots r_n}_{i_{k+1} \cdots i_n}\right)$$

where h is equal to the number of any set of interchanges necessary to bring the order $i_1, i_2, \ldots, i_n$ into the order $r_1, r_2, \ldots, r_n$, and where the summation is taken over the $n!/[k!(n-k)!]$ ways of choosing the combinations $i_1, i_2, \ldots, i_k$ from the integers $1, 2, \ldots, n$.

The *cofactor* of a_{rs} is defined as

$$A_{rs} = (-1)^{r+s}\, d\left(A\begin{matrix}1, 2, \ldots, r-1, r+1, \ldots, n\\ 1, 2, \ldots, s-1, s+1, \ldots, n\end{matrix}\right)$$

Theorem. $d(A) = \sum_{i=1}^{n} a_{ri}A_{ri}$ $\qquad d(A) = \sum_{i=1}^{n} a_{is}A_{is}$

$$\sum_{i=1}^{n} a_{ri}A_{pi} = 0 \text{ if } r \neq p \qquad \sum_{i=1}^{n} a_{is}A_{ip} = 0 \text{ if } s \neq p$$

264. SPECIAL MATRICES The *transpose* A^T of a matrix A is the matrix resulting from A by interchanging rows and columns of A; that is, if $A = [a_{rs}]$, $A^T = [a_{sr}]$.

The *adjoint* or *adjugate* of a square matrix $A = [a_{rs}]$ is the matrix

$$adj\, A = [A_{sr}] = [A_{rs}]^T$$

where A_{rs} is the cofactor of a_{rs} in A.

The *inverse* or *reciprocal* matrix A^{-1} of A exists only if $d(A) \neq 0$. A^{-1} is defined by the equation $A^{-1}A = I$ and is unique. Furthermore, $AA^{-1} = I$ and $d(A)\, A^{-1} = adj\, A$.

If $a_{ij} = a_{ji}$, $A^T = A$ and A is *symmetric.*

If $a_{ij} = -a_{ji}$ and $a_{ii} = 0$, $A^T = -A$ and A is *skew.*

If $\bar{a}_{ij}$ is the complex conjugate of a_{ij}, $\bar{A} \equiv [\bar{a}_{ij}]$ is the *complex conjugate* of $A = [a_{ij}]$.

If $\bar{A}^T = A$, A is said to be *hermitian.*

If $\bar{A}^T = -A$, A is *skew-hermitian.*

If $A^TA = I$, where I is the unit matrix, A is *orthogonal.*

If $\bar{A}^TA = I$, A is *unitary.*

265. PROPERTIES OF MATRICES Matrices obey many of the laws of Table 11.2, Chap. 11. In general, multiplication is not commutative.

$$A + B = B + A \qquad A + (B + C) = (A + B) + C$$

$$(AB)C = A(BC)$$

$$A(B + C) = AB + AC \qquad (B + C)A = BA + CA$$

$$A - B = A + (-1)B = A + (-I)B$$

$$AA^{-1} = I \qquad A^{-1}A = I$$

$$d(AB) = d(A)\, d(B) \qquad d(A^{-1}) = \frac{1}{d(A)}$$

$$d(AA^{-1}) = d(A)\,d(A^{-1}) = d(I) = 1$$

$$(A^T)^T = A \qquad (A^{-1})^{-1} = A$$

$$(A + B)^T = A^T + B^T \qquad (AB)^T = B^T A^T$$

$$(A^T)^{-1} = (A^{-1})^T$$

$$(AB)^{-1} = B^{-1}A^{-1} \qquad adj(AB) = adj\,B\; adj\,A$$

If A is square of order n

$$d(adj\,A) = (d(A))^{n-1}$$

If α and β are scalars,

$$\alpha(A + B) = \alpha A + \alpha B \qquad (\alpha + \beta)A = \alpha A + \beta A$$

$$\alpha(\beta A) = \alpha\beta A \qquad \alpha(AB) = (\alpha A)B = A(\alpha B)$$

266. ELEMENTARY MATRICES An *elementary operation* (*transformation*) *upon the rows* (*columns*) of a matrix A is a transformation of one or more of the following types:

1. The interchange of rows (columns)
2. The addition to the elements of a row (column) of h times the corresponding elements of another row (column)
3. The multiplication of the elements of a row (column) by a number k, $k \neq 0$.

An *elementary matrix* is a matrix obtained by performing any elementary operation on the identity matrix I.

Let $I_{(ij)}$ be the matrix obtained from I by interchanging ith and jth rows without disturbing remaining rows. Multiplying $I_{(ij)}$ on the left (right) of A interchanges row (column) i and row (column) j of A.

Let $H_{(ij)} = I + (h)_{(ij)}$, $(i \neq j)$, be the matrix obtained from I by inserting element h in ith row, jth column. Multiplying $H_{(ij)}$ on the left (right) of A adds to each element of ith row (jth column) of A, h times the corresponding element of jth row (ith column) of A.

Let $J_{(ii)} = I + (k)_{(ii)}$ be the matrix obtained from I by replacing element in ith row, column i, by k. Multiplying $J_{(ii)}$ on the left (right) of A multiplies each element in the ith row (column) of A by k.

$I_{(ij)}$, $H_{(ij)}$, $J_{(ii)}$ are elementary matrices.

Theorem. If D is obtained from A by a finite number of elementary transformations upon the rows and columns of A, then $D = R\,A\,S$, where R and S are the products of elementary matrices.

267. LINEAR DEPENDENCE The elements $z_1, z_2, \ldots, z_n$ (for example, matrices, the rows of matrices, polynomials, numbers, solutions to sets of equations) are said to be *linearly independent* with respect to a field F if no linear combination of them, $\sum_{i=1}^{n} a_i z_i$, with coefficients a_i in F, is zero unless $a_1 = a_2 = \cdots = a_n = 0$.

If numbers a_i not all zero exist such that $\sum_{i=1}^{n} a_i z_i = 0$, then the $z_1, z_2, \ldots, z_n$ are said to be *linearly dependent* relative to F.

268. SET OF LINEAR EQUATIONS Consider the linear equations

(1) $\sum_{s=1}^{n} a_{rs} x_s = y_r \qquad (r = 1, \ldots, m) \qquad \text{written } A\{x\} = \{y\}$

of matrix $A = [a_{rs}]$, with a_{rs} in field F. Let B be the matrix derived from A by annexing the column $\{y\}$ on the right side of A; ρ is the rank of A, and ρ' the rank of B.

Theorem. **(a)** If $\rho' > \rho$, Eq. (1) is *inconsistent*.

(b) If $\rho' = \rho$, certain ρ of the equations determine uniquely ρ of the unknowns as linear functions of the remaining $(n - \rho)$ unknowns; for all values of the latter the expressions for these ρ unknowns satisfy also the remaining $(m - \rho)$ equations.

Theorem. If Eq. (1) is *homogeneous,* i.e., if

(2) $\sum_{s=1}^{n} a_{rs} x_s = 0 \qquad (r = 1, \ldots, m) \qquad \text{written } A\{x\} = \{0\}$

(a) Equation (2) has $(n - \rho)$ linearly independent solutions in F, and every other solution is *linearly dependent* on them [that is, ρ equations may be selected from (2), whose matrix has a nonvanishing ρ-rowed determinant and these ρ equations determine ρ of the unknowns as homogeneous linear functions, with coefficients in F, of the remaining $(n - \rho)$ unknowns; for all values of the latter, the expressions for the ρ unknowns satisfy (2).]

(b) Equation (2) has solutions not all zero if and only if $\rho < n$; the unique solution $x_j = 0$, $(j = 1, \ldots, n)$, if $\rho = n$.

(c) n homogeneous linear equations (2) in n unknowns have solutions not all zero if and only if $d(A) = 0$; the unique solution $x_j = 0$, $(j = 1, \ldots, n)$, if $d(A) \neq 0$.

Remark. The finding of solutions to $A\{x\} = \{y\}$ can be facilitated by performing a sequence of elementary transformations on the rows of A, reducing A to a triangular matrix (one in which blocks of zeros appear above or below the main diagonal) or a diagonal matrix D.

If A is nonsingular, A can be reduced to the identity matrix I. The result of performing the same sequence of operations on the rows of I will be the inverse matrix A^{-1}. In other words, if $E_k E_{k-1} \cdots E_1 = E$ is the left product of the elementary matrices which reduces A to the identity matrix I, that is, if $EA = I$, then $E = A^{-1}$. Furthermore, $\{x\} = A^{-1}\{y\} = E\{y\}$ is the solution to the equation $A\{x\} = \{y\}$ in case $d(A) \neq 0$. (See Secs. 275 to 282.)

269. EQUIVALENCE The square matrix A is *equivalent* to the square matrix B if there exist nonsingular matrices P and Q such that $A = PBQ$.

Theorem. Every matrix A of rank ρ is equivalent to a diagonal matrix $\|h_1, h_2, \ldots, h_\rho, 0, \ldots, 0\|$, $h_i \neq 0$.

A is *congruent* to B if there exists a nonsingular matrix P such that $A = P^T BP$.

Theorem. Every symmetric matrix A of rank ρ with elements in a field F is congruent in F to a diagonal matrix

$$\|d_1, d_2, \ldots, d_\rho, 0, \ldots, 0\| \qquad d_i \neq 0$$

A is said to be *similar* to B if there exists a nonsingular matrix P such that $A = P^{-1}BP$.

270. BILINEAR FORMS If the variables in the bilinear form

$$(1) \qquad G \equiv \sum_{i,j=1}^{n} a_{ij}x_i y_j \qquad \text{written } G = (x)A\{y\}$$

are subjected to the nonsingular transformations

$$(2) \qquad \{x\} = P\{\xi\} \qquad \{y\} = Q\{\eta\}$$

where $P = [p_{ij}]$ and $Q = [q_{ij}]$, then G becomes

$$(3) \qquad G = (\xi)P^T AQ\{\eta\}$$

Matrix $B = P^T AQ$ is equivalent to matrix A.

271. QUADRATIC FORMS If the variables in the symmetric quadratic form

$$Q \equiv \sum_{i,j=1}^{n} a_{ij}x_i x_j \qquad \text{written } Q = (x)A\{x\} \qquad (a_{ij} = a_{ji})$$

are subjected to the nonsingular transformations

$$\{x\} = C\{y\} \qquad C = [C_{ij}]$$

then Q becomes

$$Q = (y)C^TAC\{y\}$$

Matrix $B = C^TAC$ is congruent to matrix A.

272. TRANSFORMATION OF REFERENCE FRAMES Suppose vectors $\{x\}$ and $\{y\}$, referred, respectively, to reference frames X and Y, are related by

(1) $$\{x\} = A\{y\} \qquad A = [a_{rs}]$$

Change from frame X to frame $\overline{X}$ and from frame Y to frame $\overline{Y}$ by the nonsingular transformations

$$\{x\} = P\{\bar{x}\}, \quad P = [p_{rs}] \qquad \text{and} \qquad \{y\} = Q\{\bar{y}\}, \quad Q = [q_{rs}]$$

Then

(2) $$\{\bar{x}\} = P^{-1}AQ\{\bar{y}\}$$

Matrix $B = P^{-1}AQ$ of Eq. (2) is equivalent to matrix A of Eq. (1).

273. λ-MATRICES A matrix $A = (a_{ij})$ whose elements are polynomials in λ is a *λ-matrix*. A λ-matrix is of rank ρ if ρ is the largest integer for which not all minors of order ρ are identically zero.

The *elementary λ-operations* of a matrix are the operations 1, 2, and 3 of Sec. 266, h being either a polynomial in λ or a constant.

Two matrices are *λ-equivalent* if it is possible to pass from one to the other by a finite sequence of λ-operations.

Theorem. Every λ-matrix A of order n and rank ρ is λ-equivalent to a diagonal matrix $\|E_1, E_2, \ldots, E_\rho, 0, \ldots, 0\|$, each E_i being either unity or a polynomial in λ with unity as the coefficient of the highest power of λ. Each E_{i-1} is a factor of E_i, $i = 2, 3, \ldots, \rho$. $E_1, E_2, \ldots, E_\rho$ are *invariant factors* of A.

If the $E_1, E_2, \ldots$ are regarded as complex polynomials, each E_i can be expressed in the form

$$E_i = (\lambda - \alpha)^{a_i}(\lambda - \beta)^{b_i} \ldots (\lambda - \kappa)^{k_i}$$

with $a_i < a_{i+1}$, $b_i < b_{i+1}$, . . . , $k_i < k_{i+1}$ $(i = 1, 2, \ldots, \rho - 1)$.

The *elementary divisors of A* are those factors $(\lambda - \alpha)^{a_1}, \ldots,$ $(\lambda - \kappa)^{k_\rho}$ which are not unity.

Theorem. Each one of **(a)** and **(b)** below is a necessary and sufficient condition for two square λ-matrices A and B to be λ-equivalent:

(a) A and B have the same invariant factors.

(b) There exist λ-matrices P and Q for which $|P|$ and $|Q|$ are nonzero constants and $QAP = B$.

274. EIGENVALUES AND EIGENVECTORS Let A be a square matrix of order n and $\{z\}$ an n-dimensional vector, written as a column matrix. A vector $\{z\}$ which has the same direction as vector $A\{z\}$ must satisfy

(1) $$A\{z\} = \lambda\{z\} \qquad \text{or} \qquad (A - \lambda I)\{z\} = 0$$

where λ is a scalar and I is the identity matrix of order n. If $\{z\} \neq 0$, the matrix $(A - \lambda I)$ is singular and

(2) $$\det(A - \lambda I) = 0$$

Relation (2) is a polynomial equation of degree n in λ, with roots $\lambda_1, \lambda_2, \ldots, \lambda_n$. The values of λ_i are the *eigenvalues* (or *latent roots* or *characteristic values*) of matrix A. Corresponding to each root λ_i is a vector $\{z_i\}$ called an *eigenvector.* The elements of $\{z_i\}$ can be chosen as the cofactors of any convenient row of $A - \lambda_i I$.

When the eigenvalues are distinct, the corresponding eigenvectors are linearly independent. In this case the eigenvectors can be used as a basis of the n-vector space.

Let L be the matrix whose jth column is the eigenvector $\{z_j\}$. Then

(3) $$L^{-1}AL = \operatorname{diag}(\lambda_1, \lambda_2, \ldots, \lambda_n)$$

If A is a real symmetric matrix, the eigenvectors corresponding to distinct eigenvalues are orthogonal, that is, if $\lambda_i \neq \lambda_j$, then $\{z_j\}^T\{z_i\} = 0$.

If any of the eigenvalues are multiple roots then there are, say, $(n - q)$ distinct eigenvalues and $(n - q)$ linearly independent eigenvectors.

Solution of Linear Systems of Equations

275. LINEAR SYSTEM OF EQUATIONS The linear system

$$\sum_{s=1}^{n} a_{rs}x_s = b_r \qquad r = 1, \ldots, n \tag{1}$$

of n equations in n unknowns written in matrix notation is

$$A\{x\} = \{b\} \tag{2}$$

where $A = (a_{rs})$ is the matrix of the coefficients in (1), and $\{x\}$ and $\{b\}$ are column vectors. (See Sec. 268.)

276. CRAMER'S RULE If $d(A) \equiv D \neq 0$, and D_r is the determinant of the matrix obtained from A by replacing the rth column of A by the column $\{b\}$, then the solution to (1) is

$$x_r = \frac{D_r}{D} \qquad r = 1, 2, \ldots, n \tag{3}$$

277. GAUSSIAN ELIMINATION In this method the system (1) is transformed into an equivalent *triangular* system [one in which $A = (a_{rs})$ is *upper triangular*, that is, $a_{rs} = 0$ for $r > s$] whose solution can be more easily obtained.

Elementary row operations are performed on A to carry out the elimination. These operations may be done by the use of the identity matrix I of order n, as in Sec. 266.

Suppose $a_{11} \neq 0$. Replace the first column of I by the column whose elements are $\alpha_{11}, \alpha_{21}, \ldots, \alpha_{n1}$, with $\alpha_{11} = 1$, $\alpha_{21} = -a_{21}/a_{11}$, $\alpha_{31} = -a_{31}/a_{11}, \ldots$. Call the resulting matrix $T^{(1)}$. From (2)

$$T^{(1)}A\{x\} = T^{(1)}\{b\} \qquad \text{or} \qquad A^{(1)}\{x\} = \{b^{(1)}\} \tag{4}$$

where $T^{(1)}A \equiv A^{(1)} \equiv [a_{rs}{}^{(1)}]$ is a matrix whose elements in the first column–not on the main diagonal–are all zero and $T^{(1)}\{b\} \equiv \{b^{(1)}\}$. Similarly, if $a_{22}{}^{(1)} \neq 0$, replace column two of I by the column

$\alpha_{12}, \alpha_{22}, \ldots, \alpha_{n2}$ with

$$\alpha_{12} = 0, \alpha_{22} = 1, \alpha_{32} = \frac{-a_{32}^{(1)}}{a_{22}^{(1)}}, \alpha_{42} = \frac{-a_{42}^{(1)}}{a_{22}^{(1)}}, \ldots$$

Let $T^{(2)}$ be the resulting matrix. Then

(5) $$T^{(2)}A^{(1)}\{x\} = T^{(2)}\{b^{(1)}\} \qquad \text{or} \qquad A^{(2)}\{x\} = \{b^{(2)}\}$$

where $T^{(2)}A^{(1)} \equiv A^{(2)}$ and $T^{(2)}\{b^{(1)}\} \equiv \{b^{(2)}\}$. $A^{(2)}$ is a matrix whose elements in the first and second columns below the main diagonal are all zero.

Continue the process until a system

(6) $$A^{(u)}\{x\} = \{b^{(u)}\}$$

is found in which $A^{(u)}$ is upper triangular:

(7) $$\begin{pmatrix} a_{11} & a_{12} & a_{13} & \cdots & a_{1n} \\ 0 & a_{22}^{(1)} & a_{23}^{(1)} & \cdots & a_{2n}^{(1)} \\ 0 & 0 & a_{33}^{(2)} & \cdots & a_{3n}^{(2)} \\ \cdots & \cdots & \cdots & \cdots & \cdots \\ 0 & 0 & 0 & \cdots & a_{nn}^{(n-1)} \end{pmatrix} \begin{Bmatrix} x_1 \\ x_2 \\ x_3 \\ \cdots \\ x_n \end{Bmatrix} = \begin{Bmatrix} b_1 \\ b_2^{(1)} \\ b_3^{(2)} \\ \cdots \\ b_n^{(n-1)} \end{Bmatrix}$$

The solution to (2), and (1), is found by *back substitution* to be

(8) $$\begin{cases} x_n = \dfrac{b_n^{(n-1)}}{a_{nn}^{(n-1)}} \\[2ex] x_{n-1} = \dfrac{b_{n-1}^{(n-2)} - a_{n-1,n}^{(n-2)}x_n}{a_{n-1,n-1}^{(n-2)}} \end{cases}$$

The diagonal elements $a_{11}, a_{22}^{(1)}, a_{33}^{(2)}, \ldots$ are called *pivots*.

Note: The matrix $A^{(u)}$ is the product $A^{(u)} = T^{(u)} \cdots T^{(2)}T^{(1)}A$, $d(T^{(u)}) = \cdots = d(T^{(1)}) = 1$, and $d(A^{(u)}) = d(A) = a_{11}a_{22}^{(1)}a_{33}^{(2)} \cdots a_{nn}^{(n-1)}$.

If at any stage one of the pivots vanishes, an attempt should be made to rearrange the remaining rows so as to get a nonzero pivot. If this is impossible, A is singular and (1) has no unique solution.

278. The process can be extended to reduce A to an equivalent diagonal matrix. (See Sec. 269.) When this is done and when

$d(A) \neq 0$, the solution to (1) can be read directly from (7) without back substitution.

If A is reduced to the identity matrix by the process, the product $T^{(u)} \cdots T^{(2)}T^{(1)} \equiv E = A^{-1}$ is the inverse to matrix A. (See Sec. 268.)

279. ACCURACY OF SOLUTION In dealing with systems (1), round-off errors may be introduced into the coefficients. The loss of accuracy may invalidate the solution obtained. To reduce such errors the process should be arranged (if possible) so that the pivot at any stage is larger in magnitude than any remaining element in the column.

The Gaussian elimination process requires less arithmetic than Cramer's rule.

One method for estimating the accuracy of a solution is by use of residuals. If $\{x^{(1)}\}$ is an approximate solution to $A\{x\} = \{b\}$, the smallness of the *residuals* $\{r\} = \{b\} - A\{x^{(1)}\}$ is one indication of the exactness of solution $\{x^{(1)}\}$. If the solution error is primarily due to round-off errors, residuals provide a reasonable measure of the accuracy of solution.

Systems in which small changes in the coefficients lead to large changes in the solution are *ill-conditioned.* Residuals do not provide good measures of accuracy of solution in ill-conditioned systems.

Example. Find solutions of

$$(1) \qquad \begin{cases} 2x_1 + 3x_2 - x_3 = 5 \\ 4x_1 + 4x_2 - 3x_3 = 3 \\ 2x_1 - 3x_2 + x_3 = -1 \end{cases} \qquad \text{or} \qquad A\{x\} = \{b\}$$

where

$$(2) \qquad A = \begin{pmatrix} 2 & 3 & -1 \\ 4 & 4 & -3 \\ 2 & -3 & 1 \end{pmatrix} \qquad \{b\} = \begin{Bmatrix} 5 \\ 3 \\ -1 \end{Bmatrix}$$

The addition of (-2) times the first row to the second row and (-1) times the first row to the third row is achieved by multiplying (1) on the left by

(3)
$$T^{(1)} = \begin{pmatrix} 1 & 0 & 0 \\ -2 & 1 & 0 \\ -1 & 0 & 1 \end{pmatrix}$$

Then

(4)
$$T^{(1)}A\{x\} = T^{(1)}\{b\}$$

where

$$T^{(1)}A \equiv A^{(1)} = \begin{pmatrix} 2 & 3 & -1 \\ 0 & -2 & -1 \\ 0 & -6 & 2 \end{pmatrix} \qquad T^{(1)}\{b\} = \{b^{(1)}\} = \begin{Bmatrix} 5 \\ -7 \\ -6 \end{Bmatrix}$$

Similarly, multiplication of (4) on the left by

(5)
$$T^{(2)} = \begin{pmatrix} 1 & 0 & 0 \\ 0 & 1 & 0 \\ 0 & -3 & 1 \end{pmatrix}$$

gives

(6)
$$T^{(2)}A^{(1)}\{x\} = T^{(2)}\{b^{(1)}\}$$

where

(7)
$$T^{(2)}A^{(1)} \equiv A^{(2)} = \begin{pmatrix} 2 & 3 & -1 \\ 0 & -2 & -1 \\ 0 & 0 & 5 \end{pmatrix}, \quad T^{(2)}\{b^{(1)}\} \equiv \{b^{(2)}\} = \begin{Bmatrix} 5 \\ -7 \\ 15 \end{Bmatrix}$$

In other words,

(8)
$$\begin{cases} 2x_1 + 3x_2 - x_3 = 5 \\ - 2x_2 - x_3 = -7 \\ 5x_3 = 15 \end{cases}$$

Hence $x_3 = 3$, $x_2 = 2$, $x_1 = 1$.

System (8) can be reduced to one in which the matrix $A^{(2)}$ is replaced by a diagonal matrix. This may be done by multiplying (6) on the left by $T^{(3)}$, $T^{(4)}$, $T^{(5)}$, in that order, where

$$(9)\quad T^{(3)} = \begin{pmatrix} 1 & 0 & \frac{1}{5} \\ 0 & 1 & \frac{1}{5} \\ 0 & 0 & 1 \end{pmatrix} \qquad T^{(4)} = \begin{pmatrix} 1 & \frac{3}{2} & 0 \\ 0 & 1 & 0 \\ 0 & 0 & 1 \end{pmatrix}$$

$$T^{(5)} = \begin{pmatrix} \frac{1}{2} & 0 & 0 \\ 0 & -\frac{1}{2} & 0 \\ 0 & 0 & \frac{1}{5} \end{pmatrix}$$

There results

$$(10)\qquad A^{(5)}\{x\} = \{b^{(5)}\}$$

where

$$A^{(5)} = TA^{(2)} \qquad \{b^{(5)}\} = T\{b^{(2)}\} \qquad T = T^{(5)}T^{(4)}T^{(3)}$$

$$(11)\qquad A^{(5)} = \begin{pmatrix} 1 & 0 & 0 \\ 0 & 1 & 0 \\ 0 & 0 & 1 \end{pmatrix} \qquad \{b^{(5)}\} = \begin{Bmatrix} 1 \\ 2 \\ 3 \end{Bmatrix}$$

In other words, $x_1 = 1$, $x_2 = 2$, and $x_3 = 3$.

By Sec. 268, the product

$$(12)\qquad T^{(5)}T^{(4)} \cdots T^{(1)} = \begin{pmatrix} \frac{1}{4} & 0 & \frac{1}{4} \\ \frac{1}{2} & -\frac{1}{5} & -\frac{1}{10} \\ 1 & -\frac{3}{5} & \frac{1}{5} \end{pmatrix} \equiv A^{-1}$$

is the inverse to matrix A.

The above process may be abbreviated by the use of augmented matrices. Thus,

$$\begin{pmatrix} 1 & 0 & 0 \\ -2 & 1 & 0 \\ -1 & 0 & 1 \end{pmatrix} \left(\begin{array}{ccc|c} 2 & 3 & -1 & 5 \\ 4 & 4 & -3 & 3 \\ 2 & -3 & 1 & -1 \end{array}\right) = \left(\begin{array}{ccc|c} 2 & 3 & -1 & 5 \\ 0 & -2 & -1 & -7 \\ 0 & -6 & 2 & -6 \end{array}\right)$$

$$\begin{pmatrix} 1 & 0 & 0 \\ 0 & 1 & 0 \\ 0 & -3 & 1 \end{pmatrix} \left(\begin{array}{ccc|c} 2 & 3 & -1 & 5 \\ 0 & -2 & -1 & -7 \\ 0 & -6 & 2 & -6 \end{array}\right) = \left(\begin{array}{ccc|c} 2 & 3 & -1 & 5 \\ 0 & -2 & -1 & -7 \\ 0 & 0 & 5 & 15 \end{array}\right)$$

From the right side it follows that $x_3 = 3$, $x_2 = 2$, and $x_1 = 1$.

280. INVERSE OF A MATRIX A The amount of calculation necessary to find the inverse of A from $A^{-1} = [adj\, A]/d(A)$ is considerable. (See Sec. 268.) A more efficient method is as follows.

By definition the inverse B of A is such that

$$AB = I \tag{1}$$

where I is the identity matrix of same order as A. Equation (1) states that

$$\begin{pmatrix} a_{11} & \cdots & a_{1n} \\ a_{21} & \cdots & a_{2n} \\ \cdots & \cdots & \cdots \\ a_{n1} & \cdots & a_{nn} \end{pmatrix} \begin{Bmatrix} b_{11} \\ b_{21} \\ \cdots \\ b_{n1} \end{Bmatrix} = \begin{Bmatrix} 1 \\ 0 \\ \cdots \\ 0 \end{Bmatrix}, \ldots, \tag{2}$$

$$\begin{pmatrix} a_{11} & \cdots & a_{1n} \\ a_{21} & \cdots & a_{2n} \\ \cdots & \cdots & \cdots \\ a_{n1} & \cdots & a_{nn} \end{pmatrix} \begin{Bmatrix} b_{1n} \\ b_{2n} \\ \cdots \\ b_{nn} \end{Bmatrix} = \begin{Bmatrix} 0 \\ 0 \\ \cdots \\ 1 \end{Bmatrix}$$

Each equation in (2) may be solved for the b_{ij}.

Let T be the product of elementary transformations that reduces A to an upper triangular form $A^{(1)}$. For simplicity assume no interchange of rows. Then

$$A^{(1)}B = I^{(1)} \tag{3}$$

where

$$A^{(1)} = TA = \begin{pmatrix} a_{11} & a_{12} & \cdots & a_{1n} \\ 0 & a_{22}^{(1)} & \cdots & a_{2n}^{(1)} \\ \cdots & \cdots & \cdots & \cdots \\ 0 & 0 & \cdots & a_{nn}^{(n-1)} \end{pmatrix} \tag{4}$$

$$I^{(1)} = TI = \begin{pmatrix} 1 & 0 & \cdots & 0 \\ \alpha_{21} & 1 & \cdots & 0 \\ \cdots & \cdots & \cdots & \cdots \\ \alpha_{n1} & \alpha_{n2} & \cdots & 1 \end{pmatrix}$$

From (2) and (4)

$$(5) \qquad A^{(1)}\begin{Bmatrix} b_{11} \\ b_{21} \\ \vdots \\ b_{n1} \end{Bmatrix} = \begin{Bmatrix} 1 \\ \alpha_{21} \\ \vdots \\ \alpha_{n1} \end{Bmatrix} \cdots A^{(1)}\begin{Bmatrix} b_{1n} \\ b_{2n} \\ \vdots \\ b_{nn} \end{Bmatrix} = \begin{Bmatrix} 0 \\ 0 \\ \vdots \\ 1 \end{Bmatrix}$$

From (5) the values of $b_{n1}, \ldots, b_{11}$ are found by back substitution, as are $b_{n2}, \ldots, b_{12}, \ldots$ through $b_{nn}, \ldots, b_{1n}$.

In the case where interchanges of rows of A are made, the rows of I must also be correspondingly interchanged.

The inversion of matrices should be avoided whenever possible since the amount of calculation is considerable and it is often difficult to obtain the inverse accurately.

281. FACTORIZATION OF MATRIX INTO TRIANGULAR MATRICES The matrix $A = (a_{ij})$ can be factored into the form $A = LU$, where L is *lower triangular* and U is *upper triangular*, and

$$(1) \quad L = \begin{pmatrix} l_{11} & 0 & 0 & \cdots & 0 \\ l_{21} & l_{22} & 0 & \cdots & 0 \\ \cdots & \cdots & \cdots & \cdots & \cdots \\ l_{n1} & l_{n2} & l_{n3} & \cdots & l_{nn} \end{pmatrix}$$

$$U = \begin{pmatrix} 1 & u_{12} & u_{13} & \cdots & u_{1n} \\ 0 & 1 & u_{23} & \cdots & u_{2n} \\ \cdots & \cdots & \cdots & \cdots & \cdots \\ 0 & 0 & 0 & \cdots & 1 \end{pmatrix}$$

provided all the principal minors of A are nonsingular, i.e., if

$$a_{11} \neq 0, \begin{vmatrix} a_{11} & a_{12} \\ a_{21} & a_{22} \end{vmatrix} \neq 0, \ldots, |A| \neq 0$$

The elements of L and U are found from $LU = A$. The product of the rows of L by the first column of U gives the first column of A

$$l_{11} = a_{11}, l_{21} = a_{21}, \ldots, l_{n1} = a_{n1}$$

The product of the first row of L by the second, third, . . . , nth column of U gives

$$l_{11}u_{12} = a_{12},\ l_{11}u_{13} = a_{13}, \ldots, l_{11}u_{1n} = a_{1n}$$

from which $u_{12}, u_{13}, \ldots, u_{1n}$ are found. The process is continued until all the elements of L and U are found. The order followed is: find the elements of the first column of L, then the elements of the first row of U; the elements of the second column of L, the elements of the second row of U; . . . .

In general

(2)
$$l_{ij} = a_{ij} - \sum_{k=1}^{j-1} l_{ik}u_{kj} \qquad i \geqq j$$

$$l_{ii}u_{ij} = a_{ij} - \sum_{k=1}^{j-1} l_{ik}u_{kj} \qquad i < j$$

The elements are computed in the order: $l_{11}, l_{21}, \ldots, u_{12}, u_{13}, \ldots;$ $l_{i1}, u_{1j}; l_{i2}, u_{2j}; \ldots l_{i,n-1}, u_{n-1,j}; l_{nn}$.

282. **USE OF L AND U TO SOLVE LINEAR SYSTEM** When A is nonsingular and factorable into $A = LU$ the system

(1) $$A\{x\} = \{b\} \quad \text{or} \quad LU\{x\} = \{b\}$$

can be solved as follows. Let $U\{x\} = \{z\}$. Then

(2) $$L\{z\} = \{b\}$$

or

(3) $$\begin{cases} l_{11}z_1 & = b_1 \\ l_{21}z_1 + l_{22}z_2 & = b_2 \\ \cdots\cdots\cdots\cdots\cdots\cdots & \\ l_{n1}z_1 + l_{n2}z_2 + \cdots + l_{nn}z_n & = b_n \end{cases}$$

From (3), $z_1, z_2, \ldots, z_n$ can be calculated successively. From $U\{x\} = \{z\}$

$$(4) \qquad \begin{cases} x_1 + u_{12}x_2 + u_{13}x_3 + \cdots + u_{1n}x_n = z_1 \\ \qquad\quad x_2 + u_{23}x_3 + \cdots + u_{2n}x_n = z_2 \\ \cdots\cdots\cdots\cdots\cdots\cdots\cdots\cdots\cdots \\ \qquad\qquad\qquad\qquad\qquad\qquad x_n = z_n \end{cases}$$

Back substitution gives $x_n, x_{n-1}, \ldots, x_1$, the solution to (1).

Methods based on matrix factorization are called *compact schemes.* They are useful both in hand and machine computation. Such schemes are useful in inverting matrices, since the inverses of triangular matrices are easily found. For example, if $A = LU$,

$$(5) \qquad A^{-1} = (LU)^{-1} = U^{-1}L^{-1}$$

Linear Vector Spaces

283. DEFINITIONS Let $c_1, c_2, \ldots$ be elements (*scalars*) of a field F. A *vector space* ("a linear space") V over a field F is a set of elements, called *vectors,* such that (1) any two vectors $\boldsymbol{\alpha}$ and $\boldsymbol{\beta}$ of V determine, relative to an operation "+" called "addition," a unique vector $(\boldsymbol{\alpha} + \boldsymbol{\beta})$ called the *sum,* (2) any scalar c from F and any vector $\boldsymbol{\alpha}$ from V determine a unique element $(c\boldsymbol{\alpha})$ in V, called the *scalar product,* and (3) for all vectors $\boldsymbol{\alpha}$ and $\boldsymbol{\beta}$ and all scalars c_1 and c_2,

(1) V is an abelian group under addition

(2) $c(\boldsymbol{\alpha} + \boldsymbol{\beta}) = c\boldsymbol{\alpha} + c\boldsymbol{\beta}$
$(c_1 + c_2)\boldsymbol{\alpha} = c_1\boldsymbol{\alpha} + c_2\boldsymbol{\alpha}$ } (distributive laws)

(3) $(c_1c_2)\boldsymbol{\alpha} = c_1(c_2\boldsymbol{\alpha})$
$1\boldsymbol{\alpha} = \boldsymbol{\alpha}$ where 1 is the unit of F

The identity element of the abelian group is the *null* (or *zero*) *vector* denoted by $\mathbf{O}$. $\boldsymbol{\xi} + \mathbf{O} = \mathbf{O} + \boldsymbol{\xi} = \boldsymbol{\xi}$ for all vectors $\boldsymbol{\xi}$.

284. A *subspace* S of a vector space V is a subset of V which is itself a vector space with respect to the operations of addition and scalar multiplication in V.

Example. The set $V_n(F)$ of all n-tuples over a field F is a vector space. Let a_i and b_i and c be in F and let $\boldsymbol{\alpha} = (a_1, a_2, \ldots, a_n)$

and $\boldsymbol{\beta} = (b_1, b_2, \ldots, b_n)$ be any two n-tuples. The operations of "addition" and "scalar multiplication" by elements c of F are defined as:

$$\begin{aligned} \boldsymbol{\alpha} + \boldsymbol{\beta} &= (a_1, \ldots, a_n) + (b_1, \ldots, b_n) \\ &= (a_1 + b_1, \ldots, a_n + b_n) \\ c\boldsymbol{\alpha} &= c(a_1, \ldots, a_n) = (ca_1, \ldots, ca_n) \end{aligned}$$

If $c_1, c_2, \ldots, c_m$ are in F and vectors $\boldsymbol{\xi}_1, \boldsymbol{\xi}_2, \ldots, \boldsymbol{\xi}_m$ are in V, the linear combination $c_1\boldsymbol{\xi}_1 + c_2\boldsymbol{\xi}_2 + \cdots + c_m\boldsymbol{\xi}_m$ is a unique element in V.

285. The set of all linear combinations of any set of vectors $\boldsymbol{\xi}_1, \boldsymbol{\xi}_2, \ldots, \boldsymbol{\xi}_m$ in V is a subspace of V *generated* or *spanned* by $\{\boldsymbol{\xi}_1, \ldots, \boldsymbol{\xi}_m\}$ and denoted by $\mathcal{L}\{\boldsymbol{\xi}_1, \ldots, \boldsymbol{\xi}_m\}$. The *generators* or *spanning vectors* of the subspace are $\boldsymbol{\xi}_1, \ldots, \boldsymbol{\xi}_m$.

286. UNIT VECTORS Any n-dimensional linear vector space $V_n(F)$ over the field F is spanned by n *unit vectors*

$$\boldsymbol{\epsilon}_1 = (1, 0, \ldots, 0), \boldsymbol{\epsilon}_2 = (0, 1, \ldots, 0), \ldots \\ \boldsymbol{\epsilon}_n = (0, 0, \ldots, 1)$$

Any vector $(x_1, x_2, \ldots, x_n)$ of $V_n(F)$ is a linear combination of $\boldsymbol{\epsilon}_1, \ldots, \boldsymbol{\epsilon}_n$,

$$(x_1, x_2, \ldots, x_n) = x_1\boldsymbol{\epsilon}_1 + x_2\boldsymbol{\epsilon}_2 + \cdots + x_n\boldsymbol{\epsilon}_n$$

$\boldsymbol{\epsilon}_1, \ldots, \boldsymbol{\epsilon}_n$ generates the whole of $V_n(F)$.

Theorem. There exists a unique element of V_n, called the *zero element,* denoted by $\mathbf{O}$, such that $0\boldsymbol{\xi} = \mathbf{O}$ for all vectors $\boldsymbol{\xi}$ of V_n and $c\mathbf{O} = \mathbf{O}$ for all scalars c of F. The *group inverse* of any vector $\boldsymbol{\xi}$ is $(-1)\boldsymbol{\xi}$. $x_1\boldsymbol{\xi}_1 + \cdots + x_n\boldsymbol{\xi}_n = \mathbf{O}$ if and only if $(x_1, \ldots, x_n) = (0, \ldots, 0)$, that is, if and only if $x_1 = x_2 = \cdots = x_n = 0$.

287. INDEPENDENCE The vectors $\boldsymbol{\alpha}_1, \ldots, \boldsymbol{\alpha}_m$ are *linearly independent* over field F if and only if, for all c_i in F, $c_1\boldsymbol{\alpha}_1 + c_2\boldsymbol{\alpha}_2 + \cdots + c_m\boldsymbol{\alpha}_m = \mathbf{O}$ implies $c_1 = c_2 = \cdots = c_m = 0$. Vectors which are not linearly independent are *linearly dependent.*

Theorem. The nonzero vectors $\boldsymbol{\alpha}_1, \ldots, \boldsymbol{\alpha}_m$ in a space V are linearly dependent if and only if some one of the vectors $\boldsymbol{\alpha}_k$ is a linear combination of the preceding ones.

288. BASIS A *basis* of a vector space V is a linearly independent subset which generates (spans) the whole space. If V has a finite basis, the space is finite dimensional.

Theorem. **(*a*)** Any vector in a space V is expressible as a linear combination of the vectors of a given basis in one and only one way.

(*b*) Any two bases of a finite dimensional vector space V have the same number of elements (called the *dimension* $d[V]$, or dim V, of V).

(*c*) In an n-dimensional vector space, any $(n + 1)$ vectors are linearly dependent.

(*d*) For a set A of n vectors of V to be a basis, it is sufficient that either they span V or they be linearly independent.

(*e*) If S and T are any two subspaces of V, then

$$d[S] + d[T] = d[S \cap T] + d[S \cup T]$$

289. INNER PRODUCT The *inner product* of two vectors $\boldsymbol{\xi} = (x_1, \ldots, x_n)$ and $\boldsymbol{\eta} = (y_1, \ldots, y_n)$ with real components is

$$(\boldsymbol{\xi},\boldsymbol{\eta}) = x_1y_1 + x_2y_2 + \cdots + x_ny_n$$

Properties of the inner product

(1) $(\boldsymbol{\xi} + \boldsymbol{\eta},\boldsymbol{\zeta}) = (\boldsymbol{\xi},\boldsymbol{\zeta}) + (\boldsymbol{\eta},\boldsymbol{\zeta})$ (*linear* in the left-hand factor)

(2) $(c\boldsymbol{\xi},\boldsymbol{\eta}) = c(\boldsymbol{\xi},\boldsymbol{\eta})$ (*linear* in the left-hand factor)

(3) $(\boldsymbol{\xi},\boldsymbol{\eta}) = (\boldsymbol{\eta},\boldsymbol{\xi})$ (*symmetric*)

(4) $(\boldsymbol{\xi},\boldsymbol{\xi}) > 0 \quad \text{if } \boldsymbol{\xi} \neq 0$ (*positiveness*)

Relations (1), (2), and (3) show the linearity of inner products in both factors (*bilinearity*).

Remark. If the components of the vectors $\boldsymbol{\xi}$ and $\boldsymbol{\eta}$ are complex numbers, the *inner product* is defined as

$$(\boldsymbol{\xi},\boldsymbol{\eta}) = x_1y_1^* + x_2y_2^* + \cdots + x_ny_n^*$$

where y_i^* is the complex conjugate of y_i. The *length* of $\boldsymbol{\xi}$ is $|\boldsymbol{\xi}| = (\boldsymbol{\xi},\boldsymbol{\xi})^{1/2}$.

290. EUCLIDEAN VECTOR SPACE A *Euclidean vector space* is a vector space E with real scalars, such that to any vectors $\boldsymbol{\xi}$ and $\boldsymbol{\eta}$ in E there corresponds a (real) "inner product" $(\boldsymbol{\xi},\boldsymbol{\eta})$ which has properties (1), . . . , (4).

The *length* $|\boldsymbol{\xi}|$ of a vector $\boldsymbol{\xi}$ of a Euclidean vector space E is $(\boldsymbol{\xi},\boldsymbol{\xi})^{1/2}$, the positive square root of the inner product of $\boldsymbol{\xi}$ and $\boldsymbol{\xi}$.

Theorem. In any Euclidean vector space, length has the properties

(1) $|c\boldsymbol{\xi}| = |c| \cdot |\boldsymbol{\xi}|$ (2) $|\boldsymbol{\xi}| > 0$ if $\boldsymbol{\xi} \neq 0$

(3) $|(\boldsymbol{\xi},\boldsymbol{\eta})| \leqq |\boldsymbol{\xi}| \cdot |\boldsymbol{\eta}|$ (*Schwarz inequality*)

(4) $|\boldsymbol{\xi} + \boldsymbol{\eta}| \leqq |\boldsymbol{\xi}| + |\boldsymbol{\eta}|$ (*triangle inequality*)

291. The *distance* between two elements $\boldsymbol{\xi}$ and $\boldsymbol{\eta}$ of E is $|\boldsymbol{\xi} - \boldsymbol{\eta}|$.

Theorem. Distance in Euclidean space has the properties:

(1) The distance from any point (element) to itself is zero, and any two distinct points (elements) are a positive distance apart.

(2) Distance is symmetric: $|\boldsymbol{\xi} - \boldsymbol{\eta}| = |\boldsymbol{\eta} - \boldsymbol{\xi}|$

(3) $|\boldsymbol{\xi} - \boldsymbol{\eta}| + |\boldsymbol{\eta} - \boldsymbol{\zeta}| \geqq |\boldsymbol{\xi} - \boldsymbol{\zeta}|$

Angle. The cosine of the one and only one angle θ between $0°$ and $180°$ is $(\boldsymbol{\xi},\boldsymbol{\eta})/[|\boldsymbol{\xi}| \cdot |\boldsymbol{\eta}|]$, and θ is *defined* as the *angle* between the vectors $\boldsymbol{\xi}$ and $\boldsymbol{\eta}$.

292. ORTHOGONAL Two elements $\boldsymbol{\xi}$ and $\boldsymbol{\eta}$ are *orthogonal* (written $\boldsymbol{\xi} \perp \boldsymbol{\eta}$) whenever $(\boldsymbol{\xi},\boldsymbol{\eta}) = 0$.

Theorem. If a vector is orthogonal to $\boldsymbol{\xi}_1, \ldots, \boldsymbol{\xi}_n$, it is orthogonal to every vector in the subspace spanned by $\boldsymbol{\xi}_1, \ldots, \boldsymbol{\xi}_n$.

The vectors $\boldsymbol{\alpha}_1, \ldots, \boldsymbol{\alpha}_n$ of a Euclidean vector space are *normal orthogonal* if $\boldsymbol{\alpha}_i \perp \boldsymbol{\alpha}_j$ when $i \neq j$, and $|\boldsymbol{\alpha}_i| = 1$ for all i.

Theorem. Any finite-dimensional Euclidean vector space E has a normal orthogonal basis.

Linear Transformations. Polynomials

293. LINEAR TRANSFORMATION A mapping T which assigns to each vector $\boldsymbol{\alpha}$ of a vector space V a unique vector $T\boldsymbol{\alpha}$ in a vector space W is called a *linear transformation* of V into W *if,* for any vectors $\boldsymbol{\alpha}$ and $\boldsymbol{\beta}$ in V and any real number r, the following conditions hold:

(1) $T(\boldsymbol{\alpha} + \boldsymbol{\beta}) = T\boldsymbol{\alpha} + T\boldsymbol{\beta}$ and

(2) $T(r\boldsymbol{\alpha}) = r(T\boldsymbol{\alpha})$

Conditions (1) and (2) can be combined into

(3) $T(r_1\boldsymbol{\alpha} + r_2\boldsymbol{\beta}) = r_1(T\boldsymbol{\alpha}) + r_2(T\boldsymbol{\beta})$

where r_1 and r_2 are any two real numbers.

If $V = W$, the transformation T is sometimes called an *operator* on V.

If T_1 and T_2 are linear transformations of a vector space V, $T_1 + T_2$ is defined as

$$(T_1 + T_2)\boldsymbol{\alpha} = T_1\boldsymbol{\alpha} + T_2\boldsymbol{\alpha}$$

for any vector $\boldsymbol{\alpha}$ in V. The operation $(T_1 + T_2)$ is linear.

Examples. **(*a*)** The mapping $T: V_3 \to V_2$ defined so that $T(x,y,z) = (x,z)$ is linear. This mapping can be written $(x,y,z) \xrightarrow{T} (x,z)$.

(*b*) The mapping $T: V_2 \to V_2$ where $T(x,y) = (x + 1,y)$ is not linear. This may be written $(x,y) \xrightarrow{T} (x + 1,y)$.

(*c*) Let V be the vector space of all continuous real functions. Define $T: V \to V$ so that $(Tf)x = \int_0^x f(x)\,dx$. T is linear since

$$\int_0^x [r_1f_1(x) + r_2f_2(x)]\,dx = r_1\int_0^x f_1(x)\,dx + r_2\int_0^x f_2(x)\,dx$$

294. POLYNOMIALS An expression of the form

$$p(x) = a_0 + a_1x + \cdots + a_nx^n = \sum_{i=0}^{n} a_ix^i \qquad a_n \neq 0$$

is a *polynomial* in the *indeterminate* x of degree n. It is assumed that the *coefficients* a_i belong to a field F. (See Sec. 226.)

If

$$f = \sum_{i=0}^{n} a_ix^i \qquad g = \sum_{i=0}^{k} b_ix^i \qquad k \leq n$$

then

$$f + g = \sum_{i=0}^{k} (a_i + b_i)x^i + \sum_{i=k+1}^{n} a_i x^i \qquad \text{(the } sum\text{)}$$

$$fg = \sum_{i=0}^{n} \sum_{j=0}^{k} a_i b_j x^{i+j} \qquad \text{(the } product\text{)}$$

that is,

$$fg = a_0b_0 + (a_0b_1 + a_1b_0)x + (a_0b_2 + a_1b_1 + a_2b_0)x^2 + \cdots + a_nb_kx^{k+n}$$

If $fg = fh$ and $f \neq 0$, then $g = h$.

Polynomials obey the commutative, associative, and distributive laws, that is, $fg = gf$, $f + g = g + f$; $f(gh) = (fg)h$, $f + (g + h) = (f + g) + h$; and $f(g + h) = fg + fh$.

If $g \neq 0$, there exist *unique* polynomials q and r such that $f = qg + r$, where either $r = 0$ or the degree of r is less than the degree of g. The *quotient* is q and the *remainder* r.

295. GREATEST COMMON DIVISOR Any two polynomials p and q in x have a *greatest common divisor* d, written g.c.d., such that

(1) $d|p$ and $d|q$

(2) $c|p$ and $c|q$ implies $c|d$

(3) $d = sp + tq$ where s and t are polynomials in x

Here $d|p$ means that "d divides p," etc.

Polynomials p and q are *relatively prime* if their greatest common divisors are unity and its associates. A polynomial with no proper (exact) divisors over a field is called *irreducible* (*prime*) over the field.

296. If two polynomials A and B, where

$$A = \sum_{i=0}^{n} a_i x^i \qquad \text{and} \qquad B = \sum_{i=0}^{m} b_i x^i$$

each of degree $\leqq n$, coincide for more than n distinct values of x, then $A \equiv B$, that is, $a_i \equiv b_i$ for $i = 0, 1, \ldots, n$.

297. POLYNOMIAL EQUATIONS Consider the polynomial $p(x) = \sum_{i=0}^{n} a_i x^i$ of degree n with real coefficients a_i and $a_n \neq 0$. A value of x, call it x_0, such that $p(x_0) = 0$, is a *zero* of $p(x)$, and x_0 is a *root* of $p(x) = 0$.

Theorems. **(*a*)** If $n \geqq 1$, then $p(x)$ has exactly n zeros, provided a root of multiplicity k is counted k times.

(*b*) If $x_1, x_2, \ldots, x_n$ are the zeros of $p(x)$, then

$$p(x) = a_n(x - x_1)(x - x_2) \cdots (x - x_n)$$

(*c*) If x_1 is a zero of $p(x)$ of multiplicity $k > 1$, then x_1 is a zero of $p'(x)$ of multiplicity $(k - 1)$. Here $p'(x)$ is the derivative of $p(x)$ with respect to x.

(*d*) If $z = \alpha + i\beta$ is a complex zero of $p(x)$, $i^2 = -1$, then $\bar{z} = \alpha - i\beta$ is a zero of $p(x)$; and

$$p(x) = (x^2 - 2\alpha x + \alpha^2 + \beta^2)q(x)$$

where $q(x)$ is a polynomial of degree $(n - 2)$.

(*e*) If $p(x)$ of degree n is divided by $(x - z)$, then

$$p(x) = (x - z)q + b_0$$

where $q \equiv q(x)$ is the *quotient polynomial* of degree $(n - 1)$ and b_0 is the *remainder*.

(*f*) z is a real root of $p(x)$ if and only if $p(x) = (x - z)q$, that is, if and only if $(x - z)$ *divides* $p(x)$.

15. NUMERICAL ANALYSIS

Nonlinear Equations

298. **SOLUTION OF NONLINEAR EQUATIONS** Various methods for approximating the roots of equations of the form

(1) $$f(x) = 0$$

where $f(x)$ is a function of x are given below.

299. **LINEAR ITERATION** Express (1) in the form

(2) $$x = g(x)$$

Let $x = x_0$ be an approximation of a root $x = \xi$ of (2). A second approximation x_1 of ξ is $x_1 = g(x_0)$. Successive approximations of ξ are generated by

(3) $$x_{i+1} = g(x_i) \qquad i = 0, 1, 2, \ldots$$

Many ways of expressing (1) in form (2) exist. For this process to be useful, the successive approximations x_i must converge to the solution $x = \xi$ of (2) and of (1).

Theorem 1. If I is an interval containing $x = \xi$, where ξ is a root of (1), if $g(x)$ and $g'(x)$ are continuous in I, and if $|g'(x)| \leqq K < 1$ for all x in I, then if x_0 is in I, the iteration (3) converges to ξ.

When the iteration converges, for i large enough the *error* $e_i = x_i - \xi$ is proportional to the error of the previous iteration, and $\lim_{i\to\infty} e_{i+1}/e_i = g'(\xi)$.

Stopping criteria. If accuracy ϵ is desired, the iteration might be stopped when x_i is such that

(4) $$|f(x_i)| < \epsilon \qquad \text{or} \qquad |x_{i+1} - x_i| < \epsilon \qquad \text{or} \qquad \frac{|x_{i+1} - x_i|}{|x_{i+1}|} < \epsilon$$

Example. Find positive root ξ of $f(x) \equiv x^2 - 2x - 8 = 0$. Write $x = \sqrt{2x + 8} \equiv g(x)$. Select $x_0 = 0$ as approximation of ξ. Repeated iterations $x_{i+1} = g(x_i) \equiv \sqrt{2x_i + 8}$ give $x_1 = 2.828$, $x_2 = 3.694$, $x_3 = 3.923$, $x_4 = 3.980, \ldots$. This converges to $x = 4$, a root of $f(x) = 0$.

If relation $x = (x^2 - 8)/2 \equiv g(x)$ is used and $x_0 = 6$, there results $x_0 = 6$, $x_1 = 14$, $x_2 = 94, \ldots$, which diverges. If $x_0 = 2$, $x_1 = -2$, $x_2 = -2, \ldots$, a root of $f(x) = 0$ is $x = -2$.

300. NEWTON-RAPHSON METHOD Newton's method for finding a root $x = \xi$ of $f(x)$ is based on

$$x_{i+1} = x_i - \frac{f(x_i)}{f'(x_i)} \equiv g(x_i) \tag{5}$$

to generate successive approximations of ξ.

Theorem 2. If $f(x)$, $f'(x)$, and $f''(x)$ are continuous and bounded on an interval I containing ξ, and if $f'(\xi) \neq 0$, then there exists an interval J, including $x = \xi$, such that if x_0 is in J, iteration (5) converges to ξ and $f(\xi) = 0$.

Any choice of x_0 such that

$$|g'(x_0)| = \frac{|f(x_0)f''(x_0)|}{[f'(x_0)]^2} < 1 \tag{6}$$

will lead to a convergent iteration.

Any convergent iteration such that $e_{i+1} = ke_i^2$, where k is a constant, $e_i = x_i - \xi$, is a *second-order iteration,* and the iteration *converges quadratically.*

Theorem 3. Under the assumptions of Theorem 2, if $f'''(x)$ is bounded and continuous [and hence so is $g''(x)$ on a suitably selected interval],

$$\lim_{i\to\infty} \frac{e_{i+1}}{e_i^2} = \frac{1}{2} g''(\xi) \tag{7}$$

Newton's method converges quadratically.

Example. Find $\sqrt{A}$ when $A > 0$.

Let $f(x) \equiv x^2 - A = 0$. $f'(x) = 2x$. From Eq. (5) $x_{i+1} = \frac{1}{2}[x_i + (A/x_i)]$. For example, if $A = 2$, select $x_0 = 1.4$. Then $x_1 = 1.41428$, $x_2 = 1.41421$, $x_3 = 1.41421, \ldots$, which converges rapidly.

301. GEOMETRIC INTERPRETATION Let x_i approximate a root $x = \xi$ of $f(x) = 0$. The slope of the tangent to $y = f(x)$ at $x = x_i$ is $f'(x_i)$, and x_{i+1} is the intersection of this tangent with the x axis.

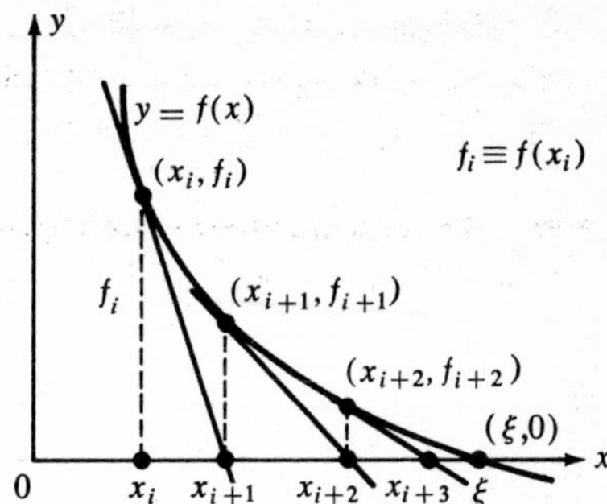

Fig. 65.
Newton-Raphson method.

Newton's method requires a knowledge of $f'(x)$ which sometimes may be difficult. If x_0 is too far away from ξ, the process may converge to another root or diverge.

302. THE SECANT METHOD Find two approximations x_0 and x_1 to root $x = \xi$ of $f(x) = 0$. From

$$x_{i+1} = \frac{x_{i-1}f(x_i) - x_i f(x_{i-1})}{f(x_i) - f(x_{i-1})} \tag{8}$$

successive approximations are generated. When the process converges, the rate of convergence is somewhat better than linear.

303. GEOMETRIC INTERPOLATION The chord passing through $(x_i, f(x_i))$ and $(x_{i-1}, f(x_{i-1}))$ intersects the x axis at $(x_{i+1}, 0)$; x_{i+1} is the next approximation of the root $x = \xi$.

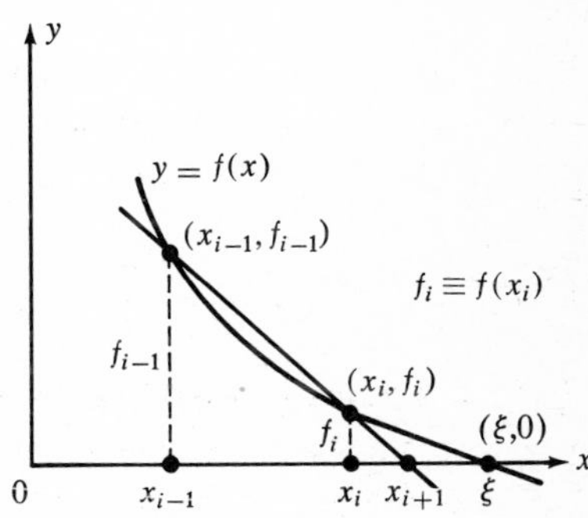

Fig. 66.
Secant method.

304. METHOD OF FALSE POSITION (REGULA FALSI) Select two approximations x_0 and x_1 of the root $x = \xi$ of $f(x) = 0$ such that $f(x_0) \cdot f(x_1) < 0$. From

$$x_2 = \frac{x_0 f(x_1) - x_1 f(x_0)}{f(x_1) - f(x_0)} \tag{9}$$

find approximation x_2. Let ϵ be a prescribed error. If $|x_2 - x_j| < \epsilon$ for $j = 0$ or $j = 1$, then accept x_2 as the answer; if not, proceed as follows. If $f(x_0) \cdot f(x_2) < 0$, then replace x_1 by x_2 in (9), leave x_0 unchanged, and compute the next approximation x_3 from (9); otherwise, replace x_0 by x_2, leave x_1 unchanged, and compute the next approximation from Eq. (9). Continue until the desired accuracy is realized.

Roots of Polynomial Equations

305. POLYNOMIAL EQUATIONS. REAL ROOTS Let

$$p(x) = a_n x^n + a_{n-1} x^{n-1} + \cdots + a_0 \qquad a_n \neq 0 \tag{1}$$

be a polynomial of degree n with real coefficients a_k. Let z be a real number. An efficient way for finding value of $p(z)$ is to generate the sequence $\{b_k\}$, $(k = n, n-1, \ldots, 1, 0)$ from

$$b_n = a_n \qquad b_{n-k} = a_{n-k} + z b_{n-k+1} \tag{2}$$

Then $b_0 = p(z)$.

If $p(x)$ is divided by $(x - z)$, then

$$p(x) = (x - z)q(x) + b_0 \tag{3}$$

where the *quotient polynomial* $q(x)$ is

$$q(x) = b_n x^{n-1} + b_{n-1} x^{n-2} + \cdots + b_1 \tag{4}$$

and b_0 is the *remainder.*

If $p(z) = 0$, then $b_0 = 0$ and the b_k, $k = 1, \ldots, n$, found from (2) are the coefficients of $q(x)$. To find the succeeding roots of (1), seek the zeros of $q(x)$.

If $p(z) = 0$, $p'(z) = q(z)$ where $p'(x)$ is the derivative of $p(x)$ with respect to x.

To find $q(z)$ generate $\{c_k\}$, $k = n, n-1, \ldots, 1$, from

$$c_n = b_n \qquad c_{n-k} = b_{n-k} + z c_{n-k+1} \tag{5}$$

Then $c_1 = q(z) = p'(z)$. [The process used in (2) and (5) is called *nested multiplication.*]

306. NEWTON'S METHOD To find real roots of $p(x) = 0$ proceed as follows. Let x_0 be an approximation to a real zero ξ of $p(x)$. Use the iteration

$$x_{i+1} = x_i - \frac{p(x_i)}{p'(x_i)} = x_i - \frac{b_0(x_i)}{c_1(x_i)} \tag{6}$$

to generate successive approximations of ξ. Here, for each approximation x_i, the corresponding values of b_0 and c_1 are found from (2) and (5), respectively, with z replaced by x_i.

If ξ is a simple zero of $p(x)$ and x_0 is sufficiently close to ξ, then (6) yields a sequence $\{x_k\}$ which converges quadratically to ξ. The reduced polynomial (4) can be used to find the remaining zeros of $p(x)$.

Method (6) may present difficulties when $p'(x)$ is small near a root and when multiple roots occur. To minimize loss of accuracy, the roots should be sought for in increasing order of magnitude.

307. POLYNOMIAL EQUATIONS. COMPLEX ROOTS If $n > 2$, Eq. (1) may be written

$$p(x) = (x^2 - \alpha x - \beta)q(x) + b_1(x - \alpha) + b_0 \tag{7}$$

where α and β are real, $q(x)$ is the *quotient*

$$q(x) = b_n x^{n-2} + b_{n-1}x^{n-3} + \cdots + b_2 \tag{8}$$

and $b_1(x - \alpha) + b_0$ is the *remainder.* The coefficients b_k, $k = 0, 1, \ldots, n$, are given by

$$\begin{cases} b_n = a_n \\ b_{n-1} = a_{n-1} + \alpha b_n \\ b_{n-2} = a_{n-2} + \alpha b_{n-1} + \beta b_n \\ \cdots\cdots\cdots\cdots\cdots\cdots \\ b_1 = a_1 + \alpha b_2 + \beta b_3 \\ b_0 = a_0 + \alpha b_1 + \beta b_2 \end{cases} \tag{9}$$

In (9) $b_1 \equiv b_1(\alpha,\beta)$ and $b_0 \equiv b_0(\alpha,\beta)$ are functions of α and β.

If $(x^2 - \alpha x - \beta)$ is a factor of $p(x)$, α and β may be found by solving

$$(10) \qquad b_1(\alpha,\beta) = 0 \qquad b_0(\alpha,\beta) = 0$$

Finding $(x^2 - \alpha x - \beta)$. To find factors of $x^2 - \alpha x - \beta$ and the corresponding roots:

(*a*) Let (α_0,β_0) be approximations of (α,β).
(*b*) Use (9) to find b_k, $k = n, n-1, \ldots, 1, 0$.
(*c*) Find c_k, $k = n, n-1, \ldots, 2, 1$, from

$$(11) \qquad \begin{cases} c_n = b_n \\ c_{n-1} = b_{n-1} + \alpha c_n \\ c_{n-2} = b_{n-2} + \alpha c_{n-1} + \beta c_n \\ \cdots\cdots\cdots\cdots\cdots\cdots \\ c_2 = b_2 + \alpha c_3 + \beta c_4 \\ c_1 = b_1 + \alpha c_2 + \beta c_3 \end{cases}$$

Note: $c_{k+1} = \partial b_k/\partial\alpha$, $k = 0, 1, \ldots, n-1$.

(*d*) Compute the corrections $\delta\alpha_0$ and $\delta\beta_0$ to α_0 and β_0, respectively, from

$$(12) \qquad \begin{cases} c_2\,\delta\alpha_0 + c_3\,\delta\beta_0 = -b_1(\alpha_0\beta_0) \\ c_1\,\delta\alpha_0 + c_2\,\delta\beta_0 = -b_0(\alpha_0,\beta_0) \end{cases}$$

New approximations (α_1,β_1) to (α,β) are

$$\alpha_1 = \alpha_0 + \delta\alpha_0 \qquad \beta_1 = \beta_0 + \delta\beta_0$$

(*e*) Use α_1 as α_0 and β_1 as β_0, and repeat steps (*a*) to (*d*) until convergence occurs, yielding α and β.

(*f*) The roots of $p(x) = 0$ corresponding to $(x^2 - \alpha x - \beta)$ are

$$(13) \; x_1 = \tfrac{1}{2}(\alpha + \sqrt{\alpha^2 + 4\beta}) \qquad x_2 = \tfrac{1}{2}(\alpha - \sqrt{\alpha^2 + 4\beta})$$

308. The above process (the *Newton-Bairstow method*) is useful for finding the complex zeros of polynomials. Its weakness lies in the difficulty sometimes experienced in finding initial approximations (α_0,β_0) adequate to realize convergence.

Interpolation and Approximation

309. Important uses for approximating functions are: (*a*) to replace complicated functions by simpler functions and (*b*) for interpolating in tables. Approximating functions in common use are polynomials, trigonometric functions, exponentials, and rational functions. Interpolating polynomials are widely used on computers.

310. **LAGRANGIAN FORM: INTERPOLATING POLYNOMIAL** Let $f(x)$ be a real function with $f(x_i) = f_i$ for $i = 0, 1, \ldots, n$. If the points (x_i, f_i) are distinct,

(1) $$p(x) = a_n x^n + a_{n-1} x^{n-1} + \cdots + a_0$$

may be made to pass through the points (x_i, f_i). The values of a_i may be found from the linear system

(2) $$\begin{cases} a_n x_0{}^n + a_{n-1} x_0{}^{n-1} + \cdots + a_0 = f_0 \\ \cdots\cdots\cdots\cdots\cdots\cdots\cdots\cdots \\ a_n x_n{}^n + a_{n-1} x_n{}^{n-1} + \cdots + a_0 = f_n \end{cases}$$

The interpolating polynomial may be written

(3) $$p(x) = \sum_{k=0}^{n} L_k(x) f_k \qquad \textit{(Lagrangian form)}$$

where

(4) $$L_k(x) = \frac{\prod\limits_{\substack{j=0 \\ j \neq k}}^{n} (x - x_j)}{\prod\limits_{\substack{j=0 \\ j \neq k}}^{n} (x_k - x_j)}$$

Form (3) is identical to (1).

311. **ERROR OF INTERPOLATING POLYNOMIAL** Since many functions exist which pass through the n points (x_i, f_i), where $f_i = f(x_i)$, little can be said about the error of approximation at a nontabular point unless more is known about $f(x)$.

Theorem. Let $f(x)$ and its derivatives of order $(n + 1)$ be continuous for all values of x in an interval I containing the $(n + 1)$ interpolating points. Construct a polynomial $p(x)$ which passes through $(x_i, f(x_i))$, $i = 0, 1, \ldots, n$. If x is in I, the error E is

$$E \equiv f(x) - p(x) = \frac{\psi(x) f^{(n+1)}(\xi)}{(n + 1)!} \tag{5}$$

where $\psi(x) = (x - x_0)(x - x_1) \cdots (x - x_n)$ and ξ is in I.

312. **LINEAR INTERPOLATION** In the case of $n = 1$, given two points (x_0, f_0) and (x_1, f_1), Eq. (3) is

$$p(x) = L_0(x) f_0 + L_1(x) f_1 \tag{6}$$

where

$$L_0(x) = \frac{x - x_1}{x_0 - x_1} \qquad L_1(x) = \frac{x - x_0}{x_1 - x_0} \tag{7}$$

Equation (6) is the usual form for *linear interpolation*, and

$$p(x) = \frac{1}{x_1 - x_0} \begin{vmatrix} f_0 & x_0 - x \\ f_1 & x_1 - x \end{vmatrix} = f_0 + \frac{f_1 - f_0}{x_1 - x_0}(x - x_0) \tag{8}$$

The error E is

$$E \equiv f(x) - p(x) = \frac{(x - x_0)(x - x_1) f''(\xi)}{2!} \tag{9}$$

where $x_0 \leqq \xi \leqq x_1$. If $|f''(x)| < M$, where M is a constant, on $x_0 < x < x_1$, then the maximum error at a point x between x_0 and x_1 is $(x_1 - x_0)^2 M/8$.

313. **QUADRATIC INTERPOLATION** In the case of $n = 2$,

$$p(x) = L_0(x) f_0 + L_1(x) f_1 + L_2(x) f_2 \tag{10}$$

where

$$L_0(x) = \frac{(x - x_1)(x - x_2)}{(x_0 - x_1)(x_0 - x_2)}$$

$$(11) \qquad L_1(x) = \frac{(x - x_0)(x - x_2)}{(x_1 - x_0)(x_1 - x_2)}$$

$$L_2(x) = \frac{(x - x_0)(x - x_1)}{(x_2 - x_0)(x_2 - x_1)}$$

The error E is

$$(12) \qquad E \equiv f(x) - p(x) = \frac{(x - x_0)(x - x_1)(x - x_2)f^{(3)}(\xi)}{3!}$$

where $x_0 \leqq \xi \leqq x_2$.

Numerical Differentiation and Integration

314. Interpolating polynomials can be used to evaluate a function at a nontabular point. Such polynomials can be used as approximations to functions with the aim of simplifying the operations of calculus.

315. NUMERICAL DIFFERENTIATION Given a set of values (x_i,f_i), $i = 0, 1, \ldots, n$, for a function $f(x)$, find the interpolating polynomial $p(x)$ passing through these points. The derivative $p'(x)$ is an approximation to $f'(x)$. Since little is usually known about the accuracy of $p(x)$ at a nontabular point, and the slopes of $p(x)$ and $f(x)$ may differ appreciably, great accuracy cannot be expected in numerical differentiation. Let I be an interval containing $x_0, \ldots, x_n$.

If $f(x)$ is continuously differentiable $(n + 2)$ times on interval I, then the error in $p'(x)$ at x_i is

$$(1) \qquad f'(x_i) - p'(x_i) = \frac{\psi'(x_i)f^{(n+1)}(\xi_i)}{(n + 1)!}$$

where ξ_i is in I and

$$(2) \qquad \psi'(x_i) = \prod_{\substack{j=0 \\ j\neq i}}^{n} (x_i - x_j)$$

Let x_i be equally spaced with spacing h. Let $s = (x - x_0)/h$. From Newton's forward-difference formula (see Sec. 330)

(3) $hf'(x) \approx \dfrac{dp}{ds} = \Delta f_0 + \dfrac{(2s-1)}{2}\Delta^2 f_0 + \cdots + \dfrac{d}{ds}\dbinom{s}{n}\Delta^n f_0$

Case n = 1. If $s = 0$, $x = x_0$, and $n = 1$, then

(4) $$f'(x_0) \approx \frac{\Delta f_0}{h} = \frac{f_1 - f_0}{h}$$

The right-hand side of (4) is the slope of chord joining (x_0, f_0) and (x_1, f_1) and is an approximation to $f'(x_0)$. The error E in this approximation of $f'(x_0)$ is $E = -(h/2)f''(\xi)$, where $x_0 \leqq \xi \leqq x_1$.

Case n = 2. If $s = 0$, $x = x_0$, and $n = 2$, then

(5) $$f'(x_0) \approx \frac{1}{2h}(-3f_0 + 4f_1 - f_2)$$

The right-hand side of (5) is the slope of a parabola passing through (x_0, f_0), (x_1, f_1), and (x_2, f_2) and is an approximation of $f'(x_0)$, with error E. $E = (h^2/3)f'''(\xi)$, $x_0 \leqq \xi \leqq x_2$.

316. Similarly, higher derivatives of $f(x)$ may be found. For example,

(6) $$f''(x_0) \approx \frac{\Delta^2 f_0}{h^2} = \frac{f_2 - 2f_1 + f_0}{h^2}$$

with error $E = -hf'''(\xi)$.

317. Differentiation formulas more accurate than those based on forward differences can be found using central differences, provided that values of $f(x)$ can be found on both sides of x_0. Examples are:

(7) $$f'(x_0) \approx \frac{f_1 - f_{-1}}{2h} \qquad E = \frac{-h^2}{6}f'''(\xi)$$

(8) $$f''(x_0) \approx \frac{f_1 - 2f_0 + f_{-1}}{h^2} \qquad E = \frac{-h^2}{12}f^{iv}(\xi)$$

318. DERIVATIVES AT TABULAR POINTS The following formulas for the derivatives f_i' may be found by differentiating three- and five-point Lagrangian interpolation formulas. Here $f_{-2} = f(x_0 - 2h)$,

$f_{-1} = f(x_0 - h)$, $f_0 = f(x_0)$, $f_1 = f(x_0 + h)$, $f_2 = f(x_0 + 2h)$. Each ξ lies between the extreme values of the abscissas involved.

Three-point:

$$f'_{-1} = \frac{1}{2h}(-3f_{-1} + 4f_0 - f_1) + \frac{h^2}{3}f'''(\xi)$$

(9) $$f'_0 = \frac{1}{2h}(-f_{-1} + f_1) - \frac{h^2}{6}f'''(\xi)$$

$$f'_1 = \frac{1}{2h}(f_{-1} - 4f_0 + 3f_1) + \frac{h^2}{3}f'''(\xi)$$

Five-point:

$$f'_{-2} = \frac{1}{12h}(-25f_{-2} + 48f_{-1} - 36f_0 + 16f_1 - 3f_2) + \frac{h^4}{5}f^{v}(\xi)$$

$$f'_{-1} = \frac{1}{12h}(-3f_{-2} - 10f_{-1} + 18f_0 - 6f_1 + f_2) - \frac{h^4}{20}f^{v}(\xi)$$

(10) $$f'_0 = \frac{1}{12h}(f_{-2} - 8f_{-1} + 8f_1 - f_2) + \frac{h^4}{30}f^{v}(\xi)$$

$$f'_1 = \frac{1}{12h}(-f_{-2} + 6f_{-1} - 18f_0 + 10f_1 + 3f_2) - \frac{h^4}{20}f^{v}(\xi)$$

$$f'_2 = \frac{1}{12h}(3f_{-2} - 16f_{-1} + 36f_0 - 48f_1 + 25f_2) + \frac{h^4}{5}f^{v}(\xi)$$

319. NUMERICAL INTEGRATION One method for calculating

(1) $$\mathrm{I} = \int_a^b f(x)\,dx$$

is to approximate $f(x)$ by a polynomial $p(x)$ and integrate $p(x)$ to obtain an approximate value of I.

Let $f(x_i) \equiv f_i$, $(i = 0, 1, \ldots, n)$, x_i being on interval $[a,b]$, $x_0 = a$, $x_n = b$. Let $p(x)$ be the polynomial of degree n which interpolates at points (x_i, f_i), $(i = 0, 1, \ldots, n)$.

If $\psi(x) = (x - x_0)(x - x_1) \cdots (x - x_n)$ does not change sign on $[a,b]$, the error E due to integration of $p(x)$ rather than $f(x)$ is

(2) $$E = \int_a^b [f(x) - p(x)]\, dx = \frac{f^{(n+1)}(\eta)}{(n+1)!} \int_a^b \psi(x)\, dx$$

where $a \leqq \eta \leqq b$.

Let $x_0, x_1, \ldots, x_n$ be equally spaced and $s = (x - x_0)/h$. Suppose $P(x)$ is the Newton forward-difference interpolating polynomial given in Sec. 330. Then

(3) $$\mathrm{I}_1 \equiv \int_{x_0}^{x_1} f(x)\, dx \approx h \int_0^1 P(s)\, ds$$

Each selection of n in $P(x)$ gives an approximation of the desired integral.

Case $n = 0$. If $n = 0$, $\mathrm{I}_1 \approx hf_0 \equiv R_1$, which is the *rectangular rule* approximation. The error in approximation R_1 is $h^2 f'(\eta_0)/2$, where $x_0 < \eta_0 < x_1$.

If $x_0 = a$, $x_1 = a + h, \ldots, x_N = a + Nh = b$, and the rectangular rule is applied over each subinterval (x_i, x_{i+1}),

(4) $$\int_a^b f(x)\, dx \approx h(f_0 + f_1 + \cdots + f_{N-1}) \equiv R_N$$

The error in R_N is $(b - a)(h/2)f'(\eta)$, $a < \eta < b$.

Case $n = 1$. If $n = 1$, $\mathrm{I}_1 \approx (h/2)(f_0 + f_1) \equiv T_1$. T_1 is the area of a trapezoid inscribed under $f(x)$ from x_0 to x_1. The error in T_1 is $(-h^3/12)f''(\eta_0)$, $x_0 < \eta_0 < x_1$. If approximation T_1 is applied over each (x_i, x_{i+1}),

(5) $$\int_a^b f(x)\, dx \approx h\left[\frac{1}{2} f_0 + f_1 + f_2 + \cdots + f_{N-1} + \frac{1}{2} f_N\right] \equiv T_N.$$

This approximation is the *trapezoidal rule.* The error in T_N is $[-(b - a)/12]h^2 f''(\eta)$, $a < \eta < b$.

Let T_N be the approximation (5) of I. A better approximation T_{2N} of I can be computed from recursion

(6) $$T_{2N} = \frac{1}{2} T_N + \frac{h}{2}\left[f\left(a + \frac{h}{2}\right) + f\left(a + \frac{3h}{2}\right) + \cdots + f\left(a + \frac{2N-1}{2} h\right)\right]$$

320. SIMPSON'S RULE From Newton's forward interpolating polynomial $P(x)$ given in Sec. 330

$$\text{(7)} \qquad \mathrm{I} = \int_a^b f(x)\,dx = S_{2N} - \frac{N}{90} h^5 f^{iv}(\eta) \qquad a < \eta < b$$

where

$$S_{2N} = \frac{h}{3}(f_0 + 4f_1 + 2f_2 + 4f_3 + \cdots + 4f_{2N-1} + f_{2N})$$

The quantity S_{2N}, called *Simpson's rule*, approximates I by the use of an *even* number $2N$ of subintervals of equal length h over $[a,b]$, $N = (b - a)/2h$. The error in S_{2N} is $[-(b - a)/180]h^4 f^{iv}(\eta)$, $a < \eta < b$. In case $N = 1$, $S_2 = (h/3)(f_0 + 4f_1 + f_2)$. The error in using S_2 is smaller than might be expected for the number of ordinates used, thus the wide use of Simpson's rule.

321. GAUSSIAN QUADRATURE The integration formulas treated above are based on the use of equally spaced points. For a fixed number of points, integration formulas of greater accuracy can be found having the form

$$\text{(8)} \qquad \int_a^b f(x)\,dx \approx \sum_{i=0}^{n} a_i f(x_i) \equiv G$$

where the arguments x_i and weights a_i are selected in such a way as to make the approximation G *exact* for all polynomials of degree less than or equal to $(2n + 1)$.

The transformation $x = [a + b + (b - a)t]/2$ transforms the interval $[a,b]$ into $[-1,1]$, and $f(x)$ into $F(t)$. Equation (8) then becomes

$$\text{(9)} \qquad \int_a^b f(x)\,dx = \frac{b-a}{2}\int_{-1}^{1} F(t)\,dt$$

where

$$\text{(10)} \qquad \int_{-1}^{1} F(t)\,dt = \sum_{i=0}^{n} A_i F(t_i)$$

The t_i are the zeros of a certain set of Legendre polynomials $P_m(t)$. All the zeros are real and distinct, and they lie on $[-1,1]$.

Tables of values for A_i and t_i have been calculated for various values of n. Examples are given below.

m	$P_m(t)$	*Zeros of* $P_m(t)$
1	$P_1(t) = t$	$t_0 = 0$
2	$P_2(t) = (3t^2 - 1)/2$	$t_1 = 1/\sqrt{3} = -t_0$
3	$P_3(t) = (5t^3 - 3t)/2$	$t_2 = \sqrt{3/5} = -t_0,\ t_1 = 0$
4	$P_4(t) = (35t^4 - 30t^2 + 3)/8$	$t_3 = -t_0,\ t_2 = -t_1$
		$t_0 = -0.86113631$
		$t_1 = -0.33998104$

n	t_i	A_i
0	$t_0 = 0.00000000$	$A_0 = 2.00000000$
1	$t_1 = -t_0 = 0.57735027$	$A_1 = A_0 = 1.00000000$
2	$t_1 = 0.00000000$	$A_1 = 0.88888889$
	$t_2 = -t_0 = 0.77459667$	$A_2 = A_0 = 0.55555556$
3	$t_2 = -t_1 = 0.33998104$	$A_2 = A_1 = 0.65214515$
	$t_3 = -t_0 = 0.86113631$	$A_3 = A_0 = 0.34785485$
4	$t_4 = -t_0 = 0.90617985$	$A_1 = A_0 = 0.23692689$
	$t_3 = -t_1 = 0.53846931$	$A_3 = A_1 = 0.47862867$
	$t_2 = 0.00000000$	$A_2 = 0.56888889$

Example. In case $n = 1$

$$\int_{-1}^{1} F(t)\,dt = A_0F(t_0) + A_1F(t_1)$$

where $A_0 = A_1 = 1$, $t_1 = -t_0$, and $t_1 = 1/\sqrt{3} = 0.57735027$. This formula is exact for all polynomials $F(t)$ whose degree is $\leqq 3$.

322. In general, Gaussian integration yields accuracy comparable to Simpson's rule for about one-half the computational work. Formulas of type (8) may present difficulties when $F(t)$ is given empirically and in some cases may present certain disadvantages in hand calculations.

Finite Difference Calculus

323. NOTATION The interpolating points x_i are equally spaced and so ordered that $x_i - x_{i-1} = h > 0$ for all i, $i = -k, -k+1, \ldots, -1, 0, 1, 2, \ldots, k$. $f(x)$ is a single-valued real function of x. $f(x_i) \equiv f_i$. $f(x_i + h/2) \equiv f_{i+1/2}$, $f(x_n + sh) \equiv f_{n+s}$.

The symbols Δ, ∇, δ, μ, and E may be considered as *operators* which transform $f(x)$ into related functions according to the definitions:

$$\Delta f(x) = f(x+h) - f(x) \qquad \textit{(forward-difference operator } \Delta)$$

$$\nabla f(x) = f(x) - f(x-h) \qquad \textit{(backward-difference operator } \nabla)$$

$$\delta f(x) = f\left(x + \frac{h}{2}\right) - f\left(x - \frac{h}{2}\right) \qquad \textit{(central-difference operator } \delta)$$

$$\mu f(x) = \frac{1}{2}\left[f\left(x + \frac{h}{2}\right) + f\left(x - \frac{h}{2}\right)\right] \qquad \textit{(averaging operator } \mu)$$

$$Ef(x) = f(x+h) \qquad \textit{(shifting operator } E)$$

324. BINOMIAL COEFFICIENTS The binomial function $\binom{x}{r}$ is defined for any real number x to be

$$\binom{x}{r} = \frac{x(x-1)\cdots(x-r+1)}{r!} \qquad r \text{ a positive integer}$$

$$\binom{x}{r} = 0 \quad \text{when } r \text{ is a negative integer} \qquad \binom{x}{0} = 1$$

$$\binom{x}{r} = 0 \quad \text{for } x = 0 \text{ and } r \neq 0 \qquad \binom{0}{0} = 1$$

If $h = 1$,

$$\Delta\binom{x}{r} = \binom{x}{r-1} \quad \text{and} \quad \Delta^k\binom{x}{r} = \binom{x}{r-k} \qquad k = 0, 1, \ldots$$

325. *Higher powers* of Δ, ∇, δ, and E are defined recursively.

$$E^k f(x) = f(x+kh) \qquad \text{for } any\ k$$

$$\Delta^k f(x) = \Delta[\Delta^{k-1} f(x)] \quad \text{and} \quad \nabla^k f(x) = \nabla[\nabla^{k-1} f(x)], \quad k = 1, 2, \ldots.$$

Examples. $\Delta^0 f(x) = f(x)$. $\Delta f(x) = (E-1)f(x)$

$$\Delta^2 f(x) = \Delta f(x+h) - \Delta f(x) = f(x+2h) - 2f(x+h) + f(x)$$

$$E^k f(x) = (1+\Delta)^k f(x) \qquad k \text{ any integer}$$

$$\nabla^k f(x) = f(x) - \binom{k}{1} f(x-h) + \binom{k}{2} f(x-2h) - \cdots + (-1)^k \binom{k}{k} f(x-kh)$$

$$\nabla^0 f(x) = f(x) \qquad \nabla f(x) = (1 - E^{-1}) f(x)$$

$$\nabla^2 f(x) = \nabla f(x) - \nabla f(x-h) = f(x) - 2f(x-h) + f(x-2h)$$

326. FORWARD DIFFERENCES Forward-difference formulas start with a set of numbers $f_0, f_1, \ldots, f_i, \ldots$. The differences $\Delta f_0 = f_1 - f_0$, $\Delta f_1 = f_2 - f_1, \ldots$; $\Delta^2 f_0 = \Delta f_1 - \Delta f_0$, $\Delta^2 f_1 = \Delta f_2 - \Delta f_1, \ldots; \ldots$ may be represented as in Table 15-1.

TABLE 15-1. FORWARD DIFFERENCES

x_{-4}	f_{-4}				
		Δf_{-4}			
x_{-3}	f_{-3}		$\Delta^2 f_{-4}$		
		Δf_{-3}		$\Delta^3 f_{-4}$	
x_{-2}	f_{-2}		$\Delta^2 f_{-3}$		$\Delta^4 f_{-4}$
		Δf_{-2}		$\Delta^3 f_{-3}$	
x_{-1}	f_{-1}		$\Delta^2 f_{-2}$		$\Delta^4 f_{-3}$
		Δf_{-1}		$\Delta^3 f_{-2}$	
x_0	f_0		$\Delta^2 f_{-1}$		$\Delta^4 f_{-2}$
		Δf_0		$\Delta^3 f_{-1}$	
x_1	f_1		$\Delta^2 f_0$		$\Delta^4 f_{-1}$
		Δf_1		$\Delta^3 f_0$	
x_2	f_2		$\Delta^2 f_1$		$\Delta^4 f_0$
		Δf_2		$\Delta^3 f_1$	
x_3	f_3		$\Delta^2 f_2$		
		Δf_3			
x_4	f_4				

327. CENTRAL-DIFFERENCE NOTATION The central-difference formulas start with a set $\ldots, f_{-i}, f_{-i+1}, \ldots, f_{-1}, f_0, f_1, \ldots, f_i, \ldots$. Central differences are represented by expressions $\delta^m f_j$. The subscript j, which is not necessarily an integer, takes the value midway between the subscripts on the two quantities whose difference is being represented. By definition

$$\delta f_j = f_{j+1/2} - f_{j-1/2}, \qquad \delta^n f_j = \delta^{n-1} f_{j+1/2} - \delta^{n-1} f_{j-1/2}, \qquad n > 1$$

$$\mu\delta^n f_j = \tfrac{1}{2}(\delta^n f_{j+1/2} + \delta^n f_{j-1/2})$$

The relation between forward- and central-difference notations is

$$\delta^n f_j = \Delta^n f_{j-n/2};\ \mu\delta^n f_j = \tfrac{1}{2}[\Delta^n f_{j-(n+1)/2} + \Delta^n f_{j-(n-1)/2}]$$

The differences may be arranged as in Table 15-2.

TABLE 15-2

x_{-4}	f_{-4}				
		$\delta f_{-7/2}$			
x_{-3}	f_{-3}		$\delta^2 f_{-3}$		
		$\delta f_{-5/2}$		$\delta^3 f_{-5/2}$	
x_{-2}	f_{-2}		$\delta^2 f_{-2}$		$\delta^4 f_{-2}$
		$\delta f_{-3/2}$		$\delta^3 f_{-3/2}$	
x_{-1}	f_{-1}		$\delta^2 f_{-1}$		$\delta^4 f_{-1}$
		$\delta f_{-1/2}$		$\delta^3 f_{-1/2}$	
x_0	f_0		$\delta^2 f_0$		$\delta^4 f_0$
		$\delta f_{1/2}$		$\delta^3 f_{1/2}$	
x_1	f_1		$\delta^2 f_1$		$\delta^4 f_1$
		$\delta f_{3/2}$		$\delta^3 f_{3/2}$	
x_2	f_2		$\delta^2 f_2$		$\delta^4 f_2$
		$\delta f_{5/2}$		$\delta^3 f_{5/2}$	
x_3	f_3		$\delta^2 f_3$		
		$\delta f_{7/2}$			
x_4	f_4				

Examples.

$\Delta f_0 = f_1 - f_0 \qquad \Delta f_{-1} = f_0 - f_{-1}$

$\delta f_{1/2} = \Delta f_0 \qquad \delta^2 f_1 = \Delta^2 f_0 \qquad \delta^3 f_{3/2} = \Delta^3 f_0 \qquad \ldots$

$\mu f_j = \tfrac{1}{2}(f_{j-1/2} + f_{j+1/2})$

$\mu\delta f_0 = \tfrac{1}{2}(\Delta f_{-1} + \Delta f_0) \qquad \delta^2 f_0 = \Delta^2 f_{-1}$

$\mu\delta^3 f_0 = \tfrac{1}{2}(\Delta^3 f_{-2} + \Delta^3 f_{-1}) \qquad \delta^4 f_0 = \Delta^4 f_{-2}$

$\delta^m f_{n/2} = \Delta^m f_{(n-m)/2}$ if the integers m and n are both odd, or both even

328. If $f(x) \equiv f$, then $f(x_i) \equiv f_i$ for $k = 1, 2, \ldots,$ and $\delta^k f(x) = \delta[\delta^{k-1} f(x)] = \delta^{k-1} f(x + \tfrac{1}{2}h) - \delta^{k-1} f(x - \tfrac{1}{2}h)$.

Examples.

$\delta^0 f(x) = f(x) \qquad \delta^{2k} f(x_i) = \Delta^{2k} f(x_{i-k}) \qquad k = 0, 1, \ldots$

$$\delta^2 f(x) = \delta[\delta f(x)] = f(x + h) - 2f(x) + f(x - h)$$

329. INTERPOLATION IN A TABLE Suppose values of $f(x)$ are given on both sides of x_0, the points x_1 ordered so that both $x_i > x_j$ if $i > j$ and $x_i - x_{i-1} = h > 0$ for all i. Let $s = (x - x_0)/h$. The tabular points are $x_s = x_0 + sh$, $s = 0, \pm 1, \ldots, \pm k$. Then $f(x_s) = f(x_0 + sh) \equiv f_s$, $s = 0, \pm 1, \ldots, \pm k$.

The forward-difference operator applied to f_s gives

$$\Delta f_s = f_{s+1} - f_s \qquad s = 0, \pm 1, \ldots, \pm k$$

Higher differences are obtained from

$$\Delta^k f(x) = \Delta^{k-1} f(x + h) - \Delta^{k-1} f(x) = (E - 1)^k f(x)$$

The differences forward from x_0 are

$$\Delta f_0 = f_1 - f_0,\ \Delta^2 f_0 = \Delta f_1 - \Delta f_0 = f_2 - 2f_1 + f_0, \ldots$$
$$\Delta^k f_0 = \Delta^{k-1} f_1 - \Delta^{k-1} f_0 = f_k - \binom{k}{1} f_{k-1} + \binom{k}{2} f_{k-2} - \cdots + (-1)^k f_0$$

The differences backward from x_0 are

$$\Delta f_{-1} = f_0 - f_{-1},\ \Delta^2 f_{-2} = \Delta f_{-1} - \Delta f_{-2} = f_0 - 2f_{-1} + f_{-2}, \ldots$$
$$\Delta^k f_{-k} = \Delta^{k-1} f_{-k+1} - \Delta^{k-1} f_{-k} = f_0 - \binom{k}{1} f_{-1} + \binom{k}{2} f_{-2} - \cdots + (-1)^k f_{-k}$$

Difference tables are used to check the smoothness of a tabulated function, to detect errors in values of the function, and to estimate the degree of a polynomial that might be used for interpolating purposes.

330. NEWTON'S FORWARD-DIFFERENCE INTERPOLATING POLYNOMIAL The polynomial of degree n which interpolates at $x_0, x_1, \ldots, x_n$ is

$$(1) \qquad P_n(s) = f_0 + \binom{s}{1} \Delta f_0 + \binom{s}{2} \Delta^2 f_0 + \cdots + \binom{s}{n} \Delta^n f_0$$

This is *Newton's forward-difference formula.* Equation (1) can be generated recursively from

$$(2) \qquad P_k(s) = P_{k-1}(s) + \binom{s}{k} \Delta^k f_0 \qquad k = 1, 2, \ldots, n$$

where $P_0 = f_0$ and $\Delta^k f_0$ are the differences forward from $x = x_0$.

Examples. The linear polynomial which interpolates x_0 and x_1 is $P_1 = f_0 + \binom{s}{1} \Delta f_0$.

The second-degree polynomial which interpolates at x_0, x_1, and x_2 is

$$P_2 = P_1 + \binom{s}{2}\Delta^2 f_0 = f_0 + \binom{s}{1}\Delta f_0 + \binom{s}{2}\Delta^2 f_0$$

331. The error E of $P_n(s)$ in (1) is

(3) $E = f(x) - P_n(s) = \binom{s}{n+1} h^{n+1} f^{n+1}(\xi) \qquad x_0 < \xi < x_n$

Formula (1) is used to interpolate values of x forward from x_0 near the beginning of a set of tabular values and to extrapolate a short distance backward from x_0.

332. NEWTON'S BACKWARD-DIFFERENCE INTERPOLATING POLYNOMIAL The polynomial of degree n which interpolates at $x_0, x_{-1}, \ldots, x_{-n}$ is *Newton's backward-difference interpolating polynomial*

(4) $P_n(s) = f_0 + \binom{s}{1}\Delta f_{-1} + \binom{s+1}{2}\Delta^2 f_{-2} + \cdots + \binom{s+n-1}{n}\Delta^n f_{-n}$

or

(4′) $P_n(s) = f_0 - \binom{-s}{1}\Delta f_{-1} + \binom{-s}{2}\Delta^2 f_{-2} - \cdots + (-1)^n\binom{-s}{n}\Delta^n f_{-n}$

since $\binom{s+k-1}{k} = (-1)^k\binom{-s}{k}$.

Equation (4) can be generated recursively from

(5) $P_k(s) = P_{k-1}(s) + \binom{s+k-1}{k}\Delta^k f_{-k} \qquad k = 1, 2, \ldots, n$

where $P_0 = f_0$ and $\Delta^k f_{-k}$ are the differences backward from x_0.

Example. The linear polynomial which interpolates x_0 and x_{-1} is $P_1(s) = f_0 + \binom{s}{1}\Delta f_{-1}$. The second-degree polynomial which interpolates x_0, x_{-1}, and x_{-2} is

$$P_2(s) = f_0 + \binom{s}{1}\Delta f_{-1} + \binom{s+1}{2}\Delta^2 f_{-2}$$

The error E of $P_n(s)$ in (4) is

(6) $E = f(x) - P_n(s) = \binom{s+n}{n+1} h^{n+1} f^{(n+1)}(\xi) \qquad x_{-k} < \xi < x_0$

333. Newton's formulas (1) and (4) have two advantages over Lagrangian formulas: (*a*) the degree of the interpolation can be increased by adding additional terms as indicated in (2) and (5), and (*b*) the error can be estimated by examining the first neglected differences.

Formula (4) is used for interpolating a value of x backward from x_0 near the end of a set of tabular values and for extrapolating values of x a short distance forward of x_0.

334. GAUSSIAN DIFFERENCE FORMULAS When data are available on both sides of a point, the following interpolating polynomials may be used

$$(1)\quad P(s) = f_0 + \binom{s}{1}\Delta f_0 + \binom{s}{2}\Delta^2 f_{-1} + \binom{s+1}{3}\Delta^3 f_{-1} + \binom{s+1}{4}\Delta^4 f_{-2} + \binom{s+2}{5}\Delta^5 f_{-2} + \cdots$$

(*Gauss' forward-difference formula*)

$$(2)\quad P(s) = f_0 + \binom{s}{1}\Delta f_{-1} + \binom{s+1}{2}\Delta^2 f_{-1} + \binom{s+1}{3}\Delta^3 f_{-2} + \binom{s+2}{4}\Delta^4 f_{-2} + \binom{s+2}{5}\Delta^5 f_{-3} + \cdots$$

(*Gauss' backward-difference formula*)

In the case where the last term used in (1) and (2) involves an even difference, say $\Delta^{2k} f_{-k}$, both (1) and (2) interpolate at $s = 0, \pm 1, \pm 2, \ldots, \pm k$.

In place of a Gaussian formula, one may always use an equivalent formula of either Stirling or Bessel, for which the coefficients are extensively tabulated.

335. STIRLING'S FORMULA When interpolations are to be done for values of x near an interior point x_0, say between $(x_0 - \frac{1}{2}h)$ and $(x_0 + \frac{1}{2}h)$, *Stirling's formula* is useful:

$$(1)\quad f_s \approx f_0 + \binom{s}{1}\mu\delta f_0 + \frac{s}{2}\binom{s}{1}\delta^2 f_0 + \binom{s+1}{3}\mu\delta^3 f_0 + \frac{s}{4}\binom{s+1}{3}\delta^4 f_0 + \cdots$$

BESSEL'S FORMULA When interpolation is needed between x_0 and x_1, *Bessel's formula* is useful:

$$f_s \approx \mu f_{1/2} + \left(s - \frac{1}{2}\right)\delta f_{1/2} + \binom{s}{2}\mu\delta^2 f_{1/2} + \frac{s - \frac{1}{2}}{3}\binom{s}{2}\delta^3 f_{1/2} + \binom{s+1}{4}\mu\delta^4 f_{1/2} + \frac{s - \frac{1}{2}}{5}\binom{s+1}{4}\delta^5 f_{1/2} + \cdots$$

16. DIFFERENTIAL EQUATIONS

Ordinary Differential Equations

336. Any equation which involves derivatives or differentials is a *differential equation.* A differential equation is of *order n* if it involves a derivative of order n and no derivative of higher order.

The *degree* of a differential equation is the greatest power to which the highest-order derivative occurs, after the equation has been made rational and integral in all of its derivatives.

Consider the ordinary differential equation of order n

$$(1) \qquad F(x, y, y', y'', \ldots, y^{(n)}) = 0$$

A *solution,* or *integral,* of (1) is a function $y = \phi(x)$ which, when $\phi(x)$, $\phi'(x)$, $\ldots$, $\phi^{(n)}(x)$ are substituted for y, y', $\ldots$, $y^{(n)}$, respectively, in (1), F is transformed into a function identically zero for all values of x.

The *general solution* of (1) is of the form

$$(2) \qquad y = \Phi(x, c_1, c_2, \ldots, c_n)$$

where $c_1, c_2, \ldots, c_n$ are n arbitrary constants.

If (2) is differentiated n times with respect to x and if the n constants c_i appearing in the resulting n equations and (2) are eliminated, the resulting differential equation is (1). (2) is the *primitive* of (1).

A solution obtained by giving the constants in (2) specific values is a *particular solution.* Solutions to (1) may exist which cannot be obtained by giving the constants c_i special values. Such solutions are called *singular solutions.*

A differential equation is said to be *solved* if the solution has been reduced to quadratures or if y has been obtained merely as an implicit function of x by $\psi(x, y, c_1, \ldots, c_n) = 0$.

337. EQUATIONS OF FIRST ORDER AND FIRST DEGREE If $f(x,y)$ is an analytic function of x and y in the neighborhood of (x_0,y_0), the equation

$$(3) \qquad \frac{dy}{dx} = f(x,y)$$

has a solution $y(x)$, where $y(x)$ is analytic in the neighborhood of x_0 and $y(x_0) = y_0$.

Methods of Solution

338. Consider the differential equation

$$M\,dx + N\,dy = 0 \tag{4}$$

where $M(x,y)$ and $N(x,y)$ are functions of x and y.

339. VARIABLES SEPARABLE If M and N are products of factors, each being a function either of x alone or of y alone, then (4) is of the form

$$A(x)P(y)\,dx + B(x)Q(y)\,dy = 0 \tag{5}$$

where A and B are functions of x alone, P and Q functions of y alone.

The general solution of (5) is

$$\int \frac{A(x)}{B(x)}\,dx + \int \frac{Q(y)}{P(y)}\,dy = c \tag{6}$$

where c is a constant.

340. EXACT EQUATIONS $M\,dx + N\,dy$ is the complete differential dU of a function $U(x,y)$ if and only if $\partial M/\partial y = \partial N/\partial x$. In this case, $M\,dx + N\,dy$ is *exact*. The general solution of (4) is $U + c = 0$.

Example. Solve $(2x - y)\,dx + (3y^2 - x)\,dy = 0$. This is exact: $M = 2x - y$, $N = 3y^2 - x$. Integrate $M\,dx$, regarding y as a constant.

$$\int M\,\partial x = x^2 - yx + f(y) \equiv \varphi(x)$$

where $f(y)$ is an unknown function of y. But $\partial\varphi/\partial y \equiv N$, and so $-x + f'(y) = 3y^2 - x$. Hence $f(y) = y^3 + c$. The general solution is $x^2 - yx + y^3 + c = 0$.

341. INTEGRATING FACTORS If (4) is not exact, look for an integrating factor $\mu(x,y)$ such that

(7) $$\mu M\,dx + \mu N\,dy \equiv dW$$

is exact.

342. HOMOGENEOUS EQUATIONS A function $f(x, y, \ldots, t)$ is *homogeneous* in $x, y, \ldots, t$ if the substitution of $\lambda x, \lambda y, \ldots, \lambda t$ for $x, y, \ldots, t$, respectively, multiplies f by λ^n, that is, if $f(\lambda x, \lambda y, \ldots, \lambda t) = \lambda^n f(x, y, \ldots, t)$. The *order of homogeneity* of f is n.

If the M and N in (4) are each of the same order of homogeneity, the substitution $y = vx$, $dy = v\,dx + x\,dv$, reduces (4) to an equation in x and v which is separable. The solution follows as in Sec. 339.

343. EQUATIONS REDUCIBLE TO HOMOGENEOUS FORM

(8) $$\frac{dy}{dx} = \frac{a_1 x + b_1 y + c_1}{a_2 x + b_2 y + c_2}$$

may be made homogeneous by the substitutions $x = x_0 + u$ and $y = y_0 + v$. If x_0 and y_0 are selected so that

$$a_1 x_0 + b_1 y_0 + c_1 = 0 \qquad a_2 x_0 + b_2 y_0 + c_2 = 0$$

then (8) becomes

(9) $$\frac{dv}{du} = \frac{a_1 u + b_1 v}{a_2 u + b_2 v}$$

which may be solved as in Sec. 342.

344. LINEAR EQUATIONS To solve

(10) $$\frac{dy}{dx} + P(x)\,y = Q(x)$$

make (10) exact by multiplying by the integrating factor $\exp\left[\int P\,dx\right]$ and integrate the resulting expression. The solution y to (10) is found from

(11) $$y \exp\left[\int P\,dx\right] = \int Q \exp\left[\int P\,dx\right] dx + c$$

If integral y_1 of (10) is known, the general integral of (10) is $y = c \exp\left[-\int P\,dx\right] + y_1$.

If y_1 and y_2 are two particular integrals of (10), the general solution is $y - y_1 = c(y_2 - y_1)$.

Example. Solve $(dy/dx) + (y/x) = x^3$. Here $P = 1/x$ and $Q = x^3$. The integrating factor is $\exp\left[\int dx/x\right] = e^{\log x} = x$. Then $x(dy/dx) + y = x^4$, $xy = (x^5/5) + c$, and $y = (x^4/5) + c/x$.

345. BERNOULLI'S EQUATION To solve

$$\frac{dy}{dx} + P(x)y = Q(x)y^n \tag{12}$$

let $z = y^{1-n}$. Then (12) reduces to

$$\frac{dz}{dx} + (1 - n)Pz = (1 - n)Q \tag{13}$$

which is solvable by the method of Sec. 344.

346. FIRST ORDER AND HIGHER DEGREE The general first-order differential equation of arbitrary degree may be written as

$$F(x,y,p) = 0 \tag{14}$$

where $p = dy/dx$, $p^2 = (dy/dx)^2$, ... Two methods for solving (14) are as follows:

Method 1. If F can be factored into

$$F = [p - f_1(x,y)][p - f_2(x,y)] \cdots [p - f_n(x,y)] = 0$$

then

$$p - f_1(x,y) = 0 \qquad p - f_2(x,y) = 0, \ldots \qquad p - f_n(x,y) = 0 \tag{15}$$

Methods for solving each equation in (15) are given in earlier sections.

Example. $(p - 2x)(p + 2y) = 0$. The integrals of $p - 2x = 0$ and $p + 2y = 0$ are, respectively,

$$y = x^2 + c \qquad \log cy = -2x$$

Method 2. If (14) is solvable for y in the form

(16) $$y = g(x,p)$$

differentiate (16) with respect to x. Then

(17) $$p = \frac{dy}{dx} = \frac{\partial g}{\partial x} + \frac{\partial g}{\partial p}\frac{dp}{\partial x}$$

which is of the first order in p.

Let a solution to (17) be

(18) $$G(x,p,c) = 0$$

Eliminate p from (16) and (18). This gives an equation of the form

(19) $$Q(x,y,c) = 0$$

which may be a solution of (16). Equation (19) should be tested in (16) to avoid the possibility of introducing extraneous factors.

347. CLAIRAUT'S EQUATION Consider

(20) $$y = px + f(p)$$

Differentiate with respect to x. Then

(21) $$[x + f'(p)]\frac{dp}{dx} = 0$$

If $dp/dx = 0$, $p = c$ and $y = cx + f(c)$ is the general solution to (20).

From the first factor in (21)

(22) $$x + f'(p) = 0 \qquad y = -pf'(p) + f(p)$$

Equations (22) are parametric solutions of (20).

348. D'ALEMBERT'S OR LAGRANGE'S EQUATION The equation

(23) $$y = xg(p) + f(p) \qquad g(p) \neq p$$

may be solved by differentiating with respect to x and treating p as the independent variable.

349. LINEAR DIFFERENTIAL EQUATIONS The general linear differential equation of order n is

$$\text{(A)} \quad X_0 \frac{d^n y}{dx^n} + X_1 \frac{d^{n-1} y}{dx^{n-1}} + \cdots + X_{n-1} \frac{dy}{dx} + X_n y = X$$

where $X_0 \neq 0, X_1, \ldots, X_n, X$ are functions of x alone or constants. Denote the left member of (A) by $P(D)y$. The equation

$$\text{(B)} \qquad P(D)y = 0$$

is the *auxiliary* or *reduced equation* corresponding to the *complete equation* (A).

Theorem. If $y_1, y_2, \ldots, y_n$ are solutions of (B) so that $P(D)y_1 = 0, \ldots, P(D)y_n = 0$, then

$$\text{(C)} \qquad y = c_1 y_1 + c_2 y_2 + \cdots + c_n y_n$$

is a solution of (B), where $c_1, \ldots, c_n$ are arbitrary constants.

The function (C) is the *complementary solution* to (A).

The solutions $y_1, \ldots, y_n$ are *linearly independent* if there exists no set of numbers $c_1, \ldots, c_n$, not all zero, such that

$$c_1 y_1 + c_2 y_2 + \cdots + c_n y_n = 0$$

Theorems. ***(a)*** A set of n solutions $y_1, \ldots, y_n$ of (B) is linearly independent when and only when $\det W \neq 0$, where $W = (y_r^{(s)})$ is a matrix of order n and $y_r^{(s)}$ is the sth derivative of y_r with respect to x, $r = 1, \ldots, n$ and $s = 0, 1, \ldots, n - 1$.

(b) If $y_1, \ldots, y_n$ are linearly independent solutions of (B) and if Y is any particular solution of (A), a general solution of (A) is

$$\text{(D)} \qquad y = c_1 y_1 + \cdots + c_n y_n + Y$$

The solutions $y_1, \ldots, y_n$ are said to form a *fundamental system of solutions* of (A).

350. LINEAR EQUATIONS WITH CONSTANT COEFFICIENTS In this case, (A) is of the form

$$\text{(I)} \quad a_0 \frac{d^n y}{dx^n} + a_1 \frac{d^{n-1} y}{dx^{n-1}} + \cdots + a_{n-1} \frac{dy}{dx} + a_n y \equiv P(D)y = X$$

where $a_0, a_1, \ldots, a_n$ are constants and X is a function of x.

351. **CASE X = 0** Let $m_1, \ldots, m_n$ be the roots of the *characteristic equation*

$$\text{(G)} \qquad a_0 m^n + a_1 m^{n-1} + \cdots + a_{n-1} m + a_n \equiv P(m) = 0$$

Then

$$\text{(F)} \qquad y = c_1 e^{m_1 x} + \cdots + c_n e^{m_n x}$$

is a solution to

$$\text{(E)} \qquad P(D)y = 0$$

If the roots of (G) are distinct, then (F) is a general solution of (E). However, if (G) factors into the form

$$(m - m_1)^\alpha (m - m_2)^\beta \cdots (m - m_r)^\rho = 0$$

then a general solution of (E) is

$$\text{(H)} \quad y = (a_1 + a_2 x + \cdots + a_\alpha x^{\alpha-1}) e^{m_1 x} + \cdots + (p_1 + p_2 x + \cdots + p_\rho x^{\rho-1}) e^{m_r x}$$

where $a_1, a_2, \ldots$ and $p_1, p_2, \ldots, p_\rho$ are arbitrary constants.

Note: The complementary solution to (I) is (F) or (H) in case (G) has multiple roots.

If $m_1 = a + bi$ and $m_2 = a - bi$ are conjugate complex roots of (G), one may write $c_1 e^{m_1 x} + c_2 e^{m_2 x}$ in the form

$$e^{ax}(A \cos bx + B \sin bx)$$

where A and B are arbitrary constants.

352. **CASE X ≠ 0** The general solution to (I) is $y = y_c + Y$, where y_c denotes the complementary solution and Y is a particular integral of (I). Methods for finding Y are given below.

353. **METHOD UNDETERMINED COEFFICIENTS** In this method a function $Y(x)$ is constructed whose form depends upon the form of X and which involves coefficients $A, B, \ldots$. Y is substituted in (I), and $A, B, \ldots$ are determined so that the result is an identity. A few rules for constructing Y are:

I. If $X = b_0x^m + b_1x^{m-1} + \cdots + b_m$ and if (G) has $m = 0$ as a root of multiplicity k, then

$$Y = x^k(A_0x^m + A_1x^{m-1} + \cdots + A_m)$$

II. If $X = be^{ax}$ and if (G) has $m = a$ as a root of multiplicity k, then $Y = Ax^ke^{ax}$.

III. If $X = b \sin ax$ or $b \cos ax$ and if $m^2 + a^2$ is a factor of (G) of multiplicity k, then

$$Y = x^k(A \sin ax + B \cos ax)$$

IV. If $X = e^{ax}\psi(x)$, where $\psi(x)$ is a function of x, let $y = ze^{ax}$ and solve for z.

V. If $X = u_1(x) + \cdots + u_m(x)$, then Y is the sum of the functions constructed for the terms $u_1(x), \ldots, u_m(x)$, respectively.

354. OPERATOR METHODS Operator methods for solving equation (I) are useful. Let $D, D^2, D^3, \ldots$ denote the symbols d/dx, d^2/dx^2, $d^3/dx^3, \ldots$, respectively. The operator D obeys the following algebraic laws. Let $f(x)$ and $g(x)$ be differentiable functions of x.

$$D[f(x) + g(x)] = Df(x) + Dg(x)$$

(1) $D[cg(x)] = c[Dg(x)]$

$$D^m[D^nf(x)] = D^n[D^mf(x)] = D^{m+n}[f(x)]$$

Equation (1) may be written $P(D)y = X$, where

$$P(D) \equiv a_0D^n + a_1D^{n-1} + \cdots + a_n$$

The following relations also hold:

$$P(D)e^{wx} = P(w)e^{wx}$$

$$De^{wx} = we^{wx} \qquad D^2e^{wx} = w^2e^{wx} \qquad \cdots$$

$$D^2 \cos wx = -w^2 \cos wx \qquad D^2 \sin wx = -w^2 \sin wx$$

(2) $P(D^2) \cos wx = P(-w^2) \cos wx$

$$P(D^2) \sin wx = P(-w^2) \sin wx$$

$$D[e^{wx}f(x)] = e^{wx}(D + w)f(x)$$

Examples of use of operators. The integral of

(3) $$(D - r)y = X(x)$$

may be written formally as $y = [1/(D - r)]\ X(x)$. By Sec. 344, the solution to (3) is

(4) $$y_s = e^{rx} \int e^{-rx}X(x)\,dx + Ce^{rx}$$

Operator $1/(D - r)$ is defined by

(5) $$\frac{1}{D - r} X(x) \equiv y_s \qquad (D - r)\left[\frac{1}{D - r} X(x)\right] = X(x)$$

The integral of

(6) $$(D - r_1)(D - r_2)y = X(x)$$

may be found as follows. Let $z = (D - r_2)y$. Then

(7) $$(D - r_1)z = X(x)$$

which has the solution

(8) $$z = \frac{1}{D - r_1} X(x) = e^{r_1 x} \int e^{-r_1 x} X(x)\, dx + c_1 e^{r_1 x}$$

Replace z by $(D - r_2)y$ and solve the resulting equation for y by the same general process.

The step-by-step process sketched above may be used to find the general solution to

(9) $$P(D) \equiv (D - r_1)(D - r_2) \cdots (D - r_n)y = X(x)$$

Another method. The solution to (9) is symbolically

(10) $$y = \frac{1}{P(D)} X(x)$$

Write $1/[P(D)]$ as the sum of partial fractions,

(11) $$\frac{1}{P(D)} = \frac{A_1}{D - r_1} + \frac{A_2}{D - r_2} + \cdots + \frac{A_n}{D - r_n}$$

The particular solution (when the roots r_i are distinct) is

(12) $$y = A_1 e^{r_1 x} \int X e^{-r_1 x}\, dx + A_2 e^{r_2 x} \int X e^{-r_2 x}\, dx + \cdots + A_n e^{r_n x} \int X e^{-r_n x}\, dx$$

When multiple roots occur, the partial fractions are expressed as in Sec. 143.

355. DAMPED SIMPLE HARMONIC MOTION Consider

(1) $$\frac{d^2x}{dt^2} + 2\rho\frac{dx}{dt} + \omega^2 x = f(t) \qquad \rho \text{ and } \omega \text{ real}$$

The solution to (1) is the sum of a particular integral x_p called the *steady-state solution* and the complementary solution x_c called the *transient solution.*

(2) $$x_c = Ae^{m_1 t} + Be^{m_2 t} \qquad \text{if } \rho^2 - \omega^2 \neq 0$$

(3) $$x_c = (A + Bt)e^{-\rho t} \qquad \text{if } \rho^2 - \omega^2 = 0$$

where A and B are arbitrary constants and m_1 and m_2 are the roots of the characteristic equation $m^2 + 2\rho m + \omega^2 = 0$.

When $\rho^2 > \omega^2$, the roots m_1 and m_2 are real, the motion is *damped*, and no vibration occurs.

When $\rho^2 = \omega^2$, the roots m_1 and m_2 are real and equal and the motion is *critically damped.*

When $\rho^2 < \omega^2$, the roots m_1 and m_2 are complex numbers and the motion is *damped oscillatory*. Then

(4) $$x_c = e^{-\rho t}[C\cos ht + D\sin ht] = e^{-\rho t}\bar{C}\cos(ht - \gamma)$$

where $\bar{C} = \sqrt{C^2 + D^2}$, $h = \sqrt{\omega^2 - \rho^2}$, and $\gamma = \tan^{-1}(D/C)$. The period of oscillation is $2\pi/h$. Here C and D are arbitrary constants of integration.

356. FORCED VIBRATION The particular solution x_p to (1) will depend on the nature of the impressed force $f(t)$. For example, if $f(t) = L\cos pt$, then

(5) $$x_p = \frac{L}{Q}\cos(pt - \beta) \qquad \text{if } \rho^2 \neq \omega^2$$

where $Q^2 = (\omega^2 - p^2)^2 + 4\rho^2 p^2$; and

(6) $$x_p = \frac{L}{\rho^2 + p^2}\cos(pt - \beta) \qquad \text{if } \rho^2 = \omega^2$$

Here $\tan\beta = 2\rho p/(\omega^2 - p^2)$.

The complete solution to (1) is $x = x_c + x_p$, where the constants of integration A, B, C, and D are determined by the initial values of x and dx/dt.

In the case where the impressed frequency p is such that the amplitude of vibration in (5) has a maximum value, *resonance* is said to occur.

357. HOMOGENEOUS LINEAR EQUATION These have the form

$$(1)\quad a_0x^n\frac{d^ny}{dx^n} + a_1x^{n-1}\frac{d^{n-1}y}{dx^{n-1}} + \cdots + a_{n-1}x\frac{dy}{dx} + a_ny = X(x)$$

Substitution of $x = e^t$ makes $x(dy/dx) = Dy$, $x^2(d^2y/dx^2) = (D^2 - D)y, \ldots$ where $D \equiv d/dt$, and it transforms (1) into a linear differential equation with constant coefficients, whose solution can be found as in Sec. 350.

358. TOTAL DIFFERENTIAL EQUATIONS A necessary and sufficient condition that

$$(1)\qquad P\,dx + Q\,dy + R\,dz = 0$$

have a solution of the form $\psi(x,y,z,c) = 0$ is that

$$(2)\quad P\left(\frac{\partial Q}{\partial z} - \frac{\partial R}{\partial y}\right) + Q\left(\frac{\partial R}{\partial x} - \frac{\partial P}{\partial z}\right) + R\left(\frac{\partial P}{\partial y} - \frac{\partial Q}{\partial x}\right) = 0$$

Here P, Q, and R are functions of x, y, and z.

359. EXACT CASE When (2) is satisfied, the solution to (1) is found as follows:

(*a*) Find an integral $V = c$ of $P\,dx + Q\,dy = 0$, treating z as a constant.

(*b*) Assume $V = c(z)$ is a solution of (1). Differentiate this, and by comparing the result with (1) find an equation for $c'(z)$.

(*c*) Integrate the resulting equation in c, obtaining a solution for c containing one arbitrary constant. $\overline{V} = c(z)$ is a solution of (1).

360. INEXACT CASE In the case where (2) is not satisfied, (1) is *nonintegrable*. To obtain a solution to (1), assume an arbitrary relation $g(x,y,z) = 0$ between x, y, and z. The relation $g = 0$ will determine (with suitable restrictions) any one of the variables in terms of the other two, say, $z = h(x,y)$. Substitute h for z in (1). A new differential equation in x and y is found. Let $w(x,y,c) = 0$ be an integral of this equation. The general solution of (1) consists of the arbitrarily chosen relation $g(x,y,z) = 0$ together with the relation $w(x,y,c) = 0$, where c is an arbitrary constant.

361. GEOMETRICAL INTERPRETATION Equation (1) may be interpreted as stating that two vectors having direction cosines proportional to $P:Q:R$ and $dx:dy:dz$, respectively, are perpendicular to each other. To solve (1) is to find geometric loci which satisfy the perpendicularity property.

If (1) is integrable, the loci corresponding to the solution of (1) consist of surfaces orthogonal to the direction $P:Q:R$ at each point P of coordinates x, y, z. Any ordinary curve drawn on the surface has this orthogonal property at each of its points.

If (1) is nonintegrable, there exists no such family of surfaces. The solution to (1) is a family of curves lying on some assumed surface having the orthogonality property to a vector of direction $P:Q:R$.

362. SIMULTANEOUS TOTAL DIFFERENTIAL EQUATIONS The system

$$\begin{cases} P_1\,dx + Q_1\,dy + R_1\,dz = 0 \\ P_2\,dx + Q_2\,dy + R_2\,dz = 0 \end{cases} \tag{1}$$

where P_1, Q_1, R_1, P_2, . . . are functions of x, y, and z, may be written

$$\frac{dx}{P} = \frac{dy}{Q} = \frac{dz}{R} \tag{2}$$

where

$$P = \begin{vmatrix} Q_1 & R_1 \\ Q_2 & R_2 \end{vmatrix} \qquad Q = \begin{vmatrix} R_1 & P_1 \\ R_2 & P_2 \end{vmatrix} \qquad R = \begin{vmatrix} P_1 & Q_1 \\ P_2 & Q_2 \end{vmatrix} \tag{3}$$

The general solution of (1) consists of two relations involving two arbitrary constants

$$\begin{cases} \varphi_1(x,y,z,c_1) = 0 \\ \varphi_2(x,y,z,c_2) = 0 \end{cases} \tag{4}$$

Methods of obtaining solutions to (2) and (1) involve approaches similar to those sketched in earlier sections.

17. LEGENDRE POLYNOMIALS. BESSEL FUNCTIONS

Legendre Polynomials

363. LEGENDRE'S EQUATION Legendre's equation is

(1) $$(1 - x^2)y'' - 2xy' + n(n + 1)y = 0$$

where n is a real constant and x and y are real variables.

(2) $$y = \sum_{k=0}^{\infty} a_k x^{m+k}$$

is a solution of (1) only if

(3) $$m(m - 1)a_0 = 0 \qquad (\textit{indicial}\ \text{equation})$$

(4) $$(m + 1)ma_1 = 0$$

(5) $$\lambda_0 \equiv (m + 2)(m + 1)a_2 - (m - n)(m + n + 1)a_0 = 0$$

(6) $$\lambda_k \equiv (m + k + 2)(m + k + 1)a_{k+2} + (n - m - k)(n + m + k + 1)a_k = 0 \qquad k = 1, 2, \ldots$$

A value of m satisfying (3) is an *index* for (1).

364. CASE $m = 0$ In this case

(7) $$(k + 2)(k + 1)a_{k+2} = -(n - k)(n + k + 1)a_k$$

The odd and even a_n form independent sets. The general solution to (1) is

(8) $$y = a_0 u(x) + a_1 v(x)$$

where a_0 and a_1 are arbitrary and

(9) $$u(x) \equiv 1 - \frac{n(n + 1)}{2!}x^2 + \frac{n(n - 2)(n + 1)(n + 3)}{4!}x^4 - \cdots$$

(10) $$v(x) \equiv x - \frac{(n - 1)(n + 2)}{3!}x^3 + \frac{(n - 1)(n - 3)(n + 2)(n + 4)}{5!}x^5 - \cdots$$

Series (8) converges uniformly in any subinterval of $[-1,1]$.

Case $m = 1$. The solution to (1) is $y = a_0u(x)$.

365. LEGENDRE POLYNOMIALS When n is an even integer, $a_0u(x)$ is a polynomial; when n is odd, so is $a_1v(x)$.

Select $a_0 = 0$, $a_1 \neq 0$ when n is odd integer; and $a_0 \neq 0$, $a_1 = 0$ when n is even. Particular solutions of (1) are:

$$P_1(x) = x$$

$$P_2(x) = \frac{3x^2 - 1}{2}$$

$$(11) \quad P_3(x) = \frac{5x^3 - 3x}{2}$$

$$P_4(x) = \frac{7 \cdot 5}{4 \cdot 2} x^4 - 2\frac{5 \cdot 3}{4 \cdot 2} x^2 + \frac{3 \cdot 1}{4 \cdot 2}$$

$$P_5(x) = \frac{9 \cdot 7}{4 \cdot 2} x^5 - 2\frac{7 \cdot 5}{4 \cdot 2} x^3 + \frac{5 \cdot 3}{4 \cdot 2} x$$

. . . .

$$(12) \quad P_n(x) = \sum_{\nu=0}^{[n/2]} (-1)^\nu \frac{1 \cdot 3 \cdot 5 \cdot \cdots \cdot (2n - 2\nu - 1)}{2^\nu \nu!(n - 2\nu)!} x^{n-2\nu}$$

where $[n/2] = n/2$ if n is even and $[n/2] = (n - 1)/2$ if n is odd. a_1, or a_0, is so determined that $P_n(1) = 1$. $P_0(x) \equiv 1$.

366. PROPERTIES

$$(13) \quad \begin{aligned} P'_{n+1} &= xP'_n + (n + 1)P_n \\ (1 - x^2)P'_n &= (n + 1)(xP_n - P_{n+1}) \\ (2n + 1)xP_n &= (n + 1)P_{n+1} + nP_{n-1} \\ (2n + 1)P_n &= P'_{n+1} - P'_{n-1} \end{aligned}$$

$P_n(x)$ are the coefficients a_n of h^n in

$$(14) \quad (1 - 2xh + h^2)^{-1/2} = \sum_{n=0}^{\infty} a_n h^n \qquad \text{(generating function)}$$

(15) $P_n(x) = \dfrac{1}{2^n n!} \dfrac{d^n}{dx^n} (x^2 - 1)^n$ (Rodrigues' formula)

$P_n(x)$ are orthogonal for $-1 \leqq x \leqq +1$,

(16) $$\int_{-1}^{+1} P_m(x)P_n(x)\,dx = \frac{2\delta_{mn}}{2n+1}$$

where $\delta_{mn} = 1$ if $m = n$ and $\delta_{mn} = 0$ if $m \neq n$.

367. ASSOCIATED LEGENDRE POLYNOMIALS

(17) $$(1 - x^2)y'' - 2xy' + \left[n(n+1) - \frac{m^2}{1 - x^2}\right]y = 0$$

has the solution $y = (1 - x^2)^{m/2}(d^m/dx^m)P_n(x)$, denoted by $P_n{}^m(x)$, and is called an *associated Legendre polynomial.* When $m > n$, $P_n{}^m(x) = 0$.

The set of functions $P_n{}^m(x)$ is orthogonal for $-1 \leqq x \leqq +1$,

(18) $$\int_{-1}^{+1} P_m{}^s(x)P_n{}^s(x)\,dx = \frac{(n+s)!}{(n-s)!}\frac{2\delta_{mn}}{(2n+1)}$$

Bessel Functions

368. BESSEL'S EQUATION is

(1) $$x^2y'' + xy' + (x^2 - n^2)y = 0$$

where n is a real constant and x and y are real variables.

(2) $$y = x^m \sum_{k=0}^{\infty} a_k x^k$$

is a solution of (1) only if

(3) $(m^2 - n^2)a_0 = 0$ (*indicial equation*)

(4) $[(m+1)^2 - n^2]a_1 = 0$

(5) $[(m+2)^2 - n^2]a_2 + a_0 = 0$

(6) $[(m+k)^2 - n^2]a_k + a_{k-2} = 0$ $k = 2, 3, \ldots$

A value of m satisfying (3) is an *index* for (1).

Consider a_0 to be arbitrary. Then by (3) $m = \pm n$.

369. CASE $m = n$ In this case

$$(7) \qquad a_1 = 0,\ a_2 = -\frac{a_0}{2(2n+2)},\ \ldots,\ a_k = -\frac{a_{k-2}}{k(2n+k)}$$

and a particular solution of (1) is

$$(8)\ y_1 = a_0 x^n \left[1 - \frac{x^2}{2(2n+2)} + \frac{x^4}{2\cdot 4(2n+2)(2n+4)} - \cdots\right]$$

provided n is not a negative integer.

370. CASE $m = -n$ In this case

$$(9) \qquad a_1 = 0,\ a_2 = \frac{1}{2(2n-2)},\ \ldots,\ a_k = \frac{a_{k-2}}{k(2n-k)}$$

and a particular solution of (1) is

$$(10)\ y_2 = a_0 x^{-n} \left[1 + \frac{x^2}{2(2n-2)} + \frac{x^4}{2\cdot 4(2n-2)(2n-4)} + \cdots\right]$$

provided n is not a positive integer. When $n = 0$, (8) and (10) are identical. If n is not an integer, (8) and (10) are independent and the general solution of (1) is

$$(11) \qquad y = c_1 y_1 + c_2 y_2$$

371. BESSEL FUNCTIONS OF THE FIRST KIND If in (8) one sets $a_0 = 1/[2^n\Gamma(n+1)]$, where $\Gamma(n+1) = n!$ when n is a positive integer, y_1 becomes

$$(12) \qquad J_n(x) = \sum_{k=0}^{\infty} \frac{(-1)^k}{k!\Gamma(n+k+1)} \left(\frac{x}{2}\right)^{n+2k}$$

(n not a negative integer)

(12) is absolutely convergent for all values of x. $J_n(x)$ is the *Bessel function of the first kind of order n.*

When n is a negative integer, $J_n(x)$ is defined by:

$$(13) \qquad J_{-n}(x) = (-1)^n J_n(x)$$

372. RECURSION RELATIONS Let $J'_n(x) \equiv dJ_n(x)/dx$.

(14)
$$\frac{d}{dx}[x^n J_n(x)] = x^n J_{n-1}(x)$$
$$xJ'_n(x) = xJ_{n-1}(x) - nJ_n(x)$$
$$\frac{d}{dx}[x^{-n} J_n(x)] = -x^{-n} J_{n+1}(x)$$
$$xJ'_n(x) = nJ_n(x) - xJ_{n+1}(x)$$
$$J_{n-1}(x) - J_{n+1}(x) = 2J'_n(x)$$
$$xJ_{n-1}(x) + xJ_{n+1}(x) = 2nJ_n(x)$$

In Eq. (10), set $a_0 = 1/[2^{-n}\Gamma(-n+1)]$. Then $y_2 = J_{-n}(x)$. When n is not an integer, the general solution of (1) is

$$y = c_1 J_n(x) + c_2 J_{-n}(x)$$

373. BESSEL FUNCTIONS OF THE SECOND KIND When n is a positive integer, the general solution of (1) is

(15)
$$y = c_1 J_n(x) + c_2 Y_n(x)$$

Here $Y_n(x)$ is the *Bessel function* of the *second kind* of order n, with

(16)
$$Y_n(x) = \frac{2}{\pi} J_n(x)\left[\log\frac{x}{2} + \gamma\right] - \frac{1}{\pi}\sum_{k=0}^{n-1}\frac{(n-k-1)!}{k!}\left(\frac{x}{2}\right)^{2k-n}$$
$$- \frac{1}{\pi}\sum_{k=0}^{\infty}\frac{(-1)^k}{k!(n+k)!}[\phi(k) + \phi(k+n)]\left(\frac{x}{2}\right)^{2k+n}$$

where

$$\phi(k) \equiv \sum_{t=1}^{k} t^{-1} \qquad \phi(0) = 0 \qquad \gamma = 0.5772157$$

The series in (16) converges for all values of x.

374. OTHER RELATIONS

(17)
$$J_n(x) = \frac{1}{\pi}\int_0^{\pi} \cos(n\phi - x\sin\phi)\, d\phi \qquad (n \text{ any integer})$$

The functions $J_n(x)$ may be *generated* from

$$e^{(x/2)[t-(1/t)]} = \sum_{n=-\infty}^{\infty} J_n(x)t^n \tag{18}$$

$$J_{1/2}(x) = \sqrt{\frac{2}{\pi x}}\sin x \qquad J_{-1/2}(x) = \sqrt{\frac{2}{\pi x}}\cos x \tag{19}$$

18. PARTIAL DIFFERENTIAL EQUATIONS

375. A *partial differential equation* is one involving partial derivatives. Explicit solutions can be written down in only a relatively few cases. The solution of a partial differential equation usually involves arbitrary functions, in much the same way that the solutions of ordinary differential equations involve arbitrary constants. In applications, the problem usually involves the determination of a particular function which satisfies the differential equation and also meets certain specific initial or boundary conditions.

376. EULER'S EQUATION The general Euler's equation

$$a\frac{\partial^2 z}{\partial x^2} + 2b\frac{\partial^2 z}{\partial x\,\partial y} + c\frac{\partial^2 z}{\partial y^2} = 0 \tag{1}$$

where a, b, and c are constants, has the general solution

$$\begin{cases} z = \varphi(x + \lambda_1 y) + \psi(x + \lambda_2 y) & \text{if } b^2 \neq ac \\ z = \varphi(x + \lambda_1 y) + (\gamma x + \delta y)\psi(x + \lambda_1 y) & \text{if } b^2 = ac \end{cases} \tag{2}$$

where λ_1 and λ_2 are the roots of $a + 2b\lambda + c\lambda^2 = 0$, φ and ψ are arbitrary functions, and γ and δ are arbitrary constants.

Examples. The general solution of $\partial^2 z/\partial x\,\partial y = 0$ is $z = \varphi(x) + \psi(y)$, φ and ψ being arbitrary functions.

The general solution of $\partial^2 u/\partial t^2 = a^2\partial^2 u/\partial x^2$ is $u = \varphi(x + at) + \psi(x - at)$, φ and ψ being arbitrary functions.

377. THE LINEAR PARTIAL EQUATION OF FIRST ORDER The equation

$$Ap + Bq = C \tag{3}$$

where A, B, and C are functions of x, y, and z and

$$p = \frac{\partial z}{\partial x} = z_x \qquad q = \frac{\partial z}{\partial y} = z_y$$

has the general solution $\Phi(u,v) = 0$. Here Φ is an arbitrary function, and $u(x,y,z) = c_1$, $v(x,y,z) = c_2$ is the general solution of the differential equations

(4) $$\frac{dx}{A} = \frac{dy}{B} = \frac{dz}{C}$$

The normal to surface $\Phi(u,v) = 0$ is perpendicular to the curves given by (4). The *characteristic curves* of (3) are $u = c_1$ and $v = c_2$.

378. LAPLACE'S EQUATION The general solution of

(5) $$\frac{\partial^2 \varphi}{\partial x^2} + \frac{\partial^2 \varphi}{\partial y^2} = 0$$

is of the form

(6) $$\varphi = f_1(x + iy) + f_2(x - iy) \qquad i^2 = -1$$

f_1 and f_2 being arbitrary functions. In practice this solution is too general, because of the difficulty of determining the functions so as to satisfy given boundary conditions. A function which satisfies (5) is a *harmonic function.*

379. A method which is quite useful depends upon assuming a particular solution to be a product of functions each of which contains only one of the variables. The combination of a number of such solutions often results in a sufficiently general solution. For example, suppose

(7) $$\varphi = X(x) \cdot Y(y)$$

is a solution of (5). Then $(X''/X) + (Y''/Y) = 0$. It follows that

(8) $$\frac{1}{X}\frac{d^2X}{dx^2} = -\omega^2 \qquad \frac{1}{Y}\frac{d^2Y}{dy^2} = \omega^2$$

where ω is a constant. The solutions to (8) are

(9) $$X = c_1 \cos \omega x + c_2 \sin \omega x \qquad Y = c_3 e^{\omega y} + c_4 e^{-\omega y}$$

where c_1, c_2, c_3, c_4 are arbitrary constants. Then (7) becomes

(10) $$\varphi \equiv \varphi(\omega) = e^{\omega y}(A_\omega \cos \omega x + B_\omega \sin \omega x) + e^{-\omega y}(C_\omega \cos \omega x + D_\omega \sin \omega x)$$

where A_ω, B_ω, C_ω, D_ω are arbitrary constants.

Generally, it may be assumed that a solution to (5) is

$$\varphi = \sum_{\omega=0}^{\infty} \varphi(\omega) \tag{11}$$

The constants in (11) must be determined so that (11) will satisfy the given (initial or boundary) conditions of the particular problem at hand.

19. FOURIER SERIES AND TRANSFORMS

380. FOURIER SERIES Let $f(t)$ be a real integrable function of the real variable t defined on $-T/2 \leqq t \leqq T/2$. The *Fourier series* generated by $f(t)$ is

$$(1) \qquad \frac{1}{2} a_0 + \sum_{n=1}^{\infty} (a_n \cos nwt + b_n \sin nwt)$$

where for $n = 0, 1, 2, \ldots,$

$$(2) \quad a_n = \frac{2}{T} \int_{-T/2}^{T/2} f(t) \cos nwt \, dt \qquad b_n = \frac{2}{T} \int_{-T/2}^{T/2} f(t) \sin nwt \, dt$$

Relations (2) are called *Euler's formulas*. The series (1) may not converge anywhere, or it may converge for only some values of t. Even if (1) converges for some value t_1, the sum of the series is not necessarily equal to $f(t_1)$. The *period* $T = 2\pi/w$. The variable t may be thought of as time (in seconds), w being the fundamental frequency (in radians per second).

Theorem. **(a)** If series (1) converges in some interval $\xi \leqq t < (\xi + T)$, then (1) represents a function of period T defined for all values of t.

(b) If (1) converges uniformly to $f(t)$ in $-T/2 \leqq t \leqq T/2$, a_n and b_n are given by (2).

(c) *Dirichlet's conditions:* If $f(t)$ is continuous except for a finite number of finite discontinuities in $-T/2 \leqq t \leqq T/2$ and if $f(t)$ has only a finite number of maxima and minima in the interval, the Fourier series generated by $f(t)$ converges to $\frac{1}{2}[f(t_0 + 0) + f(t_0 - 0)]$ for all values of t. At all points of continuity of $f(t)$, the series converges to $f(t)$.

(d) The Fourier series of $f(t)$ converges uniformly to $f(t)$ in any interval interior to an interval in which $f(t)$ is continuous and of bounded variation.*

* A function $f(t)$ is *bounded* in (a,b) if two constants h and K exist such that $h \leqq f(t) \leqq K$ for all t satisfying $a \leqq t \leqq b$. $f(t)$ is of *bounded variation* if it is the difference of two bounded increasing functions.

381. COMPLEX REPRESENTATION If the sum of (1) is $f(t)$, relation (1) can be written

(3) $$f(t) = \sum_{n=-\infty}^{\infty} F(n)e^{jnwt}$$

where

(4) $F(n) = \frac{1}{2}(a_n - jb_n)$ $\qquad n = 0, \pm 1, \pm 2, \ldots$

$j^2 = -1,\ a_{-n} = a_n,\ b_{-n} = -b_n$ $\qquad$ for $n = 1, 2, 3, \ldots$

(5) $F(n) = \dfrac{1}{T}\displaystyle\int_{-T/2}^{T/2} f(t)e^{-jnwt}\, dt$ $\qquad n = 0, \pm 1, \pm 2, \ldots$

$F(n)$ is the *Fourier transform* of $f(t)$. $F(n)$ is the *complex (line) spectrum* of $f(t)$.

382. Equation (3) may be thought of as a synthesis, Eq. (5) as an analysis. $f(t)$ is the *Fourier transform* of $F(n)$.

$$F(n) = \frac{1}{2}\sqrt{a_n^2 + b_n^2}\,\exp\left[j\tan^{-1}\left(\frac{-b_n}{a_n}\right)\right]$$

The *amplitude spectrum* of $f(t)$ is $|F(n)| = \frac{1}{2}\sqrt{a_n^2 + b_n^2}$

The *phase spectrum* of $f(t)$ is $\theta_n = \tan^{-1}\left(\dfrac{-b_n}{a_n}\right)$

$$f(t) = \sum_{n=-\infty}^{\infty} |F(n)|e^{j(nwt+\theta_n)}$$

383. The *auto-correlation function* of $f_1(t)$ is

$$\varphi_{11}(\tau) = \frac{1}{T}\int_{-T/2}^{T/2} f_1(t)f_1(t+\tau)\, dt = \sum_{n=-\infty}^{\infty} \Phi_{11}(n)e^{jnw\tau}$$

where $\Phi_{11}(n) = |F_1(n)|^2$ is the *power spectrum*.

$$\Phi_{11}(n) = \frac{1}{T}\int_{-T/2}^{T/2} \varphi_{11}(\tau)e^{-jnw\tau}\, d\tau$$

$$\frac{1}{T}\int_{-T/2}^{T/2} [f_1(t)]^2\,dt = \sum_{n=-\infty}^{\infty} |F_1(n)|^2 \qquad \text{(Parseval's formula)}$$

384. The *cross-correlation function* of two different periodic functions $f_1(t)$ and $f_2(t)$ of the same fundamental frequency is

$$\varphi_{12}(\tau) = \frac{1}{T}\int_{-T/2}^{T/2} f_1(t)f_2(t+\tau)\,dt$$

Here τ is a displacement in t.

385. The *convolution* of $f_1(t)$ and $f_2(t)$ is

$$\frac{1}{T}\int_{-T/2}^{T/2} f_1(t)f_2(t-\tau)\,dt$$

ABLE OF FINITE SINE TRANSFORMS*

The entries $f_s(n)$ give values of the finite sine transform $f_s(n)$ of the function $F(x)$.

$$f_s(n) = \int_0^{\pi} F(x) \sin nx \, dx \qquad n = 1, 2, \ldots$$

ere $F(x)$ is defined on the interval $0 < x < \pi$.

Example: If $F(x) = x/\pi$, then $f_s(n) = (-1)^{n+1}/n$, $n = 1, 2, \ldots$.

	$f_s(n)$		$F(x)$	
1.	$(-1)^{n+1} f_s(n)$		$F(\pi - x)$	
2.	$[1 - (-1)^n]/n$		1	
3.	$1/n$		$(\pi - x)/\pi$	
4.	$(-1)^{n+1}/n$		x/π	
5.	$\frac{2}{n^2} \sin \frac{n\pi}{2}$		$\begin{cases} x \\ \pi - x \end{cases}$	for $0 < x < \pi/2$ for $\pi/2 < x < \pi$
6.	$[1 - (-1)^n]/n^3$		$x(\pi - x)/2$	
7.	$(-1)^{n+1}/n^3$		$x(\pi^2 - x^2)/(6\pi)$	
8.	$[\pi^2(-1)^{n-1}/n] - 2[1 - (-1)^n]/n^3$		x^2	
9.	$\pi(-1)^n[(6/n^3) - (\pi^2/n)]$		x^3	
10.	$(\pi/n^2) \sin nc$	$0 < c < \pi$	$\begin{cases} (\pi - c)x \\ c(\pi - x) \end{cases}$	for $x \leqq c$ for $x \geqq c$
11.	$(\pi/n) \cos nc$	$0 \leqq c \leqq \pi$	$\begin{cases} -x \\ \pi - x \end{cases}$	for $x < c$ for $x > c$
12.	$n[1 - (-1)^n e^{c\pi}]/(n^2 + c^2)$		e^{cx}	
13.	$n/(n^2 + c^2)$		$\sinh c(\pi - x)/\sinh c\pi$	
14.	$n/(n^2 - k^2)$	$\lvert k \rvert \neq 0, 1, 2, \ldots$	$\sin k(\pi - x)/\sin k\pi$	
15.	0 if $n \neq m$	$\pi/2$ if $n = m$	$\sin mx$	$m = 1, 2, \ldots$
16.	$n[1 - (-1)^n \cos k\pi]/(n^2 - k^2)$		$\cos kx$	$\lvert k \rvert \neq 1, 2, \ldots$
17.	$\begin{cases} n[1 - (-1)^{n+m}]/(n^2 - m^2) & \text{if } n \neq m \\ 0 & \text{if } n = m \end{cases}$		$\cos mx$	$m = 1, 2, \ldots$
18.	$n/(n^2 - k^2)^2$	$\lvert k \rvert \neq 0, 1, 2, \ldots$	$\frac{\pi \sin kx}{2k \sin^2 k\pi} - \frac{x \cos k(\pi - x)}{2k \sin k\pi}$	
19.	b^n/n	$\lvert b \rvert \leqq 1$	$\frac{2}{\pi} \tan^{-1}\left[\frac{b \sin x}{(1 - b \cos x)}\right]$	
20.	$[1 - (-1)^n] b^n/n$	$\lvert b \rvert \leqq 1$	$\frac{2}{\pi} \tan^{-1}\left[\frac{2b \sin x}{(1 - b^2)}\right]$	

* c, k, and b are constants.

TABLE OF FINITE COSINE TRANSFORMS

The entries $f_c(n)$ give values of the finite cosine transform $f_c(n)$ of the function $F(x$

$$f_c(n) = \int_0^{\pi} F(x) \cos nx \, dx \qquad n = 0, 1, 2, \ldots$$

Here $F(x)$ is defined over the interval $0 < x < \pi$. c, k, and b are constants.

	$f_c(n)$		$F(x)$
1.	$(-1)^n f_c(n)$		$F(\pi - x)$
2.	$\begin{cases} 0 \\ \pi \end{cases}$	if $n = 1, 2, \ldots$ if $n = 0$	1
3.	$\begin{cases} (2/n) \sin (n\pi/2) \\ 0 \end{cases}$	if $n = 1, 2, \ldots$ if $n = 0$	$\begin{cases} 1 & \text{for } 0 < x < \pi/ \\ -1 & \text{for } \pi/2 < x < \end{cases}$
4.	$\begin{cases} (2/n) \sin nc \\ 2c - \pi \end{cases}$	if $n = 1, 2, \ldots$ if $n = 0$	$\begin{cases} 1 & \text{for } 0 < x < \\ -1 & \text{for } c < x < \end{cases}$
5.	$\begin{cases} [(-1)^n - 1]/n^2 \\ \pi^2/2 \end{cases}$	if $n \neq 0$ if $n = 0$	x
6.	$\begin{cases} (-1)^n/n^2 \\ \pi^2/6 \end{cases}$	if $n \neq 0$ if $n = 0$	$x^2/(2\pi)$
7.	$\begin{cases} 1/n^2 \\ 0 \end{cases}$	if $n \neq 0$ if $n = 0$	$[(\pi - x)^2/(2\pi)] - \pi/6$
8.	$\begin{cases} \dfrac{3\pi^2(-1)^n}{n^2} + \dfrac{6[1 - (-1)^n]}{n^4} \\ \pi^4/4 \end{cases}$	if $n \neq 0$ if $n = 0$	x^3
9.	$[(-1)^n e^{c\pi} - 1]/(n^2 + c^2)$		e^{cx}/c
10.	$\dfrac{(-1)^n \cos k\pi - 1}{n^2 - k^2}$	if $\lvert k \rvert \neq 0, 1, 2, \ldots$	$(\sin kx)/k$
11.	$\begin{cases} \dfrac{(-1)^{n+m} - 1}{n^2 - m^2} \\ 0 \end{cases}$	if $n \neq m$ if $n = m = 1, 2, \ldots$	$(\sin mx)/m$
12.	$1/(n^2 - k^2)$	if $\lvert k \rvert \neq 0, 1, 2, \ldots$	$-\cos k(\pi - x)/(k \sin k\pi)$
13.	$\begin{cases} 0 \\ \pi/2 \end{cases}$	if $n \neq m$ if $n = m = 1, 2, \ldots$	$\cos mx$
14.	$1/(n^2 + c^2)$		$\cosh c(\pi - x)/(c \sinh c\pi)$
15.	$\begin{cases} b^n \\ 0 \end{cases}$	if $n \neq 0$ if $n = 0, \lvert b \rvert < 1$	$\dfrac{2b(\cos x - b)}{\pi(1 - 2b \cos x + b^2)}$

20. LAPLACE TRANSFORMS

386. **LAPLACE TRANSFORM** The *Laplace transform* $\mathcal{L}(F)$ of a function $F(t)$ may be formally defined by

$$\mathcal{L}(F) = f(s) = \int_0^\infty e^{-st} F(t)\, dt \tag{1}$$

387. **DEFINITIONS** A real function $F(t)$ is *sectionally continuous* on a finite real interval $a \leqq t \leqq b$ if it is possible to subdivide the interval into a finite number of subintervals such that in each of which $F(t)$ is continuous and has finite limits as t approaches either end point of the subinterval from the interior.

Let $F(t)$ be a sectionally continuous function defined for all t such that $-\infty < t < \infty$, with $F(t) = 0$ for $t < 0$. $F(t)$ is of *exponential type*, i.e., of class E, if there exists a real number σ_F such that

$$\int_0^\infty |F(t)| e^{-\sigma_F t}\, dt \tag{2}$$

exists. In this case $F(t)$ is said to be of the *order of* $e^{\sigma_F t}$ or that $F(t)$ is $O(e^{\sigma_F t})$.

388. **PROPERTIES OF LAPLACE TRANSFORMS** In the following sections $F(t)$ is assumed to be real and to have continuous derivatives $F'(t), \ldots, F^{(n-1)}(t)$, $F^{(n)}(t)$ being sectionally continuous, which are all of class E of order $O(e^{\sigma_F t})$ in every finite interval $0 \leqq t \leqq T$. s and t are assumed to be real variables, and $s > \sigma_F$. [In a more general case, s and $f(s)$ may be complex.]

Theorem 1. **(a)** If $F(t)$ is of class E, the Laplace transform exists and the integral in (1) is uniformly convergent for $-\infty < \omega < \infty$.

(b) If c_1 and c_2 are complex numbers, then

$$\mathcal{L}(c_1 F_1 + c_2 F_2) = c_1 \mathcal{L}(F_1) + c_2 \mathcal{L}(F_2) \tag{3}$$

(c) The transform $f(s)$ in (1) is an analytic function for $\sigma > \sigma_F$; that is, all derivatives with respect to s of the complex function $f(s)$ exist if $\sigma > \sigma_F$,

$$\mathcal{L}\{F'(t)\} = s\mathcal{L}\{F(t)\} - F(+0) \qquad \text{for } s > \sigma_F \tag{4}$$

Here $F(+0)$ is the limit of $F(t)$ as t approaches zero through positive values.

(5) $\mathfrak{L}\{F^{(k)}(t)\} = s^k\mathfrak{L}(F) - s^{k-1}F(+0) - s^{k-2}F'(+0)$
$- s^{k-3}F''(+0) - \cdots - F^{(k-1)}(+0)$

for $s > \sigma_F$ and $k = 1, 2, \ldots, n$.

Theorem 2. Define

$$G(t) = D_0^{-1}F(t) = \int_0^t F(u)\,du \qquad \mathfrak{L}[F] = f(s)$$

$$D_0^{-2}(F) = D_0^{-1}\{D_0^{-1}(F)\} \qquad D_0^{-k}(F) = D_0^{-1}\{D_0^{-k+1}(F)\}$$

$$\mathfrak{L}[D_0^{-1}F] = \mathfrak{L}(G) = \frac{1}{s}\,\mathfrak{L}(F) \qquad \text{for } \sigma > \sigma_F$$

$$\mathfrak{L}[D_0^{-k}F] = \frac{1}{s^k}\,\mathfrak{L}(F)$$

$$\mathfrak{L}[F(t-a)] = e^{-as}\mathfrak{L}(F) = e^{-as}f(s) \qquad a > 0$$

[Here $F(t) = 0$ when $t < 0$.]

$$\mathfrak{L}[e^{bt}F] = f(s-b) \qquad \text{for } \sigma > \sigma_F + \text{Re } b$$

$$\mathfrak{L}[tF] = -f'(s)$$

$$\mathfrak{L}[t^nF] = (-1)^n\frac{d^n}{ds^n}\,\mathfrak{L}(F) = (-1)^n f^{(n)}(s)$$

$$\mathfrak{L}[t^{-1}F] = \int_s^\infty f(s)\,ds \qquad \text{(integration of transform)}$$

$$\mathfrak{L}\left[\frac{1}{c}\,F\left(\frac{t}{c}\right)\right] = f(cs) \qquad c > 0$$

$$\mathfrak{L}\left[\frac{1}{c}\,e^{bt/c}F\left(\frac{t}{c}\right)\right] = f(cs-b) \qquad c > 0$$

Theorem 3. **(a)** If $F(t+a) = F(t)$, that is, if $F(t)$ is *periodic*,

$$\mathfrak{L}[F(t)] = \frac{\int_0^a e^{-st}F(t)\,dt}{1 - e^{-as}} \qquad s > 0$$

(b) If $F(t+a) = -F(t)$, that is, if $F(t)$ is *antiperiodic*, then $F(t+2a) = F(t)$, and

$$\mathfrak{L}[F(t)] = \frac{\int_0^a e^{-st}F(t)\,dt}{1 + e^{-as}} \qquad s > 0$$

Theorem 4. **(a)** If $F(t)$ is antiperiodic, $f(s) = \mathfrak{L}[F(t)]$ and $F_1(t)$ is the *half-wave rectification** of $F(t)$,

$$\mathfrak{L}[F_1(t)] = \frac{f(s)}{1 - e^{-as}}$$

(b) If $F_2(t)$ is the *full-wave rectification** of $F(t)$, then

$$\mathcal{L}[F_2(t)] = f(s) \coth \frac{as}{2}$$

389. INVERSE LAPLACE TRANSFORMATION Let $\mathcal{L}\{F(t)\} = f(s)$. The *inverse Laplace transform* of $f(s)$ is $F(t) = \mathcal{L}^{-1}\{f(s)\}$.

The inverse Laplace transform is not unique. $f(s)$ determines $F(t)$ uniquely within class E functions, except at jump points. Stated another way, $f(s)$ uniquely determines $F(t)$ up to a *null function* $N(t)$, where $\int_0^t N(u)\,du = 0$ for every $t \geqq 0$. In other words, any two functions having the same transform can differ at most by a null function $N(t)$.

Theorem 5. Suppose $\mathcal{L}\{F(t)\} = f(s) = p(s)/q(s)$, where $p(s)$ and $q(s)$ are polynomials in s with no common factor and $q(s) = \prod_{i=1}^{m} (s - a_i)$, all a_i being distinct. If the degree of $p(s)$ is smaller than that of $q(s)$,

$$\mathcal{L}^{-1}[f(s)] = \mathcal{L}^{-1}\left[\frac{p(s)}{q(s)}\right] = \sum_{i=1}^{m} \frac{p(a_i)}{q'(a_i)} e^{a_i t}$$

390. CONVOLUTION Let $F(t)$ and $G(t)$ be sectionally continuous functions for $t \geqq 0$, F and G being identically zero for $t < 0$. The *convolution* (or *Faltung*) of F and G is

$$F * G = \int_0^t F(u)G(t-u)\,du$$

$$\mathcal{L}(F * G) = \mathcal{L}(F)\mathcal{L}(G) \qquad \text{for } \sigma > \sigma_0$$

Remark. In applications to differential equations the properties of transforms listed above are used to change an initial-value problem in t into an algebraic problem in the complex plane s. In general, the problem is then solved in the s plane, and a return is then made to the t plane by means of an inverse Laplace transformation.

* In half-wave rectification, the negative half-wave is eliminated entirely; in full-wave rectification, the negative half-wave is reflected relative to the t axis to the upper half of the (F,t) plane.

TABLE OF LAPLACE TRANSFORMS

Entries $f(s)$ give values of the Laplace transform $\mathcal{L}[F(t)] \equiv f(s)$ of the function $F(t)$. The entries in column $F(t)$ define $F(t)$ for $t \geqq 0$.

$$a,\ b,\ c,\ k \text{ are constants.}\quad f(s) = \int_0^\infty e^{-st}F(t)\,dt.$$

	$f(s) \equiv \mathcal{L}[F(t)]$		$F(t)$
1.	$1/s$		1
2.	$1/s^2$		t
3.	$1/s^n$	$n = 1, 2, \ldots$	$t^{n-1}/(n-1)!$
4.	$1/(s-a)$		e^{at}
5.	$1/(s-a)^2$		te^{at}
6.	$1/(s-a)^n$	$n = 1, 2, \ldots$	$t^{n-1}e^{at}/(n-1)!$
7.	$a/(s^2+a^2)$		$\sin at$
8.	$s/(s^2+a^2)$		$\cos at$
9.	$a/(s^2-a^2)$		$\sinh at$
10.	$s/(s^2-a^2)$		$\cosh at$
11.	$(s \sin b + a \cos b)/(s^2+a^2)$		$\sin (at+b)$
12.	$(s \cos b - a \sin b)/(s^2+a^2)$		$\cos (at+b)$
13.	$2as/(s^2+a^2)^2$		$t \sin at$
14.	$(s^2-a^2)/(s^2+a^2)^2$		$t \cos at$
15.	$1/\sqrt{s}$		$1/\sqrt{\pi t}$
16.	$s^{-3/2}$		$2\sqrt{t/\pi}$
17.	$s^{-[n+(1/2)]}$	$n = 1, 2, \ldots$	$2^n t^{n-(1/2)}/[1 \cdot 3 \cdot 5 \cdots (2n-1)\sqrt{\pi}]$
18.	$\Gamma(k)/s^k$	$k > 0$	t^{k-1}
19.	$\Gamma(k)/(s-a)^k$	$k > 0$	$t^{k-1}e^{at}$
20.*	$1/[(s-a)(s-b)]$		$(e^{at}-e^{bt})/(a-b)$
21.*	$s/[(s-a)(s-b)]$		$(ae^{at}-be^{bt})/(a-b)$
22.*	$1/[(s-a)(a-b)(s-c)]$		$[(b-c)e^{at}+(c-a)e^{bt} + (a-b)e^{ct}]/A$ where $A = -(a-b)(b-c)(c-a)$
23.	$1/[s(s^2+a^2)]$		$(1-\cos at)/a^2$
24.	$1/[s^2(s^2+a^2)]$		$(at-\sin at)/a^3$
25.	$1/(s^2+a^2)^2$		$(\sin at - at \cos at)/[2a^3]$

* Here a, b, and c are distinct constants.

TABLE OF LAPLACE TRANSFORMS *(continued)*

	$f(s) \equiv \mathfrak{L}[F(t)]$		$F(t)$
26.	$s/(s^2 + a^2)^2$		$(t \sin at)/[2a]$
27.	$s^2/(s^2 + a^2)^2$		$(\sin at + at \cos at)/[2a]$
28.	$s/[(s^2 + a^2)(s^2 + b^2)]$	$a^2 \neq b^2$	$(\cos at - \cos bt)/(b^2 - a^2)$
29.	$1/[(s - a)^2 + b^2]$		$(e^{at} \sin bt)/b$
30.	$(s - a)/[(s - a)^2 + b^2]$		$e^{at} \cos bt$
31.	$3a^2/(s^3 + a^3)$		$e^{-at} - e^{at/2} \cdot A$ where $A = \cos(\sqrt{3}\, at/2) - \sqrt{3} \sin(\sqrt{3}\, at/2)$
32.	$3as/(s^3 + a^3)$		$e^{-at} + e^{at/2} \cdot A$ where $A = \cos(\sqrt{3}\, at/2) + \sqrt{3} \sin(\sqrt{3}\, at/2)$
33.	$4a^3/(s^4 + 4a^4)$		$\sin at \cosh at - \cos at \sinh at$
34.	$s/(s^4 + 4a^4)$		$[1/(2a^2)] \sin at \sinh at$
35.	$1/(s^4 - a^4)$		$(\sinh at - \sin at)/[2a^3]$
36.	$s/(s^4 - a^4)$		$(\cosh at - \cos at)/[2a^2]$
37.	$8a^3s^2/(s^2 + a^2)^3$		$(1 + a^2t^2) \sin at - at \cos at$
38.*	$(1/s)[(s - 1)/s]^n$		$L_n(t) = (e^t/n!)[d^n(t^n e^{-t})/dt^n]$
39.	$\sqrt{s - a} - \sqrt{s - b}$		$(e^{bt} - e^{at})/[2\sqrt{\pi t^3}]$
40.	$s/(s - a)^{3/2}$		$e^{at}(1 + 2at)/\sqrt{\pi t}$
41.†	$1/(\sqrt{s} + a)$		$(1/\sqrt{\pi t}) - ae^{a^2 t} \operatorname{erfc}(a\sqrt{t})$
42.	$\sqrt{s}/(s + a^2)$		$(1/\sqrt{\pi t}) - (2ae^{-a^2 t}/\sqrt{\pi}) \int_0^{a\sqrt{t}} e^{\lambda^2}\, d\lambda$
43.†	$\sqrt{s}/(s - a^2)$		$(1/\sqrt{\pi t}) + ae^{a^2 t} \operatorname{erf}(a\sqrt{t})$
44.	$1/[\sqrt{s}(s + a^2)]$		$[2/(a\sqrt{\pi})]e^{-a^2 t} \int_0^{a\sqrt{t}} e^{\lambda^2}\, d\lambda$
45.	$1/[\sqrt{s}(s - a^2)]$		$(e^{a^2 t}/a) \operatorname{erf}(a\sqrt{t})$
46.	$1/[\sqrt{s}(\sqrt{s} + a)]$		$e^{a^2 t} \operatorname{erfc}(a\sqrt{t})$
47.	$(b^2 - a^2)/[(s - a^2)(\sqrt{s} + b)]$		$e^{a^2 t}[b - a \operatorname{erf}(a\sqrt{t})] - be^{b^2 t} \operatorname{erfc}(b\sqrt{t})$
48.	$(b^2 - a^2)/[\sqrt{s}(s - a^2)(\sqrt{s} + b)]$		$e^{a^2 t}[(b/a) \operatorname{erf}(a\sqrt{t}) - 1] + e^{b^2 t} \operatorname{erfc}(b\sqrt{t})$

* Here $L_n(t)$ is the *Laguerre polynomial* of degree n.

† $\operatorname{erf} y = \dfrac{2}{\sqrt{\pi}} \int_0^y e^{-u^2}\, du$; $\operatorname{erfc} y = 1 - \operatorname{erf} y$

TABLE OF LAPLACE TRANSFORMS (*continued*)

	$f(s) \equiv \mathfrak{L}[F(t)]$		$F(t)$
49.	$1/[(s+a)\sqrt{s+b}]$		$(1/\sqrt{b-a})e^{-at}\,\text{erf}\,(\sqrt{b-a}\sqrt{t})$
50.*	$1/[\sqrt{s+a}\sqrt{s+b}]$		$e^{-(a+b)t/2}\,I_0[(a-b)t/2]$
51.	$1/[(s+a)^{1/2}(s+b)^{3/2}]$		$te^{-(a+b)t/2}\cdot B$ where $B \equiv I_0[(a-b)t/2] + I_1[(a-b)t/2]$
52.	$\Gamma(k)/[(s+a)^k(s+b)^k]$	$k>0$	$\sqrt{\pi}C[t/(a-b)]^{k-(1/2)}e^{-(a+b)t/2}$ where $C \equiv I_{k-1/2}\,[(a-b)t/2]$
53.	$(\sqrt{s+2a}-\sqrt{s})/(\sqrt{s+2a}+\sqrt{s})$		$(1/t)e^{-at}I_1(at)$
54.†	$(1-s)^n/s^{[n+(1/2)]}$		$n!H_{2n}(\sqrt{t})/[(2n)!\sqrt{\pi t}]$
55.	$(1-s)^n/s^{[n+(3/2)]}$		$-n!H_{2n+1}(\sqrt{t})/[\sqrt{\pi}(2n+1)!]$
56.	$(\sqrt{s+2a}/\sqrt{s})-1$		$ae^{-at}[I_1(at)+I_0(at)]$
57.	$(\sqrt{s+a}+\sqrt{s})^{-2\nu}/(\sqrt{s}\sqrt{s+a})$	$\nu>-1$	$(1/a^\nu)e^{-at/2}I_\nu(at/2)$
58.	$(a-b)^k/(\sqrt{s+a}+\sqrt{s+b})^{2k}$	$k>0$	$(k/t)e^{-(a+b)t/2}I_k[(a-b)t/2]$
59.	$(s^2+a^2)^{-1/2}$		$J_0(at)$
60.	$(s^2+a^2)^{-k}$	$k>0$	$[\sqrt{\pi}/\Gamma(k)][t/2a]^{k-(1/2)}J_{k-1/2}(at)$
61.	$(s^2-a^2)^{-k}$	$k>0$	$[\sqrt{\pi}/\Gamma(k)][t/2a]^{k-(1/2)}I_{k-1/2}(at)$
62.	$(\sqrt{s^2+a^2}-s)^\nu/\sqrt{s^2+a^2}$	$\nu>-1$	$a^\nu J_\nu(at)$
63.	$(\sqrt{s^2+a^2}-s)^k$	$k>0$	$(ka^k/t)J_k(at)$
64.	$(s-\sqrt{s^2-a^2})^\nu/\sqrt{s^2-a^2}$	$\nu>-1$	$a^\nu I_\nu(at)$
65.	$s^{-1}e^{-k/s}$		$J_0(2\sqrt{kt})$
66.	$s^{-1/2}e^{-k/s}$		$(1/\sqrt{\pi t})\cos(2\sqrt{kt})$
67.	$s^{-1/2}e^{k/s}$		$(1/\sqrt{\pi t})\cosh(2\sqrt{kt})$
68.	$s^{-3/2}e^{-k/s}$		$(1/\sqrt{\pi k})\sin(2\sqrt{kt})$
69.	$s^{-3/2}e^{k/s}$		$(1/\sqrt{\pi k})\sinh(2\sqrt{kt})$
70.	$s^{-\mu}e^{-k/s}$	$\mu>0$	$(t/k)^{(\mu-1)/2}J_{\mu-1}(2\sqrt{kt})$
71.	$s^{-\mu}e^{k/s}$	$\mu>0$	$(t/k)^{(\mu-1)/2}I_{\mu-1}(2\sqrt{kt})$
72.	$e^{-k\sqrt{s}}$	$k>0$	$[k/(2\sqrt{\pi t^3})]\exp[-k^2/(4t)]$
73.	$(1/\sqrt{s})e^{-k\sqrt{s}}$	$k\geqq 0$	$(1/\sqrt{\pi t})\exp[-k^2/(4t)]$

* $I_n(x) \equiv i^{-n}J_n(ix)$, where J_n is *Bessel's function of the first kind.*
† $H_n(x) \equiv e^{x^2}[d^n(e^{-x^2})/dx^n]$ is the *Hermite polynomial.*

$f(s) \equiv \mathfrak{L}[F(t)]$		$F(t)$
$s^{-1}e^{-k\sqrt{s}}$	$k \geqq 0$	$\operatorname{erfc}[k/(2\sqrt{t})]$
$s^{-3/2}e^{-k\sqrt{s}}$	$k \geqq 0$	$2\sqrt{t/\pi}\exp[-k^2/(4t)] - k\operatorname{erfc}[k/(2\sqrt{t})]$
$ae^{-k\sqrt{s}}/[s(a+\sqrt{s})]$	$k \geqq 0$	$-e^{ak}e^{a^2t}\operatorname{erfc}A + \operatorname{erfc}[k/(2\sqrt{t})]$ where $A = a\sqrt{t} + k/(2\sqrt{t})$
$e^{-k\sqrt{s}}/[\sqrt{s}(a+\sqrt{s})]$	$k \geqq 0$	$e^{ak}e^{a^2t}\operatorname{erfc}A$
$e^{-k\sqrt{s(s+a)}}/\sqrt{s(s+a)}$		$\begin{cases} 0 & \text{when } 0<t<k \\ e^{-at/2}I_0(\frac{1}{2}a\sqrt{t^2-k^2}) & \text{when } t>k \end{cases}$
$e^{-k\sqrt{s^2+a^2}}/\sqrt{s^2+a^2}$		$\begin{cases} 0 & \text{when } 0<t<k \\ J_0(a\sqrt{t^2-k^2}) & \text{when } t>k \end{cases}$
$e^{-k\sqrt{s^2-a^2}}/\sqrt{s^2-a^2}$		$\begin{cases} 0 & \text{when } 0<t<k \\ I_0(a\sqrt{t^2-k^2}) & \text{when } t>k \end{cases}$
$e^{-k(\sqrt{s^2+a^2}-s)}/\sqrt{s^2+a^2}$	$k \geqq 0$	$J_0(a\sqrt{t^2+2kt})$
$e^{-ks} - e^{-k\sqrt{s^2+a^2}}$		$\begin{cases} 0 & \text{when } 0<t<k \\ (ak/\sqrt{t^2-k^2})J_1(a\sqrt{t^2-k^2}) & \text{when } t>k \end{cases}$
$e^{-k\sqrt{s^2-a^2}} - e^{-ks}$		$\begin{cases} 0 & \text{when } 0<t<k \\ (ak/\sqrt{t^2-k^2})I_1(a\sqrt{t^2-k^2}) & \text{when } t>k \end{cases}$
$[a^\nu \exp(-k\sqrt{s^2+a^2})]/[\sqrt{s^2+a^2}(\sqrt{s^2+a^2}+s)^\nu]$	$\nu > -1$	$\begin{cases} 0 & \text{when } 0<t<k \\ [(t-k)/(t+k)]^{\nu/2}J_\nu(a\sqrt{t^2-k^2}) & t>k \end{cases}$
* $(1/s)\log s$		$\Gamma'(1) - \log t,\ \Gamma'(1) = -0.5772157$
* $s^{-k}\log s$	$k > 0$	$t^{k-1}[\Gamma'(k) - \Gamma(k)\log t][\Gamma(k)]^{-2}$
* $(s-a)^{-1}\log s$	$a > 0$	$e^{at}[\log a - \operatorname{Ei}(-at)]$
* $(s^2+1)^{-1}\log s$		$\cos t\operatorname{Si} t - \sin t\operatorname{Ci} t \equiv H(t)$
* $s(s^2+1)^{-1}\log s$		$-\sin t\operatorname{Si} t - \cos t\operatorname{Ci} t$
* $s^{-1}\log(1+ks)$	$k > 0$	$-\operatorname{Ei}(-t/k)$

* In this table $\log s \equiv \log_e s \equiv \ln s$

$\operatorname{Ei} t = \int_{-\infty}^{t} r^{-1}e^{r}\,dr, \quad t<0$ (the *exponential-integral* function)

$\operatorname{Ei}(-t) = \int_{1}^{\infty} x^{-1}e^{-tx}\,dx, \quad t>0$

$\operatorname{Ci} t = -\int_{t}^{\infty} r^{-1}\cos r\,dr = -\int_{1}^{\infty} x^{-1}\cos tx\,dx, \quad t>0$ (the *cosine-integral* function)

$\operatorname{Si} t = \int_{0}^{t} r^{-1}\sin r\,dr, \quad t>0$ (the *sine-integral* function)

$H(t) = \cos t\operatorname{Si} t - \sin t\operatorname{Ci} t, \quad t>0$

TABLE OF LAPLACE TRANSFORMS (*continued*)

	$f(s) \equiv \mathfrak{L}[F(t)]$		$F(t)$	
91.	$\log[(s-a)/(s-b)]$		$t^{-1}(e^{bt}-e^{at})$	
92.	$\log[(s+a)/(s-a)]$		$2t^{-1}\sinh at$	$\text{Re}\, s > \text{Re}$
93.	$s^{-1}\log(1+k^2s^2)$		$-2\,\text{Ci}\,(t/k)$	
94.	$s^{-1}\log(s^2+a^2)$	$a>0$	$2\log a - 2\,\text{Ci}\,(at)$	
95.	$s^{-2}\log(s^2+a^2)$	$a>0$	$2a^{-1}[at\log a + \sin at - at\,\text{Ci}\,at]$	
96.	$\log[(s^2+a^2)/s^2]$		$2t^{-1}(1-\cos at)$	
97.	$\log[(s^2-a^2)/s^2]$		$2t^{-1}(1-\cosh at)$	
98.	$\cot^{-1}[(s-b)/a]$		$t^{-1}e^{bt}\sin at$	
99.	$\cot^{-1}(s/k)$		$t^{-1}\sin kt$	
100.	$(1/s)\cot^{-1}(s/k)$		$\text{Si}\,kt$	
101.	$e^{k^2s^2}\,\text{erfc}\,(ks)$	$k>0$	$[1/(k\sqrt{\pi})]\exp[-t^2/(4k^2)]$	
102.	$s^{-1}e^{k^2s^2}\,\text{erfc}\,(ks)$	$k>0$	$\text{erf}\,[t/(2k)]$	
103.	$e^{ks}\,\text{erfc}\,\sqrt{ks}$	$k>0$	$\sqrt{k}/[\pi\sqrt{t}(t+k)]$	
104.	$s^{-1/2}\,\text{erfc}\,\sqrt{ks}$		$\begin{cases}0\\(\pi t)^{-1/2}\end{cases}$	when $0<t<$ when $t>$
105.	$s^{-1/2}e^{ks}\,\text{erfc}\,\sqrt{ks}$	$k>0$	$1/\sqrt{\pi(t+k)}$	
106.	$s^{-1/2}e^{k^2/s}\,\text{erfc}\,(k/\sqrt{s})$		$(1/\sqrt{\pi t})e^{-2k\sqrt{t}}$	
107.	$\text{erf}\,(k/\sqrt{s})$		$[1/(\pi t)]\sin(2k\sqrt{t})$	
108.*	$K_0(ks)$		$\begin{cases}0\\(t^2-k^2)^{-1/2}\end{cases}$	when $0<t<$ when $t>$
109.	$K_0(k\sqrt{s})$		$[1/(2t)]\exp[-k^2/(4t)]$	
110.	$s^{-1}e^{ks}K_1(ks)$		$(1/k)\sqrt{t(t+2k)}$	
111.	$s^{-1/2}K_1(k\sqrt{s})$		$(1/k)\exp[-k^2/(4t)]$	
112.	$s^{-1/2}e^{k/s}K_0(k/s)$		$(2/\sqrt{\pi t})K_0(2\sqrt{2kt})$	
113.	$\pi e^{-ks}I_0(ks)$		$\begin{cases}[t(2k-t)]^{-1/2}\\0\end{cases}$	when $0<t<2$ when $t>2$
114.	$e^{-ks}I_1(ks)$		$\begin{cases}(k-t)/[\pi k\sqrt{t(2k-t)}]\\0\end{cases}$	when $0<t<2$ when $t>2$
115.	$-e^{as}\,\text{Ei}\,(-as)$		$(t+a)^{-1}$	$a>$
116.	$a^{-1}+se^{as}\,\text{Ei}\,(-as)$		$(t+a)^{-2}$	$a>$

* $K_n(x)$ is *Bessel's function of the second kind* for the imaginary argument.

BLE OF LAPLACE TRANSFORMS *(continued)*

	$f(s) \equiv \mathcal{L}[F(t)]$	$F(t)$
7.	$[(\pi/2) - \text{Si}\, s] \cos s + \text{Ci}\, s \sin s$ $\quad s > 0$	$(t^2 + 1)^{-1}$
		Step functions
8.	s^{-1}	$U(t) = \begin{cases} 0 & \text{when } t < 0 \\ 1 & \text{when } t \geqq 0 \end{cases}$ *(Heaviside unit function)*
9.	$s^{-1}e^{-ks}$	$U(t - k) = \begin{cases} 0 & \text{when } t < k \\ 1 & \text{when } t \geqq k \end{cases}$
0.*	1	$\delta(t)$ *(Dirac delta* or *impulse function)*
1.	e^{-as} $\quad a \geqq 0$	$\delta(t - a)$
2.	$s^k e^{-as}$ $\quad a \geqq 0$	$\delta^{(k)}(t - a)$ $\quad k = 1, 2, \ldots$
3.	$s^{-2}e^{-ks}$	$S_k(t) = \begin{cases} 0 & \text{when } t < k \\ t - k & \text{when } t \geqq k \end{cases}$
4.	$2s^{-3}e^{-ks}$	$\begin{cases} 0 & \text{when } t < k \\ (t - k)^2 & \text{when } t \geqq k \end{cases}$
5.	$\Gamma(\mu) \cdot s^{-\mu}e^{-ks}$ $\quad \mu > 0$	$\begin{cases} 0 & \text{when } t < k \\ (t - k)^{\mu - 1} & \text{when } t \geqq k \end{cases}$
6.	$s^{-1}(1 - e^{-ks})$	$\begin{cases} 1 & \text{when } 0 < t < k \\ 0 & \text{when } t > k \end{cases}$
7.†	$s^{-1}(1 - e^{-ks})^{-1}$ $= (2s)^{-1}(1 + \coth \frac{1}{2}ks)$	$1 + [t/k] = n$ when $(n - 1)k < t < nk, \quad n = 1, 2, \ldots$
8.	$s^{-1} \tanh ks$	$M(2k,t) = (-1)^{n-1}$ when $2k(n - 1) < t < 2kn$ $\quad n = 1, 2, \ldots$ *(Square-wave function)*
9.	$s^{-1}(1 + e^{-ks})^{-1}$	$[M(k,t) + 1]/2 = [1 - (-1)^n]/2$ when $(n - 1)k < t < nk, \quad n = 1, 2, \ldots$
0.	$(s \sinh ks)^{-1}$	$F(t) = 2(n - 1)$ when $(2n - 3)k < t < (2n - 1)k$ $\quad n = 1, 2, \ldots, \quad t > 0$
1.	$s^{-1} \coth ks$	$F(t) = 2n - 1$ when $2k(n - 1) < t < 2kn$ $\quad n = 1, 2, \ldots$

* See Sec. 148, formulas 439 and 442, for definitions of $\delta(t)$ and $\delta^{(k)}(t - a)$.

† When $t > 0$, $[t]$ denotes the greatest integer (0, 1, 2, . . .) that does not exceed the ımber t. For example, $[\pi] = 3 = [3]$.

TABLE OF LAPLACE TRANSFORMS (*continued*)

	$f(s) \equiv \mathcal{L}[F(t)]$	$F(t)$
132.	$(s \cosh ks)^{-1}$	$M(2k,t + 3k) + 1 = 1 + (-1)^n$ when $(2n - 3)k < t < (2n - 1)$ $n = 1, 2, \ldots, \quad t >$
133.	$s^{-1}(e^{-as} - e^{-bs})$	$F(t) = \begin{cases} 0 & \text{for } 0 < t < \\ 1 & \text{for } a < t < \\ 0 & \text{for } t > \end{cases}$
134.	$(m/s^2) - (ma/2s)[\coth (as/2) - 1]$	$F(t) = m(t - na)$ when $na < t < (n + 1)$ $n = 0, 1, 2, .$
135.	$s^{-2} \tanh (cs/2)$	$H(c,t) = \begin{cases} t & \text{when } 0 < t < \\ 2c - t & \text{when } c < t < \end{cases}$ $H(c,t + 2nc) = H(c,t) \qquad n = 1, 2, ..$ (*triangular wav*
136.	$k(s^2 + k^2)^{-1} \coth [\pi s/(2k)]$	$\lvert\sin kt\rvert$ (*full-wave rectification* of sin k
137.	$[(s^2 + 1)(1 - e^{-\pi s})]^{-1}$	$\begin{cases} \sin t & \text{when } (2n - 2)\pi < t < (2n - 1) \\ 0 & \text{when } (2n - 1)\pi < t < 2n\pi \end{cases}$ $n = 1, 2, ..$ (*half-wave rectification* of sin
138.	$[(E/s) + (m/s^2)]e^{-as}$	$F(t) = \begin{cases} 0 & \text{for } 0 < t < \\ E + m(t - a) & \text{for } t > \end{cases}$
139.	$(m/s^2)(1 - e^{-as})$	$F(t) = \begin{cases} mt & \text{for } 0 < t < \\ ma & \text{for } t > \end{cases}$
140.	$(m/s^2)(1 - 2e^{-as} + e^{-2as})$ $= (m/s^2)(1 - e^{-as})^2$	$F(t) = \begin{cases} mt & \text{for } 0 < t < \\ -m(t - 2a) & \text{for } a < t < 2 \\ 0 & \text{for } t > 2 \end{cases}$
141.	$(m/s^2)[1 - (1 + as)e^{-as}]$	$F(t) = \begin{cases} mt & \text{for } 0 < t < \\ 0 & \text{for } t > \end{cases}$

21. FUNCTIONS OF A COMPLEX VARIABLE

391. COMPLEX NUMBERS A *complex number* z is an ordered pair of real numbers (x,y) that satisfies certain rules of operation. z is written as $z = (x,y)$ or in the form $z = x + iy = x + yi$, where i is the *imaginary unit* and $i^2 = -1$. The real part x and the imaginary part y are written $\mathcal{R}(z) = x$ and $\mathcal{I}(z) = y$.

The rules of operation with complex numbers can be written formally by using the rules of ordinary algebra for real binomials to $x + iy$ and replacing i^2 by -1. (See Secs. 1 to 7, and 223 to 226.)

The number $z = x + iy$ may be represented by the point (x,y) or by the vector whose x component is x and whose y component is y (Fig. 67). The length of the vector $z = x + iy$ is the *absolute value* or the *modulus* of z and is written $|z| = \sqrt{x^2 + y^2} = \sqrt{z\bar{z}}$, where $\bar{z} = x - iy$ is the *complex conjugate* of z. $z\bar{z} = |z|^2$.

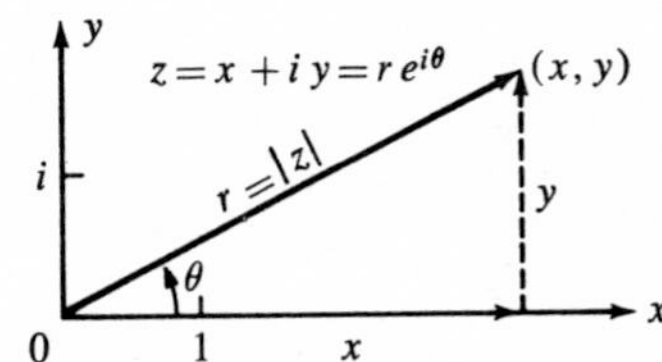

Fig. 67.

If (r,θ) are the polar coordinates of (x,y), with θ being in radians, then $|z| = r$ and

$$z = x + iy = r(\cos\theta + i\sin\theta) \equiv re^{i\theta} \tag{1}$$

The *amplitude* or *argument* of z is the angle θ and is written amp z. The amplitude θ is many-valued. The *principal value* of θ is that value for which $-\pi < \theta \leqq \pi$. The general value is $\theta + 2k\pi$, where k is an integer.

The sum $z_1 + z_2$ may be represented by the vector sum of z_1 and z_2 (Fig. 68).

$$|z_1 z_2| = |z_1||z_2| \qquad \left|\frac{z_1}{z_2}\right| = \frac{|z_1|}{|z_2|} \qquad |z^m| = |z|^m$$

$$|z_1 \pm z_2| \leqq |z_1| + |z_2| \qquad |z_1 \pm z_2| \geqq \big||z_1| - |z_2|\big|$$

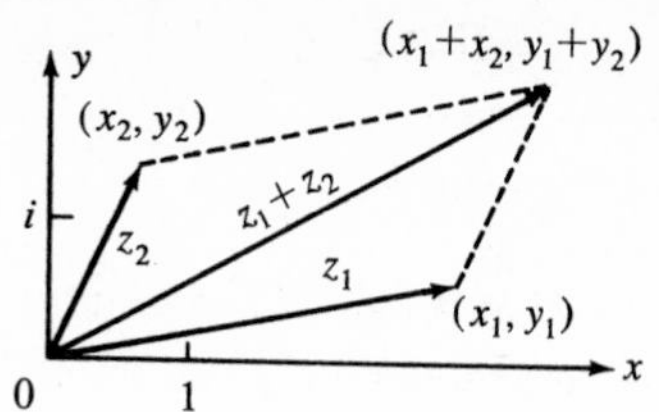

Fig. 68.

392. COMPLEX FUNCTIONS Let $z = x + iy$ be a complex variable, with $i^2 = -1$. Associate z with the point (x,y) in the complex z plane. If the complex variable $w \equiv f(z)$ has associated with it one or more values for each value of z in some subset D of all complex numbers, then w is a *complex function* of z. $w \equiv f(z)$ can be written $f(z) = u + iv$, where $u \equiv u(x,y)$ and $v \equiv v(x,y)$ are real functions of x and y.

Associate w with the point (u,v) in the complex w plane. If to each z in D there corresponds exactly one value of w, then w is a *single-valued* complex function of z.

If $f(z)$ is single-valued, if w_0 and z_0 are complex numbers, and if for each preassigned positive number ϵ there exists a positive number δ such that $|f(z) - w_0| < \epsilon$ for all z for which $0 < |z - z_0| < \delta$, then w_0 is the *limit* of $f(z)$ as $z \to z_0$ and is denoted by $\lim_{z \to z_0} f(z)$.

A single-valued function $f(z)$ is *continuous* at $z = z_0$ when and only when $f(z_0)$ exists and when for each $\epsilon > 0$ there exists a $\delta > 0$ such that

(2) $$|f(z) - f(z_0)| < \epsilon \text{ when } |z - z_0| < \delta$$

A *neighborhood* of the point z_0 is the aggregate of points z for which $|z - z_0| < \epsilon$, where ϵ is some positive number.

If for any $\epsilon > 0$ there exists a $\delta > 0$ independent of the value of z_0 in D for which (2) holds, then $f(z)$ is *uniformly continuous* over D.

393. DERIVATIVE The *derivative* of the single-valued function $f(z)$ with respect to z is defined as that function $f'(z)$ of z such that

(3) $$f'(z_0) = \lim_{h \to 0} \frac{f(z_0 + h) - f(z_0)}{h}$$

at all points z_0 where the (unique) limit exists and such that $f'(z)$ is defined for no other values of z. (The approach $h \to 0$ may be made through any complex values.)

A single-valued function $f(z)$ is *analytic* throughout a region D if it has a derivative of each point in D. (*Regular, holomorphic,* or *monogenic* are used by some writers instead of analytic.)

394. A necessary condition that $f(z) = u(x,y) + iv(x,y)$ be analytic at $z_0 = x_0 + iy_0$ is that, at z_0, both u and v satisfy the *Cauchy-Riemann* differential *equations*

$$\frac{\partial u}{\partial x} = \frac{\partial v}{\partial y} \qquad \frac{\partial u}{\partial y} = -\frac{\partial v}{\partial x} \tag{4}$$

Then

$$f'(z) = u_x + iv_x = u_x - iu_y = v_y + iv_x = v_y - iu_y \tag{3'}$$

Here $u_x = \partial u/\partial x$, $u_y = \partial u/\partial y$, ...

If each of the derivatives in (4) is continuous at all points of a region D, then a necessary and sufficient condition for $f(z) = u + iv$ to be analytic over D is that (4) hold.

A necessary condition for $u + iv$ to be analytic in D is that u and v satisfy Laplace's equation in D; that is,

$$\nabla^2 u = \frac{\partial^2 u}{\partial x^2} + \frac{\partial^2 u}{\partial y^2} = 0 \qquad \nabla^2 v = \frac{\partial^2 v}{\partial x^2} + \frac{\partial^2 v}{\partial y^2} = 0 \tag{5}$$

395. DEFINITE COMPLEX INTEGRAL Consider a continuous curve C connecting points A and B. Let $z_0, z_1, \ldots, z_n$ be n distinct points on C, with z_0 being at A and z_n at B. Let $f(z) = u(x,y) + iv(x,y)$ be single-valued and continuous everywhere on C. Consider

$$S_n = \sum_{k=1}^{n} f(\zeta_k) \cdot (z_k - z_{k-1}) \tag{6}$$

where ζ_k is any point on C between z_{k-1} and z_k. Let δ be the largest of $|z_k - z_{k-1}|$. The *complex line integral* (if it exists) of $f(z)$ along C is

$$\lim_{\delta \to 0} S_n \equiv \int_C f(z)\, dz \tag{7}$$

If C is defined by $x = \alpha(t)$ and $y = \beta(t)$, where α and β are differentiable real functions of the real variable t, then

$$(8)\quad \int_C f(z)\,dz = \int_{z_A}^{z_B} f(z)\,dz = \int_{t_A}^{t_B} (u + iv)[\alpha'(t) + i\beta'(t)]\,dt$$

Here $dz = dx + i\,dy$ and z_A and z_B are the values of z at A and B, respectively, and t_A and t_B the corresponding values of t.

396. CONTOURS Let $x = \alpha(t)$ and $y = \beta(t)$ be continuous over $a \leqq t \leqq b$ and suppose that $\alpha(t)$ and $\beta(t)$ do not assume the same pair of values for any two different values of t in $a < t < b$; then the set of points (x,y) corresponding to these values of t is called a *simple curve* or *simple contour* C. C is *closed* if $x(a) = x(b)$ and $y(a) = y(b)$.

397. CAUCHY'S INTEGRAL THEOREM If $f(z)$ is analytic everywhere within and on a simple closed curve C, then

$$\int_C f(z)\,dz = 0$$

If $f(z)$ is analytic throughout a *simply connected region* D and the path of integration lies entirely in D, then

$\int_{z_A}^{z_B} f(z)\,dz$ is independent of the path, and

$\int_{z_A}^{Z} f(z)\,dz \equiv F(Z)$ is analytic in D and $F'(Z) = f(Z)$.

$F(z) + C$, where C is a constant, is the *indefinite integral* of $f(z)$ with respect to z.

398. CAUCHY'S INTEGRAL FORMULA If $f(z)$ is analytic inside and on a simple closed curve C and if z is interior to C, then

$$(9)\qquad f(z) = \frac{1}{2\pi i}\int_C \frac{f(w)}{w - z}\,dw$$

Moreover, the derivatives $f'(z), f''(z), \ldots$ of all orders exist, and

$$(10)\quad f^{(n)}(z) = \frac{n!}{2\pi i}\int_C \frac{f(w)}{(w - z)^{n+1}}\,dw \qquad n = 1, 2, \ldots$$

399. TAYLOR'S THEOREM If $f(z)$ is analytic throughout the interior of a region D bounded by the simple closed curve C and if z and a are interior to D, then

$$(11)\quad f(z) \equiv f(a) + f'(a)(z-a) + \frac{f''(a)}{2!}(z-a)^2 + \cdots + \frac{f^{(n-1)}(a)}{(n-1)!}(z-a)^{n-1} + R_n$$

where

$$R_n = (z-a)^n P_n(z), \quad P_n(z) = \frac{1}{2\pi i}\int_C \frac{f(\omega)}{(\omega-a)^n(\omega-z)}\,d\omega$$

$P_n(z)$ is analytic throughout the interior of D.

400. TAYLOR'S SERIES The infinite series

$$(12)\quad f(z) = f(a) + f'(a)(z-a) + \frac{f''(a)}{2!}(z-a)^2 + \cdots + \frac{f^{(n)}(a)}{n!}(z-a)^n + \cdots$$

is called a *Taylor series*. When $a = 0$, (12) is a *Maclaurin series*.

401. CONSEQUENCES **(a)** Series (12) converges and is equal to $f(z)$ at every point z within any circle C' in whose interior $f(z)$ is analytic throughout.

(b) The *radius of convergence* R of (12) is the distance d from a to the nearest singular point of $f(z)$; that is, (12) converges for all z such that $|z-a| < R = d$.

(c) All derivatives $f^{(n)}(z)$, $n = 1, 2, \ldots$, of $f(z)$ can be found by differentiating term by term the right side of (12), and each of the series so obtained has R for its radius of convergence and converges uniformly in $|z-a| \leqq R_1$ where $R_1 < R$.

(d) If $|z-a| < d$, then $R_n \to 0$ as $n \to \infty$ and the series (12) converges absolutely.

(e) If R is the radius of convergence of (12) and if $|f(z)|$ never exceeds M along the circumference of the circle $|z-a| = r$, where $0 < r < R$, then $|f^{(n)}(a)| \leqq n!M/r^n$ $(n = 0, 1, 2, \ldots)$. (*Cauchy's inequality*.)

(*f*) If $f(z)$ is analytic within the circle $|z - a| < r$, then $|f(z)|$ at a point z interior to the circle never exceeds the maximum M of $|f(z)|$ on the circle's boundary.

(*g*) A necessary and sufficient condition for a function $f(z)$ to be expressible in a power series of form (12) is that it be analytic in some region.

(*h*) Suppose $f(x)$ is a real function of the real variable x, representable by a real Taylor series $\mathcal{S}(x) = \sum_{n=0}^{\infty} a_n(x - a)^n$ that is convergent within the interval $(a - R) < x < (a + R)$. The series $\mathcal{S}(z) = \sum_{n=0} a_n(z - a)^n$, obtained by replacing x by z in $\mathcal{S}(x)$, converges throughout circle C of radius R about a, and $\mathcal{S}(z)$ represents a function $f(z)$ analytic everywhere within C. When $z = x$, $f(z)$ equals $f(x)$. $f(z)$ is the only function that is analytic throughout C and that coincides with $f(x)$ on an interval of the x axis about a.

402. LIOUVILLE'S THEOREM A function $f(z)$, which is analytic at all finite points z and is bounded [e.g., $|f(z)| \leqq M$, where M is a constant, for all z], is a constant.

403. ZEROS OF ANALYTIC FUNCTIONS An analytic function $f(z)$ is said to have a *zero of order m* at $z = a$ if

$$f(a) = f'(a) = \cdots = f^{(m-1)}(a) = 0 \qquad f^{(m)}(a) \neq 0$$

404. LAURENT'S THEOREM If $f(z)$ is analytic within and on the boundaries of annulus T, which is bounded by two concentric circles C_1 and C_2 and has its center at $z = a$, then at any point z of T,

$$f(z) = \sum_{n=-\infty}^{n=+\infty} a_n(z - a)^n \tag{13}$$

where

$$a_n = \frac{1}{2\pi i} \int_\Gamma \frac{f(\omega)\, d\omega}{(\omega - a)^{n+1}} \tag{14}$$

and Γ is any circle within T.

405. SINGULAR POINTS z_0 is a *singular point* of $f(z)$ if $f(z)$ does not have a derivative at $z = z_0$, or if within every neighborhood of z_0 there are points z_1 distinct from z_0 at which $f(z)$ has no derivative.

If z_0 is a point about which there exists a neighborhood throughout which $f(z)$ is analytic, except at z_0, then z_0 is an *isolated singular point* of $f(z)$.

A singular point a is *removable* for $f(z)$ if $f(z)$ can be made analytic at a by redefining the value of $f(z)$ at $z = a$. $f(z)$ has a *pole* at $z = a$ if $f(z)$ is singular at a and the function $1/f(z)$ is analytic at a. If $z = a$ is a singular point for $f(z)$ which is neither removable nor a pole, then $f(z)$ has an *essential singularity* at $z = a$.

406. FURTHER PROPERTIES By Laurent's theorem a function $f(z)$, analytic within an annular region of outer circle C_2 and inner circle C_1, here taken to be arbitrarily small, is representable by

$$(15)\quad f(z) = \sum_{n=0}^{\infty} a_n(z-a)^n + \sum_{n=1}^{\infty} b_n(z-a)^{-n} \qquad 0 < |z-a| < R$$

(a) If all $b_n = 0$, the first series in (15) is analytic and represents $f(z)$ inside C_2, that is, for $|z - a| < R$, except possibly at $z = a$.

(b) If the second series in (15) contains only a finite number of terms, the last nonzero coefficient being b_m, with $b_n = 0$ for $n > m$, then $f(z)$ has a *pole of order* m at $z = a$. When $m = 1$, the pole is *simple*. The finite series $\sum_{n=1}^{m} b_n(z-a)^{-n}$ is called the *principal part* of $f(z)$ at $z = a$.

(c) If the second series in (15) does not terminate, it converges for all values of z except $z = a$ and the point $z = a$ is an *essential singularity* of $f(z)$.

407. SINGULARITIES AT INFINITY A function $f(z)$ has a *pole of order m* at infinity, that is, at $z = \infty$, if the function $f(1/\omega) \equiv \varphi(\omega)$ has a pole of order m at $\omega = 0$.

$f(z)$ has an essential singularity at $z = \infty$ if $\varphi(\omega)$ has an essential singularity at $\omega = 0$.

Theorems. **(a)** A function $f(z)$ analytic everywhere, including $z = \infty$, is a constant.

(*b*) If $f(z)$ has no singularities other than poles, it is a *rational function* (e.g., a function expressible as the ratio of two polynomials).

(*c*) A rational function has no singularities other than poles.

408. INTEGRAL FUNCTION A function $f(z)$ which is analytic everywhere except at $z = \infty$ is an *integral function.*

409. RESIDUES The coefficient b_1 in (15), that is, the coefficient of $1/(z - a)$, is the *residue* of $f(z)$ *relative to* $z = a$.

The residue of $f(z)$ relative to $z = \infty$ is defined to be $-b_1$.

Theorems. **(*a*)** In the neighborhood of an isolated singularity at $z = a$, an analytic function $f(z)$ may be expanded in form (15), and the residue of $f(z)$ relative to $z = a$ is

$$b_1 = \frac{1}{2\pi i} \int_C f(z)\, dz = 0 \tag{16}$$

where C is a circle with center at $z = a$, containing no other singularity of $f(z)$.

(*b*) If $z = a$ is a simple pole of $f(z)$, the residue relative to $z = a$ is $b_1 = \lim_{z \to a} (z - a) f(z)$.

(*c*) If $z = a$ is a pole of order m, the residue is

$$b_1 = \frac{1}{(m-1)!} \frac{d^{m-1}}{dz^{m-1}} [(z - a)^m f(z)]_{z=a} \tag{17}$$

(*d*) If $f(z)$ is analytic everywhere within and on a simple closed curve C except at a finite number of singularities $z_1, z_2, \ldots, z_n$ located in the interior of C, then

$$\int_C f(z)\, dz = 2\pi i \left(\sum_{i=1}^{n} R_i \right) \tag{18}$$

where R_i is the residue of $f(z)$ relative to z_i.

410. CONFORMAL MAPPING $w = f(z)$ defines a correspondence between the points of the w plane and the z plane. If z ranges over some set S_z of points in the z plane, w ranges over some other set S_w in the w plane. S_z is *mapped* upon set S_w.

The mapping is *biunique* (*one-to-one*) if $w = f(z)$ gives but one value of w for each value of z in S_z and if there exists an *inverse* function $z = F(w)$ which gives but one value of z in S_z for each w in S_w.

Theorem. A sufficient condition that $w = f(z)$ define a biunique (one-to-one) correspondence between the points of S_z and S_w is that $f(z)$ be analytic and $f'(z) \neq 0$ over S_z.

Let C_1 and C_2 be any two curves in the z plane which intersect at P_0 with angle α. Suppose $w = f(z)$ maps C_1 and C_2 into curves Γ_1 and Γ_2 on the w plane, carrying point P_0 into point Q_0 in the w plane. If the angle between Γ_1 and Γ_2 at Q_0 is also α, then the mapping is *isogonal.* If the sense of rotation of the tangents is preserved under the mapping, the transformation is *conformal.*

Theorem. The transformation $w = f(z)$ is conformal when $f(z)$ is analytic and $f'(z) \neq 0$ over S_z containing P_0.

411. Suppose P and P_0 on curve C_1 in the z plane map into Q and Q_0 on Γ_1. If Δs is the arc length P_0P on C_1 and $\Delta\sigma$ is the corresponding arc length Q_0Q on Γ_1, with corresponding chord lengths $|\Delta z|$ and $|\Delta w|$, respectively, then, if $f(z)$ is analytic in S_z,

$$\lim_{\Delta s \to 0} \frac{\Delta\sigma}{\Delta s} = \lim_{\Delta s \to 0} \left|\frac{\Delta w}{\Delta z}\right| = |f'(z_0)|$$

$\rho \equiv |f'(z_0)|$ is the *scale function* for f at z_0. ρ is the *ratio of magnification.*

412. SOME COMMON FUNCTIONS The *logarithm* of z is defined to be

$$\log z = \int_C \frac{dt}{t} = \int_{t=1}^{t=z} \frac{dt}{t}$$

where C is any path in the complex plane which does not pass through the origin, joining the point $t = 1$ to the point $t = z$.

(1) $\quad \log z = \ln z + i(2k\pi) \qquad k = 0, \pm 1, \pm 2, \ldots$

The *principal value* of the infinitely many-valued function $\log z$ is denoted by

$$\ln z \equiv \ln |z| + i\theta \qquad -\pi < \theta \leqq \pi$$

θ being the (real) principal value, in radians, of the amplitude of z.

The principal value $\ln z$ is determined as follows. A "slit" is made along the negative real axis in the z plane, and the agreement is made that the path of integration defining $\log z$ shall not cross this "cut."

If it is agreed to reach the negative real axis from above, the imaginary part of ln z is πi; if from below, $-\pi i$. (Note that ln $1 = 0$.)

$$\log z_1 + \log z_2 = \log (z_1 z_2) \tag{2}$$

in the sense that every value of either side of Eq. (2) is one of the values of the other side of (2).

413. The *exponential* $z \equiv \exp w$ is defined as the inverse function of $w = \log z$. w is single-valued and satisfies the differential equation $dw = dz/z$, $w = 0$ when $z = 1$, and

$$\exp w = 1 + \frac{w}{1!} + \frac{w^2}{2!} + \cdots + \frac{w^n}{n!} + \cdots$$

converges for $|w| < \infty$.

By definition, $e \equiv \exp 1$. Then $1 = \ln e$. $\exp w$ is often written e^w.

$$\exp w_1 \cdot \exp w_2 \equiv \exp (w_1 + w_2)$$

If $w = u + iv$, then

$$\exp (u + iv) = \exp u \cdot \exp iv$$

and

$$\exp iv = \cos v + i \sin v$$

414. Trigonometric identities for functions with real arguments can be extended without change of form to the functions with complex arguments. Thus, $\sin^2 z + \cos^2 z = 1$. Formulas for the derivatives of these functions retain the same form when the variable is complex as when it is real. Thus, $d(\sin w)/dz = \cos w(dw/dz)$. (See Secs. 72 and 108.)

415. MANY-VALUED FUNCTIONS The function $w \equiv \log z = \ln |z| + i(\theta + 2k\pi)$ in (1) is an example of an infinitely many-valued function of z. Each value of k defines a *branch* of log z. As z describes a circuit around the origin, w moves from one branch of log z to another, the origin being a branch point for each branch. In general, a point $z = a$ is called a *branch point* of a many-valued function of z, $w \equiv f(z)$, if $z = a$ is a singular point for $f(z)$ and if, as z makes a circuit about $z = a$, w moves from one branch of $f(z)$ to another.

When n is an integer, $w = z^{1/n}$ is an n-valued function of z; and if $z = re^{i\theta}$, then $w = r^{1/n}e^{i\psi}$, where $\psi \equiv (\theta + 2k\pi)/n$ and $r^{1/n}$ is the positive arithmetic nth root of the real number r.

22. STATISTICS*

416. FREQUENCY DISTRIBUTIONS Let $x_1, x_2, \ldots, x_k$ be a set of numbers (*variates*, observations, measurements) which occur with frequencies (*weighting factors*) $f_1, f_2, \ldots, f_k$, respectively, where each $f_i \geqq 0$, and $x_1 < x_2 < \cdots < x_k$. Such a set is a *frequency distribution*. The *weighted arithmetic mean*, or *weighted average*, is

$$\bar{x} = \frac{1}{n} \sum_{i=1}^{k} f_i x_i \qquad \text{where } n = \sum_{i=1}^{k} f_i \text{ is the } \textit{total weight}$$

Let L be the *least* and U the *greatest* of $x_1, \ldots, x_k$. $(U - L)$ is the *range* of the distribution. [Some writers call the interval (L, U) the range.] Other parameters in common use are listed in Table 22.1.

A plot of f_i in terms of x_i as abscissa is a *frequency diagram*. The quantity $F_j = \sum_{i=1}^{j} f_i \equiv \text{cum } f|_j$ is the *cumulative frequency* associated with item x_j. F_j is the number of items having measurements X less than or equal to x_j. A plot of F_j in terms of x_j is a *cumulative frequency graph*.

417. GROUPED DATA When the number of items $X_1, X_2, \ldots, X_n$ in the data is large, the data may be grouped by dividing the range into some number k of equal intervals of length c, called *class intervals*, these intervals being located so as to have a convenient midpoint x_i, called a *class mark* or *midvalue*. Every item in the data falling in a given class interval i is assigned a value equal to the x_i of the interval. The boundaries of class interval i are $(x_i - c/2)$ and $(x_i + c/2)$. The class intervals are ordered as $1, 2, \ldots, k$ and the corresponding *class frequencies* $f_1, \ldots, f_k$ listed. The *cumulative frequency* associated with the upper boundary of jth class interval is $F_j = \text{cum } f|_j$. The plots of f_j and F_j in terms of x_j give a *frequency histogram* and a *cumulative frequency polygon*, respectively.

* A convenient summary of theories, formulas, definitions, working rules, and tabular material useful in practical problems in probability and statistics is given in Richard S. Burington and Donald C. May, *Handbook of Probability and Statistics with Tables*, 2d ed., McGraw-Hill Book Company, New York, 1970.

TABLE 22.1. STATISTICAL PARAMETERS FOR DISCRETE FREQUENCY DISTRIBUTIONS

Measures of Location

$n = \sum_{i=1}^{k} f_i$	Total weight
$\bar{x} = \frac{1}{n} \sum_{i=1}^{k} f_i x_i$	Weighted arithmetic mean
$x_i - \bar{x}$	Deviation (or error) of x_i with respect to $\bar{x}$
$\lvert x_i - \bar{x} \rvert$	Absolute deviation (or absolute error) of x_i with respect to $\bar{x}$
$\left(\sum_{i=1}^{k} \frac{f_i x_i^2}{n} \right)^{1/2}$	Root mean square (rms) value of $x_1, \ldots, x_k$
A value of x_i having a maximum frequency f_i	Mode (Mo)
In list of ordered variates x, n in number, the *median* is the mid-value (if n odd) or the average of the two middle values (if n even)	Median

Moments

$\mu_r(x_0) = \frac{1}{n} \sum_{i=1}^{k} f_i (x_i - x_0)^r$	rth moment about $x = x_0$
$\mu_2(x_0)$	Mean square deviation from x_0
$\sqrt{\mu_2(x_0)}$	Root mean square deviation from x_0
$\nu_r = \frac{1}{n} \sum_{i=1}^{k} f_i x_i^r$	rth moment about $x = 0$
$\mu_r = \frac{1}{n} \sum_{i=1}^{k} f_i (x_i - \bar{x})^r$	rth (central) moment about weighted mean $\bar{x}$
$\sigma^2 = \mu_2 = \nu_2 - \bar{x}^2$	Variance, 2d moment about the mean
$\sigma = \sqrt{\mu_2}$	Standard deviation
$\alpha_r = \frac{1}{n} \sum_{i=1}^{k} f_i t_i^r$ where $t_i = \frac{x_i - \bar{x}}{\sigma}$	rth moment in standard t units about $t = 0$

TABLE 22.1. Continued

$\frac{1}{n}\sum_{i=1}^{k} f_i \lvert x_i - x_0 \rvert^r$	rth absolute moment about x_0
$MD\vert_0 = \frac{1}{n}\sum_{i=1}^{k} f_i \lvert x_i - x_0 \rvert$	Mean deviation from x_0
$MD = \frac{1}{n}\sum_{i=1}^{k} f_i \lvert x_i - \bar{x} \rvert$	Mean (absolute) deviation from the mean, mean absolute error (m.a.e.)

Measures of Skewness and Kurtosis

$\gamma_1 = \frac{\mu_3}{\sigma^3} = \alpha_3$	Coefficient of skewness $= \gamma_1$
$\frac{\mu_3}{2\sigma^3}$	Momental skewness
$\gamma_2 = \frac{\mu_4}{\sigma^4} - 3 = \alpha_4 - 3$	Coefficient of excess $= \gamma_2$, "flatness"
$\beta_2 = \frac{\mu_4}{\sigma^4} = \gamma_2 + 3 = \alpha_4$	Kurtosis $= \beta_2$, "flatness"

Measures of Dispersion

Consider the ordered variates, $x_1 < x_2 < \cdots < x_k$, where f_i is frequency of x_i, $\bar{x}$ is mean, and σ is standard deviation. Let $n = \sum_{i=1}^{k} f_i$, $\text{cum}\, f\vert_j = \sum_{i=1}^{j} f_i$. Plot $\text{cum}\, f\vert_j$ as a function of x. The largest x_j for which $\text{cum}\, f\vert_j \leqq p/100$ is called a pth *percentile*.

Lower (first) quartile	Q_1: the 25th percentile
Median (second) quartile	Q_2: the 50th percentile
Upper (third) quartile	Q_3: the 75th percentile
Semi-interquartile range	$Q{:}\,Q = \frac{Q_3 - Q_1}{2}$

Variance σ^2 and standard deviation σ are measures of dispersion. $V = \sigma/\bar{x}$, or $100\sigma/\bar{x}$ percent, is the *coefficient of variation.*

418. DISCRETE PROBABILITY DISTRIBUTIONS Suppose X takes on only the discrete values $x_1, x_2, \ldots, x_k$, with $x_1 < x_2 < \cdots < x_k$. Let the probability that X takes the value x_i be

$$P[X = x_i] = f(x_i) \qquad i = 1, \ldots, k$$

Then

$$\sum_{i=1}^{k} f(x_i) = 1$$

The probability that $X \leqq x'$ is

$$P[X \leqq x'] = \sum_{i=1}^{i=j} f(x_i) = F(x')$$

where x_j is the largest discrete value of X less than or equal to x'. $F(x')$ is the *cumulative discrete probability*. $F(x_k) = 1$.

The probability that $x' < X \leqq x''$ is

$$P[x' < X \leqq x''] = F(x'') - F(x')$$

The *discrete distribution* $f(x)$ is describable by parameters similar to those given in Table 22.1. The mean μ and variance σ^2 of X are, respectively,

$$\mu = \sum_{i=1}^{k} x_i f(x_i)$$

$$\sigma^2 = \sum_{i=1}^{k} (x_i - \mu)^2 f(x_i) = \sum_{i=1}^{k} x_i^2 f(x_i) - \mu^2$$

Probability distributions and sample frequency distributions are similar in form, the $f(x_i)$ corresponding to f_i/n.

Mathematical Expectation. Let $g(X)$ be a function of the discrete chance variable X, X having probability distribution $f(x)$. The expected value (mean value) of $g(X)$ is

$$E[g(X)] = \sum_{i=1}^{k} g(x_i) f(x_i)$$

The mean μ is the mathematical expectation $E[X]$ of X.

The rth moment μ_r about the mean μ is

$$E[(X - \mu)^r] = \mu_r = \sum_{i=1}^{k} (x_i - \mu)^r f(x_i)$$

The variance of X is $\mu_2 = \sigma^2$; σ is the standard deviation of X.

419. CONTINUOUS PROBABILITY DISTRIBUTION Let X be a chance variable which is certain to fall in the interval (α,β). A *continuous cumulative distribution function* $F(x)$ is a function which gives the probability that the chance variable X is less than or equal to any particular value x which may be selected, that is,

$$P[X \leqq x] = F(x)$$

$F(x)$ is called the *probability function* of X.

The probability that $x' < X \leqq x''$ is

$$P[x' < X \leqq x''] = F(x'') - F(x')$$

If the derivative $F'(x)$ of $F(x)$ with respect to x exists, then $f(x) = F'(x)$ is the *continuous probability density function* (*frequency function, distribution*), and

$$P[x' < X \leqq x''] = \int_{x'}^{x''} f(x)\,dx = F(x'') - F(x')$$

$$F(x') = \int_{\alpha}^{x'} f(x)\,dx = P[X \leqq x']$$

$$F(\beta) = \int_{\alpha}^{\beta} f(x)\,dx = 1 \qquad F(\alpha) = 0 \qquad 0 \leqq F(x) \leqq 1$$

The principal features of a continuous distribution $f(x)$ may be described by parameters similar to those given for discrete distributions.

The mean μ and variance σ^2 of X are, respectively,

$$\mu = \int_{\alpha}^{\beta} xf(x)\,dx$$

$$\sigma^2 = \int_{\alpha}^{\beta} (x - \mu)^2 f(x)\,dx = \int_{\alpha}^{\beta} x^2 f(x)\,dx - \mu^2$$

Mathematical Expectation. Let $g(X)$ be a function of the continuous variable X, X having probability distribution $f(x)$ and being certain to fall in (α,β). The *expected value* (*mean value*) of $g(X)$ is

$$E[g(X)] = \int_{\alpha}^{\beta} g(x)f(x)\,dx$$

The mean μ is the mathematical expectation $E[X]$ of X.

The rth moment μ_r of the distribution about the mean μ is

$$E[(X - \mu)]^r = \mu_r = \int_{\alpha}^{\beta} (x - \mu)^r f(x)\,dx$$

$\mu_2 = \sigma^2$ is the variance; σ, the standard deviation.

420. BINOMIAL DISTRIBUTION If X is such that the probability that X takes the value x is

(1) $$P[X = x] = f(x) = C_x^n p^x q^{n-x} \qquad x = 0, 1, \ldots, n$$

where

$$q = 1 - p \qquad \text{and} \qquad C_x^n = \frac{n!}{x!(n-x)!} \equiv \binom{n}{x}$$

then X is said to have a *binomial probability distribution.*

The probability that $X \leqq x'$, where x' is any value such that $x' \leqq n$, is

$$P[X \leqq x'] = F_B(x') = C_0^n p^0 q^n + C_1^n p^1 q^{n-1} + \cdots + C_{x^*}^n p^{x^*} q^{n-x^*}$$

where x^* is the largest integer which does not exceed x'.

The $(x + 1)$th term of the binomial expansion

$$(q + p)^n = q^n + C_1^n p q^{n-1} + \cdots + C_x^n p^x q^{n-x} + \cdots + p^n$$

is the right-hand side of (1).

Properties. The mean, variance, etc., of X having distribution (1) are

$$\text{Mean} = \mu = np \qquad \text{Variance} = \sigma^2 = npq$$

$$\text{Standard deviation} = \sigma = \sqrt{npq}$$

$$\text{Coefficient of skewness} = \gamma_1 = \frac{q - p}{\sqrt{npq}}$$

$$\text{Coefficient of excess} = \gamma_2 = \frac{1 - 6pq}{npq}$$

Mode = positive integral value (or values) of x for which

$$np - q \leqq x \leqq np + p$$

rth moment μ_r about mean μ:

$$\mu_1 = 0 \qquad \mu_2 = \sigma^2 = npq \qquad \mu_3 = npq(q - p)$$

$$\mu_4 = npq[1 + 3(n - 2)pq]$$

The normal probability distribution (Sec. 423) gives a good approximation to the binomial distribution for n sufficiently large. To fit normal curve $y = \psi(t)/\sigma$, where $\psi(t) = (1/\sqrt{2\pi}) \exp[-t^2/2]$, to binomial distribution $f(x) = C_x^n p^x q^{n-x}$, set $\mu = np$, $\sigma = \sqrt{npq}$, $t = (x - \mu)/\sigma$.

421. **REPEATED TRIALS** Let p be the probability of occurrence of an event E ("success") on each of n trials and q the probability of nonoccurrence of E ("failure"). $p + q = 1$ and $0 \leqq p \leqq 1$. The probability of x successes followed by $(n - x)$ failures is $p^x q^{n-x}$. If the order of occurrence of successes is of no consequence, the number of orders in which x successes and $(n - x)$ failures can occur is $C_x^n = n!/[x!(n - x)!]$.

The probability of *exactly* x successes (in n independent trials) irrespective of order is

$$P(X = x) = f(x) = C_x^n p^x q^{n-x}$$

The probability of *at least* x occurrences in n trials is

$$P[X \geqq x] = \sum_{i=x}^{n} C_i^n p^i q^{n-i} \qquad \text{for } x = 0, 1, \ldots, n$$

The probability of *at most* x occurrences in n trials is

$$P[X \leqq x] = \sum_{i=0}^{i=x} C_i^n p^i q^{n-i} \qquad \text{for } x = 0, 1, \ldots, n$$

The mean number of occurrences of event E is $\mu = np$; the mean number of nonoccurrences of event E is nq (see Table 18).

422. **POISSON DISTRIBUTION** If X is such that the probability that X takes the value x is

$$(1)\quad P[X = x] = f(x) = \frac{e^{-m} m^x}{x!}, \quad m > 0, \quad \text{for } x = 0, 1, 2, \ldots$$

the variable X is said to have a *Poisson distribution*. *Note:* $f(0) = e^{-m}$.

The probability that $X \geqq x'$ is

$$P[X \geqq x'] = \sum_{x=x'}^{x=\infty} \frac{e^{-m} m^x}{x!}$$

Note: $P[X \geqq 0] = 1$.

The probability that $X \leqq x^*$ is

$$P[X \leqq x^*] = \sum_{x=0}^{x=x^*} \frac{e^{-m} m^x}{x!}$$

Properties. The mean, variance, etc., of X having distribution (1) are

$$\text{Mean} = \mu = m \qquad \text{Variance} = \sigma^2 = m$$

$$\text{Standard deviation} = \sigma = \sqrt{m}$$

$$\text{Coefficient of skewness} = \gamma_1 = \frac{1}{\sqrt{m}}$$

$$\text{Coefficient of excess} = \gamma_2 = \frac{1}{m}$$

rth moment μ_r about mean μ:

$$\mu_2 = \sigma^2 = m \qquad \mu_3 = m \qquad \mu_4 = 3m^2 + m \qquad \cdots$$

rth moment ν_r about $x = 0$:

$$\nu_1 = m \qquad \nu_2 = m(m+1) \qquad \nu_3 = m(m^2 + 3m + 1)$$

$$\nu_4 = m(m^3 + 6m^2 + 7m + 1) \qquad \cdots$$

The binomial distribution $C_x^n p^x q^{n-x}$ approaches the Poisson distribution as a limit when $n \to \infty$ and $p \to 0$ such that $np = m$ is a constant. The Poisson distribution may be used to approximate the binomial distribution when n is large enough and p is small enough (e.g., when $n > 50$, $p < 0.1$, $0 < np < 10$). (See Table 19.)

423. NORMAL DISTRIBUTION If X is such that the probability that X is less than or equal to x is

$$P[X \leqq x] = F_N(x) = \frac{1}{\sqrt{2\pi}\sigma} \int_{-\infty}^{x} \exp\left[\frac{-(X-\mu)^2}{2\sigma^2}\right] dX$$

where μ and σ are constants, X is distributed normally, and $F_N(x)$ is the *cumulative normal (Gaussian) distribution.*

The normal probability density function $f_N(x)$ is

(1) $$f_N(x) = \frac{1}{\sqrt{2\pi}\sigma} \exp\left[-\frac{(x-\mu)^2}{2\sigma^2}\right]$$

The probability that X falls between x' and x^* is

$$P[x' < X \leqq x^*] = F_N(x^*) - F_N(x')$$

(See Table 18 for comprehensive table of values for the normal distribution.)

(1) may be written

$$f_N(x) = \frac{1}{\sqrt{2\pi}} \exp\left(-\frac{t^2}{2}\right) \qquad \text{where } t = \frac{x-\mu}{\sigma}$$

Properties. The mean, variance, etc., of X having distribution (1) are

$$\text{Mean} = \mu \qquad \text{Variance} = \sigma^2$$

$$\text{Standard deviation} = \sigma$$

$$F_N(-\infty) = 0 \qquad F_N(+\infty) = 1$$

$$\text{Coefficient of skewness} = \gamma_1 = 0$$

$$\text{Coefficient of excess} = \gamma_2 = 0$$

$$\text{Mode} = \mu \qquad \text{Median} = \mu$$

rth moment μ_r about mean μ:

$$\mu_1 = 0 \qquad \mu_2 = \sigma^2 \qquad \mu_3 = 0 \qquad \mu_4 = 3\sigma^4 \qquad \cdots$$

rth moment ν_r about $x = 0$:

$$\nu_1 = \mu \qquad \nu_2 = \mu^2 + \sigma^2 \qquad \nu_3 = \mu(\mu^2 + 3\sigma^2)$$

$$\nu_4 = \mu^4 + 6\mu^2\sigma^2 + 3\sigma^4 \qquad \cdots$$

Mean (absolute) deviation from mean

$$= \text{MD} = \text{m.a.e.} = \sigma\sqrt{\frac{2}{\pi}} = 0.7979\sigma$$

$$\text{Modulus of precision} = h = \frac{1}{\sigma\sqrt{2}} = 0.7071\sigma$$

$$\text{Probable error} = \text{p.e.} = 0.6745\sigma = 0.8453 \text{ (m.a.e.)}$$

Upper and lower quartiles: $Q_3 = \mu + \text{p.e.} \qquad Q_1 = \mu - \text{p.e.}$

The fractions of the areas under the curve $f_N(x)$ above x axis, plotted as a function of x between lines parallel to y axis, are:

Zone, %	*Range*
50	$\mu \pm 0.6745\sigma$
68.27	$\mu \pm \sigma$
90	$\mu \pm 1.645\sigma$
95.45	$\mu \pm 2\sigma$
99.73	$\mu \pm 3\sigma$

424. PROBABILITY OF OCCURRENCE ON CERTAIN DEVIATIONS The probability that X will differ from μ by not more than σ is

$$P[|X-\mu| \leqq \sigma] = P[-1 \leqq t \leqq 1]$$
$$= \frac{1}{\sqrt{2\pi}} \int_{t=-1}^{t=1} \exp\left(-\frac{t^2}{2}\right) dt = 0.6827$$

[Here $t = (X-\mu)/\sigma$.]

The probability that x will deviate more than $\lambda\sigma$ from the mean μ is

$$P = P[|x-u| > \lambda\sigma] = \frac{2}{\sqrt{2\pi}} \int_{\lambda}^{\infty} \exp\left(-\frac{t^2}{2}\right) dt$$

Example. Values of λ for 0.1, 1, and 5 percent *tolerance limits* on x are, respectively,

λ	3.29	2.58	1.96
P	0.001	0.01	0.05

425. FITTING A POLYNOMIAL TO DATA REPRESENTED BY A SET OF POINTS Consider the set of points (x_1, y_1), (x_2, y_2), . . . , (x_k, y_k). It is desired to find among the curves

$$y(x) = a_0 + a_1 x + \cdots + a_n x^n \qquad k > n$$

a curve which gives a "best possible fit" to the given points. By the method of least squares, a "best" system of values for the coefficients (namely, $A_0, A_1, \ldots, A_n$) is one that renders $M = \sum_{i=1}^{k} d_i^2$ a minimum. Here $d_i = y_i - y(x_i)$ is the *deviation* of y_i from the value $y(x_i)$. A curve C with coefficients $a_j = A_j$ $(j = 0, 1, \ldots, n)$ is a "best-fitting curve." C is called a *mean square regression curve.* When $n = 1$, C is a *mean square regression line.*

The coefficients of C may be found by solving the $(n+1)$ "normal" equations $\partial M/\partial a_j = 0$ $(j = 0, 1, \ldots, n)$ for a_j. The coefficients $A_0, \ldots, A_n$ of C are the roots of the simultaneous equations:

$$\begin{cases} A_0S_0 + A_1S_1 + \cdots + A_nS_n = T_0 \\ A_0S_1 + A_1S_2 + \cdots + A_nS_{n+1} = T_1 \\ \dots\dots\dots\dots\dots\dots\dots\dots \\ A_0S_n + A_1S_{n+1} + \cdots + A_nS_{2n} = T_n \end{cases}$$

where $\quad S_j = \sum_{i=1}^{i=k} x_i{}^j \qquad S_0 = k \qquad T_j = \sum_{i=1}^{i=k} x_i{}^j y_i$

426. SIMPLE (LINEAR) CORRELATION One measure of the degree of association, or correlation, between the xs and the corresponding ys in the data $(x_1, y_1), (x_2, y_2), \ldots, (x_k, y_k)$ is the ratio

$$r = \frac{\operatorname{cov}(x, y)}{\sigma_x \sigma_y}$$

where $\bar{x}$ and σ_x are the mean and standard deviation, respectively, of the xs; $\bar{y}$ and σ_y are the corresponding parameters for the ys; and

$$\operatorname{cov}(x, y) = \frac{\sum_{i=1}^{k} (x_i - \bar{x})(y_i - \bar{y})}{k} = \frac{\sum_{i=1}^{k} x_i y_i - k\bar{x}\bar{y}}{k}$$

is the covariance between the xs and ys. r is called the *correlation coefficient* (or *product moment about the mean*) of the variables x and y. *Note:* $\operatorname{var} x = \operatorname{cov}(x,x) = \sigma_x{}^2$.

427. FITTING A LINE TO POINTS A line $y(x) = a_0 + a_1 x$ may be fitted to the points $(x_1, y_1), \ldots, (x_k, y_k)$ in many ways. If the fit is made to minimize the sum of the squares of the vertical distances from the line (assuming the errors to be in the y coordinates only), the line so obtained is known as the (*mean*) *regression line of y on x* and has the equation

$$y - \bar{y} = \frac{r\sigma_y}{\sigma_x}(x - \bar{x})$$

The slope of this line is $m_1 = r\sigma_y/\sigma_x$, which is called the *regression coefficient of y on x*.

The standard deviation of the error of estimate in y is S_y, where

$S_y^2 = \sigma_y^2(1 - r^2)$. S_y^2 is the *residual variance* of y; S_y, the *standard error of estimate in* y.

If the fit is made so as to minimize the sum of the squares of the horizontal distances from the line (assuming the errors to be in the x coordinates only), the line so obtained is the (*mean*) *regression line of* x *on* y and has the equation

$$y - \bar{y} = \frac{\sigma_y}{r\sigma_x}(x - \bar{x})$$

The *regression coefficient of* x *on* y is $r\sigma_x/\sigma_y$.

S_x is the standard deviation of the errors of estimate in x and is called the *standard error of estimate in* x. The *residual variance* of x is S_x^2, where $S_x^2 = \sigma_x^2(1 - r^2)$.

When $r = \pm 1$, the lines of regression coincide and the correlation is perfect; if $r = 0$, the lines are at maximum divergence and x and y are *uncorrelated*. When $r \neq 0$, the variables are *correlated*. If $r < 0$, on the average y decreases as x increases; if $r > 0$, y increases with x.

428. SAMPLE DISTRIBUTIONS A *population* consists of a collection $\{x\}$ of the values of a variable X. A *sample* is a set of values $x_1, x_2, x_3, \ldots$ taken from population $\{x\}$. Let $f(x)$ be the frequency distribution function of X in $\{x\}$, with $\tilde{x}$ = mean of X and σ_x^2 = variance of X. In the set of all possible random samples of size n, the values $x_1, x_2, \ldots, x_n$ may be considered statistically independent. If each variable x_i has the frequency distribution function $f(x)$, the sample set $(x_1, x_2, \ldots, x_n)$ has the frequency distribution function $f(x_1)f(x_2) \cdots f(x_n)$.

For the sample $x_1, x_2, \ldots, x_n$, let the mean $\bar{x}$ and variance s^2 be, respectively,

$$\bar{x} = \sum_{i=1}^{n} \frac{x_i}{n} \qquad s^2 = \sum_{i=1}^{n} \frac{(x_i - \bar{x})^2}{n}$$

Theorem 1. An "unbiased" estimate of the population mean $\tilde{x}$ is the sample mean $\bar{x}$; of the variance σ_x^2 of a population is $s^2 n/(n - 1)$, where $s^2 = \sum_{i=1}^{n} (x_i - \bar{x})^2/n$ is the sample variance.

Theorem 2. If an infinite population of values x is normally distributed with mean $\tilde{x}$ and standard deviation σ_x, then the sampling distribution of means $\bar{x}$ is normally distributed with mean $\tilde{x}$ and standard deviation $\sigma_x/\sqrt{n}$.

The probability p that x for any one sample will fall in the interval $\tilde{x} - \delta \leqq x \leqq \tilde{x} + \delta$ is

$$p = \int_{-\beta}^{+\beta} \psi(u)\,du \qquad \text{where} \qquad \beta = \frac{\delta\sqrt{n}}{\sigma_x}$$

$$u = (\bar{x} - \tilde{x})\frac{\sqrt{n}}{\sigma_x} \qquad \text{and} \qquad \psi(u) = \frac{1}{\sqrt{2\pi}}\exp\left(-\frac{u^2}{2}\right)$$

429. GOODNESS OF FIT (CHI-SQUARE TEST) There are various statistical ways to test how well a sample distribution agrees with an assumed population distribution $f(x)$. One scheme for doing this makes use of a chi-square test. The comparison makes use of the observed and the population frequencies for a suitable set of class intervals of the variable of the distribution.

Consider a population of items falling into r mutually exclusive class intervals (*cells*). Given a sample $x_1, x_2, \ldots, x_n$ of n items having observed frequency f_i in ith cell $\left(n = \sum_{i=1}^{r} f_i\right)$, suppose the assumed theoretical frequency for the ith class is np_i. The χ^2 test is concerned with the difference between f_i and np_i for all cells. This difference is measured by

$$\chi^2 = \sum_{i=1}^{r} \frac{(f_i - np_i)^2}{np_i} = \left(\sum_{i=1}^{r} \frac{f_i^2}{np_i}\right) - n \tag{1}$$

The *hypothesis* H of the test is that the data $x_1, \ldots, x_n$ are a sample of a random variable with distribution $f(x)$.

The *purpose of the test* is to determine whether the observed data may be considered consistent with H.

The *steps in the test* are (*a*) Calculate (1). (*b*) Determine the number of degrees of freedom m. (*c*) Select a level of significance ϵ. (*d*) From tables of χ^2 distribution find a χ_0^2 such that $P[\chi^2 \geqq \chi_0^2] \approx \epsilon$ (Table 20).

The *interpretation of the test* is as follows: (*a*) If for the sample a value $\chi^2 > \chi_0^2$ is found, the sample $x_1, \ldots, x_n$ shows a significant deviation from hypothesis H, and H is rejected at the ϵ level of significance. (*b*) If for the sample a value $\chi^2 \leqq \chi_0{}^2$ is found, the sample $x_1, \ldots, x_n$ is considered consistent with hypothesis H.

In using Table 20, an estimate of the number m of degrees of freedom must be made, where $m = r - 1 - b$, r is the number of cells, and b is the number of parameters in the population distribution determined from the sample. When the population distribution is completely known, $m = r - 1$.

The chi-square test may be applied to data grouped in cells provided n is large enough (say, $n \geqq 50$), the theoretical and sample cell frequencies are not too small, (f_i and np_i each greater than, say, 5 to 10), and the constraints on the cell frequencies are linear.

Table 20 is applicable only when $m \leqq 30$. When $m > 30$, $\sqrt{2\chi^2}$ is, approximately, normally distributed, with mean $\sqrt{2m - 1}$ and unit standard deviation.

430. TEST OF ESTIMATE OF PROBABILITY OF EVENT FROM ONE SAMPLE Suppose that in a sequence of n independent trials, event E has occurred v times. It is desired to test whether this is consistent with the hypothesis that the probability of the occurrence of E is $p = 1 - q$ in each trial. There are $r = 2$ class intervals for the observations, namely, E occurred, E did not occur. From Eq. (1), Sec. 429, with $r = 2$, $f_1 = v$, $f_2 = n - v$, $f_1' = np$, $f_2' = nq$, $m = r - 1 = 1$ degree of freedom,

$$\chi^2 = \frac{(v - np)^2}{npq}$$

Example. If $n = 40$, $v = 12$, $p = 0.25$, then $\chi^2 = [12 - (0.25)(40)]^2/40(0.25)(0.75) = 0.53$. The χ^2 table (Table 20) shows that a value of χ^2 as large as 0.53 may be expected in over 40 percent of samples. Thus the sample value $v = 12$ is not unlikely if the actual probability of E is $p = 0.25$.

For further material and tables on probability and statistics see Richard S. Burington and Donald C. May, *Handbook of Probability and Statistics with Tables*, 2d ed., McGraw-Hill Book Company, New York, 1970.

PART TWO

Tables

EXPLANATION OF TABLES

Logarithms

TABLE 1. FIVE-PLACE COMMON LOGARITHMS OF NUMBERS FROM 1 TO 10,000

Definition of Logarithm. The *logarithm* x of the number N to the base b is the exponent of the power to which b must be raised to give N. That is,

$$\log_b N = x \qquad \text{or} \qquad b^x = N$$

The number N is positive, and b may be any positive number except 1.

Properties of Logarithms

(a) *The logarithm of a product is equal to the sum of the logarithms of the factors; thus*

$$\log_b MN = \log_b M + \log_b N$$

(b) *The logarithm of a quotient is equal to the logarithm of the numerator minus the logarithm of the denominator; thus*

$$\log_b \frac{M}{N} = \log_b M - \log_b N$$

(c) *The logarithm of a power of a number is equal to the logarithm of the number multiplied by the exponent of the power; thus*

$$\log_b M^p = p \log_b M$$

(d) *The logarithm of a root of a number is equal to the logarithm of the number divided by the index of the root; thus*

$$\log_b \sqrt[q]{M} = \frac{1}{q} \log_b M$$

Other properties of logarithms:

$$\log_b b = 1 \qquad \log_b \sqrt[q]{M^p} = \frac{p}{q} \log_b M$$

$$\log_b 1 = 0 \qquad \log_b N = \log_a N \log_b a = \frac{\log_a N}{\log_a b}$$

$$\log_b b^N = N \qquad b^{\log_b N} = N$$

Systems of Logarithms. There are two common systems of logarithms in use: (1) the *natural* (Napierian or hyperbolic) system which uses the base $e = 2.71828 \cdots$; (2) the *common* (Briggsian) system which uses the base 10.

We shall use the abbreviation* $\log N \equiv \log_{10} N$ in this section.

Unless otherwise stated, tables of logarithms are always tables of common logarithms.

Characteristic of a Common Logarithm of a Number. Every real positive number has a real common logarithm such that if $a < b$, $\log a < \log b$. Neither zero nor any negative number has a real logarithm.

A common logarithm, in general, consists of an integer, which is called the *characteristic*, and a decimal (usually endless), which is called the *mantissa*. The characteristic of any number may be determined from the following rules.

Rule I. The characteristic of any number greater than 1 *is one less than the number of digits before the decimal point.*

Rule II.† *The characteristic of a number less than* 1 *is found by subtracting from* 9 *the number of ciphers between the decimal point and the first significant digit and writing* -10 *after the result.*

Thus the characteristic of log 936 is 2; the characteristic of 9.36 is 0; of log 0.936 is 9 − 10; of log 0.00936 is 7 − 10.

Mantissa of a Common Logarithm of a Number. An important consequence of the use of base 10 is that the mantissa of a number is independent of the position of the decimal point. Thus 93,600, 93.600, 0.000936, all have the same mantissa. Hence in Tables of Common Logarithms only mantissas are given. This is done in Table 1. A five-place table gives the values of the mantissa correct to five places of decimals.

To Find the Logarithm of a Given Number N. By means of Rules I and II determine the characteristic. Then use Table 1 to find mantissa.

To find mantissa when the given number (neglecting decimal point) consists of four or fewer digits (exclusive of ciphers at the beginning or end), look in the column marked N for the first three significant digits

* Note, however, that $\log N \equiv \log_e N$ is used in the Table of Integrals.

† Some writers use a dash over the characteristic to indicate a negative value, for example,

$$\log .004657 = 7.66811 - 10 = \bar{3}.66811$$

and pick the column headed by the fourth digit—the mantissa is the number appearing at the intersection of this row and column. Thus to find the logarithm of 64030, first note (by Rule I) that the characteristic is 4. Next in Table 1, find 640 in column marked N and opposite it in column 3 is the desired mantissa, .80638. Hence log 64030 = 4.80638. Likewise, log 0.0064030 = 7.80638 − 10; log 0.64030 = 9.80638 − 10.

*Interpolation.** The mantissa of a number of more than four significant figures can be found approximately by assuming that the mantissa varies directly as the number in the small interval not tabulated. Thus if N has five digits (significant) and f is the fifth digit of N, the mantissa of N is

$$m = m_1 + \frac{f}{10}\left(m_2 - m_1\right)$$

where m_1 is the mantissa corresponding to the first four digits of N and m_2 is the next larger mantissa in the table. $(m_2 - m_1)$ is called a *tabular difference.* The proportional part of the difference $m_2 - m_1$ is called the *correction.* These proportional parts are printed without zeros at the right-hand side of each page as an aid to mental multiplications.

For example, find log 64034. Here $f = 4$. From the table we see $m_1 = .80638$, $m_2 = .80645$, whence

$$m = .80638 + \frac{4}{10}(.00007), \quad \log 64034 = 4 + m = 4.80641$$

To Find the Number N When Its Logarithm Is Known. (The number N whose logarithm is k is called the *antilogarithm* of k.)

Case 1. If the mantissa m is found exactly in Table 1, join the figure at the top of the column containing m to the right of the figures in the column marked N and in the same row as m, and place the decimal point according to the characteristic of the logarithm.

Case 2. If the mantissa m is not found exactly in the table, interpolate as follows: Find the next smaller mantissa m_1 to m; the first four significant digits of N correspond to the mantissa m_1, and the fifth digit f equals the nearest whole number to

$$f = 10\left(\frac{m - m_1}{m_2 - m_1}\right)$$

* A more accurate method of interpolation is given on page 297.

where m_2 is the next larger mantissa to m appearing in the table. Then locate the decimal point according to the characteristic.

The decimal point may be located by means of the following rules:

Rule III. *If the characteristic of the logarithm is positive* (*then the mantissa is not followed by* -10), *begin at the left, count digits one more than the characteristic, and place the decimal point to the right of the last digit counted.*

Rule IV. *If the characteristic is negative* (*then the mantissa will be preceded by an integer n and followed by* -10), *prefix* $(9 - n)$ *ciphers, and place the decimal point to the left of these ciphers.*

Illustrations of the Use of Logarithms.

Example 1. Given $\log x = 2.91089$, find x. The mantissa .91089 appears in the table. Join the figure 5 which appears at the top of the column to the right of the number 814 in the column N, giving the number 8145. By Rule III, the decimal point is placed to the right of 4, thus giving $x = 814.5$.

Example 2. Given $\log x = 2.34917$, find x. The mantissa $m = .34917$ does not appear in the table. The next smaller and next larger mantissas are m_1 and m_2,

$$m_1 = .34908 \qquad m = .34917 \qquad m_2 = .34928$$

The first four digits of N, corresponding to m_1, are 2234, and the fifth digit is the nearest whole number (5) to

$$10\left(\frac{m - m_1}{m_2 - m_1}\right) = 10\left(\frac{.00009}{.00020}\right) = 4.5$$

By Rule III, we locate decimal point, thus giving $x = 223.45$.

Example 3. Find $x = (396.21)(.004657)(21.21)$:

$$\begin{aligned} \log 396.21 &= 2.59792 \\ \log .004657 &= 7.66811 - 10 \\ \log 21.210 &= 1.32654 \qquad \text{(add)} \\ \hline \log x &= 11.59257 - 10, \; x = 39.135 \end{aligned}$$

Example 4. Find $x = \dfrac{396.21^*}{24.3}$:

$$\begin{aligned} \log 396.21 &= 2.59792 \\ \log 24.3 &= \underline{1.38561} \quad \text{(subtract)} \\ \log x &= 1.21231, \ x = 16.305 \end{aligned}$$

Example 5. Find $x = (3.5273)^4$:

$$\begin{aligned} \log 3.5273 &= 0.54745 \\ &\quad\ \ \underline{\qquad\ 4} \quad \text{(multiply)} \\ \log x &= 2.18980, \ x = 154.81 \end{aligned}$$

Example 6. Given $\log x = -2.23653$, find x. To convert this logarithm to one with a positive mantissa, add algebraically -2.23653 to $10.00000 - 10$. Thus

$$\begin{aligned} &10.00000 - 10 \\ &\underline{-2.23653} \\ \log x = \ &\ \ 7.76347 - 10, \text{ hence } x = 0.0058006 \end{aligned}$$

Example 7. Find $x = \sqrt[3]{.04657}$:

$$\begin{aligned} \log x &= \tfrac{1}{3} \log (.04657) \\ &= \tfrac{1}{3}(8.66811 - 10) = \tfrac{1}{3}(-1.33189) \\ \log x &= -0.44396 = 9.55604 - 10, \ x = 0.35978 \end{aligned}$$

or

$$\begin{aligned} \log x &= \tfrac{1}{3}(8.66811 - 10) = \tfrac{1}{3}(28.66811 - 30) \\ \log x &= 9.55604 - 10, \ x = 0.35978 \end{aligned}$$

Example 8. Find $x = \dfrac{1}{21.210}$:

$$\begin{aligned} \log x &= \log 1 - \log 21.210 \\ \log 1 &= 10.00000 - 10 \\ \log 21.210 &= \underline{1.32654} \qquad \text{(subtract)} \\ \log x &= 8.67346 - 10, \ x = 0.047148 \end{aligned}$$

* Some writers use *cologarithms*. The cologarithm of a number N is the negative of the logarithm of N; that is, $\text{colog } N = 10.00000 - \log N - 10$. Adding the cologarithm is equivalent to subtracting the logarithm. Thus in our example,

$$\begin{aligned} \text{colog } 24.3 &= (10.00000 - 1.38561) - 10 = 8.61439 - 10 \\ \log 396.21 - \log 24.3 &= \log 396.21 + \text{colog } 24.3 = 2.59792 + 8.61439 - 10 \\ &= 11.21231 - 10 = \text{antilog } 16.305 \end{aligned}$$

TABLE 2. COMMON LOGARITHMS OF GAMMA FUNCTIONS

Trigonometric Functions

TABLE 4. COMMON LOGARITHMS OF TRIGONOMETRIC FUNCTIONS In this table, the logarithmic values of the sine, cosine, tangent, and cotangent of angles at intervals of one minute from 0° to 90° are given. Since $\log \sec A = -\log \cos A$ and $\log \csc A = -\log \sin A$, $\log \sec A$ and $\log \csc A$ are omitted. For angles between 0° and 45°, the number of the degrees and the name of the trigonometric function are read at the top of the page and the number of minutes in the left-hand column. The corresponding information for angles between 45° and 90° is found at the bottom of the page and in the right-hand column.

The arrangement and the principles of interpolation are similar to those given in the explanation for Table 1. The -10 portion of the characteristic is not printed in the table but must be written down whenever such a logarithm is used.

Although the logarithmic values of the trigonometric functions may be interpolated to the nearest second, interpolation for the logarithms of the sine and tangent of small angles from 0° to 3° and for the logarithms of the cosine and cotangent of angles from 87° to 90° is not accurate. Table 3 should be used in these cases.

Table 3 gives

$$S = \log \sin A - \log A' \qquad \text{and} \qquad T = \log \tan A - \log A'$$

where A is the given angle and A' is the number of minutes in A, for values of A from 0° to 3°. Then

$$\log \sin A = \log A' + S \qquad \text{and} \qquad \log \tan A = \log A' + T$$

Likewise, $\log \cos A = \log (90° - A)' + S$ and $\log \text{ctn}\, A = \log (90° - A)' + T$, for values of A from 87° to 90°, the S and T corresponding to $(90° - A)'$.

For the functions of angles greater than 90°, use the relations given in Sec. 57.

Tables of proportional parts are provided in Table 4 which may be used in the interpolation of logarithmic values of the functions between 4° and 86°.

Example 1. Find log sin 21°13′26″.

Turn to the page having 21° at the top. In the row having 13′ on the left, find in the column marked "L Sin" at the top log sin 21°13′ = 9.55858 − 10. (The −10 portion of the characteristic is omitted in the table.) The tabular difference is 33 and is given in the column marked *d*. In the table of proportional parts for 33, the correction for 20″ is 11.0 and for 6″ is 3.3. These two corrections must be added to the value of log sin 21°13′. Hence log sin 21°13′26″ = 9.55872 − 10.

Example 2. Find log ctn 56°23′37″.

On the page having 56° at the bottom, and in the row having 23 on the right, find, in the column that is marked "L Ctn" at the bottom, log cot 56°23′ = 9.82270 − 10. The tabular difference is 27, given in the column marked *c.d.* In the table of proportional parts for 27, we obtain 17 as the correction for 37″. Since the logarithmic value of the cotangent decreases as the angle increases from 0° to 90°, this correction for 37″ must be subtracted from log cot 56°23′. Hence log ctn 56°23′37″ = 9.82253 − 10.

Example 3. Find the acute angle A if log tan A = 9.67341 − 10.

The tabulated value of the logarithm of the tangent just smaller than 9.67341 − 10 is 9.67327 − 10. This corresponds to 25°14′. The difference between 9.67341 − 10 and 9.67327 − 10 is 14, and the tabular difference to be used is 33. In the proportional parts table for 33, we find the largest value less than 14 is 11.0, which corresponds to 20″. The difference 14 − 11.0 = 3.0, which corresponds to 5″. Hence, 25″ is the approximate correction and A = 25°14′25″.

Example 4. Find the acute angle A if log cos A = 9.89317 − 10.

The value of the logarithm of the cosine decreases as the angle increases from 0° to 90°. Hence we must find in the column marked "L Cos" a value just larger than 9.89317 − 10. This value is 9.89324 − 10 and corresponds to 38°33′. The difference between 9.89324 − 10 and 9.89317 − 10 is 7, and the tabular difference is 10. From the table of proportional parts we see that the largest value just smaller than 7 is 6.7, which corresponds to 40″. The difference 7 − 6.7 = 0.3, which corresponds to 2″. Hence the correction is 42″, approximately. Therefore A = 38°33′42″.

Example 5. Find log sin $0°35'30''$.

Convert $35'30''$ to minutes. In Table 3, column A, we find, for 35.5 minutes, $S = 6.46372 - 10$. By Table 1, log 35.5 = 1.55023.

Hence,

$$\begin{aligned}\log \sin 0°35'30'' &= \log 35.5 + S \\ &= 1.55023 + 6.46372 - 10 = 8.01395 - 10\end{aligned}$$

TABLE 5. NATURAL TRIGONOMETRIC FUNCTIONS This table gives the values of the sine, cosine, tangent, and cotangent at intervals of one minute from 0° to 90°.

The following tables are self-explanatory: Table 6, Minutes and Seconds to Decimal Parts of a Degree; Table 7, Common Logarithms of Trigonometric Functions in Radian Measure; Table 8, Degrees, Minutes, and Seconds to Radians; Table 9, Trigonometric Functions in Radian Measure; Table 10, Radians to Degrees, Minutes, and Seconds.

Exponential and Hyperbolic Functions

TABLE 11. NATURAL LOGARITHMS OF NUMBERS This table gives the logarithms of N to the Napierian base e (= 2.71828 ···) for equidistant values of N from 0.00 to 10.09 and from 10 to 1109. For values of N greater than 1109, use the formula $\log_e 10\,N = \log_e N + \log_e 10 = \log_e N + 2.30258509$ or the formula $\log_e N = (\log_e 10)(\log_{10} N) = 2.30258509\,(\log_{10} N)$.

TABLE 12. VALUES AND COMMON LOGARITHMS OF EXPONENTIAL AND HYPERBOLIC FUNCTIONS This table gives the values of e^x, e^{-x}, $\sinh x$, $\cosh x$, $\tanh x$, and the common logarithms of e^x, $\sinh x$, and $\cosh x$ for values of x equally spaced from 0.00 to 3.00 and for certain values of x from 3.00 to 10.00. The common logarithm of e^{-x} may be found by the relation $\log_{10} e^{-x} = -\log_{10} e^x = \text{colog}\, e^x$. To find $\log_{10} \tanh x$, use $\log_{10} \tanh x = \log_{10} \sinh x - \log_{10} \cosh x$. This table may be extended indefinitely by means of Table 13, since $\log_{10} e^x = Mx$.

TABLE 13. MULTIPLES OF M AND $1/M$ The purpose of this table is to facilitate the multiplication of a number N by M and $1/M$.

This occurs whenever it is desired to change from common logarithms to natural logarithms, and conversely. Thus

$$\log_{10} x = (\log_e x)(\log_{10} e) = M \log_e x \qquad \log_e x = \frac{\log_{10} x}{M}$$

These multiples are also required in

$$\log_{10} e^x = Mx \qquad \log_e (10^n x) = \log_e x + n \frac{1}{M}$$

and in the approximate formulas

$$\log_{10} (1 \pm x) = \pm xM \qquad \text{and} \qquad 10^{\pm x} = 1 \pm \frac{1}{M} x$$

Powers, Roots, Reciprocals, Areas, and Circumferences

The following tables are self-explanatory: Table 14, Squares, Cubes, Square Roots, and Cube Roots; Table 15, Reciprocals, Circumferences, and Areas of Circles.

Statistics and Probability

Table 16, Logarithms of Factorial n; Factorials and Their Reciprocals; Table 17, Binomial Coefficients; Probability; Table 18, Probability Functions (Normal Distribution); Table 19, Summed Binomial Distribution Function; Table 20, χ^2 Distribution.

Interest and Actuarial Experience

Table 21, Amount of 1 at Compound Interest; Table 22, Present Value of 1 at Compound Interest; Table 23, Amount of an Annuity of 1; Table 24, Present Value of an Annuity of 1; Table 25, The Annuity that 1 Will Purchase; Table 26, Logarithms for Interest Computation; Table 27, American Experience Mortality Table; Constants, Equivalents, and Conversion Factors; Table 28, Decimal Equivalents of Common Fractions.

Square Roots of Certain Common Fractions

Table 29, Important Constants*; Table 30, Complete Elliptic Integrals, K and E, for Different Values of the Modulus k; Table 31, A Table of Conversion Factors (Weights and Measures).*

Four-place Tables

Table 32, Common Logarithms of Trigonometric Functions; Table 33, Natural Trigonometric Functions; Table 34, Common Logarithms of Numbers.

* All units of measurement are ultimately derived from original prototype units (e.g., units of length, mass, temperature, and time). To make measurement units more effective tools the standards that physically define such units have been changed in recent years. For this reason slight variations in relationships between units of measures are found in the literature. For details consult the various publications on Measurement Standards by the National Bureau of Standards, Washington, D.C.

LOGARITHMS

Table 1 FIVE-PLACE COMMON LOGARITHMS OF NUMBERS†

100–150

N.	0	1	2	3	4	5	6	7	8	9
100	00 000	043	087	130	173	217	260	303	346	389
101	432	475	518	561	604	647	689	732	775	817
102	860	903	945	988	*030	*072	*115	*157	*199	*242
103	01 284	326	368	410	452	494	536	578	620	662
104	703	745	787	828	870	912	953	995	*036	*078
105	02 119	160	202	243	284	325	366	407	449	490
106	531	572	612	653	694	735	776	816	857	898
107	938	979	*019	*060	*100	*141	*181	*222	*262	*302
108	03 342	383	423	463	503	543	583	623	663	703
109	743	782	822	862	902	941	981	*021	*060	*100
110	04 139	179	218	258	297	336	376	415	454	493
111	532	571	610	650	689	727	766	805	844	883
112	922	961	999	*038	*077	*115	*154	*192	*231	*269
113	05 308	346	385	423	461	500	538	576	614	652
114	690	729	767	805	843	881	918	956	994	*032
115	06 070	108	145	183	221	258	296	333	371	408
116	446	483	521	558	595	633	670	707	744	781
117	819	856	893	930	967	*004	*041	*078	*115	*151
118	07 188	225	262	298	335	372	408	445	482	518
119	555	591	628	664	700	737	773	809	846	882
120	918	954	990	*027	*063	*099	*135	*171	*207	*243
121	08 279	314	350	386	422	458	493	529	565	600
122	636	672	707	743	778	814	849	884	920	955
123	991	*026	*061	*096	*132	*167	*202	*237	*272	*307
124	09 342	377	412	447	482	517	552	587	621	656
125	691	726	760	795	830	864	899	934	968	*003
126	10 037	072	106	140	175	209	243	278	312	346
127	380	415	449	483	517	551	585	619	653	687
128	721	755	789	823	857	890	924	958	992	*025
129	11 059	093	126	160	193	227	261	294	327	361
130	394	428	461	494	528	561	594	628	661	694
131	727	760	793	826	860	893	926	959	992	*024
132	12 057	090	123	156	189	222	254	287	320	352
133	385	418	450	483	516	548	581	613	646	678
134	710	743	775	808	840	872	905	937	969	*001
135	13 033	066	098	130	162	194	226	258	290	322
136	354	386	418	450	481	513	545	577	609	640
137	672	704	735	767	799	830	862	893	925	956
138	988	*019	*051	*082	*114	*145	*176	*208	*239	*270
139	14 301	333	364	395	426	457	489	520	551	582
140	613	644	675	706	737	768	799	829	860	891
141	922	953	983	*014	*045	*076	*106	*137	*168	*198
142	15 229	259	290	320	351	381	412	442	473	503
143	534	564	594	625	655	685	715	746	776	806
144	836	866	897	927	957	987	*017	*047	*077	*107
145	16 137	167	197	227	256	286	316	346	376	406
146	435	465	495	524	554	584	613	643	673	702
147	732	761	791	820	850	879	909	938	967	997
148	17 026	056	085	114	143	173	202	231	260	289
149	319	348	377	406	435	464	493	522	551	580
150	609	638	667	696	725	754	782	811	840	869
N.	0	1	2	3	4	5	6	7	8	9

Proportional parts

	44	43	42
1	4.4	4.3	4.2
2	8.8	8.6	8.4
3	13.2	12.9	12.6
4	17.6	17.2	16.8
5	22.0	21.5	21.0
6	26.4	25.8	25.2
7	30.8	30.1	29.4
8	35.2	34.4	33.6
9	39.6	38.7	37.8

	41	40	39
1	4.1	4.0	3.9
2	8.2	8.0	7.8
3	12.3	12.0	11.7
4	16.4	16 0	15.6
5	20.5	20.0	19.5
6	24.6	24.0	23.4
7	28.7	28.0	27.3
8	32.8	32.0	31.2
9	36.9	36.0	35.1

	38	37	36
1	3.8	3.7	3.6
2	7.6	7.4	7.2
3	11.4	11.1	10.8
4	15.2	14.8	14.4
5	19.0	18.5	18.0
6	22.8	22.2	21.6
7	26.6	25.9	25.2
8	30.4	29.6	28.8
9	34.2	33.3	32.4

	35	34	33
1	3.5	3.4	3.3
2	7.0	6.8	6.6
3	10.5	10.2	9.9
4	14.0	13.6	13.2
5	17.5	17.0	16.5
6	21.0	20.4	19.8
7	24.5	23.8	23.1
8	28.0	27.2	26.4
9	31.5	30.6	29.7

	32	31	30
1	3.2	3.1	3.0
2	6.4	6.2	6.0
3	9.6	9.3	9.0
4	12.8	12.4	12.0
5	16.0	15.5	15.0
6	19.2	18.6	18.0
7	22.4	21.7	21.0
8	25.6	24.8	24.0
9	28.8	27.9	27.0

Proportional parts

.00 000–.17 869

† Entries in Table 1 are values of $\log_{10} N$ for the indicated values of N.

150–200

N.	0	1	2	3	4	5	6	7	8	9
150	17 609	638	667	696	725	754	782	811	840	869
151	898	926	955	984	*013	*041	*070	*099	*127	*156
152	18 184	213	241	270	298	327	355	384	412	441
153	469	498	526	554	583	611	639	667	696	724
154	752	780	808	837	865	893	921	949	977	*005
155	19 033	061	089	117	145	173	201	229	257	285
156	312	340	368	396	424	451	479	507	535	562
157	590	618	645	673	700	728	756	783	811	838
158	866	893	921	948	976	*003	*030	*058	*085	*112
159	20 140	167	194	222	249	276	303	330	358	385
160	412	439	466	493	520	548	575	602	629	656
161	683	710	737	763	790	817	844	871	898	925
162	952	978	*005	*032	*059	*085	*112	*139	*165	*192
163	21 219	245	272	299	325	352	378	405	431	458
164	484	511	537	564	590	617	643	669	696	722
165	748	775	801	827	854	880	906	932	958	985
166	22 011	037	063	089	115	141	167	194	220	246
167	272	298	324	350	376	401	427	453	479	505
168	531	557	583	608	634	660	686	712	737	763
169	789	814	840	866	891	917	943	968	994	*019
170	23 045	070	096	121	147	172	198	223	249	274
171	300	325	350	376	401	426	452	477	502	528
172	553	578	603	629	654	679	704	729	754	779
173	805	830	855	880	905	930	955	980	*005	*030
174	24 055	080	105	130	155	180	204	229	254	279
175	304	329	353	378	403	428	452	477	502	527
176	551	576	601	625	650	674	699	724	748	773
177	797	822	846	871	895	920	944	969	993	*018
178	25 042	066	091	115	139	164	188	212	237	261
179	285	310	334	358	382	406	431	455	479	503
180	527	551	575	600	624	648	672	696	720	744
181	768	792	816	840	864	888	912	935	959	983
182	26 007	031	055	079	102	126	150	174	198	221
183	245	269	293	316	340	364	387	411	435	458
184	482	505	529	553	576	600	623	647	670	694
185	717	741	764	788	811	834	858	881	905	928
186	951	975	998	*021	*045	*068	*091	*114	*138	*161
187	27 184	207	231	254	277	300	323	346	370	393
188	416	439	462	485	508	531	554	577	600	623
189	646	669	692	715	738	761	784	807	830	852
190	875	898	921	944	967	989	*012	*035	*058	*081
191	28 103	126	149	171	194	217	240	262	285	307
192	330	353	375	398	421	443	466	488	511	533
193	556	578	601	623	646	668	691	713	735	758
194	780	803	825	847	870	892	914	937	959	981
195	29 003	026	048	070	092	115	137	159	181	203
196	226	248	270	292	314	336	358	380	403	425
197	447	469	491	513	535	557	579	601	623	645
198	667	688	710	732	754	776	798	820	842	863
199	885	907	929	951	973	994	*016	*038	*060	*081
200	30 103	125	146	168	190	211	233	255	276	298
N.	0	1	2	3	4	5	6	7	8	9

Proportional parts

	29	28	27	26	25	24	23	22	21
1	2.9	2.8	2.7	2.6	2.5	2.4	2.3	2.2	2.1
2	5.8	5.6	5.4	5.2	5.0	4.8	4.6	4.4	4.2
3	8.7	8.4	8.1	7.8	7.5	7.2	6.9	6.6	6.3
4	11.6	11.2	10.8	10.4	10.0	9.6	9.2	8.8	8.4
5	14.5	14.0	13.5	13.0	12.5	12.0	11.5	11.0	10.5
6	17.4	16.8	16.2	15.6	15.0	14.4	13.8	13.2	12.6
7	20.3	19.6	18.9	18.2	17.5	16.8	16.1	15.4	14.7
8	23.2	22.4	21.6	20.8	20.0	19.2	18.4	17.6	16.8
9	26.1	25.2	24.3	23.4	22.5	21.6	20.7	19.8	18.9

200–250

N.	0	1	2	3	4	5	6	7	8	9
200	30 103	125	146	168	190	211	233	255	276	298
201	320	341	363	384	406	428	449	471	492	514
202	535	557	578	600	621	643	664	685	707	728
203	750	771	792	814	835	856	878	899	920	942
204	963	984	*006	*027	*048	*069	*091	*112	*133	*154
205	31 175	197	218	239	260	281	302	323	345	366
206	387	408	429	450	471	492	513	534	555	576
207	597	618	639	660	681	702	723	744	765	785
208	806	827	848	869	890	911	931	952	973	994
209	32 015	035	056	077	098	118	139	160	181	201
210	222	243	263	284	305	325	346	366	387	408
211	428	449	469	490	510	531	552	572	593	613
212	634	654	675	695	715	736	756	777	797	818
213	838	858	879	899	919	940	960	980	*001	*021
214	33 041	062	082	102	122	143	163	183	203	224
215	244	264	284	304	325	345	365	385	405	425
216	445	465	486	506	526	546	566	586	606	626
217	646	666	686	706	726	746	766	786	806	826
218	846	866	885	905	925	945	965	985	*005	*025
219	34 044	064	084	104	124	143	163	183	203	223
220	242	262	282	301	321	341	361	380	400	420
221	439	459	479	498	518	537	557	577	596	616
222	635	655	674	694	713	733	753	772	792	811
223	830	850	869	889	908	928	947	967	986	*005
224	35 025	044	064	083	102	122	141	160	180	199
225	218	238	257	276	295	315	334	353	372	392
226	411	430	449	468	488	507	526	545	564	583
227	603	622	641	660	679	698	717	736	755	774
228	793	813	832	851	870	889	908	927	946	965
229	984	*003	*021	*040	*059	*078	*097	*116	*135	*154
230	36 173	192	211	229	248	267	286	305	324	342
231	361	380	399	418	436	455	474	493	511	530
232	549	568	586	605	624	642	661	680	698	717
233	736	754	773	791	810	829	847	866	884	903
234	922	940	959	977	996	*014	*033	*051	*070	*088
235	37 107	125	144	162	181	199	218	236	254	273
236	291	310	328	346	365	383	401	420	438	457
237	475	493	511	530	548	566	585	603	621	639
238	658	676	694	712	731	749	767	785	803	822
239	840	858	876	894	912	931	949	967	985	*003
240	38 021	039	057	075	093	112	130	148	166	184
241	202	220	238	256	274	292	310	328	346	364
242	382	399	417	435	453	471	489	507	525	543
243	561	578	596	614	632	650	668	686	703	721
244	739	757	775	792	810	828	846	863	881	899
245	917	934	952	970	987	*005	*023	*041	*058	*076
246	39 094	111	129	146	164	182	199	217	235	252
247	270	287	305	322	340	358	375	393	410	428
248	445	463	480	498	515	533	550	568	585	602
249	620	637	655	672	690	707	724	742	759	777
250	794	811	829	846	863	881	898	915	933	950
N.	0	1	2	3	4	5	6	7	8	9

Proportional parts

	22	21
1	2.2	2.1
2	4.4	4.2
3	6.6	6.3
4	8.8	8.4
5	11.0	10.5
6	13.2	12.6
7	15.4	14.7
8	17.6	16.8
9	19.8	18.9

	20
1	2.0
2	4.0
3	6.0
4	8.0
5	10.0
6	12.0
7	14.0
8	16.0
9	18.0

	19
1	1.9
2	3.8
3	5.7
4	7.6
5	9.5
6	11.4
7	13.3
8	15.2
9	17.1

	18
1	1.8
2	3.6
3	5.4
4	7.2
5	9.0
6	10.8
7	12.6
8	14.4
9	16.2

	17
1	1.7
2	3.4
3	5.1
4	6.8
5	8.5
6	10.2
7	11.9
8	13.6
9	15.3

Proportional parts

.30 103–.39 950

250–300

N.	0	1	2	3	4	5	6	7	8	9
250	39 794	811	829	846	863	881	898	915	933	950
251	967	985	*002	*019	*037	*054	*071	*088	*106	*123
252	40 140	157	175	192	209	226	243	261	278	295
253	312	329	346	364	381	398	415	432	449	466
254	483	500	518	535	552	569	586	603	620	637
255	654	671	688	705	722	739	756	773	790	807
256	824	841	858	875	892	909	926	943	960	976
257	993	*010	*027	*044	*061	*078	*095	*111	*128	*145
258	41 162	179	196	212	229	246	263	280	296	313
259	330	347	363	380	397	414	430	447	464	481
260	497	514	531	547	564	581	597	614	631	647
261	664	681	697	714	731	747	764	780	797	814
262	830	847	863	880	896	913	929	946	963	979
263	996	*012	*029	*045	*062	*078	*095	*111	*127	*144
264	42 160	177	193	210	226	243	259	275	292	308
265	325	341	357	374	390	406	423	439	455	472
266	488	504	521	537	553	570	586	602	619	635
267	651	667	684	700	716	732	749	765	781	797
268	813	830	846	862	878	894	911	927	943	959
269	975	991	*008	*024	*040	*056	*072	*088	*104	*120
270	43 136	152	169	185	201	217	233	249	265	281
271	297	313	329	345	361	377	393	409	425	441
272	457	473	489	505	521	537	553	569	584	600
273	616	632	648	664	680	696	712	727	743	759
274	775	791	807	823	838	854	870	886	902	917
275	933	949	965	981	996	*012	*028	*044	*059	*075
276	44 091	107	122	138	154	170	185	201	217	232
277	248	264	279	295	311	326	342	358	373	389
278	404	420	436	451	467	483	498	514	529	545
279	560	576	592	607	623	638	654	669	685	700
280	716	731	747	762	778	793	809	824	840	855
281	871	886	902	917	932	948	963	979	994	*010
282	45 025	040	056	071	086	102	117	133	148	163
283	179	194	209	225	240	255	271	286	301	317
284	332	347	362	378	393	408	423	439	454	469
285	484	500	515	530	545	561	576	591	606	621
286	637	652	667	682	697	712	728	743	758	773
287	788	803	818	834	849	864	879	894	909	924
288	939	954	969	984	*000	*015	*030	*045	*060	*075
289	46 090	105	120	135	150	165	180	195	210	225
290	240	255	270	285	300	315	330	345	359	374
291	389	404	419	434	449	464	479	494	509	523
292	538	553	568	583	598	613	627	642	657	672
293	687	702	716	731	746	761	776	790	805	820
294	835	850	864	879	894	909	923	938	953	967
295	982	997	*012	*026	*041	*056	*070	*085	*100	*114
296	47 129	144	159	173	188	202	217	232	246	261
297	276	290	305	319	334	349	363	378	392	407
298	422	436	451	465	480	494	509	524	538	553
299	567	582	596	611	625	640	654	669	683	698
300	712	727	741	756	770	784	799	813	828	842
N.	0	1	2	3	4	5	6	7	8	9

Proportional parts

	18	17	16	15	14
1	1.8	1.7	1.6	1.5	1.4
2	3.6	3.4	3.2	3.0	2.8
3	5.4	5.1	4.8	4.5	4.2
4	7.2	6.8	6.4	6.0	5.6
5	9.0	8.5	8.0	7.5	7.0
6	10.8	10.2	9.6	9.0	8.4
7	12.6	11.9	11.2	10.5	9.8
8	14.4	13.6	12.8	12.0	11.2
9	16.2	15.3	14.4	13.5	12.6

log e = 0.43429

Proportional parts

300–350

N.	0	1	2	3	4	5	6	7	8	9
300	47 712	727	741	756	770	784	799	813	828	842
301	857	871	885	900	914	929	943	958	972	986
302	48 001	015	029	044	058	073	087	101	116	130
303	144	159	173	187	202	216	230	244	259	273
304	287	302	316	330	344	359	373	387	401	416
305	430	444	458	473	487	501	515	530	544	558
306	572	586	601	615	629	643	657	671	686	700
307	714	728	742	756	770	785	799	813	827	841
308	855	869	883	897	911	926	940	954	968	982
309	996	*010	*024	*038	*052	*066	*080	*094	*108	*122
310	49 136	150	164	178	192	206	220	234	248	262
311	276	290	304	318	332	346	360	374	388	402
312	415	429	443	457	471	485	499	513	527	541
313	554	568	582	596	610	624	638	651	665	679
314	693	707	721	734	748	762	776	790	803	817
315	831	845	859	872	886	900	914	927	941	955
316	969	982	996	*010	*024	*037	*051	*065	*079	*092
317	50 106	120	133	147	161	174	188	202	215	229
318	243	256	270	284	297	311	325	338	352	365
319	379	393	406	420	433	447	461	474	488	501
320	515	529	542	556	569	583	596	610	623	637
321	651	664	678	691	705	718	732	745	759	772
322	786	799	813	826	840	853	866	880	893	907
323	920	934	947	961	974	987	*001	*014	*028	*041
324	51 055	068	081	095	108	121	135	148	162	175
325	188	202	215	228	242	255	268	282	295	308
326	322	335	348	362	375	388	402	415	428	441
327	455	468	481	495	508	521	534	548	561	574
328	587	601	614	627	640	654	667	680	693	706
329	720	733	746	759	772	786	799	812	825	838
330	851	865	878	891	904	917	930	943	957	970
331	983	996	*009	*022	*035	*048	*061	*075	*088	*101
332	52 114	127	140	153	166	179	192	205	218	231
333	244	257	270	284	297	310	323	336	349	362
334	375	388	401	414	427	440	453	466	479	492
335	504	517	530	543	556	569	582	595	608	621
336	634	647	660	673	686	699	711	724	737	750
337	763	776	789	802	815	827	840	853	866	879
338	892	905	917	930	943	956	969	982	994	*007
339	53 020	033	046	058	071	084	097	110	122	135
340	148	161	173	186	199	212	224	237	250	263
341	275	288	301	314	326	339	352	364	377	390
342	403	415	428	441	453	466	479	491	504	517
343	529	542	555	567	580	593	605	618	631	643
344	656	668	681	694	706	719	732	744	757	769
345	782	794	807	820	832	845	857	870	882	895
346	908	920	933	945	958	970	983	995	*008	*020
347	54 033	045	058	070	083	095	108	120	133	145
348	158	170	183	195	208	220	233	245	258	270
349	283	295	307	320	332	345	357	370	382	394
350	407	419	432	444	456	469	481	494	506	518
N.	0	1	2	3	4	5	6	7	8	9

Proportional parts

	15	14	13	12
1	1.5	1.4	1.3	1.2
2	3.0	2.8	2.6	2.4
3	4.5	4.2	3.9	3.6
4	6.0	5.6	5.2	4.8
5	7.5	7.0	6.5	6.0
6	9.0	8.4	7.8	7.2
7	10.5	9.8	9.1	8.4
8	12.0	11.2	10.4	9.6
9	13.5	12.6	11.7	10.8

$\log \pi = 0.49715$

350–400

N.		0	1	2	3	4	5	6	7	8	9
350	54	407	419	432	444	456	469	481	494	506	518
351		531	543	555	568	580	593	605	617	630	642
352		654	667	679	691	704	716	728	741	753	765
353		777	790	802	814	827	839	851	864	876	888
354		900	913	925	937	949	962	974	986	998	*011
355	55	023	035	047	060	072	084	096	108	121	133
356		145	157	169	182	194	206	218	230	242	255
357		267	279	291	303	315	328	340	352	364	376
358		388	400	413	425	437	449	461	473	485	497
359		509	522	534	546	558	570	582	594	606	618
360		630	642	654	666	678	691	703	715	727	739
361		751	763	775	787	799	811	823	835	847	859
362		871	883	895	907	919	931	943	955	967	979
363		991	*003	*015	*027	*038	*050	*062	*074	*086	*098
364	56	110	122	134	146	158	170	182	194	205	217
365		229	241	253	265	277	289	301	312	324	336
366		348	360	372	384	396	407	419	431	443	455
367		467	478	490	502	514	526	538	549	561	573
368		585	597	608	620	632	644	656	667	679	691
369		703	714	726	738	750	761	773	785	797	808
370		820	832	844	855	867	879	891	902	914	926
371		937	949	961	972	984	996	*008	*019	*031	*043
372	57	054	066	078	089	101	113	124	136	148	159
373		171	183	194	206	217	229	241	252	264	276
374		287	299	310	322	334	345	357	368	380	392
375		403	415	426	438	449	461	473	484	496	507
376		519	530	542	553	565	576	588	600	611	623
377		634	646	657	669	680	692	703	715	726	738
378		749	761	772	784	795	807	818	830	841	852
379		864	875	887	898	910	921	933	944	955	967
380		978	990	*001	*013	*024	*035	*047	*058	*070	*081
381	58	092	104	115	127	138	149	161	172	184	195
382		206	218	229	240	252	263	274	286	297	309
383		320	331	343	354	365	377	388	399	410	422
384		433	444	456	467	478	490	501	512	524	535
385		546	557	569	580	591	602	614	625	636	647
386		659	670	681	692	704	715	726	737	749	760
387		771	782	794	805	816	827	838	850	861	872
388		883	894	906	917	928	939	950	961	973	984
389		995	*006	*017	*028	*040	*051	*062	*073	*084	*095
390	59	106	118	129	140	151	162	173	184	195	207
391		218	229	240	251	262	273	284	295	306	318
392		329	340	351	362	373	384	395	406	417	428
393		439	450	461	472	483	494	506	517	528	539
394		550	561	572	583	594	605	616	627	638	649
395		660	671	682	693	704	715	726	737	748	759
396		770	780	791	802	813	824	835	846	857	868
397		879	890	901	912	923	934	945	956	966	977
398		988	999	*010	*021	*032	*043	*054	*065	*076	*086
399	60	097	108	119	130	141	152	163	173	184	195
400		206	217	228	239	249	260	271	282	293	304
N.		0	1	2	3	4	5	6	7	8	9

Proportional parts

	13	12	11	10
1	1.3	1.2	1.1	1.0
2	2.6	2.4	2.2	2.0
3	3.9	3.6	3.3	3.0
4	5.2	4.8	4.4	4.0
5	6.5	6.0	5.5	5.0
6	7.8	7.2	6.6	6.0
7	9.1	8.4	7.7	7.0
8	10.4	9.6	8.8	8.0
9	11.7	10.8	9.9	9.0

Proportional parts

400–450

N.	0	1	2	3	4	5	6	7	8	9
400	60 206	217	228	239	249	260	271	282	293	304
401	314	325	336	347	358	369	379	390	401	412
402	423	433	444	455	466	477	487	498	509	520
403	531	541	552	563	574	584	595	606	617	627
404	638	649	660	670	681	692	703	713	724	735
405	746	756	767	778	788	799	810	821	831	842
406	853	863	874	885	895	906	917	927	938	949
407	959	970	981	991	*002	*013	*023	*034	*045	*055
408	61 066	077	087	098	109	119	130	140	151	162
409	172	183	194	204	215	225	236	247	257	268
410	278	289	300	310	321	331	342	352	363	374
411	384	395	405	416	426	437	448	458	469	479
412	490	500	511	521	532	542	553	563	574	584
413	595	606	616	627	637	648	658	669	679	690
414	700	711	721	731	742	752	763	773	784	794
415	805	815	826	836	847	857	868	878	888	899
416	909	920	930	941	951	962	972	982	993	*003
417	62 014	024	034	045	055	066	076	086	097	107
418	118	128	138	149	159	170	180	190	201	211
419	221	232	242	252	263	273	284	294	304	315
420	325	335	346	356	366	377	387	397	408	418
421	428	439	449	459	469	480	490	500	511	521
422	531	542	552	562	572	583	593	603	613	624
423	634	644	655	665	675	685	696	706	716	726
424	737	747	757	767	778	788	798	808	818	829
425	839	849	859	870	880	890	900	910	921	931
426	941	951	961	972	982	992	*002	*012	*022	*033
427	63 043	053	063	073	083	094	104	114	124	134
428	144	155	165	175	185	195	205	215	225	236
429	246	256	266	276	286	296	306	317	327	337
430	347	357	367	377	387	397	407	417	428	438
431	448	458	468	478	488	498	508	518	528	538
432	548	558	568	579	589	599	609	619	629	639
433	649	659	669	679	689	699	709	719	729	739
434	749	759	769	779	789	799	809	819	829	839
435	849	859	869	879	889	899	909	919	929	939
436	949	959	969	979	988	998	*008	*018	*028	*038
437	64 048	058	068	078	088	098	108	118	128	137
438	147	157	167	177	187	197	207	217	227	237
439	246	256	266	276	286	296	306	316	326	335
440	345	355	365	375	385	395	404	414	424	434
441	444	454	464	473	483	493	503	513	523	532
442	542	552	562	572	582	591	601	611	621	631
443	640	650	660	670	680	689	699	709	719	729
444	738	748	758	768	777	787	797	807	816	826
445	836	846	856	865	875	885	895	904	914	924
446	933	943	953	963	972	982	992	*002	*011	*021
447	65 031	040	050	060	070	079	089	099	108	118
448	128	137	147	157	167	176	186	196	205	215
449	225	234	244	254	263	273	283	292	302	312
450	321	331	341	350	360	369	379	389	398	408
N.	0	1	2	3	4	5	6	7	8	9

Proportional parts

	11
1	1.1
2	2.2
3	3.3
4	4.4
5	5.5
6	6.6
7	7.7
8	8.8
9	9.9

	10
1	1.0
2	2.0
3	3.0
4	4.0
5	5.0
6	6.0
7	7.0
8	8.0
9	9.0

	9
1	0.9
2	1.8
3	2.7
4	3.6
5	4.5
6	5.4
7	6.3
8	7.2
9	8.1

Proportional parts

.60 206–.65 408

450–500

N.		0	1	2	3	4	5	6	7	8	9
450	65	321	331	341	350	360	369	379	389	398	408
451		418	427	437	447	456	466	475	485	495	504
452		514	523	533	543	552	562	571	581	591	600
453		610	619	629	639	648	658	667	677	686	696
454		706	715	725	734	744	753	763	772	782	792
455		801	811	820	830	839	849	858	868	877	887
456		896	906	916	925	935	944	954	963	973	982
457		992	*001	*011	*020	*030	*039	*049	*058	*068	*077
458	66	087	096	106	115	124	134	143	153	162	172
459		181	191	200	210	219	229	238	247	257	266
460		276	285	295	304	314	323	332	342	351	361
461		370	380	389	398	408	417	427	436	445	455
462		464	474	483	492	502	511	521	530	539	549
463		558	567	577	586	596	605	614	624	633	642
464		652	661	671	680	689	699	708	717	727	736
465		745	755	764	773	783	792	801	811	820	829
466		839	848	857	867	876	885	894	904	913	922
467		932	941	950	960	969	978	987	997	*006	*015
468	67	025	034	043	052	062	071	080	089	099	108
469		117	127	136	145	154	164	173	182	191	201
470		210	219	228	237	247	256	265	274	284	293
471		302	311	321	330	339	348	357	367	376	385
472		394	403	413	422	431	440	449	459	468	477
473		486	495	504	514	523	532	541	550	560	569
474		578	587	596	605	614	624	633	642	651	660
475		669	679	688	697	706	715	724	733	742	752
476		761	770	779	788	797	806	815	825	834	843
477		852	861	870	879	888	897	906	916	925	934
478		943	952	961	970	979	988	997	*006	*015	*024
479	68	034	043	052	061	070	079	088	097	106	115
480		124	133	142	151	160	169	178	187	196	205
481		215	224	233	242	251	260	269	278	287	296
482		305	314	323	332	341	350	359	368	377	386
483		395	404	413	422	431	440	449	458	467	476
484		485	494	502	511	520	529	538	547	556	565
485		574	583	592	601	610	619	628	637	646	655
486		664	673	681	690	699	708	717	726	735	744
487		753	762	771	780	789	797	806	815	824	833
488		842	851	860	869	878	886	895	904	913	922
489		931	940	949	958	966	975	984	993	*002	*011
490	69	020	028	037	046	055	064	073	082	090	099
491		108	117	126	135	144	152	161	170	179	188
492		197	205	214	223	232	241	249	258	267	276
493		285	294	302	311	320	329	338	346	355	364
494		373	381	390	399	408	417	425	434	443	452
495		461	469	478	487	496	504	513	522	531	539
496		548	557	566	574	583	592	601	609	618	627
497		636	644	653	662	671	679	688	697	705	714
498		723	732	740	749	758	767	775	784	793	801
499		810	819	827	836	845	854	862	871	880	888
500		897	906	914	923	932	940	949	958	966	975
N.		0	1	2	3	4	5	6	7	8	9

Proportional parts

	10
1	1.0
2	2.0
3	3.0
4	4.0
5	5.0
6	6.0
7	7.0
8	8.0
9	9.0

	9
1	0.9
2	1.8
3	2.7
4	3.6
5	4.5
6	5.4
7	6.3
8	7.2
9	8.1

	8
1	0.8
2	1.6
3	2.4
4	3.2
5	4.0
6	4.8
7	5.6
8	6.4
9	7.2

Proportional parts

.65 321–.69 975

500–550

N.	0	1	2	3	4	5	6	7	8	9
500	69 897	906	914	923	932	940	949	958	966	975
501	984	992	*001	*010	*018	*027	*036	*044	*053	*062
502	70 070	079	088	096	105	114	122	131	140	148
503	157	165	174	183	191	200	209	217	226	234
504	243	252	260	269	278	286	295	303	312	321
505	329	338	346	355	364	372	381	389	398	406
506	415	424	432	441	449	458	467	475	484	492
507	501	509	518	526	535	544	552	561	569	578
508	586	595	603	612	621	629	638	646	655	663
509	672	680	689	697	706	714	723	731	740	749
510	757	766	774	783	791	800	808	817	825	834
511	842	851	859	868	876	885	893	902	910	919
512	927	935	944	952	961	969	978	986	995	*003
513	71 012	020	029	037	046	054	063	071	079	088
514	096	105	113	122	130	139	147	155	164	172
515	181	189	198	206	214	223	231	240	248	257
516	265	273	282	290	299	307	315	324	332	341
517	349	357	366	374	383	391	399	408	416	425
518	433	441	450	458	466	475	483	492	500	508
519	517	525	533	542	550	559	567	575	584	592
520	600	609	617	625	634	642	650	659	667	675
521	684	692	700	709	717	725	734	742	750	759
522	767	775	784	792	800	809	817	825	834	842
523	850	858	867	875	883	892	900	908	917	925
524	933	941	950	958	966	975	983	991	999	*008
525	72 016	024	032	041	049	057	066	074	082	090
526	099	107	115	123	132	140	148	156	165	173
527	181	189	198	206	214	222	230	239	247	255
528	263	272	280	288	296	304	313	321	329	337
529	346	354	362	370	378	387	395	403	411	419
530	428	436	444	452	460	469	477	485	493	501
531	509	518	526	534	542	550	558	567	575	583
532	591	599	607	616	624	632	640	648	656	665
533	673	681	689	697	705	713	722	730	738	746
534	754	762	770	779	787	795	803	811	819	827
535	835	843	852	860	868	876	884	892	900	908
536	916	925	933	941	949	957	965	973	981	989
537	997	*006	*014	*022	*030	*038	*046	*054	*062	*070
538	73 078	086	094	102	111	119	127	135	143	151
539	159	167	175	183	191	199	207	215	223	231
540	239	247	255	263	272	280	288	296	304	312
541	320	328	336	344	352	360	368	376	384	392
542	400	408	416	424	432	440	448	456	464	472
543	480	488	496	504	512	520	528	536	544	552
544	560	568	576	584	592	600	608	616	624	632
545	640	648	656	664	672	679	687	695	703	711
546	719	727	735	743	751	759	767	775	783	791
547	799	807	815	823	830	838	846	854	862	870
548	878	886	894	902	910	918	926	933	941	949
549	957	965	973	981	989	997	*005	*013	*020	*028
550	74 036	044	052	060	068	076	084	092	099	107
N.	0	1	2	3	4	5	6	7	8	9

Proportional parts

	9
1	0.9
2	1.8
3	2.7
4	3.6
5	4.5
6	5.4
7	6.3
8	7.2
9	8.1

	8
1	0.8
2	1.6
3	2.4
4	3.2
5	4.0
6	4.8
7	5.6
8	6.4
9	7.2

	7
1	0.7
2	1.4
3	2.1
4	2.8
5	3.5
6	4.2
7	4.9
8	5.6
9	6.3

Proportional parts

550–600

N.		0	1	2	3	4	5	6	7	8	9
550	74	036	044	052	060	068	076	084	092	099	107
551		115	123	131	139	147	155	162	170	178	186
552		194	202	210	218	225	233	241	249	257	265
553		273	280	288	296	304	312	320	327	335	343
554		351	359	367	374	382	390	398	406	414	421
555		429	437	445	453	461	468	476	484	492	500
556		507	515	523	531	539	547	554	562	570	578
557		586	593	601	609	617	624	632	640	648	656
558		663	671	679	687	695	702	710	718	726	733
559		741	749	757	764	772	780	788	796	803	811
560		819	827	834	842	850	858	865	873	881	889
561		896	904	912	920	927	935	943	950	958	966
562		974	981	989	997	*005	*012	*020	*028	*035	*043
563	75	051	059	066	074	082	089	097	105	113	120
564		128	136	143	151	159	166	174	182	189	197
565		205	213	220	228	236	243	251	259	266	274
566		282	289	297	305	312	320	328	335	343	351
567		358	366	374	381	389	397	404	412	420	427
568		435	442	450	458	465	473	481	488	496	504
569		511	519	526	534	542	549	557	565	572	580
570		587	595	603	610	618	626	633	641	648	656
571		664	671	679	686	694	702	709	717	724	732
572		740	747	755	762	770	778	785	793	800	808
573		815	823	831	838	846	853	861	868	876	884
574		891	899	906	914	921	929	937	944	952	959
575		967	974	982	989	997	*005	*012	*020	*027	*035
576	76	042	050	057	065	072	080	087	095	103	110
577		118	125	133	140	148	155	163	170	178	185
578		193	200	208	215	223	230	238	245	253	260
579		268	275	283	290	298	305	313	320	328	335
580		343	350	358	365	373	380	388	395	403	410
581		418	425	433	440	448	455	462	470	477	485
582		492	500	507	515	522	530	537	545	552	559
583		567	574	582	589	597	604	612	619	626	634
584		641	649	656	664	671	678	686	693	701	708
585		716	723	730	738	745	753	760	768	775	782
586		790	797	805	812	819	827	834	842	849	856
587		864	871	879	886	893	901	908	916	923	930
588		938	945	953	960	967	975	982	989	997	*004
589	77	012	019	026	034	041	048	056	063	070	078
590		085	093	100	107	115	122	129	137	144	151
591		159	166	173	181	188	195	203	210	217	225
592		232	240	247	254	262	269	276	283	291	298
593		305	313	320	327	335	342	349	357	364	371
594		379	386	393	401	408	415	422	430	437	444
595		452	459	466	474	481	488	495	503	510	517
596		525	532	539	546	554	561	568	576	583	590
597		597	605	612	619	627	634	641	648	656	663
598		670	677	685	692	699	706	714	721	728	735
599		743	750	757	764	772	779	786	793	801	808
600		815	822	830	837	844	851	859	866	873	880
N.		0	1	2	3	4	5	6	7	8	9

Proportional parts

	8
1	0.8
2	1.6
3	2.4
4	3.2
5	4.0
6	4.8
7	5.6
8	6.4
9	7.2

	7
1	0.7
2	1.4
3	2.1
4	2.8
5	3.5
6	4.2
7	4.9
8	5.6
9	6.3

Proportional parts

600–650

N.	0	1	2	3	4	5	6	7	8	9
600	77 815	822	830	837	844	851	859	866	873	880
601	887	895	902	909	916	924	931	938	945	952
602	960	967	974	981	988	996	*003	*010	*017	*025
603	78 032	039	046	053	061	068	075	082	089	097
604	104	111	118	125	132	140	147	154	161	168
605	176	183	190	197	204	211	219	226	233	240
606	247	254	262	269	276	283	290	297	305	312
607	319	326	333	340	347	355	362	369	376	383
608	390	398	405	412	419	426	433	440	447	455
609	462	469	476	483	490	497	504	512	519	526
610	533	540	547	554	561	569	576	583	590	597
611	604	611	618	625	633	640	647	654	661	668
612	675	682	689	696	704	711	718	725	732	739
613	746	753	760	767	774	781	789	796	803	810
614	817	824	831	838	845	852	859	866	873	880
615	888	895	902	909	916	923	930	937	944	951
616	958	965	972	979	986	993	*000	*007	*014	*021
617	79 029	036	043	050	057	064	071	078	085	092
618	099	106	113	120	127	134	141	148	155	162
619	169	176	183	190	197	204	211	218	225	232
620	239	246	253	260	267	274	281	288	295	302
621	309	316	323	330	337	344	351	358	365	372
622	379	386	393	400	407	414	421	428	435	442
623	449	456	463	470	477	484	491	498	505	511
624	518	525	532	539	546	553	560	567	574	581
625	588	595	602	609	616	623	630	637	644	650
626	657	664	671	678	685	692	699	706	713	720
627	727	734	741	748	754	761	768	775	782	789
628	796	803	810	817	824	831	837	844	851	858
629	865	872	879	886	893	900	906	913	920	927
630	934	941	948	955	962	969	975	982	989	996
631	80 003	010	017	024	030	037	044	051	058	065
632	072	079	085	092	099	106	113	120	127	134
633	140	147	154	161	168	175	182	188	195	202
634	209	216	223	229	236	243	250	257	264	271
635	277	284	291	298	305	312	318	325	332	339
636	346	353	359	366	373	380	387	393	400	407
637	414	421	428	434	441	448	455	462	468	475
638	482	489	496	502	509	516	523	530	536	543
639	550	557	564	570	577	584	591	598	604	611
640	618	625	632	638	645	652	659	665	672	679
641	686	693	699	706	713	720	726	733	740	747
642	754	760	767	774	781	787	794	801	808	814
643	821	828	835	841	848	855	862	868	875	882
644	889	895	902	909	916	922	929	936	943	949
645	956	963	969	976	983	990	996	*003	*010	*017
646	81 023	030	037	043	050	057	064	070	077	084
647	090	097	104	111	117	124	131	137	144	151
648	158	164	171	178	184	191	198	204	211	218
649	224	231	238	245	251	258	265	271	278	285
650	291	298	305	311	318	325	331	338	345	351
N.	0	1	2	3	4	5	6	7	8	9

Proportional parts

	8
1	0.8
2	1.6
3	2.4
4	3.2
5	4.0
6	4.8
7	5.6
8	6.4
9	7.2

	7
1	0.7
2	1.4
3	2.1
4	2.8
5	3.5
6	4.2
7	4.9
8	5.6
9	6.3

	6
1	0.6
2	1.2
3	1.8
4	2.4
5	3.0
6	3.6
7	4.2
8	4.8
9	5.4

Proportional parts

650–700

N.		0	1	2	3	4	5	6	7	8	9
650	81	291	298	305	311	318	325	331	338	345	351
651		358	365	371	378	385	391	398	405	411	418
652		425	431	438	445	451	458	465	471	478	485
653		491	498	505	511	518	525	531	538	544	551
654		558	564	571	578	584	591	598	604	611	617
655		624	631	637	644	651	657	664	671	677	684
656		690	697	704	710	717	723	730	737	743	750
657		757	763	770	776	783	790	796	803	809	816
658		823	829	836	842	849	856	862	869	875	882
659		889	895	902	908	915	921	928	935	941	948
660		954	961	968	974	981	987	994	*000	*007	*014
661	82	020	027	033	040	046	053	060	066	073	079
662		086	092	099	105	112	119	125	132	138	145
663		151	158	164	171	178	184	191	197	204	210
664		217	223	230	236	243	249	256	263	269	276
665		282	289	295	302	308	315	321	328	334	341
666		347	354	360	367	373	380	387	393	400	406
667		413	419	426	432	439	445	452	458	465	471
668		478	484	491	497	504	510	517	523	530	536
669		543	549	556	562	569	575	582	588	595	601
670		607	614	620	627	633	640	646	653	659	666
671		672	679	685	692	698	705	711	718	724	730
672		737	743	750	756	763	769	776	782	789	795
673		802	808	814	821	827	834	840	847	853	860
674		866	872	879	885	892	898	905	911	918	924
675		930	937	943	950	956	963	969	975	982	988
676		995	*001	*008	*014	*020	*027	*033	*040	*046	*052
677	83	059	065	072	078	085	091	097	104	110	117
678		123	129	136	142	149	155	161	168	174	181
679		187	193	200	206	213	219	225	232	238	245
680		251	257	264	270	276	283	289	296	302	308
681		315	321	327	334	340	347	353	359	366	372
682		378	385	391	398	404	410	417	423	429	436
683		442	448	455	461	467	474	480	487	493	499
684		506	512	518	525	531	537	544	550	556	563
685		569	575	582	588	594	601	607	613	620	626
686		632	639	645	651	658	664	670	677	683	689
687		696	702	708	715	721	727	734	740	746	753
688		759	765	771	778	784	790	797	803	809	816
689		822	828	835	841	847	853	860	866	872	879
690		885	891	897	904	910	916	923	929	935	942
691		948	954	960	967	973	979	985	992	998	*004
692	84	011	017	023	029	036	042	048	055	061	067
693		073	080	086	092	098	105	111	117	123	130
694		136	142	148	155	161	167	173	180	186	192
695		198	205	211	217	223	230	236	242	248	255
696		261	267	273	280	286	292	298	305	311	317
697		323	330	336	342	348	354	361	367	373	379
698		386	392	398	404	410	417	423	429	435	442
699		448	454	460	466	473	479	485	491	497	504
700		510	516	522	528	535	541	547	553	559	566
N.		0	1	2	3	4	5	6	7	8	9

Proportional parts

	7
1	0.7
2	1.4
3	2.1
4	2.8
5	3.5
6	4.2
7	4.9
8	5.6
9	6.3

	6
1	0.6
2	1.2
3	1.8
4	2.4
5	3.0
6	3.6
7	4.2
8	4.8
9	5.4

Proportional parts

700–750

N.	0	1	2	3	4	5	6	7	8	9
700	84 510	516	522	528	535	541	547	553	559	566
701	572	578	584	590	597	603	609	615	621	628
702	634	640	646	652	658	665	671	677	683	689
703	696	702	708	714	720	726	733	739	745	751
704	757	763	770	776	782	788	794	800	807	813
705	819	825	831	837	844	850	856	862	868	874
706	880	887	893	899	905	911	917	924	930	936
707	942	948	954	960	967	973	979	985	991	997
708	85 003	009	016	022	028	034	040	046	052	058
709	065	071	077	083	089	095	101	107	114	120
710	126	132	138	144	150	156	163	169	175	181
711	187	193	199	205	211	217	224	230	236	242
712	248	254	260	266	272	278	285	291	297	303
713	309	315	321	327	333	339	345	352	358	364
714	370	376	382	388	394	400	406	412	418	425
715	431	437	443	449	455	461	467	473	479	485
716	491	497	503	509	516	522	528	534	540	546
717	552	558	564	570	576	582	588	594	600	606
718	612	618	625	631	637	643	649	655	661	667
719	673	679	685	691	697	703	709	715	721	727
720	733	739	745	751	757	763	769	775	781	788
721	794	800	806	812	818	824	830	836	842	848
722	854	860	866	872	878	884	890	896	902	908
723	914	920	926	932	938	944	950	956	962	968
724	974	980	986	992	998	*004	*010	*016	*022	*028
725	86 034	040	046	052	058	064	070	076	082	088
726	094	100	106	112	118	124	130	136	141	147
727	153	159	165	171	177	183	189	195	201	207
728	213	219	225	231	237	243	249	255	261	267
729	273	279	285	291	297	303	308	314	320	326
730	332	338	344	350	356	362	368	374	380	386
731	392	398	404	410	415	421	427	433	439	445
732	451	457	463	469	475	481	487	493	499	504
733	510	516	522	528	534	540	546	552	558	564
734	570	576	581	587	593	599	605	611	617	623
735	629	635	641	646	652	658	664	670	676	682
736	688	694	700	705	711	717	723	729	735	741
737	747	753	759	764	770	776	782	788	794	800
738	806	812	817	823	829	835	841	847	853	859
739	864	870	876	882	888	894	900	906	911	917
740	923	929	935	941	947	953	958	964	970	976
741	982	988	994	999	*005	*011	*017	*023	*029	*035
742	87 040	046	052	058	064	070	075	081	087	093
743	099	105	111	116	122	128	134	140	146	151
744	157	163	169	175	181	186	192	198	204	210
745	216	221	227	233	239	245	251	256	262	268
746	274	280	286	291	297	303	309	315	320	326
747	332	338	344	349	355	361	367	373	379	384
748	390	396	402	408	413	419	425	431	437	442
749	448	454	460	466	471	477	483	489	495	500
750	506	512	518	523	529	535	541	547	552	558
N.	0	1	2	3	4	5	6	7	8	9

Proportional parts

	7
1	0.7
2	1.4
3	2.1
4	2.8
5	3.5
6	4.2
7	4.9
8	5.6
9	6.3

	6
1	0.6
2	1.2
3	1.8
4	2.4
5	3.0
6	3.6
7	4.2
8	4.8
9	5.4

	5
1	0.5
2	1.0
3	1.5
4	2.0
5	2.5
6	3.0
7	3.5
8	4.0
9	4.5

Proportional parts

750–800

N.		0	1	2	3	4	5	6	7	8	9
750	87	506	512	518	523	529	535	541	547	552	558
751		564	570	576	581	587	593	599	604	610	616
752		622	628	633	639	645	651	656	662	668	674
753		679	685	691	697	703	708	714	720	726	731
754		737	743	749	754	760	766	772	777	783	789
755		795	800	806	812	818	823	829	835	841	846
756		852	858	864	869	875	881	887	892	898	904
757		910	915	921	927	933	938	944	950	955	961
758		967	973	978	984	990	996	*001	*007	*013	*018
759	88	024	030	036	041	047	053	058	064	070	076
760		081	087	093	098	104	110	116	121	127	133
761		138	144	150	156	161	167	173	178	184	190
762		195	201	207	213	218	224	230	235	241	247
763		252	258	264	270	275	281	287	292	298	304
764		309	315	321	326	332	338	343	349	355	360
765		366	372	377	383	389	395	400	406	412	417
766		423	429	434	440	446	451	457	463	468	474
767		480	485	491	497	502	508	513	519	525	530
768		536	542	547	553	559	564	570	576	581	587
769		593	598	604	610	615	621	627	632	638	643
770		649	655	660	666	672	677	683	689	694	700
771		705	711	717	722	728	734	739	745	750	756
772		762	767	773	779	784	790	795	801	807	812
773		818	824	829	835	840	846	852	857	863	868
774		874	880	885	891	897	902	908	913	919	925
775		930	936	941	947	953	958	964	969	975	981
776		986	992	997	*003	*009	*014	*020	*025	*031	*037
777	89	042	048	053	059	064	070	076	081	087	092
778		098	104	109	115	120	126	131	137	143	148
779		154	159	165	170	176	182	187	193	198	204
780		209	215	221	226	232	237	243	248	254	260
781		265	271	276	282	287	293	298	304	310	315
782		321	326	332	337	343	348	354	360	365	371
783		376	382	387	393	398	404	409	415	421	426
784		432	437	443	448	454	459	465	470	476	481
785		487	492	498	504	509	515	520	526	531	537
786		542	548	553	559	564	570	575	581	586	592
787		597	603	609	614	620	625	631	636	642	647
788		653	658	664	669	675	680	686	691	697	702
789		708	713	719	724	730	735	741	746	752	757
790		763	768	774	779	785	790	796	801	807	812
791		818	823	829	834	840	845	851	856	862	867
792		873	878	883	889	894	900	905	911	916	922
793		927	933	938	944	949	955	960	966	971	977
794		982	988	993	998	*004	*009	*015	*020	*026	*031
795	90	037	042	048	053	059	064	069	075	080	086
796		091	097	102	108	113	119	124	129	135	140
797		146	151	157	162	168	173	179	184	189	195
798		200	206	211	217	222	227	233	238	244	249
799		255	260	266	271	276	282	287	293	298	304
800		309	314	320	325	331	336	342	347	352	358
N.		0	1	2	3	4	5	6	7	8	9

Proportional parts

	6
1	0.6
2	1.2
3	1.8
4	2.4
5	3.0
6	3.6
7	4.2
8	4.8
9	5.4

	5
1	0.5
2	1.0
3	1.5
4	2.0
5	2.5
6	3.0
7	3.5
8	4.0
9	4.5

Proportional parts

800–850

N.		0	1	2	3	4	5	6	7	8	9
800	90	309	314	320	325	331	336	342	347	352	358
801		363	369	374	380	385	390	396	401	407	412
802		417	423	428	434	439	445	450	455	461	466
803		472	477	482	488	493	499	504	509	515	520
804		526	531	536	542	547	553	558	563	569	574
805		580	585	590	596	601	607	612	617	623	628
806		634	639	644	650	655	660	666	671	677	682
807		687	693	698	703	709	714	720	725	730	736
808		741	747	752	757	763	768	773	779	784	789
809		795	800	806	811	816	822	827	832	838	843
810		849	854	859	865	870	875	881	886	891	897
811		902	907	913	918	924	929	934	940	945	950
812		956	961	966	972	977	982	988	993	998	*004
813	91	009	014	020	025	030	036	041	046	052	057
814		062	068	073	078	084	089	094	100	105	110
815		116	121	126	132	137	142	148	153	158	164
816		169	174	180	185	190	196	201	206	212	217
817		222	228	233	238	243	249	254	259	265	270
818		275	281	286	291	297	302	307	312	318	323
819		328	334	339	344	350	355	360	365	371	376
820		381	387	392	397	403	408	413	418	424	429
821		434	440	445	450	455	461	466	471	477	482
822		487	492	498	503	508	514	519	524	529	535
823		540	545	551	556	561	566	572	577	582	587
824		593	598	603	609	614	619	624	630	635	640
825		645	651	656	661	666	672	677	682	687	693
826		698	703	709	714	719	724	730	735	740	745
827		751	756	761	766	772	777	782	787	793	798
828		803	808	814	819	824	829	834	840	845	850
829		855	861	866	871	876	882	887	892	897	903
830		908	913	918	924	929	934	939	944	950	955
831		960	965	971	976	981	986	991	997	*002	*007
832	92	012	018	023	028	033	038	044	049	054	059
833		065	070	075	080	085	091	096	101	106	111
834		117	122	127	132	137	143	148	153	158	163
835		169	174	179	184	189	195	200	205	210	215
836		221	226	231	236	241	247	252	257	262	267
837		273	278	283	288	293	298	304	309	314	319
838		324	330	335	340	345	350	355	361	366	371
839		376	381	387	392	397	402	407	412	418	423
840		428	433	438	443	449	454	459	464	469	474
841		480	485	490	495	500	505	511	516	521	526
842		531	536	542	547	552	557	562	567	572	578
843		583	588	593	598	603	609	614	619	624	629
844		634	639	645	650	655	660	665	670	675	681
845		686	691	696	701	706	711	716	722	727	732
846		737	742	747	752	758	763	768	773	778	783
847		788	793	799	804	809	814	819	824	829	834
848		840	845	850	855	860	865	870	875	881	886
849		891	896	901	906	911	916	921	927	932	937
850		942	947	952	957	962	967	973	978	983	988
N.		0	1	2	3	4	5	6	7	8	9

Proportional parts

	6
1	0.6
2	1.2
3	1.8
4	2.4
5	3.0
6	3.6
7	4.2
8	4.8
9	5.4

	5
1	0.5
2	1.0
3	1.5
4	2.0
5	2.5
6	3.0
7	3.5
8	4.0
9	4.5

Proportional parts

850–900

N.	0	1	2	3	4	5	6	7	8	9
850	92 942	947	952	957	962	967	973	978	983	988
851	993	998	*003	*008	*013	*018	*024	*029	*034	*039
852	93 044	049	054	059	064	069	075	080	085	090
853	095	100	105	110	115	120	125	131	136	141
854	146	151	156	161	166	171	176	181	186	192
855	197	202	207	212	217	222	227	232	237	242
856	247	252	258	263	268	273	278	283	288	293
857	298	303	308	313	318	323	328	334	339	344
858	349	354	359	364	369	374	379	384	389	394
859	399	404	409	414	420	425	430	435	440	445
860	450	455	460	465	470	475	480	485	490	495
861	500	505	510	515	520	526	531	536	541	546
862	551	556	561	566	571	576	581	586	591	596
863	601	606	611	616	621	626	631	636	641	646
864	651	656	661	666	671	676	682	687	692	697
865	702	707	712	717	722	727	732	737	742	747
866	752	757	762	767	772	777	782	787	792	797
867	802	807	812	817	822	827	832	837	842	847
868	852	857	862	867	872	877	882	887	892	897
869	902	907	912	917	922	927	932	937	942	947
870	952	957	962	967	972	977	982	987	992	997
871	94 002	007	012	017	022	027	032	037	042	047
872	052	057	062	067	072	077	082	086	091	096
873	101	106	111	116	121	126	131	136	141	146
874	151	156	161	166	171	176	181	186	191	196
875	201	206	211	216	221	226	231	236	240	245
876	250	255	260	265	270	275	280	285	290	295
877	300	305	310	315	320	325	330	335	340	345
878	349	354	359	364	369	374	379	384	389	394
879	399	404	409	414	419	424	429	433	438	443
880	448	453	458	463	468	473	478	483	488	493
881	498	503	507	512	517	522	527	532	537	542
882	547	552	557	562	567	571	576	581	586	591
883	596	601	606	611	616	621	626	630	635	640
884	645	650	655	660	665	670	675	680	685	689
885	694	699	704	709	714	719	724	729	734	738
886	743	748	753	758	763	768	773	778	783	787
887	792	797	802	807	812	817	822	827	832	836
888	841	846	851	856	861	866	871	876	880	885
889	890	895	900	905	910	915	919	924	929	934
890	939	944	949	954	959	963	968	973	978	983
891	988	993	998	*002	*007	*012	*017	*022	*027	*032
892	95 036	041	046	051	056	061	066	071	075	080
893	085	090	095	100	105	109	114	119	124	129
894	134	139	143	148	153	158	163	168	173	177
895	182	187	192	197	202	207	211	216	221	226
896	231	236	240	245	250	255	260	265	270	274
897	279	284	289	294	299	303	308	313	318	323
898	328	332	337	342	347	352	357	361	366	371
899	376	381	386	390	395	400	405	410	415	419
900	424	429	434	439	444	448	453	458	463	468
N.	0	1	2	3	4	5	6	7	8	9

Proportional parts

	6
1	0.6
2	1.2
3	1.8
4	2.4
5	3.0
6	3.6
7	4.2
8	4.8
9	5.4

	5
1	0.5
2	1.0
3	1.5
4	2.0
5	2.5
6	3.0
7	3.5
8	4.0
9	4.5

	4
1	0.4
2	0.8
3	1.2
4	1.6
5	2.0
6	2.4
7	2.8
8	3.2
9	3.6

Proportional parts

900–950

N.		0	1	2	3	4	5	6	7	8	9
900	95	424	429	434	439	444	448	453	458	463	468
901		472	477	482	487	492	497	501	506	511	516
902		521	525	530	535	540	545	550	554	559	564
903		569	574	578	583	588	593	598	602	607	612
904		617	622	626	631	636	641	646	650	655	660
905		665	670	674	679	684	689	694	698	703	708
906		713	718	722	727	732	737	742	746	751	756
907		761	766	770	775	780	785	789	794	799	804
908		809	813	818	823	828	832	837	842	847	852
909		856	861	866	871	875	880	885	890	895	899
910		904	909	914	918	923	928	933	938	942	947
911		952	957	961	966	971	976	980	985	990	995
912		999	*004	*009	*014	*019	*023	*028	*033	*038	*042
913	96	047	052	057	061	066	071	076	080	085	090
914		095	099	104	109	114	118	123	128	133	137
915		142	147	152	156	161	166	171	175	180	185
916		190	194	199	204	209	213	218	223	227	232
917		237	242	246	251	256	261	265	270	275	280
918		284	289	294	298	303	308	313	317	322	327
919		332	336	341	346	350	355	360	365	369	374
920		379	384	388	393	398	402	407	412	417	421
921		426	431	435	440	445	450	454	459	464	468
922		473	478	483	487	492	497	501	506	511	515
923		520	525	530	534	539	544	548	553	558	562
924		567	572	577	581	586	591	595	600	605	609
925		614	619	624	628	633	638	642	647	652	656
926		661	666	670	675	680	685	689	694	699	703
927		708	713	717	722	727	731	736	741	745	750
928		755	759	764	769	774	778	783	788	792	797
929		802	806	811	816	820	825	830	834	839	844
930		848	853	858	862	867	872	876	881	886	890
931		895	900	904	909	914	918	923	928	932	937
932		942	946	951	956	960	965	970	974	979	984
933		988	993	997	*002	*007	*011	*016	*021	*025	*030
934	97	035	039	044	049	053	058	063	067	072	077
935		081	086	090	095	100	104	109	114	118	123
936		128	132	137	142	146	151	155	160	165	169
937		174	179	183	188	192	197	202	206	211	216
938		220	225	230	234	239	243	248	253	257	262
939		267	271	276	280	285	290	294	299	304	308
940		313	317	322	327	331	336	340	345	350	354
941		359	364	368	373	377	382	387	391	396	400
942		405	410	414	419	424	428	433	437	442	447
943		451	456	460	465	470	474	479	483	488	493
944		497	502	506	511	516	520	525	529	534	539
945		543	548	552	557	562	566	571	575	580	585
946		589	594	598	603	607	612	617	621	626	630
947		635	640	644	649	653	658	663	667	672	676
948		681	685	690	695	699	704	708	713	717	722
949		727	731	736	740	745	749	754	759	763	768
950		772	777	782	786	791	795	800	804	809	813

Proportional parts

	5
1	0.5
2	1.0
3	1.5
4	2.0
5	2.5
6	3.0
7	3.5
8	4.0
9	4.5

	4
1	0.4
2	0.8
3	1.2
4	1.6
5	2.0
6	2.4
7	2.8
8	3.2
9	3.6

950–1000

N.	0	1	2	3	4	5	6	7	8	9
950	97 772	777	782	786	791	795	800	804	809	813
951	818	823	827	832	836	841	845	850	855	859
952	864	868	873	877	882	886	891	896	900	905
953	909	914	918	923	928	932	937	941	946	950
954	955	959	964	968	973	978	982	987	991	996
955	98 000	005	009	014	019	023	028	032	037	041
956	046	050	055	059	064	068	073	078	082	087
957	091	096	100	105	109	114	118	123	127	132
958	137	141	146	150	155	159	164	168	173	177
959	182	186	191	195	200	204	209	214	218	223
960	227	232	236	241	245	250	254	259	263	268
961	272	277	281	286	290	295	299	304	308	313
962	318	322	327	331	336	340	345	349	354	358
963	363	367	372	376	381	385	390	394	399	403
964	408	412	417	421	426	430	435	439	444	448
965	453	457	462	466	471	475	480	484	489	493
966	498	502	507	511	516	520	525	529	534	538
967	543	547	552	556	561	565	570	574	579	583
968	588	592	597	601	605	610	614	619	623	628
969	632	637	641	646	650	655	659	664	668	673
970	677	682	686	691	695	700	704	709	713	717
971	722	726	731	735	740	744	749	753	758	762
972	767	771	776	780	784	789	793	798	802	807
973	811	816	820	825	829	834	838	843	847	851
974	856	860	865	869	874	878	883	887	892	896
975	900	905	909	914	918	923	927	932	936	941
976	945	949	954	958	963	967	972	976	981	985
977	989	994	998	*003	*007	*012	*016	*021	*025	*029
978	99 034	038	043	047	052	056	061	065	069	074
979	078	083	087	092	096	100	105	109	114	118
980	123	127	131	136	140	145	149	154	158	162
981	167	171	176	180	185	189	193	198	202	207
982	211	216	220	224	229	233	238	242	247	251
983	255	260	264	269	273	277	282	286	291	295
984	300	304	308	313	317	322	326	330	335	339
985	344	348	352	357	361	366	370	374	379	383
986	388	392	396	401	405	410	414	419	423	427
987	432	436	441	445	449	454	458	463	467	471
988	476	480	484	489	493	498	502	506	511	515
989	520	524	528	533	537	542	546	550	555	559
990	564	568	572	577	581	585	590	594	599	603
991	607	612	616	621	625	629	634	638	642	647
992	651	656	660	664	669	673	677	682	686	691
993	695	699	704	708	712	717	721	726	730	734
994	739	743	747	752	756	760	765	769	774	778
995	782	787	791	795	800	804	808	813	817	822
996	826	830	835	839	843	848	852	856	861	865
997	870	874	878	883	887	891	896	900	904	909
998	913	917	922	926	930	935	939	944	948	952
999	957	961	965	970	974	978	983	987	991	996
1000	00 000	004	009	013	017	022	026	030	035	039
N.	0	1	2	3	4	5	6	7	8	9

Proportional parts

	5
1	0.5
2	1.0
3	1.5
4	2.0
5	2.5
6	3.0
7	3.5
8	4.0
9	4.5

	4
1	0.4
2	0.8
3	1.2
4	1.6
5	2.0
6	2.4
7	2.8
8	3.2
9	3.6

Proportional parts

*Table 2** COMMON LOGARITHMS OF GAMMA FUNCTION $\Gamma(n)$

$$\Gamma(n) = \qquad x^{n-1}e^{-x}\,dx = \int_0^1 \left[\log_e \frac{1}{x}\right]^{n-1} dx$$

n	$\log_{10}\Gamma(n)+10$	n	$\log_{10}\Gamma(n)+10$	n	$\log_{10}\Gamma(n)+10$	n	$\log_{10}\Gamma(n)+10$	n	$\log_{10}\Gamma(n)+10$
1.01	9.9975	1.21	9.9617	1.41	9.9478	1.61	9.9517	1.81	9.9704
1.02	9.9951	1.22	9.9605	1.42	9.9476	1.62	9.9523	1.82	9.9717
1.03	9.9928	1.23	9.9594	1.43	9.9475	1.63	9.9529	1.83	9.9730
1.04	9.9905	1.24	9.9583	1.44	9.9473	1.64	9.9536	1.84	9.9743
1.05	9.9883	1.25	9.9573	1.45	9.9473	1.65	9.9543	1.85	9.9757
1.06	9.9862	1.26	9.9564	1.46	9.9472	1.66	9.9550	1.86	9.9771
1.07	9.9841	1.27	9.9554	1.47	9.9473	1.67	9.9558	1.87	9.9786
1.08	9.9821	1.28	9.9546	1.48	9.9473	1.68	9.9566	1.88	9.9800
1.09	9.9802	1.29	9.9538	1.49	9.9474	1.69	9.9575	1.89	9.9815
1.10	9.9783	1.30	9.9530	1.50	9.9475	1.70	9.9584	1.90	9.9831
1.11	9.9765	1.31	9.9523	1.51	9.9477	1.71	9.9593	1.91	9.9846
1.12	9.9748	1.32	9.9516	1.52	9.9479	1.72	9.9603	1.92	9.9862
1.13	9.9731	1.33	9.9510	1.53	9.9482	1.73	9.9613	1.93	9.9878
1.14	9.9715	1.34	9.9505	1.54	9.9485	1.74	9.9623	1.94	9.9895
1.15	9.9699	1.35	9.9500	1.55	9.9488	1.75	9.9633	1.95	9.9912
1.16	9.9684	1.36	9.9495	1.56	9.9492	1.76	9.9644	1.96	9.9929
1.17	9.9669	1.37	9.9491	1.57	9.9496	1.77	9.9656	1.97	9.9946
1.18	9.9655	1.38	9.9487	1.58	9.9501	1.78	9.9667	1.98	9.9964
1.19	9.9642	1.39	9.9483	1.59	9.9506	1.79	9.9679	1.99	9.9982
1.20	9.9629	1.40	9.9481	1.60	9.9511	1.80	9.9691	2.00	10.0000

$\Gamma(n+1) = n\,\Gamma(n), n > 0.\quad \Gamma(2) = \Gamma(1) = 1.$

* See page 105.

Interpolation

Let $f(x)$ be an analytic function of x. If the values of $f(x)$ are given in a able for a set of values of x separated from one another consecutively by the onstant small interval h, the differences between the successive values of the unction as tabulated are called *first tabular differences,* the differences of hese first differences, *second tabular differences,* etc. The first, second, and hird tabular differences corresponding to $x = a$ and the tabulated value of (a) are

$$\Delta_1 \equiv f(a+h) - f(a)$$
$$\Delta_2 \equiv f(a+2h) - 2f(a+h) + f(a)$$
$$\Delta_3 \equiv f(a+3h) - 3f(a+2h) + 3f(a+h) - f(a)$$

The value of $f(x)$ for $x = a + \delta$, where $\delta = kh$, $0 < k < 1$, is

$$f(a+\delta) = f(a) + k\,\Delta_1 + \frac{k(k-1)}{2!}\Delta_2 + \frac{k(k-1)(k-2)}{3!}\Delta_3 + \cdots$$

This is known as the *Gregory-Newton formula of interpolation* (see paragraph on interpolation of tables, page 269).

TRIGONOMETRIC FUNCTIONS

Table 3 AUXILIARY TABLE OF S AND T FOR A IN MINUTES (COMMON LOGARITHMS)

$$S = \log \sin A - \log A' \text{ and } T = \log \tan A - \log A'$$

$A' =$ number of minutes in A and $(90° - A)' =$ number of minutes in $90° - A$

A'	S
0 to 13	6.46373 — 10
14 to 42	72 — 10
43 to 58	71 — 10
59 to 71	6.46370 — 10
72 to 81	69 — 10
82 to 91	68 — 10
92 to 99	6.46367 — 10
100 to 107	66 — 10
108 to 115	65 — 10
116 to 121	6.46364 — 10
122 to 128	63 — 10
129 to 134	62 — 10
135 to 140	6.46361 — 10
141 to 146	60 — 10
147 to 151	59 — 10
152 to 157	6.46358 — 10
158 to 162	57 — 10
163 to 167	56 — 10
168 to 171	6.46355 — 10
172 to 176	54 — 10
177 to 181	53 — 10

A'	T
0 to 26	6.46373 — 10
27 to 39	74 — 10
40 to 48	75 — 10
49 to 56	6.46376 — 10
57 to 63	77 — 10
64 to 69	78 — 10
70 to 74	6.46379 — 10
75 to 80	80 — 10
81 to 85	81 — 10
86 to 89	6.46382 — 10
90 to 94	83 — 10
95 to 98	84 — 10
99 to 102	6.46385 — 10
103 to 106	86 — 10
107 to 110	87 — 10
111 to 113	6.46388 — 10
114 to 117	89 — 10
118 to 120	90 — 10
121 to 124	6.46391 — 10
125 to 127	92 — 10
128 to 130	93 — 10

A'	T
131 to 133	6.46394 — 10
134 to 136	95 — 10
137 to 139	96 — 10
140 to 142	6.46397 — 10
143 to 145	98 — 10
146 to 148	99 — 10
149 to 150	6.46400 — 10
151 to 153	01 — 10
154 to 156	02 — 10
157 to 158	6.46403 — 10
159 to 161	04 — 10
162 to 163	05 — 10
164 to 166	6.46406 — 10
167 to 168	07 — 10
169 to 171	08 — 10
172 to 173	6.46409 — 10
174 to 175	10 — 10
176 to 178	11 — 10
179 to 180	6.46412 — 10
181 to 182	13 — 10
183 to 184	14 — 10

For small angles, $\log \sin A = \log A' + S$ and $\log \tan A = \log A' + T$; for angles near 90°, $\log \cos A = \log (90° - A)' + S$ and $\log \text{ctn } A = \log (90° - A)' + T$.

Table 4 COMMON LOGARITHMS OF THE TRIGONOMETRIC FUNCTIONS FROM 0° TO 90°

At Intervals of One Minute to Five Decimal Places

For degrees indicated at the top of the page use the column headings at the top. For degrees indicated at the bottom of the page use the column headings at the bottom.

With degrees at the left of each block (at the top or bottom of the page) use the minute column at the left. With degrees at the right of each block (at the top or bottom of the page) use the minute column at the right.

In order to conserve space, the -10 portion of the characteristic of the logarithm is not printed in Table 4 but must be written down whenever such a logarithm is used.

The −10 portion of the characteristic of the logarithm is not printed but must be written down whenever such a logarithm is used.

0° (180°) (359°) **179°**

′	L Sin	d	L Tan	c d	L Ctn	L Cos	′
0						10.00 000	60
1	6.46 373	30103	6.46 373	30103	13.53 627	10.00 000	59
2	6.76 476	17609	6.76 476	17609	13.23 524	10.00 000	58
3	6.94 085	12494	6.94 085	12494	13.05 915	10.00 000	57
4	7.06 579	9691	7.06 579	9691	12.93 421	10.00 000	56
5	7.16 270	7918	7.16 270	7918	12.83 730	10.00 000	55
6	7.24 188	6694	7.24 188	6694	12.75 812	10.00 000	54
7	7.30 882	5800	7.30 882	5800	12.69 118	10.00 000	53
8	7.36 682	5115	7.36 682	5115	12.63 318	10.00 000	52
9	7.41 797	4576	7.41 797	4576	12.58 203	10.00 000	51
10	7.46 373	4139	7.46 373	4139	12.53 627	10.00 000	50
11	7.50 512	3779	7.50 512	3779	12.49 488	10.00 000	49
12	7.54 291	3476	7.54 291	3476	12.45 709	10.00 000	48
13	7.57 767	3218	7.57 767	3219	12.42 233	10.00 000	47
14	7.60 985	2997	7.60 986	2996	12.39 014	10.00 000	46
15	7.63 982	2802	7.63 982	2803	12.36 018	10.00 000	45
16	7.66 784	2633	7.66 785	2633	12.33 215	10.00 000	44
17	7.69 417	2483	7.69 418	2482	12.30 582	9.99 999	43
18	7.71 900	2348	7.71 900	2348	12.28 100	9.99 999	42
19	7.74 248	2227	7.74 248	2228	12.25 752	9.99 999	41
20	7.76 475	2119	7.76 476	2119	12.23 524	9.99 999	40
21	7.78 594	2021	7.78 595	2020	12.21 405	9.99 999	39
22	7.80 615	1930	7.80 615	1931	12.19 385	9.99 999	38
23	7.82 545	1848	7.82 546	1848	12.17 454	9.99 999	37
24	7.84 393	1773	7.84 394	1773	12.15 606	9.99 999	36
25	7.86 166	1704	7.86 167	1704	12.13 833	9.99 999	35
26	7.87 870	1639	7.87 871	1639	12.12 129	9.99 999	34
27	7.89 509	1579	7.89 510	1579	12.10 490	9.99 999	33
28	7.91 088	1524	7.91 089	1524	12.08 911	9.99 999	32
29	7.92 612	1472	7.92 613	1473	12.07 387	9.99 998	31
30	7.94 084	1424	7.94 086	1424	12.05 914	9.99 998	30
31	7.95 508	1379	7.95 510	1379	12.04 490	9.99 998	29
32	7.96 887	1336	7.96 889	1336	12.03 111	9.99 998	28
33	7.98 223	1297	7.98 225	1297	12.01 775	9.99 998	27
34	7.99 520	1259	7.99 522	1259	12.00 478	9.99 998	26
35	8.00 779	1223	8.00 781	1223	11.99 219	9.99 998	25
36	8.02 002	1190	8.02 004	1190	11.97 996	9.99 998	24
37	8.03 192	1158	8.03 194	1159	11.96 806	9.99 997	23
38	8.04 350	1128	8.04 353	1128	11.95 647	9.99 997	22
39	8.05 478	1100	8.05 481	1100	11.94 519	9.99 997	21
40	8.06 578	1072	8.06 581	1072	11.93 419	9.99 997	20
41	8.07 650	1046	8.07 653	1047	11.92 347	9.99 997	19
42	8.08 696	1022	8.08 700	1022	11.91 300	9.99 997	18
43	8.09 718	999	8.09 722	998	11.90 278	9.99 997	17
44	8.10 717	976	8.10 720	976	11.89 280	9.99 996	16
45	8.11 693	954	8.11 696	955	11.88 304	9.99 996	15
46	8.12 647	934	8.12 651	934	11.87 349	9.99 996	14
47	8.13 581	914	8.13 585	915	11.86 415	9.99 996	13
48	8.14 495	896	8.14 500	895	11.85 500	9.99 996	12
49	8.15 391	877	8.15 395	878	11.84 605	9.99 996	11
50	8.16 268	860	8.16 273	860	11.83 727	9.99 995	10
51	8.17 128	843	8.17 133	843	11.82 867	9.99 995	9
52	8.17 971	827	8.17 976	828	11.82 024	9.99 995	8
53	8.18 798	812	8.18 804	812	11.81 196	9.99 995	7
54	8.19 610	797	8.19 616	797	11.80 384	9.99 995	6
55	8.20 407	782	8.20 413	782	11.79 587	9.99 994	5
56	8.21 189	769	8.21 195	769	11.78 805	9.99 994	4
57	8.21 958	755	8.21 964	756	11.78 036	9.99 994	3
58	8.22 713	743	8.22 720	742	11.77 280	9.99 994	2
59	8.23 456	730	8.23 462	730	11.76 538	9.99 994	1
60	8.24 186		8.24 192		11.75 808	9.99 993	0
′	L Cos	d	L Ctn	c d	L Tan	L Sin	′

90° (270°) (269°) **89°**

For more accurate values of L sin and L tan for interpolated values of angles less than 3° (or L cos or L ctn of angles greater than 87°) use Table 3.

Table 4 COMMON LOGARITHMS OF TRIGONOMETRIC FUNCTIONS (*continued*)

The −10 portion of the characteristic of the logarithm is not printed but must be written down whenever such a logarithm is used.

1° (181°) (358°) **178°**

′	L Sin	d	L Tan	c d	L Ctn	L Cos	′
0	8.24 186	717	8.24 192	718	11.75 808	9.99 993	60
1	8.24 903	706	8.24 910	706	11.75 090	9.99 993	59
2	8.25 609	695	8.25 616	696	11.74 384	9.99 993	58
3	8.26 304	684	8.26 312	684	11.73 688	9.99 993	57
4	8.26 988	673	8.26 996	673	11.73 004	9.99 992	56
5	8.27 661	663	8.27 669	663	11.72 331	9.99 992	55
6	8.28 324	653	8.28 332	654	11.71 668	9.99 992	54
7	8.28 977	644	8.28 986	643	11.71 014	9.99 992	53
8	8.29 621	634	8.29 629	634	11.70 371	9.99 992	52
9	8.30 255	624	8.30 263	625	11.69 737	9.99 991	51
10	8.30 879	616	8.30 888	617	11.69 112	9.99 991	50
11	8.31 495	608	8.31 505	607	11.68 495	9.99 991	49
12	8.32 103	599	8.32 112	599	11.67 888	9.99 990	48
13	8.32 702	590	8.32 711	591	11.67 289	9.99 990	47
14	8.33 292	583	8.33 302	584	11.66 698	9.99 990	46
15	8.33 875	575	8.33 886	575	11.66 114	9.99 990	45
16	8.34 450	568	8.34 461	568	11.65 539	9.99 989	44
17	8.35 018	560	8.35 029	561	11.64 971	9.99 989	43
18	8.35 578	553	8.35 590	553	11.64 410	9.99 989	42
19	8.36 131	547	8.36 143	546	11.63 857	9.99 989	41
20	8.36 678	539	8.36 689	540	11.63 311	9.99 988	40
21	8.37 217	533	8.37 229	533	11.62 771	9.99 988	39
22	8.37 750	526	8.37 762	527	11.62 238	9.99 988	38
23	8.38 276	520	8.38 289	520	11.61 711	9.99 987	37
24	8.38 796	514	8.38 809	514	11.61 191	9.99 987	36
25	8.39 310	508	8.39 323	509	11.60 677	9.99 987	35
26	8.39 818	502	8.39 832	502	11.60 168	9.99 986	34
27	8.40 320	496	8.40 334	496	11.59 666	9.99 986	33
28	8.40 816	491	8.40 830	491	11.59 170	9.99 986	32
29	8.41 307	485	8.41 321	486	11.58 679	9.99 985	31
30	8.41 792	480	8.41 807	480	11.58 193	9.99 985	30
31	8.42 272	474	8.42 287	475	11.57 713	9.99 985	29
32	8.42 746	470	8.42 762	470	11.57 238	9.99 984	28
33	8.43 216	464	8.43 232	464	11.56 768	9.99 984	27
34	8.43 680	459	8.43 696	460	11.56 304	9.99 984	26
35	8.44 139	455	8.44 156	455	11.55 844	9.99 983	25
36	8.44 594	450	8.44 611	450	11.55 389	9.99 983	24
37	8.45 044	445	8.45 061	446	11.54 939	9.99 983	23
38	8.45 489	441	8.45 507	441	11.54 493	9.99 982	22
39	8.45 930	436	8.45 948	437	11.54 052	9.99 982	21
40	8.46 366	433	8.46 385	432	11.53 615	9.99 982	20
41	8.46 799	427	8.46 817	428	11.53 183	9.99 981	19
42	8.47 226	424	8.47 245	424	11.52 755	9.99 981	18
43	8.47 650	419	8.47 669	420	11.52 331	9.99 981	17
44	8.48 069	416	8.48 089	416	11.51 911	9.99 980	16
45	8.48 485	411	8.48 505	412	11.51 495	9.99 980	15
46	8.48 896	408	8.48 917	408	11.51 083	9.99 979	14
47	8.49 304	404	8.49 325	404	11.50 675	9.99 979	13
48	8.49 708	400	8.49 729	401	11.50 271	9.99 979	12
49	8.50 108	396	8.50 130	397	11.49 870	9.99 978	11
50	8.50 504	393	8.50 527	393	11.49 473	9.99 978	10
51	8.50 897	390	8.50 920	390	11.49 080	9.99 977	9
52	8.51 287	386	8.51 310	386	11.48 690	9.99 977	8
53	8.51 673	382	8.51 696	383	11.48 304	9.99 977	7
54	8.52 055	379	8.52 079	380	11.47 921	9.99 976	6
55	8.52 434	376	8.52 459	376	11.47 541	9.99 976	5
56	8.52 810	373	8.52 835	373	11.47 165	9.99 975	4
57	8.53 183	369	8.53 208	370	11.46 792	9.99 975	3
58	8.53 552	367	8.53 578	367	11.46 422	9.99 974	2
59	8.53 919	363	8.53 945	363	11.46 055	9.99 974	1
60	8.54 282		8.54 308		11.45 692	9.99 974	0
′	L Cos	d	L Ctn	c d	L Tan	L Sin	′

91° (271°) (268°) **88°**

For more accurate values of L sin and L tan for interpolated values of angles less than 3° (or L cos or L ctn of angles greater than 87°) use Table 3.

Table 4 COMMON LOGARITHMS OF TRIGONOMETRIC FUNCTIONS *(continued)*

The −10 portion of the characteristic of the logarithm is not printed but must be written down whenever such a logarithm is used.

2° (182°) (357°) 177°

′	L Sin	d	L Tan	c d	L Ctn	L Cos	′
0	8.54 282	360	8.54 308	361	11.45 692	9.99 974	60
1	8.54 642	357	8.54 669	358	11.45 331	9.99 973	59
2	8.54 999	355	8.55 027	355	11.44 973	9.99 973	58
3	8.55 354	351	8.55 382	352	11.44 618	9.99 972	57
4	8.55 705	349	8.55 734	349	11.44 266	9.99 972	56
5	8.56 054	346	8.56 083	346	11.43 917	9.99 971	55
6	8.56 400	343	8.56 429	344	11.43 571	9.99 971	54
7	8.56 743	341	8.56 773	341	11.43 227	9.99 970	53
8	8.57 084	337	8.57 114	338	11.42 886	9.99 970	52
9	8.57 421	336	8.57 452	336	11.42 548	9.99 969	51
10	8.57 757	332	8.57 788	333	11.42 212	9.99 969	50
11	8.58 089	330	8.58 121	330	11.41 879	9.99 968	49
12	8.58 419	328	8.58 451	328	11.41 549	9.99 968	48
13	8.58 747	325	8.58 779	326	11.41 221	9.99 967	47
14	8.59 072	323	8.59 105	323	11.40 895	9.99 967	46
15	8.59 395	320	8.59 428	321	11.40 572	9.99 967	45
16	8.59 715	318	8.59 749	319	11.40 251	9.99 966	44
17	8.60 033	316	8.60 068	316	11.39 932	9.99 966	43
18	8.60 349	313	8.60 384	314	11.39 616	9.99 965	42
19	8.60 662	311	8.60 698	311	11.39 302	9.99 964	41
20	8.60 973	309	8.61 009	310	11.38 991	9.99 964	40
21	8.61 282	307	8.61 319	307	11.38 681	9.99 963	39
22	8.61 589	305	8.61 626	305	11.38 374	9.99 963	38
23	8.61 894	302	8.61 931	303	11.38 069	9.99 962	37
24	8.62 196	301	8.62 234	301	11.37 766	9.99 962	36
25	8.62 497	298	8.62 535	299	11.37 465	9.99 961	35
26	8.62 795	296	8.62 834	297	11.37 166	9.99 961	34
27	8.63 091	294	8.63 131	295	11.36 869	9.99 960	33
28	8.63 385	293	8.63 426	292	11.36 574	9.99 960	32
29	8.63 678	290	8.63 718	291	11.36 282	9.99 959	31
30	8.63 968	288	8.64 009	289	11.35 991	9.99 959	30
31	8.64 256	287	8.64 298	287	11.35 702	9.99 958	29
32	8.64 543	284	8.64 585	285	11.35 415	9.99 958	28
33	8.64 827	283	8.64 870	284	11.35 130	9.99 957	27
34	8.65 110	281	8.65 154	281	11.34 846	9.99 956	26
35	8.65 391	279	8.65 435	280	11.34 565	9.99 956	25
36	8.65 670	277	8.65 715	278	11.34 285	9.99 955	24
37	8.65 947	276	8.65 993	276	11.34 007	9.99 955	23
38	8.66 223	274	8.66 269	274	11.33 731	9.99 954	22
39	8.66 497	272	8.66 543	273	11.33 457	9.99 954	21
40	8.66 769	270	8.66 816	271	11.33 184	9.99 953	20
41	8.67 039	269	8.67 087	269	11.32 913	9.99 952	19
42	8.67 308	267	8.67 356	268	11.32 644	9.99 952	18
43	8.67 575	266	8.67 624	266	11.32 376	9.99 951	17
44	8.67 841	263	8.67 890	264	11.32 110	9.99 951	16
45	8.68 104	263	8.68 154	263	11.31 846	9.99 950	15
46	8.68 367	260	8.68 417	261	11.31 583	9.99 949	14
47	8.68 627	259	8.68 678	260	11.31 322	9.99 949	13
48	8.68 886	258	8.68 938	258	11.31 062	9.99 948	12
49	8.69 144	256	8.69 196	257	11.30 804	9.99 948	11
50	8.69 400	254	8.69 453	255	11.30 547	9.99 947	10
51	8.69 654	253	8.69 708	254	11.30 292	9.99 946	9
52	8.69 907	252	8.69 962	252	11.30 038	9.99 946	8
53	8.70 159	250	8.70 214	251	11.29 786	9.99 945	7
54	8.70 409	249	8.70 465	249	11.29 535	9.99 944	6
55	8.70 658	247	8.70 714	248	11.29 286	9.99 944	5
56	8.70 905	246	8.70 962	246	11.29 038	9.99 943	4
57	8.71 151	244	8.71 208	245	11.28 792	9.99 942	3
58	8.71 395	243	8.71 453	244	11.28 547	9.99 942	2
59	8.71 638	242	8.71 697	243	11.28 303	9.99 941	1
60	8.71 880		8.71 940		11.28 060	9.99 940	0
′	L Cos	d	L Ctn	c d	L Tan	L Sin	′

92° (272°) (267°) 87°

For more accurate values of L sin and L tan for interpolated values of angles less than 3° (or L cos or L ctn of angles greater than 87°) use Table 3.

The −10 portion of the characteristic of the logarithm is not printed but must be written down whenever such a logarithm is used.

3° (183°) (356°) **176°**

′	L Sin	d	L Tan	c d	L Ctn	L Cos	′
0	8.71 880	240	8.71 940	241	11.28 060	9.99 940	60
1	8.72 120	239	8.72 181	239	11.27 819	9.99 940	59
2	8.72 359	238	8.72 420	239	11.27 580	9.99 939	58
3	8.72 597	237	8.72 659	237	11.27 341	9.99 938	57
4	8.72 834	235	8.72 896	236	11.27 104	9.99 938	56
5	8.73 069	234	8.73 132	234	11.26 868	9.99 937	55
6	8.73 303	232	8.73 366	234	11.26 634	9.99 936	54
7	8.73 535	232	8.73 600	232	11.26 400	9.99 936	53
8	8.73 767	230	8.73 832	231	11.26 168	9.99 935	52
9	8.73 997	229	8.74 063	229	11.25 937	9.99 934	51
10	8.74 226	228	8.74 292	229	11.25 708	9.99 934	50
11	8.74 454	226	8.74 521	227	11.25 479	9.99 933	49
12	8.74 680	226	8.74 748	226	11.25 252	9.99 932	48
13	8.74 906	224	8.74 974	225	11.25 026	9.99 932	47
14	8.75 130	223	8.75 199	224	11.24 801	9.99 931	46
15	8.75 353	222	8.75 423	222	11.24 577	9.99 930	45
16	8.75 575	220	8.75 645	222	11.24 355	9.99 929	44
17	8.75 795	220	8.75 867	220	11.24 133	9.99 929	43
18	8.76 015	219	8.76 087	219	11.23 913	9.99 928	42
19	8.76 234	217	8.76 306	219	11.23 694	9.99 927	41
20	8.76 451	216	8.76 525	217	11.23 475	9.99 926	40
21	8.76 667	216	8.76 742	216	11.23 258	9.99 926	39
22	8.76 883	214	8.76 958	215	11.23 042	9.99 925	38
23	8.77 097	213	8.77 173	214	11.22 827	9.99 924	37
24	8.77 310	212	8.77 387	213	11.22 613	9.99 923	36
25	8.77 522	211	8.77 600	211	11.22 400	9.99 923	35
26	8.77 733	210	8.77 811	211	11.22 189	9.99 922	34
27	8.77 943	209	8.78 022	210	11.21 978	9.99 921	33
28	8.78 152	208	8.78 232	209	11.21 768	9.99 920	32
29	8.78 360	208	8.78 441	208	11.21 559	9.99 920	31
30	8.78 568	206	8.78 649	206	11.21 351	9.99 919	30
31	8.78 774	205	8.78 855	206	11.21 145	9.99 918	29
32	8.78 979	204	8.79 061	205	11.20 939	9.99 917	28
33	8.79 183	203	8.79 266	204	11.20 734	9.99 917	27
34	8.79 386	202	8.79 470	203	11.20 530	9.99 916	26
35	8.79 588	201	8.79 673	202	11.20 327	9.99 915	25
36	8.79 789	201	8.79 875	201	11.20 125	9.99 914	24
37	8.79 990	199	8.80 076	201	11.19 924	9.99 913	23
38	8.80 189	199	8.80 277	199	11.19 723	9.99 913	22
39	8.80 388	197	8.80 476	198	11.19 524	9.99 912	21
40	8.80 585	197	8.80 674	198	11.19 326	9.99 911	20
41	8.80 782	196	8.80 872	196	11.19 128	9.99 910	19
42	8.80 978	195	8.81 068	196	11.18 932	9.99 909	18
43	8.81 173	194	8.81 264	195	11.18 736	9.99 909	17
44	8.81 367	193	8.81 459	194	11.18 541	9.99 908	16
45	8.81 560	192	8.81 653	193	11.18 347	9.99 907	15
46	8.81 752	192	8.81 846	192	11.18 154	9.99 906	14
47	8.81 944	190	8.82 038	192	11.17 962	9.99 905	13
48	8.82 134	190	8.82 230	190	11.17 770	9.99 904	12
49	8.82 324	189	8.82 420	190	11.17 580	9.99 904	11
50	8.82 513	188	8.82 610	189	11.17 390	9.99 903	10
51	8.82 701	187	8.82 799	188	11.17 201	9.99 902	9
52	8.82 888	187	8.82 987	188	11.17 013	9.99 901	8
53	8.83 075	186	8.83 175	186	11.16 825	9.99 900	7
54	8.83 261	185	8.83 361	186	11.16 639	9.99 899	6
55	8.83 446	184	8.83 547	185	11.16 453	9.99 898	5
56	8.83 630	183	8.83 732	184	11.16 268	9.99 898	4
57	8.83 813	183	8.83 916	184	11.16 084	9.99 897	3
58	8.83 996	181	8.84 100	182	11.15 900	9.99 896	2
59	8.84 177	181	8.84 282	182	11.15 718	9.99 895	1
60	8.84 358		8.84 464		11.15 536	9.99 894	0
′	L Cos	d	L Ctn	c d	L Tan	L Sin	′

Proportional parts

″	241	239	237	235	234
1	4.0	4.0	4.0	3.9	3.9
2	8.0	8.0	7.9	7.8	7.8
3	12.0	12.0	11.8	11.8	11.7
4	16.1	15.9	15.8	15.7	15.6
5	20.1	19.9	19.8	19.6	19.5
6	24.1	23.9	23.7	23.5	23.4
7	28.1	27.9	27.6	27.4	27.3
8	32.1	31.9	31.6	31.3	31.2
9	36.2	35.8	35.6	35.2	35.1
″	232	229	227	225	223
1	3.9	3.8	3.8	3.8	3.7
2	7.7	7.6	7.6	7.5	7.4
3	11.6	11.4	11.4	11.2	11.2
4	15.5	15.3	15.1	15.0	14.9
5	19.3	19.1	18.9	18.8	18.6
6	23.2	22.9	22.7	22.5	22.3
7	27.1	26.7	26.5	26.2	26.0
8	30.9	30.5	30.3	30.0	29.7
9	34.8	34.4	34.0	33.8	33.4
″	222	220	217	215	213
1	3.7	3.7	3.6	3.6	3.6
2	7.4	7.3	7.2	7.2	7.1
3	11.1	11.0	10.8	10.8	10.6
4	14.8	14.7	14.5	14.3	14.2
5	18.5	18.3	18.1	17.9	17.8
6	22.2	22.0	21.7	21.5	21.3
7	25.9	25.7	25.3	25.1	24.8
8	29.6	29.3	28.9	28.7	28.4
9	33.3	33.0	32.6	32.2	32.0
″	211	208	206	203	201
1	3.5	3.5	3.4	3.4	3.4
2	7.0	6.9	6.9	6.8	6.7
3	10.6	10.4	10.3	10.2	10.0
4	14.1	13.9	13.7	13.5	13.4
5	17.6	17.3	17.2	16.9	16.8
6	21.1	20.8	20.6	20.3	20.1
7	24.6	24.3	24.0	23.7	23.4
8	28.1	27.7	27.5	27.1	26.8
9	31.6	31.2	30.9	30.4	30.2
″	199	197	195	193	192
1	3.3	3.3	3.2	3.2	3.2
2	6.6	6.6	6.5	6.4	6.4
3	10.0	9.8	9.8	9.6	9.6
4	13.3	13.1	13.0	12.9	12.8
5	16.6	16.4	16.2	16.1	16.0
6	19.9	19.7	19.5	19.3	19.2
7	23.2	23.0	22.8	22.5	22.4
8	26.5	26.3	26.0	25.7	25.6
9	29.8	29.6	29.2	29.0	28.8
″	189	187	185	183	181
1	3.2	3.1	3.1	3.0	3.0
2	6.3	6.2	6.2	6.1	6.0
3	9.4	9.4	9.2	9.2	9.0
4	12.6	12.5	12.3	12.2	12.1
5	15.8	15.6	15.4	15.2	15.1
6	18.9	18.7	18.5	18.3	18.1
7	22.0	21.8	21.6	21.4	21.1
8	25.2	24.9	24.7	24.4	24.1
9	28.4	28.0	27.8	27.4	27.2

Proportional parts

93° (273°) (266°) **86°**

The −10 portion of the characteristic of the logarithm is not printed but must be written down whenever such a logarithm is used.

4° (184°) (355°) **175°**

′	L Sin	d	L Tan	c d	L Ctn	L Cos	′
0	8.84 358	181	8.84 464	182	11.15 536	9.99 894	60
1	8.84 539	179	8.84 646	180	11.15 354	9.99 893	59
2	8.84 718	179	8.84 826	180	11.15 174	9.99 892	58
3	8.84 897	178	8.85 006	179	11.14 994	9.99 891	57
4	8.85 075	177	8.85 185	178	11.14 815	9.99 891	56
5	8.85 252	177	8.85 363	177	11.14 637	9.99 890	55
6	8.85 429	176	8.85 540	177	11.14 460	9.99 889	54
7	8.85 605	175	8.85 717	176	11.14 283	9.99 888	53
8	8.85 780	175	8.85 893	176	11.14 107	9.99 887	52
9	8.85 955	173	8.86 069	174	11.13 931	9.99 886	51
10	8.86 128	173	8.86 243	174	11.13 757	9.99 885	50
11	8.86 301	173	8.86 417	174	11.13 583	9.99 884	49
12	8.86 474	171	8.86 591	172	11.13 409	9.99 883	48
13	8.86 645	171	8.86 763	172	11.13 237	9.99 882	47
14	8.86 816	171	8.86 935	171	11.13 065	9.99 881	46
15	8.86 987	169	8.87 106	171	11.12 894	9.99 880	45
16	8.87 156	169	8.87 277	170	11.12 723	9.99 879	44
17	8.87 325	169	8.87 447	169	11.12 553	9.99 879	43
18	8.87 494	167	8.87 616	169	11.12 384	9.99 878	42
19	8.87 661	168	8.87 785	168	11.12 215	9.99 877	41
20	8.87 829	166	8.87 953	167	11.12 047	9.99 876	40
21	8.87 995	166	8.88 120	167	11.11 880	9.99 875	39
22	8.88 161	165	8.88 287	166	11.11 713	9.99 874	38
23	8.88 326	164	8.88 453	165	11.11 547	9.99 873	37
24	8.88 490	164	8.88 618	165	11.11 382	9.99 872	36
25	8.88 654	163	8.88 783	165	11.11 217	9.99 871	35
26	8.88 817	163	8.88 948	163	11.11 052	9.99 870	34
27	8.88 980	162	8.89 111	163	11.10 889	9.99 869	33
28	8.89 142	162	8.89 274	163	11.10 726	9.99 868	32
29	8.89 304	160	8.89 437	161	11.10 563	9.99 867	31
30	8.89 464	161	8.89 598	162	11.10 402	9.99 866	30
31	8.89 625	159	8.89 760	160	11.10 240	9.99 865	29
32	8.89 784	159	8.89 920	160	11.10 080	9.99 864	28
33	8.89 943	159	8.90 080	160	11.09 920	9.99 863	27
34	8.90 102	158	8.90 240	159	11.09 760	9.99 862	26
35	8.90 260	157	8.90 399	158	11.09 601	9.99 861	25
36	8.90 417	157	8.90 557	158	11.09 443	9.99 860	24
37	8.90 574	156	8.90 715	157	11.09 285	9.99 859	23
38	8.90 730	155	8.90 872	157	11.09 128	9.99 858	22
39	8.90 885	155	8.91 029	156	11.08 971	9.99 857	21
40	8.91 040	155	8.91 185	155	11.08 815	9.99 856	20
41	8.91 195	154	8.91 340	155	11.08 660	9.99 855	19
42	8.91 349	153	8.91 495	155	11.08 505	9.99 854	18
43	8.91 502	153	8.91 650	153	11.08 350	9.99 853	17
44	8.91 655	152	8.91 803	154	11.08 197	9.99 852	16
45	8.91 807	152	8.91 957	153	11.08 043	9.99 851	15
46	8.91 959	151	8.92 110	152	11.07 890	9.99 850	14
47	8.92 110	151	8.92 262	152	11.07 738	9.99 848	13
48	8.92 261	150	8.92 414	151	11.07 586	9.99 847	12
49	8.92 411	150	8.92 565	151	11.07 435	9.99 846	11
50	8.92 561	149	8.92 716	150	11.07 284	9.99 845	10
51	8.92 710	149	8.92 866	150	11.07 134	9.99 844	9
52	8.92 859	148	8.93 016	149	11.06 984	9.99 843	8
53	8.93 007	147	8.93 165	148	11.06 835	9.99 842	7
54	8.93 154	147	8.93 313	149	11.06 687	9.99 841	6
55	8.93 301	147	8.93 462	147	11.06 538	9.99 840	5
56	8.93 448	146	8.93 609	147	11.06 391	9.99 839	4
57	8.93 594	146	8.93 756	147	11.06 244	9.99 838	3
58	8.93 740	145	8.93 903	146	11.06 097	9.99 837	2
59	8.93 885	145	8.94 049	146	11.05 951	9.99 836	1
60	8.94 030		8.94 195		11.05 805	9.99 834	0
′	L Cos	d	L Ctn	c d	L Tan	L Sin	′

Proportional parts

″	182	181	179	178	177
1	3.0	3.0	3.0	3.0	3.0
2	6.1	6.0	6.0	5.9	5.9
3	9.1	9.0	9.0	8.9	8.8
4	12.1	12.1	11.9	11.9	11.8
5	15.2	15.1	14.9	14.8	14.8
6	18.2	18.1	17.9	17.8	17.7
7	21.2	21.1	20.9	20.8	20.6
8	24.3	24.1	23.9	23.7	23.6
9	27.3	27.2	26.8	26.7	26.6
″	176	175	174	173	172
1	2.9	2.9	2.9	2.9	2.9
2	5.9	5.8	5.8	5.8	5.7
3	8.8	8.8	8.7	8.6	8.6
4	11.7	11.7	11.6	11.5	11.5
5	14.7	14.6	14.5	14.4	14.3
6	17.6	17.5	17.4	17.3	17.2
7	20.5	20.4	20.3	20.2	20.1
8	23.5	23.3	23.2	23.1	22.9
9	26.4	26.2	26.1	26.0	25.8
″	171	170	169	168	167
1	2.8	2.8	2.8	2.8	2.8
2	5.7	5.7	5.6	5.6	5.6
3	8.6	8.5	8.4	8.4	8.4
4	11.4	11.3	11.3	11.2	11.1
5	14.2	14.2	14.1	14.0	13.9
6	17.1	17.0	16.9	16.8	16.7
7	20.0	19.8	19.7	19.6	19.5
8	22.8	22.7	22.5	22.4	22.3
9	25.6	25.5	25.4	25.2	25.0
″	166	165	164	163	162
1	2.8	2.8	2.7	2.7	2.7
2	5.5	5.5	5.5	5.4	5.4
3	8.3	8.2	8.2	8.2	8.1
4	11.1	11.0	10.9	10.9	10.8
5	13.8	13.8	13.7	13.6	13.5
6	16.6	16.5	16.4	16.3	16.2
7	19.4	19.2	19.1	19.0	18.9
8	22.1	22.0	21.9	21.7	21.6
9	24.9	24.8	24.6	24.4	24.3
″	161	160	159	158	157
1	2.7	2.7	2.6	2.6	2.6
2	5.4	5.3	5.3	5.3	5.2
3	8.0	8.0	8.0	7.9	7.8
4	10.7	10.7	10.6	10.5	10.5
5	13.4	13.3	13.2	13.2	13.1
6	16.1	16.0	15.9	15.8	15.7
7	18.8	18.7	18.6	18.4	18.3
8	21.5	21.3	21.2	21.1	20.9
9	24.2	24.0	23.8	23.7	23.6
″	156	155	154	153	152
1	2.6	2.6	2.6	2.6	2.5
2	5.2	5.2	5.1	5.1	5.1
3	7.8	7.8	7.7	7.6	7.6
4	10.4	10.3	10.3	10.2	10.1
5	13.0	12.9	12.8	12.8	12.7
6	15.6	15.5	15.4	15.3	15.2
7	18.2	18.1	18.0	17.8	17.7
8	20.8	20.7	20.5	20.4	20.3
9	23.4	23.2	23.1	23.0	22.8

Proportional parts

94° (274°) (265°) **85°**

The −10 portion of the characteristic of the logarithm is not printed but must be written down whenever such a logarithm is used.

5° (185°) (354°) **174°**

′	L Sin	d	L Tan	c d	L Ctn	L Cos	′
0	8.94 030	144	8.94 195	145	11.05 805	9.99 834	60
1	8.94 174	143	8.94 340	145	11.05 660	9.99 833	59
2	8.94 317	144	8.94 485	145	11.05 515	9.99 832	58
3	8.94 461	142	8.94 630	143	11.05 370	9.99 831	57
4	8.94 603	143	8.94 773	144	11.05 227	9.99 830	56
5	8.94 746	141	8.94 917	143	11.05 083	9.99 829	55
6	8.94 887	142	8.95 060	142	11.04 940	9.99 828	54
7	8.95 029	141	8.95 202	142	11.04 798	9.99 827	53
8	8.95 170	140	8.95 344	142	11.04 656	9.99 825	52
9	8.95 310	140	8.95 486	141	11.04 514	9.99 824	51
10	8.95 450	139	8.95 627	140	11.04 373	9.99 823	50
11	8.95 589	139	8.95 767	141	11.04 233	9.99 822	49
12	8.95 728	139	8.95 908	139	11.04 092	9.99 821	48
13	8.95 867	138	8.96 047	140	11.03 953	9.99 820	47
14	8.96 005	138	8.96 187	138	11.03 813	9.99 819	46
15	8.96 143	137	8.96 325	139	11.03 675	9.99 817	45
16	8.96 280	137	8.96 464	138	11.03 536	9.99 816	44
17	8.96 417	136	8.96 602	137	11.03 398	9.99 815	43
18	8.96 553	136	8.96 739	138	11.03 261	9.99 814	42
19	8.96 689	136	8.96 877	136	11.03 123	9.99 813	41
20	8.96 825	135	8.97 013	137	11.02 987	9.99 812	40
21	8.96 960	135	8.97 150	135	11.02 850	9.99 810	39
22	8.97 095	134	8.97 285	136	11.02 715	9.99 809	38
23	8.97 229	134	8.97 421	135	11.02 579	9.99 808	37
24	8.97 363	133	8.97 556	135	11.02 444	9.99 807	36
25	8.97 496	133	8.97 691	134	11.02 309	9.99 806	35
26	8.97 629	133	8.97 825	134	11.02 175	9.99 804	34
27	8.97 762	132	8.97 959	133	11.02 041	9.99 803	33
28	8.97 894	132	8.98 092	133	11.01 908	9.99 802	32
29	8.98 026	131	8.98 225	133	11.01 775	9.99 801	31
30	8.98 157	131	8.98 358	132	11.01 642	9.99 800	30
31	8.98 288	131	8.98 490	132	11.01 510	9.99 798	29
32	8.98 419	130	8.98 622	131	11.01 378	9.99 797	28
33	8.98 549	130	8.98 753	131	11.01 247	9.99 796	27
34	8.98 679	129	8.98 884	131	11.01 116	9.99 795	26
35	8.98 808	129	8.99 015	130	11.00 985	9.99 793	25
36	8.98 937	129	8.99 145	130	11.00 855	9.99 792	24
37	8.99 066	128	8.99 275	130	11.00 725	9.99 791	23
38	8.99 194	128	8.99 405	129	11.00 595	9.99 790	22
39	8.99 322	128	8.99 534	128	11.00 466	9.99 788	21
40	8.99 450	127	8.99 662	129	11.00 338	9.99 787	20
41	8.99 577	127	8.99 791	128	11.00 209	9.99 786	19
42	8.99 704	126	8.99 919	127	11.00 081	9.99 785	18
43	8.99 830	126	9.00 046	128	10.99 954	9.99 783	17
44	8.99 956	126	9.00 174	127	10.99 826	9.99 782	16
45	9.00 082	125	9.00 301	126	10.99 699	9.99 781	15
46	9.00 207	125	9.00 427	126	10.99 573	9.99 780	14
47	9.00 332	124	9.00 553	126	10.99 447	9.99 778	13
48	9.00 456	125	9.00 679	126	10.99 321	9.99 777	12
49	9.00 581	123	9.00 805	125	10.99 195	9.99 776	11
50	9.00 704	124	9.00 930	125	10.99 070	9.99 775	10
51	9.00 828	123	9.01 055	124	10.98 945	9.99 773	9
52	9.00 951	123	9.01 179	124	10.98 821	9.99 772	8
53	9.01 074	122	9.01 303	124	10.98 697	9.99 771	7
54	9.01 196	122	9.01 427	123	10.98 573	9.99 769	6
55	9.01 318	122	9.01 550	123	10.98 450	9.99 768	5
56	9.01 440	121	9.01 673	123	10.98 327	9.99 767	4
57	9.01 561	121	9.01 796	122	10.98 204	9.99 765	3
58	9.01 682	121	9.01 918	122	10.98 082	9.99 764	2
59	9.01 803	120	9.02 040	122	10.97 960	9.99 763	1
60	9.01 923		9.02 162		10.97 838	9.99 761	0
′	L Cos	d	L Ctn	c d	L Tan	L Sin	′

Proportional parts

″	151	149	148	147	146
1	2.5	2.5	2.5	2.4	2.4
2	5.0	5.0	4.9	4.9	4.9
3	7.6	7.4	7.4	7.4	7.3
4	10.1	9.9	9.9	9.8	9.7
5	12.6	12.4	12.3	12.2	12.2
6	15.1	14.9	14.8	14.7	14.6
7	17.6	17.4	17.3	17.2	17.0
8	20.1	19.9	19.7	19.6	19.5
9	22.6	22.4	22.2	22.0	21.9

″	145	144	143	142	141
1	2.4	2.4	2.4	2.4	2.4
2	4.8	4.8	4.8	4.7	4.7
3	7.2	7.2	7.2	7.1	7.0
4	9.7	9.6	9.5	9.5	9.4
5	12.1	12.0	11.9	11.8	11.8
6	14.5	14.4	14.3	14.2	14.1
7	16.9	16.8	16.7	16.6	16.4
8	19.3	19.2	19.1	18.9	18.8
9	21.8	21.6	21.4	21.3	21.2

″	140	139	138	137	136
1	2.3	2.3	2.3	2.3	2.3
2	4.7	4.6	4.6	4.6	4.5
3	7.0	7.0	6.9	6.8	6.8
4	9.3	9.3	9.2	9.1	9.1
5	11.7	11.6	11.5	11.4	11.3
6	14.0	13.9	13.8	13.7	13.6
7	16.3	16.2	16.1	16.0	15.9
8	18.7	18.5	18.4	18.3	18.1
9	21.0	20.8	20.7	20.6	20.4

″	135	134	133	132	131
1	2.2	2.2	2.2	2.2	2.2
2	4.5	4.5	4.4	4.4	4.4
3	6.8	6.7	6.6	6.6	6.6
4	9.0	8.9	8.9	8.8	8.7
5	11.2	11.2	11.1	11.0	10.9
6	13.5	13.4	13.3	13.2	13.1
7	15.8	15.6	15.5	15.4	15.3
8	18.0	17.9	17.7	17.6	17.5
9	20.2	20.1	20.0	19.8	19.6

″	130	129	128	127	126
1	2.2	2.2	2.1	2.1	2.1
2	4.3	4.3	4.3	4.2	4.2
3	6.5	6.4	6.4	6.4	6.3
4	8.7	8.6	8.5	8.5	8.4
5	10.8	10.8	10.7	10.6	10.5
6	13.0	12.9	12.8	12.7	12.6
7	15.2	15.0	14.9	14.8	14.7
8	17.3	17.2	17.1	16.9	16.8
9	19.5	19.4	19.2	19.0	18.9

″	125	124	123	122	121
1	2.1	2.1	2.0	2.0	2.0
2	4.2	4.1	4.1	4.1	4.0
3	6.2	6.2	6.2	6.1	6.0
4	8.3	8.3	8.2	8.1	8.1
5	10.4	10.3	10.2	10.2	10.1
6	12.5	12.4	12.3	12.2	12.1
7	14.6	14.5	14.4	14.2	14.1
8	16.7	16.5	16.4	16.3	16.1
9	18.8	18.6	18.4	18.3	18.2

Proportional parts

95° (275°) (264°) **84°**

The −10 portion of the characteristic of the logarithm is not printed but must be written down whenever such a logarithm is used.

6° (186°) (353°) **173°**

′	L Sin	d	L Tan	c d	L Ctn	L Cos	′
0	9.01 923	120	9.02 162	121	10.97 838	9.99 761	60
1	9.02 043	120	9.02 283	121	10.97 717	9.99 760	59
2	9.02 163	120	9.02 404	121	10.97 596	9.99 759	58
3	9.02 283	119	9.02 525	120	10.97 475	9.99 757	57
4	9.02 402	118	9.02 645	121	10.97 355	9.99 756	56
5	9.02 520	119	9.02 766	119	10.97 234	9.99 755	55
6	9.02 639	118	9.02 885	120	10.97 115	9.99 753	54
7	9.02 757	117	9.03 005	119	10.96 995	9.99 752	53
8	9.02 874	118	9.03 124	118	10.96 876	9.99 751	52
9	9.02 992	117	9.03 242	119	10.96 758	9.99 749	51
10	9.03 109	117	9.03 361	118	10.96 639	9.99 748	50
11	9.03 226	116	9.03 479	118	10.96 521	9.99 747	49
12	9.03 342	116	9.03 597	117	10.96 403	9.99 745	48
13	9.03 458	116	9.03 714	118	10.96 286	9.99 744	47
14	9.03 574	116	9.03 832	116	10.96 168	9.99 742	46
15	9.03 690	115	9.03 948	117	10.96 052	9.99 741	45
16	9.03 805	115	9.04 065	116	10.95 935	9.99 740	44
17	9.03 920	114	9.04 181	116	10.95 819	9.99 738	43
18	9.04 034	115	9.04 297	116	10.95 703	9.99 737	42
19	9.04 149	113	9.04 413	115	10.95 587	9.99 736	41
20	9.04 262	114	9.04 528	115	10.95 472	9.99 734	40
21	9.04 376	114	9.04 643	115	10.95 357	9.99 733	39
22	9.04 490	113	9.04 758	115	10.95 242	9.99 731	38
23	9.04 603	112	9.04 873	114	10.95 127	9.99 730	37
24	9.04 715	113	9.04 987	114	10.95 013	9.99 728	36
25	9.04 828	112	9.05 101	113	10.94 899	9.99 727	35
26	9.04 940	112	9.05 214	114	10.94 786	9.99 726	34
27	9.05 052	112	9.05 328	113	10.94 672	9.99 724	33
28	9.05 164	111	9.05 441	112	10.94 559	9.99 723	32
29	9.05 275	111	9.05 553	113	10.94 447	9.99 721	31
30	9.05 386	111	9.05 666	112	10.94 334	9.99 720	30
31	9.05 497	110	9.05 778	112	10.94 222	9.99 718	29
32	9.05 607	110	9.05 890	112	10.94 110	9.99 717	28
33	9.05 717	110	9.06 002	111	10.93 998	9.99 716	27
34	9.05 827	110	9.06 113	111	10.93 887	9.99 714	26
35	9.05 937	109	9.06 224	111	10.93 776	9.99 713	25
36	9.06 046	109	9.06 335	110	10.93 665	9.99 711	24
37	9.06 155	109	9.06 445	111	10.93 555	9.99 710	23
38	9.06 264	108	9.06 556	110	10.93 444	9.99 708	22
39	9.06 372	109	9.06 666	109	10.93 334	9.99 707	21
40	9.06 481	108	9.06 775	110	10.93 225	9.99 705	20
41	9.06 589	107	9.06 885	109	10.93 115	9.99 704	19
42	9.06 696	108	9.06 994	109	10.93 006	9.99 702	18
43	9.06 804	107	9.07 103	108	10.92 897	9.99 701	17
44	9.06 911	107	9.07 211	109	10.92 789	9.99 699	16
45	9.07 018	106	9.07 320	108	10.92 680	9.99 698	15
46	9.07 124	107	9.07 428	108	10.92 572	9.99 696	14
47	9.07 231	106	9.07 536	107	10.92 464	9.99 695	13
48	9.07 337	105	9.07 643	108	10.92 357	9.99 693	12
49	9.07 442	106	9.07 751	107	10.92 249	9.99 692	11
50	9.07 548	105	9.07 858	106	10.92 142	9.99 690	10
51	9.07 653	105	9.07 964	107	10.92 036	9.99 689	9
52	9.07 758	105	9.08 071	106	10.91 929	9.99 687	8
53	9.07 863	105	9.08 177	106	10.91 823	9.99 686	7
54	9.07 968	104	9.08 283	106	10.91 717	9.99 684	6
55	9.08 072	104	9.08 389	106	10.91 611	9.99 683	5
56	9.08 176	104	9.08 495	105	10.91 505	9.99 681	4
57	9.08 280	103	9.08 600	105	10.91 400	9.99 680	3
58	9.08 383	103	9.08 705	105	10.91 295	9.99 678	2
59	9.08 486	103	9.08 810	104	10.91 190	9.99 677	1
60	9.08 589		9.08 914		10.91 086	9.99 675	0
′	L Cos	d	L Ctn	c d	L Tan	L Sin	′

Proportional parts

″	121	120	119	118
1	2.0	2.0	2.0	2.0
2	4.0	4.0	4.0	3.9
3	6.0	6.0	6.0	5.9
4	8.1	8.0	7.9	7.9
5	10.1	10.0	9.9	9.8
6	12.1	12.0	11.9	11.8
7	14.1	14.0	13.9	13.8
8	16.1	16.0	15.9	15.7
9	18.2	18.0	17.8	17.7
10	20.2	20.0	19.8	19.7
20	40.3	40.0	39.7	39.3
30	60.5	60.0	59.5	59.0
40	80.7	80.0	79.3	78.7
50	100.8	100.0	99.2	98.3

″	117	116	115	114
1	2.0	1.9	1.9	1.9
2	3.9	3.9	3.8	3.8
3	5.8	5.8	5.8	5.7
4	7.8	7.7	7.7	7.6
5	9.8	9.7	9.6	9.5
6	11.7	11.6	11.5	11.4
7	13.6	13.5	13.4	13.3
8	15.6	15.5	15.3	15.2
9	17.6	17.4	17.2	17.1
10	19.5	19.3	19.2	19.0
20	39.0	38.7	38.3	38.0
30	58.5	58.0	57.5	57.0
40	78.0	77.3	76.7	76.0
50	97.5	96.7	95.8	95.0

″	113	112	111	110
1	1.9	1.9	1.8	1.8
2	3.8	3.7	3.7	3.7
3	5.6	5.6	5.6	5.5
4	7.5	7.5	7.4	7.3
5	9.4	9.3	9.2	9.2
6	11.3	11.2	11.1	11.0
7	13.2	13.1	13.0	12.8
8	15.1	14.9	14.8	14.7
9	17.0	16.8	16.6	16.5
10	18.8	18.7	18.5	18.3
20	37.7	37.3	37.0	36.7
30	56.5	56.0	55.5	55.0
40	75.3	74.7	74.0	73.3
50	94.2	93.3	92.5	91.7

″	109	108	107	106
1	1.8	1.8	1.8	1.8
2	3.6	3.6	3.6	3.5
3	5.4	5.4	5.4	5.3
4	7.3	7.2	7.1	7.1
5	9.1	9.0	8.9	8.8
6	10.9	10.8	10.7	10.6
7	12.7	12.6	12.5	12.4
8	14.5	14.4	14.3	14.1
9	16.4	16.2	16.0	15.9
10	18.2	18.0	17.8	17.7
20	36.3	36.0	35.7	35.3
30	54.5	54.0	53.5	53.0
40	72.7	72.0	71.3	70.7
50	90.8	90.0	89.2	88.3

Proportional parts

96° (276°) (263°) **83°**

The −10 portion of the characteristic of the logarithm is not printed but must be written down whenever such a logarithm is used.

7° (187°) (352°) **172°**

′	L Sin	d	L Tan	c d	L Ctn	L Cos	′
0	9.08 589	103	9.08 914	105	10.91 086	9.99 675	60
1	9.08 692	103	9.09 019	104	10.90 981	9.99 674	59
2	9.08 795	102	9.09 123	104	10.90 877	9.99 672	58
3	9.08 897	102	9.09 227	103	10.90 773	9.99 670	57
4	9.08 999	102	9.09 330	104	10.90 670	9.99 669	56
5	9.09 101	101	9.09 434	103	10.90 566	9.99 667	55
6	9.09 202	102	9.09 537	103	10.90 463	9.99 666	54
7	9.09 304	101	9.09 640	102	10.90 360	9.99 664	53
8	9.09 405	101	9.09 742	103	10.90 258	9.99 663	52
9	9.09 506	100	9.09 845	102	10.90 155	9.99 661	51
10	9.09 606	101	9.09 947	102	10.90 053	9.99 659	50
11	9.09 707	100	9.10 049	101	10.89 951	9.99 658	49
12	9.09 807	100	9.10 150	102	10.89 850	9.99 656	48
13	9.09 907	99	9.10 252	101	10.89 748	9.99 655	47
14	9.10 006	100	9.10 353	101	10.89 647	9.99 653	46
15	9.10 106	99	9.10 454	101	10.89 546	9.99 651	45
16	9.10 205	99	9.10 555	101	10.89 445	9.99 650	44
17	9.10 304	98	9.10 656	100	10.89 344	9.99 648	43
18	9.10 402	99	9.10 756	100	10.89 244	9.99 647	42
19	9.10 501	98	9.10 856	100	10.89 144	9.99 645	41
20	9.10 599	98	9.10 956	100	10.89 044	9.99 643	40
21	9.10 697	98	9.11 056	99	10.88 944	9.99 642	39
22	9.10 795	98	9.11 155	99	10.88 845	9.99 640	38
23	9.10 893	97	9.11 254	99	10.88 746	9.99 638	37
24	9.10 990	97	9.11 353	99	10.88 647	9.99 637	36
25	9.11 087	97	9.11 452	99	10.88 548	9.99 635	35
26	9.11 184	97	9.11 551	98	10.88 449	9.99 633	34
27	9.11 281	96	9.11 649	98	10.88 351	9.99 632	33
28	9.11 377	97	9.11 747	98	10.88 253	9.99 630	32
29	9.11 474	96	9.11 845	98	10.88 155	9.99 629	31
30	9.11 570	96	9.11 943	97	10.88 057	9.99 627	30
31	9.11 666	95	9.12 040	98	10.87 960	9.99 625	29
32	9.11 761	96	9.12 138	97	10.87 862	9.99 624	28
33	9.11 857	95	9.12 235	97	10.87 765	9.99 622	27
34	9.11 952	95	9.12 332	96	10.87 668	9.99 620	26
35	9.12 047	95	9.12 428	97	10.87 572	9.99 618	25
36	9.12 142	94	9.12 525	96	10.87 475	9.99 617	24
37	9.12 236	95	9.12 621	96	10.87 379	9.99 615	23
38	9.12 331	94	9.12 717	96	10.87 283	9.99 613	22
39	9.12 425	94	9.12 813	96	10.87 187	9.99 612	21
40	9.12 519	93	9.12 909	95	10.87 091	9.99 610	20
41	9.12 612	94	9.13 004	95	10.86 996	9.99 608	19
42	9.12 706	93	9.13 099	95	10.86 901	9.99 607	18
43	9.12 799	93	9.13 194	95	10.86 806	9.99 605	17
44	9.12 892	93	9.13 289	95	10.86 711	9.99 603	16
45	9.12 985	93	9.13 384	94	10.86 616	9.99 601	15
46	9.13 078	93	9.13 478	95	10.86 522	9.99 600	14
47	9.13 171	92	9.13 573	94	10.86 427	9.99 598	13
48	9.13 263	92	9.13 667	94	10.86 333	9.99 596	12
49	9.13 355	92	9.13 761	93	10.86 239	9.99 595	11
50	9.13 447	92	9.13 854	94	10.86 146	9.99 593	10
51	9.13 539	91	9.13 948	93	10.86 052	9.99 591	9
52	9.13 630	92	9.14 041	93	10.85 959	9.99 589	8
53	9.13 722	91	9.14 134	93	10.85 866	9.99 588	7
54	9.13 813	91	9.14 227	93	10.85 773	9.99 586	6
55	9.13 904	90	9.14 320	92	10.85 680	9.99 584	5
56	9.13 994	91	9.14 412	92	10.85 588	9.99 582	4
57	9.14 085	90	9.14 504	93	10.85 496	9.99 581	3
58	9.14 175	91	9.14 597	91	10.85 403	9.99 579	2
59	9.14 266	90	9.14 688	92	10.85 312	9.99 577	1
60	9.14 356		9.14 780		10.85 220	9.99 575	0
′	L Cos	d	L Ctn	c d	L Tan	L Sin	′

Proportional parts				
″	105	104	103	102
1	1.8	1.7	1.7	1.7
2	3.5	3.5	3.4	3.4
3	5.2	5.2	5.2	5.1
4	7.0	6.9	6.9	6.8
5	8.8	8.7	8.6	8.5
6	10.5	10.4	10.3	10.2
7	12.2	12.1	12.0	11.9
8	14.0	13.9	13.7	13.6
9	15.8	15.6	15.4	15.3
10	17.5	17.3	17.2	17.0
20	35.0	34.7	34.3	34.0
30	52.5	52.0	51.5	51.0
40	70.0	69.3	68.7	68.0
50	87.5	86.7	85.8	85.0
″	101	100	99	98
1	1.7	1.7	1.6	1.6
2	3.4	3.3	3.3	3.3
3	5.0	5.0	5.0	4.9
4	6.7	6.7	6.6	6.5
5	8.4	8.3	8.2	8.2
6	10.1	10.0	9.9	9.8
7	11.8	11.7	11.6	11.4
8	13.5	13.3	13.2	13.1
9	15.2	15.0	14.8	14.7
10	16.8	16.7	16.5	16.3
20	33.7	33.3	33.0	32.7
30	50.5	50.0	49.5	49.0
40	67.3	66.7	66.0	65.3
50	84.2	83.3	82.5	81.7
″	97	96	95	94
1	1.6	1.6	1.6	1.6
2	3.2	3.2	3.2	3.1
3	4.8	4.8	4.8	4.7
4	6.5	6.4	6.3	6.3
5	8.1	8.0	7.9	7.8
6	9.7	9.6	9.5	9.4
7	11.3	11.2	11.1	11.0
8	12.9	12.8	12.7	12.5
9	14.6	14.4	14.2	14.1
10	16.2	16.0	15.8	15.7
20	32.3	32.0	31.7	31.3
30	48.5	48.0	47.5	47.0
40	64.7	64.0	63.3	62.7
50	80.8	80.0	79.2	78.3
″	93	92	91	90
1	1.6	1.5	1.5	1.5
2	3.1	3.1	3.0	3.0
3	4.6	4.6	4.6	4.5
4	6.2	6.1	6.1	6.0
5	7.8	7.7	7.6	7.5
6	9.3	9.2	9.1	9.0
7	10.8	10.7	10.6	10.5
8	12.4	12.3	12.1	12.0
9	14.0	13.8	13.6	13.5
10	15.5	15.3	15.2	15.0
20	31.0	30.7	30.3	30.0
30	46.5	46.0	45.5	45.0
40	62.0	61.3	60.7	60.0
50	77.5	76.7	75.8	75.0
Proportional parts				

97° (277°) (262°) **82°**

The −10 portion of the characteristic of the logarithm is not printed but must be written down whenever such a logarithm is used.

8° (188°) (351°) **171°**

′	L Sin	d	L Tan	c d	L Ctn	L Cos	′
0	9.14 356	89	9.14 780	92	10.85 220	9.99 575	60
1	9.14 445	90	9.14 872	91	10.85 128	9.99 574	59
2	9.14 535	89	9.14 963	91	10.85 037	9.99 572	58
3	9.14 624	90	9.15 054	91	10.84 946	9.99 570	57
4	9.14 714	89	9.15 145	91	10.84 855	9.99 568	56
5	9.14 803	88	9.15 236	91	10.84 764	9.99 566	55
6	9.14 891	89	9.15 327	90	10.84 673	9.99 565	54
7	9.14 980	89	9.15 417	91	10.84 583	9.99 563	53
8	9.15 069	88	9.15 508	90	10.84 492	9.99 561	52
9	9.15 157	88	9.15 598	90	10.84 402	9.99 559	51
10	9.15 245	88	9.15 688	89	10.84 312	9.99 557	50
11	9.15 333	88	9.15 777	90	10.84 223	9.99 556	49
12	9.15 421	87	9.15 867	89	10.84 133	9.99 554	48
13	9.15 508	88	9.15 956	90	10.84 044	9.99 552	47
14	9.15 596	87	9.16 046	89	10.83 954	9.99 550	46
15	9.15 683	87	9.16 135	89	10.83 865	9.99 548	45
16	9.15 770	87	9.16 224	88	10.83 776	9.99 546	44
17	9.15 857	87	9.16 312	89	10.83 688	9.99 545	43
18	9.15 944	86	9.16 401	88	10.83 599	9.99 543	42
19	9.16 030	86	9.16 489	88	10.83 511	9.99 541	41
20	9.16 116	87	9.16 577	88	10.83 423	9.99 539	40
21	9.16 203	86	9.16 665	88	10.83 335	9.99 537	39
22	9.16 289	85	9.16 753	88	10.83 247	9.99 535	38
23	9.16 374	86	9.16 841	87	10.83 159	9.99 533	37
24	9.16 460	85	9.16 928	88	10.83 072	9.99 532	36
25	9.16 545	86	9.17 016	87	10.82 984	9.99 530	35
26	9.16 631	85	9.17 103	87	10.82 897	9.99 528	34
27	9.16 716	85	9.17 190	87	10.82 810	9.99 526	33
28	9.16 801	85	9.17 277	86	10.82 723	9.99 524	32
29	9.16 886	84	9.17 363	87	10.82 637	9.99 522	31
30	9.16 970	85	9.17 450	86	10.82 550	9.99 520	30
31	9.17 055	84	9.17 536	86	10.82 464	9.99 518	29
32	9.17 139	84	9.17 622	86	10.82 378	9.99 517	28
33	9.17 223	84	9.17 708	86	10.82 292	9.99 515	27
34	9.17 307	84	9.17 794	86	10.82 206	9.99 513	26
35	9.17 391	83	9.17 880	85	10.82 120	9.99 511	25
36	9.17 474	84	9.17 965	86	10.82 035	9.99 509	24
37	9.17 558	83	9.18 051	85	10.81 949	9.99 507	23
38	9.17 641	83	9.18 136	85	10.81 864	9.99 505	22
39	9.17 724	83	9.18 221	85	10.81 779	9.99 503	21
40	9.17 807	83	9.18 306	85	10.81 694	9.99 501	20
41	9.17 890	83	9.18 391	84	10.81 609	9.99 499	19
42	9.17 973	82	9.18 475	85	10.81 525	9.99 497	18
43	9.18 055	82	9.18 560	84	10.81 440	9.99 495	17
44	9.18 137	83	9.18 644	84	10.81 356	9.99 494	16
45	9.18 220	82	9.18 728	84	10.81 272	9.99 492	15
46	9.18 302	81	9.18 812	84	10.81 188	9.99 490	14
47	9.18 383	82	9.18 896	83	10.81 104	9.99 488	13
48	9.18 465	82	9.18 979	84	10.81 021	9.99 486	12
49	9.18 547	81	9.19 063	83	10.80 937	9.99 484	11
50	9.18 628	81	9.19 146	83	10.80 854	9.99 482	10
51	9.18 709	81	9.19 229	83	10.80 771	9.99 480	9
52	9.18 790	81	9.19 312	83	10.80 688	9.99 478	8
53	9.18 871	81	9.19 395	83	10.80 605	9.99 476	7
54	9.18 952	81	9.19 478	83	10.80 522	9.99 474	6
55	9.19 033	80	9.19 561	82	10.80 439	9.99 472	5
56	9.19 113	80	9.19 643	82	10.80 357	9.99 470	4
57	9.19 193	80	9.19 725	82	10.80 275	9.99 468	3
58	9.19 273	80	9.19 807	82	10.80 193	9.99 466	2
59	9.19 353	80	9.19 889	82	10.80 111	9.99 464	1
60	9.19 433		9.19 971		10.80 029	9.99 462	0
′	L Cos	d	L Ctn	c d	L Tan	L Sin	′

Proportional parts

″	92	91	90
1	1.5	1.5	1.5
2	3.1	3.0	3.0
3	4.6	4.6	4.5
4	6.1	6.1	6.0
5	7.7	7.6	7.5
6	9.2	9.1	9.0
7	10.7	10.6	10.5
8	12.3	12.1	12.0
9	13.8	13.6	13.5
10	15.3	15.2	15.0
20	30.7	30.3	30.0
30	46.0	45.5	45.0
40	61.3	60.7	60.0
50	76.7	75.8	75.0

″	89	88	87
1	1.5	1.5	1.4
2	3.0	2.9	2.9
3	4.4	4.4	4.4
4	5.9	5.9	5.8
5	7.4	7.3	7.2
6	8.9	8.8	8.7
7	10.4	10.3	10.2
8	11.9	11.7	11.6
9	13.4	13.2	13.0
10	14.8	14.7	14.5
20	29.7	29.3	29.0
30	44.5	44.0	43.5
40	59.3	58.7	58.0
50	74.2	73.3	72.5

″	86	85	84
1	1.4	1.4	1.4
2	2.9	2.8	2.8
3	4.3	4.2	4.2
4	5.7	5.7	5.6
5	7.2	7.1	7.0
6	8.6	8.5	8.4
7	10.0	9.9	9.8
8	11.5	11.3	11.2
9	12.9	12.8	12.6
10	14.3	14.2	14.0
20	28.7	28.3	28.0
30	43.0	42.5	42.0
40	57.3	56.7	56.0
50	71.7	70.8	70.0

″	83	82	81
1	1.4	1.4	1.4
2	2.8	2.7	2.7
3	4.2	4.1	4.0
4	5.5	5.5	5.4
5	6.9	6.8	6.8
6	8.3	8.2	8.1
7	9.7	9.6	9.4
8	11.1	10.9	10.8
9	12.4	12.3	12.2
10	13.8	13.7	13.5
20	27.7	27.3	27.0
30	41.5	41.0	40.5
40	55.3	54.7	54.0
50	69.2	68.3	67.5

Proportional parts

98° (278°) (261°) **81°**

The −10 portion of the characteristic of the logarithm is not printed but must be written down whenever such a logarithm is used.

9° (189°) (350°) **170°**

′	L Sin	d	L Tan	c d	L Ctn	L Cos	′
0	9.19 433	80	9.19 971	82	10.80 029	9.99 462	60
1	9.19 513	79	9.20 053	81	10.79 947	9.99 460	59
2	9.19 592	80	9.20 134	82	10.79 866	9.99 458	58
3	9.19 672	79	9.20 216	81	10.79 784	9.99 456	57
4	9.19 751	79	9.20 297	81	10.79 703	9.99 454	56
5	9.19 830	79	9.20 378	81	10.79 622	9.99 452	55
6	9.19 909	79	9.20 459	81	10.79 541	9.99 450	54
7	9.19 988	79	9.20 540	81	10.79 460	9.99 448	53
8	9.20 067	78	9.20 621	80	10.79 379	9.99 446	52
9	9.20 145	78	9.20 701	81	10.79 299	9.99 444	51
10	9.20 223	79	9.20 782	80	10.79 218	9.99 442	50
11	9.20 302	78	9.20 862	80	10.79 138	9.99 440	49
12	9.20 380	78	9.20 942	80	10.79 058	9.99 438	48
13	9.20 458	77	9.21 022	80	10.78 978	9.99 436	47
14	9.20 535	78	9.21 102	80	10.78 898	9.99 434	46
15	9.20 613	78	9.21 182	79	10.78 818	9.99 432	45
16	9.20 691	77	9.21 261	80	10.78 739	9.99 429	44
17	9.20 768	77	9.21 341	79	10.78 659	9.99 427	43
18	9.20 845	77	9.21 420	79	10.78 580	9.99 425	42
19	9.20 922	77	9.21 499	79	10.78 501	9.99 423	41
20	9.20 999	77	9.21 578	79	10.78 422	9.99 421	40
21	9.21 076	77	9.21 657	79	10.78 343	9.99 419	39
22	9.21 153	76	9.21 736	78	10.78 264	9.99 417	38
23	9.21 229	77	9.21 814	79	10.78 186	9.99 415	37
24	9.21 306	76	9.21 893	78	10.78 107	9.99 413	36
25	9.21 382	76	9.21 971	78	10.78 029	9.99 411	35
26	9.21 458	76	9.22 049	78	10.77 951	9.99 409	34
27	9.21 534	76	9.22 127	78	10.77 873	9.99 407	33
28	9.21 610	75	9.22 205	78	10.77 795	9.99 404	32
29	9.21 685	76	9.22 283	78	10.77 717	9.99 402	31
30	9.21 761	75	9.22 361	77	10.77 639	9.99 400	30
31	9.21 836	76	9.22 438	78	10.77 562	9.99 398	29
32	9.21 912	75	9.22 516	77	10.77 484	9.99 396	28
33	9.21 987	75	9.22 593	77	10.77 407	9.99 394	27
34	9.22 062	75	9.22 670	77	10.77 330	9.99 392	26
35	9.22 137	74	9.22 747	77	10.77 253	9.99 390	25
36	9.22 211	75	9.22 824	77	10.77 176	9.99 388	24
37	9.22 286	75	9.22 901	76	10.77 099	9.99 385	23
38	9.22 361	74	9.22 977	77	10.77 023	9.99 383	22
39	9.22 435	74	9.23 054	76	10.76 946	9.99 381	21
40	9.22 509	74	9.23 130	76	10.76 870	9.99 379	20
41	9.22 583	74	9.23 206	77	10.76 794	9.99 377	19
42	9.22 657	74	9.23 283	76	10.76 717	9.99 375	18
43	9.22 731	74	9.23 359	76	10.76 641	9.99 372	17
44	9.22 805	73	9.23 435	75	10.76 565	9.99 370	16
45	9.22 878	74	9.23 510	76	10.76 490	9.99 368	15
46	9.22 952	73	9.23 586	75	10.76 414	9.99 366	14
47	9.23 025	73	9.23 661	76	10.76 339	9.99 364	13
48	9.23 098	73	9.23 737	75	10.76 263	9.99 362	12
49	9.23 171	73	9.23 812	75	10.76 188	9.99 359	11
50	9.23 244	73	9.23 887	75	10.76 113	9.99 357	10
51	9.23 317	73	9.23 962	75	10.76 038	9.99 355	9
52	9.23 390	72	9.24 037	75	10.75 963	9.99 353	8
53	9.23 462	73	9.24 112	74	10.75 888	9.99 351	7
54	9.23 535	72	9.24 186	75	10.75 814	9.99 348	6
55	9.23 607	72	9.24 261	74	10.75 739	9.99 346	5
56	9.23 679	73	9.24 335	75	10.75 665	9.99 344	4
57	9.23 752	71	9.24 410	74	10.75 590	9.99 342	3
58	9.23 823	72	9.24 484	74	10.75 516	9.99 340	2
59	9.23 895	72	9.24 558	74	10.75 442	9.99 337	1
60	9.23 967		9.24 632		10.75 368	9.99 335	0
′	L Cos	d	L Ctn	c d	L Tan	L Sin	′

Proportional parts

″	80	79	78	77
1	1.3	1.3	1.3	1.3
2	2.7	2.6	2.6	2.6
3	4.0	4.0	3.9	3.8
4	5.3	5.3	5.2	5.1
5	6.7	6.6	6.5	6.4
6	8.0	7.9	7.8	7.7
7	9.3	9.2	9.1	9.0
8	10.7	10.5	10.4	10.3
9	12.0	11.8	11.7	11.6
10	13.3	13.2	13.0	12.8
20	26.7	26.3	26.0	25.7
30	40.0	39.5	39.0	38.5
40	53.3	52.7	52.0	51.3
50	66.7	65.8	65.0	64.2

″	76	75	74	73
1	1.3	1.2	1.2	1.2
2	2.5	2.5	2.5	2.4
3	3.8	3.8	3.7	3.6
4	5.1	5.0	4.9	4.9
5	6.3	6.2	6.2	6.1
6	7.6	7.5	7.4	7.3
7	8.9	8.8	8.6	8.5
8	10.1	10.0	9.9	9.7
9	11.4	11.2	11.1	11.0
10	12.7	12.5	12.3	12.2
20	25.3	25.0	24.7	24.3
30	38.0	37.5	37.0	36.5
40	50.7	50.0	49.3	48.7
50	63.3	62.5	61.7	60.8

″	72	71	3	2
1	1.2	1.2	0.0	0.0
2	2.4	2.4	0.1	0.1
3	3.6	3.6	0.2	0.1
4	4.8	4.7	0.2	0.1
5	6.0	5.9	0.2	0.2
6	7.2	7.1	0.3	0.2
7	8.4	8.3	0.4	0.2
8	9.6	9.5	0.4	0.3
9	10.8	10.6	0.4	0.3
10	12.0	11.8	0.5	0.3
20	24.0	23.7	1.0	0.7
30	36.0	35.5	1.5	1.0
40	48.0	47.3	2.0	1.3
50	60.0	59.2	2.5	1.7

Proportional parts

99° (279°) (260°) **80°**

The −10 portion of the characteristic of the logarithm is not printed but must be written down whenever such a logarithm is used.

10° (190°) (349°) **169°**

′	L Sin	d	L Tan	c d	L Ctn	L Cos	d	′
0	9.23 967	72	9.24 632	74	10.75 368	9.99 335	2	60
1	9.24 039	71	9.24 706	73	10.75 294	9.99 333	2	59
2	9.24 110	71	9.24 779	74	10.75 221	9.99 331	3	58
3	9.24 181	72	9.24 853	73	10.75 147	9.99 328	2	57
4	9.24 253	71	9.24 926	74	10.75 074	9.99 326	2	56
5	9.24 324	71	9.25 000	73	10.75 000	9.99 324	2	55
6	9.24 395	71	9.25 073	73	10.74 927	9.99 322	3	54
7	9.24 466	70	9.25 146	73	10.74 854	9.99 319	2	53
8	9.24 536	71	9.25 219	73	10.74 781	9.99 317	2	52
9	9.24 607	70	9.25 292	73	10.74 708	9.99 315	2	51
10	9.24 677	71	9.25 365	72	10.74 635	9.99 313	3	50
11	9.24 748	70	9.25 437	73	10.74 563	9.99 310	2	49
12	9.24 818	70	9.25 510	72	10.74 490	9.99 308	2	48
13	9.24 888	70	9.25 582	73	10.74 418	9.99 306	2	47
14	9.24 958	70	9.25 655	72	10.74 345	9.99 304	3	46
15	9.25 028	70	9.25 727	72	10.74 273	9.99 301	2	45
16	9.25 098	70	9.25 799	72	10.74 201	9.99 299	2	44
17	9.25 168	69	9.25 871	72	10.74 129	9.99 297	3	43
18	9.25 237	70	9.25 943	72	10.74 057	9.99 294	2	42
19	9.25 307	69	9.26 015	71	10.73 985	9.99 292	2	41
20	9.25 376	69	9.26 086	72	10.73 914	9.99 290	2	40
21	9.25 445	69	9.26 158	71	10.73 842	9.99 288	3	39
22	9.25 514	69	9.26 229	72	10.73 771	9.99 285	2	38
23	9.25 583	69	9.26 301	71	10.73 699	9.99 283	2	37
24	9.25 652	69	9.26 372	71	10.73 628	9.99 281	3	36
25	9.25 721	69	9.26 443	71	10.73 557	9.99 278	2	35
26	9.25 790	68	9.26 514	71	10.73 486	9.99 276	2	34
27	9.25 858	69	9.26 585	70	10.73 415	9.99 274	3	33
28	9.25 927	68	9.26 655	71	10.73 345	9.99 271	2	32
29	9.25 995	68	9.26 726	71	10.73 274	9.99 269	2	31
30	9.26 063	68	9.26 797	70	10.73 203	9.99 267	3	30
31	9.26 131	68	9.26 867	70	10.73 133	9.99 264	2	29
32	9.26 199	68	9.26 937	71	10.73 063	9.99 262	2	28
33	9.26 267	68	9.27 008	70	10.72 992	9.99 260	3	27
34	9.26 335	68	9.27 078	70	10.72 922	9.99 257	2	26
35	9.26 403	67	9.27 148	70	10.72 852	9.99 255	3	25
36	9.26 470	68	9.27 218	70	10.72 782	9.99 252	2	24
37	9.26 538	67	9.27 288	69	10.72 712	9.99 250	2	23
38	9.26 605	67	9.27 357	70	10.72 643	9.99 248	3	22
39	9.26 672	67	9.27 427	69	10.72 573	9.99 245	2	21
40	9.26 739	67	9.27 496	70	10.72 504	9.99 243	2	20
41	9.26 806	67	9.27 566	69	10.72 434	9.99 241	3	19
42	9.26 873	67	9.27 635	69	10.72 365	9.99 238	2	18
43	9.26 940	67	9.27 704	69	10.72 296	9.99 236	3	17
44	9.27 007	66	9.27 773	69	10.72 227	9.99 233	2	16
45	9.27 073	67	9.27 842	69	10.72 158	9.99 231	2	15
46	9.27 140	66	9.27 911	69	10.72 089	9.99 229	3	14
47	9.27 206	67	9.27 980	69	10.72 020	9.99 226	2	13
48	9.27 273	66	9.28 049	68	10.71 951	9.99 224	3	12
49	9.27 339	66	9.28 117	69	10.71 883	9.99 221	2	11
50	9.27 405	66	9.28 186	68	10.71 814	9.99 219	2	10
51	9.27 471	66	9.28 254	69	10.71 746	9.99 217	3	9
52	9.27 537	65	9.28 323	68	10.71 677	9.99 214	2	8
53	9.27 602	66	9.28 391	68	10.71 609	9.99 212	3	7
54	9.27 668	66	9.28 459	68	10.71 541	9.99 209	2	6
55	9.27 734	65	9.28 527	68	10.71 473	9.99 207	3	5
56	9.27 799	65	9.28 595	67	10.71 405	9.99 204	2	4
57	9.27 864	66	9.28 662	68	10.71 338	9.99 202	2	3
58	9.27 930	65	9.28 730	68	10.71 270	9.99 200	3	2
59	9.27 995	65	9.28 798	67	10.71 202	9.99 197	2	1
60	9.28 060		9.28 865		10.71 135	9.99 195		0
′	L Cos	d	L Ctn	c d	L Tan	L Sin	d	′

Proportional parts

″	74	73	72
1	1.2	1.2	1.2
2	2.5	2.4	2.4
3	3.7	3.6	3.6
4	4.9	4.9	4.8
5	6.2	6.1	6.0
6	7.4	7.3	7.2
7	8.6	8.5	8.4
8	9.9	9.7	9.6
9	11.1	11.0	10.8
10	12.3	12.2	12.0
20	24.7	24.3	24.0
30	37.0	36.5	36.0
40	49.3	48.7	48.0
50	61.7	60.8	60.0

″	71	70	69
1	1.2	1.2	1.2
2	2.4	2.3	2.3
3	3.6	3.5	3.4
4	4.7	4.7	4.6
5	5.9	5.8	5.8
6	7.1	7.0	6.9
7	8.3	8.2	8.0
8	9.5	9.3	9.2
9	10.6	10.5	10.4
10	11.8	11.7	11.5
20	23.7	23.3	23.0
30	35.5	35.0	34.5
40	47.3	46.7	46.0
50	59.2	58.3	57.5

″	68	67	66
1	1.1	1.1	1.1
2	2.3	2.2	2.2
3	3.4	3.4	3.3
4	4.5	4.5	4.4
5	5.7	5.6	5.5
6	6.8	6.7	6.6
7	7.9	7.8	7.7
8	9.1	8.9	8.8
9	10.2	10.0	9.9
10	11.3	11.2	11.0
20	22.7	22.3	22.0
30	34.0	33.5	33.0
40	45.3	44.7	44.0
50	56.7	55.8	55.0

Proportional parts

100° (280°) (259°) **79°**

The −10 portion of the characteristic of the logarithm is not printed but must be written down whenever such a logarithm is used.

11° (191°) (348°) **168°**

′	L Sin	d	L Tan	c d	L Ctn	L Cos	d	′
0	9.28 060	65	9.28 865	68	10.71 135	9.99 195	3	60
1	9.28 125	65	9.28 933	67	10.71 067	9.99 192	2	59
2	9.28 190	64	9.29 000	67	10.71 000	9.99 190	3	58
3	9.28 254	65	9.29 067	67	10.70 933	9.99 187	2	57
4	9.28 319	65	9.29 134	67	10.70 866	9.99 185	3	56
5	9.28 384	64	9.29 201	67	10.70 799	9.99 182	2	55
6	9.28 448	64	9.29 268	67	10.70 732	9.99 180	3	54
7	9.28 512	65	9.29 335	67	10.70 665	9.99 177	2	53
8	9.28 577	64	9.29 402	66	10.70 598	9.99 175	3	52
9	9.28 641	64	9.29 468	67	10.70 532	9.99 172	2	51
10	9.28 705	64	9.29 535	66	10.70 465	9.99 170	3	50
11	9.28 769	64	9.29 601	67	10.70 399	9.99 167	2	49
12	9.28 833	63	9.29 668	66	10.70 332	9.99 165	3	48
13	9.28 896	64	9.29 734	66	10.70 266	9.99 162	2	47
14	9.28 960	64	9.29 800	66	10.70 200	9.99 160	3	46
15	9.29 024	63	9.29 866	66	10.70 134	9.99 157	2	45
16	9.29 087	63	9.29 932	66	10.70 068	9.99 155	3	44
17	9.29 150	64	9.29 998	66	10.70 002	9.99 152	2	43
18	9.29 214	63	9.30 064	66	10.69 936	9.99 150	3	42
19	9.29 277	63	9.30 130	65	10.69 870	9.99 147	2	41
20	9.29 340	63	9.30 195	66	10.69 805	9.99 145	3	40
21	9.29 403	63	9.30 261	65	10.69 739	9.99 142	2	39
22	9.29 466	63	9.30 326	65	10.69 674	9.99 140	3	38
23	9.29 529	62	9.30 391	66	10.69 609	9.99 137	2	37
24	9.29 591	63	9.30 457	65	10.69 543	9.99 135	3	36
25	9.29 654	62	9.30 522	65	10.69 478	9.99 132	2	35
26	9.29 716	63	9.30 587	65	10.69 413	9.99 130	3	34
27	9.29 779	62	9.30 652	65	10.69 348	9.99 127	3	33
28	9.29 841	62	9.30 717	65	10.69 283	9.99 124	2	32
29	9.29 903	63	9.30 782	64	10.69 218	9.99 122	3	31
30	9.29 966	62	9.30 846	65	10.69 154	9.99 119	2	30
31	9.30 028	62	9.30 911	64	10.69 089	9.99 117	3	29
32	9.30 090	61	9.30 975	65	10.69 025	9.99 114	2	28
33	9.30 151	62	9.31 040	64	10.68 960	9.99 112	3	27
34	9.30 213	62	9.31 104	64	10.68 896	9.99 109	3	26
35	9.30 275	61	9.31 168	65	10.68 832	9.99 106	2	25
36	9.30 336	62	9.31 233	64	10.68 767	9.99 104	3	24
37	9.30 398	61	9.31 297	64	10.68 703	9.99 101	2	23
38	9.30 459	62	9.31 361	64	10.68 639	9.99 099	3	22
39	9.30 521	61	9.31 425	64	10.68 575	9.99 096	3	21
40	9.30 582	61	9.31 489	63	10.68 511	9.99 093	2	20
41	9.30 643	61	9.31 552	64	10.68 448	9.99 091	3	19
42	9.30 704	61	9.31 616	63	10.68 384	9.99 088	2	18
43	9.30 765	61	9.31 679	64	10.68 321	9.99 086	3	17
44	9.30 826	61	9.31 743	63	10.68 257	9.99 083	3	16
45	9.30 887	60	9.31 806	64	10.68 194	9.99 080	2	15
46	9.30 947	61	9.31 870	63	10.68 130	9.99 078	3	14
47	9.31 008	60	9.31 933	63	10.68 067	9.99 075	3	13
48	9.31 068	61	9.31 996	63	10.68 004	9.99 072	2	12
49	9.31 129	60	9.32 059	63	10.67 941	9.99 070	3	11
50	9.31 189	61	9.32 122	63	10.67 878	9.99 067	3	10
51	9.31 250	60	9.32 185	63	10.67 815	9.99 064	2	9
52	9.31 310	60	9.32 248	63	10.67 752	9.99 062	3	8
53	9.31 370	60	9.32 311	62	10.67 689	9.99 059	3	7
54	9.31 430	60	9.32 373	63	10.67 627	9.99 056	2	6
55	9.31 490	59	9.32 436	62	10.67 564	9.99 054	3	5
56	9.31 549	60	9.32 498	63	10.67 502	9.99 051	3	4
57	9.31 609	60	9.32 561	62	10.67 439	9.99 048	2	3
58	9.31 669	59	9.32 623	62	10.67 377	9.99 046	3	2
59	9.31 728	60	9.32 685	62	10.67 315	9.99 043	3	1
60	9.31 788		9.32 747		10.67 253	9.99 040		0
′	L Cos	d	L Ctn	c d	L Tan	L Sin	d	′

Proportional parts

″	65	64	63
1	1.1	1.1	1.0
2	2.2	2.1	2.1
3	3.2	3.2	3.2
4	4.3	4.3	4.2
5	5.4	5.3	5.2
6	6.5	6.4	6.3
7	7.6	7.5	7.4
8	8.7	8.5	8.4
9	9.8	9.6	9.4
10	10.8	10.7	10.5
20	21.7	21.3	21.0
30	32.5	32.0	31.5
40	43.3	42.7	42.0
50	54.2	53.3	52.5

″	62	61	60
1	1.0	1.0	1.0
2	2.1	2.0	2.0
3	3.1	3.0	3.0
4	4.1	4.1	4.0
5	5.2	5.1	5.0
6	6.2	6.1	6.0
7	7.2	7.1	7.0
8	8.3	8.1	8.0
9	9.3	9.2	9.0
10	10.3	10.2	10.0
20	20.7	20.3	20.0
30	31.0	30.5	30.0
40	41.3	40.7	40.0
50	51.7	50.8	50.0

″	59	3	2
1	1.0	0.0	0.0
2	2.0	0.1	0.1
3	3.0	0.2	0.1
4	3.9	0.2	0.1
5	4.9	0.2	0.2
6	5.9	0.3	0.2
7	6.9	0.4	0.2
8	7.9	0.4	0.3
9	8.8	0.4	0.3
10	9.8	0.5	0.3
20	19.7	1.0	0.7
30	29.5	1.5	1.0
40	39.3	2.0	1.3
50	49.2	2.5	1.7

Proportional parts

101° (281°) (258°) **78°**

The −10 portion of the characteristic of the logarithm is not printed but must be written down whenever such a logarithm is used.

12° (192°) (347°) **167°**

′	L Sin	d	L Tan	c d	L Ctn	L Cos	d	′
0	9.31 788	59	9.32 747	63	10.67 253	9.99 040	2	60
1	9.31 847	60	9.32 810	62	10.67 190	9.99 038	3	59
2	9.31 907	59	9.32 872	61	10.67 128	9.99 035	3	58
3	9.31 966	59	9.32 933	62	10.67 067	9.99 032	2	57
4	9.32 025	59	9.32 995	62	10.67 005	9.99 030	3	56
5	9.32 084	59	9.33 057	62	10.66 943	9.99 027	3	55
6	9.32 143	59	9.33 119	61	10.66 881	9.99 024	2	54
7	9.32 202	59	9.33 180	62	10.66 820	9.99 022	3	53
8	9.32 261	58	9.33 242	61	10.66 758	9.99 019	3	52
9	9.32 319	59	9.33 303	62	10.66 697	9.99 016	3	51
10	9.32 378	59	9.33 365	61	10.66 635	9.99 013	2	50
11	9.32 437	58	9.33 426	61	10.66 574	9.99 011	3	49
12	9.32 495	58	9.33 487	61	10.66 513	9.99 008	3	48
13	9.32 553	59	9.33 548	61	10.66 452	9.99 005	3	47
14	9.32 612	58	9.33 609	61	10.66 391	9.99 002	2	46
15	9.32 670	58	9.33 670	61	10.66 330	9.99 000	3	45
16	9.32 728	58	9.33 731	61	10.66 269	9.98 997	3	44
17	9.32 786	58	9.33 792	61	10.66 208	9.98 994	3	43
18	9.32 844	58	9.33 853	60	10.66 147	9.98 991	2	42
19	9.32 902	58	9.33 913	61	10.66 087	9.98 989	3	41
20	9.32 960	58	9.33 974	60	10.66 026	9.98 986	3	40
21	9.33 018	57	9.34 034	61	10.65 966	9.98 983	3	39
22	9.33 075	58	9.34 095	60	10.65 905	9.98 980	2	38
23	9.33 133	57	9.34 155	60	10.65 845	9.98 978	3	37
24	9.33 190	58	9.34 215	61	10.65 785	9.98 975	3	36
25	9.33 248	57	9.34 276	60	10.65 724	9.98 972	3	35
26	9.33 305	57	9.34 336	60	10.65 664	9.98 969	2	34
27	9.33 362	58	9.34 396	60	10.65 604	9.98 967	3	33
28	9.33 420	57	9.34 456	60	10.65 544	9.98 964	3	32
29	9.33 477	57	9.34 516	60	10.65 484	9.98 961	3	31
30	9.33 534	57	9.34 576	59	10.65 424	9.98 958	3	30
31	9.33 591	56	9.34 635	60	10.65 365	9.98 955	2	29
32	9.33 647	57	9.34 695	60	10.65 305	9.98 953	3	28
33	9.33 704	57	9.34 755	59	10.65 245	9.98 950	3	27
34	9.33 761	57	9.34 814	60	10.65 186	9.98 947	3	26
35	9.33 818	56	9.34 874	59	10.65 126	9.98 944	3	25
36	9.33 874	57	9.34 933	59	10.65 067	9.98 941	3	24
37	9.33 931	56	9.34 992	59	10.65 008	9.98 938	2	23
38	9.33 987	56	9.35 051	60	10.64 949	9.98 936	3	22
39	9.34 043	57	9.35 111	59	10.64 889	9.98 933	3	21
40	9.34 100	56	9.35 170	59	10.64 830	9.98 930	3	20
41	9.34 156	56	9.35 229	59	10.64 771	9.98 927	3	19
42	9.34 212	56	9.35 288	59	10.64 712	9.98 924	3	18
43	9.34 268	56	9.35 347	58	10.64 653	9.98 921	2	17
44	9.34 324	56	9.35 405	59	10.64 595	9.98 919	3	16
45	9.34 380	56	9.35 464	59	10.64 536	9.98 916	3	15
46	9.34 436	55	9.35 523	58	10.64 477	9.98 913	3	14
47	9.34 491	56	9.35 581	59	10.64 419	9.98 910	3	13
48	9.34 547	55	9.35 640	58	10.64 360	9.98 907	3	12
49	9.34 602	56	9.35 698	59	10.64 302	9.98 904	3	11
50	9.34 658	55	9.35 757	58	10.64 243	9.98 901	3	10
51	9.34 713	56	9.35 815	58	10.64 185	9.98 898	2	9
52	9.34 769	55	9.35 873	58	10.64 127	9.98 896	3	8
53	9.34 824	55	9.35 931	58	10.64 069	9.98 893	3	7
54	9.34 879	55	9.35 989	58	10.64 011	9.98 890	3	6
55	9.34 934	55	9.36 047	58	10.63 953	9.98 887	3	5
56	9.34 989	55	9.36 105	58	10.63 895	9.98 884	3	4
57	9.35 044	55	9.36 163	58	10.63 837	9.98 881	3	3
58	9.35 099	55	9.36 221	58	10.63 779	9.98 878	3	2
59	9.35 154	55	9.36 279	57	10.63 721	9.98 875	3	1
60	9.35 209		9.36 336		10.63 664	9.98 872		0
′	L Cos	d	L Ctn	c d	L Tan	L Sin	d	′

Proportional parts

″	63	62	61
1	1.0	1.0	1.0
2	2.1	2.1	2.0
3	3.2	3.1	3.0
4	4.2	4.1	4.1
5	5.2	5.2	5.1
6	6.3	6.2	6.1
7	7.4	7.2	7.1
8	8.4	8.3	8.1
9	9.4	9.3	9.2
10	10.5	10.3	10.2
20	21.0	20.7	20.3
30	31.5	31.0	30.5
40	42.0	41.3	40.7
50	52.5	51.7	50.8

″	60	59	58
1	1.0	1.0	1.0
2	2.0	2.0	1.9
3	3.0	3.0	2.9
4	4.0	3.9	3.9
5	5.0	4.9	4.8
6	6.0	5.9	5.8
7	7.0	6.9	6.8
8	8.0	7.9	7.7
9	9.0	8.8	8.7
10	10.0	9.8	9.7
20	20.0	19.7	19.3
30	30.0	29.5	29.0
40	40.0	39.3	38.7
50	50.0	49.2	48.3

″	57	56	55
1	1.0	0.9	0.9
2	1.9	1.9	1.8
3	2.8	2.8	2.8
4	3.8	3.7	3.7
5	4.8	4.7	4.6
6	5.7	5.6	5.5
7	6.6	6.5	6.4
8	7.6	7.5	7.3
9	8.6	8.4	8.2
10	9.5	9.3	9.2
20	19.0	18.7	18.3
30	28.5	28.0	27.5
40	38.0	37.3	36.7
50	47.5	46.7	45.8

Proportional parts

102° (282°) (257°) **77°**

The −10 portion of the characteristic of the logarithm is not printed but must be written down whenever such a logarithm is used.

13° (193°) (346°) **166°**

′	L Sin	d	L Tan	c d	L Ctn	L Cos	d	′
0	9.35 209	54	9.36 336	58	10.63 664	9.98 872	3	60
1	9.35 263	55	9.36 394	58	10.63 606	9.98 869	2	59
2	9.35 318	55	9.36 452	57	10.63 548	9.98 867	3	58
3	9.35 373	54	9.36 509	57	10.63 491	9.98 864	3	57
4	9.35 427	54	9.36 566	58	10.63 434	9.98 861	3	56
5	9.35 481	55	9.36 624	57	10.63 376	9.98 858	3	55
6	9.35 536	54	9.36 681	57	10.63 319	9.98 855	3	54
7	9.35 590	54	9.36 738	57	10.63 262	9.98 852	3	53
8	9.35 644	54	9.36 795	57	10.63 205	9.98 849	3	52
9	9.35 698	54	9.36 852	57	10.63 148	9.98 846	3	51
10	9.35 752	54	9.36 909	57	10.63 091	9.98 843	3	50
11	9.35 806	54	9.36 966	57	10.63 034	9.98 840	3	49
12	9.35 860	54	9.37 023	57	10.62 977	9.98 837	3	48
13	9.35 914	54	9.37 080	57	10.62 920	9.98 834	3	47
14	9.35 968	54	9.37 137	56	10.62 863	9.98 831	3	46
15	9.36 022	53	9.37 193	57	10.62 807	9.98 828	3	45
16	9.36 075	54	9.37 250	56	10.62 750	9.98 825	3	44
17	9.36 129	53	9.37 306	57	10.62 694	9.98 822	3	43
18	9.36 182	54	9.37 363	56	10.62 637	9.98 819	3	42
19	9.36 236	53	9.37 419	57	10.62 581	9.98 816	3	41
20	9.36 289	53	9.37 476	56	10.62 524	9.98 813	3	40
21	9.36 342	53	9.37 532	56	10.62 468	9.98 810	3	39
22	9.36 395	54	9.37 588	56	10.62 412	9.98 807	3	38
23	9.36 449	53	9.37 644	56	10.62 356	9.98 804	3	37
24	9.36 502	53	9.37 700	56	10.62 300	9.98 801	3	36
25	9.36 555	53	9.37 756	56	10.62 244	9.98 798	3	35
26	9.36 608	52	9.37 812	56	10.62 188	9.98 795	3	34
27	9.36 660	53	9.37 868	56	10.62 132	9.98 792	3	33
28	9.36 713	53	9.37 924	56	10.62 076	9.98 789	3	32
29	9.36 766	53	9.37 980	55	10.62 020	9.98 786	3	31
30	9.36 819	52	9.38 035	56	10.61 965	9.98 783	3	30
31	9.36 871	53	9.38 091	56	10.61 909	9.98 780	3	29
32	9.36 924	52	9.38 147	55	10.61 853	9.98 777	3	28
33	9.36 976	52	9.38 202	55	10.61 798	9.98 774	3	27
34	9.37 028	53	9.38 257	56	10.61 743	9.98 771	3	26
35	9.37 081	52	9.38 313	55	10.61 687	9.98 768	3	25
36	9.37 133	52	9.38 368	55	10.61 632	9.98 765	3	24
37	9.37 185	52	9.38 423	56	10.61 577	9.98 762	3	23
38	9.37 237	52	9.38 479	55	10.61 521	9.98 759	3	22
39	9.37 289	52	9.38 534	55	10.61 466	9.98 756	3	21
40	9.37 341	52	9.38 589	55	10.61 411	9.98 753	3	20
41	9.37 393	52	9.38 644	55	10.61 356	9.98 750	4	19
42	9.37 445	52	9.38 699	55	10.61 301	9.98 746	3	18
43	9.37 497	52	9.38 754	54	10.61 246	9.98 743	3	17
44	9.37 549	51	9.38 808	55	10.61 192	9.98 740	3	16
45	9.37 600	52	9.38 863	55	10.61 137	9.98 737	3	15
46	9.37 652	51	9.38 918	54	10.61 082	9.98 734	3	14
47	9.37 703	52	9.38 972	55	10.61 028	9.98 731	3	13
48	9.37 755	51	9.39 027	55	10.60 973	9.98 728	3	12
49	9.37 806	52	9.39 082	54	10.60 918	9.98 725	3	11
50	9.37 858	51	9.39 136	54	10.60 864	9.98 722	3	10
51	9.37 909	51	9.39 190	55	10.60 810	9.98 719	4	9
52	9.37 960	51	9.39 245	54	10.60 755	9.98 715	3	8
53	9.38 011	51	9.39 299	54	10.60 701	9.98 712	3	7
54	9.38 062	51	9.39 353	54	10.60 647	9.98 709	3	6
55	9.38 113	51	9.39 407	54	10.60 593	9.98 706	3	5
56	9.38 164	51	9.39 461	54	10.60 539	9.98 703	3	4
57	9.38 215	51	9.39 515	54	10.60 485	9.98 700	3	3
58	9.38 266	51	9.39 569	54	10.60 431	9.98 697	3	2
59	9.38 317	51	9.39 623	54	10.60 377	9.98 694	4	1
60	9.38 368		9.39 677		10.60 323	9.98 690		0
′	L Cos	d	L Ctn	c d	L Tan	L Sin	d	′

Proportional parts

″	57	56	55
1	1.0	0.9	0.9
2	1.9	1.9	1.8
3	2.8	2.8	2.8
4	3.8	3.7	3.7
5	4.8	4.7	4.6
6	5.7	5.6	5.5
7	6.6	6.5	6.4
8	7.6	7.5	7.3
9	8.6	8.4	8.2
10	9.5	9.3	9.2
20	19.0	18.7	18.3
30	28.5	28.0	27.5
40	38.0	37.3	36.7
50	47.5	46.7	45.8

″	54	53	52
1	0.9	0.9	0.9
2	1.8	1.8	1.7
3	2.7	2.6	2.6
4	3.6	3.5	3.5
5	4.5	4.4	4.3
6	5.4	5.3	5.2
7	6.3	6.2	6.1
8	7.2	7.1	6.9
9	8.1	8.0	7.8
10	9.0	8.8	8.7
20	18.0	17.7	17.3
30	27.0	26.5	26.0
40	36.0	35.3	34.7
50	45.0	44.2	43.3

″	51	4	3	2
1	0.8	0.1	0.0	0.0
2	1.7	0.1	0.1	0.1
3	2.6	0.2	0.2	0.1
4	3.4	0.3	0.2	0.1
5	4.2	0.3	0.2	0.2
6	5.1	0.4	0.3	0.2
7	6.0	0.5	0.4	0.2
8	6.8	0.5	0.4	0.3
9	7.6	0.6	0.4	0.3
10	8.5	0.7	0.5	0.3
20	17.0	1.3	1.0	0.7
30	25.5	2.0	1.5	1.0
40	34.0	2.7	2.0	1.3
50	42.5	3.3	2.5	1.7

Proportional parts

103° (283°) (256°) **76°**

The −10 portion of the characteristic of the logarithm is not printed but must be written down whenever such a logarithm is used.

14° (194°) (345°) **165°**

′	L Sin	d	L Tan	c d	L Ctn	L Cos	d	′
0	9.38 368	50	9.39 677	54	10.60 323	9.98 690	3	60
1	9.38 418	51	9.39 731	54	10.60 269	9.98 687	3	59
2	9.38 469	50	9.39 785	53	10.60 215	9.98 684	3	58
3	9.38 519	51	9.39 838	54	10.60 162	9.98 681	3	57
4	9.38 570	50	9.39 892	53	10.60 108	9.98 678	3	56
5	9.38 620	50	9.39 945	54	10.60 055	9.98 675	4	55
6	9.38 670	51	9.39 999	53	10.60 001	9.98 671	3	54
7	9.38 721	50	9.40 052	54	10.59 948	9.98 668	3	53
8	9.38 771	50	9.40 106	53	10.59 894	9.98 665	3	52
9	9.38 821	50	9.40 159	53	10.59 841	9.98 662	3	51
10	9.38 871	50	9.40 212	54	10.59 788	9.98 659	3	50
11	9.38 921	50	9.40 266	53	10.59 734	9.98 656	4	49
12	9.38 971	50	9.40 319	53	10.59 681	9.98 652	3	48
13	9.39 021	50	9.40 372	53	10.59 628	9.98 649	3	47
14	9.39 071	50	9.40 425	53	10.59 575	9.98 646	3	46
15	9.39 121	49	9.40 478	53	10.59 522	9.98 643	3	45
16	9.39 170	50	9.40 531	53	10.59 469	9.98 640	4	44
17	9.39 220	50	9.40 584	52	10.59 416	9.98 636	3	43
18	9.39 270	49	9.40 636	53	10.59 364	9.98 633	3	42
19	9.39 319	50	9.40 689	53	10.59 311	9.98 630	3	41
20	9.39 369	49	9.40 742	53	10.59 258	9.98 627	4	40
21	9.39 418	49	9.40 795	52	10.59 205	9.98 623	3	39
22	9.39 467	50	9.40 847	53	10.59 153	9.98 620	3	38
23	9.39 517	49	9.40 900	52	10.59 100	9.98 617	3	37
24	9.39 566	49	9.40 952	53	10.59 048	9.98 614	4	36
25	9.39 615	49	9.41 005	52	10.58 995	9.98 610	3	35
26	9.39 664	49	9.41 057	52	10.58 943	9.98 607	3	34
27	9.39 713	49	9.41 109	52	10.58 891	9.98 604	3	33
28	9.39 762	49	9.41 161	53	10.58 839	9.98 601	4	32
29	9.39 811	49	9.41 214	52	10.58 786	9.98 597	3	31
30	9.39 860	49	9.41 266	52	10.58 734	9.98 594	3	30
31	9.39 909	49	9.41 318	52	10.58 682	9.98 591	3	29
32	9.39 958	48	9.41 370	52	10.58 630	9.98 588	4	28
33	9.40 006	49	9.41 422	52	10.58 578	9.98 584	3	27
34	9.40 055	48	9.41 474	52	10.58 526	9.98 581	3	26
35	9.40 103	49	9.41 526	52	10.58 474	9.98 578	4	25
36	9.40 152	48	9.41 578	51	10.58 422	9.98 574	3	24
37	9.40 200	49	9.41 629	52	10.58 371	9.98 571	3	23
38	9.40 249	48	9.41 681	52	10.58 319	9.98 568	3	22
39	9.40 297	49	9.41 733	51	10.58 267	9.98 565	4	21
40	9.40 346	48	9.41 784	52	10.58 216	9.98 561	3	20
41	9.40 394	48	9.41 836	51	10.58 164	9.98 558	3	19
42	9.40 442	48	9.41 887	52	10.58 113	9.98 555	4	18
43	9.40 490	48	9.41 939	51	10.58 061	9.98 551	3	17
44	9.40 538	48	9.41 990	51	10.58 010	9.98 548	3	16
45	9.40 586	48	9.42 041	52	10.57 959	9.98 545	4	15
46	9.40 634	48	9.42 093	51	10.57 907	9.98 541	3	14
47	9.40 682	48	9.42 144	51	10.57 856	9.98 538	3	13
48	9.40 730	48	9.42 195	51	10.57 805	9.98 535	4	12
49	9.40 778	47	9.42 246	51	10.57 754	9.98 531	3	11
50	9.40 825	48	9.42 297	51	10.57 703	9.98 528	3	10
51	9.40 873	48	9.42 348	51	10.57 652	9.98 525	4	9
52	9.40 921	47	9.42 399	51	10.57 601	9.98 521	3	8
53	9.40 968	48	9.42 450	51	10.57 550	9.98 518	3	7
54	9.41 016	47	9.42 501	51	10.57 499	9.98 515	4	6
55	9.41 063	48	9.42 552	51	10.57 448	9.98 511	3	5
56	9.41 111	47	9.42 603	50	10.57 397	9.98 508	3	4
57	9.41 158	47	9.42 653	51	10.57 347	9.98 505	4	3
58	9.41 205	47	9.42 704	51	10.57 296	9.98 501	3	2
59	9.41 252	48	9.42 755	50	10.57 245	9.98 498	4	1
60	9.41 300		9.42 805		10.57 195	9.98 494		0
′	L Cos	d	L Ctn	c d	L Tan	L Sin	d	′

Proportional parts

″	54	53	52
1	0.9	0.9	0.9
2	1.8	1.8	1.7
3	2.7	2.6	2.6
4	3.6	3.5	3.5
5	4.5	4.4	4.3
6	5.4	5.3	5.2
7	6.3	6.2	6.1
8	7.2	7.1	6.9
9	8.1	8.0	7.8
10	9.0	8.8	8.7
20	18.0	17.7	17.3
30	27.0	26.5	26.0
40	36.0	35.3	34.7
50	45.0	44.2	43.3

″	51	50	49
1	0.8	0.8	0.8
2	1.7	1.7	1.6
3	2.6	2.5	2.4
4	3.4	3.3	3.3
5	4.2	4.2	4.1
6	5.1	5.0	4.9
7	6.0	5.8	5.7
8	6.8	6.7	6.5
9	7.6	7.5	7.4
10	8.5	8.3	8.2
20	17.0	16.7	16.3
30	25.5	25.0	24.5
40	34.0	33.3	32.7
50	42.5	41.7	40.8

″	48	47	4	3
1	0.8	0.8	0.1	0.0
2	1.6	1.6	0.1	0.1
3	2.4	2.4	0.2	0.2
4	3.2	3.1	0.3	0.2
5	4.0	3.9	0.3	0.2
6	4.8	4.7	0.4	0.3
7	5.6	5.5	0.5	0.4
8	6.4	6.3	0.5	0.4
9	7.2	7.0	0.6	0.4
10	8.0	7.8	0.7	0.5
20	16.0	15.7	1.3	1.0
30	24.0	23.5	2.0	1.5
40	32.0	31.3	2.7	2.0
50	40.0	39.2	3.3	2.5

Proportional parts

104° (284°) (255°) **75°**

The −10 portion of the characteristic of the logarithm is not printed but must be written down whenever such a logarithm is used.

15° (195°) (344°) **164°**

′	L Sin	d	L Tan	c d	L Ctn	L Cos	d	′
0	9.41 300	47	9.42 805	51	10.57 195	9.98 494	3	60
1	9.41 347	47	9.42 856	50	10.57 144	9.98 491	3	59
2	9.41 394	47	9.42 906	51	10.57 094	9.98 488	4	58
3	9.41 441	47	9.42 957	50	10.57 043	9.98 484	3	57
4	9.41 488	47	9.43 007	50	10.56 993	9.98 481	4	56
5	9.41 535	47	9.43 057	51	10.56 943	9.98 477	3	55
6	9.41 582	46	9.43 108	50	10.56 892	9.98 474	3	54
7	9.41 628	47	9.43 158	50	10.56 842	9.98 471	4	53
8	9.41 675	47	9.43 208	50	10.56 792	9.98 467	3	52
9	9.41 722	46	9.43 258	50	10.56 742	9.98 464	4	51
10	9.41 768	47	9.43 308	50	10.56 692	9.98 460	3	50
11	9.41 815	46	9.43 358	50	10.56 642	9.98 457	4	49
12	9.41 861	47	9.43 408	50	10.56 592	9.98 453	3	48
13	9.41 908	46	9.43 458	50	10.56 542	9.98 450	3	47
14	9.41 954	47	9.43 508	50	10.56 492	9.98 447	4	46
15	9.42 001	46	9.43 558	49	10.56 442	9.98 443	3	45
16	9.42 047	46	9.43 607	50	10.56 393	9.98 440	4	44
17	9.42 093	47	9.43 657	50	10.56 343	9.98 436	3	43
18	9.42 140	46	9.43 707	49	10.56 293	9.98 433	4	42
19	9.42 186	46	9.43 756	50	10.56 244	9.98 429	3	41
20	9.42 232	46	9.43 806	49	10.56 194	9.98 426	4	40
21	9.42 278	46	9.43 855	50	10.56 145	9.98 422	3	39
22	9.42 324	46	9.43 905	49	10.56 095	9.98 419	4	38
23	9.42 370	46	9.43 954	50	10.56 046	9.98 415	3	37
24	9.42 416	45	9.44 004	49	10.55 996	9.98 412	3	36
25	9.42 461	46	9.44 053	49	10.55 947	9.98 409	4	35
26	9.42 507	46	9.44 102	49	10.55 898	9.98 405	3	34
27	9.42 553	46	9.44 151	50	10.55 849	9.98 402	4	33
28	9.42 599	45	9.44 201	49	10.55 799	9.98 398	3	32
29	9.42 644	46	9.44 250	49	10.55 750	9.98 395	4	31
30	9.42 690	45	9.44 299	49	10.55 701	9.98 391	3	30
31	9.42 735	46	9.44 348	49	10.55 652	9.98 388	4	29
32	9.42 781	45	9.44 397	49	10.55 603	9.98 384	3	28
33	9.42 826	46	9.44 446	49	10.55 554	9.98 381	4	27
34	9.42 872	45	9.44 495	49	10.55 505	9.98 377	4	26
35	9.42 917	45	9.44 544	48	10.55 456	9.98 373	3	25
36	9.42 962	46	9.44 592	49	10.55 408	9.98 370	4	24
37	9.43 008	45	9.44 641	49	10.55 359	9.98 366	3	23
38	9.43 053	45	9.44 690	48	10.55 310	9.98 363	4	22
39	9.43 098	45	9.44 738	49	10.55 262	9.98 359	3	21
40	9.43 143	45	9.44 787	49	10.55 213	9.98 356	4	20
41	9.43 188	45	9.44 836	48	10.55 164	9.98 352	3	19
42	9.43 233	45	9.44 884	49	10.55 116	9.98 349	4	18
43	9.43 278	45	9.44 933	48	10.55 067	9.98 345	3	17
44	9.43 323	44	9.44 981	48	10.55 019	9.98 342	4	16
45	9.43 367	45	9.45 029	49	10.54 971	9.98 338	4	15
46	9.43 412	45	9.45 078	48	10.54 922	9.98 334	3	14
47	9.43 457	45	9.45 126	48	10.54 874	9.98 331	4	13
48	9.43 502	44	9.45 174	48	10.54 826	9.98 327	3	12
49	9.43 546	45	9.45 222	49	10.54 778	9.98 324	4	11
50	9.43 591	44	9.45 271	48	10.54 729	9.98 320	3	10
51	9.43 635	45	9.45 319	48	10.54 681	9.98 317	4	9
52	9.43 680	44	9.45 367	48	10.54 633	9.98 313	4	8
53	9.43 724	45	9.45 415	48	10.54 585	9.98 309	3	7
54	9.43 769	44	9.45 463	48	10.54 537	9.98 306	4	6
55	9.43 813	44	9.45 511	48	10.54 489	9.98 302	3	5
56	9.43 857	44	9.45 559	47	10.54 441	9.98 299	4	4
57	9.43 901	45	9.45 606	48	10.54 394	9.98 295	4	3
58	9.43 946	44	9.45 654	48	10.54 346	9.98 291	3	2
59	9.43 990	44	9.45 702	48	10.54 298	9.98 288	4	1
60	9.44 034		9.45 750		10.54 250	9.98 284		0
′	L Cos	d	L Ctn	c d	L Tan	L Sin	d	′

105° (285°) (254°) **74°**

Proportional parts

″	51	50	49
1	0.8	0.8	0.8
2	1.7	1.7	1.6
3	2.6	2.5	2.4
4	3.4	3.3	3.3
5	4.2	4.2	4.1
6	5.1	5.0	4.9
7	6.0	5.8	5.7
8	6.8	6.7	6.5
9	7.6	7.5	7.4
10	8.5	8.3	8.2
20	17.0	16.7	16.3
30	25.5	25.0	24.5
40	34.0	33.3	32.7
50	42.5	41.7	40.8

″	48	47	46
1	0.8	0.8	0.8
2	1.6	1.6	1.5
3	2.4	2.4	2.3
4	3.2	3.1	3.1
5	4.0	3.9	3.8
6	4.8	4.7	4.6
7	5.6	5.5	5.4
8	6.4	6.3	6.1
9	7.2	7.0	6.9
10	8.0	7.8	7.7
20	16.0	15.7	15.3
30	24.0	23.5	23.0
40	32.0	31.3	30.7
50	40.0	39.2	38.3

″	45	44	4	3
1	0.8	0.7	0.1	0.0
2	1.5	1.5	0.1	0.1
3	2.2	2.2	0.2	0.2
4	3.0	2.9	0.3	0.2
5	3.8	3.7	0.3	0.2
6	4.5	4.4	0.4	0.3
7	5.2	5.1	0.5	0.4
8	6.0	5.9	0.5	0.4
9	6.8	6.6	0.6	0.4
10	7.5	7.3	0.7	0.5
20	15.0	14.7	1.3	1.0
30	22.5	22.0	2.0	1.5
40	30.0	29.3	2.7	2.0
50	37.5	36.7	3.3	2.5

Proportional parts

The −10 portion of the characteristic of the logarithm is not printed but must be written down whenever such a logarithm is used.

16° (196°) (343°) **163°**

′	L Sin	d	L Tan	c d	L Ctn	L Cos	d	′
0	9.44 034	44	9.45 750	47	10.54 250	9.98 284	3	60
1	9.44 078	44	9.45 797	48	10.54 203	9.98 281	4	59
2	9.44 122	44	9.45 845	47	10.54 155	9.98 277	4	58
3	9.44 166	44	9.45 892	48	10.54 108	9.98 273	3	57
4	9.44 210	43	9.45 940	47	10.54 060	9.98 270	4	56
5	9.44 253	44	9.45 987	48	10.54 013	9.98 266	4	55
6	9.44 297	44	9.46 035	47	10.53 965	9.98 262	3	54
7	9.44 341	44	9.46 082	48	10.53 918	9.98 259	4	53
8	9.44 385	43	9.46 130	47	10.53 870	9.98 255	4	52
9	9.44 428	44	9.46 177	47	10.53 823	9.98 251	3	51
10	9.44 472	44	9.46 224	47	10.53 776	9.98 248	4	50
11	9.44 516	43	9.46 271	48	10.53 729	9.98 244	4	49
12	9.44 559	43	9.46 319	47	10.53 681	9.98 240	3	48
13	9.44 602	44	9.46 366	47	10.53 634	9.98 237	4	47
14	9.44 646	43	9.46 413	47	10.53 587	9.98 233	4	46
15	9.44 689	44	9.46 460	47	10.53 540	9.98 229	3	45
16	9.44 733	43	9.46 507	47	10.53 493	9.98 226	4	44
17	9.44 776	43	9.46 554	47	10.53 446	9.98 222	4	43
18	9.44 819	43	9.46 601	47	10.53 399	9.98 218	3	42
19	9.44 862	43	9.46 648	46	10.53 352	9.98 215	4	41
20	9.44 905	43	9.46 694	47	10.53 306	9.98 211	4	40
21	9.44 948	44	9.46 741	47	10.53 259	9.98 207	3	39
22	9.44 992	43	9.46 788	47	10.53 212	9.98 204	4	38
23	9.45 035	42	9.46 835	46	10.53 165	9.98 200	4	37
24	9.45 077	43	9.46 881	47	10.53 119	9.98 196	4	36
25	9.45 120	43	9.46 928	47	10.53 072	9.98 192	3	35
26	9.45 163	43	9.46 975	46	10.53 025	9.98 189	4	34
27	9.45 206	43	9.47 021	47	10.52 979	9.98 185	4	33
28	9.45 249	43	9.47 068	46	10.52 932	9.98 181	4	32
29	9.45 292	42	9.47 114	46	10.52 886	9.98 177	3	31
30	9.45 334	43	9.47 160	47	10.52 840	9.98 174	4	30
31	9.45 377	42	9.47 207	46	10.52 793	9.98 170	4	29
32	9.45 419	43	9.47 253	46	10.52 747	9.98 166	4	28
33	9.45 462	42	9.47 299	47	10.52 701	9.98 162	3	27
34	9.45 504	43	9.47 346	46	10.52 654	9.98 159	4	26
35	9.45 547	42	9.47 392	46	10.52 608	9.98 155	4	25
36	9.45 589	43	9.47 438	46	10.52 562	9.98 151	4	24
37	9.45 632	42	9.47 484	46	10.52 516	9.98 147	3	23
38	9.45 674	42	9.47 530	46	10.52 470	9.98 144	4	22
39	9.45 716	42	9.47 576	46	10.52 424	9.98 140	4	21
40	9.45 758	43	9.47 622	46	10.52 378	9.98 136	4	20
41	9.45 801	42	9.47 668	46	10.52 332	9.98 132	3	19
42	9.45 843	42	9.47 714	46	10.52 286	9.98 129	4	18
43	9.45 885	42	9.47 760	46	10.52 240	9.98 125	4	17
44	9.45 927	42	9.47 806	46	10.52 194	9.98 121	4	16
45	9.45 969	42	9.47 852	45	10.52 148	9.98 117	4	15
46	9.46 011	42	9.47 897	46	10.52 103	9.98 113	3	14
47	9.46 053	42	9.47 943	46	10.52 057	9.98 110	4	13
48	9.46 095	41	9.47 989	46	10.52 011	9.98 106	4	12
49	9.46 136	42	9.48 035	45	10.51 965	9.98 102	4	11
50	9.46 178	42	9.48 080	46	10.51 920	9.98 098	4	10
51	9.46 220	42	9.48 126	45	10.51 874	9.98 094	4	9
52	9.46 262	41	9.48 171	46	10.51 829	9.98 090	3	8
53	9.46 303	42	9.48 217	45	10.51 783	9.98 087	4	7
54	9.46 345	41	9.48 262	45	10.51 738	9.98 083	4	6
55	9.46 386	42	9.48 307	46	10.51 693	9.98 079	4	5
56	9.46 428	41	9.48 353	45	10.51 647	9.98 075	4	4
57	9.46 469	42	9.48 398	45	10.51 602	9.98 071	4	3
58	9.46 511	41	9.48 443	46	10.51 557	9.98 067	4	2
59	9.46 552	42	9.48 489	45	10.51 511	9.98 063	3	1
60	9.46 594		9.48 534		10.51 466	9.98 060		0
′	L Cos	d	L Ctn	c d	L Tan	L Sin	d	′

Proportional parts

″	48	47	46
1	0.8	0.8	0.8
2	1.6	1.6	1.5
3	2.4	2.4	2.3
4	3.2	3.1	3.1
5	4.0	3.9	3.8
6	4.8	4.7	4.6
7	5.6	5.5	5.4
8	6.4	6.3	6.1
9	7.2	7.0	6.9
10	8.0	7.8	7.7
20	16.0	15.7	15.3
30	24.0	23.5	23.0
40	32.0	31.3	30.7
50	40.0	39.2	38.3

″	45	44	43
1	0.8	0.7	0.7
2	1.5	1.5	1.4
3	2.2	2.2	2.2
4	3.0	2.9	2.9
5	3.8	3.7	3.6
6	4.5	4.4	4.3
7	5.2	5.1	5.0
8	6.0	5.9	5.7
9	6.8	6.6	6.4
10	7.5	7.3	7.2
20	15.0	14.7	14.3
30	22.5	22.0	21.5
40	30.0	29.3	28.7
50	37.5	36.7	35.8

″	42	41	4	3
1	0.7	0.7	0.1	0.0
2	1.4	1.4	0.1	0.1
3	2.1	2.0	0.2	0.2
4	2.8	2.7	0.3	0.2
5	3.5	3.4	0.3	0.2
6	4.2	4.1	0.4	0.3
7	4.9	4.8	0.5	0.4
8	5.6	5.5	0.5	0.4
9	6.3	6.2	0.6	0.4
10	7.0	6.8	0.7	0.5
20	14.0	13.7	1.3	1.0
30	21.0	20.5	2.0	1.5
40	28.0	27.3	2.7	2.0
50	35.0	34.2	3.3	2.5

Proportional parts

106° (286°) (253°) **73°**

The −10 portion of the characteristic of the logarithm is not printed but must be written down whenever such a logarithm is used.

17° (197°) (342°) **162°**

′	L Sin	d	L Tan	c d	L Ctn	L Cos	d	′
0	9.46 594	41	9.48 534	45	10.51 466	9.98 060	4	60
1	9.46 635	41	9.48 579	45	10.51 421	9.98 056	4	59
2	9.46 676	41	9.48 624	45	10.51 376	9.98 052	4	58
3	9.46 717	41	9.48 669	45	10.51 331	9.98 048	4	57
4	9.46 758	42	9.48 714	45	10.51 286	9.98 044	4	56
5	9.46 800	41	9.48 759	45	10.51 241	9.98 040	4	55
6	9.46 841	41	9.48 804	45	10.51 196	9.98 036	4	54
7	9.46 882	41	9.48 849	45	10.51 151	9.98 032	3	53
8	9.46 923	41	9.48 894	45	10.51 106	9.98 029	4	52
9	9.46 964	41	9.48 939	45	10.51 061	9.98 025	4	51
10	9.47 005	40	9.48 984	45	10.51 016	9.98 021	4	50
11	9.47 045	41	9.49 029	44	10.50 971	9.98 017	4	49
12	9.47 086	41	9.49 073	45	10.50 927	9.98 013	4	48
13	9.47 127	41	9.49 118	45	10.50 882	9.98 009	4	47
14	9.47 168	41	9.49 163	44	10.50 837	9.98 005	4	46
15	9.47 209	40	9.49 207	45	10.50 793	9.98 001	4	45
16	9.47 249	41	9.49 252	44	10.50 748	9.97 997	4	44
17	9.47 290	40	9.49 296	45	10.50 704	9.97 993	4	43
18	9.47 330	41	9.49 341	44	10.50 659	9.97 989	3	42
19	9.47 371	40	9.49 385	45	10.50 615	9.97 986	4	41
20	9.47 411	41	9.49 430	44	10.50 570	9.97 982	4	40
21	9.47 452	40	9.49 474	45	10.50 526	9.97 978	4	39
22	9.47 492	41	9.49 519	44	10.50 481	9.97 974	4	38
23	9.47 533	40	9.49 563	44	10.50 437	9.97 970	4	37
24	9.47 573	40	9.49 607	45	10.50 393	9.97 966	4	36
25	9.47 613	41	9.49 652	44	10.50 348	9.97 962	4	35
26	9.47 654	40	9.49 696	44	10.50 304	9.97 958	4	34
27	9.47 694	40	9.49 740	44	10.50 260	9.97 954	4	33
28	9.47 734	40	9.49 784	44	10.50 216	9.97 950	4	32
29	9.47 774	40	9.49 828	44	10.50 172	9.97 946	4	31
30	9.47 814	40	9.49 872	44	10.50 128	9.97 942	4	30
31	9.47 854	40	9.49 916	44	10.50 084	9.97 938	4	29
32	9.47 894	40	9.49 960	44	10.50 040	9.97 934	4	28
33	9.47 934	40	9.50 004	44	10.49 996	9.97 930	4	27
34	9.47 974	40	9.50 048	44	10.49 952	9.97 926	4	26
35	9.48 014	40	9.50 092	44	10.49 908	9.97 922	4	25
36	9.48 054	40	9.50 136	44	10.49 864	9.97 918	4	24
37	9.48 094	39	9.50 180	43	10.49 820	9.97 914	4	23
38	9.48 133	40	9.50 223	44	10.49 777	9.97 910	4	22
39	9.48 173	40	9.50 267	44	10.49 733	9.97 906	4	21
40	9.48 213	39	9.50 311	44	10.49 689	9.97 902	4	20
41	9.48 252	40	9.50 355	43	10.49 645	9.97 898	4	19
42	9.48 292	40	9.50 398	44	10.49 602	9.97 894	4	18
43	9.48 332	39	9.50 442	43	10.49 558	9.97 890	4	17
44	9.48 371	40	9.50 485	44	10.49 515	9.97 886	4	16
45	9.48 411	39	9.50 529	43	10.49 471	9.97 882	4	15
46	9.48 450	40	9.50 572	44	10.49 428	9.97 878	4	14
47	9.48 490	39	9.50 616	43	10.49 384	9.97 874	4	13
48	9.48 529	39	9.50 659	44	10.49 341	9.97 870	4	12
49	9.48 568	39	9.50 703	43	10.49 297	9.97 866	5	11
50	9.48 607	40	9.50 746	43	10.49 254	9.97 861	4	10
51	9.48 647	39	9.50 789	44	10.49 211	9.97 857	4	9
52	9.48 686	39	9.50 833	43	10.49 167	9.97 853	4	8
53	9.48 725	39	9.50 876	43	10.49 124	9.97 849	4	7
54	9.48 764	39	9.50 919	43	10.49 081	9.97 845	4	6
55	9.48 803	39	9.50 962	43	10.49 038	9.97 841	4	5
56	9.48 842	39	9.51 005	43	10.48 995	9.97 837	4	4
57	9.48 881	39	9.51 048	44	10.48 952	9.97 833	4	3
58	9.48 920	39	9.51 092	43	10.48 908	9.97 829	4	2
59	9.48 959	39	9.51 135	43	10.48 865	9.97 825	4	1
60	9.48 998		9.51 178		10.48 822	9.97 821		0
′	L Cos	d	L Ctn	c d	L Tan	L Sin	d	′

Proportional parts

″	45	44	43
1	0.8	0.7	0.7
2	1.5	1.5	1.4
3	2.2	2.2	2.2
4	3.0	2.9	2.9
5	3.8	3.7	3.6
6	4.5	4.4	4.3
7	5.2	5.1	5.0
8	6.0	5.9	5.7
9	6.8	6.6	6.4
10	7.5	7.3	7.2
20	15.0	14.7	14.3
30	22.5	22.0	21.5
40	30.0	29.3	28.7
50	37.5	36.7	35.8

″	42	41	40
1	0.7	0.7	0.7
2	1.4	1.4	1.3
3	2.1	2.0	2.0
4	2.8	2.7	2.7
5	3.5	3.4	3.3
6	4.2	4.1	4.0
7	4.9	4.8	4.7
8	5.6	5.5	5.3
9	6.3	6.2	6.0
10	7.0	6.8	6.7
20	14.0	13.7	13.3
30	21.0	20.5	20.0
40	28.0	27.3	26.7
50	35.0	34.2	33.3

″	39	5	4	3
1	0.6	0.1	0.1	0.0
2	1.3	0.2	0.1	0.1
3	2.0	0.2	0.2	0.2
4	2.6	0.3	0.3	0.2
5	3.2	0.4	0.3	0.2
6	3.9	0.5	0.4	0.3
7	4.6	0.6	0.5	0.4
8	5.2	0.7	0.5	0.4
9	5.8	0.8	0.6	0.4
10	6.5	0.8	0.7	0.5
20	13.0	1.7	1.3	1.0
30	19.5	2.5	2.0	1.5
40	26.0	3.3	2.7	2.0
50	32.5	4.2	3.3	2.5

Proportional parts

107° (287°) (252°) **72°**

The −10 portion of the characteristic of the logarithm is not printed but must be written down whenever such a logarithm is used.

18° (198°) (341°) **161°**

′	L Sin	d	L Tan	c d	L Ctn	L Cos	d	′
0	9.48 998	39	9.51 178	43	10.48 822	9.97 821	4	60
1	9.49 037	39	9.51 221	43	10.48 779	9.97 817	5	59
2	9.49 076	39	9.51 264	42	10.48 736	9.97 812	4	58
3	9.49 115	38	9.51 306	43	10.48 694	9.97 808	4	57
4	9.49 153	39	9.51 349	43	10.48 651	9.97 804	4	56
5	9.49 192	39	9.51 392	43	10.48 608	9.97 800	4	55
6	9.49 231	38	9.51 435	43	10.48 565	9.97 796	4	54
7	9.49 269	39	9.51 478	42	10.48 522	9.97 792	4	53
8	9.49 308	39	9.51 520	43	10.48 480	9.97 788	4	52
9	9.49 347	38	9.51 563	43	10.48 437	9.97 784	5	51
10	9.49 385	39	9.51 606	42	10.48 394	9.97 779	4	50
11	9.49 424	38	9.51 648	43	10.48 352	9.97 775	4	49
12	9.49 462	38	9.51 691	43	10.48 309	9.97 771	4	48
13	9.49 500	39	9.51 734	42	10.48 266	9.97 767	4	47
14	9.49 539	38	9.51 776	43	10.48 224	9.97 763	4	46
15	9.49 577	38	9.51 819	42	10.48 181	9.97 759	5	45
16	9.49 615	39	9.51 861	42	10.48 139	9.97 754	4	44
17	9.49 654	38	9.51 903	43	10.48 097	9.97 750	4	43
18	9.49 692	38	9.51 946	42	10.48 054	9.97 746	4	42
19	9.49 730	38	9.51 988	43	10.48 012	9.97 742	4	41
20	9.49 768	38	9.52 031	42	10.47 969	9.97 738	4	40
21	9.49 806	38	9.52 073	42	10.47 927	9.97 734	5	39
22	9.49 844	38	9.52 115	42	10.47 885	9.97 729	4	38
23	9.49 882	38	9.52 157	43	10.47 843	9.97 725	4	37
24	9.49 920	38	9.52 200	42	10.47 800	9.97 721	4	36
25	9.49 958	38	9.52 242	42	10.47 758	9.97 717	4	35
26	9.49 996	38	9.52 284	42	10.47 716	9.97 713	5	34
27	9.50 034	38	9.52 326	42	10.47 674	9.97 708	4	33
28	9.50 072	38	9.52 368	42	10.47 632	9.97 704	4	32
29	9.50 110	38	9.52 410	42	10.47 590	9.97 700	4	31
30	9.50 148	37	9.52 452	42	10.47 548	9.97 696	5	30
31	9.50 185	38	9.52 494	42	10.47 506	9.97 691	4	29
32	9.50 223	38	9.52 536	42	10.47 464	9.97 687	4	28
33	9.50 261	37	9.52 578	42	10.47 422	9.97 683	4	27
34	9.50 298	38	9.52 620	41	10.47 380	9.97 679	5	26
35	9.50 336	38	9.52 661	42	10.47 339	9.97 674	4	25
36	9.50 374	37	9.52 703	42	10.47 297	9.97 670	4	24
37	9.50 411	38	9.52 745	42	10.47 255	9.97 666	4	23
38	9.50 449	37	9.52 787	42	10.47 213	9.97 662	5	22
39	9.50 486	37	9.52 829	41	10.47 171	9.97 657	4	21
40	9.50 523	38	9.52 870	42	10.47 130	9.97 653	4	20
41	9.50 561	37	9.52 912	41	10.47 088	9.97 649	4	19
42	9.50 598	37	9.52 953	42	10.47 047	9.97 645	5	18
43	9.50 635	38	9.52 995	42	10.47 005	9.97 640	4	17
44	9.50 673	37	9.53 037	41	10.46 963	9.97 636	4	16
45	9.50 710	37	9.53 078	42	10.46 922	9.97 632	4	15
46	9.50 747	37	9.53 120	41	10.46 880	9.97 628	5	14
47	9.50 784	37	9.53 161	41	10.46 839	9.97 623	4	13
48	9.50 821	37	9.53 202	42	10.46 798	9.97 619	4	12
49	9.50 858	38	9.53 244	41	10.46 756	9.97 615	5	11
50	9.50 896	37	9.53 285	42	10.46 715	9.97 610	4	10
51	9.50 933	37	9.53 327	41	10.46 673	9.97 606	4	9
52	9.50 970	37	9.53 368	41	10.46 632	9.97 602	5	8
53	9.51 007	36	9.53 409	41	10.46 591	9.97 597	4	7
54	9.51 043	37	9.53 450	42	10.46 550	9.97 593	4	6
55	9.51 080	37	9.53 492	41	10.46 508	9.97 589	5	5
56	9.51 117	37	9.53 533	41	10.46 467	9.97 584	4	4
57	9.51 154	37	9.53 574	41	10.46 426	9.97 580	4	3
58	9.51 191	36	9.53 615	41	10.46 385	9.97 576	5	2
59	9.51 227	37	9.53 656	41	10.46 344	9.97 571	4	1
60	9.51 264		9.53 697		10.46 303	9.97 567		0
′	L Cos	d	L Ctn	c d	L Tan	L Sin	d	′

Proportional parts

″	43	42	41
1	0.7	0.7	0.7
2	1.4	1.4	1.4
3	2.2	2.1	2.0
4	2.9	2.8	2.7
5	3.6	3.5	3.4
6	4.3	4.2	4.1
7	5.0	4.9	4.8
8	5.7	5.6	5.5
9	6.4	6.3	6.2
10	7.2	7.0	6.8
20	14.3	14.0	13.7
30	21.5	21.0	20.5
40	28.7	28.0	27.3
50	35.8	35.0	34.2

″	39	38	37
1	0.6	0.6	0.6
2	1.3	1.3	1.2
3	2.0	1.9	1.8
4	2.6	2.5	2.5
5	3.2	3.2	3.1
6	3.9	3.8	3.7
7	4.6	4.4	4.3
8	5.2	5.1	4.9
9	5.8	5.7	5.6
10	6.5	6.3	6.2
20	13.0	12.7	12.3
30	19.5	19.0	18.5
40	26.0	25.3	24.7
50	32.5	31.7	30.8

″	36	5	4
1	0.6	0.1	0.1
2	1.2	0.2	0.1
3	1.8	0.2	0.2
4	2.4	0.3	0.3
5	3.0	0.4	0.3
6	3.6	0.5	0.4
7	4.2	0.6	0.5
8	4.8	0.7	0.5
9	5.4	0.8	0.6
10	6.0	0.8	0.7
20	12.0	1.7	1.3
30	18.0	2.5	2.0
40	24.0	3.3	2.7
50	30.0	4.2	3.3

Proportional parts

108° (288°) (251°) **71°**

The −10 portion of the characteristic of the logarithm is not printed but must be written down whenever such a logarithm is used.

19° (199°) (340°) **160°**

′	L Sin	d	L Tan	c d	L Ctn	L Cos	d	′
0	9.51 264	37	9.53 697	41	10.46 303	9.97 567	4	60
1	9.51 301	37	9.53 738	41	10.46 262	9.97 563	5	59
2	9.51 338	36	9.53 779	41	10.46 221	9.97 558	4	58
3	9.51 374	37	9.53 820	41	10.46 180	9.97 554	4	57
4	9.51 411	36	9.53 861	41	10.46 139	9.97 550	5	56
5	9.51 447	37	9.53 902	41	10.46 098	9.97 545	4	55
6	9.51 484	36	9.53 943	41	10.46 057	9.97 541	5	54
7	9.51 520	37	9.53 984	41	10.46 016	9.97 536	4	53
8	9.51 557	36	9.54 025	40	10.45 975	9.97 532	4	52
9	9.51 593	36	9.54 065	41	10.45 935	9.97 528	5	51
10	9.51 629	37	9.54 106	41	10.45 894	9.97 523	4	50
11	9.51 666	36	9.54 147	40	10.45 853	9.97 519	4	49
12	9.51 702	36	9.54 187	41	10.45 813	9.97 515	5	48
13	9.51 738	36	9.54 228	41	10.45 772	9.97 510	4	47
14	9.51 774	37	9.54 269	40	10.45 731	9.97 506	5	46
15	9.51 811	36	9.54 309	41	10.45 691	9.97 501	4	45
16	9.51 847	36	9.54 350	40	10.45 650	9.97 497	5	44
17	9.51 883	36	9.54 390	41	10.45 610	9.97 492	4	43
18	9.51 919	36	9.54 431	40	10.45 569	9.97 488	4	42
19	9.51 955	36	9.54 471	41	10.45 529	9.97 484	5	41
20	9.51 991	36	9.54 512	40	10.45 488	9.97 479	4	40
21	9.52 027	36	9.54 552	41	10.45 448	9.97 475	5	39
22	9.52 063	36	9.54 593	40	10.45 407	9.97 470	4	38
23	9.52 099	36	9.54 633	40	10.45 367	9.97 466	5	37
24	9.52 135	36	9.54 673	41	10.45 327	9.97 461	4	36
25	9.52 171	36	9.54 714	40	10.45 286	9.97 457	4	35
26	9.52 207	35	9.54 754	40	10.45 246	9.97 453	5	34
27	9.52 242	36	9.54 794	41	10.45 206	9.97 448	4	33
28	9.52 278	36	9.54 835	40	10.45 165	9.97 444	5	32
29	9.52 314	36	9.54 875	40	10.45 125	9.97 439	4	31
30	9.52 350	35	9.54 915	40	10.45 085	9.97 435	5	30
31	9.52 385	36	9.54 955	40	10.45 045	9.97 430	4	29
32	9.52 421	35	9.54 995	40	10.45 005	9.97 426	5	28
33	9.52 456	36	9.55 035	40	10.44 965	9.97 421	4	27
34	9.52 492	35	9.55 075	40	10.44 925	9.97 417	5	26
35	9.52 527	36	9.55 115	40	10.44 885	9.97 412	4	25
36	9.52 563	35	9.55 155	40	10.44 845	9.97 408	5	24
37	9.52 598	36	9.55 195	40	10.44 805	9.97 403	4	23
38	9.52 634	35	9.55 235	40	10.44 765	9.97 399	5	22
39	9.52 669	36	9.55 275	40	10.44 725	9.97 394	4	21
40	9.52 705	35	9.55 315	40	10.44 685	9.97 390	5	20
41	9.52 740	35	9.55 355	40	10.44 645	9.97 385	4	19
42	9.52 775	36	9.55 395	39	10.44 605	9.97 381	5	18
43	9.52 811	35	9.55 434	40	10.44 566	9.97 376	4	17
44	9.52 846	35	9.55 474	40	10.44 526	9.97 372	5	16
45	9.52 881	35	9.55 514	40	10.44 486	9.97 367	4	15
46	9.52 916	35	9.55 554	39	10.44 446	9.97 363	5	14
47	9.52 951	35	9.55 593	40	10.44 407	9.97 358	5	13
48	9.52 986	35	9.55 633	40	10.44 367	9.97 353	4	12
49	9.53 021	35	9.55 673	39	10.44 327	9.97 349	5	11
50	9.53 056	36	9.55 712	40	10.44 288	9.97 344	4	10
51	9.53 092	34	9.55 752	39	10.44 248	9.97 340	5	9
52	9.53 126	35	9.55 791	40	10.44 209	9.97 335	4	8
53	9.53 161	35	9.55 831	39	10.44 169	9.97 331	5	7
54	9.53 196	35	9.55 870	40	10.44 130	9.97 326	4	6
55	9.53 231	35	9.55 910	39	10.44 090	9.97 322	5	5
56	9.53 266	35	9.55 949	40	10.44 051	9.97 317	5	4
57	9.53 301	35	9.55 989	39	10.44 011	9.97 312	4	3
58	9.53 336	34	9.56 028	39	10.43 972	9.97 308	5	2
59	9.53 370	35	9.56 067	40	10.43 933	9.97 303	4	1
60	9.53 405		9.56 107		10.43 893	9.97 299		0
′	L Cos	d	L Ctn	c d	L Tan	L Sin	d	′

Proportional parts

″	41	40	39
1	0.7	0.7	0.6
2	1.4	1.3	1.3
3	2.0	2.0	2.0
4	2.7	2.7	2.6
5	3.4	3.3	3.2
6	4.1	4.0	3.9
7	4.8	4.7	4.6
8	5.5	5.3	5.2
9	6.2	6.0	5.8
10	6.8	6.7	6.5
20	13.7	13.3	13.0
30	20.5	20.0	19.5
40	27.3	26.7	26.0
50	34.2	33.3	32.5

″	37	36	35
1	0.6	0.6	0.6
2	1.2	1.2	1.2
3	1.8	1.8	1.8
4	2.5	2.4	2.3
5	3.1	3.0	2.9
6	3.7	3.6	3.5
7	4.3	4.2	4.1
8	4.9	4.8	4.7
9	5.6	5.4	5.2
10	6.2	6.0	5.8
20	12.3	12.0	11.7
30	18.5	18.0	17.5
40	24.7	24.0	23.3
50	30.8	30.0	29.2

″	34	5	4
1	0.6	0.1	0.1
2	1.1	0.2	0.1
3	1.7	0.2	0.2
4	2.3	0.3	0.3
5	2.8	0.4	0.3
6	3.4	0.5	0.4
7	4.0	0.6	0.5
8	4.5	0.7	0.5
9	5.1	0.8	0.6
10	5.7	0.8	0.7
20	11.3	1.7	1.3
30	17.0	2.5	2.0
40	22.7	3.3	2.7
50	28.3	4.2	3.3

Proportional parts

109° (289°) (250°) **70°**

The −10 portion of the characteristic of the logarithm is not printed but must be written down whenever such a logarithm is used.

20° (200°) (339°) **159°**

′	L Sin	d	L Tan	c d	L Ctn	L Cos	d	′
0	9.53 405	35	9.56 107	39	10.43 893	9.97 299	5	60
1	9.53 440	35	9.56 146	39	10.43 854	9.97 294	5	59
2	9.53 475	34	9.56 185	39	10.43 815	9.97 289	4	58
3	9.53 509	35	9.56 224	40	10.43 776	9.97 285	5	57
4	9.53 544	34	9.56 264	39	10.43 736	9.97 280	4	56
5	9.53 578	35	9.56 303	39	10.43 697	9.97 276	5	55
6	9.53 613	34	9.56 342	39	10.43 658	9.97 271	5	54
7	9.53 647	35	9.56 381	39	10.43 619	9.97 266	4	53
8	9.53 682	34	9.56 420	39	10.43 580	9.97 262	5	52
9	9.53 716	35	9.56 459	39	10.43 541	9.97 257	5	51
10	9.53 751	34	9.56 498	39	10.43 502	9.97 252	4	50
11	9.53 785	34	9.56 537	39	10.43 463	9.97 248	5	49
12	9.53 819	35	9.56 576	39	10.43 424	9.97 243	5	48
13	9.53 854	34	9.56 615	39	10.43 385	9.97 238	4	47
14	9.53 888	34	9.56 654	39	10.43 346	9.97 234	5	46
15	9.53 922	35	9.56 693	39	10.43 307	9.97 229	5	45
16	9.53 957	34	9.56 732	39	10.43 268	9.97 224	4	44
17	9.53 991	34	9.56 771	39	10.43 229	9.97 220	5	43
18	9.54 025	34	9.56 810	39	10.43 190	9.97 215	5	42
19	9.54 059	34	9.56 849	38	10.43 151	9.97 210	4	41
20	9.54 093	34	9.56 887	39	10.43 113	9.97 206	5	40
21	9.54 127	34	9.56 926	39	10.43 074	9.97 201	5	39
22	9.54 161	34	9.56 965	39	10.43 035	9.97 196	4	38
23	9.54 195	34	9.57 004	38	10.42 996	9.97 192	5	37
24	9.54 229	34	9.57 042	39	10.42 958	9.97 187	5	36
25	9.54 263	34	9.57 081	39	10.42 919	9.97 182	4	35
26	9.54 297	34	9.57 120	38	10.42 880	9.97 178	5	34
27	9.54 331	34	9.57 158	39	10.42 842	9.97 173	5	33
28	9.54 365	34	9.57 197	38	10.42 803	9.97 168	5	32
29	9.54 399	34	9.57 235	39	10.42 765	9.97 163	4	31
30	9.54 433	33	9.57 274	38	10.42 726	9.97 159	5	30
31	9.54 466	34	9.57 312	39	10.42 688	9.97 154	5	29
32	9.54 500	34	9.57 351	38	10.42 649	9.97 149	4	28
33	9.54 534	33	9.57 389	39	10.42 611	9.97 145	5	27
34	9.54 567	34	9.57 428	38	10.42 572	9.97 140	5	26
35	9.54 601	34	9.57 466	38	10.42 534	9.97 135	5	25
36	9.54 635	33	9.57 504	39	10.42 496	9.97 130	4	24
37	9.54 668	34	9.57 543	38	10.42 457	9.97 126	5	23
38	9.54 702	33	9.57 581	38	10.42 419	9.97 121	5	22
39	9.54 735	34	9.57 619	39	10.42 381	9.97 116	5	21
40	9.54 769	33	9.57 658	38	10.42 342	9.97 111	4	20
41	9.54 802	34	9.57 696	38	10.42 304	9.97 107	5	19
42	9.54 836	33	9.57 734	38	10.42 266	9.97 102	5	18
43	9.54 869	34	9.57 772	38	10.42 228	9.97 097	5	17
44	9.54 903	33	9.57 810	39	10.42 190	9.97 092	5	16
45	9.54 936	33	9.57 849	38	10.42 151	9.97 087	4	15
46	9.54 969	34	9.57 887	38	10.42 113	9.97 083	5	14
47	9.55 003	33	9.57 925	38	10.42 075	9.97 078	5	13
48	9.55 036	33	9.57 963	38	10.42 037	9.97 073	5	12
49	9.55 069	33	9.58 001	38	10.41 999	9.97 068	5	11
50	9.55 102	34	9.58 039	38	10.41 961	9.97 063	4	10
51	9.55 136	33	9.58 077	38	10.41 923	9.97 059	5	9
52	9.55 169	33	9.58 115	38	10.41 885	9.97 054	5	8
53	9.55 202	33	9.58 153	38	10.41 847	9.97 049	5	7
54	9.55 235	33	9.58 191	38	10.41 809	9.97 044	5	6
55	9.55 268	33	9.58 229	38	10.41 771	9.97 039	4	5
56	9.55 301	33	9.58 267	37	10.41 733	9.97 035	5	4
57	9.55 334	33	9.58 304	38	10.41 696	9.97 030	5	3
58	9.55 367	33	9.58 342	38	10.41 658	9.97 025	5	2
59	9.55 400	33	9.58 380	38	10.41 620	9.97 020	5	1
60	9.55 433		9.58 418		10.41 582	9.97 015		0
′	L Cos	d	L Ctn	c d	L Tan	L Sin	d	′

Proportional parts

″	40	39	38
1	0.7	0.6	0.6
2	1.3	1.3	1.3
3	2.0	2.0	1.9
4	2.7	2.6	2.5
5	3.3	3.2	3.2
6	4.0	3.9	3.8
7	4.7	4.6	4.4
8	5.3	5.2	5.1
9	6.0	5.8	5.7
10	6.7	6.5	6.3
20	13.3	13.0	12.7
30	20.0	19.5	19.0
40	26.7	26.0	25.3
50	33.3	32.5	31.7

″	37	35	34
1	0.6	0.6	0.6
2	1.2	1.2	1.1
3	1.8	1.8	1.7
4	2.5	2.3	2.3
5	3.1	2.9	2.8
6	3.7	3.5	3.4
7	4.3	4.1	4.0
8	4.9	4.7	4.5
9	5.6	5.2	5.1
10	6.2	5.8	5.7
20	12.3	11.7	11.3
30	18.5	17.5	17.0
40	24.7	23.3	22.7
50	30.8	29.2	28.3

″	33	5	4
1	0.6	0.1	0.1
2	1.1	0.2	0.1
3	1.6	0.2	0.2
4	2.2	0.3	0.3
5	2.8	0.4	0.3
6	3.3	0.5	0.4
7	3.8	0.6	0.5
8	4.4	0.7	0.5
9	5.0	0.8	0.6
10	5.5	0.8	0.7
20	11.0	1.7	1.3
30	16.5	2.5	2.0
40	22.0	3.3	2.7
50	27.5	4.2	3.3

Proportional parts

110° (290°) (249°) **69°**

The −10 portion of the characteristic of the logarithm is not printed but must be written down whenever such a logarithm is used.

21° (201°) (338°) **158°**

′	L Sin	d	L Tan	c d	L Ctn	L Cos	d	′
0	9.55 433	33	9.58 418	37	10.41 582	9.97 015	5	60
1	9.55 466	33	9.58 455	38	10.41 545	9.97 010	5	59
2	9.55 499	33	9.58 493	38	10.41 507	9.97 005	4	58
3	9.55 532	32	9.58 531	38	10.41 469	9.97 001	5	57
4	9.55 564	33	9.58 569	37	10.41 431	9.96 996	5	56
5	9.55 597	33	9.58 606	38	10.41 394	9.96 991	5	55
6	9.55 630	33	9.58 644	37	10.41 356	9.96 986	5	54
7	9.55 663	32	9.58 681	38	10.41 319	9.96 981	5	53
8	9.55 695	33	9.58 719	38	10.41 281	9.96 976	5	52
9	9.55 728	33	9.58 757	37	10.41 243	9.96 971	5	51
10	9.55 761	32	9.58 794	38	10.41 206	9.96 966	4	50
11	9.55 793	33	9.58 832	37	10.41 168	9.96 962	5	49
12	9.55 826	32	9.58 869	38	10.41 131	9.96 957	5	48
13	9.55 858	33	9.58 907	37	10.41 093	9.96 952	5	47
14	9.55 891	32	9.58 944	37	10.41 056	9.96 947	5	46
15	9.55 923	33	9.58 981	38	10.41 019	9.96 942	5	45
16	9.55 956	32	9.59 019	37	10.40 981	9.96 937	5	44
17	9.55 988	33	9.59 056	38	10.40 944	9.96 932	5	43
18	9.56 021	32	9.59 094	37	10.40 906	9.96 927	5	42
19	9.56 053	32	9.59 131	37	10.40 869	9.96 922	5	41
20	9.56 085	33	9.59 168	37	10.40 832	9.96 917	5	40
21	9.56 118	32	9.59 205	38	10.40 795	9.96 912	5	39
22	9.56 150	32	9.59 243	37	10.40 757	9.96 907	4	38
23	9.56 182	33	9.59 280	37	10.40 720	9.96 903	5	37
24	9.56 215	32	9.59 317	37	10.40 683	9.96 898	5	36
25	9.56 247	32	9.59 354	37	10.40 646	9.96 893	5	35
26	9.56 279	32	9.59 391	38	10.40 609	9.96 888	5	34
27	9.56 311	32	9.59 429	37	10.40 571	9.96 883	5	33
28	9.56 343	32	9.59 466	37	10.40 534	9.96 878	5	32
29	9.56 375	33	9.59 503	37	10.40 497	9.96 873	5	31
30	9.56 408	32	9.59 540	37	10.40 460	9.96 868	5	30
31	9.56 440	32	9.59 577	37	10.40 423	9.96 863	5	29
32	9.56 472	32	9.59 614	37	10.40 386	9.96 858	5	28
33	9.56 504	32	9.59 651	37	10.40 349	9.96 853	5	27
34	9.56 536	32	9.59 688	37	10.40 312	9.96 848	5	26
35	9.56 568	31	9.59 725	37	10.40 275	9.96 843	5	25
36	9.56 599	32	9.59 762	37	10.40 238	9.96 838	5	24
37	9.56 631	32	9.59 799	36	10.40 201	9.96 833	5	23
38	9.56 663	32	9.59 835	37	10.40 165	9.96 828	5	22
39	9.56 695	32	9.59 872	37	10.40 128	9.96 823	5	21
40	9.56 727	32	9.59 909	37	10.40 091	9.96 818	5	20
41	9.56 759	31	9.59 946	37	10.40 054	9.96 813	5	19
42	9.56 790	32	9.59 983	36	10.40 017	9.96 808	5	18
43	9.56 822	32	9.60 019	37	10.39 981	9.96 803	5	17
44	9.56 854	32	9.60 056	37	10.39 944	9.96 798	5	16
45	9.56 886	31	9.60 093	37	10.39 907	9.96 793	5	15
46	9.56 917	32	9.60 130	36	10.39 870	9.96 788	5	14
47	9.56 949	31	9.60 166	37	10.39 834	9.96 783	5	13
48	9.56 980	32	9.60 203	37	10.39 797	9.96 778	6	12
49	9.57 012	32	9.60 240	36	10.39 760	9.96 772	5	11
50	9.57 044	31	9.60 276	37	10.39 724	9.96 767	5	10
51	9.57 075	32	9.60 313	36	10.39 687	9.96 762	5	9
52	9.57 107	31	9.60 349	37	10.39 651	9.96 757	5	8
53	9.57 138	31	9.60 386	36	10.39 614	9.96 752	5	7
54	9.57 169	32	9.60 422	37	10.39 578	9.96 747	5	6
55	9.57 201	31	9.60 459	36	10.39 541	9.96 742	5	5
56	9.57 232	32	9.60 495	37	10.39 505	9.96 737	5	4
57	9.57 264	31	9.60 532	36	10.39 468	9.96 732	5	3
58	9.57 295	31	9.60 568	37	10.39 432	9.96 727	5	2
59	9.57 326	32	9.60 605	36	10.39 395	9.96 722	5	1
60	9.57 358		9.60 641		10.39 359	9.96 717		0
′	L Cos	d	L Ctn	c d	L Tan	L Sin	d	′

Proportional parts

″	38	37	36
1	0.6	0.6	0.6
2	1.3	1.2	1.2
3	1.9	1.8	1.8
4	2.5	2.5	2.4
5	3.2	3.1	3.0
6	3.8	3.7	3.6
7	4.4	4.3	4.2
8	5.1	4.9	4.8
9	5.7	5.6	5.4
10	6.3	6.2	6.0
20	12.7	12.3	12.0
30	19.0	18.5	18.0
40	25.3	24.7	24.0
50	31.7	30.8	30.0

″	33	32	31
1	0.6	0.5	0.5
2	1.1	1.1	1.0
3	1.6	1.6	1.6
4	2.2	2.1	2.1
5	2.8	2.7	2.6
6	3.3	3.2	3.1
7	3.8	3.7	3.6
8	4.4	4.3	4.1
9	5.0	4.8	4.6
10	5.5	5.3	5.2
20	11.0	10.7	10.3
30	16.5	16.0	15.5
40	22.0	21.3	20.7
50	27.5	26.7	25.8

″	6	5	4
1	0.1	0.1	0.1
2	0.2	0.2	0.1
3	0.3	0.2	0.2
4	0.4	0.3	0.3
5	0.5	0.4	0.3
6	0.6	0.5	0.4
7	0.7	0.6	0.5
8	0.8	0.7	0.5
9	0.9	0.8	0.6
10	1.0	0.8	0.7
20	2.0	1.7	1.3
30	3.0	2.5	2.0
40	4.0	3.3	2.7
50	5.0	4.2	3.3

Proportional parts

111° (291°) (248°) **68°**

The −10 portion of the characteristic of the logarithm is not printed but must be written down whenever such a logarithm is used.

22° (202°) (337°) **157°**

′	L Sin	d	L Tan	c d	L Ctn	L Cos	d	′
0	9.57 358	31	9.60 641	36	10.39 359	9.96 717	6	60
1	9.57 389	31	9.60 677	37	10.39 323	9.96 711	5	59
2	9.57 420	31	9.60 714	36	10.39 286	9.96 706	5	58
3	9.57 451	31	9.60 750	36	10.39 250	9.96 701	5	57
4	9.57 482	32	9.60 786	37	10.39 214	9.96 696	5	56
5	9.57 514	31	9.60 823	36	10.39 177	9.96 691	5	55
6	9.57 545	31	9.60 859	36	10.39 141	9.96 686	5	54
7	9.57 576	31	9.60 895	36	10.39 105	9.96 681	5	53
8	9.57 607	31	9.60 931	36	10.39 069	9.96 676	6	52
9	9.57 638	31	9.60 967	37	10.39 033	9.96 670	5	51
10	9.57 669	31	9.61 004	36	10.38 996	9.96 665	5	50
11	9.57 700	31	9.61 040	36	10.38 960	9.96 660	5	49
12	9.57 731	31	9.61 076	36	10.38 924	9.96 655	5	48
13	9.57 762	31	9.61 112	36	10.38 888	9.96 650	5	47
14	9.57 793	31	9.61 148	36	10.38 852	9.96 645	5	46
15	9.57 824	31	9.61 184	36	10.38 816	9.96 640	6	45
16	9.57 855	30	9.61 220	36	10.38 780	9.96 634	5	44
17	9.57 885	31	9.61 256	36	10.38 744	9.96 629	5	43
18	9.57 916	31	9.61 292	36	10.38 708	9.96 624	5	42
19	9.57 947	31	9.61 328	36	10.38 672	9.96 619	5	41
20	9.57 978	30	9.61 364	36	10.38 636	9.96 614	6	40
21	9.58 008	31	9.61 400	36	10.38 600	9.96 608	5	39
22	9.58 039	31	9.61 436	36	10.38 564	9.96 603	5	38
23	9.58 070	31	9.61 472	36	10.38 528	9.96 598	5	37
24	9.58 101	30	9.61 508	36	10.38 492	9.96 593	5	36
25	9.58 131	31	9.61 544	35	10.38 456	9.96 588	6	35
26	9.58 162	30	9.61 579	36	10.38 421	9.96 582	5	34
27	9.58 192	31	9.61 615	36	10.38 385	9.96 577	5	33
28	9.58 223	30	9.61 651	36	10.38 349	9.96 572	5	32
29	9.58 253	31	9.61 687	35	10.38 313	9.96 567	5	31
30	9.58 284	30	9.61 722	36	10.38 278	9.96 562	6	30
31	9.58 314	31	9.61 758	36	10.38 242	9.96 556	5	29
32	9.58 345	30	9.61 794	36	10.38 206	9.96 551	5	28
33	9.58 375	31	9.61 830	35	10.38 170	9.96 546	5	27
34	9.58 406	30	9.61 865	36	10.38 135	9.96 541	6	26
35	9.58 436	31	9.61 901	35	10.38 099	9.96 535	5	25
36	9.58 467	30	9.61 936	36	10.38 064	9.96 530	5	24
37	9.58 497	30	9.61 972	36	10.38 028	9.96 525	5	23
38	9.58 527	30	9.62 008	35	10.37 992	9.96 520	6	22
39	9.58 557	31	9.62 043	36	10.37 957	9.96 514	5	21
40	9.58 588	30	9.62 079	35	10.37 921	9.96 509	5	20
41	9.58 618	30	9.62 114	36	10.37 886	9.96 504	6	19
42	9.58 648	30	9.62 150	35	10.37 850	9.96 498	5	18
43	9.58 678	31	9.62 185	36	10.37 815	9.96 493	5	17
44	9.58 709	30	9.62 221	35	10.37 779	9.96 488	5	16
45	9.58 739	30	9.62 256	36	10.37 744	9.96 483	6	15
46	9.58 769	30	9.62 292	35	10.37 708	9.96 477	5	14
47	9.58 799	30	9.62 327	35	10.37 673	9.96 472	5	13
48	9.58 829	30	9.62 362	36	10.37 638	9.96 467	6	12
49	9.58 859	30	9.62 398	35	10.37 602	9.96 461	5	11
50	9.58 889	30	9.62 433	35	10.37 567	9.96 456	5	10
51	9.58 919	30	9.62 468	36	10.37 532	9.96 451	6	9
52	9.58 949	30	9.62 504	35	10.37 496	9.96 445	5	8
53	9.58 979	30	9.62 539	35	10.37 461	9.96 440	5	7
54	9.59 009	30	9.62 574	35	10.37 426	9.96 435	6	6
55	9.59 039	30	9.62 609	36	10.37 391	9.96 429	5	5
56	9.59 069	29	9.62 645	35	10.37 355	9.96 424	5	4
57	9.59 098	30	9.62 680	35	10.37 320	9.96 419	6	3
58	9.59 128	30	9.62 715	35	10.37 285	9.96 413	5	2
59	9.59 158	30	9.62 750	35	10.37 250	9.96 408	5	1
60	9.59 188		9.62 785		10.37 215	9.96 403		0
′	L Cos	d	L Ctn	c d	L Tan	L Sin	d	′

Proportional parts

″	37	36	35
1	0.6	0.6	0.6
2	1.2	1.2	1.2
3	1.8	1.8	1.8
4	2.5	2.4	2.3
5	3.1	3.0	2.9
6	3.7	3.6	3.5
7	4.3	4.2	4.1
8	4.9	4.8	4.7
9	5.6	5.4	5.2
10	6.2	6.0	5.8
20	12.3	12.0	11.7
30	18.5	18.0	17.5
40	24.7	24.0	23.3
50	30.8	30.0	29.2

″	32	31	30
1	0.5	0.5	0.5
2	1.1	1.0	1.0
3	1.6	1.6	1.5
4	2.1	2.1	2.0
5	2.7	2.6	2.5
6	3.2	3.1	3.0
7	3.7	3.6	3.5
8	4.3	4.1	4.0
9	4.8	4.6	4.5
10	5.3	5.2	5.0
20	10.7	10.3	10.0
30	16.0	15.5	15.0
40	21.3	20.7	20.0
50	26.7	25.8	25.0

″	29	6	5
1	0.5	0.1	0.1
2	1.0	0.2	0.2
3	1.4	0.3	0.2
4	1.9	0.4	0.3
5	2.4	0.5	0.4
6	2.9	0.6	0.5
7	3.4	0.7	0.6
8	3.9	0.8	0.7
9	4.4	0.9	0.8
10	4.8	1.0	0.8
20	9.7	2.0	1.7
30	14.5	3.0	2.5
40	19.3	4.0	3.3
50	24.2	5.0	4.2

Proportional parts

112° (292°) (247°) **67°**

The −10 portion of the characteristic of the logarithm is not printed but must be written down whenever such a logarithm is used.

23° (203°) (336°) **156°**

′	L Sin	d	L Tan	c d	L Ctn	L Cos	d	′
0	9.59 188	30	9.62 785	35	10.37 215	9.96 403	6	60
1	9.59 218	29	9.62 820	35	10.37 180	9.96 397	5	59
2	9.59 247	30	9.62 855	35	10.37 145	9.96 392	5	58
3	9.59 277	30	9.62 890	36	10.37 110	9.96 387	6	57
4	9.59 307	29	9.62 926	35	10.37 074	9.96 381	5	56
5	9.59 336	30	9.62 961	35	10.37 039	9.96 376	6	55
6	9.59 366	30	9.62 996	35	10.37 004	9.96 370	5	54
7	9.59 396	29	9.63 031	35	10.36 969	9.96 365	5	53
8	9.59 425	30	9.63 066	35	10.36 934	9.96 360	6	52
9	9.59 455	29	9.63 101	34	10.36 899	9.96 354	5	51
10	9.59 484	30	9.63 135	35	10.36 865	9.96 349	6	50
11	9.59 514	29	9.63 170	35	10.36 830	9.96 343	5	49
12	9.59 543	30	9.63 205	35	10.36 795	9.96 338	5	48
13	9.59 573	29	9.63 240	35	10.36 760	9.96 333	6	47
14	9.59 602	30	9.63 275	35	10.36 725	9.96 327	5	46
15	9.59 632	29	9.63 310	35	10.36 690	9.96 322	6	45
16	9.59 661	29	9.63 345	34	10.36 655	9.96 316	5	44
17	9.59 690	30	9.63 379	35	10.36 621	9.96 311	6	43
18	9.59 720	29	9.63 414	35	10.36 586	9.96 305	5	42
19	9.59 749	29	9.63 449	35	10.36 551	9.96 300	6	41
20	9.59 778	30	9.63 484	35	10.36 516	9.96 294	5	40
21	9.59 808	29	9.63 519	34	10.36 481	9.96 289	5	39
22	9.59 837	29	9.63 553	35	10.36 447	9.96 284	6	38
23	9.59 866	29	9.63 588	35	10.36 412	9.96 278	5	37
24	9.59 895	29	9.63 623	34	10.36 377	9.96 273	6	36
25	9.59 924	30	9.63 657	35	10.36 343	9.96 267	5	35
26	9.59 954	29	9.63 692	34	10.36 308	9.96 262	6	34
27	9.59 983	29	9.63 726	35	10.36 274	9.96 256	5	33
28	9.60 012	29	9.63 761	35	10.36 239	9.96 251	6	32
29	9.60 041	29	9.63 796	34	10.36 204	9.96 245	5	31
30	9.60 070	29	9.63 830	35	10.36 170	9.96 240	6	30
31	9.60 099	29	9.63 865	34	10.36 135	9.96 234	5	29
32	9.60 128	29	9.63 899	35	10.36 101	9.96 229	6	28
33	9.60 157	29	9.63 934	34	10.36 066	9.96 223	5	27
34	9.60 186	29	9.63 968	35	10.36 032	9.96 218	6	26
35	9.60 215	29	9.64 003	34	10.35 997	9.96 212	5	25
36	9.60 244	29	9.64 037	35	10.35 963	9.96 207	6	24
37	9.60 273	29	9.64 072	34	10.35 928	9.96 201	5	23
38	9.60 302	29	9.64 106	34	10.35 894	9.96 196	6	22
39	9.60 331	28	9.64 140	35	10.35 860	9.96 190	5	21
40	9.60 359	29	9.64 175	34	10.35 825	9.96 185	6	20
41	9.60 388	29	9.64 209	34	10.35 791	9.96 179	5	19
42	9.60 417	29	9.64 243	35	10.35 757	9.96 174	6	18
43	9.60 446	28	9.64 278	34	10.35 722	9.96 168	6	17
44	9.60 474	29	9.64 312	34	10.35 688	9.96 162	5	16
45	9.60 503	29	9.64 346	35	10.35 654	9.96 157	6	15
46	9.60 532	29	9.64 381	34	10.35 619	9.96 151	5	14
47	9.60 561	28	9.64 415	34	10.35 585	9.96 146	6	13
48	9.60 589	29	9.64 449	34	10.35 551	9.96 140	5	12
49	9.60 618	28	9.64 483	34	10.35 517	9.96 135	6	11
50	9.60 646	29	9.64 517	35	10.35 483	9.96 129	6	10
51	9.60 675	29	9.64 552	34	10.35 448	9.96 123	5	9
52	9.60 704	28	9.64 586	34	10.35 414	9.96 118	6	8
53	9.60 732	29	9.64 620	34	10.35 380	9.96 112	5	7
54	9.60 761	28	9.64 654	34	10.35 346	9.96 107	6	6
55	9.60 789	29	9.64 688	34	10.35 312	9.96 101	6	5
56	9.60 818	28	9.64 722	34	10.35 278	9.96 095	5	4
57	9.60 846	29	9.64 756	34	10.35 244	9.96 090	6	3
58	9.60 875	28	9.64 790	34	10.35 210	9.96 084	5	2
59	9.60 903	28	9.64 824	34	10.35 176	9.96 079	6	1
60	9.60 931		9.64 858		10.35 142	9.96 073		0
′	L Cos	d	L Ctn	c d	L Tan	L Sin	d	′

Proportional parts

″	36	35	34
1	0.6	0.6	0.6
2	1.2	1.2	1.1
3	1.8	1.8	1.7
4	2.4	2.3	2.3
5	3.0	2.9	2.8
6	3.6	3.5	3.4
7	4.2	4.1	4.0
8	4.8	4.7	4.5
9	5.4	5.2	5.1
10	6.0	5.8	5.7
20	12.0	11.7	11.3
30	18.0	17.5	17.0
40	24.0	23.3	22.7
50	30.0	29.2	28.3

″	30	29	28
1	0.5	0.5	0.5
2	1.0	1.0	0.9
3	1.5	1.4	1.4
4	2.0	1.9	1.9
5	2.5	2.4	2.3
6	3.0	2.9	2.8
7	3.5	3.4	3.3
8	4.0	3.9	3.7
9	4.5	4.4	4.2
10	5.0	4.8	4.7
20	10.0	9.7	9.3
30	15.0	14.5	14.0
40	20.0	19.3	18.7
50	25.0	24.2	23.3

″	6	5
1	0.1	0.1
2	0.2	0.2
3	0.3	0.2
4	0.4	0.3
5	0.5	0.4
6	0.6	0.5
7	0.7	0.6
8	0.8	0.7
9	0.9	0.8
10	1.0	0.8
20	2.0	1.7
30	3.0	2.5
40	4.0	3.3
50	5.0	4.2

Proportional parts

113° (293°) (246°) **66°**

The −10 portion of the characteristic of the logarithm is not printed but must be written down whenever such a logarithm is used.

24° (204°) (335°) **155°**

′	L Sin	d	L Tan	c d	L Ctn	L Cos	d	′
0	9.60 931	29	9.64 858	34	10.35 142	9.96 073	6	60
1	9.60 960	28	9.64 892	34	10.35 108	9.96 067	5	59
2	9.60 988	28	9.64 926	34	10.35 074	9.96 062	6	58
3	9.61 016	29	9.64 960	34	10.35 040	9.96 056	6	57
4	9.61 045	28	9.64 994	34	10.35 006	9.96 050	5	56
5	9.61 073	28	9.65 028	34	10.34 972	9.96 045	6	55
6	9.61 101	28	9.65 062	34	10.34 938	9.96 039	5	54
7	9.61 129	29	9.65 096	34	10.34 904	9.96 034	6	53
8	9.61 158	28	9.65 130	34	10.34 870	9.96 028	6	52
9	9.61 186	28	9.65 164	33	10.34 836	9.96 022	5	51
10	9.61 214	28	9.65 197	34	10.34 803	9.96 017	6	50
11	9.61 242	28	9.65 231	34	10.34 769	9.96 011	6	49
12	9.61 270	28	9.65 265	34	10.34 735	9.96 005	5	48
13	9.61 298	28	9.65 299	34	10.34 701	9.96 000	6	47
14	9.61 326	28	9.65 333	33	10.34 667	9.95 994	6	46
15	9.61 354	28	9.65 366	34	10.34 634	9.95 988	6	45
16	9.61 382	29	9.65 400	34	10.34 600	9.95 982	5	44
17	9.61 411	27	9.65 434	33	10.34 566	9.95 977	6	43
18	9.61 438	28	9.65 467	34	10.34 533	9.95 971	6	42
19	9.61 466	28	9.65 501	34	10.34 499	9.95 965	5	41
20	9.61 494	28	9.65 535	33	10.34 465	9.95 960	6	40
21	9.61 522	28	9.65 568	34	10.34 432	9.95 954	6	39
22	9.61 550	28	9.65 602	34	10.34 398	9.95 948	6	38
23	9.61 578	28	9.65 636	33	10.34 364	9.95 942	5	37
24	9.61 606	28	9.65 669	34	10.34 331	9.95 937	6	36
25	9.61 634	28	9.65 703	33	10.34 297	9.95 931	6	35
26	9.61 662	27	9.65 736	34	10.34 264	9.95 925	5	34
27	9.61 689	28	9.65 770	33	10.34 230	9.95 920	6	33
28	9.61 717	28	9.65 803	34	10.34 197	9.95 914	6	32
29	9.61 745	28	9.65 837	33	10.34 163	9.95 908	6	31
30	9.61 773	27	9.65 870	34	10.34 130	9.95 902	5	30
31	9.61 800	28	9.65 904	33	10.34 096	9.95 897	6	29
32	9.61 828	28	9.65 937	34	10.34 063	9.95 891	6	28
33	9.61 856	27	9.65 971	33	10.34 029	9.95 885	6	27
34	9.61 883	28	9.66 004	34	10.33 996	9.95 879	6	26
35	9.61 911	28	9.66 038	33	10.33 962	9.95 873	5	25
36	9.61 939	27	9.66 071	33	10.33 929	9.95 868	6	24
37	9.61 966	28	9.66 104	34	10.33 896	9.95 862	6	23
38	9.61 994	27	9.66 138	33	10.33 862	9.95 856	6	22
39	9.62 021	28	9.66 171	33	10.33 829	9.95 850	6	21
40	9.62 049	27	9.66 204	34	10.33 796	9.95 844	5	20
41	9.62 076	28	9.66 238	33	10.33 762	9.95 839	6	19
42	9.62 104	27	9.66 271	33	10.33 729	9.95 833	6	18
43	9.62 131	28	9.66 304	33	10.33 696	9.95 827	6	17
44	9.62 159	27	9.66 337	34	10.33 663	9.95 821	6	16
45	9.62 186	28	9.66 371	33	10.33 629	9.95 815	5	15
46	9.62 214	27	9.66 404	33	10.33 596	9.95 810	6	14
47	9.62 241	27	9.66 437	33	10.33 563	9.95 804	6	13
48	9.62 268	28	9.66 470	33	10.33 530	9.95 798	6	12
49	9.62 296	27	9.66 503	34	10.33 497	9.95 792	6	11
50	9.62 323	27	9.66 537	33	10.33 463	9.95 786	6	10
51	9.62 350	27	9.66 570	33	10.33 430	9.95 780	5	9
52	9.62 377	28	9.66 603	33	10.33 397	9.95 775	6	8
53	9.62 405	27	9.66 636	33	10.33 364	9.95 769	6	7
54	9.62 432	27	9.66 669	33	10.33 331	9.95 763	6	6
55	9.62 459	27	9.66 702	33	10.33 298	9.95 757	6	5
56	9.62 486	27	9.66 735	33	10.33 265	9.95 751	6	4
57	9.62 513	28	9.66 768	33	10.33 232	9.95 745	6	3
58	9.62 541	27	9.66 801	33	10.33 199	9.95 739	6	2
59	9.62 568	27	9.66 834	33	10.33 166	9.95 733	5	1
60	9.62 595		9.66 867		10.33 133	9.95 728		0
′	L Cos	d	L Ctn	c d	L Tan	L Sin	d	′

114° (294°) (245°) **65°**

Proportional parts

″	34	33
1	0.6	0.6
2	1.1	1.1
3	1.7	1.6
4	2.3	2.2
5	2.8	2.8
6	3.4	3.3
7	4.0	3.8
8	4.5	4.4
9	5.1	5.0
10	5.7	5.5
20	11.3	11.0
30	17.0	16.5
40	22.7	22.0
50	28.3	27.5

″	29	28	27
1	0.5	0.5	0.4
2	1.0	0.9	0.9
3	1.4	1.4	1.4
4	1.9	1.9	1.8
5	2.4	2.3	2.2
6	2.9	2.8	2.7
7	3.4	3.3	3.2
8	3.9	3.7	3.6
9	4.4	4.2	4.0
10	4.8	4.7	4.5
20	9.7	9.3	9.0
30	14.5	14.0	13.5
40	19.3	18.7	18.0
50	24.2	23.3	22.5

″	6	5
1	0.1	0.1
2	0.2	0.2
3	0.3	0.2
4	0.4	0.3
5	0.5	0.4
6	0.6	0.5
7	0.7	0.6
8	0.8	0.7
9	0.9	0.8
10	1.0	0.8
20	2.0	1.7
30	3.0	2.5
40	4.0	3.3
50	5.0	4.2

Proportional parts

The −10 portion of the characteristic of the logarithm is not printed but must be written down whenever such a logarithm is used.

25° (205°) (334°) **154°**

′	L Sin	d	L Tan	c d	L Ctn	L Cos	d	′
0	9.62 595	27	9.66 867	33	10.33 133	9.95 728	6	60
1	9.62 622	27	9.66 900	33	10.33 100	9.95 722	6	59
2	9.62 649	27	9.66 933	33	10.33 067	9.95 716	6	58
3	9.62 676	27	9.66 966	33	10.33 034	9.95 710	6	57
4	9.62 703	27	9.66 999	33	10.33 001	9.95 704	6	56
5	9.62 730	27	9.67 032	33	10.32 968	9.95 698	6	55
6	9.62 757	27	9.67 065	33	10.32 935	9.95 692	6	54
7	9.62 784	27	9.67 098	33	10.32 902	9.95 686	6	53
8	9.62 811	27	9.67 131	32	10.32 869	9.95 680	6	52
9	9.62 838	27	9.67 163	33	10.32 837	9.95 674	6	51
10	9.62 865	27	9.67 196	33	10.32 804	9.95 668	5	50
11	9.62 892	26	9.67 229	33	10.32 771	9.95 663	6	49
12	9.62 918	27	9.67 262	33	10.32 738	9.95 657	6	48
13	9.62 945	27	9.67 295	32	10.32 705	9.95 651	6	47
14	9.62 972	27	9.67 327	33	10.32 673	9.95 645	6	46
15	9.62 999	27	9.67 360	33	10.32 640	9.95 639	6	45
16	9.63 026	26	9.67 393	33	10.32 607	9.95 633	6	44
17	9.63 052	27	9.67 426	32	10.32 574	9.95 627	6	43
18	9.63 079	27	9.67 458	33	10.32 542	9.95 621	6	42
19	9.63 106	27	9.67 491	33	10.32 509	9.95 615	6	41
20	9.63 133	26	9.67 524	32	10.32 476	9.95 609	6	40
21	9.63 159	27	9.67 556	33	10.32 444	9.95 603	6	39
22	9.63 186	27	9.67 589	33	10.32 411	9.95 597	6	38
23	9.63 213	26	9.67 622	32	10.32 378	9.95 591	6	37
24	9.63 239	27	9.67 654	33	10.32 346	9.95 585	6	36
25	9.63 266	26	9.67 687	32	10.32 313	9.95 579	6	35
26	9.63 292	27	9.67 719	33	10.32 281	9.95 573	6	34
27	9.63 319	26	9.67 752	33	10.32 248	9.95 567	6	33
28	9.63 345	27	9.67 785	32	10.32 215	9.95 561	6	32
29	9.63 372	26	9.67 817	33	10.32 183	9.95 555	6	31
30	9.63 398	27	9.67 850	32	10.32 150	9.95 549	6	30
31	9.63 425	26	9.67 882	33	10.32 118	9.95 543	6	29
32	9.63 451	27	9.67 915	32	10.32 085	9.95 537	6	28
33	9.63 478	26	9.67 947	33	10.32 053	9.95 531	6	27
34	9.63 504	27	9.67 980	32	10.32 020	9.95 525	6	26
35	9.63 531	26	9.68 012	32	10.31 988	9.95 519	6	25
36	9.63 557	26	9.68 044	33	10.31 956	9.95 513	6	24
37	9.63 583	27	9.68 077	32	10.31 923	9.95 507	7	23
38	9.63 610	26	9.68 109	33	10.31 891	9.95 500	6	22
39	9.63 636	26	9.68 142	32	10.31 858	9.95 494	6	21
40	9.63 662	27	9.68 174	32	10.31 826	9.95 488	6	20
41	9.63 689	26	9.68 206	33	10.31 794	9.95 482	6	19
42	9.63 715	26	9.68 239	32	10.31 761	9.95 476	6	18
43	9.63 741	26	9.68 271	32	10.31 729	9.95 470	6	17
44	9.63 767	27	9.68 303	33	10.31 697	9.95 464	6	16
45	9.63 794	26	9.68 336	32	10.31 664	9.95 458	6	15
46	9.63 820	26	9.68 368	32	10.31 632	9.95 452	6	14
47	9.63 846	26	9.68 400	32	10.31 600	9.95 446	6	13
48	9.63 872	26	9.68 432	33	10.31 568	9.95 440	6	12
49	9.63 898	26	9.68 465	32	10.31 535	9.95 434	7	11
50	9.63 924	26	9.68 497	32	10.31 503	9.95 427	6	10
51	9.63 950	26	9.68 529	32	10.31 471	9.95 421	6	9
52	9.63 976	26	9.68 561	32	10.31 439	9.95 415	6	8
53	9.64 002	26	9.68 593	33	10.31 407	9.95 409	6	7
54	9.64 028	26	9.68 626	32	10.31 374	9.95 403	6	6
55	9.64 054	26	9.68 658	32	10.31 342	9.95 397	6	5
56	9.64 080	26	9.68 690	32	10.31 310	9.95 391	7	4
57	9.64 106	26	9.68 722	32	10.31 278	9.95 384	6	3
58	9.64 132	26	9.68 754	32	10.31 246	9.95 378	6	2
59	9.64 158	26	9.68 786	32	10.31 214	9.95 372	6	1
60	9.64 184		9.68 818		10.31 182	9.95 366		0
′	L Cos	d	L Ctn	c d	L Tan	L Sin	d	′

Proportional parts

″	33	32
1	0.6	0.5
2	1.1	1.1
3	1.6	1.6
4	2.2	2.1
5	2.8	2.7
6	3.3	3.2
7	3.8	3.7
8	4.4	4.3
9	5.0	4.8
10	5.5	5.3
20	11.0	10.7
30	16.5	16.0
40	22.0	21.3
50	27.5	26.7

″	27	26
1	0.4	0.4
2	0.9	0.9
3	1.4	1.3
4	1.8	1.7
5	2.2	2.2
6	2.7	2.6
7	3.2	3.0
8	3.6	3.5
9	4.0	3.9
10	4.5	4.3
20	9.0	8.7
30	13.5	13.0
40	18.0	17.3
50	22.5	21.7

″	7	6	5
1	0.1	0.1	0.1
2	0.2	0.2	0.2
3	0.4	0.3	0.2
4	0.5	0.4	0.3
5	0.6	0.5	0.4
6	0.7	0.6	0.5
7	0.8	0.7	0.6
8	0.9	0.8	0.7
9	1.0	0.9	0.8
10	1.2	1.0	0.8
20	2.3	2.0	1.7
30	3.5	3.0	2.5
40	4.7	4.0	3.3
50	5.8	5.0	4.2

Proportional parts

115° (295°) (244°) **64°**

The −10 portion of the characteristic of the logarithm is not printed but must be written down whenever such a logarithm is used.

26° (206°) (333°) **153°**

′	L Sin	d	L Tan	c d	L Ctn	L Cos	d	′
0	9.64 184	26	9.68 818	32	10.31 182	9.95 366	6	60
1	9.64 210	26	9.68 850	32	10.31 150	9.95 360	6	59
2	9.64 236	26	9.68 882	32	10.31 118	9.95 354	6	58
3	9.64 262	26	9.68 914	32	10.31 086	9.95 348	7	57
4	9.64 288	25	9.68 946	32	10.31 054	9.95 341	6	56
5	9.64 313	26	9.68 978	32	10.31 022	9.95 335	6	55
6	9.64 339	26	9.69 010	32	10.30 990	9.95 329	6	54
7	9.64 365	26	9.69 042	32	10.30 958	9.95 323	6	53
8	9.64 391	26	9.69 074	32	10.30 926	9.95 317	7	52
9	9.64 417	25	9.69 106	32	10.30 894	9.95 310	6	51
10	9.64 442	26	9.69 138	32	10.30 862	9.95 304	6	50
11	9.64 468	26	9.69 170	32	10.30 830	9.95 298	6	49
12	9.64 494	25	9.69 202	32	10.30 798	9.95 292	6	48
13	9.64 519	26	9.69 234	32	10.30 766	9.95 286	7	47
14	9.64 545	26	9.69 266	32	10.30 734	9.95 279	6	46
15	9.64 571	25	9.69 298	31	10.30 702	9.95 273	6	45
16	9.64 596	26	9.69 329	32	10.30 671	9.95 267	6	44
17	9.64 622	25	9.69 361	32	10.30 639	9.95 261	7	43
18	9.64 647	26	9.69 393	32	10.30 607	9.95 254	6	42
19	9.64 673	25	9.69 425	32	10.30 575	9.95 248	6	41
20	9.64 698	26	9.69 457	31	10.30 543	9.95 242	6	40
21	9.64 724	25	9.69 488	32	10.30 512	9.95 236	7	39
22	9.64 749	26	9.69 520	32	10.30 480	9.95 229	6	38
23	9.64 775	25	9.69 552	32	10.30 448	9.95 223	6	37
24	9.64 800	26	9.69 584	31	10.30 416	9.95 217	6	36
25	9.64 826	25	9.69 615	32	10.30 385	9.95 211	7	35
26	9.64 851	26	9.69 647	32	10.30 353	9.95 204	6	34
27	9.64 877	25	9.69 679	31	10.30 321	9.95 198	6	33
28	9.64 902	25	9.69 710	32	10.30 290	9.95 192	7	32
29	9.64 927	26	9.69 742	32	10.30 258	9.95 185	6	31
30	9.64 953	25	9.69 774	31	10.30 226	9.95 179	6	30
31	9.64 978	25	9.69 805	32	10.30 195	9.95 173	6	29
32	9.65 003	26	9.69 837	31	10.30 163	9.95 167	7	28
33	9.65 029	25	9.69 868	32	10.30 132	9.95 160	6	27
34	9.65 054	25	9.69 900	32	10.30 100	9.95 154	6	26
35	9.65 079	25	9.69 932	31	10.30 068	9.95 148	7	25
36	9.65 104	26	9.69 963	32	10.30 037	9.95 141	6	24
37	9.65 130	25	9.69 995	31	10.30 005	9.95 135	6	23
38	9.65 155	25	9.70 026	32	10.29 974	9.95 129	7	22
39	9.65 180	25	9.70 058	31	10.29 942	9.95 122	6	21
40	9.65 205	25	9.70 089	32	10.29 911	9.95 116	6	20
41	9.65 230	25	9.70 121	31	10.29 879	9.95 110	7	19
42	9.65 255	26	9.70 152	32	10.29 848	9.95 103	6	18
43	9.65 281	25	9.70 184	31	10.29 816	9.95 097	7	17
44	9.65 306	25	9.70 215	32	10.29 785	9.95 090	6	16
45	9.65 331	25	9.70 247	31	10.29 753	9.95 084	6	15
46	9.65 356	25	9.70 278	31	10.29 722	9.95 078	7	14
47	9.65 381	25	9.70 309	32	10.29 691	9.95 071	6	13
48	9.65 406	25	9.70 341	31	10.29 659	9.95 065	6	12
49	9.65 431	25	9.70 372	32	10.29 628	9.95 059	7	11
50	9.65 456	25	9.70 404	31	10.29 596	9.95 052	6	10
51	9.65 481	25	9.70 435	31	10.29 565	9.95 046	7	9
52	9.65 506	25	9.70 466	32	10.29 534	9.95 039	6	8
53	9.65 531	25	9.70 498	31	10.29 502	9.95 033	6	7
54	9.65 556	24	9.70 529	31	10.29 471	9.95 027	7	6
55	9.65 580	25	9.70 560	32	10.29 440	9.95 020	6	5
56	9.65 605	25	9.70 592	31	10.29 408	9.95 014	7	4
57	9.65 630	25	9.70 623	31	10.29 377	9.95 007	6	3
58	9.65 655	25	9.70 654	31	10.29 346	9.95 001	6	2
59	9.65 680	25	9.70 685	32	10.29 315	9.94 995	7	1
60	9.65 705		9.70 717		10.29 283	9.94 988		0
′	L Cos	d	L Ctn	c d	L Tan	L Sin	d	′

Proportional parts

″	32	31
1	0.5	0.5
2	1.1	1.0
3	1.6	1.6
4	2.1	2.1
5	2.7	2.6
6	3.2	3.1
7	3.7	3.6
8	4.3	4.1
9	4.8	4.6
10	5.3	5.2
20	10.7	10.3
30	16.0	15.5
40	21.3	20.7
50	26.7	25.8

″	26	25	24
1	0.4	0.4	0.4
2	0.9	0.8	0.8
3	1.3	1.2	1.2
4	1.7	1.7	1.6
5	2.2	2.1	2.0
6	2.6	2.5	2.4
7	3.0	2.9	2.8
8	3.5	3.3	3.2
9	3.9	3.8	3.6
10	4.3	4.2	4.0
20	8.7	8.3	8.0
30	13.0	12.5	12.0
40	17.3	16.7	16.0
50	21.7	20.8	20.0

″	7	6
1	0.1	0.1
2	0.2	0.2
3	0.4	0.3
4	0.5	0.4
5	0.6	0.5
6	0.7	0.6
7	0.8	0.7
8	0.9	0.8
9	1.0	0.9
10	1.2	1.0
20	2.3	2.0
30	3.5	3.0
40	4.7	4.0
50	5.8	5.0

Proportional parts

116° (296°) (243°) **63°**

The −10 portion of the characteristic of the logarithm is not printed but must be written down whenever such a logarithm is used.

27° (207°) (332°) **152°**

′	L Sin	d	L Tan	c d	L Ctn	L Cos	d	′
0	9.65 705	24	9.70 717	31	10.29 283	9.94 988	6	60
1	9.65 729	25	9.70 748	31	10.29 252	9.94 982	7	59
2	9.65 754	25	9.70 779	31	10.29 221	9.94 975	6	58
3	9.65 779	25	9.70 810	31	10.29 190	9.94 969	7	57
4	9.65 804	24	9.70 841	32	10.29 159	9.94 962	6	56
5	9.65 828	25	9.70 873	31	10.29 127	9.94 956	7	55
6	9.65 853	25	9.70 904	31	10.29 096	9.94 949	6	54
7	9.65 878	24	9.70 935	31	10.29 065	9.94 943	7	53
8	9.65 902	25	9.70 966	31	10.29 034	9.94 936	6	52
9	9.65 927	25	9.70 997	31	10.29 003	9.94 930	7	51
10	9.65 952	24	9.71 028	31	10.28 972	9.94 923	6	50
11	9.65 976	25	9.71 059	31	10.28 941	9.94 917	6	49
12	9.66 001	24	9.71 090	31	10.28 910	9.94 911	7	48
13	9.66 025	25	9.71 121	32	10.28 879	9.94 904	6	47
14	9.66 050	25	9.71 153	31	10.28 847	9.94 898	7	46
15	9.66 075	24	9.71 184	31	10.28 816	9.94 891	6	45
16	9.66 099	25	9.71 215	31	10.28 785	9.94 885	7	44
17	9.66 124	24	9.71 246	31	10.28 754	9.94 878	7	43
18	9.66 148	25	9.71 277	31	10.28 723	9.94 871	6	42
19	9.66 173	24	9.71 308	31	10.28 692	9.94 865	7	41
20	9.66 197	24	9.71 339	31	10.28 661	9.94 858	6	40
21	9.66 221	25	9.71 370	31	10.28 630	9.94 852	7	39
22	9.66 246	24	9.71 401	30	10.28 599	9.94 845	6	38
23	9.66 270	25	9.71 431	31	10.28 569	9.94 839	7	37
24	9.66 295	24	9.71 462	31	10.28 538	9.94 832	6	36
25	9.66 319	24	9.71 493	31	10.28 507	9.94 826	7	35
26	9.66 343	25	9.71 524	31	10.28 476	9.94 819	6	34
27	9.66 368	24	9.71 555	31	10.28 445	9.94 813	7	33
28	9.66 392	24	9.71 586	31	10.28 414	9.94 806	7	32
29	9.66 416	25	9.71 617	31	10.28 383	9.94 799	6	31
30	9.66 441	24	9.71 648	31	10.28 352	9.94 793	7	30
31	9.66 465	24	9.71 679	30	10.28 321	9.94 786	6	29
32	9.66 489	24	9.71 709	31	10.28 291	9.94 780	7	28
33	9.66 513	24	9.71 740	31	10.28 260	9.94 773	6	27
34	9.66 537	25	9.71 771	31	10.28 229	9.94 767	7	26
35	9.66 562	24	9.71 802	31	10.28 198	9.94 760	7	25
36	9.66 586	24	9.71 833	30	10.28 167	9.94 753	6	24
37	9.66 610	24	9.71 863	31	10.28 137	9.94 747	7	23
38	9.66 634	24	9.71 894	31	10.28 106	9.94 740	6	22
39	9.66 658	24	9.71 925	30	10.28 075	9.94 734	7	21
40	9.66 682	24	9.71 955	31	10.28 045	9.94 727	7	20
41	9.66 706	25	9.71 986	31	10.28 014	9.94 720	6	19
42	9.66 731	24	9.72 017	31	10.27 983	9.94 714	7	18
43	9.66 755	24	9.72 048	30	10.27 952	9.94 707	7	17
44	9.66 779	24	9.72 078	31	10.27 922	9.94 700	6	16
45	9.66 803	24	9.72 109	31	10.27 891	9.94 694	7	15
46	9.66 827	24	9.72 140	30	10.27 860	9.94 687	7	14
47	9.66 851	24	9.72 170	31	10.27 830	9.94 680	6	13
48	9.66 875	24	9.72 201	30	10.27 799	9.94 674	7	12
49	9.66 899	23	9.72 231	31	10.27 769	9.94 667	7	11
50	9.66 922	24	9.72 262	31	10.27 738	9.94 660	6	10
51	9.66 946	24	9.72 293	30	10.27 707	9.94 654	7	9
52	9.66 970	24	9.72 323	31	10.27 677	9.94 647	7	8
53	9.66 994	24	9.72 354	30	10.27 646	9.94 640	6	7
54	9.67 018	24	9.72 384	31	10.27 616	9.94 634	7	6
55	9.67 042	24	9.72 415	30	10.27 585	9.94 627	7	5
56	9.67 066	24	9.72 445	31	10.27 555	9.94 620	6	4
57	9.67 090	23	9.72 476	30	10.27 524	9.94 614	7	3
58	9.67 113	24	9.72 506	31	10.27 494	9.94 607	7	2
59	9.67 137	24	9.72 537	30	10.27 463	9.94 600	7	1
60	9.67 161		9.72 567		10.27 433	9.94 593		0
′	L Cos	d	L Ctn	c d	L Tan	L Sin	d	′

117° (297°) (242°) **62°**

Proportional parts

″	32	31	30
1	0.5	0.5	0.5
2	1.1	1.0	1.0
3	1.6	1.6	1.5
4	2.1	2.1	2.0
5	2.7	2.6	2.5
6	3.2	3.1	3.0
7	3.7	3.6	3.5
8	4.3	4.1	4.0
9	4.8	4.6	4.5
10	5.3	5.2	5.0
20	10.7	10.3	10.0
30	16.0	15.5	15.0
40	21.3	20.7	20.0
50	26.7	25.8	25.0

″	25	24	23
1	0.4	0.4	0.4
2	0.8	0.8	0.8
3	1.2	1.2	1.2
4	1.7	1.6	1.5
5	2.1	2.0	1.9
6	2.5	2.4	2.3
7	2.9	2.8	2.7
8	3.3	3.2	3.1
9	3.8	3.6	3.4
10	4.2	4.0	3.8
20	8.3	8.0	7.7
30	12.5	12.0	11.5
40	16.7	16.0	15.3
50	20.8	20.0	19.2

″	7	6
1	0.1	0.1
2	0.2	0.2
3	0.4	0.3
4	0.5	0.4
5	0.6	0.5
6	0.7	0.6
7	0.8	0.7
8	0.9	0.8
9	1.0	0.9
10	1.2	1.0
20	2.3	2.0
30	3.5	3.0
40	4.7	4.0
50	5.8	5.0

Proportional parts

The −10 portion of the characteristic of the logarithm is not printed but must be written down whenever such a logarithm is used.

28° (208°) (331°) **151°**

′	L Sin	d	L Tan	c d	L Ctn	L Cos	d	′
0	9.67 161	24	9.72 567	31	10.27 433	9.94 593	6	60
1	9.67 185	23	9.72 598	30	10.27 402	9.94 587	7	59
2	9.67 208	24	9.72 628	31	10.27 372	9.94 580	7	58
3	9.67 232	24	9.72 659	30	10.27 341	9.94 573	6	57
4	9.67 256	24	9.72 689	31	10.27 311	9.94 567	7	56
5	9.67 280	23	9.72 720	30	10.27 280	9.94 560	7	55
6	9.67 303	24	9.72 750	30	10.27 250	9.94 553	7	54
7	9.67 327	23	9.72 780	31	10.27 220	9.94 546	6	53
8	9.67 350	24	9.72 811	30	10.27 189	9.94 540	7	52
9	9.67 374	24	9.72 841	31	10.27 159	9.94 533	7	51
10	9.67 398	23	9.72 872	30	10.27 128	9.94 526	7	50
11	9.67 421	24	9.72 902	30	10.27 098	9.94 519	6	49
12	9.67 445	23	9.72 932	31	10.27 068	9.94 513	7	48
13	9.67 468	24	9.72 963	30	10.27 037	9.94 506	7	47
14	9.67 492	23	9.72 993	30	10.27 007	9.94 499	7	46
15	9.67 515	24	9.73 023	31	10.26 977	9.94 492	7	45
16	9.67 539	23	9.73 054	30	10.26 946	9.94 485	6	44
17	9.67 562	24	9.73 084	30	10.26 916	9.94 479	7	43
18	9.67 586	23	9.73 114	30	10.26 886	9.94 472	7	42
19	9.67 609	24	9.73 144	31	10.26 856	9.94 465	7	41
20	9.67 633	23	9.73 175	30	10.26 825	9.94 458	7	40
21	9.67 656	24	9.73 205	30	10.26 795	9.94 451	6	39
22	9.67 680	23	9.73 235	30	10.26 765	9.94 445	7	38
23	9.67 703	23	9.73 265	30	10.26 735	9.94 438	7	37
24	9.67 726	24	9.73 295	31	10.26 705	9.94 431	7	36
25	9.67 750	23	9.73 326	30	10.26 674	9.94 424	7	35
26	9.67 773	23	9.73 356	30	10.26 644	9.94 417	7	34
27	9.67 796	24	9.73 386	30	10.26 614	9.94 410	6	33
28	9.67 820	23	9.73 416	30	10.26 584	9.94 404	7	32
29	9.67 843	23	9.73 446	30	10.26 554	9.94 397	7	31
30	9.67 866	24	9.73 476	31	10.26 524	9.94 390	7	30
31	9.67 890	23	9.73 507	30	10.26 493	9.94 383	7	29
32	9.67 913	23	9.73 537	30	10.26 463	9.94 376	7	28
33	9.67 936	23	9.73 567	30	10.26 433	9.94 369	7	27
34	9.67 959	23	9.73 597	30	10.26 403	9.94 362	7	26
35	9.67 982	24	9.73 627	30	10.26 373	9.94 355	6	25
36	9.68 006	23	9.73 657	30	10.26 343	9.94 349	7	24
37	9.68 029	23	9.73 687	30	10.26 313	9.94 342	7	23
38	9.68 052	23	9.73 717	30	10.26 283	9.94 335	7	22
39	9.68 075	23	9.73 747	30	10.26 253	9.94 328	7	21
40	9.68 098	23	9.73 777	30	10.26 223	9.94 321	7	20
41	9.68 121	23	9.73 807	30	10.26 193	9.94 314	7	19
42	9.68 144	23	9.73 837	30	10.26 163	9.94 307	7	18
43	9.68 167	23	9.73 867	30	10.26 133	9.94 300	7	17
44	9.68 190	23	9.73 897	30	10.26 103	9.94 293	7	16
45	9.68 213	24	9.73 927	30	10.26 073	9.94 286	7	15
46	9.68 237	23	9.73 957	30	10.26 043	9.94 279	6	14
47	9.68 260	23	9.73 987	30	10.26 013	9.94 273	7	13
48	9.68 283	22	9.74 017	30	10.25 983	9.94 266	7	12
49	9.68 305	23	9.74 047	30	10.25 953	9.94 259	7	11
50	9.68 328	23	9.74 077	30	10.25 923	9.94 252	7	10
51	9.68 351	23	9.74 107	30	10.25 893	9.94 245	7	9
52	9.68 374	23	9.74 137	29	10.25 863	9.94 238	7	8
53	9.68 397	23	9.74 166	30	10.25 834	9.94 231	7	7
54	9.68 420	23	9.74 196	30	10.25 804	9.94 224	7	6
55	9.68 443	23	9.74 226	30	10.25 774	9.94 217	7	5
56	9.68 466	23	9.74 256	30	10.25 744	9.94 210	7	4
57	9.68 489	23	9.74 286	30	10.25 714	9.94 203	7	3
58	9.68 512	22	9.74 316	29	10.25 684	9.94 196	7	2
59	9.68 534	23	9.74 345	30	10.25 655	9.94 189	7	1
60	9.68 557		9.74 375		10.25 625	9.94 182		0
′	L Cos	d	L Ctn	c d	L Tan	L Sin	d	′

Proportional parts

″	31	30	29
1	0.5	0.5	0.5
2	1.0	1.0	1.0
3	1.6	1.5	1.4
4	2.1	2.0	1.9
5	2.6	2.5	2.4
6	3.1	3.0	2.9
7	3.6	3.5	3.4
8	4.1	4.0	3.9
9	4.6	4.5	4.4
10	5.2	5.0	4.8
20	10.3	10.0	9.7
30	15.5	15.0	14.5
40	20.7	20.0	19.3
50	25.8	25.0	24.2

″	24	23	22
1	0.4	0.4	0.4
2	0.8	0.8	0.7
3	1.2	1.2	1.1
4	1.6	1.5	1.5
5	2.0	1.9	1.8
6	2.4	2.3	2.2
7	2.8	2.7	2.6
8	3.2	3.1	2.9
9	3.6	3.4	3.3
10	4.0	3.8	3.7
20	8.0	7.7	7.3
30	12.0	11.5	11.0
40	16.0	15.3	14.7
50	20.0	19.2	18.3

″	7	6
1	0.1	0.1
2	0.2	0.2
3	0.4	0.3
4	0.5	0.4
5	0.6	0.5
6	0.7	0.6
7	0.8	0.7
8	0.9	0.8
9	1.0	0.9
10	1.2	1.0
20	2.3	2.0
30	3.5	3.0
40	4.7	4.0
50	5.8	5.0

Proportional parts

118° (298°) (241°) **61°**

The −10 portion of the characteristic of the logarithm is not printed but must be written down whenever such a logarithm is used.

29° (209°) (330°) **150°**

′	L Sin	d	L Tan	c d	L Ctn	L Cos	d	′
0	9.68 557	23	9.74 375	30	10.25 625	9.94 182	7	60
1	9.68 580	23	9.74 405	30	10.25 595	9.94 175	7	59
2	9.68 603	22	9.74 435	30	10.25 565	9.94 168	7	58
3	9.68 625	23	9.74 465	29	10.25 535	9.94 161	7	57
4	9.68 648	23	9.74 494	30	10.25 506	9.94 154	7	56
5	9.68 671	23	9.74 524	30	10.25 476	9.94 147	7	55
6	9.68 694	22	9.74 554	29	10.25 446	9.94 140	7	54
7	9.68 716	23	9.74 583	30	10.25 417	9.94 133	7	53
8	9.68 739	23	9.74 613	30	10.25 387	9.94 126	7	52
9	9.68 762	22	9.74 643	30	10.25 357	9.94 119	7	51
10	9.68 784	23	9.74 673	29	10.25 327	9.94 112	7	50
11	9.68 807	22	9.74 702	30	10.25 298	9.94 105	7	49
12	9.68 829	23	9.74 732	30	10.25 268	9.94 098	8	48
13	9.68 852	23	9.74 762	29	10.25 238	9.94 090	7	47
14	9.68 875	22	9.74 791	30	10.25 209	9.94 083	7	46
15	9.68 897	23	9.74 821	30	10.25 179	9.94 076	7	45
16	9.68 920	22	9.74 851	29	10.25 149	9.94 069	7	44
17	9.68 942	23	9.74 880	30	10.25 120	9.94 062	7	43
18	9.68 965	22	9.74 910	29	10.25 090	9.94 055	7	42
19	9.68 987	23	9.74 939	30	10.25 061	9.94 048	7	41
20	9.69 010	22	9.74 969	29	10.25 031	9.94 041	7	40
21	9.69 032	23	9.74 998	30	10.25 002	9.94 034	7	39
22	9.69 055	22	9.75 028	30	10.24 972	9.94 027	7	38
23	9.69 077	23	9.75 058	29	10.24 942	9.94 020	8	37
24	9.69 100	22	9.75 087	30	10.24 913	9.94 012	7	36
25	9.69 122	22	9.75 117	29	10.24 883	9.94 005	7	35
26	9.69 144	23	9.75 146	30	10.24 854	9.93 998	7	34
27	9.69 167	22	9.75 176	29	10.24 824	9.93 991	7	33
28	9.69 189	23	9.75 205	30	10.24 795	9.93 984	7	32
29	9.69 212	22	9.75 235	29	10.24 765	9.93 977	7	31
30	9.69 234	22	9.75 264	30	10.24 736	9.93 970	7	30
31	9.69 256	23	9.75 294	29	10.24 706	9.93 963	8	29
32	9.69 279	22	9.75 323	30	10.24 677	9.93 955	7	28
33	9.69 301	22	9.75 353	29	10.24 647	9.93 948	7	27
34	9.69 323	22	9.75 382	29	10.24 618	9.93 941	7	26
35	9.69 345	23	9.75 411	30	10.24 589	9.93 934	7	25
36	9.69 368	22	9.75 441	29	10.24 559	9.93 927	7	24
37	9.69 390	22	9.75 470	30	10.24 530	9.93 920	8	23
38	9.69 412	22	9.75 500	29	10.24 500	9.93 912	7	22
39	9.69 434	22	9.75 529	29	10.24 471	9.93 905	7	21
40	9.69 456	23	9.75 558	30	10.24 442	9.93 898	7	20
41	9.69 479	22	9.75 588	29	10.24 412	9.93 891	7	19
42	9.69 501	22	9.75 617	30	10.24 383	9.93 884	8	18
43	9.69 523	22	9.75 647	29	10.24 353	9.93 876	7	17
44	9.69 545	22	9.75 676	29	10.24 324	9.93 869	7	16
45	9.69 567	22	9.75 705	30	10.24 295	9.93 862	7	15
46	9.69 589	22	9.75 735	29	10.24 265	9.93 855	8	14
47	9.69 611	22	9.75 764	29	10.24 236	9.93 847	7	13
48	9.69 633	22	9.75 793	29	10.24 207	9.93 840	7	12
49	9.69 655	22	9.75 822	30	10.24 178	9.93 833	7	11
50	9.69 677	22	9.75 852	29	10.24 148	9.93 826	7	10
51	9.69 699	22	9.75 881	29	10.24 119	9.93 819	8	9
52	9.69 721	22	9.75 910	29	10.24 090	9.93 811	7	8
53	9.69 743	22	9.75 939	30	10.24 061	9.93 804	7	7
54	9.69 765	22	9.75 969	29	10.24 031	9.93 797	8	6
55	9.69 787	22	9.75 998	29	10.24 002	9.93 789	7	5
56	9.69 809	22	9.76 027	29	10.23 973	9.93 782	7	4
57	9.69 831	22	9.76 056	30	10.23 944	9.93 775	7	3
58	9.69 853	22	9.76 086	29	10.23 914	9.93 768	8	2
59	9.69 875	22	9.76 115	29	10.23 885	9.93 760	7	1
60	9.69 897		9.76 144		10.23 856	9.93 753		0
′	L Cos	d	L Ctn	c d	L Tan	L Sin	d	′

Proportional parts

″	30	29	23
1	0.5	0.5	0.4
2	1.0	1.0	0.8
3	1.5	1.4	1.2
4	2.0	1.9	1.5
5	2.5	2.4	1.9
6	3.0	2.9	2.3
7	3.5	3.4	2.7
8	4.0	3.9	3.1
9	4.5	4.4	3.4
10	5.0	4.8	3.8
20	10.0	9.7	7.7
30	15.0	14.5	11.5
40	20.0	19.3	15.3
50	25.0	24.2	19.2

″	22	8	7
1	0.4	0.1	0.1
2	0.7	0.3	0.2
3	1.1	0.4	0.4
4	1.5	0.5	0.5
5	1.8	0.7	0.6
6	2.2	0.8	0.7
7	2.6	0.9	0.8
8	2.9	1.1	0.9
9	3.3	1.2	1.0
10	3.7	1.3	1.2
20	7.3	2.7	2.3
30	11.0	4.0	3.5
40	14.7	5.3	4.7
50	18.3	6.7	5.8

Proportional parts

119° (299) (240°) **60°**

The -10 portion of the characteristic of the logarithm is not printed but must be written down whenever such a logarithm is used.

30° (210°) (329°) **149°**

′	L Sin	d	L Tan	c d	L Ctn	L Cos	d	′
0	9.69 897	22	9.76 144	29	10.23 856	9.93 753	7	60
1	9.69 919	22	9.76 173	29	10.23 827	9.93 746	8	59
2	9.69 941	22	9.76 202	29	10.23 798	9.93 738	7	58
3	9.69 963	21	9.76 231	30	10.23 769	9.93 731	7	57
4	9.69 984	22	9.76 261	29	10.23 739	9.93 724	7	56
5	9.70 006	22	9.76 290	29	10.23 710	9.93 717	8	55
6	9.70 028	22	9.76 319	29	10.23 681	9.93 709	7	54
7	9.70 050	22	9.76 348	29	10.23 652	9.93 702	7	53
8	9.70 072	21	9.76 377	29	10.23 623	9.93 695	8	52
9	9.70 093	22	9.76 406	29	10.23 594	9.93 687	7	51
10	9.70 115	22	9.76 435	29	10.23 565	9.93 680	7	50
11	9.70 137	22	9.76 464	29	10.23 536	9.93 673	8	49
12	9.70 159	21	9.76 493	29	10.23 507	9.93 665	7	48
13	9.70 180	22	9.76 522	29	10.23 478	9.93 658	8	47
14	9.70 202	22	9.76 551	29	10.23 449	9.93 650	7	46
15	9.70 224	21	9.76 580	29	10.23 420	9.93 643	7	45
16	9.70 245	22	9.76 609	30	10.23 391	9.93 636	8	44
17	9.70 267	21	9.76 639	29	10.23 361	9.93 628	7	43
18	9.70 288	22	9.76 668	29	10.23 332	9.93 621	7	42
19	9.70 310	22	9.76 697	28	10.23 303	9.93 614	8	41
20	9.70 332	21	9.76 725	29	10.23 275	9.93 606	7	40
21	9.70 353	22	9.76 754	29	10.23 246	9.93 599	8	39
22	9.70 375	21	9.76 783	29	10.23 217	9.93 591	7	38
23	9.70 396	22	9.76 812	29	10.23 188	9.93 584	7	37
24	9.70 418	21	9.76 841	29	10.23 159	9.93 577	8	36
25	9.70 439	22	9.76 870	29	10.23 130	9.93 569	7	35
26	9.70 461	21	9.76 899	29	10.23 101	9.93 562	8	34
27	9.70 482	22	9.76 928	29	10.23 072	9.93 554	7	33
28	9.70 504	21	9.76 957	29	10.23 043	9.93 547	8	32
29	9.70 525	22	9.76 986	29	10.23 014	9.93 539	7	31
30	9.70 547	21	9.77 015	29	10.22 985	9.93 532	7	30
31	9.70 568	22	9.77 044	29	10.22 956	9.93 525	8	29
32	9.70 590	21	9.77 073	28	10.22 927	9.93 517	7	28
33	9.70 611	22	9.77 101	29	10.22 899	9.93 510	8	27
34	9.70 633	21	9.77 130	29	10.22 870	9.93 502	7	26
35	9.70 654	21	9.77 159	29	10.22 841	9.93 495	8	25
36	9.70 675	22	9.77 188	29	10.22 812	9.93 487	7	24
37	9.70 697	21	9.77 217	29	10.22 783	9.93 480	8	23
38	9.70 718	21	9.77 246	28	10.22 754	9.93 472	7	22
39	9.70 739	22	9.77 274	29	10.22 726	9.93 465	8	21
40	9.70 761	21	9.77 303	29	10.22 697	9.93 457	7	20
41	9.70 782	21	9.77 332	29	10.22 668	9.93 450	8	19
42	9.70 803	21	9.77 361	29	10.22 639	9.93 442	7	18
43	9.70 824	22	9.77 390	28	10.22 610	9.93 435	8	17
44	9.70 846	21	9.77 418	29	10.22 582	9.93 427	7	16
45	9.70 867	21	9.77 447	29	10.22 553	9.93 420	8	15
46	9.70 888	21	9.77 476	29	10.22 524	9.93 412	7	14
47	9.70 909	22	9.77 505	28	10.22 495	9.93 405	8	13
48	9.70 931	21	9.77 533	29	10.22 467	9.93 397	7	12
49	9.70 952	21	9.77 562	29	10.22 438	9.93 390	8	11
50	9.70 973	21	9.77 591	28	10.22 409	9.93 382	7	10
51	9.70 994	21	9.77 619	29	10.22 381	9.93 375	8	9
52	9.71 015	21	9.77 648	29	10.22 352	9.93 367	7	8
53	9.71 036	22	9.77 677	29	10.22 323	9.93 360	8	7
54	9.71 058	21	9.77 706	28	10.22 294	9.93 352	8	6
55	9.71 079	21	9.77 734	29	10.22 266	9.93 344	7	5
56	9.71 100	21	9.77 763	28	10.22 237	9.93 337	8	4
57	9.71 121	21	9.77 791	29	10.22 209	9.93 329	7	3
58	9.71 142	21	9.77 820	29	10.22 180	9.93 322	8	2
59	9.71 163	21	9.77 849	28	10.22 151	9.93 314	7	1
60	9.71 184		9.77 877		10.22 123	9.93 307		0
′	L Cos	d	L Ctn	c d	L Tan	L Sin	d	′

Proportional parts

″	30	29	28
1	0.5	0.5	0.5
2	1.0	1.0	0.9
3	1.5	1.4	1.4
4	2.0	1.9	1.9
5	2.5	2.4	2.3
6	3.0	2.9	2.8
7	3.5	3.4	3.3
8	4.0	3.9	3.7
9	4.5	4.4	4.2
10	5.0	4.8	4.7
20	10.0	9.7	9.3
30	15.0	14.5	14.0
40	20.0	19.3	18.7
50	25.0	24.2	23.3

″	22	21
1	0.4	0.4
2	0.7	0.7
3	1.1	1.0
4	1.5	1.4
5	1.8	1.8
6	2.2	2.1
7	2.6	2.4
8	2.9	2.8
9	3.3	3.2
10	3.7	3.5
20	7.3	7.0
30	11.0	10.5
40	14.7	14.0
50	18.3	17.5

″	8	7
1	0.1	0.1
2	0.3	0.2
3	0.4	0.4
4	0.5	0.5
5	0.7	0.6
6	0.8	0.7
7	0.9	0.8
8	1.1	0.9
9	1.2	1.0
10	1.3	1.2
20	2.7	2.3
30	4.0	3.5
40	5.3	4.7
50	6.7	5.8

Proportional parts

120° (300°) (239°) **59°**

The −10 portion of the characteristic of the logarithm is not printed but must be written down whenever such a logarithm is used.

31° (211°) (328°) **148°**

′	L Sin	d	L Tan	c d	L Ctn	L Cos	d	′
0	9.71 184	21	9.77 877	29	10.22 123	9.93 307	8	60
1	9.71 205	21	9.77 906	29	10.22 094	9.93 299	8	59
2	9.71 226	21	9.77 935	28	10.22 065	9.93 291	7	58
3	9.71 247	21	9.77 963	29	10.22 037	9.93 284	8	57
4	9.71 268	21	9.77 992	28	10.22 008	9.93 276	7	56
5	9.71 289	21	9.78 020	29	10.21 980	9.93 269	8	55
6	9.71 310	21	9.78 049	28	10.21 951	9.93 261	8	54
7	9.71 331	21	9.78 077	29	10.21 923	9.93 253	7	53
8	9.71 352	21	9.78 106	29	10.21 894	9.93 246	8	52
9	9.71 373	20	9.78 135	28	10.21 865	9.93 238	8	51
10	9.71 393	21	9.78 163	29	10.21 837	9.93 230	7	50
11	9.71 414	21	9.78 192	28	10.21 808	9.93 223	8	49
12	9.71 435	21	9.78 220	29	10.21 780	9.93 215	8	48
13	9.71 456	21	9.78 249	28	10.21 751	9.93 207	7	47
14	9.71 477	21	9.78 277	29	10.21 723	9.93 200	8	46
15	9.71 498	21	9.78 306	28	10.21 694	9.93 192	8	45
16	9.71 519	20	9.78 334	29	10.21 666	9.93 184	7	44
17	9.71 539	21	9.78 363	28	10.21 637	9.93 177	8	43
18	9.71 560	21	9.78 391	28	10.21 609	9.93 169	8	42
19	9.71 581	21	9.78 419	29	10.21 581	9.93 161	7	41
20	9.71 602	20	9.78 448	28	10.21 552	9.93 154	8	40
21	9.71 622	21	9.78 476	29	10.21 524	9.93 146	8	39
22	9.71 643	21	9.78 505	28	10.21 495	9.93 138	7	38
23	9.71 664	21	9.78 533	29	10.21 467	9.93 131	8	37
24	9.71 685	20	9.78 562	28	10.21 438	9.93 123	8	36
25	9.71 705	21	9.78 590	28	10.21 410	9.93 115	7	35
26	9.71 726	21	9.78 618	29	10.21 382	9.93 108	8	34
27	9.71 747	20	9.78 647	28	10.21 353	9.93 100	8	33
28	9.71 767	21	9.78 675	29	10.21 325	9.93 092	8	32
29	9.71 788	21	9.78 704	28	10.21 296	9.93 084	7	31
30	9.71 809	20	9.78 732	28	10.21 268	9.93 077	8	30
31	9.71 829	21	9.78 760	29	10.21 240	9.93 069	8	29
32	9.71 850	20	9.78 789	28	10.21 211	9.93 061	8	28
33	9.71 870	21	9.78 817	28	10.21 183	9.93 053	7	27
34	9.71 891	20	9.78 845	29	10.21 155	9.93 046	8	26
35	9.71 911	21	9.78 874	28	10.21 126	9.93 038	8	25
36	9.71 932	20	9.78 902	28	10.21 098	9.93 030	8	24
37	9.71 952	21	9.78 930	29	10.21 070	9.93 022	8	23
38	9.71 973	21	9.78 959	28	10.21 041	9.93 014	7	22
39	9.71 994	20	9.78 987	28	10.21 013	9.93 007	8	21
40	9.72 014	20	9.79 015	28	10.20 985	9.92 999	8	20
41	9.72 034	21	9.79 043	29	10.20 957	9.92 991	8	19
42	9.72 055	20	9.79 072	28	10.20 928	9.92 983	7	18
43	9.72 075	21	9.79 100	28	10.20 900	9.92 976	8	17
44	9.72 096	20	9.79 128	28	10.20 872	9.92 968	8	16
45	9.72 116	21	9.79 156	29	10.20 844	9.92 960	8	15
46	9.72 137	20	9.79 185	28	10.20 815	9.92 952	8	14
47	9.72 157	20	9.79 213	28	10.20 787	9.92 944	8	13
48	9.72 177	21	9.79 241	28	10.20 759	9.92 936	7	12
49	9.72 198	20	9.79 269	28	10.20 731	9.92 929	8	11
50	9.72 218	20	9.79 297	29	10.20 703	9.92 921	8	10
51	9.72 238	21	9.79 326	28	10.20 674	9.92 913	8	9
52	9.72 259	20	9.79 354	28	10.20 646	9.92 905	8	8
53	9.72 279	20	9.79 382	28	10.20 618	9.92 897	8	7
54	9.72 299	21	9.79 410	28	10.20 590	9.92 889	8	6
55	9.72 320	20	9.79 438	28	10.20 562	9.92 881	7	5
56	9.72 340	20	9.79 466	29	10.20 534	9.92 874	8	4
57	9.72 360	21	9.79 495	28	10.20 505	9.92 866	8	3
58	9.72 381	20	9.79 523	28	10.20 477	9.92 858	8	2
59	9.72 401	20	9.79 551	28	10.20 449	9.92 850	8	1
60	9.72 421		9.79 579		10.20 421	9.92 842		0
′	L Cos	d	L Ctn	c d	L Tan	L Sin	d	′

121° (301°) (238°) **58°**

Proportional parts

″	29	28
1	0.5	0.5
2	1.0	0.9
3	1.4	1.4
4	1.9	1.9
5	2.4	2.3
6	2.9	2.8
7	3.4	3.3
8	3.9	3.7
9	4.4	4.2
10	4.8	4.7
20	9.7	9.3
30	14.5	14.0
40	19.3	18.7
50	24.2	23.3

″	21	20
1	0.4	0.3
2	0.7	0.7
3	1.0	1.0
4	1.4	1.3
5	1.8	1.7
6	2.1	2.0
7	2.4	2.3
8	2.8	2.7
9	3.2	3.0
10	3.5	3.3
20	7.0	6.7
30	10.5	10.0
40	14.0	13.3
50	17.5	16.7

″	8	7
1	0.1	0.1
2	0.3	0.2
3	0.4	0.4
4	0.5	0.5
5	0.7	0.6
6	0.8	0.7
7	0.9	0.8
8	1.1	0.9
9	1.2	1.0
10	1.3	1.2
20	2.7	2.3
30	4.0	3.5
40	5.3	4.7
50	6.7	5.8

Proportional parts

The −10 portion of the characteristic of the logarithm is not printed but must be written down whenever such a logarithm is used.

32° (212°) (327°) **147°**

′	L Sin	d	L Tan	c d	L Ctn	L Cos	d	′
0	9.72 421	20	9.79 579	28	10.20 421	9.92 842	8	60
1	9.72 441	20	9.79 607	28	10.20 393	9.92 834	8	59
2	9.72 461	21	9.79 635	28	10.20 365	9.92 826	8	58
3	9.72 482	20	9.79 663	28	10.20 337	9.92 818	8	57
4	9.72 502	20	9.79 691	28	10.20 309	9.92 810	7	56
5	9.72 522	20	9.79 719	28	10.20 281	9.92 803	8	55
6	9.72 542	20	9.79 747	29	10.20 253	9.92 795	8	54
7	9.72 562	20	9.79 776	28	10.20 224	9.92 787	8	53
8	9.72 582	20	9.79 804	28	10.20 196	9.92 779	8	52
9	9.72 602	20	9.79 832	28	10.20 168	9.92 771	8	51
10	9.72 622	21	9.79 860	28	10.20 140	9.92 763	8	50
11	9.72 643	20	9.79 888	28	10.20 112	9.92 755	8	49
12	9.72 663	20	9.79 916	28	10.20 084	9.92 747	8	48
13	9.72 683	20	9.79 944	28	10.20 056	9.92 739	8	47
14	9.72 703	20	9.79 972	28	10.20 028	9.92 731	8	46
15	9.72 723	20	9.80 000	28	10.20 000	9.92 723	8	45
16	9.72 743	20	9.80 028	28	10.19 972	9.92 715	8	44
17	9.72 763	20	9.80 056	28	10.19 944	9.92 707	8	43
18	9.72 783	20	9.80 084	28	10.19 916	9.92 699	8	42
19	9.72 803	20	9.80 112	28	10.19 888	9.92 691	8	41
20	9.72 823	20	9.80 140	28	10.19 860	9.92 683	8	40
21	9.72 843	20	9.80 168	27	10.19 832	9.92 675	8	39
22	9.72 863	20	9.80 195	28	10.19 805	9.92 667	8	38
23	9.72 883	19	9.80 223	28	10.19 777	9.92 659	8	37
24	9.72 902	20	9.80 251	28	10.19 749	9.92 651	8	36
25	9.72 922	20	9.80 279	28	10.19 721	9.92 643	8	35
26	9.72 942	20	9.80 307	28	10.19 693	9.92 635	8	34
27	9.72 962	20	9.80 335	28	10.19 665	9.92 627	8	33
28	9.72 982	20	9.80 363	28	10.19 637	9.92 619	8	32
29	9.73 002	20	9.80 391	28	10.19 609	9.92 611	8	31
30	9.73 022	19	9.80 419	28	10.19 581	9.92 603	8	30
31	9.73 041	20	9.80 447	27	10.19 553	9.92 595	8	29
32	9.73 061	20	9.80 474	28	10.19 526	9.92 587	8	28
33	9.73 081	20	9.80 502	28	10.19 498	9.92 579	8	27
34	9.73 101	20	9.80 530	28	10.19 470	9.92 571	8	26
35	9.73 121	19	9.80 558	28	10.19 442	9.92 563	8	25
36	9.73 140	20	9.80 586	28	10.19 414	9.92 555	9	24
37	9.73 160	20	9.80 614	28	10.19 386	9.92 546	8	23
38	9.73 180	20	9.80 642	27	10.19 358	9.92 538	8	22
39	9.73 200	19	9.80 669	28	10.19 331	9.92 530	8	21
40	9.73 219	20	9.80 697	28	10.19 303	9.92 522	8	20
41	9.73 239	20	9.80 725	28	10.19 275	9.92 514	8	19
42	9.73 259	19	9.80 753	28	10.19 247	9.92 506	8	18
43	9.73 278	20	9.80 781	27	10.19 219	9.92 498	8	17
44	9.73 298	20	9.80 808	28	10.19 192	9.92 490	8	16
45	9.73 318	19	9.80 836	28	10.19 164	9.92 482	9	15
46	9.73 337	20	9.80 864	28	10.19 136	9.92 473	8	14
47	9.73 357	20	9.80 892	27	10.19 108	9.92 465	8	13
48	9.73 377	19	9.80 919	28	10.19 081	9.92 457	8	12
49	9.73 396	20	9.80 947	28	10.19 053	9.92 449	8	11
50	9.73 416	19	9.80 975	28	10.19 025	9.92 441	8	10
51	9.73 435	20	9.81 003	27	10.18 997	9.92 433	8	9
52	9.73 455	19	9.81 030	28	10.18 970	9.92 425	9	8
53	9.73 474	20	9.81 058	28	10.18 942	9.92 416	8	7
54	9.73 494	19	9.81 086	27	10.18 914	9.92 408	8	6
55	9.73 513	20	9.81 113	28	10.18 887	9.92 400	8	5
56	9.73 533	19	9.81 141	28	10.18 859	9.92 392	8	4
57	9.73 552	20	9.81 169	27	10.18 831	9.92 384	8	3
58	9.73 572	19	9.81 196	28	10.18 804	9.92 376	9	2
59	9.73 591	20	9.81 224	28	10.18 776	9.92 367	8	1
60	9.73 611		9.81 252		10.18 748	9.92 359		0
′	L Cos	d	L Ctn	c d	L Tan	L Sin	d	′

Proportional parts

″	29	28	27
1	0.5	0.5	0.4
2	1.0	0.9	0.9
3	1.4	1.4	1.4
4	1.9	1.9	1.8
5	2.4	2.3	2.2
6	2.9	2.8	2.7
7	3.4	3.3	3.2
8	3.9	3.7	3.6
9	4.4	4.2	4.0
10	4.8	4.7	4.5
20	9.7	9.3	9.0
30	14.5	14.0	13.5
40	19.3	18.7	18.0
50	24.2	23.3	22.5

″	21	20	19
1	0.4	0.3	0.3
2	0.7	0.7	0.6
3	1.0	1.0	1.0
4	1.4	1.3	1.3
5	1.8	1.7	1.6
6	2.1	2.0	1.9
7	2.4	2.3	2.2
8	2.8	2.7	2.5
9	3.2	3.0	2.8
10	3.5	3.3	3.2
20	7.0	6.7	6.3
30	10.5	10.0	9.5
40	14.0	13.3	12.7
50	17.5	16.7	15.8

″	9	8	7
1	0.2	0.1	0.1
2	0.3	0.3	0.2
3	0.4	0.4	0.4
4	0.6	0.5	0.5
5	0.8	0.7	0.6
6	0.9	0.8	0.7
7	1.0	0.9	0.8
8	1.2	1.1	0.9
9	1.4	1.2	1.0
10	1.5	1.3	1.2
20	3.0	2.7	2.3
30	4.5	4.0	3.5
40	6.0	5.3	4.7
50	7.5	6.7	5.8

Proportional parts

122° (**302°**) (237°) **57°**

Table 4 COMMON LOGARITHMS OF TRIGONOMETRIC FUNCTIONS (*continued*)

The −10 portion of the characteristic of the logarithm is not printed but must be written down whenever such a logarithm is used.

33° (213°) — (326°) **146°**

′	L Sin	d	L Tan	c d	L Ctn	L Cos	d	′
0	9.73 611	19	9.81 252	27	10.18 748	9.92 359	8	60
1	9.73 630	20	9.81 279	28	10.18 721	9.92 351	8	59
2	9.73 650	19	9.81 307	28	10.18 693	9.92 343	8	58
3	9.73 669	20	9.81 335	27	10.18 665	9.92 335	9	57
4	9.73 689	19	9.81 362	28	10.18 638	9.92 326	8	56
5	9.73 708	19	9.81 390	28	10.18 610	9.92 318	8	55
6	9.73 727	20	9.81 418	27	10.18 582	9.92 310	8	54
7	9.73 747	19	9.81 445	28	10.18 555	9.92 302	9	53
8	9.73 766	19	9.81 473	27	10.18 527	9.92 293	8	52
9	9.73 785	20	9.81 500	28	10.18 500	9.92 285	8	51
10	9.73 805	19	9.81 528	28	10.18 472	9.92 277	8	50
11	9.73 824	19	9.81 556	27	10.18 444	9.92 269	9	49
12	9.73 843	20	9.81 583	28	10.18 417	9.92 260	8	48
13	9.73 863	19	9.81 611	27	10.18 389	9.92 252	8	47
14	9.73 882	19	9.81 638	28	10.18 362	9.92 244	9	46
15	9.73 901	20	9.81 666	27	10.18 334	9.92 235	8	45
16	9.73 921	19	9.81 693	28	10.18 307	9.92 227	8	44
17	9.73 940	19	9.81 721	27	10.18 279	9.92 219	8	43
18	9.73 959	19	9.81 748	28	10.18 252	9.92 211	9	42
19	9.73 978	19	9.81 776	27	10.18 224	9.92 202	8	41
20	9.73 997	20	9.81 803	28	10.18 197	9.92 194	8	40
21	9.74 017	19	9.81 831	27	10.18 169	9.92 186	9	39
22	9.74 036	19	9.81 858	28	10.18 142	9.92 177	8	38
23	9.74 055	19	9.81 886	27	10.18 114	9.92 169	8	37
24	9.74 074	19	9.81 913	28	10.18 087	9.92 161	9	36
25	9.74 093	20	9.81 941	27	10.18 059	9.92 152	8	35
26	9.74 113	19	9.81 968	28	10.18 032	9.92 144	8	34
27	9.74 132	19	9.81 996	27	10.18 004	9.92 136	9	33
28	9.74 151	19	9.82 023	28	10.17 977	9.92 127	8	32
29	9.74 170	19	9.82 051	27	10.17 949	9.92 119	8	31
30	9.74 189	19	9.82 078	28	10.17 922	9.92 111	9	30
31	9.74 208	19	9.82 106	27	10.17 894	9.92 102	8	29
32	9.74 227	19	9.82 133	28	10.17 867	9.92 094	8	28
33	9.74 246	19	9.82 161	27	10.17 839	9.92 086	9	27
34	9.74 265	19	9.82 188	27	10.17 812	9.92 077	8	26
35	9.74 284	19	9.82 215	28	10.17 785	9.92 069	9	25
36	9.74 303	19	9.82 243	27	10.17 757	9.92 060	8	24
37	9.74 322	19	9.82 270	28	10.17 730	9.92 052	8	23
38	9.74 341	19	9.82 298	27	10.17 702	9.92 044	9	22
39	9.74 360	19	9.82 325	27	10.17 675	9.92 035	8	21
40	9.74 379	19	9.82 352	28	10.17 648	9.92 027	9	20
41	9.74 398	19	9.82 380	27	10.17 620	9.92 018	8	19
42	9.74 417	19	9.82 407	28	10.17 593	9.92 010	8	18
43	9.74 436	19	9.82 435	27	10.17 565	9.92 002	9	17
44	9.74 455	19	9.82 462	27	10.17 538	9.91 993	8	16
45	9.74 474	19	9.82 489	28	10.17 511	9.91 985	9	15
46	9.74 493	19	9.82 517	27	10.17 483	9.91 976	8	14
47	9.74 512	19	9.82 544	27	10.17 456	9.91 968	9	13
48	9.74 531	18	9.82 571	28	10.17 429	9.91 959	8	12
49	9.74 549	19	9.82 599	27	10.17 401	9.91 951	9	11
50	9.74 568	19	9.82 626	27	10.17 374	9.91 942	8	10
51	9.74 587	19	9.82 653	28	10.17 347	9.91 934	9	9
52	9.74 606	19	9.82 681	27	10.17 319	9.91 925	8	8
53	9.74 625	19	9.82 708	27	10.17 292	9.91 917	9	7
54	9.74 644	18	9.82 735	27	10.17 265	9.91 908	8	6
55	9.74 662	19	9.82 762	28	10.17 238	9.91 900	9	5
56	9.74 681	19	9.82 790	27	10.17 210	9.91 891	8	4
57	9.74 700	19	9.82 817	27	10.17 183	9.91 883	9	3
58	9.74 719	18	9.82 844	27	10.17 156	9.91 874	8	2
59	9.74 737	19	9.82 871	28	10.17 129	9.91 866	9	1
60	9.74 756		9.82 899		10.17 101	9.91 857		0
′	L Cos	d	L Ctn	c d	L Tan	L Sin	d	′

Proportional parts

″	28	27
1	0.5	0.4
2	0.9	0.9
3	1.4	1.4
4	1.9	1.8
5	2.3	2.2
6	2.8	2.7
7	3.3	3.2
8	3.7	3.6
9	4.2	4.0
10	4.7	4.5
20	9.3	9.0
30	14.0	13.5
40	18.7	18.0
50	23.3	22.5

″	20	19	18
1	0.3	0.3	0.3
2	0.7	0.6	0.6
3	1.0	1.0	0.9
4	1.3	1.3	1.2
5	1.7	1.6	1.5
6	2.0	1.9	1.8
7	2.3	2.2	2.1
8	2.7	2.5	2.4
9	3.0	2.8	2.7
10	3.3	3.2	3.0
20	6.7	6.3	6.0
30	10.0	9.5	9.0
40	13.3	12.7	12.0
50	16.7	15.8	15.0

″	9	8
1	0.2	0.1
2	0.3	0.3
3	0.4	0.4
4	0.6	0.5
5	0.8	0.7
6	0.9	0.8
7	1.0	0.9
8	1.2	1.1
9	1.4	1.2
10	1.5	1.3
20	3.0	2.7
30	4.5	4.0
40	6.0	5.3
50	7.5	6.7

Proportional parts

123° (303°) — (236°) **56°**

The −10 portion of the characteristic of the logarithm is not printed but must be written down whenever such a logarithm is used.

34° (214°) (325°) **145°**

′	L Sin	d	L Tan	c d	L Ctn	L Cos	d	′
0	9.74 756	19	9.82 899	27	10.17 101	9.91 857	8	60
1	9.74 775	19	9.82 926	27	10.17 074	9.91 849	9	59
2	9.74 794	18	9.82 953	27	10.17 047	9.91 840	8	58
3	9.74 812	19	9.82 980	28	10.17 020	9.91 832	9	57
4	9.74 831	19	9.83 008	27	10.16 992	9.91 823	8	56
5	9.74 850	18	9.83 035	27	10.16 965	9.91 815	9	55
6	9.74 868	19	9.83 062	27	10.16 938	9.91 806	8	54
7	9.74 887	19	9.83 089	28	10.16 911	9.91 798	9	53
8	9.74 906	18	9.83 117	27	10.16 883	9.91 789	8	52
9	9.74 924	19	9.83 144	27	10.16 856	9.91 781	9	51
10	9.74 943	18	9.83 171	27	10.16 829	9.91 772	9	50
11	9.74 961	19	9.83 198	27	10.16 802	9.91 763	8	49
12	9.74 980	19	9.83 225	27	10.16 775	9.91 755	9	48
13	9.74 999	18	9.83 252	28	10.16 748	9.91 746	8	47
14	9.75 017	19	9.83 280	27	10.16 720	9.91 738	9	46
15	9.75 036	18	9.83 307	27	10.16 693	9.91 729	9	45
16	9.75 054	19	9.83 334	27	10.16 666	9.91 720	8	44
17	9.75 073	18	9.83 361	27	10.16 639	9.91 712	9	43
18	9.75 091	19	9.83 388	27	10.16 612	9.91 703	8	42
19	9.75 110	18	9.83 415	27	10.16 585	9.91 695	9	41
20	9.75 128	19	9.83 442	28	10.16 558	9.91 686	9	40
21	9.75 147	18	9.83 470	27	10.16 530	9.91 677	8	39
22	9.75 165	19	9.83 497	27	10.16 503	9.91 669	9	38
23	9.75 184	18	9.83 524	27	10.16 476	9.91 660	9	37
24	9.75 202	19	9.83 551	27	10.16 449	9.91 651	8	36
25	9.75 221	18	9.83 578	27	10.16 422	9.91 643	9	35
26	9.75 239	19	9.83 605	27	10.16 395	9.91 634	9	34
27	9.75 258	18	9.83 632	27	10.16 368	9.91 625	8	33
28	9.75 276	18	9.83 659	27	10.16 341	9.91 617	9	32
29	9.75 294	19	9.83 686	27	10.16 314	9.91 608	9	31
30	9.75 313	18	9.83 713	27	10.16 287	9.91 599	8	30
31	9.75 331	19	9.83 740	28	10.16 260	9.91 591	9	29
32	9.75 350	18	9.83 768	27	10.16 232	9.91 582	9	28
33	9.75 368	18	9.83 795	27	10.16 205	9.91 573	8	27
34	9.75 386	19	9.83 822	27	10.16 178	9.91 565	9	26
35	9.75 405	18	9.83 849	27	10.16 151	9.91 556	9	25
36	9.75 423	18	9.83 876	27	10.16 124	9.91 547	9	24
37	9.75 441	18	9.83 903	27	10.16 097	9.91 538	8	23
38	9.75 459	19	9.83 930	27	10.16 070	9.91 530	9	22
39	9.75 478	18	9.83 957	27	10.16 043	9.91 521	9	21
40	9.75 496	18	9.83 984	27	10.16 016	9.91 512	8	20
41	9.75 514	19	9.84 011	27	10.15 989	9.91 504	9	19
42	9.75 533	18	9.84 038	27	10.15 962	9.91 495	9	18
43	9.75 551	18	9.84 065	27	10.15 935	9.91 486	9	17
44	9.75 569	18	9.84 092	27	10.15 908	9.91 477	8	16
45	9.75 587	18	9.84 119	27	10.15 881	9.91 469	9	15
46	9.75 605	19	9.84 146	27	10.15 854	9.91 460	9	14
47	9.75 624	18	9.84 173	27	10.15 827	9.91 451	9	13
48	9.75 642	18	9.84 200	27	10.15 800	9.91 442	9	12
49	9.75 660	18	9.84 227	27	10.15 773	9.91 433	8	11
50	9.75 678	18	9.84 254	26	10.15 746	9.91 425	9	10
51	9.75 696	18	9.84 280	27	10.15 720	9.91 416	9	9
52	9.75 714	19	9.84 307	27	10.15 693	9.91 407	9	8
53	9.75 733	18	9.84 334	27	10.15 666	9.91 398	9	7
54	9.75 751	18	9.84 361	27	10.15 639	9.91 389	8	6
55	9.75 769	18	9.84 388	27	10.15 612	9.91 381	9	5
56	9.75 787	18	9.84 415	27	10.15 585	9.91 372	9	4
57	9.75 805	18	9.84 442	27	10.15 558	9.91 363	9	3
58	9.75 823	18	9.84 469	27	10.15 531	9.91 354	9	2
59	9.75 841	18	9.84 496	27	10.15 504	9.91 345	9	1
60	9.75 859		9.84 523		10.15 477	9.91 336		0
′	L Cos	d	L Ctn	c d	L Tan	L Sin	d	′

Proportional parts

″	28	27	26
1	0.5	0.4	0.4
2	0.9	0.9	0.9
3	1.4	1.4	1.3
4	1.9	1.8	1.7
5	2.3	2.2	2.2
6	2.8	2.7	2.6
7	3.3	3.2	3.0
8	3.7	3.6	3.5
9	4.2	4.0	3.9
10	4.7	4.5	4.3
20	9.3	9.0	8.7
30	14.0	13.5	13.0
40	18.7	18.0	17.3
50	23.3	22.5	21.7

″	19	18
1	0.3	0.3
2	0.6	0.6
3	1.0	0.9
4	1.3	1.2
5	1.6	1.5
6	1.9	1.8
7	2.2	2.1
8	2.5	2.4
9	2.8	2.7
10	3.2	3.0
20	6.3	6.0
30	9.5	9.0
40	12.7	12.0
50	15.8	15.0

″	9	8
1	0.2	0.1
2	0.3	0.3
3	0.4	0.4
4	0.6	0.5
5	0.8	0.7
6	0.9	0.8
7	1.0	0.9
8	1.2	1.1
9	1.4	1.2
10	1.5	1.3
20	3.0	2.7
30	4.5	4.0
40	6.0	5.3
50	7.5	6.7

Proportional parts

124° (304°) (235°) **55°**

The −10 portion of the characteristic of the logarithm is not printed but must be written down whenever such a logarithm is used.

35° (215°) (324°) **144°**

′	L Sin	d	L Tan	c d	L Ctn	L Cos	d	′
0	9.75 859	18	9.84 523	27	10.15 477	9.91 336	8	60
1	9.75 877	18	9.84 550	26	10.15 450	9.91 328	9	59
2	9.75 895	18	9.84 576	27	10.15 424	9.91 319	9	58
3	9.75 913	18	9.84 603	27	10.15 397	9.91 310	9	57
4	9.75 931	18	9.84 630	27	10.15 370	9.91 301	9	56
5	9.75 949	18	9.84 657	27	10.15 343	9.91 292	9	55
6	9.75 967	18	9.84 684	27	10.15 316	9.91 283	9	54
7	9.75 985	18	9.84 711	27	10.15 289	9.91 274	8	53
8	9.76 003	18	9.84 738	26	10.15 262	9.91 266	9	52
9	9.76 021	18	9.84 764	27	10.15 236	9.91 257	9	51
10	9.76 039	18	9.84 791	27	10.15 209	9.91 248	9	50
11	9.76 057	18	9.84 818	27	10.15 182	9.91 239	9	49
12	9.76 075	18	9.84 845	27	10.15 155	9.91 230	9	48
13	9.76 093	18	9.84 872	27	10.15 128	9.91 221	9	47
14	9.76 111	18	9.84 899	26	10.15 101	9.91 212	9	46
15	9.76 129	17	9.84 925	27	10.15 075	9.91 203	9	45
16	9.76 146	18	9.84 952	27	10.15 048	9.91 194	9	44
17	9.76 164	18	9.84 979	27	10.15 021	9.91 185	9	43
18	9.76 182	18	9.85 006	27	10.14 994	9.91 176	9	42
19	9.76 200	18	9.85 033	26	10.14 967	9.91 167	9	41
20	9.76 218	18	9.85 059	27	10.14 941	9.91 158	9	40
21	9.76 236	17	9.85 086	27	10.14 914	9.91 149	8	39
22	9.76 253	18	9.85 113	27	10.14 887	9.91 141	9	38
23	9.76 271	18	9.85 140	26	10.14 860	9.91 132	9	37
24	9.76 289	18	9.85 166	27	10.14 834	9.91 123	9	36
25	9.76 307	17	9.85 193	27	10.14 807	9.91 114	9	35
26	9.76 324	18	9.85 220	27	10.14 780	9.91 105	9	34
27	9.76 342	18	9.85 247	26	10.14 753	9.91 096	9	33
28	9.76 360	18	9.85 273	27	10.14 727	9.91 087	9	32
29	9.76 378	17	9.85 300	27	10.14 700	9.91 078	9	31
30	9.76 395	18	9.85 327	27	10.14 673	9.91 069	9	30
31	9.76 413	18	9.85 354	26	10.14 646	9.91 060	9	29
32	9.76 431	17	9.85 380	27	10.14 620	9.91 051	9	28
33	9.76 448	18	9.85 407	27	10.14 593	9.91 042	9	27
34	9.76 466	18	9.85 434	26	10.14 566	9.91 033	10	26
35	9.76 484	17	9.85 460	27	10.14 540	9.91 023	9	25
36	9.76 501	18	9.85 487	27	10.14 513	9.91 014	9	24
37	9.76 519	18	9.85 514	26	10.14 486	9.91 005	9	23
38	9.76 537	17	9.85 540	27	10.14 460	9.90 996	9	22
39	9.76 554	18	9.85 567	27	10.14 433	9.90 987	9	21
40	9.76 572	18	9.85 594	26	10.14 406	9.90 978	9	20
41	9.76 590	17	9.85 620	27	10.14 380	9.90 969	9	19
42	9.76 607	18	9.85 647	27	10.14 353	9.90 960	9	18
43	9.76 625	17	9.85 674	26	10.14 326	9.90 951	9	17
44	9.76 642	18	9.85 700	27	10.14 300	9.90 942	9	16
45	9.76 660	17	9.85 727	27	10.14 273	9.90 933	9	15
46	9.76 677	18	9.85 754	26	10.14 246	9.90 924	9	14
47	9.76 695	17	9.85 780	27	10.14 220	9.90 915	9	13
48	9.76 712	18	9.85 807	27	10.14 193	9.90 906	10	12
49	9.76 730	17	9.85 834	26	10.14 166	9.90 896	9	11
50	9.76 747	18	9.85 860	27	10.14 140	9.90 887	9	10
51	9.76 765	17	9.85 887	26	10.14 113	9.90 878	9	9
52	9.76 782	18	9.85 913	27	10.14 087	9.90 869	9	8
53	9.76 800	17	9.85 940	27	10.14 060	9.90 860	9	7
54	9.76 817	18	9.85 967	26	10.14 033	9.90 851	9	6
55	9.76 835	17	9.85 993	27	10.14 007	9.90 842	10	5
56	9.76 852	18	9.86 020	26	10.13 980	9.90 832	9	4
57	9.76 870	17	9.86 046	27	10.13 954	9.90 823	9	3
58	9.76 887	17	9.86 073	27	10.13 927	9.90 814	9	2
59	9.76 904	18	9.86 100	26	10.13 900	9.90 805	9	1
60	9.76 922		9.86 126		10.13 874	9.90 796		0
′	L Cos	d	L Ctn	c d	L Tan	L Sin	d	′

Proportional parts

″	27	26	18
1	0.4	0.4	0.3
2	0.9	0.9	0.6
3	1.4	1.3	0.9
4	1.8	1.7	1.2
5	2.2	2.2	1.5
6	2.7	2.6	1.8
7	3.2	3.0	2.1
8	3.6	3.5	2.4
9	4.0	3.9	2.7
10	4.5	4.3	3.0
20	9.0	8.7	6.0
30	13.5	13.0	9.0
40	18.0	17.3	12.0
50	22.5	21.7	15.0

″	17	10
1	0.3	0.2
2	0.6	0.3
3	0.8	0.5
4	1.1	0.7
5	1.4	0.8
6	1.7	1.0
7	2.0	1.2
8	2.3	1.3
9	2.6	1.5
10	2.8	1.7
20	5.7	3.3
30	8.5	5.0
40	11.3	6.7
50	14.2	8.3

″	9	8
1	0.2	0.1
2	0.3	0.3
3	0.4	0.4
4	0.6	0.5
5	0.8	0.7
6	0.9	0.8
7	1.0	0.9
8	1.2	1.1
9	1.4	1.2
10	1.5	1.3
20	3.0	2.7
30	4.5	4.0
40	6.0	5.3
50	7.5	6.7

Proportional parts

125° (305°) (234°) **54°**

The −10 portion of the characteristic of the logarithm is not printed but must be written down whenever such a logarithm is used.

36° (216°) (323°) **143°**

′	L Sin	d	L Tan	c d	L Ctn	L Cos	d	′
0	9.76 922	17	9.86 126	27	10.13 874	9.90 796	9	60
1	9.76 939	18	9.86 153	26	10.13 847	9.90 787	10	59
2	9.76 957	17	9.86 179	27	10.13 821	9.90 777	9	58
3	9.76 974	17	9.86 206	26	10.13 794	9.90 768	9	57
4	9.76 991	18	9.86 232	27	10.13 768	9.90 759	9	56
5	9.77 009	17	9.86 259	26	10.13 741	9.90 750	9	55
6	9.77 026	17	9.86 285	27	10.13 715	9.90 741	10	54
7	9.77 043	18	9.86 312	26	10.13 688	9.90 731	9	53
8	9.77 061	17	9.86 338	27	10.13 662	9.90 722	9	52
9	9.77 078	17	9.86 365	27	10.13 635	9.90 713	9	51
10	9.77 095	17	9.86 392	26	10.13 608	9.90 704	10	50
11	9.77 112	18	9.86 418	27	10.13 582	9.90 694	9	49
12	9.77 130	17	9.86 445	26	10.13 555	9.90 685	9	48
13	9.77 147	17	9.86 471	27	10.13 529	9.90 676	9	47
14	9.77 164	17	9.86 498	26	10.13 502	9.90 667	10	46
15	9.77 181	18	9.86 524	27	10.13 476	9.90 657	9	45
16	9.77 199	17	9.86 551	26	10.13 449	9.90 648	9	44
17	9.77 216	17	9.86 577	26	10.13 423	9.90 639	9	43
18	9.77 233	17	9.86 603	27	10.13 397	9.90 630	10	42
19	9.77 250	18	9.86 630	26	10.13 370	9.90 620	9	41
20	9.77 268	17	9.86 656	27	10.13 344	9.90 611	9	40
21	9.77 285	17	9.86 683	26	10.13 317	9.90 602	10	39
22	9.77 302	17	9.86 709	27	10.13 291	9.90 592	9	38
23	9.77 319	17	9.86 736	26	10.13 264	9.90 583	9	37
24	9.77 336	17	9.86 762	27	10.13 238	9.90 574	9	36
25	9.77 353	17	9.86 789	26	10.13 211	9.90 565	10	35
26	9.77 370	17	9.86 815	27	10.13 185	9.90 555	9	34
27	9.77 387	18	9.86 842	26	10.13 158	9.90 546	9	33
28	9.77 405	17	9.86 868	26	10.13 132	9.90 537	10	32
29	9.77 422	17	9.86 894	27	10.13 106	9.90 527	9	31
30	9.77 439	17	9.86 921	26	10.13 079	9.90 518	9	30
31	9.77 456	17	9.86 947	27	10.13 053	9.90 509	10	29
32	9.77 473	17	9.86 974	26	10.13 026	9.90 499	9	28
33	9.77 490	17	9.87 000	27	10.13 000	9.90 490	10	27
34	9.77 507	17	9.87 027	26	10.12 973	9.90 480	9	26
35	9.77 524	17	9.87 053	26	10.12 947	9.90 471	9	25
36	9.77 541	17	9.87 079	27	10.12 921	9.90 462	10	24
37	9.77 558	17	9.87 106	26	10.12 894	9.90 452	9	23
38	9.77 575	17	9.87 132	26	10.12 868	9.90 443	9	22
39	9.77 592	17	9.87 158	27	10.12 842	9.90 434	10	21
40	9.77 609	17	9.87 185	26	10.12 815	9.90 424	9	20
41	9.77 626	17	9.87 211	27	10.12 789	9.90 415	10	19
42	9.77 643	17	9.87 238	26	10.12 762	9.90 405	9	18
43	9.77 660	17	9.87 264	26	10.12 736	9.90 396	10	17
44	9.77 677	17	9.87 290	27	10.12 710	9.90 386	9	16
45	9.77 694	17	9.87 317	26	10.12 683	9.90 377	9	15
46	9.77 711	17	9.87 343	26	10.12 657	9.90 368	10	14
47	9.77 728	16	9.87 369	27	10.12 631	9.90 358	9	13
48	9.77 744	17	9.87 396	26	10.12 604	9.90 349	10	12
49	9.77 761	17	9.87 422	26	10.12 578	9.90 339	9	11
50	9.77 778	17	9.87 448	27	10.12 552	9.90 330	10	10
51	9.77 795	17	9.87 475	26	10.12 525	9.90 320	9	9
52	9.77 812	17	9.87 501	26	10.12 499	9.90 311	10	8
53	9.77 829	17	9.87 527	27	10.12 473	9.90 301	9	7
54	9.77 846	16	9.87 554	26	10.12 446	9.90 292	10	6
55	9.77 862	17	9.87 580	26	10.12 420	9.90 282	9	5
56	9.77 879	17	9.87 606	27	10.12 394	9.90 273	10	4
57	9.77 896	17	9.87 633	26	10.12 367	9.90 263	9	3
58	9.77 913	17	9.87 659	26	10.12 341	9.90 254	10	2
59	9.77 930	16	9.87 685	26	10.12 315	9.90 244	9	1
60	9.77 946		9.87 711		10.12 289	9.90 235		0
′	L Cos	d	L Ctn	c d	L Tan	L Sin	d	′

126° (306°) (233°) **53°**

Proportional parts

″	27	26
1	0.4	0.4
2	0.9	0.9
3	1.4	1.3
4	1.8	1.7
5	2.2	2.2
6	2.7	2.6
7	3.2	3.0
8	3.6	3.5
9	4.0	3.9
10	4.5	4.3
20	9.0	8.7
30	13.5	13.0
40	18.0	17.3
50	22.5	21.7

″	18	17	16
1	0.3	0.3	0.3
2	0.6	0.6	0.5
3	0.9	0.8	0.8
4	1.2	1.1	1.1
5	1.5	1.4	1.3
6	1.8	1.7	1.6
7	2.1	2.0	1.9
8	2.4	2.3	2.1
9	2.7	2.6	2.4
10	3.0	2.8	2.7
20	6.0	5.7	5.3
30	9.0	8.5	8.0
40	12.0	11.3	10.7
50	15.0	14.2	13.3

″	10	9
1	0.2	0.2
2	0.3	0.3
3	0.5	0.4
4	0.7	0.6
5	0.8	0.8
6	1.0	0.9
7	1.2	1.0
8	1.3	1.2
9	1.5	1.4
10	1.7	1.5
20	3.3	3.0
30	5.0	4.5
40	6.7	6.0
50	8.3	7.5

Proportional parts

The −10 portion of the characteristic of the logarithm is not printed but must be written down whenever such a logarithm is used.

37° (217°) (322°) **142°**

′	L Sin	d	L Tan	c d	L Ctn	L Cos	d	′
0	9.77 946	17	9.87 711	27	10.12 289	9.90 235	10	60
1	9.77 963	17	9.87 738	26	10.12 262	9.90 225	9	59
2	9.77 980	17	9.87 764	26	10.12 236	9.90 216	10	58
3	9.77 997	16	9.87 790	27	10.12 210	9.90 206	9	57
4	9.78 013	17	9.87 817	26	10.12 183	9.90 197	10	56
5	9.78 030	17	9.87 843	26	10.12 157	9.90 187	9	55
6	9.78 047	16	9.87 869	26	10.12 131	9.90 178	10	54
7	9.78 063	17	9.87 895	27	10.12 105	9.90 168	9	53
8	9.78 080	17	9.87 922	26	10.12 078	9.90 159	10	52
9	9.78 097	16	9.87 948	26	10.12 052	9.90 149	10	51
10	9.78 113	17	9.87 974	26	10.12 026	9.90 139	9	50
11	9.78 130	17	9.88 000	27	10.12 000	9.90 130	10	49
12	9.78 147	16	9.88 027	26	10.11 973	9.90 120	9	48
13	9.78 163	17	9.88 053	26	10.11 947	9.90 111	10	47
14	9.78 180	17	9.88 079	26	10.11 921	9.90 101	10	46
15	9.78 197	16	9.88 105	26	10.11 895	9.90 091	9	45
16	9.78 213	17	9.88 131	27	10.11 869	9.90 082	10	44
17	9.78 230	16	9.88 158	26	10.11 842	9.90 072	9	43
18	9.78 246	17	9.88 184	26	10.11 816	9.90 063	10	42
19	9.78 263	17	9.88 210	26	10.11 790	9.90 053	10	41
20	9.78 280	16	9.88 236	26	10.11 764	9.90 043	9	40
21	9.78 296	17	9.88 262	27	10.11 738	9.90 034	10	39
22	9.78 313	16	9.88 289	26	10.11 711	9.90 024	10	38
23	9.78 329	17	9.88 315	26	10.11 685	9.90 014	9	37
24	9.78 346	16	9.88 341	26	10.11 659	9.90 005	10	36
25	9.78 362	17	9.88 367	26	10.11 633	9.89 995	10	35
26	9.78 379	16	9.88 393	27	10.11 607	9.89 985	9	34
27	9.78 395	17	9.88 420	26	10.11 580	9.89 976	10	33
28	9.78 412	16	9.88 446	26	10.11 554	9.89 966	10	32
29	9.78 428	17	9.88 472	26	10.11 528	9.89 956	9	31
30	9.78 445	16	9.88 498	26	10.11 502	9.89 947	10	30
31	9.78 461	17	9.88 524	26	10.11 476	9.89 937	10	29
32	9.78 478	16	9.88 550	27	10.11 450	9.89 927	9	28
33	9.78 494	16	9.88 577	26	10.11 423	9.89 918	10	27
34	9.78 510	17	9.88 603	26	10.11 397	9.89 908	10	26
35	9.78 527	16	9.88 629	26	10.11 371	9.89 898	10	25
36	9.78 543	17	9.88 655	26	10.11 345	9.89 888	9	24
37	9.78 560	16	9.88 681	26	10.11 319	9.89 879	10	23
38	9.78 576	16	9.88 707	26	10.11 293	9.89 869	10	22
39	9.78 592	17	9.88 733	26	10.11 267	9.89 859	10	21
40	9.78 609	16	9.88 759	27	10.11 241	9.89 849	9	20
41	9.78 625	17	9.88 786	26	10.11 214	9.89 840	10	19
42	9.78 642	16	9.88 812	26	10.11 188	9.89 830	10	18
43	9.78 658	16	9.88 838	26	10.11 162	9.89 820	10	17
44	9.78 674	17	9.88 864	26	10.11 136	9.89 810	9	16
45	9.78 691	16	9.88 890	26	10.11 110	9.89 801	10	15
46	9.78 707	16	9.88 916	26	10.11 084	9.89 791	10	14
47	9.78 723	16	9.88 942	26	10.11 058	9.89 781	10	13
48	9.78 739	17	9.88 968	26	10.11 032	9.89 771	10	12
49	9.78 756	16	9.88 994	26	10.11 006	9.89 761	9	11
50	9.78 772	16	9.89 020	26	10.10 980	9.89 752	10	10
51	9.78 788	17	9.89 046	27	10.10 954	9.89 742	10	9
52	9.78 805	16	9.89 073	26	10.10 927	9.89 732	10	8
53	9.78 821	16	9.89 099	26	10.10 901	9.89 722	10	7
54	9.78 837	16	9.89 125	26	10.10 875	9.89 712	10	6
55	9.78 853	16	9.89 151	26	10.10 849	9.89 702	9	5
56	9.78 869	17	9.89 177	26	10.10 823	9.89 693	10	4
57	9.78 886	16	9.89 203	26	10.10 797	9.89 683	10	3
58	9.78 902	16	9.89 229	26	10.10 771	9.89 673	10	2
59	9.78 918	16	9.89 255	26	10.10 745	9.89 663	10	1
60	9.78 934		9.89 281		10.10 719	9.89 653		0
′	L Cos	d	L Ctn	c d	L Tan	L Sin	d	′

Proportional parts

″	27	26
1	0.4	0.4
2	0.9	0.9
3	1.4	1.3
4	1.8	1.7
5	2.2	2.2
6	2.7	2.6
7	3.2	3.0
8	3.6	3.5
9	4.0	3.9
10	4.5	4.3
20	9.0	8.7
30	13.5	13.0
40	18.0	17.3
50	22.5	21.7

″	17	16
1	0.3	0.3
2	0.6	0.5
3	0.8	0.8
4	1.1	1.1
5	1.4	1.3
6	1.7	1.6
7	2.0	1.9
8	2.3	2.1
9	2.6	2.4
10	2.8	2.7
20	5.7	5.3
30	8.5	8.0
40	11.3	10.7
50	14.2	13.3

″	10	9
1	0.2	0.2
2	0.3	0.3
3	0.5	0.4
4	0.7	0.6
5	0.8	0.8
6	1.0	0.9
7	1.2	1.0
8	1.3	1.2
9	1.5	1.4
10	1.7	1.5
20	3.3	3.0
30	5.0	4.5
40	6.7	6.0
50	8.3	7.5

Proportional parts

127° (307°) (232°) **52°**

The −10 portion of the characteristic of the logarithm is not printed but must be written down whenever such a logarithm is used.

38° (218°) (321°) **141°**

′	L Sin	d	L Tan	c d	L Ctn	L Cos	d	′
0	9.78 934	16	9.89 281	26	10.10 719	9.89 653	10	60
1	9.78 950	17	9.89 307	26	10.10 693	9.89 643	10	59
2	9.78 967	16	9.89 333	26	10.10 667	9.89 633	9	58
3	9.78 983	16	9.89 359	26	10.10 641	9.89 624	10	57
4	9.78 999	16	9.89 385	26	10.10 615	9.89 614	10	56
5	9.79 015	16	9.89 411	26	10.10 589	9.89 604	10	55
6	9.79 031	16	9.89 437	26	10.10 563	9.89 594	10	54
7	9.79 047	16	9.89 463	26	10.10 537	9.89 584	10	53
8	9.79 063	16	9.89 489	26	10.10 511	9.89 574	10	52
9	9.79 079	16	9.89 515	26	10.10 485	9.89 564	10	51
10	9.79 095	16	9.89 541	26	10.10 459	9.89 554	10	50
11	9.79 111	17	9.89 567	26	10.10 433	9.89 544	10	49
12	9.79 128	16	9.89 593	26	10.10 407	9.89 534	10	48
13	9.79 144	16	9.89 619	26	10.10 381	9.89 524	10	47
14	9.79 160	16	9.89 645	26	10.10 355	9.89 514	10	46
15	9.79 176	16	9.89 671	26	10.10 329	9.89 504	9	45
16	9.79 192	16	9.89 697	26	10.10 303	9.89 495	10	44
17	9.79 208	16	9.89 723	26	10.10 277	9.89 485	10	43
18	9.79 224	16	9.89 749	26	10.10 251	9.89 475	10	42
19	9.79 240	16	9.89 775	26	10.10 225	9.89 465	10	41
20	9.79 256	16	9.89 801	26	10.10 199	9.89 455	10	40
21	9.79 272	16	9.89 827	26	10.10 173	9.89 445	10	39
22	9.79 288	16	9.89 853	26	10.10 147	9.89 435	10	38
23	9.79 304	15	9.89 879	26	10.10 121	9.89 425	10	37
24	9.79 319	16	9.89 905	26	10.10 095	9.89 415	10	36
25	9.79 335	16	9.89 931	26	10.10 069	9.89 405	10	35
26	9.79 351	16	9.89 957	26	10.10 043	9.89 395	10	34
27	9.79 367	16	9.89 983	26	10.10 017	9.89 385	10	33
28	9.79 383	16	9.90 009	26	10.09 991	9.89 375	11	32
29	9.79 399	16	9.90 035	26	10.09 965	9.89 364	10	31
30	9.79 415	16	9.90 061	25	10.09 939	9.89 354	10	30
31	9.79 431	16	9.90 086	26	10.09 914	9.89 344	10	29
32	9.79 447	16	9.90 112	26	10.09 888	9.89 334	10	28
33	9.79 463	15	9.90 138	26	10.09 862	9.89 324	10	27
34	9.79 478	16	9.90 164	26	10.09 836	9.89 314	10	26
35	9.79 494	16	9.90 190	26	10.09 810	9.89 304	10	25
36	9.79 510	16	9.90 216	26	10.09 784	9.89 294	10	24
37	9.79 526	16	9.90 242	26	10.09 758	9.89 284	10	23
38	9.79 542	16	9.90 268	26	10.09 732	9.89 274	10	22
39	9.79 558	15	9.90 294	26	10.09 706	9.89 264	10	21
40	9.79 573	16	9.90 320	26	10.09 680	9.89 254	10	20
41	9.79 589	16	9.90 346	25	10.09 654	9.89 244	11	19
42	9.79 605	16	9.90 371	26	10.09 629	9.89 233	10	18
43	9.79 621	15	9.90 397	26	10.09 603	9.89 223	10	17
44	9.79 636	16	9.90 423	26	10.09 577	9.89 213	10	16
45	9.79 652	16	9.90 449	26	10.09 551	9.89 203	10	15
46	9.79 668	16	9.90 475	26	10.09 525	9.89 193	10	14
47	9.79 684	15	9.90 501	26	10.09 499	9.89 183	10	13
48	9.79 699	16	9.90 527	26	10.09 473	9.89 173	11	12
49	9.79 715	16	9.90 553	25	10.09 447	9.89 162	10	11
50	9.79 731	15	9.90 578	26	10.09 422	9.89 152	10	10
51	9.79 746	16	9.90 604	26	10.09 396	9.89 142	10	9
52	9.79 762	16	9.90 630	26	10.09 370	9.89 132	10	8
53	9.79 778	15	9.90 656	26	10.09 344	9.89 122	10	7
54	9.79 793	16	9.90 682	26	10.09 318	9.89 112	11	6
55	9.79 809	16	9.90 708	26	10.09 292	9.89 101	10	5
56	9.79 825	15	9.90 734	25	10.09 266	9.89 091	10	4
57	9.79 840	16	9.90 759	26	10.09 241	9.89 081	10	3
58	9.79 856	16	9.90 785	26	10.09 215	9.89 071	11	2
59	9.79 872	15	9.90 811	26	10.09 189	9.89 060	10	1
60	9.79 887		9.90 837		10.09 163	9.89 050		0
′	L Cos	d	L Ctn	c d	L Tan	L Sin	d	′

Proportional parts

″	26	25
1	0.4	0.4
2	0.9	0.8
3	1.3	1.2
4	1.7	1.7
5	2.2	2.1
6	2.6	2.5
7	3.0	2.9
8	3.5	3.3
9	3.9	3.8
10	4.3	4.2
20	8.7	8.3
30	13.0	12.5
40	17.3	16.7
50	21.7	20.8

″	17	16	15
1	0.3	0.3	0.2
2	0.6	0.5	0.5
3	0.8	0.8	0.8
4	1.1	1.1	1.0
5	1.4	1.3	1.2
6	1.7	1.6	1.5
7	2.0	1.9	1.8
8	2.3	2.1	2.0
9	2.6	2.4	2.2
10	2.8	2.7	2.5
20	5.7	5.3	5.0
30	8.5	8.0	7.5
40	11.3	10.7	10.0
50	14.2	13.3	12.5

″	11	10	9
1	0.2	0.2	0.2
2	0.4	0.3	0.3
3	0.6	0.5	0.4
4	0.7	0.7	0.6
5	0.9	0.8	0.8
6	1.1	1.0	0.9
7	1.3	1.2	1.0
8	1.5	1.3	1.2
9	1.6	1.5	1.4
10	1.8	1.7	1.5
20	3.7	3.3	3.0
30	5.5	5.0	4.5
40	7.3	6.7	6.0
50	9.2	8.3	7.5

Proportional parts

128° (308°) (231°) **51°**

The −10 portion of the characteristic of the logarithm is not printed but must be written down whenever such a logarithm is used.

39° (219°) (320°) **140°**

′	L Sin	d	L Tan	c d	L Ctn	L Cos	d	′
0	9.79 887	16	9.90 837	26	10.09 163	9.89 050	10	60
1	9.79 903	15	9.90 863	26	10.09 137	9.89 040	10	59
2	9.79 918	16	9.90 889	25	10.09 111	9.89 030	10	58
3	9.79 934	16	9.90 914	26	10.09 086	9.89 020	11	57
4	9.79 950	15	9.90 940	26	10.09 060	9.89 009	10	56
5	9.79 965	16	9.90 966	26	10.09 034	9.88 999	10	55
6	9.79 981	15	9.90 992	26	10.09 008	9.88 989	11	54
7	9.79 996	16	9.91 018	25	10.08 982	9.88 978	10	53
8	9.80 012	15	9.91 043	26	10.08 957	9.88 968	10	52
9	9.80 027	16	9.91 069	26	10.08 931	9.88 958	10	51
10	9.80 043	15	9.91 095	26	10.08 905	9.88 948	11	50
11	9.80 058	16	9.91 121	26	10.08 879	9.88 937	10	49
12	9.80 074	15	9.91 147	25	10.08 853	9.88 927	10	48
13	9.80 089	16	9.91 172	26	10.08 828	9.88 917	11	47
14	9.80 105	15	9.91 198	26	10.08 802	9.88 906	10	46
15	9.80 120	16	9.91 224	26	10.08 776	9.88 896	10	45
16	9.80 136	15	9.91 250	26	10.08 750	9.88 886	11	44
17	9.80 151	15	9.91 276	25	10.08 724	9.88 875	10	43
18	9.80 166	16	9.91 301	26	10.08 699	9.88 865	10	42
19	9.80 182	15	9.91 327	26	10.08 673	9.88 855	11	41
20	9.80 197	16	9.91 353	26	10.08 647	9.88 844	10	40
21	9.80 213	15	9.91 379	25	10.08 621	9.88 834	10	39
22	9.80 228	16	9.91 404	26	10.08 596	9.88 824	11	38
23	9.80 244	15	9.91 430	26	10.08 570	9.88 813	10	37
24	9.80 259	15	9.91 456	26	10.08 544	9.88 803	10	36
25	9.80 274	16	9.91 482	25	10.08 518	9.88 793	11	35
26	9.80 290	15	9.91 507	26	10.08 493	9.88 782	10	34
27	9.80 305	15	9.91 533	26	10.08 467	9.88 772	11	33
28	9.80 320	16	9.91 559	26	10.08 441	9.88 761	10	32
29	9.80 336	15	9.91 585	25	10.08 415	9.88 751	10	31
30	9.80 351	15	9.91 610	26	10.08 390	9.88 741	11	30
31	9.80 366	16	9.91 636	26	10.08 364	9.88 730	10	29
32	9.80 382	15	9.91 662	26	10.08 338	9.88 720	11	28
33	9.80 397	15	9.91 688	25	10.08 312	9.88 709	10	27
34	9.80 412	16	9.91 713	26	10.08 287	9.88 699	11	26
35	9.80 428	15	9.91 739	26	10.08 261	9.88 688	10	25
36	9.80 443	15	9.91 765	26	10.08 235	9.88 678	10	24
37	9.80 458	15	9.91 791	25	10.08 209	9.88 668	11	23
38	9.80 473	16	9.91 816	26	10.08 184	9.88 657	10	22
39	9.80 489	15	9.91 842	26	10.08 158	9.88 647	11	21
40	9.80 504	15	9.91 868	25	10.08 132	9.88 636	10	20
41	9.80 519	15	9.91 893	26	10.08 107	9.88 626	11	19
42	9.80 534	16	9.91 919	26	10.08 081	9.88 615	10	18
43	9.80 550	15	9.91 945	26	10.08 055	9.88 605	11	17
44	9.80 565	15	9.91 971	25	10.08 029	9.88 594	10	16
45	9.80 580	15	9.91 996	26	10.08 004	9.88 584	11	15
46	9.80 595	15	9.92 022	26	10.07 978	9.88 573	10	14
47	9.80 610	15	9.92 048	25	10.07 952	9.88 563	11	13
48	9.80 625	16	9.92 073	26	10.07 927	9.88 552	10	12
49	9.80 641	15	9.92 099	26	10.07 901	9.88 542	11	11
50	9.80 656	15	9.92 125	25	10.07 875	9.88 531	10	10
51	9.80 671	15	9.92 150	26	10.07 850	9.88 521	11	9
52	9.80 686	15	9.92 176	26	10.07 824	9.88 510	11	8
53	9.80 701	15	9.92 202	25	10.07 798	9.88 499	10	7
54	9.80 716	15	9.92 227	26	10.07 773	9.88 489	11	6
55	9.80 731	15	9.92 253	26	10.07 747	9.88 478	10	5
56	9.80 746	16	9.92 279	25	10.07 721	9.88 468	11	4
57	9.80 762	15	9.92 304	26	10.07 696	9.88 457	10	3
58	9.80 777	15	9.92 330	26	10.07 670	9.88 447	11	2
59	9.80 792	15	9.92 356	25	10.07 644	9.88 436	11	1
60	9.80 807		9.92 381		10.07 619	9.88 425		0
′	L Cos	d	L Ctn	c d	L Tan	L Sin	d	′

129° (309°) (230°) **50°**

Proportional parts

″	26	25
1	0.4	0.4
2	0.9	0.8
3	1.3	1.2
4	1.7	1.7
5	2.2	2.1
6	2.6	2.5
7	3.0	2.9
8	3.5	3.3
9	3.9	3.8
10	4.3	4.2
20	8.7	8.3
30	13.0	12.5
40	17.3	16.7
50	21.7	20.8

″	16	15
1	0.3	0.2
2	0.5	0.5
3	0.8	0.8
4	1.1	1.0
5	1.3	1.2
6	1.6	1.5
7	1.9	1.8
8	2.1	2.0
9	2.4	2.2
10	2.7	2.5
20	5.3	5.0
30	8.0	7.5
40	10.7	10.0
50	13.3	12.5

″	11	10
1	0.2	0.2
2	0.4	0.3
3	0.6	0.5
4	0.7	0.7
5	0.9	0.8
6	1.1	1.0
7	1.3	1.2
8	1.5	1.3
9	1.6	1.5
10	1.8	1.7
20	3.7	3.3
30	5.5	5.0
40	7.3	6.7
50	9.2	8.3

Proportional parts

The −10 portion of the characteristic of the logarithm is not printed but must be written down whenever such a logarithm is used.

40° (220°) (319°) **139°**

′	L Sin	d	L Tan	c d	L Ctn	L Cos	d	′
0	9.80 807	15	9.92 381	26	10.07 619	9.88 425	10	60
1	9.80 822	15	9.92 407	26	10.07 593	9.88 415	11	59
2	9.80 837	15	9.92 433	25	10.07 567	9.88 404	10	58
3	9.80 852	15	9.92 458	26	10.07 542	9.88 394	11	57
4	9.80 867	15	9.92 484	26	10.07 516	9.88 383	11	56
5	9.80 882	15	9.92 510	25	10.07 490	9.88 372	10	55
6	9.80 897	15	9.92 535	26	10.07 465	9.88 362	11	54
7	9.80 912	15	9.92 561	26	10.07 439	9.88 351	11	53
8	9.80 927	15	9.92 587	25	10.07 413	9.88 340	10	52
9	9.80 942	15	9.92 612	26	10.07 388	9.88 330	11	51
10	9.80 957	15	9.92 638	25	10.07 362	9.88 319	11	50
11	9.80 972	15	9.92 663	26	10.07 337	9.88 308	10	49
12	9.80 987	15	9.92 689	26	10.07 311	9.88 298	11	48
13	9.81 002	15	9.92 715	25	10.07 285	9.88 287	11	47
14	9.81 017	15	9.92 740	26	10.07 260	9.88 276	10	46
15	9.81 032	15	9.92 766	26	10.07 234	9.88 266	11	45
16	9.81 047	14	9.92 792	25	10.07 208	9.88 255	11	44
17	9.81 061	15	9.92 817	26	10.07 183	9.88 244	10	43
18	9.81 076	15	9.92 843	25	10.07 157	9.88 234	11	42
19	9.81 091	15	9.92 868	26	10.07 132	9.88 223	11	41
20	9.81 106	15	9.92 894	26	10.07 106	9.88 212	11	40
21	9.81 121	15	9.92 920	25	10.07 080	9.88 201	10	39
22	9.81 136	15	9.92 945	26	10.07 055	9.88 191	11	38
23	9.81 151	15	9.92 971	25	10.07 029	9.88 180	11	37
24	9.81 166	14	9.92 996	26	10.07 004	9.88 169	11	36
25	9.81 180	15	9.93 022	26	10.06 978	9.88 158	10	35
26	9.81 195	15	9.93 048	25	10.06 952	9.88 148	11	34
27	9.81 210	15	9.93 073	26	10.06 927	9.88 137	11	33
28	9.81 225	15	9.93 099	25	10.06 901	9.88 126	11	32
29	9.81 240	14	9.93 124	26	10.06 876	9.88 115	10	31
30	9.81 254	15	9.93 150	25	10.06 850	9.88 105	11	30
31	9.81 269	15	9.93 175	26	10.06 825	9.88 094	11	29
32	9.81 284	15	9.93 201	26	10.06 799	9.88 083	11	28
33	9.81 299	15	9.93 227	25	10.06 773	9.88 072	11	27
34	9.81 314	14	9.93 252	26	10.06 748	9.88 061	10	26
35	9.81 328	15	9.93 278	25	10.06 722	9.88 051	11	25
36	9.81 343	15	9.93 303	26	10.06 697	9.88 040	11	24
37	9.81 358	14	9.93 329	25	10.06 671	9.88 029	11	23
38	9.81 372	15	9.93 354	26	10.06 646	9.88 018	11	22
39	9.81 387	15	9.93 380	26	10.06 620	9.88 007	11	21
40	9.81 402	15	9.93 406	25	10.06 594	9.87 996	11	20
41	9.81 417	14	9.93 431	26	10.06 569	9.87 985	10	19
42	9.81 431	15	9.93 457	25	10.06 543	9.87 975	11	18
43	9.81 446	15	9.93 482	26	10.06 518	9.87 964	11	17
44	9.81 461	14	9.93 508	25	10.06 492	9.87 953	11	16
45	9.81 475	15	9.93 533	26	10.06 467	9.87 942	11	15
46	9.81 490	15	9.93 559	25	10.06 441	9.87 931	11	14
47	9.81 505	14	9.93 584	26	10.06 416	9.87 920	11	13
48	9.81 519	15	9.93 610	26	10.06 390	9.87 909	11	12
49	9.81 534	15	9.93 636	25	10.06 364	9.87 898	11	11
50	9.81 549	14	9.93 661	26	10.06 339	9.87 887	10	10
51	9.81 563	15	9.93 687	25	10.06 313	9.87 877	11	9
52	9.81 578	14	9.93 712	26	10.06 288	9.87 866	11	8
53	9.81 592	15	9.93 738	25	10.06 262	9.87 855	11	7
54	9.81 607	15	9.93 763	26	10.06 237	9.87 844	11	6
55	9.81 622	14	9.93 789	25	10.06 211	9.87 833	11	5
56	9.81 636	15	9.93 814	26	10.06 186	9.87 822	11	4
57	9.81 651	14	9.93 840	25	10.06 160	9.87 811	11	3
58	9.81 665	15	9.93 865	26	10.06 135	9.87 800	11	2
59	9.81 680	14	9.93 891	25	10.06 109	9.87 789	11	1
60	9.81 694		9.93 916		10.06 084	9.87 778		0
′	L Cos	d	L Ctn	c d	L Tan	L Sin	d	′

Proportional parts

″	26	25
1	0.4	0.4
2	0.9	0.8
3	1.3	1.2
4	1.7	1.7
5	2.2	2.1
6	2.6	2.5
7	3.0	2.9
8	3.5	3.3
9	3.9	3.8
10	4.3	4.2
20	8.7	8.3
30	13.0	12.5
40	17.3	16.7
50	21.7	20.8

″	15	14
1	0.2	0.2
2	0.5	0.5
3	0.8	0.7
4	1.0	0.9
5	1.2	1.2
6	1.5	1.4
7	1.8	1.6
8	2.0	1.9
9	2.2	2.1
10	2.5	2.3
20	5.0	4.7
30	7.5	7.0
40	10.0	9.3
50	12.5	11.7

″	11	10
1	0.2	0.2
2	0.4	0.3
3	0.6	0.5
4	0.7	0.7
5	0.9	0.8
6	1.1	1.0
7	1.3	1.2
8	1.5	1.3
9	1.6	1.5
10	1.8	1.7
20	3.7	3.3
30	5.5	5.0
40	7.3	6.7
50	9.2	8.3

Proportional parts

130° (310°) (229°) **49°**

The −10 portion of the characteristic of the logarithm is not printed but must be written down whenever such a logarithm is used.

41° (221°) (318°) **138°**

′	L Sin	d	L Tan	c d	L Ctn	L Cos	d	′
0	9.81 694	15	9.93 916	26	10.06 084	9.87 778	11	60
1	9.81 709	14	9.93 942	25	10.06 058	9.87 767	11	59
2	9.81 723	15	9.93 967	26	10.06 033	9.87 756	11	58
3	9.81 738	14	9.93 993	25	10.06 007	9.87 745	11	57
4	9.81 752	15	9.94 018	26	10.05 982	9.87 734	11	56
5	9.81 767	14	9.94 044	25	10.05 956	9.87 723	11	55
6	9.81 781	15	9.94 069	26	10.05 931	9.87 712	11	54
7	9.81 796	14	9.94 095	25	10.05 905	9.87 701	11	53
8	9.81 810	15	9.94 120	26	10.05 880	9.87 690	11	52
9	9.81 825	14	9.94 146	25	10.05 854	9.87 679	11	51
10	9.81 839	15	9.94 171	26	10.05 829	9.87 668	11	50
11	9.81 854	14	9.94 197	25	10.05 803	9.87 657	11	49
12	9.81 868	14	9.94 222	26	10.05 778	9.87 646	11	48
13	9.81 882	15	9.94 248	25	10.05 752	9.87 635	11	47
14	9.81 897	14	9.94 273	26	10.05 727	9.87 624	11	46
15	9.81 911	15	9.94 299	25	10.05 701	9.87 613	12	45
16	9.81 926	14	9.94 324	26	10.05 676	9.87 601	11	44
17	9.81 940	15	9.94 350	25	10.05 650	9.87 590	11	43
18	9.81 955	14	9.94 375	26	10.05 625	9.87 579	11	42
19	9.81 969	14	9.94 401	25	10.05 599	9.87 568	11	41
20	9.81 983	15	9.94 426	26	10.05 574	9.87 557	11	40
21	9.81 998	14	9.94 452	25	10.05 548	9.87 546	11	39
22	9.82 012	14	9.94 477	26	10.05 523	9.87 535	11	38
23	9.82 026	15	9.94 503	25	10.05 497	9.87 524	11	37
24	9.82 041	14	9.94 528	26	10.05 472	9.87 513	12	36
25	9.82 055	14	9.94 554	25	10.05 446	9.87 501	11	35
26	9.82 069	15	9.94 579	25	10.05 421	9.87 490	11	34
27	9.82 084	14	9.94 604	26	10.05 396	9.87 479	11	33
28	9.82 098	14	9.94 630	25	10.05 370	9.87 468	11	32
29	9.82 112	14	9.94 655	26	10.05 345	9.87 457	11	31
30	9.82 126	15	9.94 681	25	10.05 319	9.87 446	12	30
31	9.82 141	14	9.94 706	26	10.05 294	9.87 434	11	29
32	9.82 155	14	9.94 732	25	10.05 268	9.87 423	11	28
33	9.82 169	15	9.94 757	26	10.05 243	9.87 412	11	27
34	9.82 184	14	9.94 783	25	10.05 217	9.87 401	11	26
35	9.82 198	14	9.94 808	26	10.05 192	9.87 390	12	25
36	9.82 212	14	9.94 834	25	10.05 166	9.87 378	11	24
37	9.82 226	14	9.94 859	25	10.05 141	9.87 367	11	23
38	9.82 240	15	9.94 884	26	10.05 116	9.87 356	11	22
39	9.82 255	14	9.94 910	25	10.05 090	9.87 345	11	21
40	9.82 269	14	9.94 935	26	10.05 065	9.87 334	12	20
41	9.82 283	14	9.94 961	25	10.05 039	9.87 322	11	19
42	9.82 297	14	9.94 986	26	10.05 014	9.87 311	11	18
43	9.82 311	15	9.95 012	25	10.04 988	9.87 300	12	17
44	9.82 326	14	9.95 037	25	10.04 963	9.87 288	11	16
45	9.82 340	14	9.95 062	26	10.04 938	9.87 277	11	15
46	9.82 354	14	9.95 088	25	10.04 912	9.87 266	11	14
47	9.82 368	14	9.95 113	26	10.04 887	9.87 255	12	13
48	9.82 382	14	9.95 139	25	10.04 861	9.87 243	11	12
49	9.82 396	14	9.95 164	26	10.04 836	9.87 232	11	11
50	9.82 410	14	9.95 190	25	10.04 810	9.87 221	12	10
51	9.82 424	15	9.95 215	25	10.04 785	9.87 209	11	9
52	9.82 439	14	9.95 240	26	10.04 760	9.87 198	11	8
53	9.82 453	14	9.95 266	25	10.04 734	9.87 187	12	7
54	9.82 467	14	9.95 291	26	10.04 709	9.87 175	11	6
55	9.82 481	14	9.95 317	25	10.04 683	9.87 164	11	5
56	9.82 495	14	9.95 342	26	10.04 658	9.87 153	12	4
57	9.82 509	14	9.95 368	25	10.04 632	9.87 141	11	3
58	9.82 523	14	9.95 393	25	10.04 607	9.87 130	11	2
59	9.82 537	14	9.95 418	26	10.04 582	9.87 119	12	1
60	9.82 551		9.95 444		10.04 556	9.87 107		0
′	L Cos	d	L Ctn	c d	L Tan	L Sin	d	′

131° (311°) (228°) **48°**

Proportional parts

″	26	25
1	0.4	0.4
2	0.9	0.8
3	1.3	1.2
4	1.7	1.7
5	2.2	2.1
6	2.6	2.5
7	3.0	2.9
8	3.5	3.3
9	3.9	3.8
10	4.3	4.2
20	8.7	8.3
30	13.0	12.5
40	17.3	16.7
50	21.7	20.8

″	15	14
1	0.2	0.2
2	0.5	0.5
3	0.8	0.7
4	1.0	0.9
5	1.2	1.2
6	1.5	1.4
7	1.8	1.6
8	2.0	1.9
9	2.2	2.1
10	2.5	2.3
20	5.0	4.7
30	7.5	7.0
40	10.0	9.3
50	12.5	11.7

″	12	11
1	0.2	0.2
2	0.4	0.4
3	0.6	0.6
4	0.8	0.7
5	1.0	0.9
6	1.2	1.1
7	1.4	1.3
8	1.6	1.5
9	1.8	1.6
10	2.0	1.8
20	4.0	3.7
30	6.0	5.5
40	8.0	7.3
50	10.0	9.2

Proportional parts

The −10 portion of the characteristic of the logarithm is not printed but must be written down whenever such a logarithm is used.

42° (222°) (317°) **137°**

′	L Sin	d	L Tan	c d	L Ctn	L Cos	d	′
0	9.82 551	14	9.95 444	25	10.04 556	9.87 107	11	60
1	9.82 565	14	9.95 469	26	10.04 531	9.87 096	11	59
2	9.82 579	14	9.95 495	25	10.04 505	9.87 085	12	58
3	9.82 593	14	9.95 520	25	10.04 480	9.87 073	11	57
4	9.82 607	14	9.95 545	26	10.04 455	9.87 062	12	56
5	9.82 621	14	9.95 571	25	10.04 429	9.87 050	11	55
6	9.82 635	14	9.95 596	26	10.04 404	9.87 039	11	54
7	9.82 649	14	9.95 622	25	10.04 378	9.87 028	12	53
8	9.82 663	14	9.95 647	25	10.04 353	9.87 016	11	52
9	9.82 677	14	9.95 672	26	10.04 328	9.87 005	12	51
10	9.82 691	14	9.95 698	25	10.04 302	9.86 993	11	50
11	9.82 705	14	9.95 723	25	10.04 277	9.86 982	12	49
12	9.82 719	14	9.95 748	26	10.04 252	9.86 970	11	48
13	9.82 733	14	9.95 774	25	10.04 226	9.86 959	12	47
14	9.82 747	14	9.95 799	26	10.04 201	9.86 947	11	46
15	9.82 761	14	9.95 825	25	10.04 175	9.86 936	12	45
16	9.82 775	13	9.95 850	25	10.04 150	9.86 924	11	44
17	9.82 788	14	9.95 875	26	10.04 125	9.86 913	11	43
18	9.82 802	14	9.95 901	25	10.04 099	9.86 902	12	42
19	9.82 816	14	9.95 926	26	10.04 074	9.86 890	11	41
20	9.82 830	14	9.95 952	25	10.04 048	9.86 879	12	40
21	9.82 844	14	9.95 977	25	10.04 023	9.86 867	12	39
22	9.82 858	14	9.96 002	26	10.03 998	9.86 855	11	38
23	9.82 872	13	9.96 028	25	10.03 972	9.86 844	12	37
24	9.82 885	14	9.96 053	25	10.03 947	9.86 832	11	36
25	9.82 899	14	9.96 078	26	10.03 922	9.86 821	12	35
26	9.82 913	14	9.96 104	25	10.03 896	9.86 809	11	34
27	9.82 927	14	9.96 129	26	10.03 871	9.86 798	12	33
28	9.82 941	14	9.96 155	25	10.03 845	9.86 786	11	32
29	9.82 955	13	9.96 180	25	10.03 820	9.86 775	12	31
30	9.82 968	14	9.96 205	26	10.03 795	9.86 763	11	30
31	9.82 982	14	9.96 231	25	10.03 769	9.86 752	12	29
32	9.82 996	14	9.96 256	25	10.03 744	9.86 740	12	28
33	9.83 010	13	9.96 281	26	10.03 719	9.86 728	11	27
34	9.83 023	14	9.96 307	25	10.03 693	9.86 717	12	26
35	9.83 037	14	9.96 332	25	10.03 668	9.86 705	11	25
36	9.83 051	14	9.96 357	26	10.03 643	9.86 694	12	24
37	9.83 065	13	9.96 383	25	10.03 617	9.86 682	12	23
38	9.83 078	14	9.96 408	25	10.03 592	9.86 670	11	22
39	9.83 092	14	9.96 433	26	10.03 567	9.86 659	12	21
40	9.83 106	14	9.96 459	25	10.03 541	9.86 647	12	20
41	9.83 120	13	9.96 484	26	10.03 516	9.86 635	11	19
42	9.83 133	14	9.96 510	25	10.03 490	9.86 624	12	18
43	9.83 147	14	9.96 535	25	10.03 465	9.86 612	12	17
44	9.83 161	13	9.96 560	26	10.03 440	9.86 600	11	16
45	9.83 174	14	9.96 586	25	10.03 414	9.86 589	12	15
46	9.83 188	14	9.96 611	25	10.03 389	9.86 577	12	14
47	9.83 202	13	9.96 636	26	10.03 364	9.86 565	11	13
48	9.83 215	14	9.96 662	25	10.03 338	9.86 554	12	12
49	9.83 229	13	9.96 687	25	10.03 313	9.86 542	12	11
50	9.83 242	14	9.96 712	26	10.03 288	9.86 530	12	10
51	9.83 256	14	9.96 738	25	10.03 262	9.86 518	11	9
52	9.83 270	13	9.96 763	25	10.03 237	9.86 507	12	8
53	9.83 283	14	9.96 788	26	10.03 212	9.86 495	12	7
54	9.83 297	13	9.96 814	25	10.03 186	9.86 483	11	6
55	9.83 310	14	9.96 839	25	10.03 161	9.86 472	12	5
56	9.83 324	14	9.96 864	26	10.03 136	9.86 460	12	4
57	9.83 338	13	9.96 890	25	10.03 110	9.86 448	12	3
58	9.83 351	14	9.96 915	25	10.03 085	9.86 436	11	2
59	9.83 365	13	9.96 940	26	10.03 060	9.86 425	12	1
60	9.83 378		9.96 966		10.03 034	9.86 413		0
′	L Cos	d	L Ctn	c d	LTan	L Sin	d	′

132° (312°) (227°) **47°**

Proportional parts

″	26	25
1	0.4	0.4
2	0.9	0.8
3	1.3	1.2
4	1.7	1.7
5	2.2	2.1
6	2.6	2.5
7	3.0	2.9
8	3.5	3.3
9	3.9	3.8
10	4.3	4.2
20	8.7	8.3
30	13.0	12.5
40	17.3	16.7
50	21.7	20.8

″	14	13
1	0.2	0.2
2	0.5	0.4
3	0.7	0.6
4	0.9	0.9
5	1.2	1.1
6	1.4	1.3
7	1.6	1.5
8	1.9	1.7
9	2.1	2.0
10	2.3	2.2
20	4.7	4.3
30	7.0	6.5
40	9.3	8.7
50	11.7	10.8

″	12	11
1	0.2	0.2
2	0.4	0.4
3	0.6	0.6
4	0.8	0.7
5	1.0	0.9
6	1.2	1.1
7	1.4	1.3
8	1.6	1.5
9	1.8	1.6
10	2.0	1.8
20	4.0	3.7
30	6.0	5.5
40	8.0	7.3
50	10.0	9.2

Proportional parts

The −10 portion of the characteristic of the logarithm is not printed but must be written down whenever such a logarithm is used.

43° (223°) (316°) **136°**

′	L Sin	d	L Tan	c d	L Ctn	L Cos	d	′
0	9.83 378	14	9.96 966	25	10.03 034	9.86 413	12	60
1	9.83 392	13	9.96 991	25	10.03 009	9.86 401	12	59
2	9.83 405	14	9.97 016	26	10.02 984	9.86 389	12	58
3	9.83 419	13	9.97 042	25	10.02 958	9.86 377	11	57
4	9.83 432	14	9.97 067	25	10.02 933	9.86 366	12	56
5	9.83 446	13	9.97 092	26	10.02 908	9.86 354	12	55
6	9.83 459	14	9.97 118	25	10.02 882	9.86 342	12	54
7	9.83 473	13	9.97 143	25	10.02 857	9.86 330	12	53
8	9.83 486	14	9.97 168	25	10.02 832	9.86 318	12	52
9	9.83 500	13	9.97 193	26	10.02 807	9.86 306	11	51
10	9.83 513	14	9.97 219	25	10.02 781	9.86 295	12	50
11	9.83 527	13	9.97 244	25	10.02 756	9.86 283	12	49
12	9.83 540	14	9.97 269	26	10.02 731	9.86 271	12	48
13	9.83 554	13	9.97 295	25	10.02 705	9.86 259	12	47
14	9.83 567	14	9.97 320	25	10.02 680	9.86 247	12	46
15	9.83 581	13	9.97 345	26	10.02 655	9.86 235	12	45
16	9.83 594	14	9.97 371	25	10.02 629	9.86 223	12	44
17	9.83 608	13	9.97 396	25	10.02 604	9.86 211	11	43
18	9.83 621	13	9.97 421	26	10.02 579	9.86 200	12	42
19	9.83 634	14	9.97 447	25	10.02 553	9.86 188	12	41
20	9.83 648	13	9.97 472	25	10.02 528	9.86 176	12	40
21	9.83 661	13	9.97 497	26	10.02 503	9.86 164	12	39
22	9.83 674	14	9.97 523	25	10.02 477	9.86 152	12	38
23	9.83 688	13	9.97 548	25	10.02 452	9.86 140	12	37
24	9.83 701	14	9.97 573	25	10.02 427	9.86 128	12	36
25	9.83 715	13	9.97 598	26	10.02 402	9.86 116	12	35
26	9.83 728	13	9.97 624	25	10.02 376	9.86 104	12	34
27	9.83 741	14	9.97 649	25	10.02 351	9.86 092	12	33
28	9.83 755	13	9.97 674	26	10.02 326	9.86 080	12	32
29	9.83 768	13	9.97 700	25	10.02 300	9.86 068	12	31
30	9.83 781	14	9.97 725	25	10.02 275	9.86 056	12	30
31	9.83 795	13	9.97 750	26	10.02 250	9.86 044	12	29
32	9.83 808	13	9.97 776	25	10.02 224	9.86 032	12	28
33	9.83 821	13	9.97 801	25	10.02 199	9.86 020	12	27
34	9.83 834	14	9.97 826	25	10.02 174	9.86 008	12	26
35	9.83 848	13	9.97 851	26	10.02 149	9.85 996	12	25
36	9.83 861	13	9.97 877	25	10.02 123	9.85 984	12	24
37	9.83 874	13	9.97 902	25	10.02 098	9.85 972	12	23
38	9.83 887	14	9.97 927	26	10.02 073	9.85 960	12	22
39	9.83 901	13	9.97 953	25	10.02 047	9.85 948	12	21
40	9.83 914	13	9.97 978	25	10.02 022	9.85 936	12	20
41	9.83 927	13	9.98 003	26	10.01 997	9.85 924	12	19
42	9.83 940	14	9.98 029	25	10.01 971	9.85 912	12	18
43	9.83 954	13	9.98 054	25	10.01 946	9.85 900	12	17
44	9.83 967	13	9.98 079	25	10.01 921	9.85 888	12	16
45	9.83 980	13	9.98 104	26	10.01 896	9.85 876	12	15
46	9.83 993	13	9.98 130	25	10.01 870	9.85 864	13	14
47	9.84 006	14	9.98 155	25	10.01 845	9.85 851	12	13
48	9.84 020	13	9.98 180	26	10.01 820	9.85 839	12	12
49	9.84 033	13	9.98 206	25	10.01 794	9.85 827	12	11
50	9.84 046	13	9.98 231	25	10.01 769	9.85 815	12	10
51	9.84 059	13	9.98 256	25	10.01 744	9.85 803	12	9
52	9.84 072	13	9.98 281	26	10.01 719	9.85 791	12	8
53	9.84 085	13	9.98 307	25	10.01 693	9.85 779	13	7
54	9.84 098	14	9.98 332	25	10.01 668	9.85 766	12	6
55	9.84 112	13	9.98 357	26	10.01 643	9.85 754	12	5
56	9.84 125	13	9.98 383	25	10.01 617	9.85 742	12	4
57	9.84 138	13	9.98 408	25	10.01 592	9.85 730	12	3
58	9.84 151	13	9.98 433	25	10.01 567	9.85 718	12	2
59	9.84 164	13	9.98 458	26	10.01 542	9.85 706	13	1
60	9.84 177		9.98 484		10.01 516	9.85 693		0
′	L Cos	d	L Ctn	c d	L Tan	L Sin	d	′

133° (313°) (226°) **46°**

Proportional parts

″	26	25
1	0.4	0.4
2	0.9	0.8
3	1.3	1.2
4	1.7	1.7
5	2.2	2.1
6	2.6	2.5
7	3.0	2.9
8	3.5	3.3
9	3.9	3.8
10	4.3	4.2
20	8.7	8.3
30	13.0	12.5
40	17.3	16.7
50	21.7	20.8

″	14	13
1	0.2	0.2
2	0.5	0.4
3	0.7	0.6
4	0.9	0.9
5	1.2	1.1
6	1.4	1.3
7	1.6	1.5
8	1.9	1.7
9	2.1	2.0
10	2.3	2.2
20	4.7	4.3
30	7.0	6.5
40	9.3	8.7
50	11.7	10.8

″	12	11
1	0.2	0.2
2	0.4	0.4
3	0.6	0.6
4	0.8	0.7
5	1.0	0.9
6	1.2	1.1
7	1.4	1.3
8	1.6	1.5
9	1.8	1.6
10	2.0	1.8
20	4.0	3.7
30	6.0	5.5
40	8.0	7.3
50	10.0	9.2

Proportional parts

The −10 portion of the characteristic of the logarithm is not printed but must be written down whenever such a logarithm is used.

44° (224°) (315°) **135°**

′	L Sin	d	L Tan	c d	L Ctn	L Cos	d	′
0	9.84 177	13	9.98 484	25	10.01 516	9.85 693	12	60
1	9.84 190	13	9.98 509	25	10.01 491	9.85 681	12	59
2	9.84 203	13	9.98 534	26	10.01 466	9.85 669	12	58
3	9.84 216	13	9.98 560	25	10.01 440	9.85 657	12	57
4	9.84 229	13	9.98 585	25	10.01 415	9.85 645	13	56
5	9.84 242	13	9.98 610	25	10.01 390	9.85 632	12	55
6	9.84 255	14	9.98 635	26	10.01 365	9.85 620	12	54
7	9.84 269	13	9.98 661	25	10.01 339	9.85 608	12	53
8	9.84 282	13	9.98 686	25	10.01 314	9.85 596	13	52
9	9.84 295	13	9.98 711	26	10.01 289	9.85 583	12	51
10	9.84 308	13	9.98 737	25	10.01 263	9.85 571	12	50
11	9.84 321	13	9.98 762	25	10.01 238	9.85 559	12	49
12	9.84 334	13	9.98 787	25	10.01 213	9.85 547	13	48
13	9.84 347	13	9.98 812	26	10.01 188	9.85 534	12	47
14	9.84 360	13	9.98 838	25	10.01 162	9.85 522	12	46
15	9.84 373	12	9.98 863	25	10.01 137	9.85 510	13	45
16	9.84 385	13	9.98 888	25	10.01 112	9.85 497	12	44
17	9.84 398	13	9.98 913	26	10.01 087	9.85 485	12	43
18	9.84 411	13	9.98 939	25	10.01 061	9.85 473	13	42
19	9.84 424	13	9.98 964	25	10.01 036	9.85 460	12	41
20	9.84 437	13	9.98 989	26	10.01 011	9.85 448	12	40
21	9.84 450	13	9.99 015	25	10.00 985	9.85 436	13	39
22	9.84 463	13	9.99 040	25	10.00 960	9.85 423	12	38
23	9.84 476	13	9.99 065	25	10.00 935	9.85 411	12	37
24	9.84 489	13	9.99 090	26	10.00 910	9.85 399	13	36
25	9.84 502	13	9.99 116	25	10.00 884	9.85 386	12	35
26	9.84 515	13	9.99 141	25	10.00 859	9.85 374	13	34
27	9.84 528	12	9.99 166	25	10.00 834	9.85 361	12	33
28	9.84 540	13	9.99 191	26	10.00 809	9.85 349	12	32
29	9.84 553	13	9.99 217	25	10.00 783	9.85 337	13	31
30	9.84 566	13	9.99 242	25	10.00 758	9.85 324	12	30
31	9.84 579	13	9.99 267	26	10.00 733	9.85 312	13	29
32	9.84 592	13	9.99 293	25	10.00 707	9.85 299	12	28
33	9.84 605	13	9.99 318	25	10.00 682	9.85 287	13	27
34	9.84 618	12	9.99 343	25	10.00 657	9.85 274	12	26
35	9.84 630	13	9.99 368	26	10.00 632	9.85 262	12	25
36	9.84 643	13	9.99 394	25	10.00 606	9.85 250	13	24
37	9.84 656	13	9.99 419	25	10.00 581	9.85 237	12	23
38	9.84 669	13	9.99 444	25	10.00 556	9.85 225	13	22
39	9.84 682	12	9.99 469	26	10.00 531	9.85 212	12	21
40	9.84 694	13	9.99 495	25	10.00 505	9.85 200	13	20
41	9.84 707	13	9.99 520	25	10.00 480	9.85 187	12	19
42	9.84 720	13	9.99 545	25	10.00 455	9.85 175	13	18
43	9.84 733	12	9.99 570	26	10.00 430	9.85 162	12	17
44	9.84 745	13	9.99 596	25	10.00 404	9.85 150	13	16
45	9.84 758	13	9.99 621	25	10.00 379	9.85 137	12	15
46	9.84 771	13	9.99 646	26	10.00 354	9.85 125	13	14
47	9.84 784	12	9.99 672	25	10.00 328	9.85 112	12	13
48	9.84 796	13	9.99 697	25	10.00 303	9.85 100	13	12
49	9.84 809	13	9.99 722	25	10.00 278	9.85 087	13	11
50	9.84 822	13	9.99 747	26	10.00 253	9.85 074	12	10
51	9.84 835	12	9.99 773	25	10.00 227	9.85 062	13	9
52	9.84 847	13	9.99 798	25	10.00 202	9.85 049	12	8
53	9.84 860	13	9.99 823	25	10.00 177	9.85 037	13	7
54	9.84 873	12	9.99 848	26	10.00 152	9.85 024	12	6
55	9.84 885	13	9.99 874	25	10.00 126	9.85 012	13	5
56	9.84 898	13	9.99 899	25	10.00 101	9.84 999	13	4
57	9.84 911	12	9.99 924	25	10.00 076	9.84 986	12	3
58	9.84 923	13	9.99 949	26	10.00 051	9.84 974	13	2
59	9.84 936	13	9.99 975	25	10.00 025	9.84 961	12	1
60	9.84 949		10.00 000		10.00 000	9.84 949		0
′	L Cos	d	L Ctn	c d	L Tan	L Sin	d	′

134° (314°) (225°) **45°**

Proportional parts

″	26	25
1	0.4	0.4
2	0.9	0.8
3	1.3	1.2
4	1.7	1.7
5	2.2	2.1
6	2.6	2.5
7	3.0	2.9
8	3.5	3.3
9	3.9	3.8
10	4.3	4.2
20	8.7	8.3
30	13.0	12.5
40	17.3	16.7
50	21.7	20.8

″	14	13	12
1	0.2	0.2	0.2
2	0.5	0.4	0.4
3	0.7	0.6	0.6
4	0.9	0.9	0.8
5	1.2	1.1	1.0
6	1.4	1.3	1.2
7	1.6	1.5	1.4
8	1.9	1.7	1.6
9	2.1	2.0	1.8
10	2.3	2.2	2.0
20	4.7	4.3	4.0
30	7.0	6.5	6.0
40	9.3	8.7	8.0
50	11.7	10.8	10.0

Proportional parts

0° (180°) (359°) **179°**

′	Sin	Tan	Ctn	Cos	′
0	.00000	.00000	———	1.0000	60
1	.00029	.00029	3437.7	1.0000	59
2	.00058	.00058	1718.9	1.0000	58
3	.00087	.00087	1145.9	1.0000	57
4	.00116	.00116	859.44	1.0000	56
5	.00145	.00145	687.55	1.0000	55
6	.00175	.00175	572.96	1.0000	54
7	.00204	.00204	491.11	1.0000	53
8	.00233	.00233	429.72	1.0000	52
9	.00262	.00262	381.97	1.0000	51
10	.00291	.00291	343.77	1.0000	50
11	.00320	.00320	312.52	.99999	49
12	.00349	.00349	286.48	.99999	48
13	.00378	.00378	264.44	.99999	47
14	.00407	.00407	245.55	.99999	46
15	.00436	.00436	229.18	.99999	45
16	.00465	.00465	214.86	.99999	44
17	.00495	.00495	202.22	.99999	43
18	.00524	.00524	190.98	.99999	42
19	.00553	.00553	180.93	.99998	41
20	.00582	.00582	171.89	.99998	40
21	.00611	.00611	163.70	.99998	39
22	.00640	.00640	156.26	.99998	38
23	.00669	.00669	149.47	.99998	37
24	.00698	.00698	143.24	.99998	36
25	.00727	.00727	137.51	.99997	35
26	.00756	.00756	132.22	.99997	34
27	.00785	.00785	127.32	.99997	33
28	.00814	.00815	122.77	.99997	32
29	.00844	.00844	118.54	.99996	31
30	.00873	.00873	114.59	.99996	30
31	.00902	.00902	110.89	.99996	29
32	.00931	.00931	107.43	.99996	28
33	.00960	.00960	104.17	.99995	27
34	.00989	.00989	101.11	.99995	26
35	.01018	.01018	98.218	.99995	25
36	.01047	.01047	95.489	.99995	24
37	.01076	.01076	92.908	.99994	23
38	.01105	.01105	90.463	.99994	22
39	.01134	.01135	88.144	.99994	21
40	.01164	.01164	85.940	.99993	20
41	.01193	.01193	83.844	.99993	19
42	.01222	.01222	81.847	.99993	18
43	.01251	.01251	79.943	.99992	17
44	.01280	.01280	78.126	.99992	16
45	.01309	.01309	76.390	.99991	15
46	.01338	.01338	74.729	.99991	14
47	.01367	.01367	73.139	.99991	13
48	.01396	.01396	71.615	.99990	12
49	.01425	.01425	70.153	.99990	11
50	.01454	.01455	68.750	.99989	10
51	.01483	.01484	67.402	.99989	9
52	.01513	.01513	66.105	.99989	8
53	.01542	.01542	64.858	.99988	7
54	.01571	.01571	63.657	.99988	6
55	.01600	.01600	62.499	.99987	5
56	.01629	.01629	61.383	.99987	4
57	.01658	.01658	60.306	.99986	3
58	.01687	.01687	59.266	.99986	2
59	.01716	.01716	58.261	.99985	1
60	.01745	.01746	57.290	.99985	0
′	Cos	Ctn	Tan	Sin	′

90° (270°) (269°) **89°**

1° (181°) (358°) **178°**

′	Sin	Tan	Ctn	Cos	′
0	.01745	.01746	57.290	.99985	60
1	.01774	.01775	56.351	.99984	59
2	.01803	.01804	55.442	.99984	58
3	.01832	.01833	54.561	.99983	57
4	.01862	.01862	53.709	.99983	56
5	.01891	.01891	52.882	.99982	55
6	.01920	.01920	52.081	.99982	54
7	.01949	.01949	51.303	.99981	53
8	.01978	.01978	50.549	.99980	52
9	.02007	.02007	49.816	.99980	51
10	.02036	.02036	49.104	.99979	50
11	.02065	.02066	48.412	.99979	49
12	.02094	.02095	47.740	.99978	48
13	.02123	.02124	47.085	.99977	47
14	.02152	.02153	46.449	.99977	46
15	.02181	.02182	45.829	.99976	45
16	.02211	.02211	45.226	.99976	44
17	.02240	.02240	44.639	.99975	43
18	.02269	.02269	44.066	.99974	42
19	.02298	.02298	43.508	.99974	41
20	.02327	.02328	42.964	.99973	40
21	.02356	.02357	42.433	.99972	39
22	.02385	.02386	41.916	.99972	38
23	.02414	.02415	41.411	.99971	37
24	.02443	.02444	40.917	.99970	36
25	.02472	.02473	40.436	.99969	35
26	.02501	.02502	39.965	.99969	34
27	.02530	.02531	39.506	.99968	33
28	.02560	.02560	39.057	.99967	32
29	.02589	.02589	38.618	.99966	31
30	.02618	.02619	38.188	.99966	30
31	.02647	.02648	37.769	.99965	29
32	.02676	.02677	37.358	.99964	28
33	.02705	.02706	36.956	.99963	27
34	.02734	.02735	36.563	.99963	26
35	.02763	.02764	36.178	.99962	25
36	.02792	.02793	35.801	.99961	24
37	.02821	.02822	35.431	.99960	23
38	.02850	.02851	35.070	.99959	22
39	.02879	.02881	34.715	.99959	21
40	.02908	.02910	34.368	.99958	20
41	.02938	.02939	34.027	.99957	19
42	.02967	.02968	33.694	.99956	18
43	.02996	.02997	33.366	.99955	17
44	.03025	.03026	33.045	.99954	16
45	.03054	.03055	32.730	.99953	15
46	.03083	.03084	32.421	.99952	14
47	.03112	.03114	32.118	.99952	13
48	.03141	.03143	31.821	.99951	12
49	.03170	.03172	31.528	.99950	11
50	.03199	.03201	31.242	.99949	10
51	.03228	.03230	30.960	.99948	9
52	.03257	.03259	30.683	.99947	8
53	.03286	.03288	30.412	.99946	7
54	.03316	.03317	30.145	.99945	6
55	.03345	.03346	29.882	.99944	5
56	.03374	.03376	29.624	.99943	4
57	.03403	.03405	29.371	.99942	3
58	.03432	.03434	29.122	.99941	2
59	.03461	.03463	28.877	.99940	1
60	.03490	.03492	28.636	.99939	0
′	Cos	Ctn	Tan	Sin	′

91° (271°) (268°) **88°**

For degrees indicated at top (bottom) of page use column headings at top (bottom). With degrees at left (right) of each block (top or bottom), use minute column at left (right). The correct sign (plus or minus) must be prefixed in accordance with Sec. 54.

2° (182°) (357°) **177°**

′	Sin	Tan	Ctn	Cos	′
0	.03490	.03492	28.636	.99939	60
1	.03519	.03521	28.399	.99938	59
2	.03548	.03550	28.166	.99937	58
3	.03577	.03579	27.937	.99936	57
4	.03606	.03609	27.712	.99935	56
5	.03635	.03638	27.490	.99934	55
6	.03664	.03667	27.271	.99933	54
7	.03693	.03696	27.057	.99932	53
8	.03723	.03725	26.845	.99931	52
9	.03752	.03754	26.637	.99930	51
10	.03781	.03783	26.432	.99929	50
11	.03810	.03812	26.230	.99927	49
12	.03839	.03842	26.031	.99926	48
13	.03868	.03871	25.835	.99925	47
14	.03897	.03900	25.642	.99924	46
15	.03926	.03929	25.452	.99923	45
16	.03955	.03958	25.264	.99922	44
17	.03984	.03987	25.080	.99921	43
18	.04013	.04016	24.898	.99919	42
19	.04042	.04046	24.719	.99918	41
20	.04071	.04075	24.542	.99917	40
21	.04100	.04104	24.368	.99916	39
22	.04129	.04133	24.196	.99915	38
23	.04159	.04162	24.026	.99913	37
24	.04188	.04191	23.859	.99912	36
25	.04217	.04220	23.695	.99911	35
26	.04246	.04250	23.532	.99910	34
27	.04275	.04279	23.372	.99909	33
28	.04304	.04308	23.214	.99907	32
29	.04333	.04337	23.058	.99906	31
30	.04362	.04366	22.904	.99905	30
31	.04391	.04395	22.752	.99904	29
32	.04420	.04424	22.602	.99902	28
33	.04449	.04454	22.454	.99901	27
34	.04478	.04483	22.308	.99900	26
35	.04507	.04512	22.164	.99898	25
36	.04536	.04541	22.022	.99897	24
37	.04565	.04570	21.881	.99896	23
38	.04594	.04599	21.743	.99894	22
39	.04623	.04628	21.606	.99893	21
40	.04653	.04658	21.470	.99892	20
41	.04682	.04687	21.337	.99890	19
42	.04711	.04716	21.205	.99889	18
43	.04740	.04745	21.075	.99888	17
44	.04769	.04774	20.946	.99886	16
45	.04798	.04803	20.819	.99885	15
46	.04827	.04833	20.693	.99883	14
47	.04856	.04862	20.569	.99882	13
48	.04885	.04891	20.446	.99881	12
49	.04914	.04920	20.325	.99879	11
50	.04943	.04949	20.206	.99878	10
51	.04972	.04978	20.087	.99876	9
52	.05001	.05007	19.970	.99875	8
53	.05030	.05037	19.855	.99873	7
54	.05059	.05066	19.740	.99872	6
55	.05088	.05095	19.627	.99870	5
56	.05117	.05124	19.516	.99869	4
57	.05146	.05153	19.405	.99867	3
58	.05175	.05182	19.296	.99866	2
59	.05205	.05212	19.188	.99864	1
60	.05234	.05241	19.081	.99863	0
′	Cos	Ctn	Tan	Sin	′

92° (272°) (267°) **87°**

3° (183°) (356°) **176°**

′	Sin	Tan	Ctn	Cos	′
0	.05234	.05241	19.081	.99863	60
1	.05263	.05270	18.976	.99861	59
2	.05292	.05299	18.871	.99860	58
3	.05321	.05328	18.768	.99858	57
4	.05350	.05357	18.666	.99857	56
5	.05379	.05387	18.564	.99855	55
6	.05408	.05416	18.464	.99854	54
7	.05437	.05445	18.366	.99852	53
8	.05466	.05474	18.268	.99851	52
9	.05495	.05503	18.171	.99849	51
10	.05524	.05533	18.075	.99847	50
11	.05553	.05562	17.980	.99846	49
12	.05582	.05591	17.886	.99844	48
13	.05611	.05620	17.793	.99842	47
14	.05640	.05649	17.702	.99841	46
15	.05669	.05678	17.611	.99839	45
16	.05698	.05708	17.521	.99838	44
17	.05727	.05737	17.431	.99836	43
18	.05756	.05766	17.343	.99834	42
19	.05785	.05795	17.256	.99833	41
20	.05814	.05824	17.169	.99831	40
21	.05844	.05854	17.084	.99829	39
22	.05873	.05883	16.999	.99827	38
23	.05902	.05912	16.915	.99826	37
24	.05931	.05941	16.832	.99824	36
25	.05960	.05970	16.750	.99822	35
26	.05989	.05999	16.668	.99821	34
27	.06018	.06029	16.587	.99819	33
28	.06047	.06058	16.507	.99817	32
29	.06076	.06087	16.428	.99815	31
30	.06105	.06116	16.350	.99813	30
31	.06134	.06145	16.272	.99812	29
32	.06163	.06175	16.195	.99810	28
33	.06192	.06204	16.119	.99808	27
34	.06221	.06233	16.043	.99806	26
35	.06250	.06262	15.969	.99804	25
36	.06279	.06291	15.895	.99803	24
37	.06308	.06321	15.821	.99801	23
38	.06337	.06350	15.748	.99799	22
39	.06366	.06379	15.676	.99797	21
40	.06395	.06408	15.605	.99795	20
41	.06424	.06438	15.534	.99793	19
42	.06453	.06467	15.464	.99792	18
43	.06482	.06496	15.394	.99790	17
44	.06511	.06525	15.325	.99788	16
45	.06540	.06554	15.257	.99786	15
46	.06569	.06584	15.189	.99784	14
47	.06598	.06613	15.122	.99782	13
48	.06627	.06642	15.056	.99780	12
49	.06656	.06671	14.990	.99778	11
50	.06685	.06700	14.924	.99776	10
51	.06714	.06730	14.860	.99774	9
52	.06743	.06759	14.795	.99772	8
53	.06773	.06788	14.732	.99770	7
54	.06802	.06817	14.669	.99768	6
55	.06831	.06847	14.606	.99766	5
56	.06860	.06876	14.544	.99764	4
57	.06889	.06905	14.482	.99762	3
58	.06918	.06934	14.421	.99760	2
59	.06947	.06963	14.361	.99758	1
60	.06976	.06993	14.301	.99756	0
′	Cos	Ctn	Tan	Sin	′

93° (273°) (266°) **86°**

4° (184°) (355°) **175°**

′	Sin	Tan	Ctn	Cos	′
0	.06976	.06993	14.301	.99756	60
1	.07005	.07022	14.241	.99754	59
2	.07034	.07051	14.182	.99752	58
3	.07063	.07080	14.124	.99750	57
4	.07092	.07110	14.065	.99748	56
5	.07121	.07139	14.008	.99746	55
6	.07150	.07168	13.951	.99744	54
7	.07179	.07197	13.894	.99742	53
8	.07208	.07227	13.838	.99740	52
9	.07237	.07256	13.782	.99738	51
10	.07266	.07285	13.727	.99736	50
11	.07295	.07314	13.672	.99734	49
12	.07324	.07344	13.617	.99731	48
13	.07353	.07373	13.563	.99729	47
14	.07382	.07402	13.510	.99727	46
15	.07411	.07431	13.457	.99725	45
16	.07440	.07461	13.404	.99723	44
17	.07469	.07490	13.352	.99721	43
18	.07498	.07519	13.300	.99719	42
19	.07527	.07548	13.248	.99716	41
20	.07556	.07578	13.197	.99714	40
21	.07585	.07607	13.146	.99712	39
22	.07614	.07636	13.096	.99710	38
23	.07643	.07665	13.046	.99708	37
24	.07672	.07695	12.996	.99705	36
25	.07701	.07724	12.947	.99703	35
26	.07730	.07753	12.898	.99701	34
27	.07759	.07782	12.850	.99699	33
28	.07788	.07812	12.801	.99696	32
29	.07817	.07841	12.754	.99694	31
30	.07846	.07870	12.706	.99692	30
31	.07875	.07899	12.659	.99689	29
32	.07904	.07929	12.612	.99687	28
33	.07933	.07958	12.566	.99685	27
34	.07962	.07987	12.520	.99683	26
35	.07991	.08017	12.474	.99680	25
36	.08020	.08046	12.429	.99678	24
37	.08049	.08075	12.384	.99676	23
38	.08078	.08104	12.339	.99673	22
39	.08107	.08134	12.295	.99671	21
40	.08136	.08163	12.251	.99668	20
41	.08165	.08192	12.207	.99666	19
42	.08194	.08221	12.163	.99664	18
43	.08223	.08251	12.120	.99661	17
44	.08252	.08280	12.077	.99659	16
45	.08281	.08309	12.035	.99657	15
46	.08310	.08339	11.992	.99654	14
47	.08339	.08368	11.950	.99652	13
48	.08368	.08397	11.909	.99649	12
49	.08397	.08427	11.867	.99647	11
50	.08426	.08456	11.826	.99644	10
51	.08455	.08485	11.785	.99642	9
52	.08484	.08514	11.745	.99639	8
53	.08513	.08544	11.705	.99637	7
54	.08542	.08573	11.664	.99635	6
55	.08571	.08602	11.625	.99632	5
56	.08600	.08632	11.585	.99630	4
57	.08629	.08661	11.546	.99627	3
58	.08658	.08690	11.507	.99625	2
59	.08687	.08720	11.468	.99622	1
60	.08716	.08749	11.430	.99619	0
′	Cos	Ctn	Tan	Sin	′

94° (274°) (265°) **85°**

5° (185°) (354°) **174°**

′	Sin	Tan	Ctn	Cos	′
0	.08716	.08749	11.430	.99619	60
1	.08745	.08778	11.392	.99617	59
2	.08774	.08807	11.354	.99614	58
3	.08803	.08837	11.316	.99612	57
4	.08831	.08866	11.279	.99609	56
5	.08860	.08895	11.242	.99607	55
6	.08889	.08925	11.205	.99604	54
7	.08918	.08954	11.168	.99602	53
8	.08947	.08983	11.132	.99599	52
9	.08976	.09013	11.095	.99596	51
10	.09005	.09042	11.059	.99594	50
11	.09034	.09071	11.024	.99591	49
12	.09063	.09101	10.988	.99588	48
13	.09092	.09130	10.953	.99586	47
14	.09121	.09159	10.918	.99583	46
15	.09150	.09189	10.883	.99580	45
16	.09179	.09218	10.848	.99578	44
17	.09208	.09247	10.814	.99575	43
18	.09237	.09277	10.780	.99572	42
19	.09266	.09306	10.746	.99570	41
20	.09295	.09335	10.712	.99567	40
21	.09324	.09365	10.678	.99564	39
22	.09353	.09394	10.645	.99562	38
23	.09382	.09423	10.612	.99559	37
24	.09411	.09453	10.579	.99556	36
25	.09440	.09482	10.546	.99553	35
26	.09469	.09511	10.514	.99551	34
27	.09498	.09541	10.481	.99548	33
28	.09527	.09570	10.449	.99545	32
29	.09556	.09600	10.417	.99542	31
30	.09585	.09629	10.385	.99540	30
31	.09614	.09658	10.354	.99537	29
32	.09642	.09688	10.322	.99534	28
33	.09671	.09717	10.291	.99531	27
34	.09700	.09746	10.260	.99528	26
35	.09729	.09776	10.229	.99526	25
36	.09758	.09805	10.199	.99523	24
37	.09787	.09834	10.168	.99520	23
38	.09816	.09864	10.138	.99517	22
39	.09845	.09893	10.108	.99514	21
40	.09874	.09923	10.078	.99511	20
41	.09903	.09952	10.048	.99508	19
42	.09932	.09981	10.019	.99506	18
43	.09961	.10011	9.9893	.99503	17
44	.09990	.10040	9.9601	.99500	16
45	.10019	.10069	9.9310	.99497	15
46	.10048	.10099	9.9021	.99494	14
47	.10077	.10128	9.8734	.99491	13
48	.10106	.10158	9.8448	.99488	12
49	.10135	.10187	9.8164	.99485	11
50	.10164	.10216	9.7882	.99482	10
51	.10192	.10246	9.7601	.99479	9
52	.10221	.10275	9.7322	.99476	8
53	.10250	.10305	9.7044	.99473	7
54	.10279	.10334	9.6768	.99470	6
55	.10308	.10363	9.6493	.99467	5
56	.10337	.10393	9.6220	.99464	4
57	.10366	.10422	9.5949	.99461	3
58	.10395	.10452	9.5679	.99458	2
59	.10424	.10481	9.5411	.99455	1
60	.10453	.10510	9.5144	.99452	0
′	Cos	Ctn	Tan	Sin	′

95° (275°) (264°) **84°**

6° (186°) (353°) **173°**

′	Sin	Tan	Ctn	Cos	′
0	.10453	.10510	9.5144	.99452	60
1	.10482	.10540	9.4878	.99449	59
2	.10511	.10569	9.4614	.99446	58
3	.10540	.10599	9.4352	.99443	57
4	.10569	.10628	9.4090	.99440	56
5	.10597	.10657	9.3831	.99437	55
6	.10626	.10687	9.3572	.99434	54
7	.10655	.10716	9.3315	.99431	53
8	.10684	.10746	9.3060	.99428	52
9	.10713	.10775	9.2806	.99424	51
10	.10742	.10805	9.2553	.99421	50
11	.10771	.10834	9.2302	.99418	49
12	.10800	.10863	9.2052	.99415	48
13	.10829	.10893	9.1803	.99412	47
14	.10858	.10922	9.1555	.99409	46
15	.10887	.10952	9.1309	.99406	45
16	.10916	.10981	9.1065	.99402	44
17	.10945	.11011	9.0821	.99399	43
18	.10973	.11040	9.0579	.99396	42
19	.11002	.11070	9.0338	.99393	41
20	.11031	.11099	9.0098	.99390	40
21	.11060	.11128	8.9860	.99386	39
22	.11089	.11158	8.9623	.99383	38
23	.11118	.11187	8.9387	.99380	37
24	.11147	.11217	8.9152	.99377	36
25	.11176	.11246	8.8919	.99374	35
26	.11205	.11276	8.8686	.99370	34
27	.11234	.11305	8.8455	.99367	33
28	.11263	.11335	8.8225	.99364	32
29	.11291	.11364	8.7996	.99360	31
30	.11320	.11394	8.7769	.99357	30
31	.11349	.11423	8.7542	.99354	29
32	.11378	.11452	8.7317	.99351	28
33	.11407	.11482	8.7093	.99347	27
34	.11436	.11511	8.6870	.99344	26
35	.11465	.11541	8.6648	.99341	25
36	.11494	.11570	8.6427	.99337	24
37	.11523	.11600	8.6208	.99334	23
38	.11552	.11629	8.5989	.99331	22
39	.11580	.11659	8.5772	.99327	21
40	.11609	.11688	8.5555	.99324	20
41	.11638	.11718	8.5340	.99320	19
42	.11667	.11747	8.5126	.99317	18
43	.11696	.11777	8.4913	.99314	17
44	.11725	.11806	8.4701	.99310	16
45	.11754	.11836	8.4490	.99307	15
46	.11783	.11865	8.4280	.99303	14
47	.11812	.11895	8.4071	.99300	13
48	.11840	.11924	8.3863	.99297	12
49	.11869	.11954	8.3656	.99293	11
50	.11898	.11983	8.3450	.99290	10
51	.11927	.12013	8.3245	.99286	9
52	.11956	.12042	8.3041	.99283	8
53	.11985	.12072	8.2838	.99279	7
54	.12014	.12101	8.2636	.99276	6
55	.12043	.12131	8.2434	.99272	5
56	.12071	.12160	8.2234	.99269	4
57	.12100	.12190	8.2035	.99265	3
58	.12129	.12219	8.1837	.99262	2
59	.12158	.12249	8.1640	.99258	1
60	.12187	.12278	8.1443	.99255	0
′	Cos	Ctn	Tan	Sin	′

96° (276°) (263°) **83°**

7° (187°) (352°) **172°**

′	Sin	Tan	Ctn	Cos	′
0	.12187	.12278	8.1443	.99255	60
1	.12216	.12308	8.1248	.99251	59
2	.12245	.12338	8.1054	.99248	58
3	.12274	.12367	8.0860	.99244	57
4	.12302	.12397	8.0667	.99240	56
5	.12331	.12426	8.0476	.99237	55
6	.12360	.12456	8.0285	.99233	54
7	.12389	.12485	8.0095	.99230	53
8	.12418	.12515	7.9906	.99226	52
9	.12447	.12544	7.9718	.99222	51
10	.12476	.12574	7.9530	.99219	50
11	.12504	.12603	7.9344	.99215	49
12	.12533	.12633	7.9158	.99211	48
13	.12562	.12662	7.8973	.99208	47
14	.12591	.12692	7.8789	.99204	46
15	.12620	.12722	7.8606	.99200	45
16	.12649	.12751	7.8424	.99197	44
17	.12678	.12781	7.8243	.99193	43
18	.12706	.12810	7.8062	.99189	42
19	.12735	.12840	7.7882	.99186	41
20	.12764	.12869	7.7704	.99182	40
21	.12793	.12899	7.7525	.99178	39
22	.12822	.12929	7.7348	.99175	38
23	.12851	.12958	7.7171	.99171	37
24	.12880	.12988	7.6996	.99167	36
25	.12908	.13017	7.6821	.99163	35
26	.12937	.13047	7.6647	.99160	34
27	.12966	.13076	7.6473	.99156	33
28	.12995	.13106	7.6301	.99152	32
29	.13024	.13136	7.6129	.99148	31
30	.13053	.13165	7.5958	.99144	30
31	.13081	.13195	7.5787	.99141	29
32	.13110	.13224	7.5618	.99137	28
33	.13139	.13254	7.5449	.99133	27
34	.13168	.13284	7.5281	.99129	26
35	.13197	.13313	7.5113	.99125	25
36	.13226	.13343	7.4947	.99122	24
37	.13254	.13372	7.4781	.99118	23
38	.13283	.13402	7.4615	.99114	22
39	.13312	.13432	7.4451	.99110	21
40	.13341	.13461	7.4287	.99106	20
41	.13370	.13491	7.4124	.99102	19
42	.13399	.13521	7.3962	.99098	18
43	.13427	.13550	7.3800	.99094	17
44	.13456	.13580	7.3639	.99091	16
45	.13485	.13609	7.3479	.99087	15
46	.13514	.13639	7.3319	.99083	14
47	.13543	.13669	7.3160	.99079	13
48	.13572	.13698	7.3002	.99075	12
49	.13600	.13728	7.2844	.99071	11
50	.13629	.13758	7.2687	.99067	10
51	.13658	.13787	7.2531	.99063	9
52	.13687	.13817	7.2375	.99059	8
53	.13716	.13846	7.2220	.99055	7
54	.13744	.13876	7.2066	.99051	6
55	.13773	.13906	7.1912	.99047	5
56	.13802	.13935	7.1759	.99043	4
57	.13831	.13965	7.1607	.99039	3
58	.13860	.13995	7.1455	.99035	2
59	.13889	.14024	7.1304	.99031	1
60	.13917	.14054	7.1154	.99027	0
′	Cos	Ctn	Tan	Sin	′

97° (277°) (262°) **82°**

8° (188°) (351°) **171°**

′	Sin	Tan	Ctn	Cos	′
0	.13917	.14054	7.1154	.99027	60
1	.13946	.14084	7.1004	.99023	59
2	.13975	.14113	7.0855	.99019	58
3	.14004	.14143	7.0706	.99015	57
4	.14033	.14173	7.0558	.99011	56
5	.14061	.14202	7.0410	.99006	55
6	.14090	.14232	7.0264	.99002	54
7	.14119	.14262	7.0117	.98998	53
8	.14148	.14291	6.9972	.98994	52
9	.14177	.14321	6.9827	.98990	51
10	.14205	.14351	6.9682	.98986	50
11	.14234	.14381	6.9538	.98982	49
12	.14263	.14410	6.9395	.98978	48
13	.14292	.14440	6.9252	.98973	47
14	.14320	.14470	6.9110	.98969	46
15	.14349	.14499	6.8969	.98965	45
16	.14378	.14529	6.8828	.98961	44
17	.14407	.14559	6.8687	.98957	43
18	.14436	.14588	6.8548	.98953	42
19	.14464	.14618	6.8408	.98948	41
20	.14493	.14648	6.8269	.98944	40
21	.14522	.14678	6.8131	.98940	39
22	.14551	.14707	6.7994	.98936	38
23	.14580	.14737	6.7856	.98931	37
24	.14608	.14767	6.7720	.98927	36
25	.14637	.14796	6.7584	.98923	35
26	.14666	.14826	6.7448	.98919	34
27	.14695	.14856	6.7313	.98914	33
28	.14723	.14886	6.7179	.98910	32
29	.14752	.14915	6.7045	.98906	31
30	.14781	.14945	6.6912	98902	30
31	.14810	.14975	6.6779	.98897	29
32	.14838	.15005	6.6646	.98893	28
33	.14867	.15034	6.6514	.98889	27
34	.14896	.15064	6.6383	.98884	26
35	.14925	.15094	6.6252	.98880	25
36	.14954	.15124	6.6122	.98876	24
37	.14982	.15153	6.5992	.98871	23
38	.15011	.15183	6.5863	.98867	22
39	.15040	.15213	6.5734	.98863	21
40	.15069	.15243	6.5606	.98858	20
41	.15097	.15272	6.5478	.98854	19
42	.15126	.15302	6.5350	.98849	18
43	.15155	.15332	6.5223	.98845	17
44	.15184	.15362	6.5097	.98841	16
45	.15212	.15391	6.4971	.98836	15
46	.15241	.15421	6.4846	.98832	14
47	.15270	.15451	6.4721	.98827	13
48	.15299	.15481	6.4596	.98823	12
49	.15327	.15511	6.4472	.98818	11
50	.15356	.15540	6.4348	.98814	10
51	.15385	.15570	6.4225	.98809	9
52	.15414	.15600	6.4103	.98805	8
53	.15442	.15630	6.3980	.98800	7
54	.15471	.15660	6.3859	.98796	6
55	.15500	.15689	6.3737	.98791	5
56	.15529	.15719	6.3617	.98787	4
57	.15557	.15749	6.3496	.98782	3
58	.15586	.15779	6.3376	.98778	2
59	.15615	.15809	6.3257	.98773	1
60	.15643	.15838	6.3138	.98769	0
′	Cos	Ctn	Tan	Sin	′

98° (278°) (261°) **81°**

9° (189°) (350°) **170°**

′	Sin	Tan	Ctn	Cos	′
0	.15643	.15838	6.3138	.98769	60
1	.15672	.15868	6.3019	.98764	59
2	.15701	.15898	6.2901	.98760	58
3	.15730	.15928	6.2783	.98755	57
4	.15758	.15958	6.2666	.98751	56
5	.15787	.15988	6.2549	.98746	55
6	.15816	.16017	6.2432	.98741	54
7	.15845	.16047	6.2316	.98737	53
8	.15873	.16077	6.2200	.98732	52
9	.15902	.16107	6.2085	.98728	51
10	.15931	.16137	6.1970	.98723	50
11	.15959	.16167	6.1856	.98718	49
12	.15988	.16196	6.1742	.98714	48
13	.16017	.16226	6.1628	.98709	47
14	.16046	.16256	6.1515	.98704	46
15	.16074	.16286	6.1402	.98700	45
16	.16103	.16316	6.1290	.98695	44
17	.16132	.16346	6.1178	.98690	43
18	.16160	.16376	6.1066	.98686	42
19	.16189	.16405	6.0955	.98681	41
20	.16218	.16435	6.0844	.98676	40
21	.16246	.16465	6.0734	.98671	39
22	.16275	.16495	6.0624	.98667	38
23	.16304	.16525	6.0514	.98662	37
24	.16333	.16555	6.0405	.98657	36
25	.16361	.16585	6.0296	.98652	35
26	.16390	.16615	6.0188	.98648	34
27	.16419	.16645	6.0080	.98643	33
28	.16447	.16674	5.9972	.98638	32
29	.16476	.16704	5.9865	.98633	31
30	.16505	.16734	5.9758	.98629	30
31	.16533	.16764	5.9651	.98624	29
32	.16562	.16794	5.9545	.98619	28
33	.16591	.16824	5.9439	.98614	27
34	.16620	.16854	5.9333	.98609	26
35	.16648	.16884	5.9228	.98604	25
36	.16677	.16914	5.9124	.98600	24
37	.16706	.16944	5.9019	.98595	23
38	.16734	.16974	5.8915	.98590	22
39	.16763	.17004	5.8811	.98585	21
40	.16792	.17033	5.8708	.98580	20
41	.16820	.17063	5.8605	.98575	19
42	.16849	.17093	5.8502	.98570	18
43	.16878	.17123	5.8400	.98565	17
44	.16906	.17153	5.8298	.98561	16
45	.16935	.17183	5.8197	.98556	15
46	.16964	.17213	5.8095	.98551	14
47	.16992	.17243	5.7994	.98546	13
48	.17021	.17273	5.7894	.98541	12
49	.17050	.17303	5.7794	.98536	11
50	.17078	.17333	5.7694	.98531	10
51	.17107	.17363	5.7594	.98526	9
52	.17136	.17393	5.7495	.98521	8
53	.17164	.17423	5.7396	.98516	7
54	.17193	.17453	5.7297	.98511	6
55	.17222	.17483	5.7199	.98506	5
56	.17250	.17513	5.7101	.98501	4
57	.17279	.17543	5.7004	.98496	3
58	.17308	.17573	5.6906	.98491	2
59	.17336	.17603	5.6809	.98486	1
60	.17365	.17633	5.6713	.98481	0
′	Cos	Ctn	Tan	Sin	′

99° (279°) (260°) **80°**

10° (190°) (349°) **169°**

′	Sin	Tan	Ctn	Cos	′
0	.17365	.17633	5.6713	.98481	60
1	.17393	.17663	5.6617	.98476	59
2	.17422	.17693	5.6521	.98471	58
3	.17451	.17723	5.6425	.98466	57
4	.17479	.17753	5.6329	.98461	56
5	.17508	.17783	5.6234	.98455	55
6	.17537	.17813	5.6140	.98450	54
7	.17565	.17843	5.6045	.98445	53
8	.17594	.17873	5.5951	.98440	52
9	.17623	.17903	5.5857	.98435	51
10	.17651	.17933	5.5764	.98430	50
11	.17680	.17963	5.5671	.98425	49
12	.17708	.17993	5.5578	.98420	48
13	.17737	.18023	5.5485	.98414	47
14	.17766	.18053	5.5393	.98409	46
15	.17794	.18083	5.5301	.98404	45
16	.17823	.18113	5.5209	.98399	44
17	.17852	.18143	5.5118	.98394	43
18	.17880	.18173	5.5026	.98389	42
19	.17909	.18203	5.4936	.98383	41
20	.17937	.18233	5.4845	.98378	40
21	.17966	.18263	5.4755	.98373	39
22	.17995	.18293	5.4665	.98368	38
23	.18023	.18323	5.4575	.98362	37
24	.18052	.18353	5.4486	.98357	36
25	.18081	.18384	5.4397	.98352	35
26	.18109	.18414	5.4308	.98347	34
27	.18138	.18444	5.4219	.98341	33
28	.18166	.18474	5.4131	.98336	32
29	.18195	.18504	5.4043	.98331	31
30	.18224	.18534	5.3955	.98325	30
31	.18252	.18564	5.3868	.98320	29
32	.18281	.18594	5.3781	.98315	28
33	.18309	.18624	5.3694	.98310	27
34	.18338	.18654	5.3607	.98304	26
35	.18367	.18684	5.3521	.98299	25
36	.18395	.18714	5.3435	.98294	24
37	.18424	.18745	5.3349	.98288	23
38	.18452	.18775	5.3263	.98283	22
39	.18481	.18805	5.3178	.98277	21
40	.18509	.18835	5.3093	.98272	20
41	.18538	.18865	5.3008	.98267	19
42	.18567	.18895	5.2924	.98261	18
43	.18595	.18925	5.2839	.98256	17
44	.18624	.18955	5.2755	.98250	16
45	.18652	.18986	5.2672	.98245	15
46	.18681	.19016	5.2588	.98240	14
47	.18710	.19046	5.2505	.98234	13
48	.18738	.19076	5.2422	.98229	12
49	.18767	.19106	5.2339	.98223	11
50	.18795	.19136	5.2257	.98218	10
51	.18824	.19166	5.2174	.98212	9
52	.18852	.19197	5.2092	.98207	8
53	.18881	.19227	5.2011	.98201	7
54	.18910	.19257	5.1929	.98196	6
55	.18938	.19287	5.1848	.98190	5
56	.18967	.19317	5.1767	.98185	4
57	.18995	.19347	5.1686	.98179	3
58	.19024	.19378	5.1606	.98174	2
59	.19052	.19408	5.1526	.98168	1
60	.19081	.19438	5.1446	.98163	0
′	Cos	Ctn	Tan	Sin	′

100° (280°) (259°) **79°**

11° (191°) (348°) **168°**

′	Sin	Tan	Ctn	Cos	′
0	.19081	.19438	5.1446	.98163	60
1	.19109	.19468	5.1366	.98157	59
2	.19138	.19498	5.1286	.98152	58
3	.19167	.19529	5.1207	.98146	57
4	.19195	.19559	5.1128	.98140	56
5	.19224	.19589	5.1049	.98135	55
6	.19252	.19619	5.0970	.98129	54
7	.19281	.19649	5.0892	.98124	53
8	.19309	.19680	5.0814	.98118	52
9	.19338	.19710	5.0736	.98112	51
10	.19366	.19740	5.0658	.98107	50
11	.19395	.19770	5.0581	.98101	49
12	.19423	.19801	5.0504	.98096	48
13	.19452	.19831	5.0427	.98090	47
14	.19481	.19861	5.0350	.98084	46
15	.19509	.19891	5.0273	.98079	45
16	.19538	.19921	5.0197	.98073	44
17	.19566	.19952	5.0121	.98067	43
18	.19595	.19982	5.0045	.98061	42
19	.19623	.20012	4.9969	.98056	41
20	.19652	.20042	4.9894	.98050	40
21	.19680	.20073	4.9819	.98044	39
22	.19709	.20103	4.9744	.98039	38
23	.19737	.20133	4.9669	.98033	37
24	.19766	.20164	4.9594	.98027	36
25	.19794	.20194	4.9520	.98021	35
26	.19823	.20224	4.9446	.98016	34
27	.19851	.20254	4.9372	.98010	33
28	.19880	.20285	4.9298	.98004	32
29	.19908	.20315	4.9225	.97998	31
30	.19937	.20345	4.9152	.97992	30
31	.19965	.20376	4.9078	.97987	29
32	.19994	.20406	4.9006	.97981	28
33	.20022	.20436	4.8933	.97975	27
34	.20051	.20466	4.8860	.97969	26
35	.20079	.20497	4.8788	.97963	25
36	.20108	.20527	4.8716	.97958	24
37	.20136	.20557	4.8644	.97952	23
38	.20165	.20588	4.8573	.97946	22
39	.20193	.20618	4.8501	.97940	21
40	.20222	.20648	4.8430	.97934	20
41	.20250	.20679	4.8359	.97928	19
42	.20279	.20709	4.8288	.97922	18
43	.20307	.20739	4.8218	.97916	17
44	.20336	.20770	4.8147	.97910	16
45	.20364	.20800	4.8077	.97905	15
46	.20393	.20830	4.8007	.97899	14
47	.20421	.20861	4.7937	.97893	13
48	.20450	.20891	4.7867	.97887	12
49	.20478	.20921	4.7798	.97881	11
50	.20507	.20952	4.7729	.97875	10
51	.20535	.20982	4.7659	.97869	9
52	.20563	.21013	4.7591	.97863	8
53	.20592	.21043	4.7522	.97857	7
54	.20620	.21073	4.7453	.97851	6
55	.20649	.21104	4.7385	.97845	5
56	.20677	.21134	4.7317	.97839	4
57	.20706	.21164	4.7249	.97833	3
58	.20734	.21195	4.7181	.97827	2
59	.20763	.21225	4.7114	.97821	1
60	.20791	.21256	4.7046	.97815	0
′	Cos	Ctn	Tan	Sin	′

101° (281°) (258°) **78°**

12° (192°) (347°) **167°**

′	Sin	Tan	Ctn	Cos	′
0	.20791	.21256	4.7046	.97815	60
1	.20820	.21286	4.6979	.97809	59
2	.20848	.21316	4.6912	.97803	58
3	.20877	.21347	4.6845	.97797	57
4	.20905	.21377	4.6779	.97791	56
5	.20933	.21408	4.6712	.97784	55
6	.20962	.21438	4.6646	.97778	54
7	.20990	.21469	4.6580	.97772	53
8	.21019	.21499	4.6514	.97766	52
9	.21047	.21529	4.6448	.97760	51
10	.21076	.21560	4.6382	.97754	50
11	.21104	.21590	4.6317	.97748	49
12	.21132	.21621	4.6252	.97742	48
13	.21161	.21651	4.6187	.97735	47
14	.21189	.21682	4.6122	.97729	46
15	.21218	.21712	4.6057	.97723	45
16	.21246	.21743	4.5993	.97717	44
17	.21275	.21773	4.5928	.97711	43
18	.21303	.21804	4.5864	.97705	42
19	.21331	.21834	4.5800	.97698	41
20	.21360	.21864	4.5736	.97692	40
21	.21388	.21895	4.5673	.97686	39
22	.21417	.21925	4.5609	.97680	38
23	.21445	.21956	4.5546	.97673	37
24	.21474	.21986	4.5483	.97667	36
25	.21502	.22017	4.5420	.97661	35
26	.21530	.22047	4.5357	.97655	34
27	.21559	.22078	4.5294	.97648	33
28	.21587	.22108	4.5232	.97642	32
29	.21616	.22139	4.5169	.97636	31
30	.21644	.22169	4.5107	.97630	30
31	.21672	.22200	4.5045	.97623	29
32	.21701	.22231	4.4983	.97617	28
33	.21729	.22261	4.4922	.97611	27
34	.21758	.22292	4.4860	.97604	26
35	.21786	.22322	4.4799	.97598	25
36	.21814	.22353	4.4737	.97592	24
37	.21843	.22383	4.4676	.97585	23
38	.21871	.22414	4.4615	.97579	22
39	.21899	.22444	4.4555	.97573	21
40	.21928	.22475	4.4494	.97566	20
41	.21956	.22505	4.4434	.97560	19
42	.21985	.22536	4.4373	.97553	18
43	.22013	.22567	4.4313	.97547	17
44	.22041	.22597	4.4253	.97541	16
45	.22070	.22628	4.4194	.97534	15
46	.22098	.22658	4.4134	.97528	14
47	.22126	.22689	4.4075	.97521	13
48	.22155	.22719	4.4015	.97515	12
49	.22183	.22750	4.3956	.97508	11
50	.22212	.22781	4.3897	.97502	10
51	.22240	.22811	4.3838	.97496	9
52	.22268	.22842	4.3779	.97489	8
53	.22297	.22872	4.3721	.97483	7
54	.22325	.22903	4.3662	.97476	6
55	.22353	.22934	4.3604	.97470	5
56	.22382	.22964	4.3546	.97463	4
57	.22410	.22995	4.3488	.97457	3
58	.22438	.23026	4.3430	.97450	2
59	.22467	.23056	4.3372	.97444	1
60	.22495	.23087	4.3315	.97437	0
′	Cos	Ctn	Tan	Sin	′

102° (282°) (257°) **77°**

13° (193°) (346°) **166°**

′	Sin	Tan	Ctn	Cos	′
0	.22495	.23087	4.3315	.97437	60
1	.22523	.23117	4.3257	.97430	59
2	.22552	.23148	4.3200	.97424	58
3	.22580	.23179	4.3143	.97417	57
4	.22608	.23209	4.3086	.97411	56
5	.22637	.23240	4.3029	.97404	55
6	.22665	.23271	4.2972	.97398	54
7	.22693	.23301	4.2916	.97391	53
8	.22722	.23332	4.2859	.97384	52
9	.22750	.23363	4.2803	.97378	51
10	.22778	.23393	4.2747	.97371	50
11	.22807	.23424	4.2691	.97365	49
12	.22835	.23455	4.2635	.97358	48
13	.22863	.23485	4.2580	.97351	47
14	.22892	.23516	4.2524	.97345	46
15	.22920	.23547	4.2468	.97338	45
16	.22948	.23578	4.2413	.97331	44
17	.22977	.23608	4.2358	.97325	43
18	.23005	.23639	4.2303	.97318	42
19	.23033	.23670	4.2248	.97311	41
20	.23062	.23700	4.2193	.97304	40
21	.23090	.23731	4.2139	.97298	39
22	.23118	.23762	4.2084	.97291	38
23	.23146	.23793	4.2030	.97284	37
24	.23175	.23823	4.1976	.97278	36
25	.23203	.23854	4.1922	.97271	35
26	.23231	.23885	4.1868	.97264	34
27	.23260	.23916	4.1814	.97257	33
28	.23288	.23946	4.1760	.97251	32
29	.23316	.23977	4.1706	.97244	31
30	.23345	.24008	4.1653	.97237	30
31	.23373	.24039	4.1600	.97230	29
32	.23401	.24069	4.1547	.97223	28
33	.23429	.24100	4.1493	.97217	27
34	.23458	.24131	4.1441	.97210	26
35	.23486	.24162	4.1388	.97203	25
36	.23514	.24193	4.1335	.97196	24
37	.23542	.24223	4.1282	.97189	23
38	.23571	.24254	4.1230	.97182	22
39	.23599	.24285	4.1178	.97176	21
40	.23627	.24316	4.1126	.97169	20
41	.23656	.24347	4.1074	.97162	19
42	.23684	.24377	4.1022	.97155	18
43	.23712	.24408	4.0970	.97148	17
44	.23740	.24439	4.0918	.97141	16
45	.23769	.24470	4.0867	.97134	15
46	.23797	.24501	4.0815	.97127	14
47	.23825	.24532	4.0764	.97120	13
48	.23853	.24562	4.0713	.97113	12
49	.23882	.24593	4.0662	.97106	11
50	.23910	.24624	4.0611	.97100	10
51	.23938	.24655	4.0560	.97093	9
52	.23966	.24686	4.0509	.97086	8
53	.23995	.24717	4.0459	.97079	7
54	.24023	.24747	4.0408	.97072	6
55	.24051	.24778	4.0358	.97065	5
56	.24079	.24809	4.0308	.97058	4
57	.24108	.24840	4.0257	.97051	3
58	.24136	.24871	4.0207	.97044	2
59	.24164	.24902	4.0158	.97037	1
60	.24192	.24933	4.0108	.97030	0
′	Cos	Ctn	Tan	Sin	′

103° (283°) (256°) **76°**

14° (194°) (345°) **165°**

′	Sin	Tan	Ctn	Cos	′
0	.24192	.24933	4.0108	.97030	60
1	.24220	.24964	4.0058	.97023	59
2	.24249	.24995	4.0009	.97015	58
3	.24277	.25026	3.9959	.97008	57
4	.24305	.25056	3.9910	.97001	56
5	.24333	.25087	3.9861	.96994	55
6	.24362	.25118	3.9812	.96987	54
7	.24390	.25149	3.9763	.96980	53
8	.24418	.25180	3.9714	.96973	52
9	.24446	.25211	3.9665	.96966	51
10	.24474	.25242	3.9617	.96959	50
11	.24503	.25273	3.9568	.96952	49
12	.24531	.25304	3.9520	.96945	48
13	.24559	.25335	3.9471	.96937	47
14	.24587	.25366	3.9423	.96930	46
15	.24615	.25397	3.9375	.96923	45
16	.24644	.25428	3.9327	.96916	44
17	.24672	.25459	3.9279	.96909	43
18	.24700	.25490	3.9232	.96902	42
19	.24728	.25521	3.9184	.96894	41
20	.24756	.25552	3.9136	.96887	40
21	.24784	.25583	3.9089	.96880	39
22	.24813	.25614	3.9042	.96873	38
23	.24841	.25645	3.8995	.96866	37
24	.24869	.25676	3.8947	.96858	36
25	.24897	.25707	3.8900	.96851	35
26	.24925	.25738	3.8854	.96844	34
27	.24954	.25769	3.8807	.96837	33
28	.24982	.25800	3.8760	.96829	32
29	.25010	.25831	3.8714	.96822	31
30	.25038	.25862	3.8667	.96815	30
31	.25066	.25893	3.8621	.96807	29
32	.25094	.25924	3.8575	.96800	28
33	.25122	.25955	3.8528	.96793	27
34	.25151	.25986	3.8482	.96786	26
35	.25179	.26017	3.8436	.96778	25
36	.25207	.26048	3.8391	.96771	24
37	.25235	.26079	3.8345	.96764	23
38	.25263	.26110	3.8299	.96756	22
39	.25291	.26141	3.8254	.96749	21
40	.25320	.26172	3.8208	.96742	20
41	.25348	.26203	3.8163	.96734	19
42	.25376	.26235	3.8118	.96727	18
43	.25404	.26266	3.8073	.96719	17
44	.25432	.26297	3.8028	.96712	16
45	.25460	.26328	3.7983	.96705	15
46	.25488	.26359	3.7938	.96697	14
47	.25516	.26390	3.7893	.96690	13
48	.25545	.26421	3.7848	.96682	12
49	.25573	.26452	3.7804	.96675	11
50	.25601	.26483	3.7760	.96667	10
51	.25629	.26515	3.7715	.96660	9
52	.25657	.26546	3.7671	.96653	8
53	.25685	.26577	3.7627	.96645	7
54	.25713	.26608	3.7583	.96638	6
55	.25741	.26639	3.7539	.96630	5
56	.25769	.26670	3.7495	.96623	4
57	.25798	.26701	3.7451	.96615	3
58	.25826	.26733	3.7408	.96608	2
59	.25854	.26764	3.7364	.96600	1
60	.25882	.26795	3.7321	.96593	0
′	Cos	Ctn	Tan	Sin	′

104° (284°) (255°) **75°**

15° (195°) (344°) **164°**

′	Sin	Tan	Ctn	Cos	′
0	.25882	.26795	3.7321	.96593	60
1	.25910	.26826	3.7277	.96585	59
2	.25938	.26857	3.7234	.96578	58
3	.25966	.26888	3.7191	.96570	57
4	.25994	.26920	3.7148	.96562	56
5	.26022	.26951	3.7105	.96555	55
6	.26050	.26982	3.7062	.96547	54
7	.26079	.27013	3.7019	.96540	53
8	.26107	.27044	3.6976	.96532	52
9	.26135	.27076	3.6933	.96524	51
10	.26163	.27107	3.6891	.96517	50
11	.26191	.27138	3.6848	.96509	49
12	.26219	.27169	3.6806	.96502	48
13	.26247	.27201	3.6764	.96494	47
14	.26275	.27232	3.6722	.96486	46
15	.26303	.27263	3.6680	.96479	45
16	.26331	.27294	3.6638	.96471	44
17	.26359	.27326	3.6596	.96463	43
18	.26387	.27357	3.6554	.96456	42
19	.26415	.27388	3.6512	.96448	41
20	.26443	.27419	3.6470	.96440	40
21	.26471	.27451	3.6429	.96433	39
22	.26500	.27482	3.6387	.96425	38
23	.26528	.27513	3.6346	.96417	37
24	.26556	.27545	3.6305	.96410	36
25	.26584	.27576	3.6264	.96402	35
26	.26612	.27607	3.6222	.96394	34
27	.26640	.27638	3.6181	.96386	33
28	.26668	.27670	3.6140	.96379	32
29	.26696	.27701	3.6100	.96371	31
30	.26724	.27732	3.6059	.96363	30
31	.26752	.27764	3.6018	.96355	29
32	.26780	.27795	3.5978	.96347	28
33	.26808	.27826	3.5937	.96340	27
34	.26836	.27858	3.5897	.96332	26
35	.26864	.27889	3.5856	.96324	25
36	.26892	.27921	3.5816	.96316	24
37	.26920	.27952	3.5776	.96308	23
38	.26948	.27983	3.5736	.96301	22
39	.26976	.28015	3.5696	.96293	21
40	.27004	.28046	3.5656	.96285	20
41	.27032	.28077	3.5616	.96277	19
42	.27060	.28109	3.5576	.96269	18
43	.27088	.28140	3.5536	.96261	17
44	.27116	.28172	3.5497	.96253	16
45	.27144	.28203	3.5457	.96246	15
46	.27172	.28234	3.5418	.96238	14
47	.27200	.28266	3.5379	.96230	13
48	.27228	.28297	3.5339	.96222	12
49	.27256	.28329	3.5300	.96214	11
50	.27284	.28360	3.5261	.96206	10
51	.27312	.28391	3.5222	.96198	9
52	.27340	.28423	3.5183	.96190	8
53	.27368	.28454	3.5144	.96182	7
54	.27396	.28486	3.5105	.96174	6
55	.27424	.28517	3.5067	.96166	5
56	.27452	.28549	3.5028	.96158	4
57	.27480	.28580	3.4989	.96150	3
58	.27508	.28612	3.4951	.96142	2
59	.27536	.28643	3.4912	.96134	1
60	.27564	.28675	3.4874	.96126	0
′	Cos	Ctn	Tan	Sin	′

105° (285°) (254°) **74°**

16° (196°) (343°) **163°**

′	Sin	Tan	Ctn	Cos	′
0	.27564	.28675	3.4874	.96126	60
1	.27592	.28706	3.4836	.96118	59
2	.27620	.28738	3.4798	.96110	58
3	.27648	.28769	3.4760	.96102	57
4	.27676	.28801	3.4722	.96094	56
5	.27704	.28832	3.4684	.96086	55
6	.27731	.28864	3.4646	.96078	54
7	.27759	.28895	3.4608	.96070	53
8	.27787	.28927	3.4570	.96062	52
9	.27815	.28958	3.4533	.96054	51
10	.27843	.28990	3.4495	.96046	50
11	.27871	.29021	3.4458	.96037	49
12	.27899	.29053	3.4420	.96029	48
13	.27927	.29084	3.4383	.96021	47
14	.27955	.29116	3.4346	.96013	46
15	.27983	.29147	3.4308	.96005	45
16	.28011	.29179	3.4271	.95997	44
17	.28039	.29210	3.4234	.95989	43
18	.28067	.29242	3.4197	.95981	42
19	.28095	.29274	3.4160	.95972	41
20	.28123	.29305	3.4124	.95964	40
21	.28150	.29337	3.4087	.95956	39
22	.28178	.29368	3.4050	.95948	38
23	.28206	.29400	3.4014	.95940	37
24	.28234	.29432	3.3977	.95931	36
25	.28262	.29463	3.3941	.95923	35
26	.28290	.29495	3.3904	.95915	34
27	.28318	.29526	3.3868	.95907	33
28	.28346	.29558	3.3832	.95898	32
29	.28374	.29590	3.3796	.95890	31
30	.28402	.29621	3.3759	.95882	30
31	.28429	.29653	3.3723	.95874	29
32	.28457	.29685	3.3687	.95865	28
33	.28485	.29716	3.3652	.95857	27
34	.28513	.29748	3.3616	.95849	26
35	.28541	.29780	3.3580	.95841	25
36	.28569	.29811	3.3544	.95832	24
37	.28597	.29843	3.3509	.95824	23
38	.28625	.29875	3.3473	.95816	22
39	.28652	.29906	3.3438	.95807	21
40	.28680	.29938	3.3402	.95799	20
41	.28708	.29970	3.3367	.95791	19
42	.28736	.30001	3.3332	.95782	18
43	.28764	.30033	3.3297	.95774	17
44	.28792	.30065	3.3261	.95766	16
45	.28820	.30097	3.3226	.95757	15
46	.28847	.30128	3.3191	.95749	14
47	.28875	.30160	3.3156	.95740	13
48	.28903	.30192	3.3122	.95732	12
49	.28931	.30224	3.3087	.95724	11
50	.28959	.30255	3.3052	.95715	10
51	.28987	.30287	3.3017	.95707	9
52	.29015	.30319	3.2983	.95698	8
53	.29042	.30351	3.2948	.95690	7
54	.29070	.30382	3.2914	.95681	6
55	.29098	.30414	3.2879	.95673	5
56	.29126	.30446	3.2845	.95664	4
57	.29154	.30478	3.2811	.95656	3
58	.29182	.30509	3.2777	.95647	2
59	.29209	.30541	3.2743	.95639	1
60	.29237	.30573	3.2709	.95630	0
′	Cos	Ctn	Tan	Sin	′

106° (286°) (253°) **73°**

17° (197°) (342°) **162°**

′	Sin	Tan	Ctn	Cos	′
0	.29237	.30573	3.2709	.95630	60
1	.29265	.30605	3.2675	.95622	59
2	.29293	.30637	3.2641	.95613	58
3	.29321	.30669	3.2607	.95605	57
4	.29348	.30700	3.2573	.95596	56
5	.29376	.30732	3.2539	.95588	55
6	.29404	.30764	3.2506	.95579	54
7	.29432	.30796	3.2472	.95571	53
8	.29460	.30828	3.2438	.95562	52
9	.29487	.30860	3.2405	.95554	51
10	.29515	.30891	3.2371	.95545	50
11	.29543	.30923	3.2338	.95536	49
12	.29571	.30955	3.2305	.95528	48
13	.29599	.30987	3.2272	.95519	47
14	.29626	.31019	3.2238	.95511	46
15	.29654	.31051	3.2205	.95502	45
16	.29682	.31083	3.2172	.95493	44
17	.29710	.31115	3.2139	.95485	43
18	.29737	.31147	3.2106	.95476	42
19	.29765	.31178	3.2073	.95467	41
20	.29793	.31210	3.2041	.95459	40
21	.29821	.31242	3.2008	.95450	39
22	.29849	.31274	3.1975	.95441	38
23	.29876	.31306	3.1943	.95433	37
24	.29904	.31338	3.1910	.95424	36
25	.29932	.31370	3.1878	.95415	35
26	.29960	.31402	3.1845	.95407	34
27	.29987	.31434	3.1813	.95398	33
28	.30015	.31466	3.1780	.95389	32
29	.30043	.31498	3.1748	.95380	31
30	.30071	.31530	3.1716	.95372	30
31	.30098	.31562	3.1684	.95363	29
32	.30126	.31594	3.1652	.95354	28
33	.30154	.31626	3.1620	.95345	27
34	.30182	.31658	3.1588	.95337	26
35	.30209	.31690	3.1556	.95328	25
36	.30237	.31722	3.1524	.95319	24
37	.30265	.31754	3.1492	.95310	23
38	.30292	.31786	3.1460	.95301	22
39	.30320	.31818	3.1429	.95293	21
40	.30348	.31850	3.1397	.95284	20
41	.30376	.31882	3.1366	.95275	19
42	.30403	.31914	3.1334	.95266	18
43	.30431	.31946	3.1303	.95257	17
44	.30459	.31978	3.1271	.95248	16
45	.30486	.32010	3.1240	.95240	15
46	.30514	.32042	3.1209	.95231	14
47	.30542	.32074	3.1178	.95222	13
48	.30570	.32106	3.1146	.95213	12
49	.30597	.32139	3.1115	.95204	11
50	.30625	.32171	3.1084	.95195	10
51	.30653	.32203	3.1053	.95186	9
52	.30680	.32235	3.1022	.95177	8
53	.30708	.32267	3.0991	.95168	7
54	.30736	.32299	3.0961	.95159	6
55	.30763	.32331	3.0930	.95150	5
56	.30791	.32363	3.0899	.95142	4
57	.30819	.32396	3.0868	.95133	3
58	.30846	.32428	3.0838	.95124	2
59	.30874	.32460	3.0807	.95115	1
60	.30902	.32492	3.0777	.95106	0
′	Cos	Ctn	Tan	Sin	′

107° (287°) (252°) **72°**

18° (198°) (341°) **161°**

′	Sin	Tan	Ctn	Cos	′
0	.30902	.32492	3.0777	.95106	60
1	.30929	.32524	3.0746	.95097	59
2	.30957	.32556	3.0716	.95088	58
3	.30985	.32588	3.0686	.95079	57
4	.31012	.32621	3.0655	.95070	56
5	.31040	.32653	3.0625	.95061	55
6	.31068	.32685	3.0595	.95052	54
7	.31095	.32717	3.0565	.95043	53
8	.31123	.32749	3.0535	.95033	52
9	.31151	.32782	3.0505	.95024	51
10	.31178	.32814	3.0475	.95015	50
11	.31206	.32846	3.0445	.95006	49
12	.31233	.32878	3.0415	.94997	48
13	.31261	.32911	3.0385	.94988	47
14	.31289	.32943	3.0356	.94979	46
15	.31316	.32975	3.0326	.94970	45
16	.31344	.33007	3.0296	.94961	44
17	.31372	.33040	3.0267	.94952	43
18	.31399	.33072	3.0237	.94943	42
19	.31427	.33104	3.0208	.94933	41
20	.31454	.33136	3.0178	.94924	40
21	.31482	.33169	3.0149	.94915	39
22	.31510	.33201	3.0120	.94906	38
23	.31537	.33233	3.0090	.94897	37
24	.31565	.33266	3.0061	.94888	36
25	.31593	.33298	3.0032	.94878	35
26	.31620	.33330	3.0003	.94869	34
27	.31648	.33363	2.9974	.94860	33
28	.31675	.33395	2.9945	.94851	32
29	.31703	.33427	2.9916	.94842	31
30	.31730	.33460	2.9887	.94832	30
31	.31758	.33492	2.9858	.94823	29
32	.31786	.33524	2.9829	.94814	28
33	.31813	.33557	2.9800	.94805	27
34	.31841	.33589	2.9772	.94795	26
35	.31868	.33621	2.9743	.94786	25
36	.31896	.33654	2.9714	.94777	24
37	.31923	.33686	2.9686	.94768	23
38	.31951	.33718	2.9657	.94758	22
39	.31979	.33751	2.9629	.94749	21
40	.32006	.33783	2.9600	.94740	20
41	.32034	.33816	2.9572	.94730	19
42	.32061	.33848	2.9544	.94721	18
43	.32089	.33881	2.9515	.94712	17
44	.32116	.33913	2.9487	.94702	16
45	.32144	.33945	2.9459	.94693	15
46	.32171	.33978	2.9431	.94684	14
47	.32199	.34010	2.9403	.94674	13
48	.32227	.34043	2.9375	.94665	12
49	.32254	.34075	2.9347	.94656	11
50	.32282	.34108	2.9319	.94646	10
51	.32309	.34140	2.9291	.94637	9
52	.32337	.34173	2.9263	.94627	8
53	.32364	.34205	2.9235	.94618	7
54	.32392	.34238	2.9208	.94609	6
55	.32419	.34270	2.9180	.94599	5
56	.32447	.34303	2.9152	.94590	4
57	.32474	.34335	2.9125	.94580	3
58	.32502	.34368	2.9097	.94571	2
59	.32529	.34400	2.9070	.94561	1
60	.32557	.34433	2.9042	.94552	0
′	Cos	Ctn	Tan	Sin	′

108° (288°) (251°) **71°**

19° (199°) (340°) **160°**

′	Sin	Tan	Ctn	Cos	′
0	.32557	.34433	2.9042	.94552	60
1	.32584	.34465	2.9015	.94542	59
2	.32612	.34498	2.8987	.94533	58
3	.32639	.34530	2.8960	.94523	57
4	.32667	.34563	2.8933	.94514	56
5	.32694	.34596	2.8905	.94504	55
6	.32722	.34628	2.8878	.94495	54
7	.32749	.34661	2.8851	.94485	53
8	.32777	.34693	2.8824	.94476	52
9	.32804	.34726	2.8797	.94466	51
10	.32832	.34758	2.8770	.94457	50
11	.32859	.34791	2.8743	.94447	49
12	.32887	.34824	2.8716	.94438	48
13	.32914	.34856	2.8689	.94428	47
14	.32942	.34889	2.8662	.94418	46
15	.32969	.34922	2.8636	.94409	45
16	.32997	.34954	2.8609	.94399	44
17	.33024	.34987	2.8582	.94390	43
18	.33051	.35020	2.8556	.94380	42
19	.33079	.35052	2.8529	.94370	41
20	.33106	.35085	2.8502	.94361	40
21	.33134	.35118	2.8476	.94351	39
22	.33161	.35150	2.8449	.94342	38
23	.33189	.35183	2.8423	.94332	37
24	.33216	.35216	2.8397	.94322	36
25	.33244	.35248	2.8370	.94313	35
26	.33271	.35281	2.8344	.94303	34
27	.33298	.35314	2.8318	.94293	33
28	.33326	.35346	2.8291	.94284	32
29	.33353	.35379	2.8265	.94274	31
30	.33381	.35412	2.8239	.94264	30
31	.33408	.35445	2.8213	.94254	29
32	.33436	.35477	2.8187	.94245	28
33	.33463	.35510	2.8161	.94235	27
34	.33490	.35543	2.8135	.94225	26
35	.33518	.35576	2.8109	.94215	25
36	.33545	.35608	2.8083	.94206	24
37	.33573	.35641	2.8057	.94196	23
38	.33600	.35674	2.8032	.94186	22
39	.33627	.35707	2.8006	.94176	21
40	.33655	.35740	2.7980	.94167	20
41	.33682	.35772	2.7955	.94157	19
42	.33710	.35805	2.7929	.94147	18
43	.33737	.35838	2.7903	.94137	17
44	.33764	.35871	2.7878	.94127	16
45	.33792	.35904	2.7852	.94118	15
46	.33819	.35937	2.7827	.94108	14
47	.33846	.35969	2.7801	.94098	13
48	.33874	.36002	2.7776	.94088	12
49	.33901	.36035	2.7751	.94078	11
50	.33929	.36068	2.7725	.94068	10
51	.33956	.36101	2.7700	.94058	9
52	.33983	.36134	2.7675	.94049	8
53	.34011	.36167	2.7650	.94039	7
54	.34038	.36199	2.7625	.94029	6
55	.34065	.36232	2.7600	.94019	5
56	.34093	.36265	2.7575	.94009	4
57	.34120	.36298	2.7550	.93999	3
58	.34147	.36331	2.7525	.93989	2
59	.34175	.36364	2.7500	.93979	1
60	.34202	.36397	2.7475	.93969	0
′	Cos	Ctn	Tan	Sin	′

109° (289°) (250°) **70°**

20° (200°) (339°) **159°**

′	Sin	Tan	Ctn	Cos	′
0	.34202	.36397	2.7475	.93969	60
1	.34229	.36430	2.7450	.93959	59
2	.34257	.36463	2.7425	.93949	58
3	.34284	.36496	2.7400	.93939	57
4	.34311	.36529	2.7376	.93929	56
5	.34339	.36562	2.7351	.93919	55
6	.34366	.36595	2.7326	.93909	54
7	.34393	.36628	2.7302	.93899	53
8	.34421	.36661	2.7277	.93889	52
9	.34448	.36694	2.7253	.93879	51
10	.34475	.36727	2.7228	.93869	50
11	.34503	.36760	2.7204	.93859	49
12	.34530	.36793	2.7179	.93849	48
13	.34557	.36826	2.7155	.93839	47
14	.34584	.36859	2.7130	.93829	46
15	.34612	.36892	2.7106	.93819	45
16	.34639	.36925	2.7082	.93809	44
17	.34666	.36958	2.7058	.93799	43
18	.34694	.36991	2.7034	.93789	42
19	.34721	.37024	2.7009	.93779	41
20	.34748	.37057	2.6985	.93769	40
21	.34775	.37090	2.6961	.93759	39
22	.34803	.37123	2.6937	.93748	38
23	.34830	.37157	2.6913	.93738	37
24	.34857	.37190	2.6889	.93728	36
25	.34884	.37223	2.6865	.93718	35
26	.34912	.37256	2.6841	.93708	34
27	.34939	.37289	2.6818	.93698	33
28	.34966	.37322	2.6794	.93688	32
29	.34993	.37355	2.6770	.93677	31
30	.35021	.37388	2.6746	.93667	30
31	.35048	.37422	2.6723	.93657	29
32	.35075	.37455	2.6699	.93647	28
33	.35102	.37488	2.6675	.93637	27
34	.35130	.37521	2.6652	.93626	26
35	.35157	.37554	2.6628	.93616	25
36	.35184	.37588	2.6605	.93606	24
37	.35211	.37621	2.6581	.93596	23
38	.35239	.37654	2.6558	.93585	22
39	.35266	.37687	2.6534	.93575	21
40	.35293	.37720	2.6511	.93565	20
41	.35320	.37754	2.6488	.93555	19
42	.35347	.37787	2.6464	.93544	18
43	.35375	.37820	2.6441	.93534	17
44	.35402	.37853	2.6418	.93524	16
45	.35429	.37887	2.6395	.93514	15
46	.35456	.37920	2.6371	.93503	14
47	.35484	.37953	2.6348	.93493	13
48	.35511	.37986	2.6325	.93483	12
49	.35538	.38020	2.6302	.93472	11
50	.35565	.38053	2.6279	.93462	10
51	.35592	.38086	2.6256	.93452	9
52	.35619	.38120	2.6233	.93441	8
53	.35647	.38153	2.6210	.93431	7
54	.35674	.38186	2.6187	.93420	6
55	.35701	.38220	2.6165	.93410	5
56	.35728	.38253	2.6142	.93400	4
57	.35755	.38286	2.6119	.93389	3
58	.35782	.38320	2.6096	.93379	2
59	.35810	.38353	2.6074	.93368	1
60	.35837	.38386	2.6051	.93358	0
′	Cos	Ctn	Tan	Sin	′

110° (290°) (249°) **69°**

21° (201°) (338°) **158°**

′	Sin	Tan	Ctn	Cos	′
0	.35837	.38386	2.6051	.93358	60
1	.35864	.38420	2.6028	.93348	59
2	.35891	.38453	2.6006	.93337	58
3	.35918	.38487	2.5983	.93327	57
4	.35945	.38520	2.5961	.93316	56
5	.35973	.38553	2.5938	.93306	55
6	.36000	.38587	2.5916	.93295	54
7	.36027	.38620	2.5893	.93285	53
8	.36054	.38654	2.5871	.93274	52
9	.36081	.38687	2.5848	.93264	51
10	.36108	.38721	2.5826	.93253	50
11	.36135	.38754	2.5804	.93243	49
12	.36162	.38787	2.5782	.93232	48
13	.36190	.38821	2.5759	.93222	47
14	.36217	.38854	2.5737	.93211	46
15	.36244	.38888	2.5715	.93201	45
16	.36271	.38921	2.5693	.93190	44
17	.36298	.38955	2.5671	.93180	43
18	.36325	.38988	2.5649	.93169	42
19	.36352	.39022	2.5627	.93159	41
20	.36379	.39055	2.5605	.93148	40
21	.36406	.39089	2.5583	.93137	39
22	.36434	.39122	2.5561	.93127	38
23	.36461	.39156	2.5539	.93116	37
24	.36488	.39190	2.5517	.93106	36
25	.36515	.39223	2.5495	.93095	35
26	.36542	.39257	2.5473	.93084	34
27	.36569	.39290	2.5452	.93074	33
28	.36596	.39324	2.5430	.93063	32
29	.36623	.39357	2.5408	.93052	31
30	.36650	.39391	2.5386	.93042	30
31	.36677	.39425	2.5365	.93031	29
32	.36704	.39458	2.5343	.93020	28
33	.36731	.39492	2.5322	.93010	27
34	.36758	.39526	2.5300	.92999	26
35	.36785	.39559	2.5279	.92988	25
36	.36812	.39593	2.5257	.92978	24
37	.36839	.39626	2.5236	.92967	23
38	.36867	.39660	2.5214	.92956	22
39	.36894	.39694	2.5193	.92945	21
40	.36921	.39727	2.5172	.92935	20
41	.36948	.39761	2.5150	.92924	19
42	.36975	.39795	2.5129	.92913	18
43	.37002	.39829	2.5108	.92902	17
44	.37029	.39862	2.5086	.92892	16
45	.37056	.39896	2.5065	.92881	15
46	.37083	.39930	2.5044	.92870	14
47	.37110	.39963	2.5023	.92859	13
48	.37137	.39997	2.5002	.92849	12
49	.37164	.40031	2.4981	.92838	11
50	.37191	.40065	2.4960	.92827	10
51	.37218	.40098	2.4939	.92816	9
52	.37245	.40132	2.4918	.92805	8
53	.37272	.40166	2.4897	.92794	7
54	.37299	.40200	2.4876	.92784	6
55	.37326	.40234	2.4855	.92773	5
56	.37353	.40267	2.4834	.92762	4
57	.37380	.40301	2.4813	.92751	3
58	.37407	.40335	2.4792	.92740	2
59	.37434	.40369	2.4772	.92729	1
60	.37461	.40403	2.4751	.92718	0
′	Cos	Ctn	Tan	Sin	′

111° (291°) (248°) **68°**

22° (202°) (337°) **157°**

′	Sin	Tan	Ctn	Cos	′
0	.37461	.40403	2.4751	.92718	60
1	.37488	.40436	2.4730	.92707	59
2	.37515	.40470	2.4709	.92697	58
3	.37542	.40504	2.4689	.92686	57
4	.37569	.40538	2.4668	.92675	56
5	.37595	.40572	2.4648	.92664	55
6	.37622	.40606	2.4627	.92653	54
7	.37649	.40640	2.4606	.92642	53
8	.37676	.40674	2.4586	.92631	52
9	.37703	.40707	2.4566	.92620	51
10	.37730	.40741	2.4545	.92609	50
11	.37757	.40775	2.4525	.92598	49
12	.37784	.40809	2.4504	.92587	48
13	.37811	.40843	2.4484	.92576	47
14	.37838	.40877	2.4464	.92565	46
15	.37865	.40911	2.4443	.92554	45
16	.37892	.40945	2.4423	.92543	44
17	.37919	.40979	2.4403	.92532	43
18	.37946	.41013	2.4383	.92521	42
19	.37973	.41047	2.4362	.92510	41
20	.37999	.41081	2.4342	.92499	40
21	.38026	.41115	2.4322	.92488	39
22	.38053	.41149	2.4302	.92477	38
23	.38080	.41183	2.4282	.92466	37
24	.38107	.41217	2.4262	.92455	36
25	.38134	.41251	2.4242	.92444	35
26	.38161	.41285	2.4222	.92432	34
27	.38188	.41319	2.4202	.92421	33
28	.38215	.41353	2.4182	.92410	32
29	.38241	.41387	2.4162	.92399	31
30	.38268	.41421	2.4142	.92388	30
31	.38295	.41455	2.4122	.92377	29
32	.38322	.41490	2.4102	.92366	28
33	.38349	.41524	2.4083	.92355	27
34	.38376	.41558	2.4063	.92343	26
35	.38403	.41592	2.4043	.92332	25
36	.38430	.41626	2.4023	.92321	24
37	.38456	.41660	2.4004	.92310	23
38	.38483	.41694	2.3984	.92299	22
39	.38510	.41728	2.3964	.92287	21
40	.38537	.41763	2.3945	.92276	20
41	.38564	.41797	2.3925	.92265	19
42	.38591	.41831	2.3906	.92254	18
43	.38617	.41865	2.3886	.92243	17
44	.38644	.41899	2.3867	.92231	16
45	.38671	.41933	2.3847	.92220	15
46	.38698	.41968	2.3828	.92209	14
47	.38725	.42002	2.3808	.92198	13
48	.38752	.42036	2.3789	.92186	12
49	.38778	.42070	2.3770	.92175	11
50	.38805	.42105	2.3750	.92164	10
51	.38832	.42139	2.3731	.92152	9
52	.38859	.42173	2.3712	.92141	8
53	.38886	.42207	2.3693	.92130	7
54	.38912	.42242	2.3673	.92119	6
55	.38939	.42276	2.3654	.92107	5
56	.38966	.42310	2.3635	.92096	4
57	.38993	.42345	2.3616	.92085	3
58	.39020	.42379	2.3597	.92073	2
59	.39046	.42413	2.3578	.92062	1
60	.39073	.42447	2.3559	.92050	0
′	Cos	Ctn	Tan	Sin	′

112° (292°) (247°) **67°**

23° (203°) (336°) **156°**

′	Sin	Tan	Ctn	Cos	′
0	.39073	.42447	2.3559	.92050	60
1	.39100	.42482	2.3539	.92039	59
2	.39127	.42516	2.3520	.92028	58
3	.39153	.42551	2.3501	.92016	57
4	.39180	.42585	2.3483	.92005	56
5	.39207	.42619	2.3464	.91994	55
6	.39234	.42654	2.3445	.91982	54
7	.39260	.42688	2.3426	.91971	53
8	.39287	.42722	2.3407	.91959	52
9	.39314	.42757	2.3388	.91948	51
10	.39341	.42791	2.3369	.91936	50
11	.39367	.42826	2.3351	.91925	49
12	.39394	.42860	2.3332	.91914	48
13	.39421	.42894	2.3313	.91902	47
14	.39448	.42929	2.3294	.91891	46
15	.39474	.42963	2.3276	.91879	45
16	.39501	.42998	2.3257	.91868	44
17	.39528	.43032	2.3238	.91856	43
18	.39555	.43067	2.3220	.91845	42
19	.39581	.43101	2.3201	.91833	41
20	.39608	.43136	2.3183	.91822	40
21	.39635	.43170	2.3164	.91810	39
22	.39661	.43205	2.3146	.91799	38
23	.39688	.43239	2.3127	.91787	37
24	.39715	.43274	2.3109	.91775	36
25	.39741	.43308	2.3090	.91764	35
26	.39768	.43343	2.3072	.91752	34
27	.39795	.43378	2.3053	.91741	33
28	.39822	.43412	2.3035	.91729	32
29	.39848	.43447	2.3017	.91718	31
30	.39875	.43481	2.2998	.91706	30
31	.39902	.43516	2.2980	.91694	29
32	.39928	.43550	2.2962	.91683	28
33	.39955	.43585	2.2944	.91671	27
34	.39982	.43620	2.2925	.91660	26
35	.40008	.43654	2.2907	.91648	25
36	.40035	.43689	2.2889	.91636	24
37	.40062	.43724	2.2871	.91625	23
38	.40088	.43758	2.2853	.91613	22
39	.40115	.43793	2.2835	.91601	21
40	.40141	.43828	2.2817	.91590	20
41	.40168	.43862	2.2799	.91578	19
42	.40195	.43897	2.2781	.91566	18
43	.40221	.43932	2.2763	.91555	17
44	.40248	.43966	2.2745	.91543	16
45	.40275	.44001	2.2727	.91531	15
46	.40301	.44036	2.2709	.91519	14
47	.40328	.44071	2.2691	.91508	13
48	.40355	.44105	2.2673	.91496	12
49	.40381	.44140	2.2655	.91484	11
50	.40408	.44175	2.2637	.91472	10
51	.40434	.44210	2.2620	.91461	9
52	.40461	.44244	2.2602	.91449	8
53	.40488	.44279	2.2584	.91437	7
54	.40514	.44314	2.2566	.91425	6
55	.40541	.44349	2.2549	.91414	5
56	.40567	.44384	2.2531	.91402	4
57	.40594	.44418	2.2513	.91390	3
58	.40621	.44453	2.2496	.91378	2
59	.40647	.44488	2.2478	.91366	1
60	.40674	.44523	2.2460	.91355	0
′	Cos	Ctn	Tan	Sin	′

113° (293°) (246°) **66°**

24° (204°) (335°) **155°**

′	Sin	Tan	Ctn	Cos	′
0	.40674	.44523	2.2460	.91355	60
1	.40700	.44558	2.2443	.91343	59
2	.40727	.44593	2.2425	.91331	58
3	.40753	.44627	2.2408	.91319	57
4	.40780	.44662	2.2390	.91307	56
5	.40806	.44697	2.2373	.91295	55
6	.40833	.44732	2.2355	.91283	54
7	.40860	.44767	2.2338	.91272	53
8	.40886	.44802	2.2320	.91260	52
9	.40913	.44837	2.2303	.91248	51
10	.40939	.44872	2.2286	.91236	50
11	.40966	.44907	2.2268	.91224	49
12	.40992	.44942	2.2251	.91212	48
13	.41019	.44977	2.2234	.91200	47
14	.41045	.45012	2.2216	.91188	46
15	.41072	.45047	2.2199	.91176	45
16	.41098	.45082	2.2182	.91164	44
17	.41125	.45117	2.2165	.91152	43
18	.41151	.45152	2.2148	.91140	42
19	.41178	.45187	2.2130	.91128	41
20	.41204	.45222	2.2113	.91116	40
21	.41231	.45257	2.2096	.91104	39
22	.41257	.45292	2.2079	.91092	38
23	.41284	.45327	2.2062	.91080	37
24	.41310	.45362	2.2045	.91068	36
25	.41337	.45397	2.2028	.91056	35
26	.41363	.45432	2.2011	.91044	34
27	.41390	.45467	2.1994	.91032	33
28	.41416	.45502	2.1977	.91020	32
29	.41443	.45538	2.1960	.91008	31
30	.41469	.45573	2.1943	.90996	30
31	.41496	.45608	2.1926	.90984	29
32	.41522	.45643	2.1909	.90972	28
33	.41549	.45678	2.1892	.90960	27
34	.41575	.45713	2.1876	.90948	26
35	.41602	.45748	2.1859	.90936	25
36	.41628	.45784	2.1842	.90924	24
37	.41655	.45819	2.1825	.90911	23
38	.41681	.45854	2.1808	.90899	22
39	.41707	.45889	2.1792	.90887	21
40	.41734	.45924	2.1775	.90875	20
41	.41760	.45960	2.1758	.90863	19
42	.41787	.45995	2.1742	.90851	18
43	.41813	.46030	2.1725	.90839	17
44	.41840	.46065	2.1708	.90826	16
45	.41866	.46101	2.1692	.90814	15
46	.41892	.46136	2.1675	.90802	14
47	.41919	.46171	2.1659	.90790	13
48	.41945	.46206	2.1642	.90778	12
49	.41972	.46242	2.1625	.90766	11
50	.41998	.46277	2.1609	.90753	10
51	.42024	.46312	2.1592	.90741	9
52	.42051	.46348	2.1576	.90729	8
53	.42077	.46383	2.1560	.90717	7
54	.42104	.46418	2.1543	.90704	6
55	.42130	.46454	2.1527	.90692	5
56	.42156	.46489	2.1510	.90680	4
57	.42183	.46525	2.1494	.90668	3
58	.42209	.46560	2.1478	.90655	2
59	.42235	.46595	2.1461	.90643	1
60	.42262	.46631	2.1445	.90631	0
′	Cos	Ctn	Tan	Sin	′

114° (294°) (245°) **65°**

25° (205°) (334°) **154°**

′	Sin	Tan	Ctn	Cos	′
0	.42262	.46631	2.1445	.90631	60
1	.42288	.46666	2.1429	.90618	59
2	.42315	.46702	2.1413	.90606	58
3	.42341	.46737	2.1396	.90594	57
4	.42367	.46772	2.1380	.90582	56
5	.42394	.46808	2.1364	.90569	55
6	.42420	.46843	2.1348	.90557	54
7	.42446	.46879	2.1332	.90545	53
8	.42473	.46914	2.1315	.90532	52
9	.42499	.46950	2.1299	.90520	51
10	.42525	.46985	2.1283	.90507	50
11	.42552	.47021	2.1267	.90495	49
12	.42578	.47056	2.1251	.90483	48
13	.42604	.47092	2.1235	.90470	47
14	.42631	.47128	2.1219	.90458	46
15	.42657	.47163	2.1203	.90446	45
16	.42683	.47199	2.1187	.90433	44
17	.42709	.47234	2.1171	.90421	43
18	.42736	.47270	2.1155	.90408	42
19	.42762	.47305	2.1139	.90396	41
20	.42788	.47341	2.1123	.90383	40
21	.42815	.47377	2.1107	.90371	39
22	.42841	.47412	2.1092	.90358	38
23	.42867	.47448	2.1076	.90346	37
24	.42894	.47483	2.1060	.90334	36
25	.42920	.47519	2.1044	.90321	35
26	.42946	.47555	2.1028	.90309	34
27	.42972	.47590	2.1013	.90296	33
28	.42999	.47626	2.0997	.90284	32
29	.43025	.47662	2.0981	.90271	31
30	.43051	.47698	2.0965	.90259	30
31	.43077	.47733	2.0950	.90246	29
32	.43104	.47769	2.0934	.90233	28
33	.43130	.47805	2.0918	.90221	27
34	.43156	.47840	2.0903	.90208	26
35	.43182	.47876	2.0887	.90196	25
36	.43209	.47912	2.0872	.90183	24
37	.43235	.47948	2.0856	.90171	23
38	.43261	.47984	2.0840	.90158	22
39	.43287	.48019	2.0825	.90146	21
40	.43313	.48055	2.0809	.90133	20
41	.43340	.48091	2.0794	.90120	19
42	.43366	.48127	2.0778	.90108	18
43	.43392	.48163	2.0763	.90095	17
44	.43418	.48198	2.0748	.90082	16
45	.43445	.48234	2.0732	.90070	15
46	.43471	.48270	2.0717	.90057	14
47	.43497	.48306	2.0701	.90045	13
48	.43523	.48342	2.0686	.90032	12
49	.43549	.48378	2.0671	.90019	11
50	.43575	.48414	2.0655	.90007	10
51	.43602	.48450	2.0640	.89994	9
52	.43628	.48486	2.0625	.89981	8
53	.43654	.48521	2.0609	.89968	7
54	.43680	.48557	2.0594	.89956	6
55	.43706	.48593	2.0579	.89943	5
56	.43733	.48629	2.0564	.89930	4
57	.43759	.48665	2.0549	.89918	3
58	.43785	.48701	2.0533	.89905	2
59	.43811	.48737	2.0518	.89892	1
60	.43837	.48773	2.0503	.89879	0
′	Cos	Ctn	Tan	Sin	′

115° (295°) (244°) **64°**

26° (206°) (333°) **153°**

′	Sin	Tan	Ctn	Cos	′
0	.43837	.48773	2.0503	.89879	60
1	.43863	.48809	2.0488	.89867	59
2	.43889	.48845	2.0473	.89854	58
3	.43916	.48881	2.0458	.89841	57
4	.43942	.48917	2.0443	.89828	56
5	.43968	.48953	2.0428	.89816	55
6	.43994	.48989	2.0413	.89803	54
7	.44020	.49026	2.0398	.89790	53
8	.44046	.49062	2.0383	.89777	52
9	.44072	.49098	2.0368	.89764	51
10	.44098	.49134	2.0353	.89752	50
11	.44124	.49170	2.0338	.89739	49
12	.44151	.49206	2.0323	.89726	48
13	.44177	.49242	2.0308	.89713	47
14	.44203	.49278	2.0293	.89700	46
15	.44229	.49315	2.0278	.89687	45
16	.44255	.49351	2.0263	.89674	44
17	.44281	.49387	2.0248	.89662	43
18	.44307	.49423	2.0233	.89649	42
19	.44333	.49459	2.0219	.89636	41
20	.44359	.49495	2.0204	.89623	40
21	.44385	.49532	2.0189	.89610	39
22	.44411	.49568	2.0174	.89597	38
23	.44437	.49604	2.0160	.89584	37
24	.44464	.49640	2.0145	.89571	36
25	.44490	.49677	2.0130	.89558	35
26	.44516	.49713	2.0115	.89545	34
27	.44542	.49749	2.0101	.89532	33
28	.44568	.49786	2.0086	.89519	32
29	.44594	.49822	2.0072	.89506	31
30	.44620	.49858	2.0057	.89493	30
31	.44646	.49894	2.0042	.89480	29
32	.44672	.49931	2.0028	.89467	28
33	.44698	.49967	2.0013	.89454	27
34	.44724	.50004	1.9999	.89441	26
35	.44750	.50040	1.9984	.89428	25
36	.44776	.50076	1.9970	.89415	24
37	.44802	.50113	1.9955	.89402	23
38	.44828	.50149	1.9941	.89389	22
39	.44854	.50185	1.9926	.89376	21
40	.44880	.50222	1.9912	.89363	20
41	.44906	.50258	1.9897	.89350	19
42	.44932	.50295	1.9883	.89337	18
43	.44958	.50331	1.9868	.89324	17
44	.44984	.50368	1.9854	.89311	16
45	.45010	.50404	1.9840	.89298	15
46	.45036	.50441	1.9825	.89285	14
47	.45062	.50477	1.9811	.89272	13
48	.45088	.50514	1.9797	.89259	12
49	.45114	.50550	1.9782	.89245	11
50	.45140	.50587	1.9768	.89232	10
51	.45166	.50623	1.9754	.89219	9
52	.45192	.50660	1.9740	.89206	8
53	.45218	.50696	1.9725	.89193	7
54	.45243	.50733	1.9711	.89180	6
55	.45269	.50769	1.9697	.89167	5
56	.45295	.50806	1.9683	.89153	4
57	.45321	.50843	1.9669	.89140	3
58	.45347	.50879	1.9654	.89127	2
59	.45373	.50916	1.9640	.89114	1
60	.45399	.50953	1.9626	.89101	0
′	Cos	Ctn	Tan	Sin	′

116° (296°) (243°) **63°**

27° (207°) (332°) **152°**

′	Sin	Tan	Ctn	Cos	′
0	.45399	.50953	1.9626	.89101	60
1	.45425	.50989	1.9612	.89087	59
2	.45451	.51026	1.9598	.89074	58
3	.45477	.51063	1.9584	.89061	57
4	.45503	.51099	1.9570	.89048	56
5	.45529	.51136	1.9556	.89035	55
6	.45554	.51173	1.9542	.89021	54
7	.45580	.51209	1.9528	.89008	53
8	.45606	.51246	1.9514	.88995	52
9	.45632	.51283	1.9500	.88981	51
10	.45658	.51319	1.9486	.88968	50
11	.45684	.51356	1.9472	.88955	49
12	.45710	.51393	1.9458	.88942	48
13	.45736	.51430	1.9444	.88928	47
14	.45762	.51467	1.9430	.88915	46
15	.45787	.51503	1.9416	.88902	45
16	.45813	.51540	1.9402	.88888	44
17	.45839	.51577	1.9388	.88875	43
18	.45865	.51614	1.9375	.88862	42
19	.45891	.51651	1.9361	.88848	41
20	.45917	.51688	1.9347	.88835	40
21	.45942	.51724	1.9333	.88822	39
22	.45968	.51761	1.9319	.88808	38
23	.45994	.51798	1.9306	.88795	37
24	.46020	.51835	1.9292	.88782	36
25	.46046	.51872	1.9278	.88768	35
26	.46072	.51909	1.9265	.88755	34
27	.46097	.51946	1.9251	.88741	33
28	.46123	.51983	1.9237	.88728	32
29	.46149	.52020	1.9223	.88715	31
30	.46175	.52057	1.9210	.88701	30
31	.46201	.52094	1.9196	.88688	29
32	.46226	.52131	1.9183	.88674	28
33	.46252	.52168	1.9169	.88661	27
34	.46278	.52205	1.9155	.88647	26
35	.46304	.52242	1.9142	.88634	25
36	.46330	.52279	1.9128	.88620	24
37	.46355	.52316	1.9115	.88607	23
38	.46381	.52353	1.9101	.88593	22
39	.46407	.52390	1.9088	.88580	21
40	.46433	.52427	1.9074	.88566	20
41	.46458	.52464	1.9061	.88553	19
42	.46484	.52501	1.9047	.88539	18
43	.46510	.52538	1.9034	.88526	17
44	.46536	.52575	1.9020	.88512	16
45	.46561	.52613	1.9007	.88499	15
46	.46587	.52650	1.8993	.88485	14
47	.46613	.52687	1.8980	.88472	13
48	.46639	.52724	1.8967	.88458	12
49	.46664	.52761	1.8953	.88445	11
50	.46690	.52798	1.8940	.88431	10
51	.46716	.52836	1.8927	.88417	9
52	.46742	.52873	1.8913	.88404	8
53	.46767	.52910	1.8900	.88390	7
54	.46793	.52947	1.8887	.88377	6
55	.46819	.52985	1.8873	.88363	5
56	.46844	.53022	1.8860	.88349	4
57	.46870	.53059	1.8847	.88336	3
58	.46896	.53096	1.8834	.88322	2
59	.46921	.53134	1.8820	.88308	1
60	.46947	.53171	1.8807	.88295	0
′	Cos	Ctn	Tan	Sin	′

117° (297°) (242°) **62°**

28° (208°) (331°) **151°**

′	Sin	Tan	Ctn	Cos	′
0	.46947	.53171	1.8807	.88295	60
1	.46973	.53208	1.8794	.88281	59
2	.46999	.53246	1.8781	.88267	58
3	.47024	.53283	1.8768	.88254	57
4	.47050	.53320	1.8755	.88240	56
5	.47076	.53358	1.8741	.88226	55
6	.47101	.53395	1.8728	.88213	54
7	.47127	.53432	1.8715	.88199	53
8	.47153	.53470	1.8702	.88185	52
9	.47178	.53507	1.8689	.88172	51
10	.47204	.53545	1.8676	.88158	50
11	.47229	.53582	1.8663	.88144	49
12	.47255	.53620	1.8650	.88130	48
13	.47281	.53657	1.8637	.88117	47
14	.47306	.53694	1.8624	.88103	46
15	.47332	.53732	1.8611	.88089	45
16	.47358	.53769	1.8598	.88075	44
17	.47383	.53807	1.8585	.88062	43
18	.47409	.53844	1.8572	.88048	42
19	.47434	.53882	1.8559	.88034	41
20	.47460	.53920	1.8546	.88020	40
21	.47486	.53957	1.8533	.88006	39
22	.47511	.53995	1.8520	.87993	38
23	.47537	.54032	1.8507	.87979	37
24	.47562	.54070	1.8495	.87965	36
25	.47588	.54107	1.8482	.87951	35
26	.47614	.54145	1.8469	.87937	34
27	.47639	.54183	1.8456	.87923	33
28	.47665	.54220	1.8443	.87909	32
29	.47690	.54258	1.8430	.87896	31
30	.47716	.54296	1.8418	.87882	30
31	.47741	.54333	1.8405	.87868	29
32	.47767	.54371	1.8392	.87854	28
33	.47793	.54409	1.8379	.87840	27
34	.47818	.54446	1.8367	.87826	26
35	.47844	.54484	1.8354	.87812	25
36	.47869	.54522	1.8341	.87798	24
37	.47895	.54560	1.8329	.87784	23
38	.47920	.54597	1.8316	.87770	22
39	.47946	.54635	1.8303	.87756	21
40	.47971	.54673	1.8291	.87743	20
41	.47997	.54711	1.8278	.87729	19
42	.48022	.54748	1.8265	.87715	18
43	.48048	.54786	1.8253	.87701	17
44	.48073	.54824	1.8240	.87687	16
45	.48099	.54862	1.8228	.87673	15
46	.48124	.54900	1.8215	.87659	14
47	.48150	.54938	1.8202	.87645	13
48	.48175	.54975	1.8190	.87631	12
49	.48201	.55013	1.8177	.87617	11
50	.48226	.55051	1.8165	.87603	10
51	.48252	.55089	1.8152	.87589	9
52	.48277	.55127	1.8140	.87575	8
53	.48303	.55165	1.8127	.87561	7
54	.48328	.55203	1.8115	.87546	6
55	.48354	.55241	1.8103	.87532	5
56	.48379	.55279	1.8090	.87518	4
57	.48405	.55317	1.8078	.87504	3
58	.48430	.55355	1.8065	.87490	2
59	.48456	.55393	1.8053	.87476	1
60	.48481	.55431	1.8040	.87462	0
′	Cos	Ctn	Tan	Sin	′

118° (298°) (241°) **61°**

29° (209°) (330°) **150°**

′	Sin	Tan	Ctn	Cos	′
0	.48481	.55431	1.8040	.87462	60
1	.48506	.55469	1.8028	.87448	59
2	.48532	.55507	1.8016	.87434	58
3	.48557	.55545	1.8003	.87420	57
4	.48583	.55583	1.7991	.87406	56
5	.48608	.55621	1.7979	.87391	55
6	.48634	.55659	1.7966	.87377	54
7	.48659	.55697	1.7954	.87363	53
8	.48684	.55736	1.7942	.87349	52
9	.48710	.55774	1.7930	.87335	51
10	.48735	.55812	1.7917	.87321	50
11	.48761	.55850	1.7905	.87306	49
12	.48786	.55888	1.7893	.87292	48
13	.48811	.55926	1.7881	.87278	47
14	.48837	.55964	1.7868	.87264	46
15	.48862	.56003	1.7856	.87250	45
16	.48888	.56041	1.7844	.87235	44
17	.48913	.56079	1.7832	.87221	43
18	.48938	.56117	1.7820	.87207	42
19	.48964	.56156	1.7808	.87193	41
20	.48989	.56194	1.7796	.87178	40
21	.49014	.56232	1.7783	.87164	39
22	.49040	.56270	1.7771	.87150	38
23	.49065	.56309	1.7759	.87136	37
24	.49090	.56347	1.7747	.87121	36
25	.49116	.56385	1.7735	.87107	35
26	.49141	.56424	1.7723	.87093	34
27	.49166	.56462	1.7711	.87079	33
28	.49192	.56501	1.7699	.87064	32
29	.49217	.56539	1.7687	.87050	31
30	.49242	.56577	1.7675	.87036	30
31	.49268	.56616	1.7663	.87021	29
32	.49293	.56654	1.7651	.87007	28
33	.49318	.56693	1.7639	.86993	27
34	.49344	.56731	1.7627	.86978	26
35	.49369	.56769	1.7615	.86964	25
36	.49394	.56808	1.7603	.86949	24
37	.49419	.56846	1.7591	.86935	23
38	.49445	.56885	1.7579	.86921	22
39	.49470	.56923	1.7567	.86906	21
40	.49495	.56962	1.7556	.86892	20
41	.49521	.57000	1.7544	.86878	19
42	.49546	.57039	1.7532	.86863	18
43	.49571	.57078	1.7520	.86849	17
44	.49596	.57116	1.7508	.86834	16
45	.49622	.57155	1.7496	.86820	15
46	.49647	.57193	1.7485	.86805	14
47	.49672	.57232	1.7473	.86791	13
48	.49697	.57271	1.7461	.86777	12
49	.49723	.57309	1.7449	.86762	11
50	.49748	.57348	1.7437	.86748	10
51	.49773	.57386	1.7426	.86733	9
52	.49798	.57425	1.7414	.86719	8
53	.49824	.57464	1.7402	.86704	7
54	.49849	.57503	1.7391	.86690	6
55	.49874	.57541	1.7379	.86675	5
56	.49899	.57580	1.7367	.86661	4
57	.49924	.57619	1.7355	.86646	3
58	.49950	.57657	1.7344	.86632	2
59	.49975	.57696	1.7332	.86617	1
60	.50000	.57735	1.7321	.86603	0
′	Cos	Ctn	Tan	Sin	′

119° (299°) (240°) **60°**

30° (210°) (329°) **149°**

′	Sin	Tan	Ctn	Cos	′
0	.50000	.57735	1.7321	.86603	60
1	.50025	.57774	1.7309	.86588	59
2	.50050	.57813	1.7297	.86573	58
3	.50076	.57851	1.7286	.86559	57
4	.50101	.57890	1.7274	.86544	56
5	.50126	.57929	1.7262	.86530	55
6	.50151	.57968	1.7251	.86515	54
7	.50176	.58007	1.7239	.86501	53
8	.50201	.58046	1.7228	.86486	52
9	.50227	.58085	1.7216	.86471	51
10	.50252	.58124	1.7205	.86457	50
11	.50277	.58162	1.7193	.86442	49
12	.50302	.58201	1.7182	.86427	48
13	.50327	.58240	1.7170	.86413	47
14	.50352	.58279	1.7159	.86398	46
15	.50377	.58318	1.7147	.86384	45
16	.50403	.58357	1.7136	.86369	44
17	.50428	.58396	1.7124	.86354	43
18	.50453	.58435	1.7113	.86340	42
19	.50478	.58474	1.7102	.86325	41
20	.50503	.58513	1.7090	.86310	40
21	.50528	.58552	1.7079	.86295	39
22	.50553	.58591	1.7067	.86281	38
23	.50578	.58631	1.7056	.86266	37
24	.50603	.58670	1.7045	.86251	36
25	.50628	.58709	1.7033	.86237	35
26	.50654	.58748	1.7022	.86222	34
27	.50679	.58787	1.7011	.86207	33
28	.50704	.58826	1.6999	.86192	32
29	.50729	.58865	1.6988	.86178	31
30	.50754	.58905	1.6977	.86163	30
31	.50779	.58944	1.6965	.86148	29
32	.50804	.58983	1.6954	.86133	28
33	.50829	.59022	1.6943	.86119	27
34	.50854	.59061	1.6932	.86104	26
35	.50879	.59101	1.6920	.86089	25
36	.50904	.59140	1.6909	.86074	24
37	.50929	.59179	1.6898	.86059	23
38	.50954	.59218	1.6887	.86045	22
39	.50979	.59258	1.6875	.86030	21
40	.51004	.59297	1.6864	.86015	20
41	.51029	.59336	1.6853	.86000	19
42	.51054	.59376	1.6842	.85985	18
43	.51079	.59415	1.6831	.85970	17
44	.51104	.59454	1.6820	.85956	16
45	.51129	.59494	1.6808	.85941	15
46	.51154	.59533	1.6797	.85926	14
47	.51179	.59573	1.6786	.85911	13
48	.51204	.59612	1.6775	.85896	12
49	.51229	.59651	1.6764	.85881	11
50	.51254	.59691	1.6753	.85866	10
51	.51279	.59730	1.6742	.85851	9
52	.51304	.59770	1.6731	.85836	8
53	.51329	.59809	1.6720	.85821	7
54	.51354	.59849	1.6709	.85806	6
55	.51379	.59888	1.6698	.85792	5
56	.51404	.59928	1.6687	.85777	4
57	.51429	.59967	1.6676	.85762	3
58	.51454	.60007	1.6665	.85747	2
59	.51479	.60046	1.6654	.85732	1
60	.51504	.60086	1.6643	.85717	0
′	Cos	Ctn	Tan	Sin	′

120° (300°) (239°) **59°**

31° (211°) (328°) **148°**

′	Sin	Tan	Ctn	Cos	′
0	.51504	.60086	1.6643	.85717	60
1	.51529	.60126	1.6632	.85702	59
2	.51554	.60165	1.6621	.85687	58
3	.51579	.60205	1.6610	.85672	57
4	.51604	.60245	1.6599	.85657	56
5	.51628	.60284	1.6588	.85642	55
6	.51653	.60324	1.6577	.85627	54
7	.51678	.60364	1.6566	.85612	53
8	.51703	.60403	1.6555	.85597	52
9	.51728	.60443	1.6545	.85582	51
10	.51753	.60483	1.6534	.85567	50
11	.51778	.60522	1.6523	.85551	49
12	.51803	.60562	1.6512	.85536	48
13	.51828	.60602	1.6501	.85521	47
14	.51852	.60642	1.6490	.85506	46
15	.51877	.60681	1.6479	.85491	45
16	.51902	.60721	1.6469	.85476	44
17	.51927	.60761	1.6458	.85461	43
18	.51952	.60801	1.6447	.85446	42
19	.51977	.60841	1.6436	.85431	41
20	.52002	.60881	1.6426	.85416	40
21	.52026	.60921	1.6415	.85401	39
22	.52051	.60960	1.6404	.85385	38
23	.52076	.61000	1.6393	.85370	37
24	.52101	.61040	1.6383	.85355	36
25	.52126	.61080	1.6372	.85340	35
26	.52151	.61120	1.6361	.85325	34
27	.52175	.61160	1.6351	.85310	33
28	.52200	.61200	1.6340	.85294	32
29	.52225	.61240	1.6329	.85279	31
30	.52250	.61280	1.6319	.85264	30
31	.52275	.61320	1.6308	.85249	29
32	.52299	.61360	1.6297	.85234	28
33	.52324	.61400	1.6287	.85218	27
34	.52349	.61440	1.6276	.85203	26
35	.52374	.61480	1.6265	.85188	25
36	.52399	.61520	1.6255	.85173	24
37	.52423	.61561	1.6244	.85157	23
38	.52448	.61601	1.6234	.85142	22
39	.52473	.61641	1.6223	.85127	21
40	.52498	.61681	1.6212	.85112	20
41	.52522	.61721	1.6202	.85096	19
42	.52547	.61761	1.6191	.85081	18
43	.52572	.61801	1.6181	.85066	17
44	.52597	.61842	1.6170	.85051	16
45	.52621	.61882	1.6160	.85035	15
46	.52646	.61922	1.6149	.85020	14
47	.52671	.61962	1.6139	.85005	13
48	.52696	.62003	1.6128	.84989	12
49	.52720	.62043	1.6118	.84974	11
50	.52745	.62083	1.6107	.84959	10
51	.52770	.62124	1.6097	.84943	9
52	.52794	.62164	1.6087	.84928	8
53	.52819	.62204	1.6076	.84913	7
54	.52844	.62245	1.6066	.84897	6
55	.52869	.62285	1.6055	.84882	5
56	.52893	.62325	1.6045	.84866	4
57	.52918	.62366	1.6034	.84851	3
58	.52943	.62406	1.6024	.84836	2
59	.52967	.62446	1.6014	.84820	1
60	.52992	.62487	1.6003	.84805	0
′	Cos	Ctn	Tan	Sin	′

121° (301°) (238°) **58°**

32° (212°) (327°) **147°**

′	Sin	Tan	Ctn	Cos	′
0	.52992	.62487	1.6003	.84805	60
1	.53017	.62527	1.5993	.84789	59
2	.53041	.62568	1.5983	.84774	58
3	.53066	.62608	1.5972	.84759	57
4	.53091	.62649	1.5962	.84743	56
5	.53115	.62689	1.5952	.84728	55
6	.53140	.62730	1.5941	.84712	54
7	.53164	.62770	1.5931	.84697	53
8	.53189	.62811	1.5921	.84681	52
9	.53214	.62852	1.5911	.84666	51
10	.53238	.62892	1.5900	.84650	50
11	.53263	.62933	1.5890	.84635	49
12	.53288	.62973	1.5880	.84619	48
13	.53312	.63014	1.5869	.84604	47
14	.53337	.63055	1.5859	.84588	46
15	.53361	.63095	1.5849	.84573	45
16	.53386	.63136	1.5839	.84557	44
17	.53411	.63177	1.5829	.84542	43
18	.53435	.63217	1.5818	.84526	42
19	.53460	.63258	1.5808	.84511	41
20	.53484	.63299	1.5798	.84495	40
21	.53509	.63340	1.5788	.84480	39
22	.53534	.63380	1.5778	.84464	38
23	.53558	.63421	1.5768	.84448	37
24	.53583	.63462	1.5757	.84433	36
25	.53607	.63503	1.5747	.84417	35
26	.53632	.63544	1.5737	.84402	34
27	.53656	.63584	1.5727	.84386	33
28	.53681	.63625	1.5717	.84370	32
29	.53705	.63666	1.5707	.84355	31
30	.53730	.63707	1.5697	.84339	30
31	.53754	.63748	1.5687	.84324	29
32	.53779	.63789	1.5677	.84308	28
33	.53804	.63830	1.5667	.84292	27
34	.53828	.63871	1.5657	.84277	26
35	.53853	.63912	1.5647	.84261	25
36	.53877	.63953	1.5637	.84245	24
37	.53902	.63994	1.5627	.84230	23
38	.53926	.64035	1.5617	.84214	22
39	.53951	.64076	1.5607	.84198	21
40	.53975	.64117	1.5597	.84182	20
41	.54000	.64158	1.5587	.84167	19
42	.54024	.64199	1.5577	.84151	18
43	.54049	.64240	1.5567	.84135	17
44	.54073	.64281	1.5557	.84120	16
45	.54097	.64322	1.5547	.84104	15
46	.54122	.64363	1.5537	.84088	14
47	.54146	.64404	1.5527	.84072	13
48	.54171	.64446	1.5517	.84057	12
49	.54195	.64487	1.5507	.84041	11
50	.54220	.64528	1.5497	.84025	10
51	.54244	.64569	1.5487	.84009	9
52	.54269	.64610	1.5477	.83994	8
53	.54293	.64652	1.5468	.83978	7
54	.54317	.64693	1.5458	.83962	6
55	.54342	.64734	1.5448	.83946	5
56	.54366	.64775	1.5438	.83930	4
57	.54391	.64817	1.5428	.83915	3
58	.54415	.64858	1.5418	.83899	2
59	.54440	.64899	1.5408	.83883	1
60	.54464	.64941	1.5399	.83867	0
′	Cos	Ctn	Tan	Sin	′

122° (302°) (237°) **57°**

33° (213°) (326°) **146°**

′	Sin	Tan	Ctn	Cos	′
0	.54464	.64941	1.5399	.83867	60
1	.54488	.64982	1.5389	.83851	59
2	.54513	.65024	1.5379	.83835	58
3	.54537	.65065	1.5369	.83819	57
4	.54561	.65106	1.5359	.83804	56
5	.54586	.65148	1.5350	.83788	55
6	.54610	.65189	1.5340	.83772	54
7	.54635	.65231	1.5330	.83756	53
8	.54659	.65272	1.5320	.83740	52
9	.54683	.65314	1.5311	.83724	51
10	.54708	.65355	1.5301	.83708	50
11	.54732	.65397	1.5291	.83692	49
12	.54756	.65438	1.5282	.83676	48
13	.54781	.65480	1.5272	.83660	47
14	.54805	.65521	1.5262	.83645	46
15	.54829	.65563	1.5253	.83629	45
16	.54854	.65604	1.5243	.83613	44
17	.54878	.65646	1.5233	.83597	43
18	.54902	.65688	1.5224	.83581	42
19	.54927	.65729	1.5214	.83565	41
20	.54951	.65771	1.5204	.83549	40
21	.54975	.65813	1.5195	.83533	39
22	.54999	.65854	1.5185	.83517	38
23	.55024	.65896	1.5175	.83501	37
24	.55048	.65938	1.5166	.83485	36
25	.55072	.65980	1.5156	.83469	35
26	.55097	.66021	1.5147	.83453	34
27	.55121	.66063	1.5137	.83437	33
28	.55145	.66105	1.5127	.83421	32
29	.55169	.66147	1.5118	.83405	31
30	.55194	.66189	1.5108	.83389	30
31	.55218	.66230	1.5099	.83373	29
32	.55242	.66272	1.5089	.83356	28
33	.55266	.66314	1.5080	.83340	27
34	.55291	.66356	1.5070	.83324	26
35	.55315	.66398	1.5061	.83308	25
36	.55339	.66440	1.5051	.83292	24
37	.55363	.66482	1.5042	.83276	23
38	.55388	.66524	1.5032	.83260	22
39	.55412	.66566	1.5023	.83244	21
40	.55436	.66608	1.5013	.83228	20
41	.55460	.66650	1.5004	.83212	19
42	.55484	.66692	1.4994	.83195	18
43	.55509	.66734	1.4985	.83179	17
44	.55533	.66776	1.4975	.83163	16
45	.55557	.66818	1.4966	.83147	15
46	.55581	.66860	1.4957	.83131	14
47	.55605	.66902	1.4947	.83115	13
48	.55630	.66944	1.4938	.83098	12
49	.55654	.66986	1.4928	.83082	11
50	.55678	.67028	1.4919	.83066	10
51	.55702	.67071	1.4910	.83050	9
52	.55726	.67113	1.4900	.83034	8
53	.55750	.67155	1.4891	.83017	7
54	.55775	.67197	1.4882	.83001	6
55	.55799	.67239	1.4872	.82985	5
56	.55823	.67282	1.4863	.82969	4
57	.55847	.67324	1.4854	.82953	3
58	.55871	.67366	1.4844	.82936	2
59	.55895	.67409	1.4835	.82920	1
60	.55919	.67451	1.4826	.82904	0
′	Cos	Ctn	Tan	Sin	′

123° (303°) (236°) **56°**

34° (214°) (325°) **145°**

′	Sin	Tan	Ctn	Cos	′
0	.55919	.67451	1.4826	.82904	60
1	.55943	.67493	1.4816	.82887	59
2	.55968	.67536	1.4807	.82871	58
3	.55992	.67578	1.4798	.82855	57
4	.56016	.67620	1.4788	.82839	56
5	.56040	.67663	1.4779	.82822	55
6	.56064	.67705	1.4770	.82806	54
7	.56088	.67748	1.4761	.82790	53
8	.56112	.67790	1.4751	.82773	52
9	.56136	.67832	1.4742	.82757	51
10	.56160	.67875	1.4733	.82741	50
11	.56184	.67917	1.4724	.82724	49
12	.56208	.67960	1.4715	.82708	48
13	.56232	.68002	1.4705	.82692	47
14	.56256	.68045	1.4696	.82675	46
15	.56280	.68088	1.4687	.82659	45
16	.56305	.68130	1.4678	.82643	44
17	.56329	.68173	1.4669	.82626	43
18	.56353	.68215	1.4659	.82610	42
19	.56377	.68258	1.4650	.82593	41
20	.56401	.68301	1.4641	.82577	40
21	.56425	.68343	1.4632	.82561	39
22	.56449	.68386	1.4623	.82544	38
23	.56473	.68429	1.4614	.82528	37
24	.56497	.68471	1.4605	.82511	36
25	.56521	.68514	1.4596	.82495	35
26	.56545	.68557	1.4586	.82478	34
27	.56569	.68600	1.4577	.82462	33
28	.56593	.68642	1.4568	.82446	32
29	.56617	.68685	1.4559	.82429	31
30	.56641	.68728	1.4550	.82413	30
31	.56665	.68771	1.4541	.82396	29
32	.56689	.68814	1.4532	.82380	28
33	.56713	.68857	1.4523	.82363	27
34	.56736	.68900	1.4514	.82347	26
35	.56760	.68942	1.4505	.82330	25
36	.56784	.68985	1.4496	.82314	24
37	.56808	.69028	1.4487	.82297	23
38	.56832	.69071	1.4478	.82281	22
39	.56856	.69114	1.4469	.82264	21
40	.56880	.69157	1.4460	.82248	20
41	.56904	.69200	1.4451	.82231	19
42	.56928	.69243	1.4442	.82214	18
43	.56952	.69286	1.4433	.82198	17
44	.56976	.69329	1.4424	.82181	16
45	.57000	.69372	1.4415	.82165	15
46	.57024	.69416	1.4406	.82148	14
47	.57047	.69459	1.4397	.82132	13
48	.57071	.69502	1.4388	.82115	12
49	.57095	.69545	1.4379	.82098	11
50	.57119	.69588	1.4370	.82082	10
51	.57143	.69631	1.4361	.82065	9
52	.57167	.69675	1.4352	.82048	8
53	.57191	.69718	1.4344	.82032	7
54	.57215	.69761	1.4335	.82015	6
55	.57238	.69804	1.4326	.81999	5
56	.57262	.69847	1.4317	.81982	4
57	.57286	.69891	1.4308	.81965	3
58	.57310	.69934	1.4299	.81949	2
59	.57334	.69977	1.4290	.81932	1
60	.57358	.70021	1.4281	.81915	0
′	Cos	Ctn	Tan	Sin	′

124° (304°) (235°) **55°**

35° (215°) (324°) **144°**

′	Sin	Tan	Ctn	Cos	′
0	.57358	.70021	1.4281	.81915	60
1	.57381	.70064	1.4273	.81899	59
2	.57405	.70107	1.4264	.81882	58
3	.57429	.70151	1.4255	.81865	57
4	.57453	.70194	1.4246	.81848	56
5	.57477	.70238	1.4237	.81832	55
6	.57501	.70281	1.4229	.81815	54
7	.57524	.70325	1.4220	.81798	53
8	.57548	.70368	1.4211	.81782	52
9	.57572	.70412	1.4202	.81765	51
10	.57596	.70455	1.4193	.81748	50
11	.57619	.70499	1.4185	.81731	49
12	.57643	.70542	1.4176	.81714	48
13	.57667	.70586	1.4167	.81698	47
14	.57691	.70629	1.4158	.81681	46
15	.57715	.70673	1.4150	.81664	45
16	.57738	.70717	1.4141	.81647	44
17	.57762	.70760	1.4132	.81631	43
18	.57786	.70804	1.4124	.81614	42
19	.57810	.70848	1.4115	.81597	41
20	.57833	.70891	1.4106	.81580	40
21	.57857	.70935	1.4097	.81563	39
22	.57881	.70979	1.4089	.81546	38
23	.57904	.71023	1.4080	.81530	37
24	.57928	.71066	1.4071	.81513	36
25	.57952	.71110	1.4063	.81496	35
26	.57976	.71154	1.4054	.81479	34
27	.57999	.71198	1.4045	.81462	33
28	.58023	.71242	1.4037	.81445	32
29	.58047	.71285	1.4028	.81428	31
30	.58070	.71329	1.4019	.81412	30
31	.58094	.71373	1.4011	.81395	29
32	.58118	.71417	1.4002	.81378	28
33	.58141	.71461	1.3994	.81361	27
34	.58165	.71505	1.3985	.81344	26
35	.58189	.71549	1.3976	.81327	25
36	.58212	.71593	1.3968	.81310	24
37	.58236	.71637	1.3959	.81293	23
38	.58260	.71681	1.3951	.81276	22
39	.58283	.71725	1.3942	.81259	21
40	.58307	.71769	1.3934	.81242	20
41	.58330	.71813	1.3925	.81225	19
42	.58354	.71857	1.3916	.81208	18
43	.58378	.71901	1.3908	.81191	17
44	.58401	.71946	1.3899	.81174	16
45	.58425	.71990	1.3891	.81157	15
46	.58449	.72034	1.3882	.81140	14
47	.58472	.72078	1.3874	.81123	13
48	.58496	.72122	1.3865	.81106	12
49	.58519	.72167	1.3857	.81089	11
50	.58543	.72211	1.3848	.81072	10
51	.58567	.72255	1.3840	.81055	9
52	.58590	.72299	1.3831	.81038	8
53	.58614	.72344	1.3823	.81021	7
54	.58637	.72388	1.3814	.81004	6
55	.58661	.72432	1.3806	.80987	5
56	.58684	.72477	1.3798	.80970	4
57	.58708	.72521	1.3789	.80953	3
58	.58731	.72565	1.3781	.80936	2
59	.58755	.72610	1.3772	.80919	1
60	.58779	.72654	1.3764	.80902	0
′	Cos	Ctn	Tan	Sin	′

125° (305°) (234°) **54°**

36° (216°)				**(323°) 143°**	
′	Sin	Tan	Ctn	Cos	′
0	.58779	.72654	1.3764	.80902	60
1	.58802	.72699	1.3755	.80885	59
2	.58826	.72743	1.3747	.80867	58
3	.58849	.72788	1.3739	.80850	57
4	.58873	.72832	1.3730	.80833	56
5	.58896	.72877	1.3722	.80816	55
6	.58920	.72921	1.3713	.80799	54
7	.58943	.72966	1.3705	.80782	53
8	.58967	.73010	1.3697	.80765	52
9	.58990	.73055	1.3688	.80748	51
10	.59014	.73100	1.3680	.80730	50
11	.59037	.73144	1.3672	.80713	49
12	.59061	.73189	1.3663	.80696	48
13	.59084	.73234	1.3655	.80679	47
14	.59108	.73278	1.3647	.80662	46
15	.59131	.73323	1.3638	.80644	45
16	.59154	.73368	1.3630	.80627	44
17	.59178	.73413	1.3622	.80610	43
18	.59201	.73457	1.3613	.80593	42
19	.59225	.73502	1.3605	.80576	41
20	.59248	.73547	1.3597	.80558	40
21	.59272	.73592	1.3588	.80541	39
22	.59295	.73637	1.3580	.80524	38
23	.59318	.73681	1.3572	.80507	37
24	.59342	.73726	1.3564	.80489	36
25	.59365	.73771	1.3555	.80472	35
26	.59389	.73816	1.3547	.80455	34
27	.59412	.73861	1.3539	.80438	33
28	.59436	.73906	1.3531	.80420	32
29	.59459	.73951	1.3522	.80403	31
30	.59482	.73996	1.3514	.80386	30
31	.59506	.74041	1.3506	.80368	29
32	.59529	.74086	1.3498	.80351	28
33	.59552	.74131	1.3490	.80334	27
34	.59576	.74176	1.3481	.80316	26
35	.59599	.74221	1.3473	.80299	25
36	.59622	.74267	1.3465	.80282	24
37	.59646	.74312	1.3457	.80264	23
38	.59669	.74357	1.3449	.80247	22
39	.59693	.74402	1.3440	.80230	21
40	.59716	.74447	1.3432	.80212	20
41	.59739	.74492	1.3424	.80195	19
42	.59763	.74538	1.3416	.80178	18
43	.59786	.74583	1.3408	.80160	17
44	.59809	.74628	1.3400	.80143	16
45	.59832	.74674	1.3392	.80125	15
46	.59856	.74719	1.3384	.80108	14
47	.59879	.74764	1.3375	.80091	13
48	.59902	.74810	1.3367	.80073	12
49	.59926	.74855	1.3359	.80056	11
50	.59949	.74900	1.3351	.80038	10
51	.59972	.74946	1.3343	.80021	9
52	.59995	.74991	1.3335	.80003	8
53	.60019	.75037	1.3327	.79986	7
54	.60042	.75082	1.3319	.79968	6
55	.60065	.75128	1.3311	.79951	5
56	.60089	.75173	1.3303	.79934	4
57	.60112	.75219	1.3295	.79916	3
58	.60135	.75264	1.3287	.79899	2
59	.60158	.75310	1.3278	.79881	1
60	.60182	.75355	1.3270	.79864	0
′	Cos	Ctn	Tan	Sin	′
126° (306°)				**(233°) 53°**	

37° (217°)				**(322°) 142°**	
′	Sin	Tan	Ctn	Cos	′
0	.60182	.75355	1.3270	.79864	60
1	.60205	.75401	1.3262	.79846	59
2	.60228	.75447	1.3254	.79829	58
3	.60251	.75492	1.3246	.79811	57
4	.60274	.75538	1.3238	.79793	56
5	.60298	.75584	1.3230	.79776	55
6	.60321	.75629	1.3222	.79758	54
7	.60344	.75675	1.3214	.79741	53
8	.60367	.75721	1.3206	.79723	52
9	.60390	.75767	1.3198	.79706	51
10	.60414	.75812	1.3190	.79688	50
11	.60437	.75858	1.3182	.79671	49
12	.60460	.75904	1.3175	.79653	48
13	.60483	.75950	1.3167	.79635	47
14	.60506	.75996	1.3159	.79618	46
15	.60529	.76042	1.3151	.79600	45
16	.60553	.76088	1.3143	.79583	44
17	.60576	.76134	1.3135	.79565	43
18	.60599	.76180	1.3127	.79547	42
19	.60622	.76226	1.3119	.79530	41
20	.60645	.76272	1.3111	.79512	40
21	.60668	.76318	1.3103	.79494	39
22	.60691	.76364	1.3095	.79477	38
23	.60714	.76410	1.3087	.79459	37
24	.60738	.76456	1.3079	.79441	36
25	.60761	.76502	1.3072	.79424	35
26	.60784	.76548	1.3064	.79406	34
27	.60807	.76594	1.3056	.79388	33
28	.60830	.76640	1.3048	.79371	32
29	.60853	.76686	1.3040	.79353	31
30	.60876	.76733	1.3032	.79335	30
31	.60899	.76779	1.3024	.79318	29
32	.60922	.76825	1.3017	.79300	28
33	.60945	.76871	1.3009	.79282	27
34	.60968	.76918	1.3001	.79264	26
35	.60991	.76964	1.2993	.79247	25
36	.61015	.77010	1.2985	.79229	24
37	.61038	.77057	1.2977	.79211	23
38	.61061	.77103	1.2970	.79193	22
39	.61084	.77149	1.2962	.79176	21
40	.61107	.77196	1.2954	.79158	20
41	.61130	.77242	1.2946	.79140	19
42	.61153	.77289	1.2938	.79122	18
43	.61176	.77335	1.2931	.79105	17
44	.61199	.77382	1.2923	.79087	16
45	.61222	.77428	1.2915	.79069	15
46	.61245	.77475	1.2907	.79051	14
47	.61268	.77521	1.2900	.79033	13
48	.61291	.77568	1.2892	.79016	12
49	.61314	.77615	1.2884	.78998	11
50	.61337	.77661	1.2876	.78980	10
51	.61360	.77708	1.2869	.78962	9
52	.61383	.77754	1.2861	.78944	8
53	.61406	.77801	1.2853	.78926	7
54	.61429	.77848	1.2846	.78908	6
55	.61451	.77895	1.2838	.78891	5
56	.61474	.77941	1.2830	.78873	4
57	.61497	.77988	1.2822	.78855	3
58	.61520	.78035	1.2815	.78837	2
59	.61543	.78082	1.2807	.78819	1
60	.61566	.78129	1.2799	.78801	0
′	Cos	Ctn	Tan	Sin	′
127° (307°)				**(232°) 52°**	

38° (218°) (321°) **141°**

′	Sin	Tan	Ctn	Cos	′
0	.61566	.78129	1.2799	.78801	60
1	.61589	.78175	1.2792	.78783	59
2	.61612	.78222	1.2784	.78765	58
3	.61635	.78269	1.2776	.78747	57
4	.61658	.78316	1.2769	.78729	56
5	.61681	.78363	1.2761	.78711	55
6	.61704	.78410	1.2753	.78694	54
7	.61726	.78457	1.2746	.78676	53
8	.61749	.78504	1.2738	.78658	52
9	.61772	.78551	1.2731	.78640	51
10	.61795	.78598	1.2723	.78622	50
11	.61818	.78645	1.2715	.78604	49
12	.61841	.78692	1.2708	.78586	48
13	.61864	.78739	1.2700	.78568	47
14	.61887	.78786	1.2693	.78550	46
15	.61909	.78834	1.2685	.78532	45
16	.61932	.78881	1.2677	.78514	44
17	.61955	.78928	1.2670	.78496	43
18	.61978	.78975	1.2662	.78478	42
19	.62001	.79022	1.2655	.78460	41
20	.62024	.79070	1.2647	.78442	40
21	.62046	.79117	1.2640	.78424	39
22	.62069	.79164	1.2632	.78405	38
23	.62092	.79212	1.2624	.78387	37
24	.62115	.79259	1.2617	.78369	36
25	.62138	.79306	1.2609	.78351	35
26	.62160	.79354	1.2602	.78333	34
27	.62183	.79401	1.2594	.78315	33
28	.62206	.79449	1.2587	.78297	32
29	.62229	.79496	1.2579	.78279	31
30	.62251	.79544	1.2572	.78261	30
31	.62274	.79591	1.2564	.78243	29
32	.62297	.79639	1.2557	.78225	28
33	.62320	.79686	1.2549	.78206	27
34	.62342	.79734	1.2542	.78188	26
35	.62365	.79781	1.2534	.78170	25
36	.62388	.79829	1.2527	.78152	24
37	.62411	.79877	1.2519	.78134	23
38	.62433	.79924	1.2512	.78116	22
39	.62456	.79972	1.2504	.78098	21
40	.62479	.80020	1.2497	.78079	20
41	.62502	.80067	1.2489	.78061	19
42	.62524	.80115	1.2482	.78043	18
43	.62547	.80163	1.2475	.78025	17
44	.62570	.80211	1.2467	.78007	16
45	.62592	.80258	1.2460	.77988	15
46	.62615	.80306	1.2452	.77970	14
47	.62638	.80354	1.2445	.77952	13
48	.62660	.80402	1.2437	.77934	12
49	.62683	.80450	1.2430	.77916	11
50	.62706	.80498	1.2423	.77897	10
51	.62728	.80546	1.2415	.77879	9
52	.62751	.80594	1.2408	.77861	8
53	.62774	.80642	1.2401	.77843	7
54	.62796	.80690	1.2393	.77824	6
55	.62819	.80738	1.2386	.77806	5
56	.62842	.80786	1.2378	.77788	4
57	.62864	.80834	1.2371	.77769	3
58	.62887	.80882	1.2364	.77751	2
59	.62909	.80930	1.2356	.77733	1
60	.62932	.80978	1.2349	.77715	0
′	Cos	Ctn	Tan	Sin	′

128° (308°) (231°) **51°**

39° (219°) (320°) **140°**

′	Sin	Tan	Ctn	Cos	′
0	.62932	.80978	1.2349	.77715	60
1	.62955	.81027	1.2342	.77696	59
2	.62977	.81075	1.2334	.77678	58
3	.63000	.81123	1.2327	.77660	57
4	.63022	.81171	1.2320	.77641	56
5	.63045	.81220	1.2312	.77623	55
6	.63068	.81268	1.2305	.77605	54
7	.63090	.81316	1.2298	.77586	53
8	.63113	.81364	1.2290	.77568	52
9	.63135	.81413	1.2283	.77550	51
10	.63158	.81461	1.2276	.77531	50
11	.63180	.81510	1.2268	.77513	49
12	.63203	.81558	1.2261	.77494	48
13	.63225	.81606	1.2254	.77476	47
14	.63248	.81655	1.2247	.77458	46
15	.63271	.81703	1.2239	.77439	45
16	.63293	.81752	1.2232	.77421	44
17	.63316	.81800	1.2225	.77402	43
18	.63338	.81849	1.2218	.77384	42
19	.63361	.81898	1.2210	.77366	41
20	.63383	.81946	1.2203	.77347	40
21	.63406	.81995	1.2196	.77329	39
22	.63428	.82044	1.2189	.77310	38
23	.63451	.82092	1.2181	.77292	37
24	.63473	.82141	1.2174	.77273	36
25	.63496	.82190	1.2167	.77255	35
26	.63518	.82238	1.2160	.77236	34
27	.63540	.82287	1.2153	.77218	33
28	.63563	.82336	1.2145	.77199	32
29	.63585	.82385	1.2138	.77181	31
30	.63608	.82434	1.2131	.77162	30
31	.63630	.82483	1.2124	.77144	29
32	.63653	.82531	1.2117	.77125	28
33	.63675	.82580	1.2109	.77107	27
34	.63698	.82629	1.2102	.77088	26
35	.63720	.82678	1.2095	.77070	25
36	.63742	.82727	1.2088	.77051	24
37	.63765	.82776	1.2081	.77033	23
38	.63787	.82825	1.2074	.77014	22
39	.63810	.82874	1.2066	.76996	21
40	.63832	.82923	1.2059	.76977	20
41	.63854	.82972	1.2052	.76959	19
42	.63877	.83022	1.2045	.76940	18
43	.63899	.83071	1.2038	.76921	17
44	.63922	.83120	1.2031	.76903	16
45	.63944	.83169	1.2024	.76884	15
46	.63966	.83218	1.2017	.76866	14
47	.63989	.83268	1.2009	.76847	13
48	.64011	.83317	1.2002	.76828	12
49	.64033	.83366	1.1995	.76810	11
50	.64056	.83415	1.1988	.76791	10
51	.64078	.83465	1.1981	.76772	9
52	.64100	.83514	1.1974	.76754	8
53	.64123	.83564	1.1967	.76735	7
54	.64145	.83613	1.1960	.76717	6
55	.64167	.83662	1.1953	.76698	5
56	.64190	.83712	1.1946	.76679	4
57	.64212	.83761	1.1939	.76661	3
58	.64234	.83811	1.1932	.76642	2
59	.64256	.83860	1.1925	.76623	1
60	.64279	.83910	1.1918	.76604	0
′	Cos	Ctn	Tan	Sin	′

129° (309°) (230°) **50°**

40° (220°) (319°) **139°**

′	Sin	Tan	Ctn	Cos	′
0	.64279	.83910	1.1918	.76604	60
1	.64301	.83960	1.1910	.76586	59
2	.64323	.84009	1.1903	.76567	58
3	.64346	.84059	1.1896	.76548	57
4	.64368	.84108	1.1889	.76530	56
5	.64390	.84158	1.1882	.76511	55
6	.64412	.84208	1.1875	.76492	54
7	.64435	.84258	1.1868	.76473	53
8	.64457	.84307	1.1861	.76455	52
9	.64479	.84357	1.1854	.76436	51
10	.64501	.84407	1.1847	.76417	50
11	.64524	.84457	1.1840	.76398	49
12	.64546	.84507	1.1833	.76380	48
13	.64568	.84556	1.1826	.76361	47
14	.64590	.84606	1.1819	.76342	46
15	.64612	.84656	1.1812	.76323	45
16	.64635	.84706	1.1806	.76304	44
17	.64657	.84756	1.1799	.76286	43
18	.64679	.84806	1.1792	.76267	42
19	.64701	.84856	1.1785	.76248	41
20	.64723	.84906	1.1778	.76229	40
21	.64746	.84956	1.1771	.76210	39
22	.64768	.85006	1.1764	.76192	38
23	.64790	.85057	1.1757	.76173	37
24	.64812	.85107	1.1750	.76154	36
25	.64834	.85157	1.1743	.76135	35
26	.64856	.85207	1.1736	.76116	34
27	.64878	.85257	1.1729	.76097	33
28	.64901	.85308	1.1722	.76078	32
29	.64923	.85358	1.1715	.76059	31
30	.64945	.85408	1.1708	.76041	30
31	.64967	.85458	1.1702	.76022	29
32	.64989	.85509	1.1695	.76003	28
33	.65011	.85559	1.1688	.75984	27
34	.65033	.85609	1.1681	.75965	26
35	.65055	.85660	1.1674	.75946	25
36	.65077	.85710	1.1667	.75927	24
37	.65100	.85761	1.1660	.75908	23
38	.65122	.85811	1.1653	.75889	22
39	.65144	.85862	1.1647	.75870	21
40	.65166	.85912	1.1640	.75851	20
41	.65188	.85963	1.1633	.75832	19
42	.65210	.86014	1.1626	.75813	18
43	.65232	.86064	1.1619	.75794	17
44	.65254	.86115	1.1612	.75775	16
45	.65276	.86166	1.1606	.75756	15
46	.65298	.86216	1.1599	.75738	14
47	.65320	.86267	1.1592	.75719	13
48	.65342	.86318	1.1585	.75700	12
49	.65364	.86368	1.1578	.75680	11
50	.65386	.86419	1.1571	.75661	10
51	.65408	.86470	1.1565	.75642	9
52	.65430	.86521	1.1558	.75623	8
53	.65452	.86572	1.1551	.75604	7
54	.65474	.86623	1.1544	.75585	6
55	.65496	.86674	1.1538	.75566	5
56	.65518	.86725	1.1531	.75547	4
57	.65540	.86776	1.1524	.75528	3
58	.65562	.86827	1.1517	.75509	2
59	.65584	.86878	1.1510	.75490	1
60	.65606	.86929	1.1504	.75471	0
′	Cos	Ctn	Tan	Sin	′

130° (310°) (229°) **49°**

41° (221°) (318°) **138°**

′	Sin	Tan	Ctn	Cos	′
0	.65606	.86929	1.1504	.75471	60
1	.65628	.86980	1.1497	.75452	59
2	.65650	.87031	1.1490	.75433	58
3	.65672	.87082	1.1483	.75414	57
4	.65694	.87133	1.1477	.75395	56
5	.65716	.87184	1.1470	.75375	55
6	.65738	.87236	1.1463	.75356	54
7	.65759	.87287	1.1456	.75337	53
8	.65781	.87338	1.1450	.75318	52
9	.65803	.87389	1.1443	.75299	51
10	.65825	.87441	1.1436	.75280	50
11	.65847	.87492	1.1430	.75261	49
12	.65869	.87543	1.1423	.75241	48
13	.65891	.87595	1.1416	.75222	47
14	.65913	.87646	1.1410	.75203	46
15	.65935	.87698	1.1403	.75184	45
16	.65956	.87749	1.1396	.75165	44
17	.65978	.87801	1.1389	.75146	43
18	.66000	.87852	1.1383	.75126	42
19	.66022	.87904	1.1376	.75107	41
20	.66044	.87955	1.1369	.75088	40
21	.66066	.88007	1.1363	.75069	39
22	.66088	.88059	1.1356	.75050	38
23	.66109	.88110	1.1349	.75030	37
24	.66131	.88162	1.1343	.75011	36
25	.66153	.88214	1.1336	.74992	35
26	.66175	.88265	1.1329	.74973	34
27	.66197	.88317	1.1323	.74953	33
28	.66218	.88369	1.1316	.74934	32
29	.66240	.88421	1.1310	.74915	31
30	.66262	.88473	1.1303	.74896	30
31	.66284	.88524	1.1296	.74876	29
32	.66306	.88576	1.1290	.74857	28
33	.66327	.88628	1.1283	.74838	27
34	.66349	.88680	1.1276	.74818	26
35	.66371	.88732	1.1270	.74799	25
36	.66393	.88784	1.1263	.74780	24
37	.66414	.88836	1.1257	.74760	23
38	.66436	.88888	1.1250	.74741	22
39	.66458	.88940	1.1243	.74722	21
40	.66480	.88992	1.1237	.74703	20
41	.66501	.89045	1.1230	.74683	19
42	.66523	.89097	1.1224	.74664	18
43	.66545	.89149	1.1217	.74644	17
44	.66566	.89201	1.1211	.74625	16
45	.66588	.89253	1.1204	.74606	15
46	.66610	.89306	1.1197	.74586	14
47	.66632	.89358	1.1191	.74567	13
48	.66653	.89410	1.1184	.74548	12
49	.66675	.89463	1.1178	.74528	11
50	.66697	.89515	1.1171	.74509	10
51	.66718	.89567	1.1165	.74489	9
52	.66740	.89620	1.1158	.74470	8
53	.66762	.89672	1.1152	.74451	7
54	.66783	.89725	1.1145	.74431	6
55	.66805	.89777	1.1139	.74412	5
56	.66827	.89830	1.1132	.74392	4
57	.66848	.89883	1.1126	.74373	3
58	.66870	.89935	1.1119	.74353	2
59	.66891	.89988	1.1113	.74334	1
60	.66913	.90040	1.1106	.74314	0
′	Cos	Ctn	Tan	Sin	′

131° (311°) (228°) **48°**

42° (222°) (317°) **137°**

′	Sin	Tan	Ctn	Cos	′
0	.66913	.90040	1.1106	.74314	60
1	.66935	.90093	1.1100	.74295	59
2	.66956	.90146	1.1093	.74276	58
3	.66978	.90199	1.1087	.74256	57
4	.66999	.90251	1.1080	.74237	56
5	.67021	.90304	1.1074	.74217	55
6	.67043	.90357	1.1067	.74198	54
7	.67064	.90410	1.1061	.74178	53
8	.67086	.90463	1.1054	.74159	52
9	.67107	.90516	1.1048	.74139	51
10	.67129	.90569	1.1041	.74120	50
11	.67151	.90621	1.1035	.74100	49
12	.67172	.90674	1.1028	.74080	48
13	.67194	.90727	1.1022	.74061	47
14	.67215	.90781	1.1016	.74041	46
15	.67237	.90834	1.1009	.74022	45
16	.67258	.90887	1.1003	.74002	44
17	.67280	.90940	1.0996	.73983	43
18	.67301	.90993	1.0990	.73963	42
19	.67323	.91046	1.0983	.73944	41
20	.67344	.91099	1.0977	.73924	40
21	.67366	.91153	1.0971	.73904	39
22	.67387	.91206	1.0964	.73885	38
23	.67409	.91259	1.0958	.73865	37
24	.67430	.91313	1.0951	.73846	36
25	.67452	.91366	1.0945	.73826	35
26	.67473	.91419	1.0939	.73806	34
27	.67495	.91473	1.0932	.73787	33
28	.67516	.91526	1.0926	.73767	32
29	.67538	.91580	1.0919	.73747	31
30	.67559	.91633	1.0913	.73728	30
31	.67580	.91687	1.0907	.73708	29
32	.67602	.91740	1.0900	.73688	28
33	.67623	.91794	1.0894	.73669	27
34	.67645	.91847	1.0888	.73649	26
35	.67666	.91901	1.0881	.73629	25
36	.67688	.91955	1.0875	.73610	24
37	.67709	.92008	1.0869	.73590	23
38	.67730	.92062	1.0862	.73570	22
39	.67752	.92116	1.0856	.73551	21
40	.67773	.92170	1.0850	.73531	20
41	.67795	.92224	1.0843	.73511	19
42	.67816	.92277	1.0837	.73491	18
43	.67837	.92331	1.0831	.73472	17
44	.67859	.92385	1.0824	.73452	16
45	.67880	.92439	1.0818	.73432	15
46	.67901	.92493	1.0812	.73413	14
47	.67923	.92547	1.0805	.73393	13
48	.67944	.92601	1.0799	.73373	12
49	.67965	.92655	1.0793	.73353	11
50	.67987	.92709	1.0786	.73333	10
51	.68008	.92763	1.0780	.73314	9
52	.68029	.92817	1.0774	.73294	8
53	.68051	.92872	1.0768	.73274	7
54	.68072	.92926	1.0761	.73254	6
55	.68093	.92980	1.0755	.73234	5
56	.68115	.93034	1.0749	.73215	4
57	.68136	.93088	1.0742	.73195	3
58	.68157	.93143	1.0736	.73175	2
59	.68179	.93197	1.0730	.73155	1
60	.68200	.93252	1.0724	.73135	0
′	Cos	Ctn	Tan	Sin	′

132° (312°) (227°) **47°**

43° (223°) (316°) **136°**

′	Sin	Tan	Ctn	Cos	′
0	.68200	.93252	1.0724	.73135	60
1	.68221	.93306	1.0717	.73116	59
2	.68242	.93360	1.0711	.73096	58
3	.68264	.93415	1.0705	.73076	57
4	.68285	.93469	1.0699	.73056	56
5	.68306	.93524	1.0692	.73036	55
6	.68327	.93578	1.0686	.73016	54
7	.68349	.93633	1.0680	.72996	53
8	.68370	.93688	1.0674	.72976	52
9	.68391	.93742	1.0668	.72957	51
10	.68412	.93797	1.0661	.72937	50
11	.68434	.93852	1.0655	.72917	49
12	.68455	.93906	1.0649	.72897	48
13	.68476	.93961	1.0643	.72877	47
14	.68497	.94016	1.0637	.72857	46
15	.68518	.94071	1.0630	.72837	45
16	.68539	.94125	1.0624	.72817	44
17	.68561	.94180	1.0618	.72797	43
18	.68582	.94235	1.0612	.72777	42
19	.68603	.94290	1.0606	.72757	41
20	.68624	.94345	1.0599	.72737	40
21	.68645	.94400	1.0593	.72717	39
22	.68666	.94455	1.0587	.72697	38
23	.68688	.94510	1.0581	.72677	37
24	.68709	.94565	1.0575	.72657	36
25	.68730	.94620	1.0569	.72637	35
26	.68751	.94676	1.0562	.72617	34
27	.68772	.94731	1.0556	.72597	33
28	.68793	.94786	1.0550	.72577	32
29	.68814	.94841	1.0544	.72557	31
30	.68835	.94896	1.0538	.72537	30
31	.68857	.94952	1.0532	.72517	29
32	.68878	.95007	1.0526	.72497	28
33	.68899	.95062	1.0519	.72477	27
34	.68920	.95118	1.0513	.72457	26
35	.68941	.95173	1.0507	.72437	25
36	.68962	.95229	1.0501	.72417	24
37	.68983	.95284	1.0495	.72397	23
38	.69004	.95340	1.0489	.72377	22
39	.69025	.95395	1.0483	.72357	21
40	.69046	.95451	1.0477	.72337	20
41	.69067	.95506	1.0470	.72317	19
42	.69088	.95562	1.0464	.72297	18
43	.69109	.95618	1.0458	.72277	17
44	.69130	.95673	1.0452	.72257	16
45	.69151	.95729	1.0446	.72236	15
46	.69172	.95785	1.0440	.72216	14
47	.69193	.95841	1.0434	.72196	13
48	.69214	.95897	1.0428	.72176	12
49	.69235	.95952	1.0422	.72156	11
50	.69256	.96008	1.0416	.72136	10
51	.69277	.96064	1.0410	.72116	9
52	.69298	.96120	1.0404	.72095	8
53	.69319	.96176	1.0398	.72075	7
54	.69340	.96232	1.0392	.72055	6
55	.69361	.96288	1.0385	.72035	5
56	.69382	.96344	1.0379	.72015	4
57	.69403	.96400	1.0373	.71995	3
58	.69424	.96457	1.0367	.71974	2
59	.69445	.96513	1.0361	.71954	1
60	.69466	.96569	1.0355	.71934	0
′	Cos	Ctn	Tan	Sin	′

133° (313°) (226°) **46°**

44° (224°) (315°) **135°**

′	Sin	Tan	Ctn	Cos	′
0	.69466	.96569	1.0355	.71934	60
1	.69487	.96625	1.0349	.71914	59
2	.69508	.96681	1.0343	.71894	58
3	.69529	.96738	1.0337	.71873	57
4	.69549	.96794	1.0331	.71853	56
5	.69570	.96850	1.0325	.71833	55
6	.69591	.96907	1.0319	.71813	54
7	.69612	.96963	1.0313	.71792	53
8	.69633	.97020	1.0307	.71772	52
9	.69654	.97076	1.0301	.71752	51
10	.69675	.97133	1.0295	.71732	50
11	.69696	.97189	1.0289	.71711	49
12	.69717	.97246	1.0283	.71691	48
13	.69737	.97302	1.0277	.71671	47
14	.69758	.97359	1.0271	.71650	46
15	.69779	.97416	1.0265	.71630	45
16	.69800	.97472	1.0259	.71610	44
17	.69821	.97529	1.0253	.71590	43
18	.69842	.97586	1.0247	.71569	42
19	.69862	.97643	1.0241	.71549	41
20	.69883	.97700	1.0235	.71529	40
21	.69904	.97756	1.0230	.71508	39
22	.69925	.97813	1.0224	.71488	38
23	.69946	.97870	1.0218	.71468	37
24	.69966	.97927	1.0212	.71447	36
25	.69987	.97984	1.0206	.71427	35
26	.70008	.98041	1.0200	.71407	34
27	.70029	.98098	1.0194	.71386	33
28	.70049	.98155	1.0188	.71366	32
29	.70070	.98213	1.0182	.71345	31
30	.70091	.98270	1.0176	.71325	30
31	.70112	.98327	1.0170	.71305	29
32	.70132	.98384	1.0164	.71284	28
33	.70153	.98441	1.0158	.71264	27
34	.70174	.98499	1.0152	.71243	26
35	.70195	.98556	1.0147	.71223	25
36	.70215	.98613	1.0141	.71203	24
37	.70236	.98671	1.0135	.71182	23
38	.70257	.98728	1.0129	.71162	22
39	.70277	.98786	1.0123	.71141	21
40	.70298	.98843	1.0117	.71121	20
41	.70319	.98901	1.0111	.71100	19
42	.70339	.98958	1.0105	.71080	18
43	.70360	.99016	1.0099	.71059	17
44	.70381	.99073	1.0094	.71039	16
45	.70401	.99131	1.0088	.71019	15
46	.70422	.99189	1.0082	.70998	14
47	.70443	.99247	1.0076	.70978	13
48	.70463	.99304	1.0070	.70957	12
49	.70484	.99362	1.0064	.70937	11
50	.70505	.99420	1.0058	.70916	10
51	.70525	.99478	1.0052	.70896	9
52	.70546	.99536	1.0047	.70875	8
53	.70567	.99594	1.0041	.70855	7
54	.70587	.99652	1.0035	.70834	6
55	.70608	.99710	1.0029	.70813	5
56	.70628	.99768	1.0023	.70793	4
57	.70649	.99826	1.0017	.70772	3
58	.70670	.99884	1.0012	.70752	2
59	.70690	.99942	1.0006	.70731	1
60	.70711	1.0000	1.0000	.70711	0
′	Cos	Ctn	Tan	Sin	′

134° (314°) (225°) **45°**

Table 6 MINUTES AND SECONDS TO DECIMAL PARTS OF A DEGREE

MINUTES AND SECONDS TO DECIMAL PARTS OF A DEGREE

Min.	Degrees	Sec.	Degrees
0	0.00000	0	0.00000
1	.01667	1	.00028
2	.03333	2	.00055
3	.05000	3	.00083
4	.06667	4	.00111
5	.08333	5	.00139
6	.10000	6	.00167
7	.11667	7	.00194
8	.13333	8	.00222
9	.15000	9	.00250
10	0.16667	10	0.00278
11	.18333	11	.00305
12	.20000	12	.00333
13	.21667	13	.00361
14	.23333	14	.00389
15	.25000	15	.00417
16	.26667	16	.00444
17	.28333	17	.00472
18	.30000	18	.00500
19	.31667	19	.00527
20	0.33333	20	0.00556
21	.35000	21	.00583
22	.36667	22	.00611
23	.38333	23	.00639
24	.40000	24	.00667
25	.41667	25	.00694
26	.43333	26	.00722
27	.45000	27	.00750
28	.46667	28	.00778
29	.48333	29	.00805
30	0.50000	30	0.00833
31	.51667	31	.00861
32	.53333	32	.00889
33	.55000	33	.00916
34	.56667	34	.00944
35	.58333	35	.00972
36	.60000	36	.01000
37	.61667	37	.01028
38	.63333	38	.01055
39	.65000	39	.01083
40	0.66667	40	0.01111
41	.68333	41	.01139
42	.70000	42	.01167
43	.71667	43	.01194
44	.73333	44	.01222
45	.75000	45	.01250
46	.76667	46	.01278
47	.78333	47	.01305
48	.80000	48	.01333
49	.81667	49	.01361
50	0.83333	50	0.01389
51	.85000	51	.01416
52	.86667	52	.01444
53	.88333	53	.01472
54	.90000	54	.01500
55	.91667	55	.01527
56	.93333	56	.01555
57	.95000	57	.01583
58	.96667	58	.01611
59	.98333	59	.01639
60	1.00000	60	0.01667

DECIMAL PARTS OF A DEGREE TO MINUTES AND SECONDS

Deg.	′	″	Deg.	′	″
0.00	0	00	0.60	36	00
.01	0	36	.61	36	36
.02	1	12	.62	37	12
.03	1	48	.63	37	48
.04	2	24	.64	38	24
.05	3	00	.65	39	00
.06	3	36	.66	39	36
.07	4	12	.67	40	12
.08	4	48	.68	40	48
.09	5	24	.69	41	24
0.10	6	00	0.70	42	00
.11	6	36	.71	42	36
.12	7	12	.72	43	12
.13	7	48	.73	43	48
.14	8	24	.74	44	24
.15	9	00	.75	45	00
.16	9	36	.76	45	36
.17	10	12	.77	46	12
.18	10	48	.78	46	48
.19	11	24	.79	47	24
0.20	12	00	0.80	48	00
.21	12	36	.81	48	36
.22	13	12	.82	49	12
.23	13	48	.83	49	48
.24	14	24	.84	50	24
.25	15	00	.85	51	00
.26	15	36	.86	51	36
.27	16	12	.87	52	12
.28	16	48	.88	52	48
.29	17	24	.89	53	24
0.30	18	00	0.90	54	00
.31	18	36	.91	54	36
.32	19	12	.92	55	12
.33	19	48	.93	55	48
.34	20	24	.94	56	24
.35	21	00	.95	57	00
.36	21	36	.96	57	36
.37	22	12	.97	58	12
.38	22	48	.98	58	48
.39	23	24	.99	59	24
0.40	24	00	1.00	60	00
.41	24	36			
.42	25	12			
.43	25	48			
.44	26	24			
.45	27	00			
.46	27	36			
.47	28	12			
.48	28	48			
.49	29	24			
0.50	30	00			
.51	30	36			
.52	31	12			
.53	31	48			
.54	32	24			
.55	33	00			
.56	33	36			
.57	34	12			
.58	34	48			
.59	35	24			
0.60	36				

Deg.	Sec.
0.000	0.0
.001	3.6
.002	7.2
.003	10.8
.004	14.4
.005	18.0
.006	21.6
.007	25.2
.008	28.8
.009	32.4
0.010	36.0

Table 7 COMMON LOGARITHMS OF TRIGONOMETRIC FUNCTIONS IN RADIAN MEASURE

Rad	L Sin	L Tan	L Ctn	L Cos
.00	-------	-------	-------	10.0000
.01	8.0000	8.0000	12.0000	.0000
.02	.3010	.3011	11.6989	9.9999
.03	.4771	.4773	.5227	.9998
.04	.6019	.6023	.3977	.9997
.05	8.6988	8.6993	11.3007	9.9995
.06	.7779	.7787	.2213	.9992
.07	.8447	.8458	.1542	.9989
.08	.9026	.9040	.0960	.9986
.09	.9537	.9554	.0446	.9982
.10	8.9993	9.0015	10.9985	9.9978
.11	9.0405	.0431	.9569	.9974
.12	.0781	.0813	.9187	.9969
.13	.1127	.1164	.8836	.9963
.14	.1447	.1490	.8510	.9957
.15	9.1745	9.1794	10.8206	9.9951
.16	.2023	.2078	.7922	.9944
.17	.2284	.2347	.7653	.9937
.18	.2529	.2600	.7400	.9929
.19	.2761	.2840	.7160	.9921
.20	9.2981	9.3069	10.6931	9.9913
.21	.3190	.3287	.6713	.9904
.22	.3389	.3495	.6505	.9894
.23	.3579	.3695	.6305	.9884
.24	.3760	.3887	.6113	.9874
.25	9.3934	9.4071	10.5929	9.9863
.26	.4101	.4249	.5751	.9852
.27	.4261	.4421	.5579	.9840
.28	.4415	.4587	.5413	.9827
.29	.4563	.4748	.5252	.9815
.30	9.4706	9.4904	10.5096	9.9802
.31	.4844	.5056	.4944	.9788
.32	.4977	.5203	.4797	.9774
.33	.5106	.5347	.4653	.9759
.34	.5231	.5487	.4513	.9744
.35	9.5352	9.5623	10.4377	9.9728
.36	.5469	.5757	.4243	.9712
.37	.5582	.5887	.4113	.9696
.38	.5693	.6014	.3986	.9679
.39	.5800	.6139	.3861	.9661
.40	9.5904	9.6261	10.3739	9.9643
.41	.6005	.6381	.3619	.9624
.42	.6104	.6499	.3501	.9605
.43	.6200	.6615	.3385	.9585
.44	.6293	.6728	.3272	.9565
.45	9.6385	9.6840	10.3160	9.9545
.46	.6473	.6950	.3050	.9523
.47	.6560	.7058	.2942	.9502
.48	.6644	.7165	.2835	.9479
.49	.6727	.7270	.2730	.9456
.50	9.6807	9.7374	10.2626	9.9433

Rad	L Sin	L Tan	L Ctn	L Cos
.50	9.6807	9.7374	10.2626	9.9433
.51	.6886	.7477	.2523	.9409
.52	.6963	.7578	.2422	.9384
.53	.7037	.7678	.2322	.9359
.54	.7111	.7777	.2223	.9333
.55	9.7182	9.7875	10.2125	9.9307
.56	.7252	.7972	.2028	.9280
.57	.7321	.8068	.1932	.9253
.58	.7388	.8164	.1836	.9224
.59	.7454	.8258	.1742	.9196
.60	9.7518	9.8351	10.1649	9.9166
.61	.7581	.8444	.1556	.9136
.62	.7642	.8536	.1464	.9106
.63	.7702	.8628	.1372	.9074
.64	.7761	.8719	.1281	.9042
.65	9.7819	9.8809	10.1191	9.9010
.66	.7875	.8899	.1101	.8976
.67	.7931	.8989	.1011	.8942
.68	.7985	.9078	.0922	.8907
.69	.8038	.9166	.0834	.8872
.70	9.8090	9.9255	10.0745	9.8836
.71	.8141	.9343	.0657	.8799
.72	.8191	.9430	.0570	.8761
.73	.8240	.9518	.0482	.8723
.74	.8288	.9605	.0395	.8683
.75	9.8336	9.9692	10.0308	9.8643
.76	.8382	.9779	.0221	.8602
.77	.8427	.9866	.0134	.8561
.78	.8471	9.9953	10.0047	.8518
.79	.8515	10.0040	9.9960	.8475
.80	9.8557	10.0127	9.9873	9.8431
.81	.8599	.0214	.9786	.8385
.82	.8640	.0301	.9699	.8339
.83	.8680	.0388	.9612	.8292
.84	.8719	.0475	.9525	.8244
.85	9.8758	10.0563	9.9437	9.8195
.86	.8796	.0650	.9350	.8145
.87	.8833	.0738	.9262	.8094
.88	.8869	.0827	.9173	.8042
.89	.8905	.0915	.9085	.7989
.90	9.8939	10.1004	9.8996	9.7935
.91	.8974	.1094	.8906	.7880
.92	.9007	.1184	.8816	.7823
.93	.9040	.1274	.8726	.7766
.94	.9072	.1365	.8635	.7707
.95	9.9103	10.1456	9.8544	9.7647
.96	.9134	.1548	.8452	.7585
.97	.9164	.1641	.8359	.7523
.98	.9193	.1735	.8265	.7459
.99	.9222	.1829	.8171	.7393
1.00	9.9250	10.1924	9.8076	9.7326

The — 10 portion of the characteristic of the logarithm is not printed in Table 7 but must be written down whenever such a logarithm is used.

Rad	L Sin	L Tan	L Ctn	L Cos
1.00	9.9250	10.1924	9.8076	9.7326
1.01	.9278	.2020	.7980	.7258
1.02	.9305	.2117	.7883	.7188
1.03	.9331	.2215	.7785	.7117
1.04	.9357	.2314	.7686	.7043
1.05	9.9382	10.2414	9.7586	9.6969
1.06	.9407	.2515	.7485	.6892
1.07	.9431	.2617	.7383	.6814
1.08	.9454	.2721	.7279	.6733
1.09	.9477	.2826	.7174	.6651
1.10	9.9500	10.2933	9.7067	9.6567
1.11	.9522	.3041	.6959	.6480
1.12	.9543	.3151	.6849	.6392
1.13	.9564	.3263	.6737	.6301
1.14	.9584	.3376	.6624	.6208
1.15	9.9604	10.3492	9.6508	9.6112
1.16	.9623	.3609	.6391	.6013
1.17	.9641	.3729	.6271	.5912
1.18	.9660	.3851	.6149	.5808
1.19	.9677	.3976	.6024	.5701
1.20	9.9694	10.4103	9.5897	9.5591
1.21	.9711	.4233	.5767	.5478
1.22	.9727	.4366	.5634	.5361
1.23	.9743	.4502	.5498	.5241
1.24	.9758	.4642	.5358	.5116
1.25	9.9773	10.4785	9.5215	9.4988
1.26	.9787	.4932	.5068	.4855
1.27	.9800	.5083	.4917	.4717
1.28	.9814	.5239	.4761	.4575
1.29	.9826	.5400	.4600	.4427
1.30	9.9839	10.5566	9.4434	9.4273

Rad	L Sin	L Tan	L Ctn	L Cos
1.30	9.9839	10.5566	9.4434	9.4273
1.31	.9851	.5737	.4263	.4114
1.32	.9862	.5914	.4086	.3948
1.33	.9873	.6098	.3902	.3774
1.34	.9883	.6290	.3710	.3594
1.35	9.9893	10.6489	9.3511	9.3405
1.36	.9903	.6696	.3304	.3206
1.37	.9912	.6914	.3086	.2998
1.38	.9920	.7141	.2859	.2779
1.39	.9929	.7380	.2620	.2548
1.40	9.9936	10.7633	9.2367	9.2304
1.41	.9944	.7900	.2100	.2044
1.42	.9950	.8183	.1817	.1767
1.43	.9957	.8485	.1515	.1472
1.44	.9963	.8809	.1191	.1154
1.45	9.9968	10.9158	9.0842	9.0810
1.46	.9973	.9537	.0463	.0436
1.47	.9978	.9951	.0049	.0027
1.48	.9982	11.0407	8.9593	8.9575
1.49	.9986	.0917	.9083	.9069
1.50	9.9989	11.1493	8.8507	8.8496
1.51	.9992	.2156	.7844	.7836
1.52	.9994	.2938	.7062	.7056
1.53	.9996	.3891	.6109	.6105
1.54	.9998	.5114	.4886	.4884
1.55	9.9999	11.6820	8.3180	8.3180
1.56	10.0000	11.9667	8.0333	8.0333
1.57	10.0000	13.0989	6.9011	6.9011
1.58	10.0000	12.0360*	7.9640*	7.9640*
1.59	9.9999	11.7166*	8.2834*	8.2834*
1.60	9.9998	11.5344*	8.4656*	8.4654*

* The tangent, cotangent, and cosine of these angles are negative.

π radians = 180° π = 3.14159 26536

1 radian = 57°17′44″ .80625 = 57° .29577 95131

1° = 0.01745 32925 19943 radian = 60′ = 3600″

Table 8 DEGREES, MINUTES, AND SECONDS TO RADIANS

Deg	Rad	Min	Rad	Sec	Rad
1	0.01745 33	1	0.00029 09	1	0.00000 48
2	0.03490 66	2	0.00058 18	2	0.00000 97
3	0.05235 99	3	0.00087 27	3	0.00001 45
4	0.06981 32	4	0.00116 36	4	0.00001 94
5	0.08726 65	5	0.00145 44	5	0.00002 42
6	0.10471 98	6	0.00174 53	6	0.00002 91
7	0.12217 30	7	0.00203 62	7	0.00003 39
8	0.13962 63	8	0.00232 71	8	0.00003 88
9	0.15707 96	9	0.00261 80	9	0.00004 36
10	0.17453 29	10	0.00290 89	10	0.00004 85
20	0.34906 59	20	0.00581 78	20	0.00009 70
30	0.52359 88	30	0.00872 66	30	0.00014 54
40	0.69813 17	40	0.01163 55	40	0.00019 39
50	0.87266 46	50	0.01454 44	50	0.00024 24
60	1.04719 76	60	0.01745 33	60	0.00029 09
70	1.22173 05				
80	1.39626 34				
90	1.57079 63				

Table 9 TRIGONOMETRIC FUNCTIONS IN RADIAN MEASURE

Rad	Sin	Tan	Ctn	Cos
.00	.0000	.0000	......	1.0000
.01	.0100	.0100	99.997	1.0000
.02	.0200	.0200	49.993	.9998
.03	.0300	.0300	33.323	.9996
.04	.0400	.0400	24.987	.9992
.05	.0500	.0500	19.983	.9988
.06	.0600	.0601	16.647	.9982
.07	.0699	.0701	14.262	.9976
.08	.0799	.0802	12.473	.9968
.09	.0899	.0902	11.081	.9960
.10	.0998	.1003	9.967	.9950
.11	.1098	.1104	9.054	.9940
.12	.1197	.1206	8.293	.9928
.13	.1296	.1307	7.649	.9916
.14	.1395	.1409	7.096	.9902
.15	.1494	.1511	6.617	.9888
.16	.1593	.1614	6.197	.9872
.17	.1692	.1717	5.826	.9856
.18	.1790	.1820	5.495	.9838
.19	.1889	.1923	5.200	.9820
.20	.1987	.2027	4.933	.9801
.21	.2085	.2131	4.692	.9780
.22	.2182	.2236	4.472	.9759
.23	.2280	.2341	4.271	.9737
.24	.2377	.2447	4.086	.9713
.25	.2474	.2553	3.916	.9689
.26	.2571	.2660	3.759	.9664
.27	.2667	.2768	3.613	.9638
.28	.2764	.2876	3.478	.9611
.29	.2860	.2984	3.351	.9582
.30	.2955	.3093	3.233	.9553
.31	.3051	.3203	3.122	.9523
.32	.3146	.3314	3.018	.9492
.33	.3240	.3425	2.920	.9460
.34	.3335	.3537	2.827	.9428
.35	.3429	.3650	2.740	.9394
.36	.3523	.3764	2.657	.9359
.37	.3616	.3879	2.578	.9323
.38	.3709	.3994	2.504	.9287
.39	.3802	.4111	2.433	.9249
.40	.3894	.4228	2.365	.9211
.41	.3986	.4346	2.301	.9171
.42	.4078	.4466	2.239	.9131
.43	.4169	.4586	2.180	.9090
.44	.4259	.4708	2.124	.9048
.45	.4350	.4831	2.070	.9004
.46	.4439	.4954	2.018	.8961
.47	.4529	.5080	1.969	.8916
.48	.4618	.5206	1.921	.8870
.49	.4706	.5334	1.875	.8823
.50	.4794	.5463	1.830	.8776

Rad	Sin	Tan	Ctn	Cos
.50	.4794	.5463	1.830	.8776
.51	.4882	.5594	1.788	.8727
.52	.4969	.5726	1.747	.8678
.53	.5055	.5859	1.707	.8628
.54	.5141	.5994	1.668	.8577
.55	.5227	.6131	1.631	.8525
.56	.5312	.6269	1.595	.8473
.57	.5396	.6410	1.560	.8419
.58	.5480	.6552	1.526	.8365
.59	.5564	.6696	1.494	.8309
.60	.5646	.6841	1.462	.8253
.61	.5729	.6989	1.431	.8196
.62	.5810	.7139	1.401	.8139
.63	.5891	.7291	1.372	.8080
.64	.5972	.7445	1.343	.8021
.65	.6052	.7602	1.315	.7961
.66	.6131	.7761	1.288	.7900
.67	.6210	.7923	1.262	.7838
.68	.6288	.8087	1.237	.7776
.69	.6365	.8253	1.212	.7712
.70	.6442	.8423	1.187	.7648
.71	.6518	.8595	1.163	.7584
.72	.6594	.8771	1.140	.7518
.73	.6669	.8949	1.117	.7452
.74	.6743	.9131	1.095	.7385
.75	.6816	.9316	1.073	.7317
.76	.6889	.9505	1.052	.7248
.77	.6961	.9697	1.031	.7179
.78	.7033	.9893	1.011	.7109
.79	.7104	1.009	.9908	.7038
.80	.7174	1.030	.9712	.6967
.81	.7243	1.050	.9520	.6895
.82	.7311	1.072	.9331	.6822
.83	.7379	1.093	.9146	.6749
.84	.7446	1.116	.8964	.6675
.85	.7513	1.138	.8785	.6600
.86	.7578	1.162	.8609	.6524
.87	.7643	1.185	.8437	.6448
.88	.7707	1.210	.8267	.6372
.89	.7771	1.235	.8100	.6294
.90	.7833	1.260	.7936	.6216
.91	.7895	1.286	.7774	.6137
.92	.7956	1.313	.7615	.6058
.93	.8016	1.341	.7458	.5978
.94	.8076	1.369	.7303	.5898
.95	.8134	1.398	.7151	.5817
.96	.8192	1.428	.7001	.5735
.97	.8249	1.459	.6853	.5653
.98	.8305	1.491	.6707	.5570
.99	.8360	1.524	.6563	.5487
1.00	.8415	1.557	.6421	.5403

π radians = 180° $\qquad \pi = 3.14159\ 26536$

1 radian = 57°17′44″.80625 = 57°.29577 95131

1° = 0.01745 32925 19943 radian = 60′ = 3600″

Rad	Sin	Tan	Ctn	Cos
1.00	.8415	1.557	.6421	.5403
1.01	.8468	1.592	.6281	.5319
1.02	.8521	1.628	.6142	.5234
1.03	.8573	1.665	.6005	.5148
1.04	.8624	1.704	.5870	.5062
1.05	.8674	1.743	.5736	.4976
1.06	.8724	1.784	.5604	.4889
1.07	.8772	1.827	.5473	.4801
1.08	.8820	1.871	.5344	.4713
1.09	.8866	1.917	.5216	.4625
1.10	.8912	1.965	.5090	.4536
1.11	.8957	2.014	.4964	.4447
1.12	.9001	2.066	.4840	.4357
1.13	.9044	2.120	.4718	.4267
1.14	.9086	2.176	.4596	.4176
1.15	.9128	2.234	.4475	.4085
1.16	.9168	2.296	.4356	.3993
1.17	.9208	2.360	.4237	.3902
1.18	.9246	2.427	.4120	.3809
1.19	.9284	2.498	.4003	.3717
1.20	.9320	2.572	.3888	.3624
1.21	.9356	2.650	.3773	.3530
1.22	.9391	2.733	.3659	.3436
1.23	.9425	2.820	.3546	.3342
1.24	.9458	2.912	.3434	.3248
1.25	.9490	3.010	.3323	.3153
1.26	.9521	3.113	.3212	.3058
1.27	.9551	3.224	.3102	.2963
1.28	.9580	3.341	.2993	.2867
1.29	.9608	3.467	.2884	.2771
1.30	.9636	3.602	.2776	.2675

Rad	Sin	Tan	Ctn	Cos
1.30	.9636	3.602	.2776	.2675
1.31	.9662	3.747	.2669	.2579
1.32	.9687	3.903	.2562	.2482
1.33	.9711	4.072	.2456	.2385
1.34	.9735	4.256	.2350	.2288
1.35	.9757	4.455	.2245	.2190
1.36	.9779	4.673	.2140	.2092
1.37	.9799	4.913	.2035	.1994
1.38	.9819	5.177	.1931	.1896
1.39	.9837	5.471	.1828	.1798
1.40	.9854	5.798	.1725	.1700
1.41	.9871	6.165	.1622	.1601
1.42	.9887	6.581	.1519	.1502
1.43	.9901	7.055	.1417	.1403
1.44	.9915	7.602	.1315	.1304
1.45	.9927	8.238	.1214	.1205
1.46	.9939	8.989	.1113	.1106
1.47	.9949	9.887	.1011	.1006
1.48	.9959	10.983	.0910	.0907
1.49	.9967	12.350	.0810	.0807
1.50	.9975	14.101	.0709	.0707
1.51	.9982	16.428	.0609	.0608
1.52	.9987	19.670	.0508	.0508
1.53	.9992	24.498	.0408	.0408
1.54	.9995	32.461	.0308	.0308
1.55	.9998	48.078	.0208	.0208
1.56	.9999	92.620	.0108	.0108
1.57	1.0000	1255.8	.0008	.0008
1.58	1.0000	−108.65	−.0092	−.0092
1.59	.9998	−52.067	−.0192	−.0192
1.60	.9996	−34.233	−.0292	−.0292

Table 10 RADIANS TO DEGREES, MINUTES, AND SECONDS

Rad		Rad		Rad		Rad		Rad	
1	57°17′44″.8	.1	5°43′46″.5	.01	0°34′22″.6	.001	0° 3′26″.3	.0001	0°0′20″.6
2	114°35′29″.6	.2	11°27′33″.0	.02	1° 8′45″.3	.002	0° 6′52″.5	.0002	0°0′41″.3
3	171°53′14″.4	.3	17°11′19″.4	.03	1°43′07″.9	.003	0°10′18″.8	.0003	0°1′01″.9
4	229°10′59″.2	.4	22°55′05″.9	.04	2°17′30″.6	.004	0°13′45″.1	.0004	0°1′22″.5
5	286°28′44″.0	.5	28°38′52″.4	.05	2°51′53″.2	.005	0°17′11″.3	.0005	0°1′43″.1
6	343°46′28″.8	.6	34°22′38″.9	.06	3°26′15″.9	.006	0°20′37″.6	.0006	0°2′03″.8
7	401° 4′13″.6	.7	40° 6′25″.4	.07	4° 0′38″.5	.007	0°24′03″.9	.0007	0°2′24″.4
8	458°21′58″.4	.8	45°50′11″.8	.08	4°35′01″.2	.008	0°27′30″.1	.0008	0°2′45″.0
9	515°39′43″.3	.9	51°33′58″.3	.09	5° 9′23″.8	.009	0°30′56″.4	.0009	0°3′05″.6

RADIANS TO DEGREES

Rad	Degrees	Rad	Degrees	Rad	Degrees
1	57.2958	4	229.1831	7	401.0705
2	114.5916	5	286.4789	8	458.3662
3	171.8873	6	343.7747	9	515.6620

EXPONENTIAL AND HYPERBOLIC FUNCTIONS

0.00–5.99†

N	0	1	2	3	4	5	6	7	8	9
0.0	Take tabular value −10	5.395	6.088	6.493	6.781	7.004	7.187	7.341	7.474	7.592
0.1	7.697	7.793	7.880	7.960	8.034	8.103	8.167	8.228	8.285	8.339
0.2	8.391	8.439	8.486	8.530	8.573	8.614	8.653	8.691	8.727	8.762
0.3	8.796	8.829	8.861	8.891	8.921	8.950	8.978	9.006	9.032	9.058
0.4	9.084	9.108	9.132	9.156	9.179	9.201	9.223	9.245	9.266	9.287
0.5	9.307	9.327	9.346	9.365	9.384	9.402	9.420	9.438	9.455	9.472
0.6	9.489	9.506	9.522	9.538	9.554	9.569	9.584	9.600	9.614	9.629
0.7	9.643	9.658	9.671	9.685	9.699	9.712	9.726	9.739	9.752	9.764
0.8	9.777	9.789	9.802	9.814	9.826	9.837	9.849	9.861	9.872	9.883
0.9	9.895	9.906	9.917	9.927	9.938	9.949	9.959	9.970	9.980	9.990
1.0	0.0 0000	0995	1980	2956	3922	4879	5827	6766	7696	8618
1.1	9531	*0436	*1333	*2222	*3103	*3976	*4842	*5700	*6551	*7395
1.2	0.1 8232	9062	9885	*0701	*1511	*2314	*3111	*3902	*4686	*5464
1.3	0.2 6236	7003	7763	8518	9267	*0010	*0748	*1481	*2208	*2930
1.4	0.3 3647	4359	5066	5767	6464	7156	7844	8526	9204	9878
1.5	0.4 0547	1211	1871	2527	3178	3825	4469	5108	5742	6373
1.6	7000	7623	8243	8858	9470	*0078	*0682	*1282	*1879	*2473
1.7	0.5 3063	3649	4232	4812	5389	5962	6531	7098	7661	8222
1.8	8779	9333	9884	*0432	*0977	*1519	*2058	*2594	*3127	*3658
1.9	0.6 4185	4710	5233	5752	6269	6783	7294	7803	8310	8813
2.0	9315	9813	*0310	*0804	*1295	*1784	*2271	*2755	*3237	*3716
2.1	0.7 4194	4669	5142	5612	6081	6547	7011	7473	7932	8390
2.2	8846	9299	9751	*0200	*0648	*1093	*1536	*1978	*2418	*2855
2.3	0.8 3291	3725	4157	4587	5015	5442	5866	6289	6710	7129
2.4	7547	7963	8377	8789	9200	9609	*0016	*0422	*0826	*1228
2.5	0.9 1629	2028	2426	2822	3216	3609	4001	4391	4779	5166
2.6	5551	5935	6317	6698	7078	7456	7833	8208	8582	8954
2.7	9325	9695	*0063	*0430	*0796	*1160	*1523	*1885	*2245	*2604
2.8	1.0 2962	3318	3674	4028	4380	4732	5082	5431	5779	6126
2.9	6471	6815	7158	7500	7841	8181	8519	8856	9192	9527
3.0	9861	*0194	*0526	*0856	*1186	*1514	*1841	*2168	*2493	*2817
3.1	1.1 3140	3462	3783	4103	4422	4740	5057	5373	5688	6002
3.2	6315	6627	6938	7248	7557	7865	8173	8479	8784	9089
3.3	9392	9695	9996	*0297	*0597	*0896	*1194	*1491	*1788	*2083
3.4	1.2 2378	2671	2964	3256	3547	3837	4127	4415	4703	4990
3.5	5276	5562	5846	6130	6413	6695	6976	7257	7536	7815
3.6	8093	8371	8647	8923	9198	9473	9746	*0019	*0291	*0563
3.7	1.3 0833	1103	1372	1641	1909	2176	2442	2708	2972	3237
3.8	3500	3763	4025	4286	4547	4807	5067	5325	5584	5841
3.9	6098	6354	6609	6864	7118	7372	7624	7877	8128	8379
4.0	8629	8879	9128	9377	9624	9872	*0118	*0364	*0610	*0854
4.1	1.4 1099	1342	1585	1828	2070	2311	2552	2792	3031	3270
4.2	3508	3746	3984	4220	4456	4692	4927	5161	5395	5629
4.3	5862	6094	6326	6557	6787	7018	7247	7476	7705	7933
4.4	8160	8387	8614	8840	9065	9290	9515	9739	9962	*0185
4.5	1.5 0408	0630	0851	1072	1293	1513	1732	1951	2170	2388
4.6	2606	2823	3039	3256	3471	3687	3902	4116	4330	4543
4.7	4756	4969	5181	5393	5604	5814	6025	6235	6444	6653
4.8	6862	7070	7277	7485	7691	7898	8104	8309	8515	8719
4.9	8924	9127	9331	9534	9737	9939	*0141	*0342	*0543	*0744
5.0	1.6 0944	1144	1343	1542	1741	1939	2137	2334	2531	2728
5.1	2924	3120	3315	3511	3705	3900	4094	4287	4481	4673
5.2	4866	5058	5250	5441	5632	5823	6013	6203	6393	6582
5.3	6771	6959	7147	7335	7523	7710	7896	8083	8269	8455
5.4	8640	8825	9010	9194	9378	9562	9745	9928	*0111	*0293
5.5	1.7 0475	0656	0838	1019	1199	1380	1560	1740	1919	2098
5.6	2277	2455	2633	2811	2988	3166	3342	3519	3695	3871
5.7	4047	4222	4397	4572	4746	4920	5094	5267	5440	5613
5.8	5786	5958	6130	6302	6473	6644	6815	6985	7156	7326
5.9	7495	7665	7834	8002	8171	8339	8507	8675	8842	9009
N	0	1	2	3	4	5	6	7	8	9

Base $e = 2.718 \cdots$ $\log_e 0.10 = 7.69741\ 49070 - 10$

† Entries in Table 11 are values of $\log_e N$ for the indicated values of N.

6.00–10.09

N	0	1	2	3	4	5	6	7	8	9
6.0	1.7 9176	9342	9509	9675	9840	*0006	*0171	*0336	*0500	*0665
6.1	1.8 0829	0993	1156	1319	1482	1645	1808	1970	2132	2294
6.2	2455	2616	2777	2938	3098	3258	3418	3578	3737	3896
6.3	4055	4214	4372	4530	4688	4845	5003	5160	5317	5473
6.4	5630	5786	5942	6097	6253	6408	6563	6718	6872	7026
6.5	7180	7334	7487	7641	7794	7947	8099	8251	8403	8555
6.6	8707	8858	9010	9160	9311	9462	9612	9762	9912	*0061
6.7	1.9 0211	0360	0509	0658	0806	0954	1102	1250	1398	1545
6.8	1692	1839	1986	2132	2279	2425	2571	2716	2862	3007
6.9	3152	3297	3442	3586	3730	3874	4018	4162	4305	4448
7.0	4591	4734	4876	5019	5161	5303	5445	5586	5727	5869
7.1	6009	6150	6291	6431	6571	6711	6851	6991	7130	7269
7.2	7408	7547	7685	7824	7962	8100	8238	8376	8513	8650
7.3	8787	8924	9061	9198	9334	9470	9606	9742	9877	*0013
7.4	2.0 0148	0283	0418	0553	0687	0821	0956	1089	1223	1357
7.5	1490	1624	1757	1890	2022	2155	2287	2419	2551	2683
7.6	2815	2946	3078	3209	3340	3471	3601	3732	3862	3992
7.7	4122	4252	4381	4511	4640	4769	4898	5027	5156	5284
7.8	5412	5540	5668	5796	5924	6051	6179	6306	6433	6560
7.9	6686	6813	6939	7065	7191	7317	7443	7568	7694	7819
8.0	7944	8069	8194	8318	8443	8567	8691	8815	8939	9063
8.1	9186	9310	9433	9556	9679	9802	9924	*0047	*0169	*0291
8.2	2.1 0413	0535	0657	0779	0900	1021	1142	1263	1384	1505
8.3	1626	1746	1866	1986	2106	2226	2346	2465	2585	2704
8.4	2823	2942	3061	3180	3298	3417	3535	3653	3771	3889
8.5	4007	4124	4242	4359	4476	4593	4710	4827	4943	5060
8.6	5176	5292	5409	5524	5640	5756	5871	5987	6102	6217
8.7	6332	6447	6562	6677	6791	6905	7020	7134	7248	7361
8.8	7475	7589	7702	7816	7929	8042	8155	8267	8380	8493
8.9	8605	8717	8830	8942	9054	9165	9277	9389	9500	9611
9.0	9722	9834	9944	*0055	*0166	*0276	*0387	*0497	*0607	*0717
9.1	2.2 0827	0937	1047	1157	1266	1375	1485	1594	1703	1812
9.2	1920	2029	2138	2246	2354	2462	2570	2678	2786	2894
9.3	3001	3109	3216	3324	3431	3538	3645	3751	3858	3965
9.4	4071	4177	4284	4390	4496	4601	4707	4813	4918	5024
9.5	5129	5234	5339	5444	5549	5654	5759	5863	5968	6072
9.6	6176	6280	6384	6488	6592	6696	6799	6903	7006	7109
9.7	7213	7316	7419	7521	7624	7727	7829	7932	8034	8136
9.8	8238	8340	8442	8544	8646	8747	8849	8950	9051	9152
9.9	9253	9354	9455	9556	9657	9757	9858	9958	*0058	*0158
10.0	2.3 0259	0358	0458	0558	0658	0757	0857	0956	1055	1154
N	0	1	2	3	4	5	6	7	8	9

Table 11 NATURAL LOGARITHMS OF NUMBERS (*continued*)

10–99

N	0	1	2	3	4	5	6	7	8	9
1	2.30259	39790	48491	56495	63906	70805	77259	83321	89037	94444
2	99573	*04452	*09104	*13549	*17805	*21888	*25810	*29584	*33220	*36730
3	3.40120	43399	46574	49651	52636	55535	58352	61092	63759	66356
4	68888	71357	73767	76120	78419	80666	82864	85015	87120	89182
5	91202	93183	95124	97029	98898	*00733	*02535	*04305	*06044	*07754
6	4.09434	11087	12713	14313	15888	17439	18965	20469	21951	23411
7	24850	26268	27667	29046	30407	31749	33073	34381	35671	36945
8	38203	39445	40672	41884	43082	44265	45435	46591	47734	48864
9	49981	51086	52179	53260	54329	55388	56435	57471	58497	59512

Examples. $\log_e 9.12 = 2.21047$ $\qquad \log_e 51 = 3.93183$

$\log_e 10 = 2.30258\ 50930$

100–609

N	0	1	2	3	4	5	6	7	8	9
10	4.6 0517	1512	2497	3473	4439	5396	6344	7283	8213	9135
11	4.7 0048	0953	1850	2739	3620	4493	5359	6217	7068	7912
12	8749	9579	*0402	*1218	*2028	*2831	*3628	*4419	*5203	*5981
13	4.8 6753	7520	8280	9035	9784	*0527	*1265	*1998	*2725	*3447
14	4.9 4164	4876	5583	6284	6981	7673	8361	9043	9721	*0395
15	5.0 1064	1728	2388	3044	3695	4343	4986	5625	6260	6890
16	7517	8140	8760	9375	9987	*0595	*1199	*1799	*2396	*2990
17	5.1 3580	4166	4749	5329	5906	6479	7048	7615	8178	8739
18	9296	9850	*0401	*0949	*1494	*2036	*2575	*3111	*3644	*4175
19	5.2 4702	5227	5750	6269	6786	7300	7811	8320	8827	9330
20	9832	*0330	*0827	*1321	*1812	*2301	*2788	*3272	*3754	*4233
21	5.3 4711	5186	5659	6129	6598	7064	7528	7990	8450	8907
22	9363	9816	*0268	*0717	*1165	*1610	*2053	*2495	*2935	*3372
23	5.4 3808	4242	4674	5104	5532	5959	6383	6806	7227	7646
24	8064	8480	8894	9306	9717	*0126	*0533	*0939	*1343	*1745
25	5.5 2146	2545	2943	3339	3733	4126	4518	4908	5296	5683
26	6068	6452	6834	7215	7595	7973	8350	8725	9099	9471
27	9842	*0212	*0580	*0947	*1313	*1677	*2040	*2402	*2762	*3121
28	5.6 3479	3835	4191	4545	4897	5249	5599	5948	6296	6643
29	6988	7332	7675	8017	8358	8698	9036	9373	9709	*0044
30	5.7 0378	0711	1043	1373	1703	2031	2359	2685	3010	3334
31	3657	3979	4300	4620	4939	5257	5574	5890	6205	6519
32	6832	7144	7455	7765	8074	8383	8690	8996	9301	9606
33	9909	*0212	*0513	*0814	*1114	*1413	*1711	*2008	*2305	*2600
34	5.8 2895	3188	3481	3773	4064	4354	4644	4932	5220	5507
35	5793	6079	6363	6647	6930	7212	7493	7774	8053	8332
36	8610	8888	9164	9440	9715	9990	*0263	*0536	*0808	*1080
37	5.9 1350	1620	1889	2158	2426	2693	2959	3225	3489	3754
38	4017	4280	4542	4803	5064	5324	5584	5842	6101	6358
39	6615	6871	7126	7381	7635	7889	8141	8394	8645	8896
40	9146	9396	9645	9894	*0141	*0389	*0635	*0881	*1127	*1372
41	6.0 1616	1859	2102	2345	2587	2828	3069	3309	3548	3787
42	4025	4263	4501	4737	4973	5209	5444	5678	5912	6146
43	6379	6611	6843	7074	7304	7535	7764	7993	8222	8450
44	8677	8904	9131	9357	9582	9807	*0032	*0256	*0479	*0702
45	6.1 0925	1147	1368	1589	1810	2030	2249	2468	2687	2905
46	3123	3340	3556	3773	3988	4204	4419	4633	4847	5060
47	5273	5486	5698	5910	6121	6331	6542	6752	6961	7170
48	7379	7587	7794	8002	8208	8415	8621	8826	9032	9236
49	9441	9644	9848	*0051	*0254	*0456	*0658	*0859	*1060	*1261
50	6.2 1461	1661	1860	2059	2258	2456	2654	2851	3048	3245
51	3441	3637	3832	4028	4222	4417	4611	4804	4998	5190
52	5383	5575	5767	5958	6149	6340	6530	6720	6910	7099
53	7288	7476	7664	7852	8040	8227	8413	8600	8786	8972
54	9157	9342	9527	9711	9895	*0079	*0262	*0445	*0628	*0810
55	6.3 0992	1173	1355	1536	1716	1897	2077	2257	2436	2615
56	2794	2972	3150	3328	3505	3683	3859	4036	4212	4388
57	4564	4739	4914	5089	5263	5437	5611	5784	5957	6130
58	6303	6475	6647	6819	6990	7161	7332	7502	7673	7843
59	8012	8182	8351	8519	8688	8856	9024	9192	9359	9526
60	9693	9859	*0026	*0192	*0357	*0523	*0688	*0853	*1017	*1182
N	0	1	2	3	4	5	6	7	8	9

Example. $\log_e 447 = 6.10256$

$$\log_e 100 = 4.60517\,01860$$

600–1109

N	0	1	2	3	4	5	6	7	8	9
60	6.3 9693	9859	*0026	*0192	*0357	*0523	*0688	*0853	*1017	*1182
61	6.4 1346	1510	1673	1836	1999	2162	2325	2487	2649	2811
62	2972	3133	3294	3455	3615	3775	3935	4095	4254	4413
63	4572	4731	4889	5047	5205	5362	5520	5677	5834	5990
64	6147	6303	6459	6614	6770	6925	7080	7235	7389	7543
65	7697	7851	8004	8158	8311	8464	8616	8768	8920	9072
66	9224	9375	9527	9677	9828	9979	*0129	*0279	*0429	*0578
67	6.5 0728	0877	1026	1175	1323	1471	1619	1767	1915	2062
68	2209	2356	2503	2649	2796	2942	3088	3233	3379	3524
69	3669	3814	3959	4103	4247	4391	4535	4679	4822	4965
70	5108	5251	5393	5536	5678	5820	5962	6103	6244	6386
71	6526	6667	6808	6948	7088	7228	7368	7508	7647	7786
72	7925	8064	8203	8341	8479	8617	8755	8893	9030	9167
73	9304	9441	9578	9715	9851	9987	*0123	*0259	*0394	*0530
74	6.6 0665	0800	0935	1070	1204	1338	1473	1607	1740	1874
75	2007	2141	2274	2407	2539	2672	2804	2936	3068	3200
76	3332	3463	3595	3726	3857	3988	4118	4249	4379	4509
77	4639	4769	4898	5028	5157	5286	5415	5544	5673	5801
78	5929	6058	6185	6313	6441	6568	6696	6823	6950	7077
79	7203	7330	7456	7582	7708	7834	7960	8085	8211	8336
80	8461	8586	8711	8835	8960	9084	9208	9332	9456	9580
81	9703	9827	9950	*0073	*0196	*0319	*0441	*0564	*0686	*0808
82	6.7 0930	1052	1174	1296	1417	1538	1659	1780	1901	2022
83	2143	2263	2383	2503	2623	2743	2863	2982	3102	3221
84	3340	3459	3578	3697	3815	3934	4052	4170	4288	4406
85	4524	4641	4759	4876	4993	5110	5227	5344	5460	5577
86	5693	5809	5926	6041	6157	6273	6388	6504	6619	6734
87	6849	6964	7079	7194	7308	7422	7537	7651	7765	7878
88	7992	8106	8219	8333	8446	8559	8672	8784	8897	9010
89	9122	9234	9347	9459	9571	9682	9794	9906	*0017	*0128
90	6.8 0239	0351	0461	0572	0683	0793	0904	1014	1124	1235
91	1344	1454	1564	1674	1783	1892	2002	2111	2220	2329
92	2437	2546	2655	2763	2871	2979	3087	3195	3303	3411
93	3518	3626	3733	3841	3948	4055	4162	4268	4375	4482
94	4588	4694	4801	4907	5013	5118	5224	5330	5435	5541
95	5646	5751	5857	5961	6066	6171	6276	6380	6485	6589
96	6693	6797	6901	7005	7109	7213	7316	7420	7523	7626
97	7730	7833	7936	8038	8141	8244	8346	8449	8551	8653
98	8755	8857	8959	9061	9163	9264	9366	9467	9568	9669
99	9770	9871	9972	*0073	*0174	*0274	*0375	*0475	*0575	*0675
100	6.9 0776	0875	0975	1075	1175	1274	1374	1473	1572	1672
101	1771	1870	1968	2067	2166	2264	2363	2461	2560	2658
102	2756	2854	2952	3049	3147	3245	3342	3440	3537	3634
103	3731	3828	3925	4022	4119	4216	4312	4409	4505	4601
104	4698	4794	4890	4986	5081	5177	5273	5368	5464	5559
105	5655	5750	5845	5940	6035	6130	6224	6319	6414	6508
106	6602	6697	6791	6885	6979	7073	7167	7261	7354	7448
107	7541	7635	7728	7821	7915	8008	8101	8193	8286	8379
108	8472	8564	8657	8749	8841	8934	9026	9118	9210	9302
109	9393	9485	9577	9668	9760	9851	9942	*0033	*0125	*0216
110	7.0 0307	0397	0488	0579	0670	0760	0851	0941	1031	1121
N	0	1	2	3	4	5	6	7	8	9

$$\log_e 1000 = 6.90775\ 52790$$

To find the logarithm of a number which is 10 (or 1/10) times a number whose logarithm is given, add to (or subtract from) the given logarithm the logarithm of 10.

Example. $\log_e 932 = 6.83733$.

Table 12 VALUES AND COMMON LOGARITHMS OF EXPONENTIAL AND HYPERBOLIC FUNCTIONS

x	e^x		e^{-x}	Sinh x		Cosh x		Tanh x
	Value	Log_{10}	Value	Value	Log_{10}	Value	Log_{10}	Value
0.00	1.0000	.00000	1.00000	0.0000	$-\infty$	1.0000	.00000	.00000
0.01	1.0101	.00434	0.99005	0.0100	$\bar{2}.00001$	1.0001	.00002	.01000
0.02	1.0202	.00869	.98020	0.0200	$\bar{2}.30106$	1.0002	.00009	.02000
0.03	1.0305	.01303	.97045	0.0300	$\bar{2}.47719$	1.0005	.00020	.02999
0.04	1.0408	.01737	.96079	0.0400	$\bar{2}.60218$	1.0008	.00035	.03998
0.05	1.0513	.02171	.95123	0.0500	$\bar{2}.69915$	1.0013	.00054	.04996
0.06	1.0618	.02606	.94176	0.0600	$\bar{2}.77841$	1.0018	.00078	.05993
0.07	1.0725	.03040	.93239	0.0701	$\bar{2}.84545$	1.0025	.00106	.06989
0.08	1.0833	.03474	.92312	0.0801	$\bar{2}.90355$	1.0032	.00139	.07983
0.09	1.0942	.03909	.91393	0.0901	$\bar{2}.95483$	1.0041	.00176	.08976
0.10	1.1052	.04343	.90484	0.1002	$\bar{1}.00072$	1.0050	.00217	.09967
0.11	1.1163	.04777	.89583	0.1102	$\bar{1}.04227$	1.0061	.00262	.10956
0.12	1.1275	.05212	.88692	0.1203	$\bar{1}.08022$	1.0072	.00312	.11943
0.13	1.1388	.05646	.87809	0.1304	$\bar{1}.11517$	1.0085	.00366	.12927
0.14	1.1503	.06080	.86936	0.1405	$\bar{1}.14755$	1.0098	.00424	.13909
0.15	1.1618	.06514	.86071	0.1506	$\bar{1}.17772$	1.0113	.00487	.14889
0.16	1.1735	.06949	.85214	0.1607	$\bar{1}.20597$	1.0128	.00554	.15865
0.17	1.1853	.07383	.84366	0.1708	$\bar{1}.23254$	1.0145	.00625	.16838
0.18	1.1972	.07817	.83527	0.1810	$\bar{1}.25762$	1.0162	.00700	.17808
0.19	1.2092	.08252	.82696	0.1911	$\bar{1}.28136$	1.0181	.00779	.18775
0.20	1.2214	.08686	.81873	0.2013	$\bar{1}.30392$	1.0201	.00863	.19738
0.21	1.2337	.09120	.81058	0.2115	$\bar{1}.32541$	1.0221	.00951	.20697
0.22	1.2461	.09554	.80252	0.2218	$\bar{1}.34592$	1.0243	.01043	.21652
0.23	1.2586	.09989	.79453	0.2320	$\bar{1}.36555$	1.0266	.01139	.22603
0.24	1.2712	.10423	.78663	0.2423	$\bar{1}.38437$	1.0289	.01239	.23550
0.25	1.2840	.10857	.77880	0.2526	$\bar{1}.40245$	1.0314	.01343	.24492
0.26	1.2969	.11292	.77105	0.2629	$\bar{1}.41986$	1.0340	.01452	.25430
0.27	1.3100	.11726	.76338	0.2733	$\bar{1}.43663$	1.0367	.01564	.26362
0.28	1.3231	.12160	.75578	0.2837	$\bar{1}.45282$	1.0395	.01681	.27291
0.29	1.3364	.12595	.74826	0.2941	$\bar{1}.46847$	1.0423	.01801	.28213
0.30	1.3499	.13029	.74082	0.3045	$\bar{1}.48362$	1.0453	.01926	.29131
0.31	1.3634	.13463	.73345	0.3150	$\bar{1}.49830$	1.0484	.02054	.30044
0.32	1.3771	.13897	.72615	0.3255	$\bar{1}.51254$	1.0516	.02187	.30951
0.33	1.3910	.14332	.71892	0.3360	$\bar{1}.52637$	1.0549	.02323	.31852
0.34	1.4049	.14766	.71177	0.3466	$\bar{1}.53981$	1.0584	.02463	.32748
0.35	1.4191	.15200	.70469	0.3572	$\bar{1}.55290$	1.0619	.02607	.33638
0.36	1.4333	.15635	.69768	0.3678	$\bar{1}.56564$	1.0655	.02755	.34521
0.37	1.4477	.16069	.69073	0.3785	$\bar{1}.57807$	1.0692	.02907	.35399
0.38	1.4623	.16503	.68386	0.3892	$\bar{1}.59019$	1.0731	.03063	.36271
0.39	1.4770	.16937	.67706	0.4000	$\bar{1}.60202$	1.0770	.03222	.37136
0.40	1.4918	.17372	.67032	0.4108	$\bar{1}.61358$	1.0811	.03385	.37995
0.41	1.5068	.17806	.66365	0.4216	$\bar{1}.62488$	1.0852	.03552	.38847
0.42	1.5220	.18240	.65705	0.4325	$\bar{1}.63594$	1.0895	.03723	.39693
0.43	1.5373	.18675	.65051	0.4434	$\bar{1}.64677$	1.0939	.03897	.40532
0.44	1.5527	.19109	.64404	0.4543	$\bar{1}.65738$	1.0984	.04075	.41364
0.45	1.5683	.19543	.63763	0.4653	$\bar{1}.66777$	1.1030	.04256	.42190
0.46	1.5841	.19978	.63128	0.4764	$\bar{1}.67797$	1.1077	.04441	.43008
0.47	1.6000	.20412	.62500	0.4875	$\bar{1}.68797$	1.1125	.04630	.43820
0.48	1.6161	.20846	.61878	0.4986	$\bar{1}.69779$	1.1174	.04822	.44624
0.49	1.6323	.21280	.61263	0.5098	$\bar{1}.70744$	1.1225	.05018	.45422
0.50	1.6487	.21715	.60653	0.5211	$\bar{1}.71692$	1.1276	.05217	.46212

Table 12 VALUES AND COMMON LOGARITHMS OF EXPONENTIAL AND HYPERBOLIC FUNCTIONS (*continued*)

x	e^x Value	e^x Log_{10}	e^{-x} Value	Sinh x Value	Sinh x Log_{10}	Cosh x Value	Cosh x Log_{10}	Tanh x Value
0.50	1.6487	.21715	.60653	0.5211	$\bar{1}.71692$	1.1276	.05217	.46212
0.51	1.6653	.22149	.60050	0.5324	$\bar{1}.72624$	1.1329	.05419	.46995
0.52	1.6820	.22583	.59452	0.5438	$\bar{1}.73540$	1.1383	.05625	.47770
0.53	1.6989	.23018	.58860	0.5552	$\bar{1}.74442$	1.1438	.05834	.48538
0.54	1.7160	.23452	.58275	0.5666	$\bar{1}.75330$	1.1494	.06046	.49299
0.55	1.7333	.23886	.57695	0.5782	$\bar{1}.76204$	1.1551	.06262	.50052
0.56	1.7507	.24320	.57121	0.5897	$\bar{1}.77065$	1.1609	.06481	.50798
0.57	1.7683	.24755	.56553	0.6014	$\bar{1}.77914$	1.1669	.06703	.51536
0.58	1.7860	.25189	.55990	0.6131	$\bar{1}.78751$	1.1730	.06929	.52267
0.59	1.8040	.25623	.55433	0.6248	$\bar{1}.79576$	1.1792	.07157	.52990
0.60	1.8221	.26058	.54881	0.6367	$\bar{1}.80390$	1.1855	.07389	.53705
0.61	1.8404	.26492	.54335	0.6485	$\bar{1}.81194$	1.1919	.07624	.54413
0.62	1.8589	.26926	.53794	0.6605	$\bar{1}.81987$	1.1984	.07861	.55113
0.63	1.8776	.27361	.53259	0.6725	$\bar{1}.82770$	1.2051	.08102	.55805
0.64	1.8965	.27795	.52729	0.6846	$\bar{1}.83543$	1.2119	.08346	.56490
0.65	1.9155	.28229	.52205	0.6967	$\bar{1}.84308$	1.2188	.08593	.57167
0.66	1.9348	.28664	.51685	0.7090	$\bar{1}.85063$	1.2258	.08843	.57836
0.67	1.9542	.29098	.51171	0.7213	$\bar{1}.85809$	1.2330	.09095	.58498
0.68	1.9739	.29532	.50662	0.7336	$\bar{1}.86548$	1.2402	.09351	.59152
0.69	1.9937	.29966	.50158	0.7461	$\bar{1}.87278$	1.2476	.09609	.59798
0.70	2.0138	.30401	.49659	0.7586	$\bar{1}.88000$	1.2552	.09870	.60437
0.71	2.0340	.30835	.49164	0.7712	$\bar{1}.88715$	1.2628	.10134	.61068
0.72	2.0544	.31269	.48675	0.7838	$\bar{1}.89423$	1.2706	.10401	.61691
0.73	2.0751	.31703	.48191	0.7966	$\bar{1}.90123$	1.2785	.10670	.62307
0.74	2.0959	.32138	.47711	0.8094	$\bar{1}.90817$	1.2865	.10942	.62915
0.75	2.1170	.32572	.47237	0.8223	$\bar{1}.91504$	1.2947	.11216	.63515
0.76	2.1383	.33006	.46767	0.8353	$\bar{1}.92185$	1.3030	.11493	.64108
0.77	2.1598	.33441	.46301	0.8484	$\bar{1}.92859$	1.3114	.11773	.64693
0.78	2.1815	.33875	.45841	0.8615	$\bar{1}.93527$	1.3199	.12055	.65271
0.79	2.2034	.34309	.45384	0.8748	$\bar{1}.94190$	1.3286	.12340	.65841
0.80	2.2255	.34744	.44933	0.8881	$\bar{1}.94846$	1.3374	.12627	.66404
0.81	2.2479	.35178	.44486	0.9015	$\bar{1}.95498$	1.3464	.12917	.66959
0.82	2.2705	.35612	.44043	0.9150	$\bar{1}.96144$	1.3555	.13209	.67507
0.83	2.2933	.36046	.43605	0.9286	$\bar{1}.96784$	1.3647	.13503	.68048
0.84	2.3164	.36481	.43171	0.9423	$\bar{1}.97420$	1.3740	.13800	.68581
0.85	2.3396	.36915	.42741	0.9561	$\bar{1}.98051$	1.3835	.14099	.69107
0.86	2.3632	.37349	.42316	0.9700	$\bar{1}.98677$	1.3932	.14400	.69626
0.87	2.3869	.37784	.41895	0.9840	$\bar{1}.99299$	1.4029	.14704	.70137
0.88	2.4109	.38218	.41478	0.9981	$\bar{1}.99916$	1.4128	.15009	.70642
0.89	2.4351	.38652	.41066	1.0122	0.00528	1.4229	.15317	.71139
0.90	2.4596	.39087	.40657	1.0265	0.01137	1.4331	.15627	.71630
0.91	2.4843	.39521	.40252	1.0409	.01741	1.4434	.15939	.72113
0.92	2.5093	.39955	.39852	1.0554	.02341	1.4539	.16254	.72590
0.93	2.5345	.40389	.39455	1.0700	.02937	1.4645	.16570	.73059
0.94	2.5600	.40824	.39063	1.0847	.03530	1.4753	.16888	.73522
0.95	2.5857	.41258	.38674	1.0995	.04119	1.4862	.17208	.73978
0.96	2.6117	.41692	.38289	1.1144	.04704	1.4973	.17531	.74428
0.97	2.6379	.42127	.37908	1.1294	.05286	1.5085	.17855	.74870
0.98	2.6645	.42561	.37531	1.1446	.05864	1.5199	.18181	.75307
0.99	2.6912	.42995	.37158	1.1598	.06439	1.5314	.18509	.75736
1.00	2.7183	.43429	.36788	1.1752	.07011	1.5431	.18839	.76159

Table 12 VALUES AND COMMON LOGARITHMS OF EXPONENTIAL AND HYPERBOLIC FUNCTIONS (*continued*)

x	e^x		e^{-x}	Sinh x		Cosh x		Tanh x
	Value	Log_{10}	Value	Value	Log_{10}	Value	Log_{10}	Value
1.00	2.7183	.43429	.36788	1.1752	.07011	1.5431	.18839	.76159
1.01	2.7456	.43864	.36422	1.1907	.07580	1.5549	.19171	.76576
1.02	2.7732	.44298	.36060	1.2063	.08146	1.5669	.19504	.76987
1.03	2.8011	.44732	.35701	1.2220	.08708	1.5790	.19839	.77391
1.04	2.8292	.45167	.35345	1.2379	.09268	1.5913	.20176	.77789
1.05	2.8577	.45601	.34994	1.2539	.09825	1.6038	.20515	.78181
1.06	2.8864	.46035	.34646	1.2700	.10379	1.6164	.20855	.78566
1.07	2.9154	.46470	.34301	1.2862	.10930	1.6292	.21197	.78946
1.08	2.9447	.46904	.33960	1.3025	.11479	1.6421	.21541	.79320
1.09	2.9743	.47338	.33622	1.3190	.12025	1.6552	.21886	.79688
1.10	3.0042	.47772	.33287	1.3356	.12569	1.6685	.22233	.80050
1.11	3.0344	.48207	.32956	1.3524	.13111	1.6820	.22582	.80406
1.12	3.0649	.48641	.32628	1.3693	.13649	1.6956	.22931	.80757
1.13	3.0957	.49075	.32303	1.3863	.14186	1.7093	.23283	.81102
1.14	3.1268	.49510	.31982	1.4035	.14720	1.7233	.23636	.81441
1.15	3.1582	.49944	.31664	1.4208	.15253	1.7374	.23990	.81775
1.16	3.1899	.50378	.31349	1.4382	.15783	1.7517	.24346	.82104
1.17	3.2220	.50812	.31037	1.4558	.16311	1.7662	.24703	.82427
1.18	3.2544	.51247	.30728	1.4735	.16836	1.7808	.25062	.82745
1.19	3.2871	.51681	.30422	1.4914	.17360	1.7957	.25422	.83058
1.20	3.3201	.52115	.30119	1.5095	.17882	1.8107	.25784	.83365
1.21	3.3535	.52550	.29820	1.5276	.18402	1.8258	.26146	.83668
1.22	3.3872	.52984	.29523	1.5460	.18920	1.8412	.26510	.83965
1.23	3.4212	.53418	.29229	1.5645	.19437	1.8568	.26876	.84258
1.24	3.4556	.53853	.28938	1.5831	.19951	1.8725	.27242	.84546
1.25	3.4903	.54287	.28650	1.6019	.20464	1.8884	.27610	.84828
1.26	3.5254	.54721	.28365	1.6209	.20975	1.9045	.27979	.85106
1.27	3.5609	.55155	.28083	1.6400	.21485	1.9208	.28349	.85380
1.28	3.5966	.55590	.27804	1.6593	.21993	1.9373	.28721	.85648
1.29	3.6328	.56024	.27527	1.6788	.22499	1.9540	.29093	.85913
1.30	3.6693	.56458	.27253	1.6984	.23004	1.9709	.29467	.86172
1.31	3.7062	.56893	.26982	1.7182	.23507	1.9880	.29842	.86428
1.32	3.7434	.57327	.26714	1.7381	.24009	2.0053	.30217	.86678
1.33	3.7810	.57761	.26448	1.7583	.24509	2.0228	.30594	.86925
1.34	3.8190	.58195	.26185	1.7786	.25008	2.0404	.30972	.87167
1.35	3.8574	.58630	.25924	1.7991	.25505	2.0583	.31352	.87405
1.36	3.8962	.59064	.25666	1.8198	.26002	2.0764	.31732	.87639
1.37	3.9354	.59498	.25411	1.8406	.26496	2.0947	.32113	.87869
1.38	3.9749	.59933	.25158	1.8617	.26990	2.1132	.32495	.88095
1.39	4.0149	.60367	.24908	1.8829	.27482	2.1320	.32878	.88317
1.40	4.0552	.60801	.24660	1.9043	.27974	2.1509	.33262	.88535
1.41	4.0960	.61236	.24414	1.9259	.28464	2.1700	.33647	.88749
1.42	4.1371	.61670	.24171	1.9477	.28952	2.1894	.34033	.88960
1.43	4.1787	.62104	.23931	1.9697	.29440	2.2090	.34420	.89167
1.44	4.2207	.62538	.23693	1.9919	.29926	2.2288	.34807	.89370
1.45	4.2631	.62973	.23457	2.0143	.30412	2.2488	.35196	.89569
1.46	4.3060	.63407	.23224	2.0369	.30896	2.2691	.35585	.89765
1.47	4.3492	.63841	.22993	2.0597	.31379	2.2896	.35976	.89958
1.48	4.3929	.64276	.22764	2.0827	.31862	2.3103	.36367	.90147
1.49	4.4371	.64710	.22537	2.1059	.32343	2.3312	.36759	.90332
1.50	4.4817	.65144	.22313	2.1293	.32823	2.3524	.37151	.90515

x	e^x		e^{-x}	Sinh x		Cosh x		Tanh x
	Value	Log_{10}	Value	Value	Log_{10}	Value	Log_{10}	Value
1.50	4.4817	.65144	.22313	2.1293	.32823	2.3524	.37151	.90515
1.51	4.5267	.65578	.22091	2.1529	.33303	2.3738	.37545	.90694
1.52	4.5722	.66013	.21871	2.1768	.33781	2.3955	.37939	.90870
1.53	4.6182	.66447	.21654	2.2008	.34258	2.4174	.38334	.91042
1.54	4.6646	.66881	.21438	2.2251	.34735	2.4395	.38730	.91212
1.55	4.7115	.67316	.21225	2.2496	.35211	2.4619	.39126	.91379
1.56	4.7588	.67750	.21014	2.2743	.35686	2.4845	.39524	.91542
1.57	4.8066	.68184	.20805	2.2993	.36160	2.5073	.39921	.91703
1.58	4.8550	.68619	.20598	2.3245	.36633	2.5305	.40320	.91860
1.59	4.9037	.69053	.20393	2.3499	.37105	2.5538	.40719	.92015
1.60	4.9530	.69487	.20190	2.3756	.37577	2.5775	.41119	.92167
1.61	5.0028	.69921	.19989	2.4015	.38048	2.6013	.41520	.92316
1.62	5.0531	.70356	.19790	2.4276	.38518	2.6255	.41921	.92462
1.63	5.1039	.70790	.19593	2.4540	.38987	2.6499	.42323	.92606
1.64	5.1552	.71224	.19398	2.4806	.39456	2.6746	.42725	.92747
1.65	5.2070	.71659	.19205	2.5075	.39923	2.6995	.43129	.92886
1.66	5.2593	.72093	.19014	2.5346	.40391	2.7247	.43532	.93022
1.67	5.3122	.72527	.18825	2.5620	.40857	2.7502	.43937	.93155
1.68	5.3656	.72961	.18637	2.5896	.41323	2.7760	.44341	.93286
1.69	5.4195	.73396	.18452	2.6175	.41788	2.8020	.44747	.93415
1.70	5.4739	.73830	.18268	2.6456	.42253	2.8283	.45153	.93541
1.71	5.5290	.74264	.18087	2.6740	.42717	2.8549	.45559	.93665
1.72	5.5845	.74699	.17907	2.7027	.43180	2.8818	.45966	.93786
1.73	5.6407	.75133	.17728	2.7317	.43643	2.9090	.46374	.93906
1.74	5.6973	.75567	.17552	2.7609	.44105	2.9364	.46782	.94023
1.75	5.7546	.76002	.17377	2.7904	.44567	2.9642	.47191	.94138
1.76	5.8124	.76436	.17204	2.8202	.45028	2.9922	.47600	.94250
1.77	5.8709	.76870	.17033	2.8503	.45488	3.0206	.48009	.94361
1.78	5.9299	.77304	.16864	2.8806	.45948	3.0492	.48419	.94470
1.79	5.9895	.77739	.16696	2.9112	.46408	3.0782	.48830	.94576
1.80	6.0496	.78173	.16530	2.9422	.46867	3.1075	.49241	.94681
1.81	6.1104	.78607	.16365	2.9734	.47325	3.1371	.49652	.94783
1.82	6.1719	.79042	.16203	3.0049	.47783	3.1669	.50064	.94884
1.83	6.2339	.79476	.16041	3.0367	.48241	3.1972	.50476	.94983
1.84	6.2965	.79910	.15882	3.0689	.48698	3.2277	.50889	.95080
1.85	6.3598	.80344	.15724	3.1013	.49154	3.2585	.51302	.95175
1.86	6.4237	.80779	.15567	3.1340	.49610	3.2897	.51716	.95268
1.87	6.4883	.81213	.15412	3.1671	.50066	3.3212	.52130	.95359
1.88	6.5535	.81647	.15259	3.2005	.50521	3.3530	.52544	.95449
1.89	6.6194	.82082	.15107	3.2341	.50976	3.3852	.52959	.95537
1.90	6.6859	.82516	.14957	3.2682	.51430	3.4177	.53374	.95624
1.91	6.7531	.82950	.14808	3.3025	.51884	3.4506	.53789	.95709
1.92	6.8210	.83385	.14661	3.3372	.52338	3.4838	.54205	.95792
1.93	6.8895	.83819	.14515	3.3722	.52791	3.5173	.54621	.95873
1.94	6.9588	.84253	.14370	3.4075	.53244	3.5512	.55038	.95953
1.95	7.0287	.84687	.14227	3.4432	.53696	3.5855	.55455	.96032
1.96	7.0993	.85122	.14086	3.4792	.54148	3.6201	.55872	.96109
1.97	7.1707	.85556	.13946	3.5156	.54600	3.6551	.56290	.96185
1.98	7.2427	.85990	.13807	3.5523	.55051	3.6904	.56707	.96259
1.99	7.3155	.86425	.13670	3.5894	.55502	3.7261	.57126	.96331
2.00	7.3891	.86859	.13534	3.6269	.55953	3.7622	.57544	.96403

x	e^x Value	e^x Log_{10}	e^{-x} Value	Sinh x Value	Sinh x Log_{10}	Cosh x Value	Cosh x Log_{10}	Tanh x Value
2.00	7.3891	.86859	.13534	3.6269	.55953	3.7622	.57544	.96403
2.01	7.4633	.87293	.13399	3.6647	.56403	3.7987	.57963	.96473
2.02	7.5383	.87727	.13266	3.7028	.56853	3.8355	.58382	.96541
2.03	7.6141	.88162	.13134	3.7414	.57303	3.8727	.58802	.96609
2.04	7.6906	.88596	.13003	3.7803	.57753	3.9103	.59221	.96675
2.05	7.7679	.89030	.12873	3.8196	.58202	3.9483	.59641	.96740
2.06	7.8460	.89465	.12745	3.8593	.58650	3.9867	.60061	.96803
2.07	7.9248	.89899	.12619	3.8993	.59099	4.0255	.60482	.96865
2.08	8.0045	.90333	.12493	3.9398	.59547	4.0647	.60903	.96926
2.09	8.0849	.90768	.12369	3.9806	.59995	4.1043	.61324	.96986
2.10	8.1662	.91202	.12246	4.0219	.60443	4.1443	.61745	.97045
2.11	8.2482	.91636	.12124	4.0635	.60890	4.1847	.62167	.97103
2.12	8.3311	.92070	.12003	4.1056	.61337	4.2256	.62589	.97159
2.13	8.4149	.92505	.11884	4.1480	.61784	4.2669	.63011	.97215
2.14	8.4994	.92939	.11765	4.1909	.62231	4.3085	.63433	.97269
2.15	8.5849	.93373	.11648	4.2342	.62677	4.3507	.63856	.97323
2.16	8.6711	.93808	.11533	4.2779	.63123	4.3932	.64278	.97375
2.17	8.7583	.94242	.11418	4.3221	.63569	4.4362	.64701	.97426
2.18	8.8463	.94676	.11304	4.3666	.64015	4.4797	.65125	.97477
2.19	8.9352	.95110	.11192	4.4116	.64460	4.5236	.65548	.97526
2.20	9.0250	.95545	.11080	4.4571	.64905	4.5679	.65972	.97574
2.21	9.1157	.95979	.10970	4.5030	.65350	4.6127	.66396	.97622
2.22	9.2073	.96413	.10861	4.5494	.65795	4.6580	.66820	.97668
2.23	9.2999	.96848	.10753	4.5962	.66240	4.7037	.67244	.97714
2.24	9.3933	.97282	.10646	4.6434	.66684	4.7499	.67668	.97759
2.25	9.4877	.97716	.10540	4.6912	.67128	4.7966	.68093	.97803
2.26	9.5831	.98151	.10435	4.7394	.67572	4.8437	.68518	.97846
2.27	9.6794	.98585	.10331	4.7880	.68016	4.8914	.68943	.97888
2.28	9.7767	.99019	.10228	4.8372	.68459	4.9395	.69368	.97929
2.29	9.8749	.99453	.10127	4.8868	.68903	4.9881	.69794	.97970
2.30	9.9742	0.99888	.10026	4.9370	.69346	5.0372	.70219	.98010
2.31	10.074	1.00322	.09926	4.9876	.69789	5.0868	.70645	.98049
2.32	10.176	1.00756	.09827	5.0387	.70232	5.1370	.71071	.98087
2.33	10.278	1.01191	.09730	5.0903	.70675	5.1876	.71497	.98124
2.34	10.381	1.01625	.09633	5.1425	.71117	5.2388	.71923	.98161
2.35	10.486	1.02059	.09537	5.1951	.71559	5.2905	.72349	.98197
2.36	10.591	1.02493	.09442	5.2483	.72002	5.3427	.72776	.98233
2.37	10.697	1.02928	.09348	5.3020	.72444	5.3954	.73203	.98267
2.38	10.805	1.03362	.09255	5.3562	.72885	5.4487	.73630	.98301
2.39	10.913	1.03796	.09163	5.4109	.73327	5.5026	.74056	.98335
2.40	11.023	1.04231	.09072	5.4662	.73769	5.5569	.74484	.98367
2.41	11.134	1.04665	.08982	5.5221	.74210	5.6119	.74911	.98400
2.42	11.246	1.05099	.08892	5.5785	.74652	5.6674	.75338	.98431
2.43	11.359	1.05534	.08804	5.6354	.75093	5.7235	.75766	.98462
2.44	11.473	1.05968	.08716	5.6929	75534	5.7801	.76194	.98492
2.45	11.588	1.06402	.08629	5.7510	.75975	5.8373	.76621	.98522
2.46	11.705	1.06836	.08543	5.8097	.76415	5.8951	.77049	.98551
2.47	11.822	1.07271	.08458	5.8689	.76856	5.9535	.77477	.98579
2.48	11.941	1.07705	.08374	5.9288	.77296	6.0125	.77906	.98607
2.49	12.061	1.08139	.08291	5.9892	.77737	6.0721	.78334	.98635
2.50	12.182	1.08574	.08208	6.0502	.78177	6.1323	.78762	.98661

Table 12 VALUES AND COMMON LOGARITHMS OF EXPONENTIAL AND HYPERBOLIC FUNCTIONS (*continued*)

x	e^x		e^{-x}	Sinh x		Cosh x		Tanh x
	Value	Log_{10}	Value	Value	Log_{10}	Value	Log_{10}	Value
2.50	12.182	1.08574	.08208	6.0502	.78177	6.1323	.78762	.98661
2.51	12.305	1.09008	.08127	6.1118	.78617	6.1931	.79191	.98688
2.52	12.429	1.09442	.08046	6.1741	.79057	6.2545	.79619	.98714
2.53	12.554	1.09877	.07966	6.2369	.79497	6.3166	.80048	.98739
2.54	12.680	1.10311	.07887	6.3004	.79937	6.3793	.80477	.98764
2.55	12.807	1.10745	.07808	6.3645	.80377	6.4426	.80906	.98788
2.56	12.936	1.11179	.07730	6.4293	.80816	6.5066	.81335	.98812
2.57	13.066	1.11614	.07654	6.4946	.81256	6.5712	.81764	.98835
2.58	13.197	1.12048	.07577	6.5607	.81695	6.6365	.82194	.98858
2.59	13.330	1.12482	.07502	6.6274	.82134	6.7024	.82623	.98881
2.60	13.464	1.12917	.07427	6.6947	.82573	6.7690	.83052	.98903
2.61	13.599	1.13351	.07353	6.7628	.83012	6.8363	.83482	.98924
2.62	13.736	1.13785	.07280	6.8315	.83451	6.9043	.83912	.98946
2.63	13.874	1.14219	.07208	6.9008	.83890	6.9729	.84341	.98966
2.64	14.013	1.14654	.07136	6.9709	.84329	7.0423	.84771	.98987
2.65	14.154	1.15088	.07065	7.0417	.84768	7.1123	.85201	.99007
2.66	14.296	1.15522	.06995	7.1132	.85206	7.1831	.85631	.99026
2.67	14.440	1.15957	.06925	7.1854	.85645	7.2546	.86061	.99045
2.68	14.585	1.16391	.06856	7.2583	.86083	7.3268	.86492	.99064
2.69	14.732	1.16825	.06788	7.3319	.86522	7.3998	.86922	.99083
2.70	14.880	1.17260	.06721	7.4063	.86960	7.4735	.87352	.99101
2.71	15.029	1.17694	.06654	7.4814	.87398	7.5479	.87783	.99118
2.72	15.180	1.18128	.06587	7.5572	.87836	7.6231	.88213	.99136
2.73	15.333	1.18562	.06522	7.6338	.88274	7.6991	.88644	.99153
2.74	15.487	1.18997	.06457	7.7112	.88712	7.7758	.89074	.99170
2.75	15.643	1.19431	.06393	7.7894	.89150	7.8533	.89505	.99186
2.76	15.800	1.19865	.06329	7.8683	.89588	7.9316	.89936	.99202
2.77	15.959	1.20300	.06266	7.9480	.90026	8.0106	.90367	.99218
2.78	16.119	1.20734	.06204	8.0285	.90463	8.0905	.90798	.99233
2.79	16.281	1.21168	.06142	8.1098	.90901	8.1712	.91229	.99248
2.80	16.445	1.21602	.06081	8.1919	.91339	8.2527	.91660	.99263
2.81	16.610	1.22037	.06020	8.2749	.91776	8.3351	.92091	.99278
2.82	16.777	1.22471	.05961	8.3586	.92213	8.4182	.92522	.99292
2.83	16.945	1.22905	.05901	8.4432	.92651	8.5022	.92953	.99306
2.84	17.116	1.23340	.05843	8.5287	.93088	8.5871	.93385	.99320
2.85	17.288	1.23774	.05784	8.6150	.93525	8.6728	.93816	.99333
2.86	17.462	1.24208	.05727	8.7021	.93963	8.7594	.94247	.99346
2.87	17.637	1.24643	.05670	8.7902	.94400	8.8469	.94679	.99359
2.88	17.814	1.25077	.05613	8.8791	.94837	8.9352	.95110	.99372
2.89	17.993	1.25511	.05558	8.9689	.95274	9.0244	.95542	.99384
2.90	18.174	1.25945	.05502	9.0596	.95711	9.1146	.95974	.99396
2.91	18.357	1.26380	.05448	9.1512	.96148	9.2056	.96405	.99408
2.92	18.541	1.26814	.05393	9.2437	.96584	9.2976	.96837	.99420
2.93	18.728	1.27248	.05340	9.3371	.97021	9.3905	.97269	.99431
2.94	18.916	1.27683	.05287	9.4315	.97458	9.4844	.97701	.99443
2.95	19.106	1.28117	.05234	9.5268	.97895	9.5791	.98133	.99454
2.96	19.298	1.28551	.05182	9.6231	.98331	9.6749	.98565	.99464
2.97	19.492	1.28985	.05130	9.7203	.98768	9.7716	.98997	.99475
2.98	19.688	1.29420	.05079	9.8185	.99205	9.8693	.99429	.99485
2.99	19.886	1.29854	.05029	9.9177	0.99641	9.9680	0.99861	.99496
3.00	20.086	1.30288	.04979	10.018	1.00078	10.068	1.00293	.99505

Table 12 VALUES AND COMMON LOGARITHMS OF EXPONENTIAL AND HYPERBOLIC FUNCTIONS *(continued)*

x	e^x Value	e^x Log10	e^{-x} Value	Sinh x Value	Sinh x Log10	Cosh x Value	Cosh x Log10	Tanh x Value
3.00	20.086	1.30288	.04979	10.018	1.00078	10.068	1.00293	.99505
3.05	21.115	1.32460	.04736	10.534	1.02259	10.581	1.02454	.99552
3.10	22.198	1.34631	.04505	11.076	1.04440	11.122	1.04616	.99595
3.15	23.336	1.36803	.04285	11.647	1.06620	11.690	1.06779	.99633
3.20	24.533	1.38974	.04076	12.246	1.08799	12.287	1.08943	.99668
3.25	25.790	1.41146	.03877	12.876	1.10977	12.915	1.11108	.99700
3.30	27.113	1.43317	.03688	13.538	1.13155	13.575	1.13273	.99728
3.35	28.503	1.45489	.03508	14.234	1.15332	14.269	1.15439	.99754
3.40	29.964	1.47660	.03337	14.965	1.17509	14.999	1.17605	.99777
3.45	31.500	1.49832	.03175	15.734	1.19685	15.766	1.19772	.99799
3.50	33.115	1.52003	.03020	16.543	1.21860	16.573	1.21940	.99818
3.55	34.813	1.54175	.02872	17.392	1.24036	17.421	1.24107	.99835
3.60	36.598	1.56346	.02732	18.286	1.26211	18.313	1.26275	.99851
3.65	38.475	1.58517	.02599	19.224	1.28385	19.250	1.28444	.99865
3.70	40.447	1.60689	.02472	20.211	1.30559	20.236	1.30612	.99878
3.75	42.521	1.62860	.02352	21.249	1.32733	21.272	1.32781	.99889
3.80	44.701	1.65032	.02237	22.339	1.34907	22.362	1.34951	.99900
3.85	46.993	1.67203	.02128	23.486	1.37081	23.507	1.37120	.99909
3.90	49.402	1.69375	.02024	24.691	1.39254	24.711	1.39290	.99918
3.95	51.935	1.71546	.01925	25.958	1.41427	25.977	1.41459	.99926
4.00	54.598	1.73718	.01832	27.290	1.43600	27.308	1.43629	.99933
4.10	60.340	1.78061	.01657	30.162	1.47946	30.178	1.47970	.99945
4.20	66.686	1.82404	.01500	33.336	1.52291	33.351	1.52310	.99955
4.30	73.700	1.86747	.01357	36.843	1.56636	36.857	1.56652	.99963
4.40	81.451	1.91090	.01227	40.719	1.60980	40.732	1.60993	.99970
4.50	90.017	1.95433	.01111	45.003	1.65324	45.014	1.65335	.99975
4.60	99.484	1.99775	.01005	49.737	1.69668	49.747	1.69677	.99980
4.70	109.95	2.04118	.00910	54.969	1.74012	54.978	1.74019	.99983
4.80	121.51	2.08461	.00823	60.751	1.78355	60.759	1.78361	.99986
4.90	134.29	2.12804	.00745	67.141	1.82699	67.149	1.82704	.99989
5.00	148.41	2.17147	.00674	74.203	1.87042	74.210	1.87046	.99991
5.10	164.02	2.21490	.00610	82.008	1.91389	82.014	1.91389	.99993
5.20	181.27	2.25833	.00552	90.633	1.95729	90.639	1.95731	.99994
5.30	200.34	2.30176	.00499	100.17	2.00074	100.17	2.00074	.99995
5.40	221.41	2.34519	.00452	110.70	2.04415	110.71	2.04417	.99996
5.50	244.69	2.38862	.00409	122.34	2.08758	122.35	2.08760	.99997
5.60	270.43	2.43205	.00370	135.21	2.13101	135.22	2.13103	.99997
5.70	298.87	2.47548	.00335	149.43	2.17444	149.44	2.17445	.99998
5.80	330.30	2.51891	.00303	165.15	2.21787	165.15	2.21788	.99998
5.90	365.04	2.56234	.00274	182.52	2.26130	182.52	2.26131	.99998
6.00	403.43	2.60577	.00248	201.71	2.30473	201.72	2.30474	.99999
6.25	518.01	2.71434	.00193	259.01	2.41331	259.01	2.41331	.99999
6.50	665.14	2.82291	.00150	332.57	2.52188	332.57	2.52189	1.0000
6.75	854.06	2.93149	.00117	427.03	2.63046	427.03	2.63046	1.0000
7.00	1096.6	3.04006	.00091	548.32	2.73903	548.32	2.73903	1.0000
7.50	1808.0	3.25721	.00055	904.02	2.95618	904.02	2.95618	1.0000
8.00	2981.0	3.47436	.00034	1490.5	3.17333	1490.5	3.17333	1.0000
8.50	4914.8	3.69150	.00020	2457.4	3.39047	2457.4	3.39047	1.0000
9.00	8103.1	3.90865	.00012	4051.5	3.60762	4051.5	3.60762	1.0000
9.50	13360.	4.12580	.00007	6679.9	3.82477	6679.9	3.82477	1.0000
10.00	22026.	4.34294	.00005	11013.	4.04191	11013.	4.04191	1.0000

Table 13 MULTIPLES OF M AND OF $1/M$

N	N·M	N	N·M	N	N÷M	N	N÷M
0	0.00000 000	50	21.71472 410	0	0.00000 000	50	115.12925 465
1	0.43429 448	51	22.14901 858	1	2.30258 509	51	117.43183 974
2	0.86858 896	52	22.58331 306	2	4.60517 019	52	119.73442 484
3	1.30288 345	53	23.01760 754	3	6.90775 528	53	122.03700 993
4	1.73717 793	54	23.45190 202	4	9.21034 037	54	124.33959 502
5	2.17147 241	55	23.88619 650	5	11.51292 546	55	126.64218 011
6	2.60576 689	56	24.32049 099	6	13.81551 056	56	128.94476 521
7	3.04006 137	57	24.75478 547	7	16.11809 565	57	131.24735 030
8	3.47435 586	58	25.18907 995	8	18.42068 074	58	133.54993 539
9	3.90865 034	59	25.62337 443	9	20.72326 584	59	135.85252 049
10	4.34294 482	60	26.05766 891	10	23.02585 093	60	138.15510 558
11	4.77723 930	61	26.49196 340	11	25.32843 602	61	140.45769 067
12	5.21153 378	62	26.92625 788	12	27.63102 112	62	142.76027 577
13	5.64582 826	63	27.36055 236	13	29.93360 621	63	145.06286 086
14	6.08012 275	64	27.79484 684	14	32.23619 130	64	147.36544 595
15	6.51441 723	65	28.22914 132	15	34.53877 639	65	149.66803 104
16	6.94871 171	66	28.66343 581	16	36.84136 149	66	151.97061 614
17	7.38300 619	67	29.09773 029	17	39.14394 658	67	154.27320 123
18	7.81730 067	68	29.53202 477	18	41.44653 167	68	156.57578 632
19	8.25159 516	69	29.96631 925	19	43.74911 677	69	158.87837 142
20	8.68588 964	70	30.40061 373	20	46.05170 186	70	161.18095 651
21	9.12018 412	71	30.83490 822	21	48.35428 695	71	163.48354 160
22	9.55447 860	72	31.26920 270	22	50.65687 205	72	165.78612 670
23	9.98877 308	73	31.70349 718	23	52.95945 714	73	168.08871 179
24	10.42306 757	74	32.13779 166	24	55.26204 223	74	170.39129 688
25	10.85736 205	75	32.57208 614	25	57.56462 732	75	172.69388 197
26	11.29165 653	76	33.00638 062	26	59.86721 242	76	174.99646 707
27	11.72595 101	77	33.44067 511	27	62.16979 751	77	177.29905 216
28	12.16024 549	78	33.87496 959	28	64.47238 260	78	179.60163 725
29	12.59453 998	79	34.30926 407	29	66.77496 770	79	181.90422 235
30	13.02883 446	80	34.74355 855	30	69.07755 279	80	184.20680 744
31	13.46312 894	81	35.17785 303	31	71.38013 788	81	186.50939 253
32	13.89742 342	82	35.61214 752	32	73.68272 298	82	188.81197 763
33	14.33171 790	83	36.04644 200	33	75.98530 807	83	191.11456 272
34	14.76601 238	84	36.48073 648	34	78.28789 316	84	193.41714 781
35	15.20030 687	85	36.91503 096	35	80.59047 825	85	195.71973 290
36	15.63460 135	86	37.34932 544	36	82.89306 335	86	198.02231 800
37	16.06889 583	87	37.78361 993	37	85.19564 844	87	200.32490 309
38	16.50319 031	88	38.21791 441	38	87.49823 353	88	202.62748 818
39	16.93748 479	89	38.65220 889	39	89.80081 863	89	204.93007 328
40	17.37177 928	90	39.08650 337	40	92.10340 372	90	207.23265 837
41	17.80607 376	91	39.52079 785	41	94.40598 881	91	209.53524 346
42	18.24036 824	92	39.95509 234	42	96.70857 391	92	211.83782 856
43	18.67466 272	93	40.38938 682	43	99.01115 900	93	214.14041 365
44	19.10895 720	94	40.82368 130	44	101.31374 409	94	216.44299 874
45	19.54325 169	95	41.25797 578	45	103.61632 918	95	218.74558 383
46	19.97754 617	96	41.69227 026	46	105.91891 428	96	221.04816 893
47	20.41184 065	97	42.12656 474	47	108.22149 937	97	223.35075 402
48	20.84613 513	98	42.56085 923	48	110.52408 446	98	225.65333 911
49	21.28042 961	99	42.99515 371	49	112.82666 956	99	227.95592 421
50	21.71472 410	100	43.42944 819	50	115.12925 465	100	230.25850 930

$M = \log_{10} e = .43429\ 44819\ 03251\ 82765$

$\frac{1}{M} = \log_e 10 = 2.30258\ 50929\ 94045\ 68402$

$\log_{10} n = \log_e n \log_{10} e = M \log_e n$

$\log_e n = \log_{10} n \log_e 10 = \frac{1}{M} \log_{10} n$

$\log_{10} e^x = x \log_{10} e = xM$

$\log_e (10^n x) = \log_e x + n \frac{1}{M}$

POWERS, ROOTS, RECIPROCALS, AREAS, AND CIRCUMFERENCES

Table 14 SQUARES, CUBES, SQUARE ROOTS, AND CUBE ROOTS

n	n^2	n^3	$\sqrt{n}$	$\sqrt{10n}$	$\sqrt[3]{n}$	$\sqrt[3]{10n}$	$\sqrt[3]{100n}$
1	1	1	1.000 000	3.162 278	1.000 000	2.154 435	4.641 589
2	4	8	1.414 214	4.472 136	1.259 921	2.714 418	5.848 035
3	9	27	1.732 051	5.477 226	1.442 250	3.107 233	6.694 330
4	16	64	2.000 000	6.324 555	1.587 401	3.419 952	7.368 063
5	25	125	2.236 068	7.071 068	1.709 976	3.684 031	7.937 005
6	36	216	2.449 490	7.745 967	1.817 121	3.914 868	8.434 327
7	49	343	2.645 751	8.366 600	1.912 931	4.121 285	8.879 040
8	64	512	2.828 427	8.944 272	2.000 000	4.308 869	9.283 178
9	81	729	3.000 000	9.486 833	2.080 084	4.481 405	9.654 894
10	100	1 000	3.162 278	10.000 00	2.154 435	4.641 589	10.000 00
11	121	1 331	3.316 625	10.488 09	2.223 980	4.791 420	10.322 80
12	144	1 728	3.464 102	10.954 45	2.289 428	4.932 424	10.626 59
13	169	2 197	3.605 551	11.401 75	2.351 335	5.065 797	10.913 93
14	196	2 744	3.741 657	11.832 16	2.410 142	5.192 494	11.186 89
15	225	3 375	3.872 983	12.247 45	2.466 212	5.313 293	11.447 14
16	256	4 096	4.000 000	12.649 11	2.519 842	5.428 835	11.696 07
17	289	4 913	4.123 106	13.038 40	2.571 282	5.539 658	11.934 83
18	324	5 832	4.242 641	13.416 41	2.620 741	5.646 216	12.164 40
19	361	6 859	4.358 899	13.784 05	2.668 402	5.748 897	12.385 62
20	400	8 000	4.472 136	14.142 14	2.714 418	5.848 035	12.599 21
21	441	9 261	4.582 576	14.491 38	2.758 924	5.943 922	12.805 79
22	484	10 648	4.690 416	14.832 40	2.802 039	6.036 811	13.005 91
23	529	12 167	4.795 832	15.165 75	2.843 867	6.126 926	13.200 06
24	576	13 824	4.898 979	15.491 93	2.884 499	6.214 465	13.388 66
25	625	15 625	5.000 000	15.811 39	2.924 018	6.299 605	13.572 09
26	676	17 576	5.099 020	16.124 52	2.962 496	6.382 504	13.750 69
27	729	19 683	5.196 152	16.431 68	3.000 000	6.463 304	13.924 77
28	784	21 952	5.291 503	16.733 20	3.036 589	6.542 133	14.094 60
29	841	24 389	5.385 165	17.029 39	3.072 317	6.619 106	14.260 43
30	900	27 000	5.477 226	17.320 51	3.107 233	6.694 330	14.422 50
31	961	29 791	5.567 764	17.606 82	3.141 381	6.767 899	14.581 00
32	1 024	32 768	5.656 854	17.888 54	3.174 802	6.839 904	14.736 13
33	1 089	35 937	5.744 563	18.165 90	3.207 534	6.910 423	14.888 06
34	1 156	39 304	5.830 952	18.439 09	3.239 612	6.979 532	15.036 95
35	1 225	42 875	5.916 080	18.708 29	3.271 066	7.047 299	15.182 94
36	1 296	46 656	6.000 000	18.973 67	3.301 927	7.113 787	15.326 19
37	1 369	50 653	6.082 763	19.235 38	3.332 222	7.179 054	15.466 80
38	1 444	54 872	6.164 414	19.493 59	3.361 975	7.243 156	15.604 91
39	1 521	59 319	6.244 998	19.748 42	3.391 211	7.306 144	15.740 61
40	1 600	64 000	6.324 555	20.000 00	3.419 952	7.368 063	15.874 01
41	1 681	68 921	6.403 124	20.248 46	3.448 217	7.428 959	16.005 21
42	1 764	74 088	6.480 741	20.493 90	3.476 027	7.488 872	16.134 29
43	1 849	79 507	6.557 439	20.736 44	3.503 398	7.547 842	16.261 33
44	1 936	85 184	6.633 250	20.976 18	3.530 348	7.605 905	16.386 43
45	2 025	91 125	6.708 204	21.213 20	3.556 893	7.663 094	16.509 64
46	2 116	97 336	6.782 330	21.447 61	3.583 048	7.719 443	16.631 03
47	2 209	103 823	6.855 655	21.679 48	3.608 826	7.774 980	16.750 69
48	2 304	110 592	6.928 203	21.908 90	3.634 241	7.829 735	16.868 65
49	2 401	117 649	7.000 000	22.135 94	3.659 306	7.883 735	16.984 99

n	n^2	n^3	$\sqrt{n}$	$\sqrt{10n}$	$\sqrt[3]{n}$	$\sqrt[3]{10n}$	$\sqrt[3]{100n}$
50	2 500	125 000	7.071 068	22.360 68	3.684 031	7.937 005	17.099 76
51	2 601	132 651	7.141 428	22.583 18	3.708 430	7.989 570	17.213 01
52	2 704	140 608	7.211 103	22.803 51	3.732 511	8.041 452	17.324 78
53	2 809	148 877	7.280 110	23.021 73	3.756 286	8.092 672	17.435 13
54	2 916	157 464	7.348 469	23.237 90	3.779 763	8.143 253	17.544 11
55	3 025	166 375	7.416 198	23.452 08	3.802 952	8.193 213	17.651 74
56	3 136	175 616	7.483 315	23.664 32	3.825 862	8.242 571	17.758 08
57	3 249	185 193	7.549 834	23.874 67	3.848 501	8.291 344	17.863 16
58	3 364	195 112	7.615 773	24.083 19	3.870 877	8.339 551	17.967 02
59	3 481	205 379	7.681 146	24.289 92	3.892 996	8.387 207	18.069 69
60	3 600	216 000	7.745 967	24.494 90	3.914 868	8.434 327	18.171 21
61	3 721	226 981	7.810 250	24.698 18	3.936 497	8.480 926	18.271 60
62	3 844	238 328	7.874 008	24.899 80	3.957 892	8.527 019	18.370 91
63	3 969	250 047	7.937 254	25.099 80	3.979 057	8.572 619	18.469 15
64	4 096	262 144	8.000 000	25.298 22	4.000 000	8.617 739	18.566 36
65	4 225	274 625	8.062 258	25.495 10	4.020 726	8.662 391	18.662 56
66	4 356	287 496	8.124 038	25.690 47	4.041 240	8.706 588	18.757 77
67	4 489	300 763	8.185 353	25.884 36	4.061 548	8.750 340	18.852 04
68	4 624	314 432	8.246 211	26.076 81	4.081 655	8.793 659	18.945 36
69	4 761	328 509	8.306 624	26.267 85	4.101 566	8.836 556	19.037 78
70	4 900	343 000	8.366 600	26.457 51	4.121 285	8.879 040	19.129 31
71	5 041	357 911	8.426 150	26.645 83	4.140 818	8.921 121	19.219 97
72	5 184	373 248	8.485 281	26.832 82	4.160 168	8.962 809	19.309 79
73	5 329	389 017	8.544 004	27.018 51	4.179 339	9.004 113	19.398 77
74	5 476	405 224	8.602 325	27.202 94	4.198 336	9.045 042	19.486 95
75	5 625	421 875	8.660 254	27.386 13	4.217 163	9.085 603	19.574 34
76	5 776	438 976	8.717 798	27.568 10	4.235 824	9.125 805	19.660 95
77	5 929	456 533	8.774 964	27.748 87	4.254 321	9.165 656	19.746 81
78	6 084	474 552	8.831 761	27.928 48	4.272 659	9.205 164	19.831 92
79	6 241	493 039	8.888 194	28.106 94	4.290 840	9.244 335	19.916 32
80	6 400	512 000	8.944 272	28.284 27	4.308 869	9.283 178	20.000 00
81	6 561	531 441	9.000 000	28.460 50	4.326 749	9.321 698	20.082 99
82	6 724	551 368	9.055 385	28.635 64	4.344 481	9.359 902	20.165 30
83	6 889	571 787	9.110 434	28.809 72	4.362 071	9.397 796	20.246 94
84	7 056	592 704	9.165 151	28.982 75	4.379 519	9.435 388	20.327 93
85	7 225	614 125	9.219 544	29.154 76	4.396 830	9.472 682	20.408 28
86	7 396	636 056	9.273 618	29.325 76	4.414 005	9.509 685	20.488 00
87	7 569	658 503	9.327 379	29.495 76	4.431 048	9.546 403	20.567 10
88	7 744	681 472	9.380 832	29.664 79	4.447 960	9.582 840	20.645 60
89	7 921	704 969	9.433 981	29.832 87	4.464 745	9.619 002	20.723 51
90	8 100	729 000	9.486 833	30.000 00	4.481 405	9.654 894	20.800 84
91	8 281	753 571	9.539 392	30.166 21	4.497 941	9.690 521	20.877 59
92	8 464	778 688	9.591 663	30.331 50	4.514 357	9.725 888	20.953 79
93	8 649	804 357	9.643 651	30.495 90	4.530 655	9.761 000	21.029 44
94	8 836	830 584	9.695 360	30.659 42	4.546 836	9.795 861	21.104 54
95	9 025	857 375	9.746 794	30.822 07	4.562 903	9.830 476	21.179 12
96	9 216	884 736	9.797 959	30.983 87	4.578 857	9.864 848	21.253 17
97	9 409	912 673	9.848 858	31.144 82	4.594 701	9.898 983	21.326 71
98	9 604	941 192	9.899 495	31.304 95	4.610 436	9.932 884	21.399 75
99	9 801	970 299	9.949 874	31.464 27	4.626 065	9.966 555	21.472 29

n	n^2	n^3	$\sqrt{n}$	$\sqrt{10n}$	$\sqrt[3]{n}$	$\sqrt[3]{10n}$	$\sqrt[3]{100n}$
100	10 000	1 000 000	10.000 00	31.622 78	4.641 589	10.000 00	21.544 35
101	10 201	1 030 301	10.049 88	31.780 50	4.657 010	10.033 22	21.615 92
102	10 404	1 061 208	10.099 50	31.937 44	4.672 329	10.066 23	21.687 03
103	10 609	1 092 727	10.148 89	32.093 61	4.687 548	10.099 02	21.757 67
104	10 816	1 124 864	10.198 04	32.249 03	4.702 669	10.131 59	21.827 86
105	11 025	1 157 625	10.246 95	32.403 70	4.717 694	10.163 96	21.897 60
106	11 236	1 191 016	10.295 63	32.557 64	4.732 623	10.196 13	21.966 89
107	11 449	1 225 043	10.344 08	32.710 85	4.747 459	10.228 09	22.035 75
108	11 664	1 259 712	10.392 30	32.863 35	4.762 203	10.259 86	22.104 19
109	11 881	1 295 029	10.440 31	33.015 15	4.776 856	10.291 42	22.172 20
110	12 100	1 331 000	10.488 09	33.166 25	4.791 420	10.322 80	22.239 80
111	12 321	1 367 631	10.535 65	33.316 66	4.805 896	10.353 99	22.306 99
112	12 544	1 404 928	10.583 01	33.466 40	4.820 285	10.384 99	22.373 78
113	12 769	1 442 897	10.630 15	33.615 47	4.834 588	10.415 80	22.440 17
114	12 996	1 481 544	10.677 08	33.763 89	4.848 808	10.446 44	22.506 17
115	13 225	1 520 875	10.723 81	33.911 65	4.862 944	10.476 90	22.571 79
116	13 456	1 560 896	10.770 33	34.058 77	4.876 999	10.507 18	22.637 02
117	13 689	1 601 613	10.816 65	34.205 26	4.890 973	10.537 28	22.701 89
118	13 924	1 643 032	10.862 78	34.351 13	4.904 868	10.567 22	22.766 38
119	14 161	1 685 159	10.908 71	34.496 38	4.918 685	10.596 99	22.830 51
120	14 400	1 728 000	10.954 45	34.641 02	4.932 424	10.626 59	22.894 28
121	14 641	1 771 561	11.000 00	34.785 05	4.946 087	10.656 02	22.957 70
122	14 884	1 815 848	11.045 36	34.928 50	4.959 676	10.685 30	23.020 78
123	15 129	1 860 867	11.090 54	35.071 36	4.973 190	10.714 41	23.083 50
124	15 376	1 906 624	11.135 53	35.213 63	4.986 631	10.743 37	23.145 89
125	15 625	1 953 125	11.180 34	35.355 34	5.000 000	10.772 17	23.207 94
126	15 876	2 000 376	11.224 97	35.496 48	5.013 298	10.800 82	23.269 67
127	16 129	2 048 383	11.269 43	35.637 06	5.026 526	10.829 32	23.331 07
128	16 384	2 097 152	11.313 71	35.777 09	5.039 684	10.857 67	23.392 14
129	16 641	2 146 689	11.357 82	35.916 57	5.052 774	10.885 87	23.452 90
130	16 900	2 197 000	11.401 75	36.055 51	5.065 797	10.913 93	23.513 35
131	17 161	2 248 091	11.445 52	36.193 92	5.078 753	10.941 84	23.573 48
132	17 424	2 299 968	11.489 13	36.331 80	5.091 643	10.969 61	23.633 32
133	17 689	2 352 637	11.532 56	36.469 17	5.104 469	10.997 24	23.692 85
134	17 956	2 406 104	11.575 84	36.606 01	5.117 230	11.024 74	23.752 08
135	18 225	2 460 375	11.618 95	36.742 35	5.129 928	11.052 09	23.811 02
136	18 496	2 515 456	11.661 90	36.878 18	5.142 563	11.079 32	23.869 66
137	18 769	2 571 353	11.704 70	37.013 51	5.155 137	11.106 41	23.928 03
138	19 044	2 628 072	11.747 34	37.148 35	5.167 649	11.133 36	23.986 10
139	19 321	2 685 619	11.789 83	37.282 70	5.180 101	11.160 19	24.043 90
140	19 600	2 744 000	11.832 16	37.416 57	5.192 494	11.186 89	24.101 42
141	19 881	2 803 221	11.874 34	37.549 97	5.204 828	11.213 46	24.158 67
142	20 164	2 863 288	11.916 38	37.682 89	5.217 103	11.239 91	24.215 65
143	20 449	2 924 207	11.958 26	37.815 34	5.229 322	11.266 23	24.272 36
144	20 736	2 985 984	12.000 00	37.947 33	5.241 483	11.292 43	24.328 81
145	21 025	3 048 625	12.041 59	38.078 87	5.253 588	11.318 51	24.384 99
146	21 316	3 112 136	12.083 05	38.209 95	5.265 637	11.344 47	24.440 92
147	21 609	3 176 523	12.124 36	38.340 58	5.277 632	11.370 31	24.496 60
148	21 904	3 241 792	12.165 53	38.470 77	5.289 572	11.396 04	24.552 02
149	22 201	3 307 949	12.206 56	38.600 52	5.301 459	11.421 65	24.607 19

n	n^2	n^3	$\sqrt{n}$	$\sqrt{10n}$	$\sqrt[3]{n}$	$\sqrt[3]{10n}$	$\sqrt[3]{100n}$
150	22 500	3 375 000	12.247 45	38.729 83	5.313 293	11.447 14	24.662 12
151	22 801	3 442 951	12.288 21	38.858 72	5.325 074	11.472 52	24.716 80
152	23 104	3 511 808	12.328 83	38.987 18	5.336 803	11.497 79	24.771 25
153	23 409	3 581 577	12.369 32	39.115 21	5.348 481	11.522 95	24.825 45
154	23 716	3 652 264	12.409 67	39.242 83	5.360 108	11.548 00	24.879 42
155	24 025	3 723 875	12.449 90	39.370 04	5.371 685	11.572 95	24.933 15
156	24 336	3 796 416	12.490 00	39.496 84	5.383 213	11.597 78	24.986 66
157	24 649	3 869 893	12.529 96	39.623 23	5.394 691	11.622 51	25.039 94
158	24 964	3 944 312	12.569 81	39.749 21	5.406 120	11.647 13	25.092 99
159	25 281	4 019 679	12.609 52	39.874 80	5.417 502	11.671 65	25.145 81
160	25 600	4 096 000	12.649 11	40.000 00	5.428 835	11.696 07	25.198 42
161	25 921	4 173 281	12.688 58	40.124 81	5.440 122	11.720 39	25.250 81
162	26 244	4 251 528	12.727 92	40.249 22	5.451 362	11.744 60	25.302 98
163	26 569	4 330 747	12.767 15	40.373 26	5.462 556	11.768 72	25.354 94
164	26 896	4 410 944	12.806 25	40.496 91	5.473 704	11.792 74	25.406 68
165	27 225	4 492 125	12.845 23	40.620 19	5.484 807	11.816 66	25.458 22
166	27 556	4 574 296	12.884 10	40.743 10	5.495 865	11.840 48	25.509 54
167	27 889	4 657 463	12.922 85	40.865 63	5.506 878	11.864 21	25.560 67
168	28 224	4 741 632	12.961 48	40.987 80	5.517 848	11.887 84	25.611 58
169	28 561	4 826 809	13.000 00	41.109 61	5.528 775	11.911 38	25.662 30
170	28 900	4 913 000	13.038 40	41.231 06	5.539 658	11.934 83	25.712 82
171	29 241	5 000 211	13.076 70	41.352 15	5.550 499	11.958 19	25.763 13
172	29 584	5 088 448	13.114 88	41.472 88	5.561 298	11.981 45	25.813 26
173	29 929	5 177 717	13.152 95	41.593 27	5.572 055	12.004 63	25.863 19
174	30 276	5 268 024	13.190 91	41.713 31	5.582 770	12.027 71	25.912 92
175	30 625	5 359 375	13.228 76	41.833 00	5.593 445	12.050 71	25.962 47
176	30 976	5 451 776	13.266 50	41.952 35	5.604 079	12.073 62	26.011 83
177	31 329	5 545 233	13.304 13	42.071 37	5.614 672	12.096 45	26.061 00
178	31 684	5 639 752	13.341 66	42.190 05	5.625 226	12.119 18	26.109 99
179	32 041	5 735 339	13.379 09	42.308 39	5.635 741	12.141 84	26.158 79
180	32 400	5 832 000	13.416 41	42.426 41	5.646 216	12.164 40	26.207 41
181	32 761	5 929 741	13.453 62	42.544 09	5.656 653	12.186 89	26.255 86
182	33 124	6 028 568	13.490 74	42.661 46	5.667 051	12.209 29	26.304 12
183	33 489	6 128 487	13.527 75	42.778 50	5.677 411	12.231 61	26.352 21
184	33 856	6 229 504	13.564 66	42.895 22	5.687 734	12.253 85	26.400 12
185	34 225	6 331 625	13.601 47	43.011 63	5.698 019	12.276 01	26.447 86
186	34 596	6 434 856	13.638 18	43.127 72	5.708 267	12.298 09	26.495 43
187	34 969	6 539 203	13.674 79	43.243 50	5.718 479	12.320 09	26.542 83
188	35 344	6 644 672	13.711 31	43.358 97	5.728 654	12.342 01	26.590 06
189	35 721	6 751 269	13.747 73	43.474 13	5.738 794	12.363 86	26.637 12
190	36 100	6 859 000	13.784 05	43.588 99	5.748 897	12.385 62	26.684 02
191	36 481	6 967 871	13.820 27	43.703 55	5.758 965	12.407 31	26.730 75
192	36 864	7 077 888	13.856 41	43.817 80	5.768 998	12.428 93	26.777 32
193	37 249	7 189 057	13.892 44	43.931 77	5.778 997	12.450 47	26.823 73
194	37 636	7 301 384	13.928 39	44.045 43	5.788 960	12.471 94	26.869 97
195	38 025	7 414 875	13.964 24	44.158 80	5.798 890	12.493 33	26.916 06
196	38 416	7 529 536	14.000 00	44.271 89	5.808 786	12.514 65	26.961 99
197	38 809	7 645 373	14.035 67	44.384 68	5.818 648	12.535 90	27.007 77
198	39 204	7 762 392	14.071 25	44.497 19	5.828 477	12.557 07	27.053 39
199	39 601	7 880 599	14.106 74	44.609 42	5.838 272	12.578 18	27.098 86

n	n^2	n^3	$\sqrt{n}$	$\sqrt{10n}$	$\sqrt[3]{n}$	$\sqrt[3]{10n}$	$\sqrt[3]{100n}$
200	40 000	8 000 000	14.142 14	44.721 36	5.848 035	12.599 21	27.144 18
201	40 401	8 120 601	14.177 45	44.833 02	5.857 766	12.620 17	27.189 34
202	40 804	8 242 408	14.212 67	44.944 41	5.867 464	12.641 07	27.234 36
203	41 209	8 365 427	14.247 81	45.055 52	5.877 131	12.661 89	27.279 22
204	41 616	8 489 664	14.282 86	45.166 36	5.886 765	12.682 65	27.323 94
205	42 025	8 615 125	14.317 82	45.276 93	5.896 369	12.703 34	27.368 52
206	42 436	8 741 816	14.352 70	45.387 22	5.905 941	12.723 96	27.412 95
207	42 849	8 869 743	14.387 49	45.497 25	5.915 482	12.744 52	27.457 23
208	43 264	8 998 912	14.422 21	45.607 02	5.924 992	12.765 01	27.501 38
209	43 681	9 129 329	14.456 83	45.716 52	5.934 472	12.785 43	27.545 38
210	44 100	9 261 000	14.491 38	45.825 76	5.943 922	12.805 79	27.589 24
211	44 521	9 393 931	14.525 84	45.934 74	5.953 342	12.826 09	27.632 96
212	44 944	9 528 128	14.560 22	46.043 46	5.962 732	12.846 32	27.676 55
213	45 369	9 663 597	14.594 52	46.151 92	5.972 093	12.866 48	27.720 00
214	45 796	9 800 344	14.628 74	46.260 13	5.981 424	12.886 59	27.763 31
215	46 225	9 938 375	14.662 88	46.368 09	5.990 726	12.906 63	27.806 49
216	46 656	10 077 696	14.696 94	46.475 80	6.000 000	12.926 61	27.849 53
217	47 089	10 218 313	14.730 92	46.583 26	6.009 245	12.946 53	27.892 44
218	47 524	10 360 232	14.764 82	46.690 47	6.018 462	12.966 38	27.935 22
219	47 961	10 503 459	14.798 65	46.797 44	6.027 650	12.986 18	27.977 87
220	48 400	10 648 000	14.832 40	46.904 16	6.036 811	13.005 91	28.020 39
221	48 841	10 793 861	14.866 07	47.010 64	6.045 944	13.025 59	28.062 78
222	49 284	10 941 048	14.899 66	47.116 88	6.055 049	13.045 21	28.105 05
223	49 729	11 089 567	14.933 18	47.222 88	6.064 127	13.064 77	28.147 18
224	50 176	11 239 424	14.966 63	47.328 64	6.073 178	13.084 27	28.189 19
225	50 625	11 390 625	15.000 00	47.434 16	6.082 202	13.103 71	28.231 08
226	51 076	11 543 176	15.033 30	47.539 46	6.091 199	13.123 09	28.272 84
227	51 529	11 697 083	15.066 52	47.644 52	6.100 170	13.142 42	28.314 48
228	51 984	11 852 352	15.099 67	47.749 35	6.109 115	13.161 69	28.356 00
229	52 441	12 008 989	15.132 75	47.853 94	6.118 033	13.180 90	28.397 39
230	52 900	12 167 000	15.165 75	47.958 32	6.126 926	13.200 06	28.438 67
231	53 361	12 326 391	15.198 68	48.062 46	6.135 792	13.219 16	28.479 83
232	53 824	12 487 168	15.231 55	48.166 38	6.144 634	13.238 21	28.520 86
233	54 289	12 649 337	15.264 34	48.270 07	6.153 449	13.257 21	28.561 78
234	54 756	12 812 904	15.297 06	48.373 55	6.162 240	13.276 14	28.602 59
235	55 225	12 977 875	15.329 71	48.476 80	6.171 006	13.295 03	28.643 27
236	55 696	13 144 256	15.362 29	48.579 83	6.179 747	13.313 86	28.683 84
237	56 169	13 312 053	15.394 80	48.682 65	6.188 463	13.332 64	28.724 30
238	56 644	13 481 272	15.427 25	48.785 24	6.197 154	13.351 36	28.764 64
239	57 121	13 651 919	15.459 62	48.887 63	6.205 822	13.370 04	28.804 87
240	57 600	13 824 000	15.491 93	48.989 79	6.214 465	13.388 66	28.844 99
241	58 081	13 997 521	15.524 17	49.091 75	6.223 084	13.407 23	28.885 00
242	58 564	14 172 488	15.556 35	49.193 50	6.231 680	13.425 75	28.924 89
243	59 049	14 348 907	15.588 46	49.295 03	6.240 251	13.444 21	28.964 68
244	59 536	14 526 784	15.620 50	49.396 36	6.248 800	13.462 63	29.004 36
245	60 025	14 706 125	15.652 48	49.497 47	6.257 325	13.481 00	29.043 93
246	60 516	14 886 936	15.684 39	49.598 39	6.265 827	13.499 31	29.083 39
247	61 009	15 069 223	15.716 23	49.699 09	6.274 305	13.517 58	29.122 75
248	61 504	15 252 992	15.748 02	49.799 60	6.282 761	13.535 80	29.161 99
249	62 001	15 438 249	15.779 73	49.899 90	6.291 195	13.553 97	29.201 14

n	n^2	n^3	$\sqrt{n}$	$\sqrt{10n}$	$\sqrt[3]{n}$	$\sqrt[3]{10n}$	$\sqrt[3]{100n}$
250	62 500	15 625 000	15.811 39	50.000 00	6.299 605	13.572 09	29.240 18
251	63 001	15 813 251	15.842 98	50.099 90	6.307 994	13.590 16	29.279 11
252	63 504	16 003 008	15.874 51	50.199 60	6.316 360	13.608 18	29.317 94
253	64 009	16 194 277	15.905 97	50.299 11	6.324 704	13.626 16	29.356 67
254	64 516	16 387 064	15.937 38	50.398 41	6.333 026	13.644 09	29.395 30
255	65 025	16 581 375	15.968 72	50.497 52	6.341 326	13.661 97	29.433 83
256	65 536	16 777 216	16.000 00	50.596 44	6.349 604	13.679 81	29.472 25
257	66 049	16 974 593	16.031 22	50.695 17	6.357 861	13.697 60	29.510 58
258	66 564	17 173 512	16.062 38	50.793 70	6.366 097	13.715 34	29.548 80
259	67 081	17 373 979	16.093 48	50.892 04	6.374 311	13.733 04	29.586 93
260	67 600	17 576 000	16.124 52	50.990 20	6.382 504	13.750 69	29.624 96
261	68 121	17 779 581	16.155 49	51.088 16	6.390 677	13.768 30	29.662 89
262	68 644	17 984 728	16.186 41	51.185 94	6.398 828	13.785 86	29.700 73
263	69 169	18 191 447	16.217 27	51.283 53	6.406 959	13.803 37	29.738 47
264	69 696	18 399 744	16.248 08	51.380 93	6.415 069	13.820 85	29.776 11
265	70 225	18 609 625	16.278 82	51.478 15	6.423 158	13.838 28	29.813 66
266	70 756	18 821 096	16.309 51	51.575 19	6.431 228	13.855 66	29.851 11
267	71 289	19 034 163	16.340 13	51.672 04	6.439 277	13.873 00	29.888 47
268	71 824	19 248 832	16.370 71	51.768 72	6.447 306	13.890 30	29.925 74
269	72 361	19 465 109	16.401 22	51.865 21	6.455 315	13.907 55	29.962 92
270	72 900	19 683 000	16.431 68	51.961 52	6.463 304	13.924 77	30.000 00
271	73 441	19 902 511	16.462 08	52.057 66	6.471 274	13.941 94	30.036 99
272	73 984	20 123 648	16.492 42	52.153 62	6.479 224	13.959 06	30.073 89
273	74 529	20 346 417	16.522 71	52.249 40	6.487 154	13.976 15	30.110 70
274	75 076	20 570 824	16.552 95	52.345 01	6.495 065	13.993 19	30.147 42
275	75 625	20 796 875	16.583 12	52.440 44	6.502 957	14.010 20	30.184 05
276	76 176	21 024 576	16.613 25	52.535 70	6.510 830	14.027 16	30.220 60
277	76 729	21 253 933	16.643 32	52.630 79	6.518 684	14.044 08	30.257 05
278	77 284	21 484 952	16.673 33	52.725 71	6.526 519	14.060 96	30.293 42
279	77 841	21 717 639	16.703 29	52.820 45	6.534 335	14.077 80	30.329 70
280	78 400	21 952 000	16.733 20	52.915 03	6.542 133	14.094 60	30.365 89
281	78 961	22 188 041	16.763 05	53.009 43	6.549 912	14.111 36	30.402 00
282	79 524	22 425 768	16.792 86	53.103 67	6.557 672	14.128 08	30.438 02
283	80 089	22 665 187	16.822 60	53.197 74	6.565 414	14.144 76	30.473 95
284	80 656	22 906 304	16.852 30	53.291 65	6.573 138	14.161 40	30.509 81
285	81 225	23 149 125	16.881 94	53.385 39	6.580 844	14.178 00	30.545 57
286	81 796	23 393 656	16.911 53	53.478 97	6.588 532	14.194 56	30.581 26
287	82 369	23 639 903	16.941 07	53.572 38	6.596 202	14.211 09	30.616 86
288	82 944	23 887 872	16.970 56	53.665 63	6.603 854	14.227 57	30.652 38
289	83 521	24 137 569	17.000 00	53.758 72	6.611 489	14.244 02	30.687 81
290	84 100	24 389 000	17.029 39	53.851 65	6.619 106	14.260 43	30.723 17
291	84 681	24 642 171	17.058 72	53.944 42	6.626 705	14.276 80	30.758 44
292	85 264	24 897 088	17.088 01	54.037 02	6.634 287	14.293 14	30.793 63
293	85 849	25 153 757	17.117 24	54.129 47	6.641 852	14.309 44	30.828 75
294	86 436	25 412 184	17.146 43	54.221 77	6.649 400	14.325 70	30.863 78
295	87 025	25 672 375	17.175 56	54.313 90	6.656 930	14.341 92	30.898 73
296	87 616	25 934 336	17.204 65	54.405 88	6.664 444	14.358 11	30.933 61
297	88 209	26 198 073	17.233 69	54.497 71	6.671 940	14.374 26	30.968 40
298	88 804	26 463 592	17.262 68	54.589 38	6.679 420	14.390 37	31.003 12
299	89 401	26 730 899	17.291 62	54.680 89	6.686 883	14.406 45	31.037 76

n	n^2	n^3	$\sqrt{n}$	$\sqrt{10n}$	$\sqrt[3]{n}$	$\sqrt[3]{10n}$	$\sqrt[3]{100n}$
300	90 000	27 000 000	17.320 51	54.772 26	6.694 330	14.422 50	31.072 33
301	90 601	27 270 901	17.349 35	54.863 47	6.701 759	14.438 50	31.106 81
302	91 204	27 543 608	17.378 15	54.954 53	6.709 173	14.454 47	31.141 22
303	91 809	27 818 127	17.406 90	55.045 44	6.716 570	14.470 41	31.175 56
304	92 416	28 094 464	17.435 60	55.136 20	6.723 951	14.486 31	31.209 82
305	93 025	28 372 625	17.464 25	55.226 81	6.731 315	14.502 18	31.244 00
306	93 636	28 652 616	17.492 86	55.317 27	6.738 664	14.518 01	31.278 11
307	94 249	28 934 443	17.521 42	55.407 58	6.745 997	14.533 81	31.312 14
308	94 864	29 218 112	17.549 93	55.497 75	6.753 313	14.549 57	31.346 10
309	95 481	29 503 629	17.578 40	55.587 77	6.760 614	14.565 30	31.379 99
310	96 100	29 791 000	17.606 82	55.677 64	6.767 899	14.581 00	31.413 81
311	96 721	30 080 231	17.635 19	55.767 37	6.775 169	14.596 66	31.447 55
312	97 344	30 371 328	17.663 52	55.856 96	6.782 423	14.612 29	31.481 22
313	97 969	30 664 297	17.691 81	55.946 40	6.789 661	14.627 88	31.514 82
314	98 596	30 959 144	17.720 05	56.035 70	6.796 884	14.643 44	31.548 34
315	99 225	31 255 875	17.748 24	56.124 86	6.804 092	14.658 97	31.581 80
316	99 856	31 554 496	17.776 39	56.213 88	6.811 285	14.674 47	31.615 18
317	100 489	31 855 013	17.804 49	56.302 75	6.818 462	14.689 93	31.648 50
318	101 124	32 157 432	17.832 55	56.391 49	6.825 624	14.705 36	31.681 74
319	101 761	32 461 759	17.860 57	56.480 08	6.832 771	14.720 76	31.714 92
320	102 400	32 768 000	17.888 54	56.568 54	6.839 904	14.736 13	31.748 02
321	103 041	33 076 161	17.916 47	56.656 86	6.847 021	14.751 46	31.781 06
322	103 684	33 386 248	17.944 36	56.745 04	6.854 124	14.766 76	31.814 03
323	104 329	33 698 267	17.972 20	56.833 09	6.861 212	14.782 03	31.846 93
324	104 976	34 012 224	18.000 00	56.921 00	6.868 285	14.797 27	31.879 76
325	105 625	34 328 125	18.027 76	57.008 77	6.875 344	14.812 48	31.912 52
326	106 276	34 645 976	18.055 47	57.096 41	6.882 389	14.827 66	31.945 22
327	106 929	34 965 783	18.083 14	57.183 91	6.889 419	14.842 80	31.977 85
328	107 584	35 287 552	18.110 77	57.271 28	6.896 434	14.857 92	32.010 41
329	108 241	35 611 289	18.138 36	57.358 52	6.903 436	14.873 00	32.042 91
330	108 900	35 937 000	18.165 90	57.445 63	6.910 423	14.888 06	32.075 34
331	109 561	36 264 691	18.193 41	57.532 60	6.917 396	14.903 08	32.107 71
332	110 224	36 594 368	18.220 87	57.619 44	6.924 356	14.918 07	32.140 01
333	110 889	36 926 037	18.248 29	57.706 15	6.931 301	14.933 03	32.172 25
334	111 556	37 259 704	18.275 67	57.792 73	6.938 232	14.947 97	32.204 42
335	112 225	37 595 375	18.303 01	57.879 18	6.945 150	14.962 87	32.236 53
336	112 896	37 933 056	18.330 30	57.965 51	6.952 053	14.977 74	32.268 57
337	113 569	38 272 753	18.357 56	58.051 70	6.958 943	14.992 59	32.300 55
338	114 244	38 614 472	18.384 78	58.137 77	6.965 820	15.007 40	32.332 47
339	114 921	38 958 219	18.411 95	58.223 71	6.972 683	15.022 19	32.364 33
340	115 600	39 304 000	18.439 09	58.309 52	6.979 532	15.036 95	32.396 12
341	116 281	39 651 821	18.466 19	58.395 21	6.986 368	15.051 67	32.427 85
342	116 964	40 001 688	18.493 24	58.480 77	6.993 191	15.066 37	32.459 52
343	117 649	40 353 607	18.520 26	58.566 20	7.000 000	15.081 04	32.491 12
344	118 336	40 707 584	18.547 24	58.651 51	7.006 796	15.095 68	32.522 67
345	119 025	41 063 625	18.574 18	58.736 70	7.013 579	15.110 30	32.554 15
346	119 716	41 421 736	18.601 08	58.821 76	7.020 349	15.124 88	32.585 57
347	120 409	41 781 923	18.627 94	58.906 71	7.027 106	15.139 44	32.616 94
348	121 104	42 144 192	18.654 76	58.991 52	7.033 850	15.153 97	32.648 24
349	121 801	42 508 549	18.681 54	59.076 22	7.040 581	15.168 47	32.679 48

n	n^2	n^3	$\sqrt{n}$	$\sqrt{10n}$	$\sqrt[3]{n}$	$\sqrt[3]{10n}$	$\sqrt[3]{100n}$
350	122 500	42 875 000	18.708 29	59.160 80	7.047 299	15.182 94	32.710 66
351	123 201	43 243 551	18.734 99	59.245 25	7.054 004	15.197 39	32.741 79
352	123 904	43 614 208	18.761 66	59.329 59	7.060 697	15.211 81	32.772 85
353	124 609	43 986 977	18.788 29	59.413 80	7.067 377	15.226 20	32.803 86
354	125 316	44 361 864	18.814 89	59.497 90	7.074 044	15.240 57	32.834 80
355	126 025	44 738 875	18.841 44	59.581 88	7.080 699	15.254 90	32.865 69
356	126 736	45 118 016	18.867 96	59.665 74	7.087 341	15.269 21	32.896 52
357	127 449	45 499 293	18.894 44	59.749 48	7.093 971	15.283 50	32.927 30
358	128 164	45 882 712	18.920 89	59.833 10	7.100 588	15.297 75	32.958 01
359	128 881	46 268 279	18.947 30	59.916 61	7.107 194	15.311 98	32.988 67
360	129 600	46 656 000	18.973 67	60.000 00	7.113 787	15.326 19	33.019 27
361	130 321	47 045 881	19.000 00	60.083 28	7.120 367	15.340 37	33.049 82
362	131 044	47 437 928	19.026 30	60.166 44	7.126 936	15.354 52	33.080 31
363	131 769	47 832 147	19.052 56	60.249 48	7.133 492	15.368 64	33.110 74
364	132 496	48 228 544	19.078 78	60.332 41	7.140 037	15.382 74	33.141 12
365	133 225	48 627 125	19.104 97	60.415 23	7.146 569	15.396 82	33.171 44
366	133 956	49 027 896	19.131 13	60.497 93	7.153 090	15.410 87	33.201 70
367	134 689	49 430 863	19.157 24	60.580 52	7.159 599	15.424 89	33.231 91
368	135 424	49 836 032	19.183 33	60.663 00	7.166 096	15.438 89	33.262 07
369	136 161	50 243 409	19.209 37	60.745 37	7.172 581	15.452 86	33.292 17
370	136 900	50 653 000	19.235 38	60.827 63	7.179 054	15.466 80	33.322 22
371	137 641	51 064 811	19.261 36	60.909 77	7.185 516	15.480 73	33.352 21
372	138 384	51 478 848	19.287 30	60.991 80	7.191 966	15.494 62	33.382 15
373	139 129	51 895 117	19.313 21	61.073 73	7.198 405	15.508 49	33.412 04
374	139 876	52 313 624	19.339 08	61.155 54	7.204 832	15.522 34	33.441 87
375	140 625	52 734 375	19.364 92	61.237 24	7.211 248	15.536 16	33.471 65
376	141 376	53 157 376	19.390 72	61.318 84	7.217 652	15.549 96	33.501 37
377	142 129	53 582 633	19.416 49	61.400 33	7.224 045	15.563 73	33.531 05
378	142 884	54 010 152	19.442 22	61.481 70	7.230 427	15.577 48	33.560 67
379	143 641	54 439 939	19.467 92	61.562 98	7.236 797	15.591 21	33.590 24
380	144 400	54 872 000	19.493 59	61.644 14	7.243 156	15.604 91	33.619 75
381	145 161	55 306 341	19.519 22	61.725 20	7.249 505	15.618 58	33.649 22
382	145 924	55 742 968	19.544 82	61.806 15	7.255 842	15.632 24	33.678 63
383	146 689	56 181 887	19.570 39	61.886 99	7.262 167	15.645 87	33.708 00
384	147 456	56 623 104	19.595 92	61.967 73	7.268 482	15.659 47	33.737 31
385	148 225	57 066 625	19.621 42	62.048 37	7.274 786	15.673 05	33.766 57
386	148 996	57 512 456	19.646 88	62.128 90	7.281 079	15.686 61	33.795 78
387	149 769	57 960 603	19.672 32	62.209 32	7.287 362	15.700 14	33.824 94
388	150 544	58 411 072	19.697 72	62.289 65	7.293 633	15.713 66	33.854 05
389	151 321	58 863 869	19.723 08	62.369 86	7.299 894	15.727 14	33.883 10
390	152 100	59 319 000	19.748 42	62.449 98	7.306 144	15.740 61	33.912 11
391	152 881	59 776 471	19.773 72	62.529 99	7.312 383	15.754 05	33.941 07
392	153 664	60 236 288	19.798 99	62.609 90	7.318 611	15.767 47	33.969 99
393	154 449	60 698 457	19.824 23	62.689 71	7.324 829	15.780 87	33.998 85
394	155 236	61 162 984	19.849 43	62.769 42	7.331 037	15.794 24	34.027 66
395	156 025	61 629 875	19.874 61	62.849 03	7.337 234	15.807 59	34.056 42
396	156 816	62 099 136	19.899 75	62.928 53	7.343 420	15.820 92	34.085 14
397	157 609	62 570 773	19.924 86	63.007 94	7.349 597	15.834 23	34.113 81
398	158 404	63 044 792	19.949 94	63.087 24	7.355 762	15.847 51	34.142 42
399	159 201	63 521 199	19.974 98	63.166 45	7.361 918	15.860 77	34.171 00

n	n^2	n^3	$\sqrt{n}$	$\sqrt{10n}$	$\sqrt[3]{n}$	$\sqrt[3]{10n}$	$\sqrt[3]{100n}$
400	160 000	64 000 000	20.000 00	63.245 55	7.368 063	15.874 01	34.199 52
401	160 801	64 481 201	20.024 98	63.324 56	7.374 198	15.887 23	34.227 99
402	161 604	64 964 808	20.049 94	63.403 47	7.380 323	15.900 42	34.256 42
403	162 409	65 450 827	20.074 86	63.482 28	7.386 437	15.913 60	34.284 80
404	163 216	65 939 264	20.099 75	63.560 99	7.392 542	15.926 75	34.313 14
405	164 025	66 430 125	20.124 61	63.639 61	7.398 636	15.939 88	34.341 43
406	164 836	66 923 416	20.149 44	63.718 13	7.404 721	15.952 99	34.369 67
407	165 649	67 419 143	20.174 24	63.796 55	7.410 795	15.966 07	34.397 86
408	166 464	67 917 312	20.199 01	63.874 88	7.416 860	15.979 14	34.426 01
409	167 281	68 417 929	20.223 75	63.953 11	7.422 914	15.992 18	34.454 12
410	168 100	68 921 000	20.248 46	64.031 24	7.428 959	16.005 21	34.482 17
411	168 921	69 426 531	20.273 13	64.109 28	7.434 994	16.018 21	34.510 18
412	169 744	69 934 528	20.297 78	64.187 23	7.441 019	16.031 19	34.538 15
413	170 569	70 444 997	20.322 40	64.265 08	7.447 034	16.044 15	34.566 07
414	171 396	70 957 944	20.346 99	64.342 83	7.453 040	16.057 09	34.593 95
415	172 225	71 473 375	20.371 55	64.420 49	7.459 036	16.070 01	34.621 78
416	173 056	71 991 296	20.396 08	64.498 06	7.465 022	16.082 90	34.649 56
417	173 889	72 511 713	20.420 58	64.575 54	7.470 999	16.095 78	34.677 31
418	174 724	73 034 632	20.445 05	64.652 92	7.476 966	16.108 64	34.705 00
419	175 561	73 560 059	20.469 49	64.730 21	7.482 924	16.121 47	34.732 66
420	176 400	74 088 000	20.493 90	64.807 41	7.488 872	16.134 29	34.760 27
421	177 241	74 618 461	20.518 28	64.884 51	7.494 811	16.147 08	34.787 83
422	178 084	75 151 448	20.542 64	64.961 53	7.500 741	16.159 86	34.815 35
423	178 929	75 686 967	20.566 96	65.038 45	7.506 661	16.172 61	34.842 83
424	179 776	76 225 024	20.591 26	65.115 28	7.512 572	16.185 34	34.870 27
425	180 625	76 765 625	20.615 53	65.192 02	7.518 473	16.198 06	34.897 66
426	181 476	77 308 776	20.639 77	65.268 68	7.524 365	16.210 75	34.925 01
427	182 329	77 854 483	20.663 98	65.345 24	7.530 248	16.223 43	34.952 32
428	183 184	78 402 752	20.688 16	65.421 71	7.536 122	16.236 08	34.979 58
429	184 041	78 953 589	20.712 32	65.498 09	7.541 987	16.248 72	35.006 80
430	184 900	79 507 000	20.736 44	65.574 39	7.547 842	16.261 33	35.033 98
431	185 761	80 062 991	20.760 54	65.650 59	7.553 689	16.273 93	35.061 12
432	186 624	80 621 568	20.784 61	65.726 71	7.559 526	16.286 51	35.088 21
433	187 489	81 182 737	20.808 65	65.802 74	7.565 355	16.299 06	35.115 27
434	188 356	81 746 504	20.832 67	65.878 68	7.571 174	16.311 60	35.142 28
435	189 225	82 312 875	20.856 65	65.954 53	7.576 985	16.324 12	35.169 25
436	190 096	82 881 856	20.880 61	66.030 30	7.582 787	16.336 62	35.196 18
437	190 969	83 453 453	20.904 54	66.105 98	7.588 579	16.349 10	35.223 07
438	191 844	84 027 672	20.928 45	66.181 57	7.594 363	16.361 56	35.249 91
439	192 721	84 604 519	20.952 33	66.257 08	7.600 139	16.374 00	35.276 72
440	193 600	85 184 000	20.976 18	66.332 50	7.605 905	16.386 43	35.303 48
441	194 481	85 766 121	21.000 00	66.407 83	7.611 663	16.398 83	35.330 21
442	195 364	86 350 888	21.023 80	66.483 08	7.617 412	16.411 22	35.356 89
443	196 249	86 938 307	21.047 57	66.558 25	7.623 152	16.423 58	35.383 54
444	197 136	87 528 384	21.071 31	66.633 32	7.628 884	16.435 93	35.410 14
445	198 025	88 121 125	21.095 02	66.708 32	7.634 607	16.448 26	35.436 71
446	198 916	88 716 536	21.118 71	66.783 23	7.640 321	16.460 57	35.463 23
447	199 809	89 314 623	21.142 37	66.858 06	7.646 027	16.472 87	35.489 71
448	200 704	89 915 392	21.166 01	66.932 80	7.651 725	16.485 14	35.516 16
449	201 601	90 518 849	21.189 62	67.007 46	7.657 414	16.497 40	35.542 57

n	n^2	n^3	$\sqrt{n}$	$\sqrt{10n}$	$\sqrt[3]{n}$	$\sqrt[3]{10n}$	$\sqrt[3]{100n}$
450	202 500	91 125 000	21.213 20	67.082 04	7.663 094	16.509 64	35.568 93
451	203 401	91 733 851	21.236 76	67.156 53	7.668 766	16.521 86	35.595 26
452	204 304	92 345 408	21.260 29	67.230 95	7.674 430	16.534 06	35.621 55
453	205 209	92 959 677	21.283 80	67.305 27	7.680 086	16.546 24	35.647 80
454	206 116	93 576 664	21.307 28	67.379 52	7.685 733	16.558 41	35.674 01
455	207 025	94 196 375	21.330 73	67.453 69	7.691 372	16.570 56	35.700 18
456	207 936	94 818 816	21.354 16	67.527 77	7.697 002	16.582 69	35.726 32
457	208 849	95 443 993	21.377 56	67.601 78	7.702 625	16.594 80	35.752 42
458	209 764	96 071 912	21.400 93	67.675 70	7.708 239	16.606 90	35.778 48
459	210 681	96 702 579	21.424 29	67.749 54	7.713 845	16.618 97	35.804 50
460	211 600	97 336 000	21.447 61	67.823 30	7.719 443	16.631 03	35.830 48
461	212 521	97 972 181	21.470 91	67.896 98	7.725 032	16.643 08	35.856 42
462	213 444	98 611 128	21.494 19	67.970 58	7.730 614	16.655 10	35.882 33
463	214 369	99 252 847	21.517 43	68.044 10	7.736 188	16.667 11	35.908 20
464	215 296	99 897 344	21.540 66	68.117 55	7.741 753	16.679 10	35.934 04
465	216 225	100 544 625	21.563 86	68.190 91	7.747 311	16.691 08	35.959 83
466	217 156	101 194 696	21.587 03	68.264 19	7.752 861	16.703 03	35.985 59
467	218 089	101 847 563	21.610 18	68.337 40	7.758 402	16.714 97	36.011 31
468	219 024	102 503 232	21.633 31	68.410 53	7.763 936	16.726 89	36.037 00
469	219 961	103 161 709	21.656 41	68.483 57	7.769 462	16.738 80	36.062 65
470	220 900	103 823 000	21.679 48	68.556 55	7.774 980	16.750 69	36.088 26
471	221 841	104 487 111	21.702 53	68.629 44	7.780 490	16.762 56	36.113 84
472	222 784	105 154 048	21.725 56	68.702 26	7.785 993	16.774 41	36.139 38
473	223 729	105 823 817	21.748 56	68.775 00	7.791 488	16.786 25	36.164 88
474	224 676	106 496 424	21.771 54	68.847 66	7.796 975	16.798 07	36.190 35
475	225 625	107 171 875	21.794 49	68.920 24	7.802 454	16.809 88	36.215 78
476	226 576	107 850 176	21.817 42	68.992 75	7.807 925	16.821 67	36.241 18
477	227 529	108 531 333	21.840 33	69.065 19	7.813 389	16.833 44	36.266 54
478	228 484	109 215 352	21.863 21	69.137 54	7.818 846	16.845 19	36.291 87
479	229 441	109 902 239	21.886 07	69.209 83	7.824 294	16.856 93	36.317 16
480	230 400	110 592 000	21.908 90	69.282 03	7.829 735	16.868 65	36.342 41
481	231 361	111 284 641	21.931 71	69.354 16	7.835 169	16.880 36	36.367 63
482	232 324	111 980 168	21.954 50	69.426 22	7.840 595	16.892 05	36.392 82
483	233 289	112 678 587	21.977 26	69.498 20	7.846 013	16.903 72	36.417 97
484	234 256	113 379 904	22.000 00	69.570 11	7.851 424	16.915 38	36.443 08
485	235 225	114 084 125	22.022 72	69.641 94	7.856 828	16.927 02	36.468 17
486	236 196	114 791 256	22.045 41	69.713 70	7.862 224	16.938 65	36.493 21
487	237 169	115 501 303	22.068 08	69.785 39	7.867 613	16.950 26	36.518 22
488	238 144	116 214 272	22.090 72	69.857 00	7.872 994	16.961 85	36.543 20
489	239 121	116 930 169	22.113 34	69.928 53	7.878 368	16.973 43	36.568 15
490	240 100	117 649 000	22.135 94	70.000 00	7.883 735	16.984 99	36.593 06
491	241 081	118 370 771	22.158 52	70.071 39	7.889 095	16.996 54	36.617 93
492	242 064	119 095 488	22.181 07	70.142 71	7.894 447	17.008 07	36.642 78
493	243 049	119 823 157	22.203 60	70.213 96	7.899 792	17.019 59	36.667 58
494	244 036	120 553 784	22.226 11	70.285 13	7.905 129	17.031 08	36.692 36
495	245 025	121 287 375	22.248 60	70.356 24	7.910 460	17.042 57	36.717 10
496	246 016	122 023 936	22.271 06	70.427 27	7.915 783	17.054 04	36.741 81
497	247 009	122 763 473	22.293 50	70.498 23	7.921 099	17.065 49	36.766 49
498	248 004	123 505 992	22.315 91	70.569 12	7.926 408	17.076 93	36.791 13
499	249 001	124 251 499	22.338 31	70.639 93	7.931 710	17.088 35	36.815 74

n	n^2	n^3	$\sqrt{n}$	$\sqrt{10n}$	$\sqrt[3]{n}$	$\sqrt[3]{10n}$	$\sqrt[3]{100n}$
500	250 000	125 000 000	22.360 68	70.710 68	7.937 005	17.099 76	36.840 31
501	251 001	125 751 501	22.383 03	70.781 35	7.942 293	17.111 15	36.864 86
502	252 004	126 506 008	22.405 36	70.851 96	7.947 574	17.122 53	36.889 37
503	253 009	127 263 527	22.427 66	70.922 49	7.952 848	17.133 89	36.913 85
504	254 016	128 024 064	22.449 94	70.992 96	7.958 114	17.145 24	36.938 30
505	255 025	128 787 625	22.472 21	71.063 35	7.963 374	17.156 57	36.962 71
506	256 036	129 554 216	22.494 44	71.133 68	7.968 627	17.167 89	36.987 09
507	257 049	130 323 843	22.516 66	71.203 93	7.973 873	17.179 19	37.011 44
508	258 064	131 096 512	22.538 86	71.274 12	7.979 112	17.190 48	37.035 76
509	259 081	131 872 229	22.561 03	71.344 24	7.984 344	17.201 75	37.060 04
510	260 100	132 651 000	22.583 18	71.414 28	7.989 570	17.213 01	37.084 30
511	261 121	133 432 831	22.605 31	71.484 26	7.994 788	17.224 25	37.108 52
512	262 144	134 217 728	22.627 42	71.554 18	8.000 000	17.235 48	37.132 71
513	263 169	135 005 697	22.649 50	71.624 02	8.005 205	17.246 69	37.156 87
514	264 196	135 796 744	22.671 57	71.693 79	8.010 403	17.257 89	37.181 00
515	265 225	136 590 875	22.693 61	71.763 50	8.015 595	17.269 08	37.205 09
516	266 256	137 388 096	22.715 63	71.833 14	8.020 779	17.280 25	37.229 16
517	267 289	138 188 413	22.737 63	71.902 71	8.025 957	17.291 40	37.253 19
518	268 324	138 991 832	22.759 61	71.972 22	8.031 129	17.302 54	37.277 20
519	269 361	139 798 359	22.781 57	72.041 65	8.036 293	17.313 67	37.301 17
520	270 400	140 608 000	22.803 51	72.111 03	8.041 452	17.324 78	37.325 11
521	271 441	141 420 761	22.825 42	72.180 33	8.046 603	17.335 88	37.349 02
522	272 484	142 236 648	22.847 32	72.249 57	8.051 748	17.346 96	37.372 90
523	273 529	143 055 667	22.869 19	72.318 74	8.056 886	17.358 04	37.396 75
524	274 576	143 877 824	22.891 05	72.387 84	8.062 018	17.369 09	37.420 57
525	275 625	144 703 125	22.912 88	72.456 88	8.067 143	17.380 13	37.444 36
526	276 676	145 531 576	22.934 69	72.525 86	8.072 262	17.391 16	37.468 12
527	277 729	146 363 183	22.956 48	72.594 77	8.077 374	17.402 18	37.491 85
528	278 784	147 197 952	22.978 25	72.663 61	8.082 480	17.413 18	37.515 55
529	279 841	148 035 889	23.000 00	72.732 39	8.087 579	17.424 16	37.539 22
530	280 900	148 877 000	23.021 73	72.801 10	8.092 672	17.435 13	37.562 86
531	281 961	149 721 291	23.043 44	72.869 75	8.097 759	17.446 09	37.586 47
532	283 024	150 568 768	23.065 13	72.938 33	8.102 839	17.457 04	37.610 05
533	284 089	151 419 437	23.086 79	73.006 85	8.107 913	17.467 97	37.633 60
534	285 156	152 273 304	23.108 44	73.075 30	8.112 980	17.478 89	37.657 12
535	286 225	153 130 375	23.130 07	73.143 69	8.118 041	17.489 79	37.680 61
536	287 296	153 990 656	23.151 67	73.212 02	8.123 096	17.500 68	37.704 07
537	288 369	154 854 153	23.173 26	73.280 28	8.128 145	17.511 56	37.727 51
538	289 444	155 720 872	23.194 83	73.348 48	8.133 187	17.522 42	37.750 91
539	290 521	156 590 819	23.216 37	73.416 62	8.138 223	17.533 27	37.774 29
540	291 600	157 464 000	23.237 90	73.484 69	8.143 253	17.544 11	37.797 63
541	292 681	158 340 421	23.259 41	73.552 70	8.148 276	17.554 93	37.820 95
542	293 764	159 220 088	23.280 89	73.620 65	8.153 294	17.565 74	37.844 24
543	294 849	160 103 007	23.302 36	73.688 53	8.158 305	17.576 54	37.867 50
544	295 936	160 989 184	23.323 81	73.756 36	8.163 310	17.587 32	37.890 73
545	297 025	161 878 625	23.345 24	73.824 12	8.168 309	17.598 09	37.913 93
546	298 116	162 771 336	23.366 64	73.891 81	8.173 302	17.608 85	37.937 11
547	299 209	163 667 323	23.388 03	73.959 45	8.178 289	17.619 59	37.960 25
548	300 304	164 566 592	23.409 40	74.027 02	8.183 269	17.630 32	37.983 37
549	301 401	165 469 149	23.430 75	74.094 53	8.188 244	17.641 04	38.006 46

n	n^2	n^3	$\sqrt{n}$	$\sqrt{10n}$	$\sqrt[3]{n}$	$\sqrt[3]{10n}$	$\sqrt[3]{100n}$
550	302 500	166 375 000	23.452 08	74.161 98	8.193 213	17.651 74	38.029 52
551	303 601	167 284 151	23.473 39	74.229 37	8.198 175	17.662 43	38.052 56
552	304 704	168 196 608	23.494 68	74.296 70	8.203 132	17.673 11	38.075 57
553	305 809	169 112 377	23.515 95	74.363 97	8.208 082	17.683 78	38.098 54
554	306 916	170 031 464	23.537 20	74.431 18	8.213 027	17.694 43	38.121 49
555	308 025	170 953 875	23.558 44	74.498 32	8.217 966	17.705 07	38.144 42
556	309 136	171 879 616	23.579 65	74.565 41	8.222 899	17.715 70	38.167 31
557	310 249	172 808 693	23.600 85	74.632 43	8.227 825	17.726 31	38.190 18
558	311 364	173 741 112	23.622 02	74.699 40	8.232 746	17.736 91	38.213 02
559	312 481	174 676 879	23.643 18	74.766 30	8.237 661	17.747 50	38.235 84
560	313 600	175 616 000	23.664 32	74.833 15	8.242 571	17.758 08	38.258 62
561	314 721	176 558 481	23.685 44	74.899 93	8.247 474	17.768 64	38.281 38
562	315 844	177 504 328	23.706 54	74.966 66	8.252 372	17.779 20	38.304 12
563	316 969	178 453 547	23.727 62	75.033 33	8.257 263	17.789 73	38.326 82
564	318 096	179 406 144	23.748 68	75.099 93	8.262 149	17.800 26	38.349 50
565	319 225	180 362 125	23.769 73	75.166 48	8.267 029	17.810 77	38.372 15
566	320 356	181 321 496	23.790 75	75.232 97	8.271 904	17.821 28	38.394 78
567	321 489	182 284 263	23.811 76	75.299 40	8.276 773	17.831 77	38.417 37
568	322 624	183 250 432	23.832 75	75.365 77	8.281 635	17.842 24	38.439 95
569	323 761	184 220 009	23.853 72	75.432 09	8.286 493	17.852 71	38.462 49
570	324 900	185 193 000	23.874 67	75.498 34	8.291 344	17.863 16	38.485 01
571	326 041	186 169 411	23.895 61	75.564 54	8.296 190	17.873 60	38.507 50
572	327 184	187 149 248	23.916 52	75.630 68	8.301 031	17.884 03	38.529 97
573	328 329	188 132 517	23.937 42	75.696 76	8.305 865	17.894 44	38.552 41
574	329 476	189 119 224	23.958 30	75.762 79	8.310 694	17.904 85	38.574 82
575	330 625	190 109 375	23.979 16	75.828 75	8.315 517	17.915 24	38.597 21
576	331 776	191 102 976	24.000 00	75.894 66	8.320 335	17.925 62	38.619 58
577	332 929	192 100 033	24.020 82	75.960 52	8.325 148	17.935 99	38.641 91
578	334 084	193 100 552	24.041 63	76.026 31	8.329 954	17.946 34	38.664 22
579	335 241	194 104 539	24.062 42	76.092 05	8.334 755	17.956 69	38.686 51
580	336 400	195 112 000	24.083 19	76.157 73	8.339 551	17.967 02	38.708 77
581	337 561	196 122 941	24.103 94	76.223 36	8.344 341	17.977 34	38.731 00
582	338 724	197 137 368	24.124 68	76.288 92	8.349 126	17.987 65	38.753 21
583	339 889	198 155 287	24.145 39	76.354 44	8.353 905	17.997 94	38.775 39
584	341 056	199 176 704	24.166 09	76.419 89	8.358 678	18.008 23	38.797 55
585	342 225	200 201 625	24.186 77	76.485 29	8.363 447	18.018 50	38.819 68
586	343 396	201 230 056	24.207 44	76.550 64	8.368 209	18.028 76	38.841 79
587	344 569	202 262 003	24.228 08	76.615 93	8.372 967	18.039 01	38.863 87
588	345 744	203 297 472	24.248 71	76.681 16	8.377 719	18.049 25	38.885 93
589	346 921	204 336 469	24.269 32	76.746 34	8.382 465	18.059 47	38.907 96
590	348 100	205 379 000	24.289 92	76.811 46	8.387 207	18.069 69	38.929 96
591	349 281	206 425 071	24.310 49	76.876 52	8.391 942	18.079 89	38.951 95
592	350 464	207 474 688	24.331 05	76.941 54	8.396 673	18.090 08	38.973 90
593	351 649	208 527 857	24.351 59	77.006 49	8.401 398	18.100 26	38.995 84
594	352 836	209 584 584	24.372 12	77.071 40	8.406 118	18.110 43	39.017 74
595	354 025	210 644 875	24.392 62	77.136 24	8.410 833	18.120 59	39.039 63
596	355 216	211 708 736	24.413 11	77.201 04	8.415 542	18.130 74	39.061 49
597	356 409	212 776 173	24.433 58	77.265 78	8.420 246	18.140 87	39.083 32
598	357 604	213 847 192	24.454 04	77.330 46	8.424 945	18.150 99	39.105 13
599	358 801	214 921 799	24.474 48	77.395 09	8.429 638	18.161 11	39.126 92

n	n^2	n^3	$\sqrt{n}$	$\sqrt{10n}$	$\sqrt[3]{n}$	$\sqrt[3]{10n}$	$\sqrt[3]{100n}$
600	360 000	216 000 000	24.494 90	77.459 67	8.434 327	18.171 21	39.148 68
601	361 201	217 081 801	24.515 30	77.524 19	8.439 010	18.181 30	39.170 41
602	362 404	218 167 208	24.535 69	77.588 66	8.443 688	18.191 37	39.192 13
603	363 609	219 256 227	24.556 06	77.653 07	8.448 361	18.201 44	39.213 82
604	364 816	220 348 864	24.576 41	77.717 44	8.453 028	18.211 50	39.235 48
605	366 025	221 445 125	24.596 75	77.781 75	8.457 691	18.221 54	39.257 12
606	367 236	222 545 016	24.617 07	77.846 00	8.462 348	18.231 58	39.278 74
607	368 449	223 648 543	24.637 37	77.910 20	8.467 000	18.241 60	39.300 33
608	369 664	224 755 712	24.657 66	77.974 35	8.471 647	18.251 61	39.321 90
609	370 881	225 866 529	24.677 93	78.038 45	8.476 289	18.261 61	39.343 45
610	372 100	226 981 000	24.698 18	78.102 50	8.480 926	18.271 60	39.364 97
611	373 321	228 099 131	24.718 41	78.166 49	8.485 558	18.281 58	39.386 47
612	374 544	229 220 928	24.738 63	78.230 43	8.490 185	18.291 55	39.407 95
613	375 769	230 346 397	24.758 84	78.294 32	8.494 807	18.301 51	39.429 40
614	376 996	231 475 544	24.779 02	78.358 15	8.499 423	18.311 45	39.450 83
615	378 225	232 608 375	24.799 19	78.421 94	8.504 035	18.321 39	39.472 23
616	379 456	233 744 896	24.819 35	78.485 67	8.508 642	18.331 31	39.493 62
617	380 689	234 885 113	24.839 48	78.549 35	8.513 243	18.341 23	39.514 98
618	381 924	236 029 032	24.859 61	78.612 98	8.517 840	18.351 13	39.536 31
619	383 161	237 176 659	24.879 71	78.676 55	8.522 432	18.361 02	39.557 63
620	384 400	238 328 000	24.899 80	78.740 08	8.527 019	18.370 91	39.578 92
621	385 641	239 483 061	24.919 87	78.803 55	8.531 601	18.380 78	39.600 18
622	386 884	240 641 848	24.939 93	78.866 98	8.536 178	18.390 64	39.621 43
623	388 129	241 804 367	24.959 97	78.930 35	8.540 750	18.400 49	39.642 65
624	389 376	242 970 624	24.979 99	78.993 67	8.545 317	18.410 33	39.663 85
625	390 625	244 140 625	25.000 00	79.056 94	8.549 880	18.420 16	39.685 03
626	391 876	245 314 376	25.019 99	79.120 16	8.554 437	18.429 98	39.706 18
627	393 129	246 491 883	25.039 97	79.183 33	8.558 990	18.439 78	39.727 31
628	394 384	247 673 152	25.059 93	79.246 45	8.563 538	18.449 58	39.748 42
629	395 641	248 858 189	25.079 87	79.309 52	8.568 081	18.459 37	39.769 51
630	396 900	250 047 000	25.099 80	79.372 54	8.572 619	18.469 15	39.790 57
631	398 161	251 239 591	25.119 71	79.435 51	8.577 152	18.478 91	39.811 61
632	399 424	252 435 968	25.139 61	79.498 43	8.581 681	18.488 67	39.832 63
633	400 689	253 636 137	25.159 49	79.561 30	8.586 205	18 498 42	39.853 63
634	401 956	254 840 104	25.179 36	79.624 12	8.590 724	18.508 15	39.874 61
635	403 225	256 047 875	25.199 21	79.686 89	8.595 238	18.517 88	39.895 56
636	404 496	257 259 456	25.219 04	79.749 61	8.599 748	18.527 59	39.916 49
637	405 769	258 474 853	25.238 86	79.812 28	8.604 252	18.537 30	39.937 40
638	407 044	259 694 072	25.258 66	79.874 90	8.608 753	18.547 00	39.958 29
639	408 321	260 917 119	25.278 45	79.937 48	8.613 248	18.556 68	39.979 16
640	409 600	262 144 000	25.298 22	80.000 00	8.617 739	18.566 36	40.000 00
641	410 881	263 374 721	25.317 98	80.062 48	8.622 225	18.576 02	40.020 82
642	412 164	264 609 288	25.337 72	80.124 90	8.626 706	18.585 68	40.041 62
643	413 449	265 847 707	25.357 44	80.187 28	8.631 183	18.595 32	40.062 40
644	414 736	267 089 984	25.377 16	80.249 61	8.635 655	18.604 95	40.083 16
645	416 025	268 336 125	25.396 85	80.311 89	8.640 123	18.614 58	40.103 90
646	417 316	269 586 136	25.416 53	80.374 13	8.644 585	18.624 19	40.124 61
647	418 609	270 840 023	25.436 19	80.436 31	8.649 044	18.633 80	40.145 30
648	419 904	272 097 792	25.455 84	80.498 45	8.653 497	18.643 40	40.165 98
649	421 201	273 359 449	25.475 48	80.560 54	8.657 947	18.652 98	40.186 63

n	n^2	n^3	$\sqrt{n}$	$\sqrt{10n}$	$\sqrt[3]{n}$	$\sqrt[3]{10n}$	$\sqrt[3]{100n}$
650	422 500	274 625 000	25.495 10	80.622 58	8.662 391	18.662 56	40.207 26
651	423 801	275 894 451	25.514 70	80.684 57	8.666 831	18.672 12	40.227 87
652	425 104	277 167 808	25.534 29	80.746 52	8.671 266	18.681 68	40.248 45
653	426 409	278 445 077	25.553 86	80.808 42	8.675 697	18.691 22	40.269 02
654	427 716	279 726 264	25.573 42	80.870 27	8.680 124	18.700 76	40.289 57
655	429 025	281 011 375	25.592 97	80.932 07	8.684 546	18.710 29	40.310 09
656	430 336	282 300 416	25.612 50	80.993 83	8.688 963	18.719 80	40.330 59
657	431 649	283 593 393	25.632 01	81.055 54	8.693 376	18.729 31	40.351 08
658	432 964	284 890 312	25.651 51	81.117 20	8.697 784	18.738 81	40.371 54
659	434 281	286 191 179	25.671 00	81.178 81	8.702 188	18.748 30	40.391 98
660	435 600	287 496 000	25.690 47	81.240 38	8.706 588	18.757 77	40.412 40
661	436 921	288 804 781	25.709 92	81.301 91	8.710 983	18.767 24	40.432 80
662	438 244	290 117 528	25.729 36	81.363 38	8.715 373	18.776 70	40.453 18
663	439 569	291 434 247	25.748 79	81.424 81	8.719 760	18.786 15	40.473 54
664	440 896	292 754 944	25.768 20	81.486 20	8.724 141	18.795 59	40.493 88
665	442 225	294 079 625	25.787 59	81.547 53	8.728 519	18.805 02	40.514 20
666	443 556	295 408 296	25.806 98	81.608 82	8.732 892	18.814 44	40.534 49
667	444 889	296 740 963	25.826 34	81.670 07	8.737 260	18.823 86	40.554 77
668	446 224	298 077 632	25.845 70	81.731 27	8.741 625	18.833 26	40.575 03
669	447 561	299 418 309	25.865 03	81.792 42	8.745 985	18.842 65	40.595 26
670	448 900	300 763 000	25.884 36	81.853 53	8.750 340	18.852 04	40.615 48
671	450 241	302 111 711	25.903 67	81.914 59	8.754 691	18.861 41	40.635 68
672	451 584	303 464 448	25.922 96	81.975 61	8.759 038	18.870 78	40.655 85
673	452 929	304 821 217	25.942 24	82.036 58	8.763 381	18.880 13	40.676 01
674	454 276	306 182 024	25.961 51	82.097 50	8.767 719	18.889 48	40.696 15
675	455 625	307 546 875	25.980 76	82.158 38	8.772 053	18.898 82	40.716 26
676	456 976	308 915 776	26.000 00	82.219 22	8.776 383	18.908 14	40.736 36
677	458 329	310 288 733	26.019 22	82.280 01	8.780 708	18.917 46	40.756 44
678	459 684	311 665 752	26.038 43	82.340 76	8.785 030	18.926 77	40.776 50
679	461 041	313 046 839	26.057 63	82.401 46	8.789 347	18.936 07	40.796 53
680	462 400	314 432 000	26.076 81	82.462 11	8.793 659	18.945 36	40.816 55
681	463 761	315 821 241	26.095 98	82.522 72	8.797 968	18.954 65	40.836 55
682	465 124	317 214 568	26.115 13	82.583 29	8.802 272	18.963 92	40.856 53
683	466 489	318 611 987	26.134 27	82.643 81	8.806 572	18.973 18	40.876 49
684	467 856	320 013 504	26.153 39	82.704 29	8.810 868	18.982 44	40.896 43
685	469 225	321 419 125	26.172 50	82.764 73	8.815 160	18.991 69	40.916 35
686	470 596	322 828 856	26.191 60	82.825 12	8.819 447	19.000 92	40.936 25
687	471 969	324 242 703	26.210 68	82.885 46	8.823 731	19.010 15	40.956 13
688	473 344	325 660 672	26.229 75	82.945 77	8.828 010	19.019 37	40.975 99
689	474 721	327 082 769	26.248 81	83.006 02	8.832 285	19.028 58	40.995 84
690	476 100	328 509 000	26.267 85	83.066 24	8.836 556	19.037 78	41.015 66
691	477 481	329 939 371	26.286 88	83.126 41	8.840 823	19.046 98	41.035 46
692	478 864	331 373 888	26.305 89	83.186 54	8.845 085	19.056 16	41.055 25
693	480 249	332 812 557	26.324 89	83.246 62	8.849 344	19.065 33	41.075 02
694	481 636	334 255 384	26.343 88	83.306 66	8.853 599	19.074 50	41.094 76
695	483 025	335 702 375	26.362 85	83.366 66	8.857 849	19.083 66	41.114 49
696	484 416	337 153 536	26.381 81	83.426 61	8.862 095	19.092 81	41.134 20
697	485 809	338 608 873	26.400 76	83.486 53	8.866 338	19.101 95	41.153 89
698	487 204	340 068 392	26.419 69	83.546 39	8.870 576	19.111 08	41.173 57
699	488 601	341 532 099	26.438 61	83.606 22	8.874 810	19.120 20	41.193 22

n	n^2	n^3	$\sqrt{n}$	$\sqrt{10n}$	$\sqrt[3]{n}$	$\sqrt[3]{10n}$	$\sqrt[3]{100n}$
700	490 000	343 000 000	26.457 51	83.666 00	8.879 040	19.129 31	41.212 85
701	491 401	344 472 101	26.476 40	83.725 74	8.883 266	19.138 42	41.232 47
702	492 804	345 948 408	26.495 28	83.785 44	8.887 488	19.147 51	41.252 07
703	494 209	347 428 927	26.514 15	83.845 10	8.891 706	19.156 60	41.271 64
704	495 616	348 913 664	26.533 00	83.904 71	8.895 920	19.165 68	41.291 20
705	497 025	350 402 625	26.551 84	83.964 28	8.900 130	19.174 75	41.310 75
706	498 436	351 895 816	26.570 66	84.023 81	8.904 337	19.183 81	41.330 27
707	499 849	353 393 243	26.589 47	84.083 29	8.908 539	19.192 86	41.349 77
708	501 264	354 894 912	26.608 27	84.142 74	8.912 737	19.201 91	41.369 26
709	502 681	356 400 829	26.627 05	84.202 14	8.916 931	19.210 95	41.388 73
710	504 100	357 911 000	26.645 83	84.261 50	8.921 121	19.219 97	41.408 18
711	505 521	359 425 431	26.664 58	84.320 82	8.925 308	19.228 99	41.427 61
712	506 944	360 944 128	26.683 33	84.380 09	8.929 490	19.238 00	41.447 02
713	508 369	362 467 097	26.702 06	84.439 33	8.933 669	19.247 01	41.466 42
714	509 796	363 994 344	26.720 78	84.498 52	8.937 843	19.256 00	41.485 79
715	511 225	365 525 875	26.739 48	84.557 67	8.942 014	19.264 99	41.505 15
716	512 656	367 061 696	26.758 18	84.616 78	8.946 181	19.273 96	41.524 49
717	514 089	368 601 813	26.776 86	84.675 85	8.950 344	19.282 93	41.543 82
718	515 524	370 146 232	26.795 52	84.734 88	8.954 503	19.291 89	41.563 12
719	516 961	371 694 959	26.814 18	84.793 87	8.958 658	19.300 84	41.582 41
720	518 400	373 248 000	26.832 82	84.852 81	8.962 809	19.309 79	41.601 68
721	519 841	374 805 361	26.851 44	84.911 72	8.966 957	19.318 72	41.620 93
722	521 284	376 367 048	26.870 06	84.970 58	8.971 101	19.327 65	41.640 16
723	522 729	377 933 067	26.888 66	85.029 41	8.975 241	19.336 57	41.659 38
724	524 176	379 503 424	26.907 25	85.088 19	8.979 377	19.345 48	41.678 57
725	525 625	381 078 125	26.925 82	85.146 93	8.983 509	19.354 38	41.697 75
726	527 076	382 657 176	26.944 39	85.205 63	8.987 637	19.363 28	41.716 92
727	528 529	384 240 583	26.962 94	85.264 29	8.991 762	19.372 16	41.736 06
728	529 984	385 828 352	26.981 48	85.322 92	8.995 883	19.381 04	41.755 19
729	531 441	387 420 489	27.000 00	85.381 50	9.000 000	19.389 91	41.774 30
730	532 900	389 017 000	27.018 51	85.440 04	9.004 113	19.398 77	41.793 39
731	534 361	390 617 891	27.037 01	85.498 54	9.008 223	19.407 63	41.812 47
732	535 824	392 223 168	27.055 50	85.557 00	9.012 329	19.416 47	41.831 52
733	537 289	393 832 837	27.073 97	85.615 42	9.016 431	19.425 31	41.850 56
734	538 756	395 446 904	27.092 43	85.673 80	9.020 529	19.434 14	41.869 59
735	540 225	397 065 375	27.110 88	85.732 14	9.024 624	19.442 96	41.888 59
736	541 696	398 688 256	27.129 32	85.790 44	9.028 715	19.451 78	41.907 58
737	543 169	400 315 553	27.147 74	85.848 70	9.032 802	19.460 58	41.926 55
738	544 644	401 947 272	27.166 16	85.906 93	9.036 886	19.469 38	41.945 51
739	546 121	403 583 419	27.184 55	85.965 11	9.040 966	19.478 17	41.964 44
740	547 600	405 224 000	27.202 94	86.023 25	9.045 042	19.486 95	41.983 36
741	549 081	406 869 021	27.221 32	86.081 36	9.049 114	19.495 73	42.002 27
742	550 564	408 518 488	27.239 68	86.139 42	9.053 183	19.504 49	42.021 15
743	552 049	410 172 407	27.258 03	86.197 45	9.057 248	19.513 25	42.040 02
744	553 536	411 830 784	27.276 36	86.255 43	9.061 310	19.522 00	42.058 87
745	555 025	413 493 625	27.294 69	86.313 38	9.065 368	19.530 74	42.077 71
746	556 516	415 160 936	27.313 00	86.371 29	9.069 422	19.539 48	42.096 53
747	558 009	416 832 723	27.331 30	86.429 16	9.073 473	19.548 20	42.115 33
748	559 504	418 508 992	27.349 59	86.486 99	9.077 520	19.556 92	42.134 11
749	561 001	420 189 749	27.367 86	86.544 79	9.081 563	19.565 63	42.152 88

n	n^2	n^3	$\sqrt{n}$	$\sqrt{10n}$	$\sqrt[3]{n}$	$\sqrt[3]{10n}$	$\sqrt[3]{100n}$
750	562 500	421 875 000	27.386 13	86.602 54	9.085 603	19.574 34	42.171 63
751	564 001	423 564 751	27.404 38	86.660 26	9.089 639	19.583 03	42.190 37
752	565 504	425 259 008	27.422 62	86.717 93	9.093 672	19.591 72	42.209 09
753	567 009	426 957 777	27.440 85	86.775 57	9.097 701	19.600 40	42.227 79
754	568 516	428 661 064	27.459 06	86.833 17	9.101 727	19.609 08	42.246 47
755	570 025	430 368 875	27.477 26	86.890 74	9.105 748	19.617 74	42.265 14
756	571 536	432 081 216	27.495 45	86.948 26	9.109 767	19.626 40	42.283 79
757	573 049	433 798 093	27.513 63	87.005 75	9.113 782	19.635 05	42.302 43
758	574 564	435 519 512	27.531 80	87.063 20	9.117 793	19.643 69	42.321 05
759	576 081	437 245 479	27.549 95	87.120 61	9.121 801	19.652 32	42.339 65
760	577 600	438 976 000	27.568 10	87.177 98	9.125 805	19.660 95	42.358 24
761	579 121	440 711 081	27.586 23	87.235 31	9.129 806	19.669 57	42.376 81
762	580 644	442 450 728	27.604 35	87.292 61	9.133 803	19.678 18	42.395 36
763	582 169	444 194 947	27.622 45	87.349 87	9.137 797	19.686 79	42.413 90
764	583 696	445 943 744	27.640 55	87.407 09	9.141 787	19.695 38	42.432 42
765	585 225	447 697 125	27.658 63	87.464 28	9.145 774	19.703 97	42.450 92
766	586 756	449 455 096	27.676 71	87.521 43	9.149 758	19.712 56	42.469 41
767	588 289	451 217 663	27.694 76	87.578 54	9.153 738	19.721 13	42.487 89
768	589 824	452 984 832	27.712 81	87.635 61	9.157 714	19.729 70	42.506 34
769	591 361	454 756 609	27.730 85	87.692 65	9.161 687	19.738 26	42.524 78
770	592 900	456 533 000	27.748 87	87.749 64	9.165 656	19.746 81	42.543 21
771	594 441	458 314 011	27.766 89	87.806 61	9.169 623	19.755 35	42.561 62
772	595 984	460 099 648	27.784 89	87.863 53	9.173 585	19.763 89	42.580 01
773	597 529	461 889 917	27.802 88	87.920 42	9.177 544	19.772 42	42.598 39
774	599 076	463 684 824	27.820 86	87.977 27	9.181 500	19.780 94	42.616 75
775	600 625	465 484 375	27.838 82	88.034 08	9.185 453	19.789 46	42.635 09
776	602 176	467 288 576	27.856 78	88.090 86	9.189 402	19.797 97	42.653 42
777	603 729	469 097 433	27.874 72	88.147 60	9.193 347	19.806 47	42.671 74
778	605 284	470 910 952	27.892 65	88.204 31	9.197 290	19.814 96	42.690 04
779	606 841	472 729 139	27.910 57	88.260 98	9.201 229	19.823 45	42.708 32
780	608 400	474 552 000	27.928 48	88.317 61	9.205 164	19.831 92	42.726 59
781	609 961	476 379 541	27.946 38	88.374 20	9.209 096	19.840 40	42.744 84
782	611 524	478 211 768	27.964 26	88.430 76	9.213 025	19.848 86	42.763 07
783	613 089	480 048 687	27.982 14	88.487 29	9.216 950	19.857 32	42.781 29
784	614 656	481 890 304	28.000 00	88.543 77	9.220 873	19.865 77	42.799 50
785	616 225	483 736 625	28.017 85	88.600 23	9.224 791	19.874 21	42.817 69
786	617 796	485 587 656	28.035 69	88.656 64	9.228 707	19.882 65	42.835 86
787	619 369	487 443 403	28.053 52	88.713 02	9.232 619	19.891 07	42.854 02
788	620 944	489 303 872	28.071 34	88.769 36	9.236 528	19.899 50	42.872 16
789	622 521	491 169 069	28.089 14	88.825 67	9.240 433	19.907 91	42.890 29
790	624 100	493 039 000	28.106 94	88.881 94	9.244 335	19.916 32	42.908 40
791	625 681	494 913 671	28.124 72	88.938 18	9.248 234	19.924 72	42.926 50
792	627 264	496 793 088	28.142 49	88.994 38	9.252 130	19.933 11	42.944 58
793	628 849	498 677 257	28.160 26	89.050 55	9.256 022	19.941 50	42.962 65
794	630 436	500 566 184	28.178 01	89.106 68	9.259 911	19.949 87	42.980 70
795	632 025	502 459 875	28.195 74	89.162 77	9.263 797	19.958 25	42.998 74
796	633 616	504 358 336	28.213 47	89.218 83	9.267 680	19.966 61	43.016 76
797	635 209	506 261 573	28.231 19	89.274 86	9.271 559	19.974 97	43.034 77
798	636 804	508 169 592	28.248 89	89.330 85	9.275 435	19.983 32	43.052 76
799	638 401	510 082 399	28.266 59	89.386 80	9.279 308	19.991 66	43.070 73

n	n^2	n^3	$\sqrt{n}$	$\sqrt{10n}$	$\sqrt[3]{n}$	$\sqrt[3]{10n}$	$\sqrt[3]{100n}$
800	640 000	512 000 000	28.284 27	89.442 72	9.283 178	20.000 00	43.088 69
801	641 601	513 922 401	28.301 94	89.498 60	9.287 044	20.008 33	43.106 64
802	643 204	515 849 608	28.319 60	89.554 45	9.290 907	20.016 65	43.124 57
803	644 809	517 781 627	28.337 25	89.610 27	9.294 767	20.024 97	43.142 49
804	646 416	519 718 464	28.354 89	89.666 05	9.298 624	20.033 28	43.160 39
805	648 025	521 660 125	28.372 52	89.721 79	9.302 477	20.041 58	43.178 28
806	649 636	523 606 616	28.390 14	89.777 50	9.306 328	20.049 88	43.196 15
807	651 249	525 557 943	28.407 75	89.833 18	9.310 175	20.058 16	43.214 00
808	652 864	527 514 112	28.425 34	89.888 82	9.314 019	20.066 45	43.231 85
809	654 481	529 475 129	28.442 93	89.944 43	9.317 860	20.074 72	43.249 67
810	656 100	531 441 000	28.460 50	90.000 00	9.321 698	20.082 99	43.267 49
811	657 721	533 411 731	28.478 06	90.055 54	9.325 532	20.091 25	43.285 29
812	659 344	535 387 328	28.495 61	90.111 04	9.329 363	20.099 50	43.303 07
813	660 969	537 367 797	28.513 15	90.166 51	9.333 192	20.107 75	43.320 84
814	662 596	539 353 144	28.530 69	90.221 95	9.337 017	20.115 99	43.338 59
815	664 225	541 343 375	28.548 20	90.277 35	9.340 839	20.124 23	43.356 33
816	665 856	543 338 496	28.565 71	90.332 72	9.344 657	20.132 45	43.374 06
817	667 489	545 338 513	28.583 21	90.388 05	9.348 473	20.140 67	43.391 77
818	669 124	547 343 432	28.600 70	90.443 35	9.352 286	20.148 89	43.409 47
819	670 761	549 353 259	28.618 18	90.498 62	9.356 095	20.157 10	43.427 15
820	672 400	551 368 000	28.635 64	90.553 85	9.359 902	20.165 30	43.444 81
821	674 041	553 387 661	28.653 10	90.609 05	9.363 705	20.173 49	43.462 47
822	675 684	555 412 248	28.670 54	90.664 22	9.367 505	20.181 68	43.480 11
823	677 329	557 441 767	28.687 98	90.719 35	9.371 302	20.189 86	43.497 73
824	678 976	559 476 224	28.705 40	90.774 45	9.375 096	20.198 03	43.515 34
825	680 625	561 515 625	28.722 81	90.829 51	9.378 887	20.206 20	43.532 94
826	682 276	563 559 976	28.740 22	90.884 54	9.382 675	20.214 36	43.550 52
827	683 929	565 609 283	28.757 61	90.939 54	9.386 460	20.222 52	43.568 09
828	685 584	567 663 552	28.774 99	90.994 51	9.390 242	20.230 66	43.585 64
829	687 241	569 722 789	28.792 36	91.049 44	9.394 021	20.238 80	43.603 18
830	688 900	571 787 000	28.809 72	91.104 34	9.397 796	20.246 94	43.620 71
831	690 561	573 856 191	28.827 07	91.159 20	9.401 569	20.255 07	43.638 22
832	692 224	575 930 368	28.844 41	91.214 03	9.405 339	20.263 19	43.655 72
833	693 889	578 009 537	28.861 74	91.268 83	9.409 105	20.271 30	43.673 20
834	695 556	580 093 704	28.879 06	91.323 60	9.412 869	20.279 41	43.690 67
835	697 225	582 182 875	28.896 37	91.378 33	9.416 630	20.287 51	43.708 12
836	698 896	584 277 056	28.913 66	91.433 04	9.420 387	20.295 61	43.725 56
837	700 569	586 376 253	28.930 95	91.487 70	9.424 142	20.303 70	43.742 99
838	702 244	588 480 472	28.948 23	91.542 34	9.427 894	20.311 78	43.760 41
839	703 921	590 589 719	28.965 50	91.596 94	9.431 642	20.319 86	43.777 81
840	705 600	592 704 000	28.982 75	91.651 51	9.435 388	20.327 93	43.795 19
841	707 281	594 823 321	29.000 00	91.706 05	9.439 131	20.335 99	43.812 56
842	708 964	596 947 688	29.017 24	91.760 56	9.442 870	20.344 05	43.829 92
843	710 649	599 077 107	29.034 46	91.815 03	9.446 607	20.352 10	43.847 27
844	712 336	601 211 584	29.051 68	91.869 47	9.450 341	20.360 14	43.864 60
845	714 025	603 351 125	29.068 88	91.923 88	9.454 072	20.368 18	43.881 91
846	715 716	605 495 736	29.086 08	91.978 26	9.457 800	20.376 21	43.899 22
847	717 409	607 645 423	29.103 26	92.032 60	9.461 525	20.384 24	43.916 51
848	719 104	609 800 192	29.120 44	92.086 92	9.465 247	20.392 26	43.933 78
849	720 801	611 960 049	29.137 60	92.141 20	9.468 966	20.400 27	43.951 05

n	n^2	n^3	$\sqrt{n}$	$\sqrt{10n}$	$\sqrt[3]{n}$	$\sqrt[3]{10n}$	$\sqrt[3]{100n}$
850	722 500	614 125 000	29.154 76	92.195 44	9.472 682	20.408 28	43.968 30
851	724 201	616 295 051	29.171 90	92.249 66	9.476 396	20.416 28	43.985 53
852	725 904	618 470 208	29.189 04	92.303 85	9.480 106	20.424 27	44.002 75
853	727 609	620 650 477	29.206 16	92.358 00	9.483 814	20.432 26	44.019 96
854	729 316	622 835 864	29.223 28	92.412 12	9.487 518	20.440 24	44.037 16
855	731 025	625 026 375	29.240 38	92.466 21	9.491 220	20.448 21	44.054 34
856	732 736	627 222 016	29.257 48	92.520 27	9.494 919	20.456 18	44.071 51
857	734 449	629 422 793	29.274 56	92.574 29	9.498 615	20.464 15	44.088 66
858	736 164	631 628 712	29.291 64	92.628 29	9.502 308	20.472 10	44.105 81
859	737 881	633 839 779	29.308 70	92.682 25	9.505 998	20.480 05	44.122 93
860	739 600	636 056 000	29.325 76	92.736 18	9.509 685	20.488 00	44.140 05
861	741 321	638 277 381	29.342 80	92.790 09	9.513 370	20.495 93	44.157 15
862	743 044	640 503 928	29.359 84	92.843 96	9.517 052	20.503 87	44.174 24
863	744 769	642 735 647	29.376 86	92.897 79	9.520 730	20.511 79	44.191 32
864	746 496	644 972 544	29.393 88	92.951 60	9.524 406	20.519 71	44.208 38
865	748 225	647 214 625	29.410 88	93.005 38	9.528 079	20.527 62	44.225 43
866	749 956	649 461 896	29.427 88	93.059 12	9.531 750	20.535 53	44.242 46
867	751 689	651 714 363	29.444 86	93.112 83	9.535 417	20.543 43	44.259 49
868	753 424	653 972 032	29.461 84	93.166 52	9.539 082	20.551 33	44.276 50
869	755 161	656 234 909	29.478 81	93.220 17	9.542 744	20.559 22	44.293 49
870	756 900	658 503 000	29.495 76	93.273 79	9.546 403	20.567 10	44.310 48
871	758 641	660 776 311	29.512 71	93.327 38	9.550 059	20.574 98	44.327 45
872	760 384	663 054 848	29.529 65	93.380 94	9.553 712	20.582 85	44.344 40
873	762 129	665 338 617	29.546 57	93.434 47	9.557 363	20.590 71	44.361 35
874	763 876	667 627 624	29.563 49	93.487 97	9.561 011	20.598 57	44.378 28
875	765 625	669 921 875	29.580 40	93.541 43	9.564 656	20.606 43	44.395 20
876	767 376	672 221 376	29.597 30	93.594 87	9.568 298	20.614 27	44.412 11
877	769 129	674 526 133	29.614 19	93.648 28	9.571 938	20.622 11	44.429 00
878	770 884	676 836 152	29.631 06	93.701 65	9.575 574	20.629 95	44.445 88
879	772 641	679 151 439	29.647 93	93.755 00	9.579 208	20.637 78	44.462 75
880	774 400	681 472 000	29.664 79	93.808 32	9.582 840	20.645 60	44.479 60
881	776 161	683 797 841	29.681 64	93.861 60	9.586 468	20.653 42	44.496 44
882	777 924	686 128 968	29.698 48	93.914 86	9.590 094	20.661 23	44.513 27
883	779 689	688 465 387	29.715 32	93.968 08	9.593 717	20.669 04	44.530 09
884	781 456	690 807 104	29.732 14	94.021 27	9.597 337	20.676 84	44.546 89
885	783 225	693 154 125	29.748 95	94.074 44	9.600 955	20.684 63	44.563 68
886	784 996	695 506 456	29.765 75	94.127 57	9.604 570	20.692 42	44.580 46
887	786 769	697 864 103	29.782 55	94.180 68	9.608 182	20.700 20	44.597 23
888	788 544	700 227 072	29.799 33	94.233 75	9.611 791	20.707 98	44.613 98
889	790 321	702 595 369	29.816 10	94.286 80	9.615 398	20.715 75	44.630 72
890	792 100	704 969 000	29.832 87	94.339 81	9.619 002	20.723 51	44.647 45
891	793 881	707 347 971	29.849 62	94.392 80	9.622 603	20.731 27	44.664 17
892	795 664	709 732 288	29.866 37	94.445 75	9.626 202	20.739 02	44.680 87
893	797 449	712 121 957	29.883 11	94.498 68	9.629 797	20.746 77	44.697 56
894	799 236	714 516 984	29.899 83	94.551 57	9.633 391	20.754 51	44.714 24
895	801 025	716 917 375	29.916 55	94.604 44	9.636 981	20.762 25	44.730 90
896	802 816	719 323 136	29.933 26	94.657 28	9.640 569	20.769 98	44.747 56
897	804 609	721 734 273	29.949 96	94.710 08	9.644 154	20.777 70	44.764 20
898	806 404	724 150 792	29.966 65	94.762 86	9.647 737	20.785 42	44.780 83
899	808 201	726 572 699	29.983 33	94.815 61	9.651 317	20.793 13	44.797 44

n	n^2	n^3	$\sqrt{n}$	$\sqrt{10n}$	$\sqrt[3]{n}$	$\sqrt[3]{10n}$	$\sqrt[3]{100n}$
900	810 000	729 000 000	30.000 00	94.868 33	9.654 894	20.800 84	44.814 05
901	811 801	731 432 701	30.016 66	94.921 02	9.658 468	20.808 54	44.830 64
902	813 604	733 870 808	30.033 31	94.973 68	9.662 040	20.816 23	44.847 22
903	815 409	736 314 327	30.049 96	95.026 31	9.665 610	20.823 92	44.863 79
904	817 216	738 763 264	30.066 59	95.078 91	9.669 176	20.831 61	44.880 34
905	819 025	741 217 625	30.083 22	95.131 49	9.672 740	20.839 29	44.896 88
906	820 836	743 677 416	30.099 83	95.184 03	9.676 302	20.846 96	44.913 41
907	822 649	746 142 643	30.116 44	95.236 55	9.679 860	20.854 63	44.929 93
908	824 464	748 613 312	30.133 04	95.289 03	9.683 417	20.862 29	44.946 44
909	826 281	751 089 429	30.149 63	95.341 49	9.686 970	20.869 94	44.962 93
910	828 100	753 571 000	30.166 21	95.393 92	9.690 521	20.877 59	44.979 41
911	829 921	756 058 031	30.182 78	95.446 32	9.694 069	20.885 24	44.995 88
912	831 744	758 550 528	30.199 34	95.498 69	9.697 615	20.892 88	45.012 34
913	833 569	761 048 497	30.215 89	95.551 03	9.701 158	20.900 51	45.028 79
914	835 396	763 551 944	30.232 43	95.603 35	9.704 699	20.908 14	45.045 22
915	837 225	766 060 875	30.248 97	95.655 63	9.708 237	20.915 76	45.061 64
916	839 056	768 575 296	30.265 49	95.707 89	9.711 772	20.923 38	45.078 05
917	840 889	771 095 213	30.282 01	95.760 12	9.715 305	20.930 99	45.094 45
918	842 724	773 620 632	30.298 51	95.812 32	9.718 835	20.938 60	45.110 84
919	844 561	776 151 559	30.315 01	95.864 49	9.722 363	20.946 20	45.127 21
920	846 400	778 688 000	30.331 50	95.916 63	9.725 888	20.953 79	45.143 57
921	848 241	781 229 961	30.347 98	95.968 74	9.729 411	20.961 38	45.159 92
922	850 084	783 777 448	30.364 45	96.020 83	9.732 931	20.968 96	45.176 26
923	851 929	786 330 467	30.380 92	96.072 89	9.736 448	20.976 54	45.192 59
924	853 776	788 889 024	30.397 37	96.124 92	9.739 963	20.984 11	45.208 91
925	855 625	791 453 125	30.413 81	96.176 92	9.743 476	20.991 68	45.225 21
926	857 476	794 022 776	30.430 25	96.228 89	9.746 986	20.999 24	45.241 50
927	859 329	796 597 983	30.446 67	96.280 84	9.750 493	21.006 80	45.257 78
928	861 184	799 178 752	30.463 09	96.332 76	9.753 998	21.014 35	45.274 05
929	863 041	801 765 089	30.479 50	96.384 65	9.757 500	21.021 90	45.290 30
930	864 900	804 357 000	30.495 90	96.436 51	9.761 000	21.029 44	45.306 55
931	866 761	806 954 491	30.512 29	96.488 34	9.764 497	21.036 97	45.322 78
932	868 624	809 557 568	30.528 68	96.540 15	9.767 992	21.044 50	45.339 00
933	870 489	812 166 237	30.545 05	96.591 93	9.771 485	21.052 03	45.355 21
934	872 356	814 780 504	30.561 41	96.643 68	9.774 974	21.059 54	45.371 41
935	874 225	817 400 375	30.577 77	96.695 40	9.778 462	21.067 06	45.387 60
936	876 096	820 025 856	30.594 12	96.747 09	9.781 946	21.074 56	45.403 77
937	877 969	822 656 953	30.610 46	96.798 76	9.785 429	21.082 07	45.419 94
938	879 844	825 293 672	30.626 79	96.850 40	9.788 909	21.089 56	45.436 09
939	881 721	827 936 019	30.643 11	96.902 01	9.792 386	21.097 06	45.452 23
940	883 600	830 584 000	30.659 42	96.953 60	9.795 861	21.104 54	45.468 36
941	885 481	833 237 621	30.675 72	97.005 15	9.799 334	21.112 02	45.484 48
942	887 364	835 896 888	30.692 02	97.056 68	9.802 804	21.119 50	45.500 58
943	889 249	838 561 807	30.708 31	97.108 19	9.806 271	21.126 97	45.516 68
944	891 136	841 232 384	30.724 58	97.159 66	9.809 736	21.134 44	45.532 76
945	893 025	843 908 625	30.740 85	97.211 11	9.813 199	21.141 90	45.548 83
946	894 916	846 590 536	30.757 11	97.262 53	9.816 659	21.149 35	45.564 90
947	896 809	849 278 123	30.773 37	97.313 93	9.820 117	21.156 80	45.580 95
948	898 704	851 971 392	30.789 61	97.365 29	9.823 572	21.164 24	45.596 98
949	900 601	854 670 349	30.805 84	97.416 63	9.827 025	21.171 68	45.613 01

n	n^2	n^3	$\sqrt{n}$	$\sqrt{10n}$	$\sqrt[3]{n}$	$\sqrt[3]{10n}$	$\sqrt[3]{100n}$
950	902 500	857 375 000	30.822 07	97.467 94	9.830 476	21.179 12	45.629 03
951	904 401	860 085 351	30.838 29	97.519 23	9.833 924	21.186 55	45.645 03
952	906 304	862 801 408	30.854 50	97.570 49	9.837 369	21.193 97	45.661 02
953	908 209	865 523 177	30.870 70	97.621 72	9.840 813	21.201 39	45.677 01
954	910 116	868 250 664	30.886 89	97.672 92	9.844 254	21.208 80	45.692 98
955	912 025	870 983 875	30.903 07	97.724 10	9.847 692	21.216 21	45.708 94
956	913 936	873 722 816	30.919 25	97.775 25	9.851 128	21.223 61	45.724 89
957	915 849	876 467 493	30.935 42	97.826 38	9.854 562	21.231 01	45.740 82
958	917 764	879 217 912	30.951 58	97.877 47	9.857 993	21.238 40	45.756 75
959	919 681	881 974 079	30.967 73	97.928 55	9.861 422	21.245 79	45.772 67
960	921 600	884 736 000	30.983 87	97.979 59	9.864 848	21.253 17	45.788 57
961	923 521	887 503 681	31.000 00	98.030 61	9.868 272	21.260 55	45.804 46
962	925 444	890 277 128	31.016 12	98.081 60	9.871 694	21.267 92	45.820 35
963	927 369	893 056 347	31.032 24	98.132 56	9.875 113	21.275 29	45.836 22
964	929 296	895 841 344	31.048 35	98.183 50	9.878 530	21.282 65	45.852 08
965	931 225	898 632 125	31.064 45	98.234 41	9.881 945	21.290 01	45.867 93
966	933 156	901 428 696	31.080 54	98.285 30	9.885 357	21.297 36	45.883 76
967	935 089	904 231 063	31.096 62	98.336 16	9.888 767	21.304 70	45.899 59
968	937 024	907 039 232	31.112 70	98.386 99	9.892 175	21.312 04	45.915 41
969	938 961	909 853 209	31.128 76	98.437 80	9.895 580	21.319 38	45.931 21
970	940 900	912 673 000	31.144 82	98.488 58	9.898 983	21.326 71	45.947 01
971	942 841	915 498 611	31.160 87	98.539 33	9.902 384	21.334 04	45.962 79
972	944 784	918 330 048	31.176 91	98.590 06	9.905 782	21.341 36	45.978 57
973	946 729	921 167 317	31.192 95	98.640 76	9.909 178	21.348 68	45.994 33
974	948 676	924 010 424	31.208 97	98.691 44	9.912 571	21.355 99	46.010 08
975	950 625	926 859 375	31.224 99	98.742 09	9.915 962	21.363 29	46.025 82
976	952 576	929 714 176	31.241 00	98.792 71	9.919 351	21.370 59	46.041 55
977	954 529	932 574 833	31.257 00	98.843 31	9.922 738	21.377 89	46.057 27
978	956 484	935 441 352	31.272 99	98.893 88	9.926 122	21.385 18	46.072 98
979	958 441	938 313 739	31.288 98	98.944 43	9.929 504	21.392 47	46.088 68
980	960 400	941 192 000	31.304 95	98.994 95	9.932 884	21.399 75	46.104 36
981	962 361	944 076 141	31.320 92	99.045 44	9.936 261	21.407 03	46.120 04
982	964 324	946 966 168	31.336 88	99.095 91	9.939 636	21.414 30	46.135 71
983	966 289	949 862 087	31.352 83	99.146 36	9.943 009	21.421 56	46.151 36
984	968 256	952 763 904	31.368 77	99.196 77	9.946 380	21.428 83	46.167 00
985	970 225	955 671 625	31.384 71	99.247 17	9.949 748	21.436 08	46.182 64
986	972 196	958 585 256	31.400 64	99.297 53	9.953 114	21.443 33	46.198 26
987	974 169	961 504 803	31.416 56	99.347 87	9.956 478	21.450 58	46.213 87
988	976 144	964 430 272	31.432 47	99.398 19	9.959 839	21.457 82	46.229 48
989	978 121	967 361 669	31.448 37	99.448 48	9.963 198	21.465 06	46.245 07
990	980 100	970 299 000	31.464 27	99.498 74	9.966 555	21.472 29	46.260 65
991	982 081	973 242 271	31.480 15	99.548 98	9.969 910	21.479 52	46.276 22
992	984 064	976 191 488	31.496 03	99.599 20	9.973 262	21.486 74	46.291 78
993	986 049	979 146 657	31.511 90	99.649 39	9.976 612	21.493 96	46.307 33
994	988 036	982 107 784	31.527 77	99.699 55	9.979 960	21.501 17	46.322 87
995	990 025	985 074 875	31.543 62	99.749 69	9.983 305	21.508 38	46.338 40
996	992 016	988 047 936	31.559 47	99.799 80	9.986 649	21.515 58	46.353 92
997	994 009	991 026 973	31.575 31	99.849 89	9.989 990	21.522 78	46.369 43
998	996 004	994 011 992	31.591 14	99.899 95	9.993 329	21.529 97	46.384 92
999	998 001	997 002 999	31.606 96	99.949 99	9.996 666	21.537 16	46.400 41

Table 15 RECIPROCALS, CIRCUMFERENCES, AND AREAS OF CIRCLES

n	$1000/n$	Circum. of circle πn	Area of circle $\pi n^2/4$	n	$1000/n$	Circum. of circle πn	Area of circle $\pi n^2/4$
				50	20.000 00	157.079 6	1 963.495
1	1000.000	3.141 593	.785 3982	51	19.607 84	160.221 2	2 042.821
2	500.000 0	6.283 185	3.141 593	52	19.230 77	163.362 8	2 123.717
3	333.333 3	9.424 778	7.068 583	53	18.867 92	166.504 4	2 206.183
4	250.000 0	12.566 37	12.566 37	54	18.518 52	169.646 0	2 290.221
5	200.000 0	15.707 96	19.634 95	55	18.181 82	172.787 6	2 375.829
6	166.666 7	18.849 56	28.274 33	56	17.857 14	175.929 2	2 463.009
7	142.857 1	21.991 15	38.484 51	57	17.543 86	179.070 8	2 551.759
8	125.000 0	25.132 74	50.265 48	58	17.241 38	182.212 4	2 642.079
9	111.111 1	28.274 33	63.617 25	59	16.949 15	185.354 0	2 733.971
10	100.000 0	31.415 93	78.539 82	60	16.666 67	188.495 6	2 827.433
11	90.909 09	34.557 52	95.033 18	61	16.393 44	191.637 2	2 922.467
12	83.333 33	37.699 11	113.097 3	62	16.129 03	194.778 7	3 019.071
13	76.923 08	40.840 70	132.732 3	63	15.873 02	197.920 3	3 117.245
14	71.428 57	43.982 30	153.938 0	64	15.625 00	201.061 9	3 216.991
15	66.666 67	47.123 89	176.714 6	65	15.384 62	204.203 5	3 318.307
16	62.500 00	50.265 48	201.061 9	66	15.151 52	207.345 1	3 421.194
17	58.823 53	53.407 08	226.980 1	67	14.925 37	210.486 7	3 525.652
18	55.555 56	56.548 67	254.469 0	68	14.705 88	213.628 3	3 631.681
19	52.631 58	59.690 26	283.528 7	69	14.492 75	216.769 9	3 739.281
20	50.000 00	62.831 85	314.159 3	70	14.285 71	219.911 5	3 848.451
21	47.619 05	65.973 45	346.360 6	71	14.084 51	223.053 1	3 959.192
22	45.454 55	69.115 04	380.132 7	72	13.888 89	226.194 7	4 071.504
23	43.478 26	72.256 63	415.475 6	73	13.698 63	229.336 3	4 185.387
24	41.666 67	75.398 22	452.389 3	74	13.513 51	232.477 9	4 300.840
25	40.000 00	78.539 82	490.873 9	75	13.333 33	235.619 4	4 417.865
26	38.461 54	81.681 41	530.929 2	76	13.157 89	238.761 0	4 536.460
27	37.037 04	84.823 00	572.555 3	77	12.987 01	241.902 6	4 656.626
28	35.714 29	87.964 59	615.752 2	78	12.820 51	245.044 2	4 778.362
29	34.482 76	91.106 19	660.519 9	79	12.658 23	248.185 8	4 901.670
30	33.333 33	94.247 78	706.858 3	80	12.500 00	251.327 4	5 026.548
31	32.258 06	97.389 37	754.767 6	81	12.345 68	254.469 0	5 152.997
32	31.250 00	100.531 0	804.247 7	82	12.195 12	257.610 6	5 281.017
33	30.303 03	103.672 6	855.298 6	83	12.048 19	260.752 2	5 410.608
34	29.411 76	106.814 2	907.920 3	84	11.904 76	263.893 8	5 541.769
35	28.571 43	109.955 7	962.112 8	85	11.764 71	267.035 4	5 674.502
36	27.777 78	113.097 3	1 017.876	86	11.627 91	270.177 0	5 808.805
37	27.027 03	116.238 9	1 075.210	87	11.494 25	273.318 6	5 944.679
38	26.315 79	119.380 5	1 134.115	88	11.363 64	276.460 2	6 082.123
39	25.641 03	122.522 1	1 194.591	89	11.235 96	279.601 7	6 221.139
40	25.000 00	125.663 7	1 256.637	90	11.111 11	282.743 3	6 361.725
41	24.390 24	128.805 3	1 320.254	91	10.989 01	285.884 9	6 503.882
42	23.809 52	131.946 9	1 385.442	92	10.869 57	289.026 5	6 647.610
43	23.255 81	135.088 5	1 452.201	93	10.752 69	292.168 1	6 792.909
44	22.727 27	138.230 1	1 520.531	94	10.638 30	295.309 7	6 939.778
45	22.222 22	141.371 7	1 590.431	95	10.526 32	298.451 3	7 088.218
46	21.739 13	144.513 3	1 661.903	96	10.416 67	301.592 9	7 238.229
47	21.276 60	147.654 9	1 734.945	97	10.309 28	304.734 5	7 389.811
48	20.833 33	150.796 4	1 809.557	98	10.204 08	307.876 1	7 542.964
49	20.408 16	153.938 0	1 885.741	99	10.101 01	311.017 7	7 697.687

Table 15 RECIPROCALS, CIRCUMFERENCES, AND AREAS OF CIRCLES (*continued*)

n	$1000/n$	Circum. of circle πn	Area of circle $\pi n^2/4$	n	$1000/n$	Circum. of circle πn	Area of circle $\pi n^2/4$
100	10.000 000	314.159 3	7 853.982	150	6.666 667	471.238 9	17 671.46
101	9.900 990	317.300 9	8 011.847	151	6.622 517	474.380 5	17 907.86
102	9.803 922	320.442 5	8 171.282	152	6.578 947	477.522 1	18 145.84
103	9.708 738	323.584 0	8 332.289	153	6.535 948	480.663 7	18 385.39
104	9.615 385	326.725 6	8 494.867	154	6.493 506	483.805 3	18 626.50
105	9.523 810	329.867 2	8 659.015	155	6.451 613	486.946 9	18 869.19
106	9.433 962	333.008 8	8 824.734	156	6.410 256	490.088 5	19 113.45
107	9.345 794	336.150 4	8 992.024	157	6.369 427	493.230 0	19 359.28
108	9.259 259	339.292 0	9 160.884	158	6.329 114	496.371 6	19 606.68
109	9.174 312	342.433 6	9 331.316	159	6.289 308	499.513 2	19 855.65
110	9.090 909	345.575 2	9 503.318	160	6.250 000	502.654 8	20 106.19
111	9.009 009	348.716 8	9 676.891	161	6.211 180	505.796 4	20 358.31
112	8.928 571	351.858 4	9 852.035	162	6.172 840	508.938 0	20 611.99
113	8.849 558	355.000 0	10 028.75	163	6.134 969	512.079 6	20 867.24
114	8.771 930	358.141 6	10 207.03	164	6.097 561	515.221 2	21 124.07
115	8.695 652	361.283 2	10 386.89	165	6.060 606	518.362 8	21 382.46
116	8.620 690	364.424 7	10 568.32	166	6.024 096	521.504 4	21 642.43
117	8.547 009	367.566 3	10 751.32	167	5.988 024	524.646 0	21 903.97
118	8.474 576	370.707 9	10 935.88	168	5.952 381	527.787 6	22 167.08
119	8.403 361	373.849 5	11 122.02	169	5.917 160	530.929 2	22 431.76
120	8.333 333	376.991 1	11 309.73	170	5.882 353	534.070 8	22 698.01
121	8.264 463	380.132 7	11 499.01	171	5.847 953	537.212 3	22 965.83
122	8.196 721	383.274 3	11 689.87	172	5.813 953	540.353 9	23 235.22
123	8.130 081	386.415 9	11 882.29	173	5.780 347	543.495 5	23 506.18
124	8.064 516	389.557 5	12 076.28	174	5.747 126	546.637 1	23 778.71
125	8.000 000	392.699 1	12 271.85	175	5.714 286	549.778 7	24 052.82
126	7.936 508	395.840 7	12 468.98	176	5.681 818	552.920 3	24 328.49
127	7.874 016	398.982 3	12 667.69	177	5.649 718	556.061 9	24 605.74
128	7.812 500	402.123 9	12 867.96	178	5.617 978	559.203 5	24 884.56
129	7.751 938	405.265 5	13 069.81	179	5.586 592	562.345 1	25 164.94
130	7.692 308	408.407 0	13 273.23	180	5.555 556	565.486 7	25 446.90
131	7.633 588	411.548 6	13 478.22	181	5.524 862	568.628 3	25 730.43
132	7.575 758	414.690 2	13 684.78	182	5.494 505	571.769 9	26 015.53
133	7.518 797	417.831 8	13 892.91	183	5.464 481	574.911 5	26 302.20
134	7.462 687	420.973 4	14 102.61	184	5.434 783	578.053 0	26 590.44
135	7.407 407	424.115 0	14 313.88	185	5.405 405	581.194 6	26 880.25
136	7.352 941	427.256 6	14 526.72	186	5.376 344	584.336 2	27 171.63
137	7.299 270	430.398 2	14 741.14	187	5.347 594	587.477 8	27 464.59
138	7.246 377	433.539 8	14 957.12	188	5.319 149	590.619 4	27 759.11
139	7.194 245	436.681 4	15 174.68	189	5.291 005	593.761 0	28 055.21
140	7.142 857	439.823 0	15 393.80	190	5.263 158	596.902 6	28 352.87
141	7.092 199	442.964 6	15 614.50	191	5.235 602	600.044 2	28 652.11
142	7.042 254	446.106 2	15 836.77	192	5.208 333	603.185 8	28 952.92
143	6.993 007	449.247 7	16 060.61	193	5.181 347	606.327 4	29 255.30
144	6.944 444	452.389 3	16 286.02	194	5.154 639	609.469 0	29 559.25
145	6.896 552	455.530 9	16 513.00	195	5.128 205	612.610 6	29 864.77
146	6.849 315	458.672 5	16 741.55	196	5.102 041	615.752 2	30 171.86
147	6.802 721	461.814 1	16 971.67	197	5.076 142	618.893 8	30 480.52
148	6.756 757	464.955 7	17 203.36	198	5.050 505	622.035 3	30 790.75
149	6.711 409	468.097 3	17 436.62	199	5.025 126	625.176 9	31 102.55

Table 15 RECIPROCALS, CIRCUMFERENCES, AND AREAS OF CIRCLES (*continued*)

n	$1000/n$	Circum. of circle πn	Area of circle $\pi n^2/4$	n	$1000/n$	Circum. of circle πn	Area of circle $\pi n^2/4$
200	5.000 000	628.318 5	31 415.93	250	4.000 000	785.398 2	49 087.39
201	4.975 124	631.460 1	31 730.87	251	3.984 064	788.539 8	49 480.87
202	4.950 495	634.601 7	32 047.39	252	3.968 254	791.681 3	49 875.92
203	4.926 108	637.743 3	32 365.47	253	3.952 569	794.822 9	50 272.55
204	4.901 961	640.884 9	32 685.13	254	3.937 008	797.964 5	50 670.75
205	4.878 049	644.026 5	33 006.36	255	3.921 569	801.106 1	51 070.52
206	4.854 369	647.168 1	33 329.16	256	3.906 250	804.247 7	51 471.85
207	4.830 918	650.309 7	33 653.53	257	3.891 051	807.389 3	51 874.76
208	4.807 692	653.451 3	33 979.47	258	3.875 969	810.530 9	52 279.24
209	4.784 689	656.592 9	34 306.98	259	3.861 004	813.672 5	52 685.29
210	4.761 905	659.734 5	34 636.06	260	3.846 154	816.814 1	53 092.92
211	4.739 336	662.876 0	34 966.71	261	3.831 418	819.955 7	53 502.11
212	4.716 981	666.017 6	35 298.94	262	3.816 794	823.097 3	53 912.87
213	4.694 836	669.159 2	35 632.73	263	3.802 281	826.238 9	54 325.21
214	4.672 897	672.300 8	35 968.09	264	3.787 879	829.380 5	54 739.11
215	4.651 163	675.442 4	36 305.03	265	3.773 585	832.522 1	55 154.59
216	4.629 630	678.584 0	36 643.54	266	3.759 398	835.663 6	55 571.63
217	4.608 295	681.725 6	36 983.61	267	3.745 318	838.805 2	55 990.25
218	4.587 156	684.867 2	37 325.26	268	3.731 343	841.946 8	56 410.44
219	4.566 210	688.008 8	37 668.48	269	3.717 472	845.088 4	56 832.20
220	4.545 455	691.150 4	38 013.27	270	3.703 704	848.230 0	57 255.53
221	4.524 887	694.292 0	38 359.63	271	3.690 037	851.371 6	57 680.43
222	4.504 505	697.433 6	38 707.56	272	3.676 471	854.513 2	58 106.90
223	4.484 305	700.575 2	39 057.07	273	3.663 004	857.654 8	58 534.94
224	4.464 286	703.716 8	39 408.14	274	3.649 635	860.796 4	58 964.55
225	4.444 444	706.858 3	39 760.78	275	3.636 364	863.938 0	59 395.74
226	4.424 779	709.999 9	40 115.00	276	3.623 188	867.079 6	59 828.49
227	4.405 286	713.141 5	40 470.78	277	3.610 108	870.221 2	60 262.82
228	4.385 965	716.283 1	40 828.14	278	3.597 122	873.362 8	60 698.71
229	4.366 812	719.424 7	41 187.07	279	3.584 229	876.504 4	61 136.18
230	4.347 826	722.566 3	41 547.56	280	3.571 429	879.645 9	61 575.22
231	4.329 004	725.707 9	41 909.63	281	3.558 719	882.787 5	62 015.82
232	4.310 345	728.849 5	42 273.27	282	3.546 099	885.929 1	62 458.00
233	4.291 845	731.991 1	42 638.48	283	3.533 569	889.070 7	62 901.75
234	4.273 504	735.132 7	43 005.26	284	3.521 127	892.212 3	63 347.07
235	4.255 319	738.274 3	43 373.61	285	3.508 772	895.353 9	63 793.97
236	4.237 288	741.415 9	43 743.54	286	3.496 503	898.495 5	64 242.43
237	4.219 409	744.557 5	44 115.03	287	3.484 321	901.637 1	64 692.46
238	4.201 681	747.699 1	44 488.09	288	3.472 222	904.778 7	65 144.07
239	4.184 100	750.840 6	44 862.73	289	3.460 208	907.920 3	65 597.24
240	4.166 667	753.982 2	45 238.93	290	3.448 276	911.061 9	66 051.99
241	4.149 378	757.123 8	45 616.71	291	3.436 426	914.203 5	66 508.30
242	4.132 231	760.265 4	45 996.06	292	3.424 658	917.345 1	66 966.19
243	4.115 226	763.407 0	46 376.98	293	3.412 969	920.486 6	67 425.65
244	4.098 361	766.548 6	46 759.47	294	3.401 361	923.628 2	67 886.68
245	4.081 633	769.690 2	47 143.52	295	3.389 831	926.769 8	68 349.28
246	4.065 041	772.831 8	47 529.16	296	3.378 378	929.911 4	68 813.45
247	4.048 583	775.973 4	47 916.36	297	3.367 003	933.053 0	69 279.19
248	4.032 258	779.115 0	48 305.13	298	3.355 705	936.194 6	69 746.50
249	4.016 064	782.256 6	48 695.47	299	3.344 482	939.336 2	70 215.38

Table 15 RECIPROCALS, CIRCUMFERENCES, AND AREAS OF CIRCLES (*continued*)

n	$1000/n$	Circum. of circle πn	Area of circle $\pi n^2/4$	n	$1000/n$	Circum. of circle πn	Area of circle $\pi n^2/4$
300	3.333 333	942.477 8	70 685.83	350	2.857 143	1 099.557	96 211.28
301	3.322 259	945.619 4	71 157.86	351	2.849 003	1 102.699	96 761.84
302	3.311 258	948.761 0	71 631.45	352	2.840 909	1 105.841	97 313.97
303	3.300 330	951.902 6	72 106.62	353	2.832 861	1 108.982	97 867.68
304	3.289 474	955.044 2	72 583.36	354	2.824 859	1 112.124	98 422.96
305	3.278 689	958.185 8	73 061.66	355	2.816 901	1 115.265	98 979.80
306	3.267 974	961.327 4	73 541.54	356	2.808 989	1 118.407	99 538.22
307	3.257 329	964.468 9	74 022.99	357	2.801 120	1 121.549	100 098.2
308	3.246 753	967.610 5	74 506.01	358	2.793 296	1 124.690	100 659.8
309	3.236 246	970.752 1	74 990.60	359	2.785 515	1 127.832	101 222.9
310	3.225 806	973.893 7	75 476.76	360	2.777 778	1 130.973	101 787.6
311	3.215 434	977.035 3	75 964.50	361	2.770 083	1 134.115	102 353.9
312	3.205 128	980.176 9	76 453.80	362	2.762 431	1 137.257	102 921.7
313	3.194 888	983.318 5	76 944.67	363	2.754 821	1 140.398	103 491.1
314	3.184 713	986.460 1	77 437.12	364	2.747 253	1 143.540	104 062.1
315	3.174 603	989.601 7	77 931.13	365	2.739 726	1 146.681	104 634.7
316	3.164 557	992.743 3	78 426.72	366	2.732 240	1 149.823	105 208.8
317	3.154 574	995.884 9	78 923.88	367	2.724 796	1 152.965	105 784.5
318	3.144 654	999.026 5	79 422.60	368	2.717 391	1 156.106	106 361.8
319	3.134 796	1 002.168	79 922.90	369	2.710 027	1 159.248	106 940.6
320	3.125 000	1 005.310	80 424.77	370	2.702 703	1 162.389	107 521.0
321	3.115 265	1 008.451	80 928.21	371	2.695 418	1 165.531	108 103.0
322	3.105 590	1 011.593	81 433.22	372	2.688 172	1 168.672	108 686.5
323	3.095 975	1 014.734	81 939.80	373	2.680 965	1 171.814	109 271.7
324	3.086 420	1 017.876	82 447.96	374	2.673 797	1 174.956	109 858.4
325	3.076 923	1 021.018	82 957.68	375	2.666 667	1 178.097	110 446.6
326	3.067 485	1 024.159	83 468.97	376	2.659 574	1 181.239	111 036.5
327	3.058 104	1 027.301	83 981.84	377	2.652 520	1 184.380	111 627.9
328	3.048 780	1 030.442	84 496.28	378	2.645 503	1 187.522	112 220.8
329	3.039 514	1 033.584	85 012.28	379	2.638 522	1 190.664	112 815.4
330	3.030 303	1 036.726	85 529.86	380	2.631 579	1 193.805	113 411.5
331	3.021 148	1 039.867	86 049.01	381	2.624 672	1 196.947	114 009.2
332	3.012 048	1 043.009	86 569.73	382	2.617 801	1 200.088	114 608.4
333	3.003 003	1 046.150	87 092.02	383	2.610 966	1 203.230	115 209.3
334	2.994 012	1 049.292	87 615.88	384	2.604 167	1 206.372	115 811.7
335	2.985 075	1 052.434	88 141.31	385	2.597 403	1 209.513	116 415.6
336	2.976 190	1 055.575	88 668.31	386	2.590 674	1 212.655	117 021.2
337	2.967 359	1 058.717	89 196.88	387	2.583 979	1 215.796	117 628.3
338	2.958 580	1 061.858	89 727.03	388	2.577 320	1 218.938	118 237.0
339	2.949 853	1 065.000	90 258.74	389	2.570 694	1 222.080	118 847.2
340	2.941 176	1 068.142	90 792.03	390	2.564 103	1 225.221	119 459.1
341	2.932 551	1 071.283	91 326.88	391	2.557 545	1 228.363	120 072.5
342	2.923 977	1 074.425	91 863.31	392	2.551 020	1 231.504	120 687.4
343	2.915 452	1 077.566	92 401.31	393	2.544 529	1 234.646	121 304.0
344	2.906 977	1 080.708	92 940.88	394	2.538 071	1 237.788	121 922.1
345	2.898 551	1 083.849	93 482.02	395	2.531 646	1 240.929	122 541.7
346	2.890 173	1 086.991	94 024.73	396	2.525 253	1 244.071	123 163.0
347	2.881 844	1 090.133	94 569.01	397	2.518 892	1 247.212	123 785.8
348	2.873 563	1 093.274	95 114.86	398	2.512 563	1 250.354	124 410.2
349	2.865 330	1 096.416	95 662.28	399	2.506 266	1 253.495	125 036.2

Table 15 RECIPROCALS, CIRCUMFERENCES, AND AREAS OF CIRCLES (*continued*)

n	$1000/n$	Circum. of circle πn	Area of circle $\pi n^2/4$	n	$1000/n$	Circum. of circle πn	Area of circle $\pi n^2/4$
400	2.500 000	1 256.637	125 663.7	450	2.222 222	1 413.717	159 043.1
401	2.493 766	1 259.779	126 292.8	451	2.217 295	1 416.858	159 750.8
402	2.487 562	1 262.920	126 923.5	452	2.212 389	1 420.000	160 460.0
403	2.481 390	1 266.062	127 555.7	453	2.207 506	1 423.141	161 170.8
404	2.475 248	1 269.203	128 189.5	454	2.202 643	1 426.283	161 883.1
405	2.469 136	1 272.345	128 824.9	455	2.197 802	1 429.425	162 597.1
406	2.463 054	1 275.487	129 461.9	456	2.192 982	1 432.566	163 312.6
407	2.457 002	1 278.628	130 100.4	457	2.188 184	1 435.708	164 029.6
408	2.450 980	1 281.770	130 740.5	458	2.183 406	1 438.849	164 748.3
409	2.444 988	1 284.911	131 382.2	459	2.178 649	1 441.991	165 468.5
410	2.439 024	1 288.053	132 025.4	460	2.173 913	1 445.133	166 190.3
411	2.433 090	1 291.195	132 670.2	461	2.169 197	1 448.274	166 913.6
412	2.427 184	1 294.336	133 316.6	462	2.164 502	1 451.416	167 638.5
413	2.421 308	1 297.478	133 964.6	463	2.159 827	1 454.557	168 365.0
414	2.415 459	1 300.619	134 614.1	464	2.155 172	1 457.699	169 093.1
415	2.409 639	1 303.761	135 265.2	465	2.150 538	1 460.841	169 822.7
416	2.403 846	1 306.903	135 917.9	466	2.145 923	1 463.982	170 553.9
417	2.398 082	1 310.044	136 572.1	467	2.141 328	1 467.124	171 286.7
418	2.392 344	1 313.186	137 227.9	468	2.136 752	1 470.265	172 021.0
419	2.386 635	1 316.327	137 885.3	469	2.132 196	1 473.407	172 757.0
420	2.380 952	1 319.469	138 544.2	470	2.127 660	1 476.549	173 494.5
421	2.375 297	1 322.611	139 204.8	471	2.123 142	1 479.690	174 233.5
422	2.369 668	1 325.752	139 866.8	472	2.118 644	1 482.832	174 974.1
423	2.364 066	1 328.894	140 530.5	473	2.114 165	1 485.973	175 716.3
424	2.358 491	1 332.035	141 195.7	474	2.109 705	1 489.115	176 460.1
425	2.352 941	1 335.177	141 862.5	475	2.105 263	1 492.257	177 205.5
426	2.347 418	1 338.318	142 530.9	476	2.100 840	1 495.398	177 952.4
427	2.341 920	1 341.460	143 200.9	477	2.096 436	1 498.540	178 700.9
428	2.336 449	1 344.602	143 872.4	478	2.092 050	1 501.681	179 450.9
429	2.331 002	1 347.743	144 545.5	479	2.087 683	1 504.823	180 202.5
430	2.325 581	1 350.885	145 220.1	480	2.083 333	1 507.964	180 955.7
431	2.320 186	1 354.026	145 896.3	481	2.079 002	1 511.106	181 710.5
432	2.314 815	1 357.168	146 574.1	482	2.074 689	1 514.248	182 466.8
433	2.309 469	1 360.310	147 253.5	483	2.070 393	1 517.389	183 224.8
434	2.304 147	1 363.451	147 934.5	484	2.066 116	1 520.531	183 984.2
435	2.298 851	1 366.593	148 617.0	485	2.061 856	1 523.672	184 745.3
436	2.293 578	1 369.734	149 301.0	486	2.057 613	1 526.814	185 507.9
437	2.288 330	1 372.876	149 986.7	487	2.053 388	1 529.956	186 272.1
438	2.283 105	1 376.018	150 673.9	488	2.049 180	1 533.097	187 037.9
439	2.277 904	1 379.159	151 362.7	489	2.044 990	1 536.239	187 805.2
440	2.272 727	1 382.301	152 053.1	490	2.040 816	1 539.380	188 574.1
441	2.267 574	1 385.442	152 745.0	491	2.036 660	1 542.522	189 344.6
442	2.262 443	1 388.584	153 438.5	492	2.032 520	1 545.664	190 116.6
443	2.257 336	1 391.726	154 133.6	493	2.028 398	1 548.805	190 890.2
444	2.252 252	1 394.867	154 830.3	494	2.024 291	1 551.947	191 665.4
445	2.247 191	1 398.009	155 528.5	495	2.020 202	1 555.088	192 442.2
446	2.242 152	1 401.150	156 228.3	496	2.016 129	1 558.230	193 220.5
447	2.237 136	1 404.292	156 929.6	497	2.012 072	1 561.372	194 000.4
448	2.232 143	1 407.434	157 632.6	498	2.008 032	1 564.513	194 781.9
449	2.227 171	1 410.575	158 337.1	499	2.004 008	1 567.655	195 564.9

Table 15 RECIPROCALS, CIRCUMFERENCES, AND AREAS OF CIRCLES (*continued*)

n	$1000/n$	Circum. of circle πn	Area of circle $\pi n^2/4$
500	2.000 000	1 570.796	196 349.5
501	1.996 008	1 573.938	197 135.7
502	1.992 032	1 577.080	197 923.5
503	1.988 072	1 580.221	198 712.8
504	1.984 127	1 583.363	199 503.7
505	1.980 198	1 586.504	200 296.2
506	1.976 285	1 589.646	201 090.2
507	1.972 387	1 592.787	201 885.8
508	1.968 504	1 595.929	202 683.0
509	1.964 637	1 599.071	203 481.7
510	1.960 784	1 602.212	204 282.1
511	1.956 947	1 605.354	205 084.0
512	1.953 125	1 608.495	205 887.4
513	1.949 318	1 611.637	206 692.4
514	1.945 525	1 614.779	207 499.1
515	1.941 748	1 617.920	208 307.2
516	1.937 984	1 621.062	209 117.0
517	1.934 236	1 624.203	209 928.3
518	1.930 502	1 627.345	210 741.2
519	1.926 782	1 630.487	211 555.6
520	1.923 077	1 633.628	212 371.7
521	1.919 386	1 636.770	213 189.3
522	1.915 709	1 639.911	214 008.4
523	1.912 046	1 643.053	214 829.2
524	1.908 397	1 646.195	215 651.5
525	1.904 762	1 649.336	216 475.4
526	1.901 141	1 652.478	217 300.8
527	1.897 533	1 655.619	218 127.8
528	1.893 939	1 658.761	218 956.4
529	1.890 359	1 661.903	219 786.6
530	1.886 792	1 665.044	220 618.3
531	1.883 239	1 668.186	221 451.7
532	1.879 699	1 671.327	222 286.5
533	1.876 173	1 674.469	223 123.0
534	1.872 659	1 677.610	223 961.0
535	1.869 159	1 680.752	224 800.6
536	1.865 672	1 683.894	225 641.8
537	1.862 197	1 687.035	226 484.5
538	1.858 736	1 690.177	227 328.8
539	1.855 288	1 693.318	228 174.7
540	1.851 852	1 696.460	229 022.1
541	1.848 429	1 699.602	229 871.1
542	1.845 018	1 702.743	230 721.7
543	1.841 621	1 705.885	231 573.9
544	1.838 235	1 709.026	232 427.6
545	1.834 862	1 712.168	233 282.9
546	1.831 502	1 715.310	234 139.8
547	1.828 154	1 718.451	234 998.2
548	1.824 818	1 721.593	235 858.2
549	1.821 494	1 724.734	236 719.8
550	1.818 182	1 727.876	237 582.9
551	1.814 882	1 731.018	238 447.7
552	1.811 594	1 734.159	239 314.0
553	1.808 318	1 737.301	240 181.8
554	1.805 054	1 740.442	241 051.3
555	1.801 802	1 743.584	241 922.3
556	1.798 561	1 746.726	242 794.8
557	1.795 332	1 749.867	243 669.0
558	1.792 115	1 753.009	244 544.7
559	1.788 909	1 756.150	245 422.0
560	1.785 714	1 759.292	246 300.9
561	1.782 531	1 762.433	247 181.3
562	1.779 359	1 765.575	248 063.3
563	1.776 199	1 768.717	248 946.9
564	1.773 050	1 771.858	249 832.0
565	1.769 912	1 775.000	250 718.7
566	1.766 784	1 778.141	251 607.0
567	1.763 668	1 781.283	252 496.9
568	1.760 563	1 784.425	253 388.3
569	1.757 469	1 787.566	254 281.3
570	1.754 386	1 790.708	255 175.9
571	1.751 313	1 793.849	256 072.0
572	1.748 252	1 796.991	256 969.7
573	1.745 201	1 800.133	257 869.0
574	1.742 160	1 803.274	258 769.8
575	1.739 130	1 806.416	259 672.3
576	1.736 111	1 809.557	260 576.3
577	1.733 102	1 812.699	261 481.8
578	1.730 104	1 815.841	262 389.0
579	1.727 116	1 818.982	263 297.7
580	1.724 138	1 822.124	264 207.9
581	1.721 170	1 825.265	265 119.8
582	1.718 213	1 828.407	266 033.2
583	1.715 266	1 831.549	266 948.2
584	1.712 329	1 834.690	267 864.8
585	1.709 402	1 837.832	268 782.9
586	1.706 485	1 840.973	269 702.6
587	1.703 578	1 844.115	270 623.9
588	1.700 680	1 847.256	271 546.7
589	1.697 793	1 850.398	272 471.1
590	1.694 915	1 853.540	273 397.1
591	1.692 047	1 856.681	274 324.7
592	1.689 189	1 859.823	275 253.8
593	1.686 341	1 862.964	276 184.5
594	1.683 502	1 866.106	277 116.7
595	1.680 672	1 869.248	278 050.6
596	1.677 852	1 872.389	278 986.0
597	1.675 042	1 875.531	279 923.0
598	1.672 241	1 878.672	280 861.5
599	1.669 449	1 881.814	281 801.6

Table 15 RECIPROCALS, CIRCUMFERENCES, AND AREAS OF CIRCLES (*continued*)

n	$1000/n$	Circum. of circle πn	Area of circle $\pi n^2/4$	n	$1000/n$	Circum. of circle πn	Area of circle $\pi n^2/4$
600	1.666 667	1 884.956	282 743.3	650	1.538 462	2 042.035	331 830.7
601	1.663 894	1 888.097	283 686.6	651	1.536 098	2 045.177	332 852.5
602	1.661 130	1 891.239	284 631.4	652	1.533 742	2 048.318	333 875.9
603	1.658 375	1 894.380	285 577.8	653	1.531 394	2 051.460	334 900.8
604	1.655 629	1 897.522	286 525.8	654	1.529 052	2.054.602	335 927.4
605	1.652 893	1 900.664	287 475.4	655	1.526 718	2 057.743	336 955.4
606	1.650 165	1 903.805	288 426.5	656	1.524 390	2 060.885	337 985.1
607	1.647 446	1 906.947	289 379.2	657	1.522 070	2 064.026	339 016.3
608	1.644 737	1 910.088	290 333.4	658	1.519 757	2 067.168	340 049.1
609	1.642 036	1 913.230	291 289.3	659	1.517 451	2 070.310	341 083.5
610	1.639 344	1 916.372	292 246.7	660	1.515 152	2 073.451	342 119.4
611	1.636 661	1 919.513	293 205.6	661	1.512 859	2 076.593	343 157.0
612	1.633 987	1 922.655	294 166.2	662	1.510 574	2 079.734	344 196.0
613	1.631 321	1 925.796	295 128.3	663	1.508 296	2 082.876	345 236.7
614	1.628 664	1 928.938	296 092.0	664	1.506 024	2 086.018	346 278.9
615	1.626 016	1 932.079	297 057.2	665	1.503 759	2 089.159	347 322.7
616	1.623 377	1 935.221	298 024.0	666	1.501 502	2 092.301	348 368.1
617	1.620 746	1 938.363	298 992.4	667	1.499 250	2 095.442	349 415.0
618	1.618 123	1 941.504	299 962.4	668	1.497 006	2 098.584	350 463.5
619	1.615 509	1 944.646	300 933.9	669	1.494 768	2 101.725	351 513.6
620	1.612 903	1 947.787	301 907.1	670	1.492 537	2 104.867	352 565.2
621	1.610 306	1 950.929	302 881.7	671	1.490 313	2 108.009	353 618.5
622	1.607 717	1 954.071	303 858.0	672	1.488 095	2 111.150	354 673.2
623	1.605 136	1 957.212	304 835.8	673	1.485 884	2 114.292	355 729.6
624	1.602 564	1 960.354	305 815.2	674	1.483 680	2 117.433	356 787.5
625	1.600 000	1 963.495	306 796.2	675	1.481 481	2 120.575	357 847.0
626	1.597 444	1 966.637	307 778.7	676	1.479 290	2 123.717	358 908.1
627	1.594 896	1 969.779	308 762.8	677	1.477 105	2 126.858	359 970.8
628	1.592 357	1 972.920	309 748.5	678	1.474 926	2 130.000	361 035.0
629	1.589 825	1 976.062	310 735.7	679	1.472 754	2 133.141	362 100.8
630	1.587 302	1 979.203	311 724.5	680	1.470 588	2 136.283	363 168.1
631	1.584 786	1 982.345	312 714.9	681	1.468 429	2 139.425	364 237.0
632	1.582 278	1 985.487	313 706.9	682	1.466 276	2 142.566	365 307.5
633	1.579 779	1 988.628	314 700.4	683	1.464 129	2 145.708	366 379.6
634	1.577 287	1 991.770	315 695.5	684	1.461 988	2 148.849	367 453.2
635	1.574 803	1 994.911	316 692.2	685	1.459 854	2 151.991	368 528.5
636	1.572 327	1 998.053	317 690.4	686	1.457 726	2 155.133	369 605.2
637	1.569 859	2 001.195	318 690.2	687	1.455 604	2 158.274	370 683.6
638	1.567 398	2 004.336	319 691.6	688	1.453 488	2 161.416	371 763.5
639	1.564 945	2 007.478	320 694.6	689	1.451 379	2 164.557	372 845.0
640	1.562 500	2 010.619	321 699.1	690	1.449 275	2 167.699	373 928.1
641	1.560 062	2 013.761	322 705.2	691	1.447 178	2 170.841	375 012.7
642	1.557 632	2 016.902	323 712.8	692	1.445 087	2 173.982	376 098.9
643	1.555 210	2 020.044	324 722.1	693	1.443 001	2 177.124	377 186.7
644	1.552 795	2 023.186	325 732.9	694	1.440 922	2 180.265	378 276.0
645	1.550 388	2 026.327	326 745.3	695	1.438 849	2 183.407	379 366.9
646	1.547 988	2 029.469	327 759.2	696	1.436 782	2 186.548	380 459.4
647	1.545 595	2 032.610	328 774.7	697	1.434 720	2 189.690	381 553.5
648	1.543 210	2 035.752	329 791.8	698	1.432 665	2 192.832	382 649.1
649	1.540 832	2 038.894	330 810.5	699	1.430 615	2 195.973	383 746.3

Table 15 RECIPROCALS, CIRCUMFERENCES, AND AREAS OF CIRCLES (*continued*)

n	$1000/n$	Circum. of circle πn	Area of circle $\pi n^2/4$	n	$1000/n$	Circum. of circle πn	Area of circle $\pi n^2/4$
700	1.428 571	2 199.115	384 845.1	750	1.333 333	2 356.194	441 786.5
701	1.426 534	2 202.256	385 945.4	751	1.331 558	2 359.336	442 965.3
702	1.424 501	2 205.398	387 047.4	752	1.329 787	2 362.478	444 145.8
703	1.422 475	2 208.540	388 150.8	753	1.328 021	2 365.619	445 327.8
704	1.420 455	2 211.681	389 255.9	754	1.326 260	2 368.761	446 511.4
705	1.418 440	2 214.823	390 362.5	755	1.324 503	2 371.902	447 696.6
706	1.416 431	2 217.964	391 470.7	756	1.322 751	2 375.044	448 883.3
707	1.414 427	2 221.106	392 580.5	757	1.321 004	2 378.186	450 071.6
708	1.412 429	2 224.248	393 691.8	758	1.319 261	2 381.327	451 261.5
709	1.410 437	2 227.389	394 804.7	759	1.317 523	2 384.469	452 453.0
710	1.408 451	2 230.531	395 919.2	760	1.315 789	2 387.610	453 646.0
711	1.406 470	2 233.672	397 035.3	761	1.314 060	2 390.752	454 840.6
712	1.404 494	2 236.814	398 152.9	762	1.312 336	2 393.894	456 036.7
713	1.402 525	2 239.956	399 272.1	763	1.310 616	2 397.035	457 234.5
714	1.400 560	2 243.097	400 392.8	764	1.308 901	2 400.177	458 433.8
715	1.398 601	2 246.239	401 515.2	765	1.307 190	2 403.318	459 634.6
716	1.396 648	2.249.380	402 639.1	766	1.305 483	2 406.460	460 837.1
717	1.394 700	2 252.522	403 764.6	767	1.303 781	2 409.602	462 041.1
718	1.392 758	2 255.664	404 891.6	768	1.302 083	2 412.743	463 246.7
719	1.390 821	2 258.805	406 020.2	769	1.300 390	2 415.885	464 453.8
720	1.388 889	2 261.947	407 150.4	770	1.298 701	2 419.026	465 662.6
721	1.386 963	2 265.088	408 282.2	771	1.297 017	2 422.168	466 872.9
722	1.385 042	2 268.230	409 415.5	772	1.295 337	2 425.310	468 084.7
723	1.383 126	2 271.371	410 550.4	773	1.293 661	2 428.451	469 298.2
724	1.381 215	2 274.513	411 686.9	774	1.291 990	2 431.593	470 513.2
725	1.379 310	2 277.655	412 824.9	775	1.290 323	2 434.734	471 729.8
726	1.377 410	2 280.796	413 964.5	776	1.288 660	2 437.876	472 947.9
727	1.375 516	2 283.938	415 105.7	777	1.287 001	2 441.017	474 167.6
728	1.373 626	2 287.079	416 248.5	778	1.285 347	2 444.159	475 388.9
729	1.371 742	2 290.221	417 392.8	779	1.283 697	2 447.301	476 611.8
730	1.369 863	2 293.363	418 538.7	780	1.282 051	2 450.442	477 836.2
731	1.367 989	2 296.504	419 686.1	781	1.280 410	2 453.584	479 062.2
732	1.366 120	2 299.646	420 835.2	782	1.278 772	2 456.725	480 289.8
733	1.364 256	2 302.787	421 985.8	783	1.277 139	2 459.867	481 519.0
734	1.362 398	2 305.929	423 138.0	784	1.275 510	2 463.009	482 749.7
735	1.360 544	2 309.071	424 291.7	785	1.273 885	2 466.150	483 982.0
736	1.358 696	2 312.212	425 447.0	786	1.272 265	2 469.292	485 215.8
737	1.356 852	2 315.354	426 603.9	787	1.270 648	2 472.433	486 451.3
738	1.355 014	2 318.495	427 762.4	788	1.269 036	2 475.575	487 688.3
739	1.353 180	2 321.637	428 922.4	789	1.267 427	2 478.717	488 926.9
740	1.351 351	2 324.779	430 084.0	790	1.265 823	2 481.858	490 167.0
741	1.349 528	2 327.920	431 247.2	791	1.264 223	2 485.000	491 408.7
742	1.347 709	2 331.062	432 412.0	792	1.262 626	2 488.141	492 652.0
743	1.345 895	2 334.203	433 578.3	793	1.261 034	2 491.283	493 896.8
744	1.344 086	2 337.345	434 746.2	794	1.259 446	2 494.425	495 143.3
745	1.342 282	2 340.487	435 915.6	795	1.257 862	2 497.566	496 391.3
746	1.340 483	2 343.628	437 086.6	796	1.256 281	2 500.708	497 640.8
747	1.338 688	2 346.770	438 259.2	797	1.254 705	2 503.849	498 892.0
748	1.336 898	2 349.911	439 433.4	798	1.253 133	2 506.991	500 144.7
749	1.335 113	2 353.053	440 609.2	799	1.251 564	2 510.133	501 399.0

STATISTICS AND PROBABILITY

Table 16 LOGARITHMS OF FACTORIAL *n*

n	log $n!$	n	log $n!$	n	log $n!$	n	log $n!$
		50	64.48307	100	157.97000	150	262.75689
1	0.00000	51	66.19064	101	159.97432	151	264.93587
2	0.30103	52	67.90665	102	161.98293	152	267.11771
3	0.77815	53	69.63092	103	163.99576	153	269.30241
4	1.38021	54	71.36332	104	166.01280	154	271.48993
5	2.07918	55	73.10368	105	168.03399	155	273.68026
6	2.85733	56	74.85187	106	170.05929	156	275.87338
7	3.70243	57	76.60774	107	172.08867	157	278.06928
8	4.60552	58	78.37117	108	174.12210	158	280.26794
9	5.55976	59	80.14202	109	176.15952	159	282.46934
10	6.55976	60	81.92017	110	178.20092	160	284.67346
11	7.60116	61	83.70550	111	180.24624	161	286.88028
12	8.68034	62	85.49790	112	182.29546	162	289.08980
13	9.79428	63	87.29724	113	184.34854	163	291.30198
14	10.94041	64	89.10342	114	186.40544	164	293.51683
15	12.11650	65	90.91633	115	188.46614	165	295.73431
16	13.32062	66	92.73587	116	190.53060	166	297.95442
17	14.55107	67	94.56195	117	192.59878	167	300.17714
18	15.80634	68	96.39446	118	194.67067	168	302.40245
19	17.08509	69	98.23331	119	196.74621	169	304.63033
20	18.38612	70	100.07840	120	198.82539	170	306.86078
21	19.70834	71	101.92966	121	200.90818	171	309.09378
22	21.05077	72	103.78700	122	202.99454	172	311.32931
23	22.41249	73	105.65032	123	205.08444	173	313.56735
24	23.79271	74	107.51955	124	207.17787	174	315.80790
25	25.19065	75	109.39461	125	209.27478	175	318.05094
26	26.60562	76	111.27543	126	211.37515	176	320.29645
27	28.03698	77	113.16192	127	213.47895	177	322.54443
28	29.48414	78	115.05401	128	215.58616	178	324.79485
29	30.94654	79	116.95164	129	217.69675	179	327.04770
30	32.42366	80	118.85473	130	219.81069	180	329.30297
31	33.91502	81	120.76321	131	221.92796	181	331.56065
32	35.42017	82	122.67703	132	224.04854	182	333.82072
33	36.93869	83	124.59610	133	226.17239	183	336.08317
34	38.47016	84	126.52038	134	228.29949	184	338.34799
35	40.01423	85	128.44980	135	230.42983	185	340.61516
36	41.57054	86	130.38430	136	232.56337	186	342.88468
37	43.13874	87	132.32382	137	234.70009	187	345.15652
38	44.71852	88	134.26830	138	236.83997	188	347.43067
39	46.30959	89	136.21769	139	238.98298	189	349.70714
40	47.91165	90	138.17194	140	241.12911	190	351.98589
41	49.52443	91	140.13098	141	243.27833	191	354.26692
42	51.14768	92	142.09476	142	245.43062	192	356.55022
43	52.78115	93	144.06325	143	247.58595	193	358.83578
44	54.42460	94	146.03638	144	249.74432	194	361.12358
45	56.07781	95	148.01410	145	251.90568	195	363.41362
46	57.74057	96	149.99637	146	254.07004	196	365.70587
47	59.41267	97	151.98314	147	256.23735	197	368.00034
48	61.09391	98	153.97437	148	258.40762	198	370.29701
49	62.78410	99	155.97000	149	260.58080	199	372.59586

FACTORIALS AND THEIR RECIPROCALS

n	$n!$	n	$n!$
1	1	11	39916800
2	2	12	479001600
3	6	13	6227020800
4	24	14	87178291200
5	120	15	1307674368000
6	720	16	20922789888000
7	5040	17	355687428096000
8	40320	18	6402373705728000
9	362880	19	121645100408832000
10	3628800	20	2432902008176640000

n	$1/n!$	n	$1/n!$
1	1.	11	$.25052 \times 10^{-7}$
2	0.5	12	$.20877 \times 10^{-8}$
3	.16667	13	$.16059 \times 10^{-9}$
4	$.41667 \times 10^{-1}$	14	$.11471 \times 10^{-10}$
5	$.83333 \times 10^{-2}$	15	$.76472 \times 10^{-12}$
6	$.13889 \times 10^{-2}$	16	$.47795 \times 10^{-13}$
7	$.19841 \times 10^{-3}$	17	$.28115 \times 10^{-14}$
8	$.24802 \times 10^{-4}$	18	$.15619 \times 10^{-15}$
9	$.27557 \times 10^{-5}$	19	$.82206 \times 10^{-17}$
10	$.27557 \times 10^{-6}$	20	$.41103 \times 10^{-18}$

Table 17 BINOMIAL COEFFICIENTS

n	$\binom{n}{0}$	$\binom{n}{1}$	$\binom{n}{2}$	$\binom{n}{3}$	$\binom{n}{4}$	$\binom{n}{5}$	$\binom{n}{6}$	$\binom{n}{7}$	$\binom{n}{8}$	$\binom{n}{9}$	$\binom{n}{10}$
0	1										
1	1	1									
2	1	2	1								
3	1	3	3	1							
4	1	4	6	4	1						
5	1	5	10	10	5	1					
6	1	6	15	20	15	6	1				
7	1	7	21	35	35	21	7	1			
8	1	8	28	56	70	56	28	8	1		
9	1	9	36	84	126	126	84	36	9	1	
10	1	10	45	120	210	252	210	120	45	10	1
11	1	11	55	165	330	462	462	330	165	55	11
12	1	12	66	220	495	792	924	792	495	220	66
13	1	13	78	286	715	1287	1716	1716	1287	715	286
14	1	14	91	364	1001	2002	3003	3432	3003	2002	1001
15	1	15	105	455	1365	3003	5005	6435	6435	5005	3003
16	1	16	120	560	1820	4368	8008	11440	12870	11440	8008
17	1	17	136	680	2380	6188	12376	19448	24310	24310	19448
18	1	18	153	816	3060	8568	18564	31824	43758	48620	43758
19	1	19	171	969	3876	11628	27132	50388	75582	92378	92378
20	1	20	190	1140	4845	15504	38760	77520	125970	167960	184756

$$ {}_nC_m = \binom{n}{m} = \frac{n!}{(n-m)!m!} = \binom{n}{n-m} \qquad \binom{n}{0} = 1 $$

$$ (p+q)^n = p^n + \binom{n}{1}p^{n-1}q + \cdots + \binom{n}{s}p^s q^t + \cdots + q^n \qquad s+t=n $$

PROBABILITY

Let p be the probability that an event E will happen in a single trial and $q = 1 - p$ the probability that the event will fail in a single trial. The probability that E will happen exactly r times in n trials is $\binom{n}{r}p^r q^{n-r}$. The probability that E will occur at least r times in n trials is $\sum_{i=r}^{i=n}\binom{n}{i}p^i q^{n-i}$; at most r times in n trials is $\sum_{i=0}^{i=r}\binom{n}{i}p^i q^{n-i}$.

In a *binomial* (*Bernoulli*) *distribution* $(q+p)^n$, the mean number $\bar{x}$ of favorable events is np; the mean number of unfavorable events is nq, where $q + p = 1$; the *standard deviation* is $\sigma = \sqrt{pqn}$.

If a variable y is *normally* distributed (*Gaussian distribution*), the probability P that y will lie between $y = y_1$ and $y = y_2$ is

$$ P = \int_{y_1}^{y_2} p(y)\,dy \qquad \text{where } p(y) = \frac{1}{\sigma\sqrt{2\pi}}e^{-(y-a)^2/2\sigma^2} $$

a is the *mean* of the distribution, and σ is the *standard deviation* (*root mean square*) of the distribution. If $x = (y-a)/\sigma$, $P = \int_{x_1}^{x_2}\Phi(x)\,dx$, where $\Phi(x) = \frac{1}{\sqrt{2\pi}}e^{-x^2/2}$ is the *normal function*, and x_i is the value of x when $y = y_i$. (See Chap. 22.)

For further details see Richard S. Burington and Donald C. May, *Handbook of Probability and Statistics with Tables*, 2d ed., McGraw-Hill Book Company, New York, 1970.

Table 18 PROBABILITY FUNCTIONS (NORMAL DISTRIBUTION)

x	$\frac{1}{2}(1+\alpha)$	$\Phi(x)$	$\Phi^{(2)}(x)$	$\Phi^{(3)}(x)$	$\Phi^{(4)}(x)$	x	$\frac{1}{2}(1+\alpha)$	$\Phi(x)$	$\Phi^{(2)}(x)$	$\Phi^{(3)}(x)$	$\Phi^{(4)}(x)$
0.00	.5000	.3989	−.3989	.0000	1.1968	0.50	.6915	.3521	−.2641	.4841	.5501
0.01	.5040	.3989	−.3989	.0120	1.1965	0.51	.6950	.3503	−.2592	.4895	.5279
0.02	.5080	.3989	−.3987	.0239	1.1956	0.52	.6985	.3485	−.2543	.4947	.5056
0.03	.5120	.3988	−.3984	.0359	1.1941	0.53	.7019	.3467	−.2493	.4996	.4831
0.04	.5160	.3986	−.3980	.0478	1.1920	0.54	.7054	.3448	−.2443	.5043	.4605
0.05	.5199	.3984	−.3975	.0597	1.1894	0.55	.7088	.3429	−.2392	.5088	.4378
0.06	.5239	.3982	−.3968	.0716	1.1861	0.56	.7123	.3410	−.2341	.5131	.4150
0.07	.5279	.3980	−.3960	.0834	1.1822	0.57	.7157	.3391	−.2289	.5171	.3921
0.08	.5319	.3977	−.3951	.0952	1.1778	0.58	.7190	.3372	−.2238	.5209	.3691
0.09	.5359	.3973	−.3941	.1070	1.1727	0.59	.7224	.3352	−.2185	.5245	.3461
0.10	.5398	.3970	−.3930	.1187	1.1671	0.60	.7257	.3332	−.2133	.5278	.3231
0.11	.5438	.3965	−.3917	.1303	1.1609	0.61	.7291	.3312	−.2080	.5309	.3000
0.12	.5478	.3961	−.3904	.1419	1.1541	0.62	.7324	.3292	−.2027	.5338	.2770
0.13	.5517	.3956	−.3889	.1534	1.1468	0.63	.7357	.3271	−.1973	.5365	.2539
0.14	.5557	.3951	−.3873	.1648	1.1389	0.64	.7389	.3251	−.1919	.5389	.2309
0.15	.5596	.3945	−.3856	.1762	1.1304	0.65	.7422	.3230	−.1865	.5411	.2078
0.16	.5636	.3939	−.3838	.1874	1.1214	0.66	.7454	.3209	−.1811	.5431	.1849
0.17	.5675	.3932	−.3819	.1986	1.1118	0.67	.7486	.3187	−.1757	.5448	.1620
0.18	.5714	.3925	−.3798	.2097	1.1017	0.68	.7517	.3166	−.1702	.5463	.1391
0.19	.5753	.3918	−.3777	.2206	1.0911	0.69	.7549	.3144	−.1647	.5476	.1164
0.20	.5793	.3910	−.3754	.2315	1.0799	0.70	.7580	.3123	−.1593	.5486	.0937
0.21	.5832	.3902	−.3730	.2422	1.0682	0.71	.7611	.3101	−.1538	.5495	.0712
0.22	.5871	.3894	−.3706	.2529	1.0560	0.72	.7642	.3079	−.1483	.5501	.0487
0.23	.5910	.3885	−.3680	.2634	1.0434	0.73	.7673	.3056	−.1428	.5504	.0265
0.24	.5948	.3876	−.3653	.2737	1.0302	0.74	.7704	.3034	−.1373	.5506	.0043
0.25	.5987	.3867	−.3625	.2840	1.0165	0.75	.7734	.3011	−.1318	.5505	−.0176
0.26	.6026	.3857	−.3596	.2941	1.0024	0.76	.7764	.2989	−.1262	.5502	−.0394
0.27	.6064	.3847	−.3566	.3040	0.9878	0.77	.7794	.2966	−.1207	.5497	−.0611
0.28	.6103	.3836	−.3535	.3138	0.9727	0.78	.7823	.2943	−.1153	.5490	−.0825
0.29	.6141	.3825	−.3504	.3235	0.9572	0.79	.7852	.2920	−.1098	.5481	−.1037
0.30	.6179	.3814	−.3471	.3330	0.9413	0.80	.7881	.2897	−.1043	.5469	−.1247
0.31	.6217	.3802	−.3437	.3423	0.9250	0.81	.7910	.2874	−.0988	.5456	−.1455
0.32	.6255	.3790	−.3402	.3515	0.9082	0.82	.7939	.2850	−.0934	.5440	−.1660
0.33	.6293	.3778	−.3367	.3605	0.8910	0.83	.7967	.2827	−.0880	.5423	−.1862
0.34	.6331	.3765	−.3330	.3693	0.8735	0.84	.7995	.2803	−.0825	.5403	−.2063
0.35	.6368	.3752	−.3293	.3779	0.8556	0.85	.8023	.2780	−.0771	.5381	−.2260
0.36	.6406	.3739	−.3255	.3864	0.8373	0.86	.8051	.2756	−.0718	.5358	−.2455
0.37	.6443	.3725	−.3216	.3947	0.8186	0.87	.8078	.2732	−.0664	.5332	−.2646
0.38	.6480	.3712	−.3176	.4028	0.7996	0.88	.8106	.2709	−.0611	.5305	−.2835
0.39	.6517	.3697	−.3135	.4107	0.7803	0.89	.8133	.2685	−.0558	.5276	−.3021
0.40	.6554	.3683	−.3094	.4184	0.7607	0.90	.8159	.2661	−.0506	.5245	−.3203
0.41	.6591	.3668	−.3059	.4259	0.7408	0.91	.8186	.2637	−.0453	.5212	−.3383
0.42	.6628	.3653	−.3008	.4332	0.7206	0.92	.8212	.2613	−.0401	.5177	−.3559
0.43	.6664	.3637	−.2965	.4403	0.7001	0.93	.8238	.2589	−.0350	.5140	−.3731
0.44	.6700	.3621	−.2920	.4472	0.6793	0.94	.8264	.2565	−.0299	.5102	−.3901
0.45	.6736	.3605	−.2875	.4539	0.6583	0.95	.8289	.2541	−.0248	.5062	−.4066
0.46	.6772	.3589	−.2830	.4603	0.6371	0.96	.8315	.2516	−.0197	.5021	−.4228
0.47	.6808	.3572	−.2783	.4666	0.6156	0.97	.8340	.2492	−.0147	.4978	−.4387
0.48	.6844	.3555	−.2736	.4727	0.5940	0.98	.8365	.2468	−.0098	.4933	−.4541
0.49	.6879	.3538	−.2689	.4785	0.5721	0.99	.8389	.2444	−.0049	.4887	−.4692

$$\tfrac{1}{2}(1+\alpha) = \int_{-\infty}^{x} \Phi(x)\,dx = \text{area under } \Phi(x) \text{ from } -\infty \text{ to } x$$

$$\alpha = \int_{-x}^{x} \Phi(x)\,dx \qquad \Phi(x) = \frac{1}{\sqrt{2\pi}} e^{-x^2/2} = \text{normal function}$$

$$\Phi^{(2)}(x) = (x^2 - 1)\,\Phi(x) = \text{second derivative of } \Phi(x)$$
$$\Phi^{(3)}(x) = (3x - x^3)\,\Phi(x) = \text{third derivative of } \Phi(x)$$
$$\Phi^{(4)}(x) = (x^4 - 6x^2 + 3)\,\Phi(x) = \text{fourth derivative of } \Phi(x)$$

x	$\frac{1}{2}(1+\alpha)$	$\Phi(x)$	$\Phi^{(2)}(x)$	$\Phi^{(3)}(x)$	$\Phi^{(4)}(x)$	x	$\frac{1}{2}(1+\alpha)$	$\Phi(x)$	$\Phi^{(2)}(x)$	$\Phi^{(3)}(x)$	$\Phi^{(4)}(x)$
1.00	.8413	.2420	.0000	.4839	−.4839	1.50	.9332	.1295	.1619	.1457	−.7043
1.01	.8438	.2396	.0048	.4790	−.4983	1.51	.9345	.1276	.1633	.1387	−.6994
1.02	.8461	.2371	.0096	.4740	−.5122	1.52	.9357	.1257	.1647	.1317	−.6942
1.03	.8485	.2347	.0143	.4688	−.5257	1.53	.9370	.1238	.1660	.1248	−.6888
1.04	.8508	.2323	.0190	.4635	−.5389	1.54	.9382	.1219	.1672	.1180	−.6831
1.05	.8531	.2299	.0236	.4580	−.5516	1.55	.9394	.1200	.1683	.1111	−.6772
1.06	.8554	.2275	.0281	.4524	−.5639	1.56	.9406	.1182	.1694	.1044	−.6710
1.07	.8577	.2251	.0326	.4467	−.5758	1.57	.9418	.1163	.1704	.0977	−.6646
1.08	.8599	.2227	.0371	.4409	−.5873	1.58	.9429	.1145	.1714	.0911	−.6580
1.09	.8621	.2203	.0414	.4350	−.5984	1.59	.9441	.1127	.1722	.0846	−.6511
1.10	.8643	.2179	.0458	.4290	−.6091	1.60	.9452	.1109	.1730	.0781	−.6441
1.11	.8665	.2155	.0500	.4228	−.6193	1.61	.9463	.1092	.1738	.0717	−.6368
1.12	.8686	.2131	.0542	.4166	−.6292	1.62	.9474	.1074	.1745	.0654	−.6293
1.13	.8708	.2107	.0583	.4102	−.6386	1.63	.9484	.1057	.1751	.0591	−.6216
1.14	.8729	.2083	.0624	.4038	−.6476	1.64	.9495	.1040	.1757	.0529	−.6138
1.15	.8749	.2059	.0664	.3973	−.6561	1.65	.9505	.1023	.1762	.0468	−.6057
1.16	.8770	.2036	.0704	.3907	−.6643	1.66	.9515	.1006	.1766	.0408	−.5975
1.17	.8790	.2012	.0742	.3840	−.6720	1.67	.9525	.0989	.1770	.0349	−.5891
1.18	.8810	.1989	.0780	.3772	−.6792	1.68	.9535	.0973	.1773	.0290	−.5806
1.19	.8830	.1965	.0818	.3704	−.6861	1.69	.9545	.0957	.1776	.0233	−.5720
1.20	.8849	.1942	.0854	.3635	−.6926	1.70	.9554	.0940	.1778	.0176	−.5632
1.21	.8869	.1919	.0890	.3566	−.6986	1.71	.9564	.0925	.1779	.0120	−.5542
1.22	.8888	.1895	.0926	.3496	−.7042	1.72	.9573	.0909	.1780	.0065	−.5452
1.23	.8907	.1872	.0960	.3425	−.7094	1.73	.9582	.0893	.1780	.0011	−.5360
1.24	.8925	.1849	.0994	.3354	−.7141	1.74	.9591	.0878	.1780	−.0042	−.5267
1.25	.8944	.1826	.1027	.3282	−.7185	1.75	.9599	.0863	.1780	−.0094	−.5173
1.26	.8962	.1804	.1060	.3210	−.7224	1.76	.9608	.0848	.1778	−.0146	−.5079
1.27	.8980	.1781	.1092	.3138	−.7259	1.77	.9616	.0833	.1777	−.0196	−.4983
1.28	.8997	.1758	.1123	.3065	−.7291	1.78	.9625	.0818	.1774	−.0245	−.4887
1.29	.9015	.1736	.1153	.2992	−.7318	1.79	.9633	.0804	.1772	−.0294	−.4789
1.30	.9032	.1714	.1182	.2918	−.7341	1.80	.9641	.0790	.1769	−.0341	−.4692
1.31	.9049	.1691	.1211	.2845	−.7361	1.81	.9649	.0775	.1765	−.0388	−.4593
1.32	.9066	.1669	.1239	.2771	−.7376	1.82	.9656	.0761	.1761	−.0433	−.4494
1.33	.9082	.1647	.1267	.2697	−.7388	1.83	.9664	.0748	.1756	−.0477	−.4395
1.34	.9099	.1626	.1293	.2624	−.7395	1.84	.9671	.0734	.1751	−.0521	−.4295
1.35	.9115	.1604	.1319	.2550	−.7399	1.85	.9678	.0721	.1746	−.0563	−.4195
1.36	.9131	.1582	.1344	.2476	−.7400	1.86	.9686	.0707	.1740	−.0605	−.4095
1.37	.9147	.1561	.1369	.2402	−.7396	1.87	.9693	.0694	.1734	−.0645	−.3995
1.38	.9162	.1539	.1392	.2328	−.7389	1.88	.9699	.0681	.1727	−.0685	−.3894
1.39	.9177	.1518	.1415	.2254	−.7378	1.89	.9706	.0669	.1720	−.0723	−.3793
1.40	.9192	.1497	.1437	.2180	−.7364	1.90	.9713	.0656	.1713	−.0761	−.3693
1.41	.9207	.1476	.1459	.2107	−.7347	1.91	.9719	.0644	.1705	−.0797	−.3592
1.42	.9222	.1456	.1480	.2033	−.7326	1.92	.9726	.0632	.1697	−.0832	−.3492
1.43	.9236	.1435	.1500	.1960	−.7301	1.93	.9732	.0620	.1688	−.0867	−.3392
1.44	.9251	.1415	.1519	.1887	−.7274	1.94	.9738	.0608	.1679	−.0900	−.3292
1.45	.9265	.1394	.1537	.1815	−.7243	1.95	.9744	.0596	.1670	−.0933	−.3192
1.46	.9279	.1374	.1555	.1742	−.7209	1.96	.9750	.0584	.1661	−.0964	−.3093
1.47	.9292	.1354	.1572	.1670	−.7172	1.97	.9756	.0573	.1651	−.0994	−.2994
1.48	.9306	.1334	.1588	.1599	−.7132	1.98	.9761	.0562	.1641	−.1024	−.2895
1.49	.9319	.1315	.1604	.1528	−.7089	1.99	.9767	.0551	.1630	−.1052	−.2797

The sum of those terms of $(p+q)^n \equiv \sum_{t=0}^{n} \binom{n}{t} p^{n-t}q^t$, $p+q=1$, in which t ranges from a to b inclusive, a and b being integers ($a \leqq t \leqq b$), is (if n is large enough) approximately

$$\int_{x_1}^{x_2} \Phi(x)\,dx + \left[\frac{q-p}{6\sigma}\Phi^{(2)}(x) + \frac{1}{24}\left(\frac{1}{\sigma^2} - \frac{6}{n}\right)\Phi^{(3)}(x)\right]_{x_1}^{x_2}$$

where $x_1 = (a - \frac{1}{2} - nq)/\sigma$, $x_2 = (b + \frac{1}{2} - nq)/\sigma$, and $\sigma = \sqrt{npq}$.

x	$\frac{1}{2}(1+\alpha)$	$\Phi(x)$	$\Phi^{(2)}(x)$	$\Phi^{(3)}(x)$	$\Phi^{(4)}(x)$	x	$\frac{1}{2}(1+\alpha)$	$\Phi(x)$	$\Phi^{(2)}(x)$	$\Phi^{(3)}(x)$	$\Phi^{(4)}(x)$
2.00	.9772	.0540	.1620	−.1080	−.2700	2.50	.9938	.0175	.0920	−.1424	.0800
2.01	.9778	.0529	.1609	−.1106	−.2603	2.51	.9940	.0171	.0906	−.1416	.0836
2.02	.9783	.0519	.1598	−.1132	−.2506	2.52	.9941	.0167	.0892	−.1408	.0871
2.03	.9788	.0508	.1586	−.1157	−.2411	2.53	.9943	.0163	.0878	−.1399	.0905
2.04	.9793	.0498	.1575	−.1180	−.2316	2.54	.9945	.0158	.0864	−.1389	.0937
2.05	.9798	.0488	.1563	−.1203	−.2222	2.55	.9946	.0154	.0850	−.1380	.0968
2.06	.9803	.0478	.1550	−.1225	−.2129	2.56	.9948	.0151	.0836	−.1370	.0998
2.07	.9808	.0468	.1538	−.1245	−.2036	2.57	.9949	.0147	.0823	−.1360	.1027
2.08	.9812	.0459	.1526	−.1265	−.1945	2.58	.9951	.0143	.0809	−.1350	.1054
2.09	.9817	.0449	.1513	−.1284	−.1854	2.59	.9952	.0139	.0796	−.1339	.1080
2.10	.9821	.0440	.1500	−.1302	−.1765	2.60	.9953	.0136	.0782	−.1328	.1105
2.11	.9826	.0431	.1487	−.1320	−.1676	2.61	.9955	.0132	.0769	−.1317	.1129
2.12	.9830	.0422	.1474	−.1336	−.1588	2.62	.9956	.0129	.0756	−.1305	.1152
2.13	.9834	.0413	.1460	−.1351	−.1502	2.63	.9957	.0126	.0743	−.1294	.1173
2.14	.9838	.0404	.1446	−.1366	−.1416	2.64	.9959	.0122	.0730	−.1282	.1194
2.15	.9842	.0395	.1433	−.1380	−.1332	2.65	.9960	.0119	.0717	−.1270	.1213
2.16	.9846	.0387	.1419	−.1393	−.1249	2.66	.9961	.0116	.0705	−.1258	.1231
2.17	.9850	.0379	.1405	−.1405	−.1167	2.67	.9962	.0113	.0692	−.1245	.1248
2.18	.9854	.0371	.1391	−.1416	−.1086	2.68	.9963	.0110	.0680	−.1233	.1264
2.19	.9857	.0363	.1377	−.1426	−.1006	2.69	.9964	.0107	.0668	−.1220	.1279
2.20	.9861	.0355	.1362	−.1436	−.0927	2.70	.9965	.0104	.0656	−.1207	.1293
2.21	.9864	.0347	.1348	−.1445	−.0850	2.71	.9966	.0101	.0644	−.1194	.1306
2.22	.9868	.0339	.1333	−.1453	−.0774	2.72	.9967	.0099	.0632	−.1181	.1317
2.23	.9871	.0332	.1319	−.1460	−.0700	2.73	.9968	.0096	.0620	−.1168	.1328
2.24	.9875	.0325	.1304	−.1467	−.0626	2.74	.9969	.0093	.0608	−.1154	.1338
2.25	.9878	.0317	.1289	−.1473	−.0554	2.75	.9970	.0091	.0597	−.1141	.1347
2.26	.9881	.0310	.1275	−.1478	−.0484	2.76	.9971	.0088	.0585	−.1127	.1356
2.27	.9884	.0303	.1260	−.1483	−.0414	2.77	.9972	.0086	.0574	−.1114	.1363
2.28	.9887	.0297	.1245	−.1486	−.0346	2.78	.9973	.0084	.0563	−.1100	.1369
2.29	.9890	.0290	.1230	−.1490	−.0279	2.79	.9974	.0081	.0552	−.1087	.1375
2.30	.9893	.0283	.1215	−.1492	−.0214	2.80	.9974	.0079	.0541	−.1073	.1379
2.31	.9896	.0277	.1200	−.1494	−.0150	2.81	.9975	.0077	.0531	−.1059	.1383
2.32	.9898	.0270	.1185	−.1495	−.0088	2.82	.9976	.0075	.0520	−.1045	.1386
2.33	.9901	.0264	.1170	−.1496	−.0027	2.83	.9977	.0073	.0510	−.1031	.1389
2.34	.9904	.0258	.1155	−.1496	.0033	2.84	.9977	.0071	.0500	−.1017	.1390
2.35	.9906	.0252	.1141	−.1495	.0092	2.85	.9978	.0069	.0490	−.1003	.1391
2.36	.9909	.0246	.1126	−.1494	.0149	2.86	.9979	.0067	.0480	−.0990	.1391
2.37	.9911	.0241	.1111	−.1492	.0204	2.87	.9979	.0065	.0470	−.0976	.1391
2.38	.9913	.0235	.1096	−.1490	.0258	2.88	.9980	.0063	.0460	−.0962	.1389
2.39	.9916	.0229	.1081	−.1487	.0311	2.89	.9981	.0061	.0451	−.0948	.1388
2.40	.9918	.0224	.1066	−.1483	.0362	2.90	.9981	.0060	.0441	−.0934	.1385
2.41	.9920	.0219	.1051	−.1480	.0412	2.91	.9982	.0058	.0432	−.0920	.1382
2.42	.9922	.0213	.1036	−.1475	.0461	2.92	.9982	.0056	.0423	−.0906	.1378
2.43	.9925	.0208	.1022	−.1470	.0508	2.93	.9983	.0055	.0414	−.0893	.1374
2.44	.9927	.0203	.1007	−.1465	.0554	2.94	.9984	.0053	.0405	−.0879	.1369
2.45	.9929	.0198	.0992	−.1459	.0598	2.95	.9984	.0051	.0396	−.0865	.1364
2.46	.9931	.0194	.0978	−.1453	.0641	2.96	.9985	.0050	.0388	−.0852	.1358
2.47	.9932	.0189	.0963	−.1446	.0683	2.97	.9985	.0048	.0379	−.0838	.1352
2.48	.9934	.0184	.0949	−.1439	.0723	2.98	.9986	.0047	.0371	−.0825	.1345
2.49	.9936	.0180	.0935	−.1432	.0762	2.99	.9986	.0046	.0363	−.0811	.1337

The sum of the first $(t+1)$ terms of $(p+q)^n \equiv \sum_{t=0}^{n} \binom{n}{t} p^{n-t}q^t$, $p+q=1$, is approximately

$$\int_x^\infty \Phi(x)\,dx + \frac{q-p}{6\sigma}\Phi^{(2)}(x) - \frac{1}{24}\left(\frac{1}{\sigma^2} - \frac{6}{n}\right)\Phi^{(3)}(x)$$

where $x = (s - \frac{1}{2} - np)/\sigma$, $s = n - t$. The sum of the last $(s+1)$ terms is approximately

$$\int_x^\infty \Phi(x)\,dx - \frac{q-p}{6\sigma}\Phi^{(2)}(x) - \frac{1}{24}\left(\frac{1}{\sigma^2} - \frac{6}{n}\right)\Phi^{(3)}(x)$$

where $x = (t - \frac{1}{2} - nq)/\sigma$, $t = n - s$, and $\sigma = \sqrt{npq}$.

x	$\frac{1}{2}(1+\alpha)$	$\Phi(x)$	$\Phi^{(2)}(x)$	$\Phi^{(3)}$	$\Phi^{(4)}(x)$	x	$\frac{1}{2}(1+\alpha)$	$\Phi(x)$	$\Phi^{(2)}(x)$	$\Phi^{(3)}(x)$	$\Phi^{(4)}(x)$
3.00	.9987	.0044	.0355	−.0798	.1330	3.50	.9998	.0009	.0098	−.0283	.0694
3.01	.9987	.0043	.0347	−.0785	.1321	3.51	.9998	.0008	.0095	−.0276	.0681
3.02	.9987	.0042	.0339	−.0771	.1313	3.52	.9998	.0008	.0093	−.0269	.0669
3.03	.9988	.0040	.0331	−.0758	.1304	3.53	.9998	.0008	.0090	−.0262	.0656
3.04	.9988	.0039	.0324	−.0745	.1294	3.54	.9998	.0008	.0087	−.0256	.0643
3.05	.9989	.0038	.0316	−.0732	.1285	3.55	.9998	.0007	.0085	−.0249	.0631
3.06	.9989	.0037	.0309	−.0720	.1275	3.56	.9998	.0007	.0082	−.0243	.0618
3.07	.9989	.0036	.0302	−.0707	.1264	3.57	.9998	.0007	.0080	−.0237	.0606
3.08	.9990	.0035	.0295	−.0694	.1254	3.58	.9998	.0007	.0078	−.0231	.0594
3.09	.9990	.0034	.0288	−.0682	.1243	3.59	.9998	.0006	.0075	−.0225	.0582
3.10	.9990	.0033	.0281	−.0669	.1231	3.60	.9998	.0006	.0073	−.0219	.0570
3.11	.9991	.0032	.0275	−.0657	.1220	3.61	.9998	.0006	.0071	−.0214	.0559
3.12	.9991	.0031	.0268	−.0645	.1208	3.62	.9999	.0006	.0069	−.0208	.0547
3.13	.9991	.0030	.0262	−.0633	.1196	3.63	.9999	.0005	.0067	−.0203	.0536
3.14	.9992	.0029	.0256	−.0621	.1184	3.64	.9999	.0005	.0065	−.0198	.0524
3.15	.9992	.0028	.0249	−.0609	.1171	3.65	.9999	.0005	.0063	−.0192	.0513
3.16	.9992	.0027	.0243	−.0598	.1159	3.66	.9999	.0005	.0061	−.0187	.0502
3.17	.9992	.0026	.0237	−.0586	.1146	3.67	.9999	.0005	.0059	−.0182	.0492
3.18	.9993	.0025	.0232	−.0575	.1133	3.68	.9999	.0005	.0057	−.0177	.0481
3.19	.9993	.0025	.0226	−.0564	.1120	3.69	.9999	.0004	.0056	−.0173	.0470
3.20	.9993	.0024	.0220	−.0552	.1107	3.70	.9999	.0004	.0054	−.0168	.0460
3.21	.9993	.0023	.0215	−.0541	.1093	3.71	.9999	.0004	.0052	−.0164	.0450
3.22	.9994	.0022	.0210	−.0531	.1080	3.72	.9999	.0004	.0051	−.0159	.0440
3.23	.9994	.0022	.0204	−.0520	.1066	3.73	.9999	.0004	.0049	−.0155	.0430
3.24	.9994	.0021	.0199	−.0509	.1053	3.74	.9999	.0004	.0048	−.0150	.0420
3.25	.9994	.0020	.0194	−.0499	.1039	3.75	.9999	.0004	.0046	−.0146	.0410
3.26	.9994	.0020	.0189	−.0488	.1025	3.76	.9999	.0003	.0045	−.0142	.0401
3.27	.9995	.0019	.0184	−.0478	.1011	3.77	.9999	.0003	.0043	−.0138	.0392
3.28	.9995	.0018	.0180	−.0468	.0997	3.78	.9999	.0003	.0042	−.0134	.0382
3.29	.9995	.0018	.0175	−.0458	.0983	3.79	.9999	.0003	.0041	−.0131	.0373
3.30	.9995	.0017	.0170	−.0449	.0969	3.80	.9999	.0003	.0039	−.0127	.0365
3.31	.9995	.0017	.0166	−.0439	.0955	3.81	.9999	.0003	.0038	−.0123	.0356
3.32	.9995	.0016	.0162	−.0429	.0941	3.82	.9999	.0003	.0037	−.0120	.0347
3.33	.9996	.0016	.0157	−.0420	.0927	3.83	.9999	.0003	.0036	−.0116	.0339
3.34	.9996	.0015	.0153	−.0411	.0913	3.84	.9999	.0003	.0034	−.0113	.0331
3.35	.9996	.0015	.0149	−.0402	.0899	3.85	.9999	.0002	.0033	−.0110	.0323
3.36	.9996	.0014	.0145	−.0393	.0885	3.86	.9999	.0002	.0032	−.0107	.0315
3.37	.9996	.0014	.0141	−.0384	.0871	3.87	.9999	.0002	.0031	−.0104	.0307
3.38	.9996	.0013	.0138	−.0376	.0857	3.88	.9999	.0002	.0030	−.0100	.0299
3.39	.9997	.0013	.0134	−.0367	.0843	3.89	.9999	.0002	.0029	−.0098	.0292
3.40	.9997	.0012	.0130	−.0359	.0829	3.90	1.0000	.0002	.0028	−.0095	.0284
3.41	.9997	.0012	.0127	−.0350	.0815	3.91	1.0000	.0002	.0027	−.0092	.0277
3.42	.9997	.0012	.0123	−.0342	.0801	3.92	1.0000	.0002	.0026	−.0089	.0270
3.43	.9997	.0011	.0120	−.0334	.0788	3.93	1.0000	.0002	.0026	−.0086	.0263
3.44	.9997	.0011	.0116	−.0327	.0774	3.94	1.0000	.0002	.0025	−.0084	.0256
3.45	.9997	.0010	.0113	−.0319	.0761	3.95	1.0000	.0002	.0024	−.0081	.0250
3.46	.9997	.0010	.0110	−.0311	.0747	3.96	1.0000	.0002	.0023	−.0079	.0243
3.47	.9997	.0010	.0107	−.0304	.0734	3.97	1.0000	.0002	.0022	−.0076	.0237
3.48	.9997	.0009	.0104	−.0297	.0721	3.98	1.0000	.0001	.0022	−.0074	.0230
3.49	.9998	.0009	.0101	−.0290	.0707	3.99	1.0000	.0001	.0021	−.0072	.0224

4.00	1.0000	.0001	.0020	−.0070	.0218	4.50	1.0000	.0000	.0003	−.0012	.0047
4.05	1.0000	.0001	.0017	−.0059	.0190	4.55	1.0000	.0000	.0003	−.0010	.0039
4.10	1.0000	.0001	.0014	−.0051	.0165	4.60	1.0000	.0000	.0002	−.0009	.0033
4.15	1.0000	.0001	.0012	−.0043	.0143	4.65	1.0000	.0000	.0002	−.0007	.0027
4.20	1.0000	.0001	.0010	−.0036	.0123	4.70	1.0000	.0000	.0001	−.0006	.0023
4.25	1.0000	.0000	.0008	−.0031	.0105	4.75	1.0000	.0000	.0001	−.0005	.0019
4.30	1.0000	.0000	.0007	−.0026	.0090	4.80	1.0000	.0000	.0001	−.0004	.0016
4.35	1.0000	.0000	.0006	−.0022	.0077	4.85	1.0000	.0000	.0001	−.0003	.0013
4.40	1.0000	.0000	.0005	−.0018	.0065	4.90	1.0000	.0000	.0001	−.0003	.0011
4.45	1.0000	.0000	.0004	−.0015	.0055	4.95	1.0000	.0000	.0000	−.0002	.0009

Table 19 SUMMED POISSON DISTRIBUTION FUNCTION*

$$\sum_{x=x'}^{x=\infty} \frac{e^{-m}m^x}{x!}$$

(See Chapter 22)

Entries in the table are values of $\sum_{x=x'}^{x=\infty} e^{-m}m^x/x!$ for the indicated values of x' and m.

x' \ m	0.1	0.2	0.3	0.4	0.5	0.6	0.7	0.8	0.9	1.0
0	1.0000	1.0000	1.0000	1.0000	1.0000	1.0000	1.0000	1.0000	1.0000	1.0000
1	.0952	.1813	.2592	.3297	.3935	.4512	.5034	.5507	.5934	.6321
2	.0047	.0175	.0369	.0616	.0902	.1219	.1558	.1912	.2275	.2642
3	.0002	.0011	.0036	.0079	.0144	.0231	.0341	.0474	.0629	.0803
4	.0000	.0001	.0003	.0008	.0018	.0034	.0058	.0091	.0135	.0190
5	.0000	.0000	.0000	.0001	.0002	.0004	.0008	.0014	.0023	.0037
6	.0000	.0000	.0000	.0000	.0000	.0000	.0001	.0002	.0003	.0006
7	.0000	.0000	.0000	.0000	.0000	.0000	.0000	.0000	.0000	.0001

x' \ m	1.1	1.2	1.3	1.4	1.5	1.6	1.7	1.8	1.9	2.0
0	1.0000	1.0000	1.0000	1.0000	1.0000	1.0000	1.0000	1.0000	1.0000	1.0000
1	.6671	.6988	.7275	.7534	.7769	.7981	.8173	.8347	.8504	.8647
2	.3010	.3374	.3732	.4082	.4422	.4751	.5068	.5372	.5663	.5940
3	.0996	.1205	.1429	.1665	.1912	.2166	.2428	.2694	.2963	.3233
4	.0257	.0338	.0431	.0537	.0656	.0788	.0932	.1087	.1253	.1429
5	.0054	.0077	.0107	.0143	.0186	.0237	.0296	.0364	.0441	.0527
6	.0010	.0015	.0022	.0032	.0045	.0060	.0080	.0104	.0132	.0166
7	.0001	.0003	.0004	.0006	.0009	.0013	.0019	.0026	.0034	.0045
8	.0000	.0000	.0001	.0001	.0002	.0003	.0004	.0006	.0008	.0011
9	.0000	.0000	.0000	.0000	.0000	.0000	.0001	.0001	.0002	.0002

x' \ m	2.1	2.2	2.3	2.4	2.5	2.6	2.7	2.8	2.9	3.0
0	1.0000	1.0000	1.0000	1.0000	1.0000	1.0000	1.0000	1.0000	1.0000	1.0000
1	.8775	.8892	.8997	.9093	.9179	.9257	.9328	.9392	.9450	.9502
2	.6204	.6454	.6691	.6916	.7127	.7326	.7513	.7689	.7854	.8009
3	.3504	.3773	.4040	.4303	.4562	.4816	.5064	.5305	.5540	.5768
4	.1614	.1806	.2007	.2213	.2424	.2640	.2859	.3081	.3304	.3528
5	.0621	.0725	.0838	.0959	.1088	.1226	.1371	.1523	.1682	.1847
6	.0204	.0249	.0300	.0357	.0420	.0490	.0567	.0651	.0742	.0839
7	.0059	.0075	.0094	.0116	.0142	.0172	.0206	.0244	.0287	.0335
8	.0015	.0020	.0026	.0033	.0042	.0053	.0066	.0081	.0099	.0119
9	.0003	.0005	.0006	.0009	.0011	.0015	.0019	.0024	.0031	.0038
10	.0001	.0001	.0001	.0002	.0003	.0004	.0005	.0007	.0009	.0011
11	.0000	.0000	.0000	.0000	.0001	.0001	.0001	.0002	.0002	.0003
12	.0000	.0000	.0000	.0000	.0000	.0000	.0000	.0000	.0001	.0001

x' \ m	3.1	3.2	3.3	3.4	3.5	3.6	3.7	3.8	3.9	4.0
0	1.0000	1.0000	1.0000	1.0000	1.0000	1.0000	1.0000	1.0000	1.0000	1.0000
1	.9550	.9592	.9631	.9666	.9698	.9727	.9753	.9776	.9798	.9817
2	.8153	.8288	.8414	.8532	.8641	.8743	.8838	.8926	.9008	.9084
3	.5988	.6201	.6406	.6603	.6792	.6973	.7146	.7311	.7469	.7619
4	.3752	.3975	.4197	.4416	.4634	.4848	.5058	.5265	.5468	.5665
5	.2018	.2194	.2374	.2558	.2746	.2936	.3128	.3322	.3516	.3712
6	.0943	.1054	.1171	.1295	.1424	.1559	.1699	.1844	.1994	.2149
7	.0388	.0446	.0510	.0579	.0653	.0733	.0818	.0909	.1005	.1107
8	.0142	.0168	.0198	.0231	.0267	.0308	.0352	.0401	.0454	.0511
9	.0047	.0057	.0069	.0083	.0099	.0117	.0137	.0160	.0185	.0214
10	.0014	.0018	.0022	.0027	.0033	.0040	.0048	.0058	.0069	.0081
11	.0004	.0005	.0006	.0008	.0010	.0013	.0016	.0019	.0023	.0028
12	.0001	.0001	.0002	.0002	.0003	.0004	.0005	.0006	.0007	.0009
13	.0000	.0000	.0000	.0001	.0001	.0001	.0001	.0002	.0002	.0003
14	.0000	.0000	.0000	.0000	.0000	.0000	.0000	.0000	.0001	.0001

For extensive tables of $\sum_{x=x'}^{x=\infty} e^{-m}m^x/x!$ see E. C. Molina, *Poisson's Exponential Binomial Limit*, D. Van Nostrand, New York, 1942.

*From Richard S. Burington and Donald C. May, *Handbook of Probability and Statistics with Tables*, 2d ed., McGraw-Hill Book Company, 1970, pp. 363–366.

$$\sum_{x=x'}^{x=\infty} \frac{e^{-m}m^x}{x!}$$

x'	m = 4.1	4.2	4.3	4.4	4.5	4.6	4.7	4.8	4.9	5.0
0	1.0000	1.0000	1.0000	1.0000	1.0000	1.0000	1.0000	1.0000	1.0000	1.0000
1	.9834	.9850	.9864	.9877	.9889	.9899	.9909	.9918	.9926	.9933
2	.9155	.9220	.9281	.9337	.9389	.9437	.9482	.9523	.9561	.9596
3	.7762	.7898	.8026	.8149	.8264	.8374	.8477	.8575	.8667	.8753
4	.5858	.6046	.6228	.6406	.6577	.6743	.6903	.7058	.7207	.7350
5	.3907	.4102	.4296	.4488	.4679	.4868	.5054	.5237	.5418	.5595
6	.2307	.2469	.2633	.2801	.2971	.3142	.3316	.3490	.3665	.3840
7	.1214	.1325	.1442	.1564	.1689	.1820	.1954	.2092	.2233	.2378
8	.0573	.0639	.0710	.0786	.0866	.0951	.1040	.1133	.1231	.1334
9	.0245	.0279	.0317	.0358	.0403	.0451	.0503	.0558	.0618	.0681
10	.0095	.0111	.0129	.0149	.0171	.0195	.0222	.0251	.0283	.0318
11	.0034	.0041	.0048	.0057	.0067	.0078	.0090	.0104	.0120	.0137
12	.0011	.0014	.0017	.0020	.0024	.0029	.0034	.0040	.0047	.0055
13	.0003	.0004	.0005	.0007	.0008	.0010	.0012	.0014	.0017	.0020
14	.0001	.0001	.0002	.0002	.0003	.0003	.0004	.0005	.0006	.0007
15	.0000	.0000	.0000	.0001	.0001	.0001	.0001	.0001	.0002	.0002
16	.0000	.0000	.0000	.0000	.0000	.0000	.0000	.0000	.0001	.0001

x'	m = 5.1	5.2	5.3	5.4	5.5	5.6	5.7	5.8	5.9	6.0
0	1.0000	1.0000	1.0000	1.0000	1.0000	1.0000	1.0000	1.0000	1.0000	1.0000
1	.9939	.9945	.9950	.9955	.9959	.9963	.9967	.9970	.9973	.9975
2	.9628	.9658	.9686	.9711	.9734	.9756	.9776	.9794	.9811	.9826
3	.8835	.8912	.8984	.9052	.9116	.9176	.9232	.9285	.9334	.9380
4	.7487	.7619	.7746	.7867	.7983	.8094	.8200	.8300	.8396	.8488
5	.5769	.5939	.6105	.6267	.6425	.6579	.6728	.6873	.7013	.7149
6	.4016	.4191	.4365	.4539	.4711	.4881	.5050	.5217	.5381	.5543
7	.2526	.2676	.2829	.2983	.3140	.3297	.3456	.3616	.3776	.3937
8	.1440	.1551	.1665	.1783	.1905	.2030	.2159	.2290	.2424	.2560
9	.0748	.0819	.0894	.0974	.1056	.1143	.1234	.1328	.1426	.1528
10	.0356	.0397	.0441	.0488	.0538	.0591	.0648	.0708	.0772	.0839
11	.0156	.0177	.0200	.0225	.0253	.0282	.0314	.0349	.0386	.0426
12	.0063	.0073	.0084	.0096	.0110	.0125	.0141	.0160	.0179	.0201
13	.0024	.0028	.0033	.0038	.0045	.0051	.0059	.0068	.0078	.0088
14	.0008	.0010	.0012	.0014	.0017	.0020	.0023	.0027	.0031	.0036
15	.0003	.0003	.0004	.0005	.0006	.0007	.0009	.0010	.0012	.0014
16	.0001	.0001	.0001	.0002	.0002	.0002	.0003	.0004	.0004	.0005
17	.0000	.0000	.0000	.0001	.0001	.0001	.0001	.0001	.0001	.0002
18	.0000	.0000	.0000	.0000	.0000	.0000	.0000	.0000	.0000	.0001

x'	m = 6.1	6.2	6.3	6.4	6.5	6.6	6.7	6.8	6.9	7.0
0	1.0000	1.0000	1.0000	1.0000	1.0000	1.0000	1.0000	1.0000	1.0000	1.0000
1	.9978	.9980	.9982	.9983	.9985	.9986	.9988	.9989	.9990	.9991
2	.9841	.9854	.9866	.9877	.9887	.9897	.9905	.9913	.9920	.9927
3	.9423	.9464	.9502	.9537	.9570	.9600	.9629	.9656	.9680	.9704
4	.8575	.8658	.8736	.8811	.8882	.8948	.9012	.9072	.9129	.9182
5	.7281	.7408	.7531	.7649	.7763	.7873	.7978	.8080	.8177	.8270
6	.5702	.5859	.6012	.6163	.6310	.6453	.6594	.6730	.6863	.6993
7	.4098	.4258	.4418	.4577	.4735	.4892	.5047	.5201	.5353	.5503
8	.2699	.2840	.2983	.3127	.3272	.3419	.3567	.3715	.3864	.4013
9	.1633	.1741	.1852	.1967	.2084	.2204	.2327	.2452	.2580	.2709
10	.0910	.0984	.1061	.1142	.1226	.1314	.1404	.1498	.1505	.1695
11	.0469	.0514	.0563	.0614	.0668	.0726	.0786	.0849	.0916	.0985
12	.0224	.0250	.0277	.0307	.0339	.0373	.0409	.0448	.0490	.0534
13	.0100	.0113	.0127	.0143	.0160	.0179	.0199	.0221	.0245	.0270
14	.0042	.0048	.0055	.0063	.0071	.0080	.0091	.0102	.0115	.0128
15	.0016	.0019	.0022	.0026	.0030	.0034	.0039	.0044	.0050	.0057
16	.0006	.0007	.0008	.0010	.0012	.0014	.0016	.0018	.0021	.0024
17	.0002	.0003	.0003	.0004	.0004	.0005	.0006	.0007	.0008	.0010
18	.0001	.0001	.0001	.0001	.0002	.0002	.0002	.0003	.0003	.0004
19	.0000	.0000	.0000	.0000	.0001	.0001	.0001	.0001	.0001	.0001

$$\sum_{x=x'}^{x=\infty} \frac{e^{-m}m^x}{x!}$$

	m									
x'	7.1	7.2	7.3	7.4	7.5	7.6	7.7	7.8	7.9	8.0
0	1.0000	1.0000	1.0000	1.0000	1.0000	1.0000	1.0000	1.0000	1.0000	1.0000
1	.9992	.9993	.9993	.9994	.9994	.9995	.9995	.9996	.9996	.9997
2	.9933	.9939	.9944	.9949	.9953	.9957	.9961	.9964	.9967	.9970
3	.9725	.9745	.9764	.9781	.9797	.9812	.9826	.9839	.9851	.9862
4	.9233	.9281	.9326	.9368	.9409	.9446	.9482	.9515	.9547	.9576
5	.8359	.8445	.8527	.8605	.8679	.8751	.8819	.8883	.8945	.9004
6	.7119	.7241	.7360	.7474	.7586	.7693	.7797	.7897	.7994	.8088
7	.5651	.5796	.5940	.6080	.6218	.6354	.6486	.6616	.6743	.6866
8	.4162	.4311	.4459	.4607	.4754	.4900	.5044	.5188	.5330	.5470
9	.2840	.2973	.3108	.3243	.3380	.3518	.3657	.3796	.3935	.4075
10	.1798	.1904	.2012	.2123	.2236	.2351	.2469	.2589	.2710	.2834
11	.1058	.1133	.1212	.1293	.1378	.1465	.1555	.1648	.1743	.1841
12	.0580	.0629	.0681	.0735	.0792	.0852	.0915	.0980	.1048	.1119
13	.0297	.0327	.0358	.0391	.0427	.0464	.0504	.0546	.0591	.0638
14	.0143	.0159	.0176	.0195	.0216	.0238	.0261	.0286	.0313	.0342
15	.0065	.0073	.0082	.0092	.0103	.0114	.0127	.0141	.0156	.0173
16	.0028	.0031	.0036	.0041	.0046	.0052	.0059	.0066	.0074	.0082
17	.0011	.0013	.0015	.0017	.0020	.0022	.0026	.0029	.0033	.0037
18	.0004	.0005	.0006	.0007	.0008	.0009	.0011	.0012	.0014	.0016
19	.0002	.0002	.0002	.0003	.0003	.0004	.0004	.0005	.0006	.0006
20	.0001	.0001	.0001	.0001	.0001	.0001	.0002	.0002	.0002	.0003
21	.0000	.0000	.0000	.0000	.0000	.0000	.0001	.0001	.0001	.0001

	m									
x'	8.1	8.2	8.3	8.4	8.5	8.6	8.7	8.8	8.9	9.0
0	1.0000	1.0000	1.0000	1.0000	1.0000	1.0000	1.0000	1.0000	1.0000	1.0000
1	.9997	.9997	.9998	.9998	.9998	.9998	.9998	.9998	.9999	.9999
2	.9972	.9975	.9977	.9979	.9981	.9982	.9984	.9985	.9987	.9988
3	.9873	.9882	.9891	.9900	.9907	.9914	.9921	.9927	.9932	.9938
4	.9604	.9630	.9654	.9677	.9699	.9719	.9738	.9756	.9772	.9788
5	.9060	.9113	.9163	.9211	.9256	.9299	.9340	.9379	.9416	.9450
6	.8178	.8264	.8347	.8427	.8504	.8578	.8648	.8716	.8781	.8843
7	.6987	.7104	.7219	.7330	.7438	.7543	.7645	.7744	.7840	.7932
8	.5609	.5746	.5881	.6013	.6144	.6272	.6398	.6522	.6643	.6761
9	.4214	.4353	.4493	.4631	.4769	.4906	.5042	.5177	.5311	.5443
10	.2959	.3085	.3212	.3341	.3470	.3600	.3731	.3863	.3994	.4126
11	.1942	.2045	.2150	.2257	.2366	.2478	.2591	.2706	.2822	.2940
12	.1193	.1269	.1348	.1429	.1513	.1600	.1689	.1780	.1874	.1970
13	.0687	.0739	.0793	.0850	.0909	.0971	.1035	.1102	.1171	.1242
14	.0372	.0405	.0439	.0476	.0514	.0555	.0597	.0642	.0689	.0739
15	.0190	.0209	.0229	.0251	.0274	.0299	.0325	.0353	.0383	.0415
16	.0092	.0102	.0113	.0125	.0138	.0152	.0168	.0184	.0202	.0220
17	.0042	.0047	.0053	.0059	.0066	.0074	.0082	.0091	.0101	.0111
18	.0018	.0021	.0023	.0027	.0030	.0034	.0038	.0043	.0048	.0053
19	.0008	.0009	.0010	.0011	.0013	.0015	.0017	.0019	.0022	.0024
20	.0003	.0003	.0004	.0005	.0005	.0006	.0007	.0008	.0009	.0011
21	.0001	.0001	.0002	.0002	.0002	.0002	.0003	.0003	.0004	.0004
22	.0000	.0000	.0001	.0001	.0001	.0001	.0001	.0001	.0002	.0002
23	.0000	.0000	.0000	.0000	.0000	.0000	.0000	.0000	.0001	.0001

	m									
x'	9.1	9.2	9.3	9.4	9.5	9.6	9.7	9.8	9.9	10
0	1.0000	1.0000	1.0000	1.0000	1.0000	1.0000	1.0000	1.0000	1.0000	1.0000
1	.9999	.9999	.9999	.9999	.9999	.9999	.9999	.9999	1.0000	1.0000
2	.9989	.9990	.9991	.9991	.9992	.9993	.9993	.9994	.9995	.9995
3	.9942	.9947	.9951	.9955	.9958	.9962	.9965	.9967	.9970	.9972
4	.9802	.9816	.9828	.9840	.9851	.9862	.9871	.9880	.9889	.9897
5	.9483	.9514	.9544	.9571	.9597	.9622	.9645	.9667	.9688	.9707
6	.8902	.8959	.9014	.9065	.9115	.9162	.9207	.9250	.9290	.9329
7	.8022	.8108	.8192	.8273	.8351	.8426	.8498	.8567	.8634	.8699
8	.6877	.6990	.7101	.7208	.7313	.7416	.7515	.7612	.7706	.7798
9	.5574	.5704	.5832	.5958	.6082	.6204	.6324	.6442	.6558	.6672

$$\sum_{x=x'}^{x=\infty} \frac{e^{-m}m^x}{x!}$$

x'	m 9.1	9.2	9.3	9.4	9.5	9.6	9.7	9.8	9.9	10
10	.4258	.4389	.4521	.4651	.4782	.4911	.5040	.5168	.5295	.5421
11	.3059	.3180	.3301	.3424	.3547	.3671	.3795	.3920	.4045	.4170
12	.2068	.2168	.2270	.2374	.2480	.2588	.2697	.2807	.2919	.3032
13	.1316	.1393	.1471	.1552	.1636	.1721	.1809	.1899	.1991	.2084
14	.0790	.0844	.0900	.0958	.1019	.1081	.1147	.1214	.1284	.1355
15	.0448	.0483	.0520	.0559	.0600	.0643	.0688	.0735	.0784	.0835
16	.0240	.0262	.0285	.0309	.0335	.0362	.0391	.0421	.0454	.0487
17	.0122	.0135	.0148	.0162	.0177	.0194	.0211	.0230	.0249	.0270
18	.0059	.0066	.0073	.0081	.0089	.0098	.0108	.0119	.0130	.0143
19	.0027	.0031	.0034	.0038	.0043	.0048	.0053	.0059	.0065	.0072
20	.0012	.0014	.0015	.0017	.0020	.0022	.0025	.0028	.0031	.0035
21	.0005	.0006	.0007	.0008	.0009	.0010	.0011	.0013	.0014	.0016
22	.0002	.0002	.0003	.0003	.0004	.0004	.0005	.0005	.0006	.0007
23	.0001	.0001	.0001	.0001	.0001	.0002	.0002	.0002	.0003	.0003
24	.0000	.0000	.0000	.0000	.0001	.0001	.0001	.0001	.0001	.0001

x'	m 11	12	13	14	15	16	17	18	19	20
0	1.0000	1.0000	1.0000	1.0000	1.0000	1.0000	1.0000	1.0000	1.0000	1.0000
1	1.0000	1.0000	1.0000	1.0000	1.0000	1.0000	1.0000	1.0000	1.0000	1.0000
2	.9998	.9999	1.0000	1.0000	1.0000	1.0000	1.0000	1.0000	1.0000	1.0000
3	.9988	.9995	.9998	.9999	1.0000	1.0000	1.0000	1.0000	1.0000	1.0000
4	.9951	.9977	.9990	.9995	.9998	.9999	1.0000	1.0000	1.0000	1.0000
5	.9849	.9924	.9963	.9982	.9991	.9996	.9998	.9999	1.0000	1.0000
6	.9625	.9797	.9893	.9945	.9972	.9986	.9993	.9997	.9998	.9999
7	.9214	.9542	.9741	.9858	.9924	.9960	.9979	.9990	.9995	.9997
8	.8568	.9105	.9460	.9684	.9820	.9900	.9946	.9971	.9985	.9992
9	.7680	.8450	.9002	.9379	.9626	.9780	.9874	.9929	.9961	.9979
10	.6595	.7576	.8342	.8906	.9301	.9567	.9739	.9846	.9911	.9950
11	.5401	.6528	.7483	.8243	.8815	.9226	.9509	.9696	.9817	.9892
12	.4207	.5384	.6468	.7400	.8152	.8730	.9153	.9451	.9653	.9786
13	.3113	.4240	.5369	.6415	.7324	.8069	.8650	.9083	.9394	.9610
14	.2187	.3185	.4270	.5356	.6368	.7255	.7991	.8574	.9016	.9339
15	.1460	.2280	.3249	.4296	.5343	.6325	.7192	.7919	.8503	.8951
16	.0926	.1556	.2364	.3306	.4319	.5333	.6285	.7133	.7852	.8435
17	.0559	.1013	.1645	.2441	.3359	.4340	.5323	.6250	.7080	.7789
18	.0322	.0630	.1095	.1728	.2511	.3407	.4360	.5314	.6216	.7030
19	.0177	.0374	.0698	.1174	.1805	.2577	.3450	.4378	.5305	.6186
20	.0093	.0213	.0427	.0765	.1248	.1878	.2637	.3491	.4394	.5297
21	.0047	.0116	.0250	.0479	.0830	.1318	.1945	.2693	.3528	.4409
22	.0023	.0061	.0141	.0288	.0531	.0892	.1385	.2009	.2745	.3563
23	.0010	.0030	.0076	.0167	.0327	.0582	.0953	.1449	.2069	.2794
24	.0005	.0015	.0040	.0093	.0195	.0367	.0633	.1011	.1510	.2125
25	.0002	.0007	.0020	.0050	.0112	.0223	.0406	.0683	.1067	.1568
26	.0001	.0003	.0010	.0026	.0062	.0131	.0252	.0446	.0731	.1122
27	.0000	.0001	.0005	.0013	.0033	.0075	.0152	.0282	.0486	.0779
28	.0000	.0001	.0002	.0006	.0017	.0041	.0088	.0173	.0313	.0525
29	.0000	.0000	.0001	.0003	.0009	.0022	.0050	.0103	.0195	.0343
30	.0000	.0000	.0000	.0001	.0004	.0011	.0027	.0059	.0118	.0218
31	.0000	.0000	.0000	.0001	.0002	.0006	.0014	.0033	.0070	.0135
32	.0000	.0000	.0000	.0000	.0001	.0003	.0007	.0018	.0040	.0081
33	.0000	.0000	.0000	.0000	.0000	.0001	.0004	.0010	.0022	.0047
34	.0000	.0000	.0000	.0000	.0000	.0001	.0002	.0005	.0012	.0027
35	.0000	.0000	.0000	.0000	.0000	.0000	.0001	.0002	.0006	.0015
36	.0000	.0000	.0000	.0000	.0000	.0000	.0000	.0001	.0003	.0008
37	.0000	.0000	.0000	.0000	.0000	.0000	.0000	.0001	.0002	.0004
38	.0000	.0000	.0000	.0000	.0000	.0000	.0000	.0000	.0001	.0002
39	.0000	.0000	.0000	.0000	.0000	.0000	.0000	.0000	.0000	.0001
40	.0000	.0000	.0000	.0000	.0000	.0000	.0000	.0000	.0000	.0001

Entries in this table give values of χ_0^2 for the indicated values of m and ϵ such that

$$\int_{\omega=\chi_0^2}^{\omega=\infty} f_m(\omega)\, d\omega = \epsilon$$

where $$f_m(\omega) = \left[\frac{\omega^{(m/2)-1}}{2^{m/2}\Gamma(m/2)}\right]\exp\left(-\frac{\omega}{2}\right) \qquad \omega > 0$$

Degrees of freedom, m	ϵ = .99	.98	.95	.90	.80	.70	.50
	χ_0^2						
1	0.000	0.001	0.004	0.016	0.064	0.148	0.455
2	0.020	0.040	0.103	0.211	0.446	0.713	1.386
3	0.115	0.185	0.352	0.584	1.005	1.424	2.366
4	0.297	0.429	0.711	1.064	1.649	2.195	3.357
5	0.554	0.752	1.145	1.610	2.343	3.000	4.351
6	0.872	1.134	1.635	2.204	3.070	3.828	5.348
7	1.239	1.564	2.167	2.833	3.822	4.671	6.346
8	1.646	2.032	2.733	3.490	4.594	5.527	7.344
9	2.088	2.532	3.325	4.168	5.380	6.393	8.343
10	2.558	3.059	3.940	4.865	6.179	7.267	9.342
11	3.053	3.609	4.575	5.578	6.989	8.148	10.341
12	3.571	4.178	5.226	6.304	7.807	9.034	11.340
13	4.107	4.765	5.892	7.042	8.634	9.926	12.340
14	4.660	5.368	6.571	7.790	9.467	10.821	13.339
15	5.229	5.985	7.261	8.547	10.307	11.721	14.339
16	5.812	6.614	7.962	9.312	11.152	12.624	15.338
17	6.408	7.255	8.672	10.085	12.002	13.531	16.338
18	7.015	7.906	9.390	10.865	12.857	14.440	17.338
19	7.633	8.567	10.117	11.651	13.716	15.352	18.338
20	8.260	9.237	10.851	12.443	14.578	16.266	19.337
21	8.897	9.915	11.591	13.240	15.445	17.182	20.337
22	9.542	10.600	12.338	14.041	16.314	18.101	21.337
23	10.196	11.293	13.091	14.848	17.187	19.021	22.337
24	10.856	11.992	13.848	15.659	18.062	19.943	23.337
25	11.524	12.697	14.611	16.473	18.940	20.867	24.337
26	12.198	13.409	15.379	17.292	19.820	21.792	25.336
27	12.879	14.125	16.151	18.114	20.703	22.719	26.336
28	13.565	14.847	16.928	18.939	21.588	23.647	27.336
29	14.256	15.574	17.708	19.768	22.475	24.577	28.336
30	14.953	16.306	18.493	20.599	23.364	25.508	29.336

Table 20 is abridged from Table III of Fisher: *Statistical Methods for Research Workers,* published by Oliver and Boyd Ltd., Edinburgh, by permission of the author and publishers.

Degrees of freedom, *m*	.30	.20	.10	ϵ .05	.02	.01	.001
				χ_0^2			
1	1.074	1.642	2.706	3.841	5.412	6.635	10.827
2	2.408	3.219	4.605	5.991	7.824	9.210	13.815
3	3.665	4.642	6.251	7.815	9.837	11.345	16.268
4	4.878	5.989	7.779	9.488	11.668	13.277	18.465
5	6.064	7.289	9.236	11.070	13.388	15.086	20.517
6	7.231	8.558	10.645	12.592	15.033	16.812	22.457
7	8.383	9.803	12.017	14.067	16.622	18.475	24.322
8	9.524	11.030	13.362	15.507	18.168	20.090	26.125
9	10.656	12.242	14.684	16.919	19.679	21.666	27.877
10	11.781	13.442	15.987	18.307	21.161	23.209	29.588
11	12.899	14.631	17.275	19.675	22.618	24.725	31.264
12	14.011	15.812	18.549	21.026	24.054	26.217	32.909
13	15.119	16.985	19.812	22.362	25.472	27.688	34.528
14	16.222	18.151	21.064	23.685	26.873	29.141	36.123
15	17.322	19.311	22.307	24.996	28.259	30.578	37.697
16	18.418	20.465	23.542	26.296	29.633	32.000	39.252
17	19.511	21.615	24.769	27.587	30.995	33.409	40.790
18	20.601	22.760	25.989	28.869	32.346	34.805	42.312
19	21.689	23.900	27.204	30.144	33.687	36.191	43.820
20	22.775	25.038	28.412	31.410	35.020	37.566	45.315
21	23.858	26.171	29.615	32.671	36.343	38.932	46.797
22	24.939	27.301	30.813	33.924	37.659	40.289	48.268
23	26.018	28.429	32.007	35.172	38.968	41.638	49.728
24	27.096	29.553	33.196	36.415	40.270	42.980	51.179
25	28.172	30.675	34.382	37.652	41.566	44.314	52.620
26	29.246	31.795	35.563	38.885	42.856	45.642	54.052
27	30.319	32.912	36.741	40.113	44.140	46.963	55.476
28	31.391	34.027	37.916	41.337	45.419	48.278	56.893
29	32.461	35.139	39.087	42.557	46.693	49.588	58.302
30	33.530	36.250	40.256	43.773	47.962	50.892	59.703

INTEREST AND ACTUARIAL EXPERIENCE

Table 21 AMOUNT OF 1 AT COMPOUND INTEREST*

n	1%	2%	$2\frac{1}{2}$%	3%	$3\frac{1}{2}$%	4%	$4\frac{1}{2}$%	5%	$5\frac{1}{2}$%
1	1.0100	1.0200	1.0250	1.0300	1.0350	1.0400	1.0450	1.0500	1.0550
2	1.0201	1.0404	1.0506	1.0609	1.0712	1.0816	1.0920	1.1025	1.1130
3	1.0303	1.0612	1.0769	1.0927	1.1087	1.1249	1.1412	1.1576	1.1742
4	1.0406	1.0824	1.1038	1.1255	1.1475	1.1699	1.1925	1.2155	1.2388
5	1.0510	1.1041	1.1314	1.1593	1.1877	1.2167	1.2462	1.2763	1.3070
6	1.0615	1.1262	1.1597	1.1941	1.2293	1.2653	1.3023	1.3401	1.3788
7	1.0721	1.1487	1.1887	1.2299	1.2723	1.3159	1.3609	1.4071	1.4547
8	1.0829	1.1717	1.2184	1.2668	1.3168	1.3686	1.4221	1.4775	1.5347
9	1.0937	1.1951	1.2489	1.3048	1.3629	1.4233	1.4861	1.5513	1.6191
10	1.1046	1.2190	1.2801	1.3439	1.4106	1.4802	1.5530	1.6289	1.7081
11	1.1157	1.2434	1.3121	1.3842	1.4600	1.5395	1.6229	1.7103	1.8021
12	1.1268	1.2682	1.3449	1.4258	1.5111	1.6010	1.6959	1.7959	1.9012
13	1.1381	1.2936	1.3785	1.4685	1.5640	1.6651	1.7722	1.8856	2.0058
14	1.1495	1.3195	1.4130	1.5126	1.6187	1.7317	1.8519	1.9799	2.1161
15	1.1610	1.3459	1.4483	1.5580	1.6753	1.8009	1.9353	2.0789	2.2325
16	1.1726	1.3728	1.4845	1.6047	1.7340	1.8730	2.0224	2.1829	2.3553
17	1.1843	1.4002	1.5216	1.6528	1.7947	1.9479	2.1134	2.2920	2.4848
18	1.1961	1.4282	1.5597	1.7024	1.8575	2.0258	2.2085	2.4066	2.6215
19	1.2081	1.4568	1.5987	1.7535	1.9225	2.1068	2.3079	2.5270	2.7656
20	1.2202	1.4859	1.6386	1.8061	1.9898	2.1911	2.4117	2.6533	2.9178
21	1.2324	1.5157	1.6796	1.8603	2.0594	2.2788	2.5202	2.7860	3.0782
22	1.2447	1.5460	1.7216	1.9161	2.1315	2.3699	2.6337	2.9253	3.2475
23	1.2572	1.5769	1.7646	1.9736	2.2061	2.4647	2.7522	3.0715	3.4262
24	1.2697	1.6084	1.8087	2.0328	2.2833	2.5633	2.8760	3.2251	3.6146
25	1.2824	1.6406	1.8539	2.0938	2.3632	2.6658	3.0054	3.3864	3.8134
26	1.2953	1.6734	1.9003	2.1566	2.4460	2.7725	3.1407	3.5557	4.0231
27	1.3082	1.7069	1.9478	2.2213	2.5316	2.8834	3.2820	3.7335	4.2444
28	1.3213	1.7410	1.9965	2.2879	2.6202	2.9987	3.4297	3.9201	4.4778
29	1.3345	1.7758	2.0464	2.3566	2.7119	3.1187	3.5840	4.1161	4.7241
30	1.3478	1.8114	2.0976	2.4273	2.8068	3.2434	3.7453	4.3219	4.9840
31	1.3613	1.8476	2.1500	2.5001	2.9050	3.3731	3.9139	4.5380	5.2581
32	1.3749	1.8845	2.2038	2.5751	3.0067	3.5081	4.0900	4.7649	5.5473
33	1.3887	1.9222	2.2589	2.6523	3.1119	3.6484	4.2740	5.0032	5.8524
34	1.4026	1.9607	2.3153	2.7319	3.2209	3.7943	4.4664	5.2533	6.1742
35	1.4166	1.9999	2.3732	2.8139	3.3336	3.9461	4.6673	5.5160	6.5138
36	1.4308	2.0399	2.4325	2.8983	3.4503	4.1039	4.8774	5.7918	6.8721
37	1.4451	2.0807	2.4933	2.9852	3.5710	4.2681	5.0969	6.0814	7.2500
38	1.4595	2.1223	2.5557	3.0748	3.6960	4.4388	5.3262	6.3855	7.6488
39	1.4741	2.1647	2.6196	3.1670	3.8254	4.6164	5.5659	6.7048	8.0695
40	1.4889	2.2080	2.6851	3.2620	3.9593	4.8010	5.8164	7.0400	8.5133
41	1.5038	2.2522	2.7522	3.3599	4.0978	4.9931	6.0781	7.3920	8.9815
42	1.5188	2.2972	2.8210	3.4607	4.2413	5.1928	6.3516	7.7616	9.4755
43	1.5340	2.3432	2.8915	3.5645	4.3897	5.4005	6.6374	8.1497	9.9967
44	1.5493	2.3901	2.9638	3.6715	4.5433	5.6165	6.9361	8.5572	10.5465
45	1.5648	2.4379	3.0379	3.7816	4.7024	5.8412	7.2482	8.9850	11.1266
46	1.5805	2.4866	3.1139	3.8950	4.8669	6.0748	7.5744	9.4343	11.7385
47	1.5963	2.5363	3.1917	4.0119	5.0373	6.3178	7.9153	9.9060	12.3841
48	1.6122	2.5871	3.2715	4.1323	5.2136	6.5705	8.2715	10.4013	13.0653
49	1.6283	2.6388	3.3533	4.2562	5.3961	6.8333	8.6437	10.9213	13.7838
50	1.6446	2.6916	3.4371	4.3839	5.5849	7.1067	9.0326	11.4674	14.5420

$$s = (1 + r)^n$$

* See Sec. 18. A principal of 1 placed at a rate of interest r (expressed as decimal), compounded annually, accumulates to an amount $(1 + r)^n$ at the end of n years. In this table the rate r is expressed in percent.

n	6%	6½%	7%	7½%	8%	8½%	9%	9½%	10%
1	1.0600	1.0650	1.0700	1.0750	1.0800	1.0850	1.0900	1.0950	1.1000
2	1.1236	1.1342	1.1449	1.1556	1.1664	1.1772	1.1881	1.1990	1.2100
3	1.1910	1.2079	1.2250	1.2423	1.2597	1.2773	1.2950	1.3129	1.3310
4	1.2625	1.2865	1.3108	1.3355	1.3605	1.3859	1.4116	1.4377	1.4641
5	1.3382	1.3701	1.4026	1.4356	1.4693	1.5037	1.5386	1.5742	1.6105
6	1.4185	1.4591	1.5007	1.5433	1.5869	1.6315	1.6771	1.7238	1.7716
7	1.5036	1.5540	1.6058	1.6590	1.7138	1.7701	1.8280	1.8876	1.9487
8	1.5938	1.6550	1.7182	1.7835	1.8509	1.9206	1.9926	2.0669	2.1436
9	1.6895	1.7626	1.8385	1.9172	1.9990	2.0839	2.1719	2.2632	2.3579
10	1.7908	1.8771	1.9672	2.0610	2.1589	2.2610	2.3674	2.4782	2.5937
11	1.8983	1.9992	2.1049	2.2156	2.3316	2.4532	2.5804	2.7137	2.8531
12	2.0122	2.1291	2.2522	2.3818	2.5182	2.6617	2.8127	2.9715	3.1384
13	2.1329	2.2675	2.4098	2.5604	2.7196	2.8879	3.0658	3.2537	3.4523
14	2.2609	2.4149	2.5785	2.7524	2.9372	3.1334	3.3417	3.5629	3.7975
15	2.3966	2.5718	2.7590	2.9589	3.1722	3.3997	3.6425	3.9013	4.1772
16	2.5404	2.7390	2.9522	3.1808	3.4259	3.6887	3.9703	4.2719	4.5950
17	2.6928	2.9170	3.1588	3.4194	3.7000	4.0023	4.3276	4.6778	5.0545
18	2.8543	3.1067	3.3799	3.6758	3.9960	4.3425	4.7171	5.1222	5.5599
19	3.0256	3.3086	3.6165	3.9515	4.3157	4.7116	5.1417	5.6088	6.1159
20	3.2071	3.5236	3.8697	4.2479	4.6610	5.1120	5.6044	6.1416	6.7275
21	3.3996	3.7527	4.1406	4.5664	5.0338	5.5466	6.1088	6.7251	7.4002
22	3.6035	3.9966	4.4304	4.9089	5.4365	6.0180	6.6586	7.3639	8.1403
23	3.8197	4.2564	4.7405	5.2771	5.8715	6.5296	7.2579	8.0635	8.9543
24	4.0489	4.5330	5.0724	5.6729	6.3412	7.0846	7.9111	8.8296	9.8497
25	4.2919	4.8277	5.4274	6.0983	6.8485	7.6868	8.6231	9.6684	10.8347
26	4.5494	5.1415	5.8074	6.5557	7.3964	8.3401	9.3992	10.5869	11.9182
27	4.8223	5.4757	6.2139	7.0474	7.9881	9.0490	10.2451	11.5926	13.1100
28	5.1117	5.8316	6.6488	7.5759	8.6271	9.8182	11.1671	12.6939	14.4210
29	5.4184	6.2107	7.1143	8.1441	9.3173	10.6528	12.1722	13.8998	15.8631
30	5.7435	6.6144	7.6123	8.7550	10.0627	11.5583	13.2677	15.2203	17.4494
31	6.0881	7.0443	8.1451	9.4116	10.8677	12.5407	14.4618	16.6662	19.1943
32	6.4534	7.5022	8.7153	10.1174	11.7371	13.6067	15.7633	18.2495	21.1138
33	6.8406	7.9898	9.3253	10.8763	12.6760	14.7632	17.1820	19.9832	23.2252
34	7.2510	8.5092	9.9781	11.6920	13.6901	16.0181	18.7284	21.8816	25.5477
35	7.6861	9.0623	10.6766	12.5689	14.7853	17.3796	20.4140	23.9604	28.1024
36	8.1473	9.6513	11.4239	13.5115	15.9682	18.8569	22.2512	26.2366	30.9127
37	8.6361	10.2786	12.2236	14.5249	17.2456	20.4597	24.2538	28.7291	34.0039
38	9.1543	10.9467	13.0793	15.6143	18.6253	22.1988	26.4367	31.4584	37.4043
39	9.7035	11.6583	13.9948	16.7853	20.1153	24.0857	28.8160	34.4469	41.1448
40	10.2857	12.4161	14.9745	18.0442	21.7245	26.1330	31.4094	37.7194	45.2593
41	10.9029	13.2231	16.0227	19.3976	23.4625	28.3543	34.2363	41.3027	49.7852
42	11.5570	14.0826	17.1443	20.8524	25.3395	30.7644	37.3175	45.2265	54.7637
43	12.2505	14.9980	18.3444	22.4163	27.3666	33.3794	40.6761	49.5230	60.2401
44	12.9855	15.9729	19.6285	24.0975	29.5560	36.2167	44.3370	54.2277	66.2641
45	13.7646	17.0111	21.0025	25.9048	31.9204	39.2951	48.3273	59.3793	72.8905
46	14.5905	18.1168	22.4726	27.8477	34.4741	42.6352	52.6767	65.0204	80.1795
47	15.4659	19.2944	24.0457	29.9363	37.2320	46.2592	57.4176	71.1973	88.1975
48	16.3939	20.5485	25.7289	32.1815	40.2106	50.1912	62.5852	77.9611	97.0172
49	17.3775	21.8842	27.5299	34.5951	43.4274	54.4574	68.2179	85.3674	106.7190
50	18.4202	23.3067	29.4570	37.1897	46.9016	59.0863	74.3575	93.4773	117.3909

$$s = (1 + r)^n$$

* See Sec. 18. A principal of 1 placed at a rate of interest r (expressed as decimal), compounded annually, accumulates to an amount $(1 + r)^n$ at the end of n years. In this table the rate r is expressed in percent.

Table 22 PRESENT VALUE OF 1 AT COMPOUND INTEREST*

n	1%	2%	$2\frac{1}{2}$%	3%	$3\frac{1}{2}$%	4%	$4\frac{1}{2}$%	5%	$5\frac{1}{2}$%
1	.99010	.98039	.97561	.97087	.96618	.96154	.95694	.95238	.94787
2	.98030	.96117	.95181	.94260	.93351	.92456	.91573	.90703	.89845
3	.97059	.94232	.92860	.91514	.90194	.88900	.87630	.86384	.85161
4	.96098	.92385	.90595	.88849	.87144	.85480	.83856	.82270	.80722
5	.95147	.90573	.88385	.86261	.84197	.82193	.80245	.78353	.76513
6	.94205	.88797	.86230	.83748	.81350	.79031	.76790	.74622	.72525
7	.93272	.87056	.84127	.81309	.78599	.75992	.73483	.71068	.68744
8	.92348	.85349	.82075	.78941	.75941	.73069	.70319	.67684	.65160
9	.91434	.83676	.80073	.76642	.73373	.70259	.67290	.64461	.61763
10	.90529	.82035	.78120	.74409	.70892	.67556	.64393	.61391	.58543
11	.89632	.80426	.76214	.72242	.68495	.64958	.61620	.58468	.55491
12	.88745	.78849	.74356	.70138	.66178	.62460	.58966	.55684	.52598
13	.87866	.77303	.72542	.68095	.63940	.60057	.56427	.53032	.49856
14	.86996	.75788	.70773	.66112	.61778	.57748	.53997	.50507	.47257
15	.86135	.74301	.69047	.64186	.59689	.55526	.51672	.48102	.44793
16	.85282	.72845	.67362	.62317	.57671	.53391	.49447	.45811	.42458
17	.84438	.71416	.65720	.60502	.55720	.51337	.47318	.43630	.40245
18	.83602	.70016	.64117	.58739	.53836	.49363	.45280	.41552	.38147
19	.82774	.68643	.62553	.57029	.52016	.47464	.43330	.39573	.36158
20	.81954	.67297	.61027	.55368	.50257	.45639	.41464	.37689	.34273
21	.81143	.65978	.59539	.53755	.48557	.43883	.39679	.35894	.32486
22	.80340	.64684	.58086	.52189	.46915	.42196	.37970	.34185	.30793
23	.79544	.63416	.56670	.50669	.45329	.40573	.36335	.32557	.29187
24	.78757	.62172	.55288	.49193	.43796	.39012	.34770	.31007	.27666
25	.77977	.60953	.53939	.47761	.42315	.37512	.33273	.29530	.26223
26	.77205	.59758	.52623	.46369	.40884	.36069	.31840	.28124	.24856
27	.76440	.58586	.51340	.45019	.39501	.34682	.30469	.26785	.23560
28	.75684	.57437	.50088	.43708	.38165	.33348	.29157	.25509	.22332
29	.74934	.56311	.48866	.42435	.36875	.32065	.27902	.24295	.21168
30	.74192	.55207	.47674	.41199	.35628	.30832	.26700	.23138	.20064
31	.73458	.54125	.46511	.39999	.34423	.29646	.25550	.22036	.19018
32	.72730	.53063	.45377	.38834	.33259	.28506	.24450	.20987	.18027
33	.72010	.52023	.44270	.37703	.32134	.27409	.23397	.19987	.17087
34	.71297	.51003	.43191	.36604	.31048	.26355	.22390	.19035	.16196
35	.70591	.50003	.42137	.35538	.29998	.25342	.21425	.18129	.15352
36	.69892	.49022	.41109	.34503	.28983	.24367	.20503	.17266	.14552
37	.69200	.48061	.40107	.33498	.28003	.23430	.19620	.16444	.13793
38	.68515	.47119	.39128	.32523	.27056	.22529	.18775	.15661	.13074
39	.67837	.46195	.38174	.31575	.26141	.21662	.17967	.14915	.12392
40	.67165	.45289	.37243	.30656	.25257	.20829	.17193	.14205	.11746
41	.66500	.44401	.36335	.29763	.24403	.20028	.16453	.13528	.11134
42	.65842	.43530	.35448	.28896	.23578	.19257	.15744	.12884	.10554
43	.65190	.42677	.34584	.28054	.22781	.18517	.15066	.12270	.10003
44	.64545	.41840	.33740	.27237	.22010	.17805	.14417	.11686	.09482
45	.63905	.41020	.32917	.26444	.21266	.17120	.13796	.11130	.08988
46	.63273	.40215	.32115	.25674	.20547	.16461	.13202	.10600	.08519
47	.62646	.39427	.31331	.24926	.19852	.15828	.12634	.10095	.08075
48	.62026	.38654	.30567	.24200	.19181	.15219	.12090	.09614	.07654
49	.61412	.37896	.29822	.23495	.18532	.14634	.11569	.09156	.07255
50	.60804	.37153	.29094	.22811	.17905	.14071	.11071	.08720	.06877

$$v^n = (1 + r)^{-n}$$

* See Sec. 20. The present quantity which in n years will accumulate to 1 at the rate of interest r (expressed as decimal), compounded annually, is $(1 + r)^{-n}$. In this table the rate r is expressed in percent.

n	6%	6½%	7%	7½%	8%	8½%	9%	9½%	10%
1	.94340	.93897	.93458	.93023	.92593	.92166	.91743	.91324	.90909
2	.89000	.88166	.87344	.86533	.85734	.84946	.84168	.83401	.82645
3	.83962	.82785	.81630	.80496	.79383	.78291	.77218	.76165	.75131
4	.79209	.77732	.76290	.74880	.73503	.72157	.70843	.69557	.68301
5	.74726	.72988	.71299	.69656	.68058	.66505	.64993	.63523	.62092
6	.70496	.68533	.66634	.64796	.63017	.61295	.59627	.58012	.56447
7	.66506	.64351	.62275	.60275	.58349	.56493	.54703	.52979	.51316
8	.62741	.60423	.58201	.56070	.54027	.52067	.50187	.48382	.46651
9	.59190	.56735	.54393	.52158	.50025	.47988	.46043	.44185	.42410
10	.55839	.53273	.50835	.48519	.46319	.44229	.42241	.40351	.38554
11	.52679	.50021	.47509	.45134	.42888	.40764	.38753	.36851	.35049
12	.49697	.46968	.44401	.41985	.39711	.37570	.35553	.33654	.31863
13	.46884	.44102	.41496	.39056	.36770	.34627	.32618	.30734	.28966
14	.44230	.41410	.38782	.36331	.34046	.31914	.29925	.28067	.26333
15	.41727	.38883	.36245	.33797	.31524	.29414	.27454	.25632	.23939
16	.39365	.36510	.33873	.31439	.29189	.27110	.25187	.23409	.21763
17	.37136	.34281	.31657	.29245	.27027	.24986	.23107	.21378	.19784
18	.35034	.32189	.29586	.27205	.25025	.23028	.21199	.19523	.17986
19	.33051	.30224	.27651	.25307	.23171	.21224	.19449	.17829	.16351
20	.31180	.28380	.25842	.23541	.21455	.19562	.17843	.16282	.14864
21	.29416	.26648	.24151	.21899	.19866	.18029	.16370	.14870	.13513
22	.27751	.25021	.22571	.20371	.18394	.16617	.15018	.13580	.12285
23	.26180	.23494	.21095	.18950	.17032	.15315	.13778	.12402	.11168
24	.24698	.22060	.19715	.17628	.15770	.14115	.12640	.11326	.10153
25	.23300	.20714	.18425	.16398	.14602	.13009	.11597	.10343	.09230
26	.21981	.19450	.17220	.15254	.13520	.11990	.10639	.09446	.08391
27	.20737	.18263	.16093	.14190	.12519	.11051	.09761	.08626	.07628
28	.19563	.17148	.15040	.13200	.11591	.10185	.08955	.07878	.06934
29	.18456	.16101	.14056	.12279	.10733	.09387	.08215	.07194	.06304
30	.17411	.15119	.13137	.11422	.09938	.08652	.07537	.06570	.05731
31	.16425	.14196	.12277	.10625	.09202	.07974	.06915	.06000	.05210
32	.15496	.13329	.11474	.09884	.08520	.07349	.06344	.05480	.04736
33	.14619	.12516	.10723	.09194	.07889	.06774	.05820	.05004	.04306
34	.13791	.11752	.10022	.08553	.07305	.06243	.05339	.04570	.03914
35	.13011	.11035	.09366	.07956	.06763	.05754	.04899	.04174	.03558
36	.12274	.10361	.08754	.07401	.06262	.05303	.04494	.03811	.03235
37	.11579	.09729	.08181	.06885	.05799	.04888	.04123	.03481	.02941
38	.10924	.09135	.07646	.06404	.05369	.04505	.03783	.03179	.02673
39	.10306	.08578	.07146	.05958	.04971	.04152	.03470	.02903	.02430
40	.09722	.08054	.06678	.05542	.04603	.03827	.03184	.02651	.02209
41	.09172	.07563	.06241	.05155	.04262	.03527	.02921	.02421	.02009
42	.08653	.07101	.05833	.04796	.03946	.03250	.02680	.02211	.01826
43	.08163	.06668	.05451	.04461	.03654	.02996	.02458	.02019	.01660
44	.07701	.06261	.05095	.04150	.03383	.02761	.02255	.01844	.01509
45	.07265	.05879	.04761	.03860	.03133	.02545	.02069	.01684	.01372
46	.06854	.05520	.04450	.03591	.02901	.02345	.01898	.01538	.01247
47	.06466	.05183	.04159	.03340	.02686	.02162	.01742	.01405	.01134
48	.06100	.04867	.03887	.03107	.02487	.01992	.01598	.01283	.01031
49	.05755	.04570	.03632	.02891	.02303	.01836	.01466	.01171	.00937
50	.05429	.04291	.03395	.02689	.02132	.01692	.01345	.01070	.00852

$$v^n = (1 + r)^{-n}$$

* See Sec. 20. The present quantity which in n years will accumulate to 1 at the rate of interest r (expressed as decimal), compounded annually, is $(1 + r)^{-n}$. In this table the rate r is expressed in percent.

Table 23 AMOUNT OF AN ANNUITY* OF 1

n	1%	2%	2½%	3%	3½%	4%	4½%	5%	5½%
1	1.0000	1.0000	1.0000	1.0000	1.0000	1.0000	1.0000	1.0000	1.0000
2	2.0100	2.0200	2.0250	2.0300	2.0350	2.0400	2.0450	2.0500	2.0550
3	3.0301	3.0604	3.0756	3.0909	3.1062	3.1216	3.1370	3.1525	3.1680
4	4.0604	4.1216	4.1525	4.1836	4.2149	4.2465	4.2782	4.3101	4.3423
5	5.1010	5.2040	5.2563	5.3091	5.3625	5.4163	5.4707	5.5256	5.5811
6	6.1520	6.3081	6.3877	6.4684	6.5502	6.6330	6.7169	6.8019	6.8880
7	7.2135	7.4343	7.5474	7.6625	7.7794	7.8983	8.0192	8.1420	8.2669
8	8.2857	8.5830	8.7361	8.8923	9.0517	9.2142	9.3800	9.5491	9.7216
9	9.3685	9.7546	9.9545	10.1591	10.3685	10.5828	10.8021	11.0266	11.2563
10	10.4622	10.9497	11.2034	11.4639	11.7314	12.0061	12.2882	12.5779	12.8754
11	11.5668	12.1687	12.4835	12.8078	13.1420	13.4864	13.8412	14.2068	14.5835
12	12.6825	13.4121	13.7956	14.1920	14.6020	15.0258	15.4640	15.9171	16.3856
13	13.8093	14.6803	15.1404	15.6178	16.1130	16.6268	17.1599	17.7130	18.2868
14	14.9474	15.9739	16.5190	17.0863	17.6770	18.2919	18.9321	19.5986	20.2926
15	16.0969	17.2934	17.9319	18.5989	19.2957	20.0236	20.7841	21.5786	22.4087
16	17.2579	18.6393	19.3802	20.1569	20.9710	21.8245	22.7193	23.6575	24.6411
17	18.4304	20.0121	20.8647	21.7616	22.7050	23.6975	24.7417	25.8404	26.9964
18	19.6147	21.4123	22.3863	23.4144	24.4997	25.6454	26.8551	28.1324	29.4812
19	20.8109	22.8406	23.9460	25.1169	26.3572	27.6712	29.0636	30.5390	32.1027
20	22.0190	24.2974	25.5447	26.8704	28.2797	29.7781	31.3714	33.0660	34.8683
21	23.2392	25.7833	27.1833	28.6765	30.2695	31.9692	33.7831	35.7193	37.7861
22	24.4716	27.2990	28.8629	30.5368	32.3289	34.2480	36.3034	38.5052	40.8643
23	25.7163	28.8450	30.5844	32.4529	34.4604	36.6179	38.9370	41.4305	44.1118
24	26.9735	30.4219	32.3490	34.4265	36.6665	39.0826	41.6892	44.5020	47.5380
25	28.2432	32.0303	34.1578	36.4593	38.9499	41.6459	44.5652	47.7271	51.1526
26	29.5256	33.6709	36.0117	38.5530	41.3131	44.3117	47.5706	51.1135	54.9660
27	30.8209	35.3443	37.9120	40.7096	43.7591	47.0842	50.7113	54.6691	58.9891
28	32.1291	37.0512	39.8598	42.9309	46.2906	49.9676	53.9933	58.4026	63.2335
29	33.4504	38.7922	41.8563	45.2189	48.9108	52.9663	57.4230	62.3227	67.7114
30	34.7849	40.5681	43.9027	47.5754	51.6227	56.0849	61.0071	66.4388	72.4355
31	36.1327	42.3794	46.0003	50.0027	54.4295	59.3283	64.7524	70.7608	77.4194
32	37.4941	44.2270	48.1503	52.5028	57.3345	62.7015	68.6662	75.2988	82.6775
33	38.8690	46.1116	50.3540	55.0778	60.3412	66.2095	72.7562	80.0638	88.2248
34	40.2577	48.0338	52.6129	57.7302	63.4532	69.8579	77.0303	85.0670	94.0771
35	41.6603	49.9945	54.9282	60.4621	66.6740	73.6522	81.4966	90.3203	100.2514
36	43.0769	51.9944	57.3014	63.2759	70.0076	77.5983	86.1640	95.8363	106.7652
37	44.5076	54.0343	59.7339	66.1742	73.4579	81.7022	91.0413	101.6281	113.6373
38	45.9527	56.1149	62.2273	69.1594	77.0289	85.9703	96.1382	107,7095	120.8873
39	47.4123	58.2372	64.7830	72.2342	80.7249	90.4091	101.4644	114.0950	128.5361
40	48.8864	60.4020	67.4026	75.4013	84.5503	95.0255	107.0303	120.7998	136.6056
41	50.3752	62.6100	70.0876	78.6633	88.5095	99.8265	112.8467	127.8398	145.1189
42	51.8790	64.8622	72.8398	82.0232	92.6074	104.8196	118.9248	135.2318	154.1005
43	53.3978	67.1595	75.6608	85.4839	96.8486	110.0124	125.2764	142.9933	163.5760
44	54.9318	69.5027	78.5523	89.0484	101.2383	115.4129	131.9138	151.1430	173.5727
45	56.4811	71.8927	81.5161	92.7199	105.7817	121.0294	138.8500	159.7002	184.1192
46	58.0459	74.3306	84.5540	96.5015	110.4840	126.8706	146.0982	168.6852	195.2457
47	59.6263	76.8172	87.6679	100.3965	115.3510	132.9454	153.6726	178.1194	206.9842
48	61.2226	79.3535	90.8596	104.4084	120.3883	139.2632	161.5879	188.0254	219.3684
49	62.8348	81.9406	94.1311	108.5406	125.6018	145.8337	169.8594	198.4267	232.4336
50	64.4632	84.5794	97.4843	112.7969	130.9979	152.6671	178.5030	209.3480	246.2175

$$s_{\overline{n|}} \text{ at } r = \frac{(1+r)^n - 1}{r}$$

* See Sec. 21. If 1 is deposited at the end of each successive year (beginning one year hence) and if interest at rate r (expressed as decimal), compounded annually, is paid on the accumulated deposit at the end of each year, the total amount accumulated at the end of n years is $[(1+r)^n - 1]/r$. In this table the rate r is expressed in percent.

n	6%	6½%	7%	7½%	8%	8½%	9%	9½%	10%
1	1.0000	1.0000	1.0000	1.0000	1.0000	1.0000	1.0000	1.0000	1.0000
2	2.0600	2.0650	2.0700	2.0750	2.0800	2.0850	2.0900	2.0950	2.1000
3	3.1836	3.1992	3.2149	3.2306	3.2464	3.2622	3.2781	3.2940	3.3100
4	4.3746	4.4072	4.4399	4.4729	4.5061	4.5395	4.5731	4.6070	4.6410
5	5.6371	5.6936	5.7507	5.8084	5.8666	5.9254	5.9847	6.0446	6.1051
6	6.9753	7.0637	7.1533	7.2440	7.3359	7.4290	7.5233	7.6189	7.7156
7	8.3938	8.5229	8.6540	8.7873	8.9228	9.0605	9.2004	9.3426	9.4872
8	9.8975	10.0769	10.2598	10.4464	10.6366	10.8306	11.0285	11.2302	11.4359
9	11.4913	11.7319	11.9780	12.2298	12.4876	12.7512	13.0210	13.2971	13.5795
10	13.1808	13.4944	13.8164	14.1471	14.4866	14.8351	15.1929	15.5603	15.9374
11	14.9716	15.3716	15.7836	16.2081	16.6455	17.0961	17.5603	18.0385	18.5312
12	16.8699	17.3707	17.8885	18.4237	18.9771	19.5492	20.1407	20.7522	21.3843
13	18.8821	19.4998	20.1406	20.8055	21.4953	22.2109	22.9534	23.7236	24.5227
14	21.0151	21.7673	22.5505	23.3659	24.2149	25.0989	26.0192	26.9774	27.9750
15	23.2760	24.1822	25.1290	26.1184	27.1521	28.2323	29.3609	30.5402	31.7725
16	25.6725	26.7540	27.8881	29.0772	30.3243	31.6320	33.0034	34.4416	35.9497
17	28.2129	29.4930	30.8402	32.2580	33.7502	35.3207	36.9737	38.7135	40.5447
18	30.9057	32.4101	33.9990	35.6774	37.4502	39.3230	41.3013	43.3913	45.5992
19	33.7600	35.5167	37.3790	39.3532	41.4463	43.6654	46.0185	48.5135	51.1591
20	36.7856	38.8253	40.9955	43.3047	45.7620	48.3770	51.1601	54.1222	57.2750
21	39.9927	42.3490	44.8652	47.5525	50.4229	53.4891	56.7645	60.2638	64.0025
22	43.3923	46.1016	49.0057	52.1190	55.4568	59.0356	62.8733	66.9889	71.4027
23	46.9958	50.0982	53.4361	57.0279	60.8933	65.0537	69.5319	74.3529	79.5430
24	50.8156	54.3546	58.1767	62.3050	66.7648	71.5832	76.7898	82.4164	88.4973
25	54.8645	58.8877	63.2490	67.9779	73.1059	78.6678	84.7009	91.2459	98.3471
26	59.1564	63.7154	68.6765	74.0762	79.9544	86.3546	93.3240	100.9143	109.1818
27	63.7058	68.8569	74.4838	80.6319	87.3508	94.6947	102.7231	111.5012	121.0999
28	68.5281	74.3326	80.6977	87.6793	95.3388	103.7437	112.9682	123.0938	134.2099
29	73.6398	80.1642	87.3465	95.2553	103.9659	113.5620	124.1354	135.7877	148.6309
30	79.0582	86.3749	94.4608	103.3994	113.2832	124.2147	136.3075	149.6875	164.4940
31	84.8017	92.9892	102.0730	112.1544	123.3459	135.7730	149.5752	164.9078	181.9434
32	90.8898	100.0335	110.2182	121.5659	134.2135	148.3137	164.0370	181.5741	201.1378
33	97.3432	107.5357	118.9334	131.6834	145.9506	161.9203	179.8003	199.8236	222.2515
34	104.1838	115.5255	128.2588	142.5596	158.6267	176.6836	196.9823	219.8068	245.4767
35	111.4348	124.0347	138.2369	154.2516	172.3168	192.7017	215.7108	241.6885	271.0244
36	119.1209	133.0969	148.9135	166.8205	187.1021	210.0813	236.1247	265.6489	299.1268
37	127.2681	142.7482	160.3374	180.3320	203.0703	228.9382	258.3759	291.8855	330.0395
38	135.9042	153.0269	172.5610	194.8569	220.3159	249.3980	282.6298	320.6147	364.0434
39	145.0585	163.9736	185.6403	210.4712	238.9412	271.5968	309.0665	352.0731	401.4478
40	154.7620	175.6319	199.6351	227.2565	259.0565	295.6825	337.8824	386.5200	442.5926
41	165.0477	188.0480	214.6096	245.3008	280.7810	321.8156	369.2919	424.2394	487.8518
42	175.9505	201.2711	230.6322	264.6983	304.2435	350.1699	403.5281	465.5421	537.6370
43	187.5076	215.3537	247.7765	285.5507	329.5830	380.9343	440.8457	510.7686	592.4007
44	199.7580	230.3517	266.1209	307.9670	356.9496	414.3137	481.5218	560.2917	652.6408
45	212.7435	246.3246	285.7493	332.0645	386.5056	450.5304	525.8587	614.5194	718.9048
46	226.5081	263.3357	306.7518	357.9694	418.4261	489.8255	574.1860	673.8987	791.7953
47	241.0986	281.4525	329.2244	385.8171	452.9002	532.4606	626.8628	738.9191	871.9749
48	256.5645	300.7469	353.2701	415.7533	490.1322	578.7198	684.2804	810.1164	960.1723
49	272.9584	321.2955	378.9990	447.9348	530.3427	628.9110	746.8656	888.0775	1057.1896
50	290.3359	343.1797	406.5289	482.5299	573.7702	683.3684	815.0836	973.4448	1163.9085

$$s_{\overline{n|}} \text{ at } r = \frac{(1+r)^n - 1}{r}$$

* See Sec. 21. If 1 is deposited at the end of each successive year (beginning one year hence) and if interest at rate r (expressed as decimal), compounded annually, is paid on the accumulated deposit at the end of each year, the total amount accumulated at the end of n years is $[(1+r)^n - 1]/r$. In this table the rate r is expressed in percent.

Table 24 PRESENT VALUE OF AN ANNUITY* OF 1

n	1%	2%	2½%	3%	3½%	4%	4½%	5%	5½%
1	0.9901	.9804	.9756	.9709	.9662	.9615	.9569	.9524	.9479
2	1.9704	1.9416	1.9274	1.9135	1.8997	1.8861	1.8727	1.8594	1.8463
3	2.9410	2.8839	2.8560	2.8286	2.8016	2.7751	2.7490	2.7232	2.6979
4	3.9020	3.8077	3.7620	3.7171	3.6731	3.6299	3.5875	3.5460	3.5052
5	4.8534	4.7135	4.6458	4.5797	4.5151	4.4518	4.3900	4.3295	4.2703
6	5.7955	5.6014	5.5081	5.4172	5.3286	5.2421	5.1579	5.0757	4.9955
7	6.7282	6.4720	6.3494	6.2303	6.1145	6.0021	5.8927	5.7864	5.6830
8	7.6517	7.3255	7.1701	7.0197	6.8740	6.7327	6.5959	6.4632	6.3346
9	8.5660	8.1622	7.9709	7.7861	7.6077	7.4353	7.2688	7.1078	6.9522
10	9.4713	8.9826	8.7521	8.5302	8.3166	8.1109	7.9127	7.7217	7.5376
11	10.3676	9.7868	9.5142	9.2526	9.0016	8.7605	8.5289	8.3064	8.0925
12	11.2551	10.5753	10.2578	9.9540	9.6633	9.3851	9.1186	8.8633	8.6185
13	12.1337	11.3484	10.9832	10.6350	10.3027	9.9856	9.6829	9.3936	9.1171
14	13.0037	12.1062	11.6909	11.2961	10.9205	10.5631	10.2228	9.8986	9.5896
15	13.8651	12.8493	12.3814	11.9379	11.5174	11.1184	10.7395	10.3797	10.0376
16	14.7179	13.5777	13.0550	12.5611	12.0941	11.6523	11.2340	10.8378	10.4622
17	15.5623	14.2919	13.7122	13.1661	12.6513	12.1657	11.7072	11.2741	10.8646
18	16.3983	14.9920	14.3534	13.7535	13.1897	12.6593	12.1600	11.6896	11.2461
19	17.2260	15.6785	14.9789	14.3238	13.7098	13.1339	12.5933	12.0853	11.6077
20	18.0456	16.3514	15.5892	14.8775	14.2124	13.5903	13.0079	12.4622	11.9504
21	18.8570	17.0112	16.1845	15.4150	14.6980	14.0292	13.4047	12.8212	12.2752
22	19.6604	17.6580	16.7654	15.9369	15.1671	14.4511	13.7844	13.1630	12.5832
23	20.4558	18.2922	17.3321	16.4436	15.6204	14.8568	14.1478	13.4886	12.8750
24	21.2434	18.9139	17.8850	16.9355	16.0584	15.2470	14.4955	13.7986	13.1517
25	22.0232	19.5235	18.4244	17.4131	16.4815	15.6221	14.8282	14.0939	13.4139
26	22.7952	20.1210	18.9506	17.8768	16.8904	15.9828	15.1466	14.3752	13.6625
27	23.5596	20.7069	19.4640	18.3270	17.2854	16.3296	15.4513	14.6430	13.8981
28	24.3164	21.2813	19.9649	18.7641	17.6670	16.6631	15.7429	14.8981	14.1214
29	25.0658	21.8444	20.4535	19.1885	18.0358	16.9837	16.0219	15.1411	14.3331
30	25.8077	22.3965	20.9303	19.6004	18.3920	17.2920	16.2889	15.3725	14.5337
31	26.5423	22.9377	21.3954	20.0004	18.7363	17.5885	16.5444	15.5928	14.7239
32	27.2696	23.4683	21.8492	20.3888	19.0689	17.8736	16.7889	15.8027	14.9042
33	27.9897	23.9886	22.2919	20.7658	19.3902	18.1476	17.0229	16.0025	15.0751
34	28.7027	24.4986	22.7238	21.1318	19.7007	18.4112	17.2468	16.1929	15.2370
35	29.4086	24.9986	23.1452	21.4872	20.0007	18.6646	17.4610	16.3742	15.3906
36	30.1075	25.4888	23.5563	21.8323	20.2905	18.9083	17.6660	16.5469	15.5361
37	30.7995	25.9695	23.9573	22.1672	20.5705	19.1426	17.8622	16.7113	15.6740
38	31.4847	26.4406	24.3486	22.4925	20.8411	19.3679	18.0500	16.8679	15.8047
39	32.1630	26.9026	24.7303	22.8082	21.1025	19.5845	18.2297	17.0170	15.9287
40	32.8347	27.3555	25.1028	23.1148	21.3551	19.7928	18.4016	17.1591	16.0461
41	33.4997	27.7995	25.4661	23.4124	21.5991	19.9931	18.5661	17.2944	16.1575
42	34.1581	28.2348	25.8206	23.7014	21.8349	20.1856	18.7235	17.4232	16.2630
43	34.8100	28.6616	26.1664	23.9819	22.0627	20.3708	18.8742	17.5459	16.3630
44	35.4555	29.0800	26.5038	24.2543	22.2828	20.5488	19.0184	17.6628	16.4578
45	36.0945	29.4902	26.8330	24.5187	22.4955	20.7200	19.1563	17.7741	16.5477
46	36.7272	29.8923	27.1542	24.7754	22.7009	20.8847	19.2884	17.8801	16.6329
47	37.3537	30.2866	27.4675	25.0247	22.8994	21.0429	19.4147	17.9810	16.7137
48	37.9740	30.6731	27.7732	25.2667	23.0912	21.1951	19.5356	18.0772	16.7902
49	38.5881	31.0521	28.0714	25.5017	23.2766	21.3415	19.6513	18.1687	16.8628
50	39.1961	31.4236	28.3623	25.7298	23.4556	21.4822	19.7620	18.2559	16.9315

$$a_{\overline{n}|} \text{ at } r = \frac{1 - (1 + r)^{-n}}{r}$$

* See Sec. 21. The total present amount which will supply an annuity of 1 at the end of each year for n years, beginning one year hence (assuming that in successive years the amount not yet paid out earns interest at rate r, compounded annually), is $[1 - (1 + r)^{-n}]/r$. In this table the rate r is expressed in percent.

n	6%	6½%	7%	7½%	8%	8½%	9%	9½%	10%
1	.9434	.9390	.9346	.9302	.9259	.9217	.9174	.9132	.9091
2	1.8334	1.8206	1.8080	1.7956	1.7833	1.7711	1.7591	1.7473	1.7355
3	2.6730	2.6485	2.6243	2.6005	2.5771	2.5540	2.5313	2.5089	2.4869
4	3.4651	3.4258	3.3872	3.3493	3.3121	3.2756	3.2397	3.2045	3.1699
5	4.2124	4.1557	4.1002	4.0459	3.9927	3.9406	3.8897	3.8397	3.7908
6	4.9173	4.8410	4.7665	4.6938	4.6229	4.5536	4.4859	4.4198	4.3553
7	5.5824	5.4845	5.3893	5.2966	5.2064	5.1185	5.0330	4.9496	4.8684
8	6.2098	6.0887	5.9713	5.8573	5.7466	5.6392	5.5348	5.4334	5.3349
9	6.8017	6.6561	6.5152	6.3789	6.2469	6.1191	5.9952	5.8753	5.7590
10	7.3601	7.1888	7.0236	6.8641	6.7101	6.5613	6.4177	6.2788	6.1446
11	7.8869	7.6890	7.4987	7.3154	7.1390	6.9690	6.8052	6.6473	6.4951
12	8.3838	8.1587	7.9427	7.7353	7.5361	7.3447	7.1607	6.9838	6.8137
13	8.8527	8.5997	8.3576	8.1258	7.9038	7.6910	7.4869	7.2912	7.1034
14	9.2950	9.0138	8.7455	8.4892	8.2442	8.0101	7.7862	7.5719	7.3667
15	9.7122	9.4027	9.1079	8.8271	8.5595	8.3042	8.0607	7.8282	7.6061
16	10.1059	9.7678	9.4466	9.1415	8.8514	8.5753	8.3126	8.0623	7.8237
17	10.4773	10.1106	9.7632	9.4340	9.1216	8.8252	8.5436	8.2760	8.0216
18	10.8276	10.4325	10.0591	9.7060	9.3719	9.0555	8.7556	8.4713	8.2014
19	11.1581	10.7347	10.3356	9.9591	9.6036	9.2677	8.9501	8.6496	8.3649
20	11.4699	11.0185	10.5940	10.1945	9.8181	9.4633	9.1285	8.8124	8.5136
21	11.7641	11.2850	10.8355	10.4135	10.0168	9.6436	9.2922	8.9611	8.6487
22	12.0416	11.5352	11.0612	10.6172	10.2007	9.8098	9.4424	9.0969	8.7715
23	12.3034	11.7701	11.2722	10.8067	10.3711	9.9629	9.5802	9.2209	8.8832
24	12.5504	11.9907	11.4693	10.9830	10.5288	10.1041	9.7066	9.3341	8.9847
25	12.7834	12.1979	11.6536	11.1469	10.6748	10.2342	9.8226	9.4376	9.0770
26	13.0032	12.3924	11.8258	11.2995	10.8100	10.3541	9.9290	9.5320	9.1609
27	13.2105	12.5750	11.9867	11.4414	10.9352	10.4646	10.0266	9.6183	9.2372
28	13.4062	12.7465	12.1371	11.5734	11.0511	10.5665	10.1161	9.6971	9.3066
29	13.5907	12.9075	12.2777	11.6962	11.1584	10.6603	10.1983	9.7690	9.3696
30	13.7648	13.0587	12.4090	11.8104	11.2578	10.7468	10.2737	9.8347	9.4269
31	13.9291	13.2006	12.5318	11.9166	11.3498	10.8266	10.3428	9.8947	9.4790
32	14.0840	13.3339	12.6465	12.0155	11.4350	10.9001	10.4062	9.9495	9.5264
33	14.2302	13.4591	12.7538	12.1074	11.5139	10.9678	10.4644	9.9996	9.5694
34	14.3681	13.5766	12.8540	12.1929	11.5869	11.0302	10.5178	10.0453	9.6086
35	14.4982	13.6870	12.9477	12.2725	11.6546	11.0878	10.5668	10.0870	9.6442
36	14.6210	13.7906	13.0352	12.3465	11.7172	11.1408	10.6118	10.1251	9.6765
37	14.7368	13.8879	13.1170	12.4154	11.7752	11.1897	10.6530	10.1599	9.7059
38	14.8460	13.9792	13.1935	12.4794	11.8289	11.2347	10.6908	10.1917	9.7327
39	14.9491	14.0650	13.2649	12.5390	11.8786	11.2763	10.7255	10.2207	9.7570
40	15.0463	14.1455	13.3317	12.5944	11.9246	11.3145	10.7574	10.2472	9.7790
41	15.1380	14.2212	13.3941	12.6460	11.9672	11.3498	10.7866	10.2715	9.7991
42	15.2245	14.2922	13.4524	12.6939	12.0067	11.3823	10.8134	10.2936	9.8174
43	15.3062	14.3588	13.5070	12.7385	12.0432	11.4123	10.8380	10.3138	9.8340
44	15.3832	14.4214	13.5579	12.7800	12.0771	11.4399	10.8605	10.3322	9.8491
45	15.4558	14.4802	13.6055	12.8186	12.1084	11.4653	10.8812	10.3490	9.8628
46	15.5244	14.5354	13.6500	12.8545	12.1374	11.4888	10.9002	10.3644	9.8753
47	15.5890	14.5873	13.6916	12.8879	12.1643	11.5104	10.9176	10.3785	9.8866
48	15.6500	14.6359	13.7305	12.9190	12.1891	11.5303	10.9336	10.3913	9.8969
49	15.7076	14.6816	13.7668	12.9479	12.2122	11.5487	10.9482	10.4030	9.9063
50	15.7619	14.7245	13.8007	12.9748	12.2335	11.5656	10.9617	10.4137	9.9148

$$a_{\overline{n}|} \text{ at } r = \frac{1 - (1 + r)^{-n}}{r}$$

* See Sec. 21. The total present amount which will supply an annuity of 1 at the end of each year for n years, beginning one year hence (assuming that in successive years the amount not yet paid out earns interest at rate r, compounded annually), is $[1 - (1 + r)^{-n}]/r$. In this table the rate r is expressed in percent.

Table 25 THE ANNUITY* THAT 1 WILL PURCHASE

n	1%	2%	2½%	3%	3½%	4%	4½%	5%	5½%
1	1.01000	1.02000	1.02500	1.03000	1.03500	1.04000	1.04500	1.05000	1.05500
2	0.50751	0.51505	0.51883	0.52261	0.52640	0.53020	0.53400	0.53780	0.54162
3	.34002	.34675	.35014	.35353	.35693	.36035	.36377	.36721	.37065
4	.25628	.26262	.26582	.26903	.27225	.27549	.27874	.28201	.28529
5	.20604	.21216	.21525	.21835	.22148	.22463	.22779	.23097	.23418
6	.17255	.17853	.18155	.18460	.18767	.19076	.19388	.19702	.20018
7	.14863	.15451	.15750	.16051	.16354	.16661	.16970	.17282	.17596
8	.13069	.13651	.13947	.14246	.14548	.14853	.15161	.15472	.15786
9	.11674	.12252	.12546	.12843	.13145	.13449	.13757	.14069	.14384
10	.10558	.11133	.11426	.11723	.12024	.12329	.12638	.12950	.13267
11	.09645	.10218	.10511	.10808	.11109	.11415	.11725	.12039	.12357
12	.08885	.09456	.09749	.10046	.10348	.10655	.10967	.11283	.11603
13	.08241	.08812	.09105	.09403	.09706	.10014	.10328	.10646	.10968
14	.07690	.08260	.08554	.08853	.09157	.09467	.09782	.10102	.10428
15	.07212	.07783	.08077	.08377	.08683	.08994	.09311	.09634	.09963
16	.06794	.07365	.07660	.07961	.08268	.08582	.08902	.09227	.09558
17	.06426	.06997	.07293	.07595	.07904	.08220	.08542	.08870	.09205
18	.06098	.06670	.06967	.07271	.07582	.07899	.08224	.08555	.08892
19	.05805	.06378	.06676	.06981	.07294	.07614	.07941	.08275	.08615
20	.05542	.06116	.06415	.06722	.07036	.07358	.07688	.08024	.08368
21	.05303	.05878	.06179	.06487	.06804	.07128	.07460	.07800	.08146
22	.05086	.05663	.05965	.06275	.06593	.06920	.07255	.07597	.07947
23	.04889	.05467	.05770	.06081	.06402	.06731	.07068	.07414	.07767
24	.04707	.05287	.05591	.05905	.06227	.06559	.06899	.07247	.07604
25	.04541	.05122	.05428	.05743	.06067	.06401	.06744	.07095	.07455
26	.04387	.04970	.05277	.05594	.05921	.06257	.06602	.06956	.07319
27	.04245	.04829	.05138	.05456	.05785	.06124	.06472	.06829	.07195
28	.04112	.04699	.05009	.05329	.05660	.06001	.06352	.06712	.07081
29	.03990	.04578	.04889	.05211	.05545	.05888	.06241	.06605	.06977
30	.03875	.04465	.04778	.05102	.05437	.05783	.06139	.06505	.06880
31	.03768	.04360	.04674	.05000	.05337	.05686	.06044	.06413	.06792
32	.03667	.04261	.04577	.04905	.05244	.05595	.05956	.06328	.06710
33	.03573	.04169	.04486	.04816	.05157	.05510	.05874	.06249	.06633
34	.03484	.04082	.04401	.04732	.05076	.05431	.05798	.06176	.06563
35	.03400	.04000	.04321	.04654	.05000	.05358	.05727	.06107	.06497
36	.03321	.03923	.04245	.04580	.04928	.05289	.05661	.06043	.06437
37	.03247	.03851	.04174	.04511	.04861	.05224	.05598	.05984	.06380
38	.03176	.03782	.04107	.04446	.04798	.05163	.05540	.05928	.06327
39	.03109	.03717	.04044	.04384	.04739	.05106	.05486	.05876	.06278
40	.03046	.03656	.03984	.04326	.04683	.05052	.05434	.05828	.06232
41	.02985	.03597	.03927	.04271	.04630	.05002	.05386	.05782	.06189
42	.02928	.03542	.03873	.04219	.04580	.04954	.05341	.05739	.06149
43	.02873	.03489	.03822	.04170	.04533	.04909	.05298	.05699	.06111
44	.02820	.03439	.03773	.04123	.04488	.04866	.05258	.05662	.06076
45	.02771	.03391	.03727	.04079	.04445	.04826	.05220	.05626	.06043
46	.02723	.03345	.03683	.04036	.04405	.04788	.05184	.05593	.06012
47	.02677	.03302	.03641	.03996	.04367	.04752	.05151	.05561	.05983
48	.02633	.03260	.03601	.03958	.04331	.04718	.05119	.05532	.05956
49	.02591	.03220	.03562	.03921	.04296	.04686	.05089	.05504	.05930
50	.02551	.03182	.03526	.03887	.04263	.04655	.05060	.05478	.05906

$$a_{\overline{n}|}^{-1} \text{ at } r = \frac{r}{1 - (1 + r)^{-n}}$$

* See Sec. 21. One (1) now will purchase an annuity $r/[1 - (1 + r)^{-n}]$ at the end of each year for n years, beginning one year hence (assuming that in successive years the amount not yet paid out earns interest at rate r, expressed as decimal, compounded annually). In this table the rate r is expressed in percent.

The total amount accumulated at the end of n years is 1 if an annuity $s_{\overline{n}|}^{-1}$ is deposited at the end of each successive year (beginning one year hence), and interest at rate r, compounded annually, is paid on the accumulated deposit at the end of each year.

$$s_{\overline{n}|}^{-1} = a_{\overline{n}|}^{-1} - r$$

n	6%	$6\frac{1}{2}$%	7%	$7\frac{1}{2}$%	8%	$8\frac{1}{2}$%	9%	$9\frac{1}{2}$%	10%
1	1.06000	1.06500	1.07000	1.07500	1.08000	1.08500	1.09000	1.09500	1.10000
2	0.54544	0.54926	0.55309	0.55693	0.56077	0.56462	0.56847	0.57233	0.57619
3	.37411	.37758	.38105	.38454	.38803	.39154	.39505	.39858	.40211
4	.28859	.29190	.29523	.29857	.30192	.30529	.30867	.31206	.31547
5	.23740	.24063	.24389	.24716	.25046	.25377	.25709	.26044	.26380
6	.20336	.20657	.20980	.21304	.21632	.21961	.22292	.22625	.22961
7	.17914	.18233	.18555	.18880	.19207	.19537	.19869	.20204	.20540
8	.16104	.16424	.16747	.17073	.17401	.17733	.18067	.18404	.18744
9	.14702	.15024	.15349	.15677	.16008	.16342	.16680	.17020	.17364
10	.13587	.13910	.14238	.14568	.14903	.15241	.15582	.15927	.16274
11	.12679	.13006	.13336	.13670	.14008	.14349	.14695	.15044	.15396
12	.11928	.12257	.12590	.12928	.13270	.13615	.13965	.14319	.14676
13	.11296	.11628	.11965	.12306	.12652	.13002	.13357	.13715	.14078
14	.10758	.11094	.11434	.11780	.12130	.12484	.12843	.13207	.13575
15	.10296	.10635	.10979	.11329	.11683	.12042	.12406	.12774	.13147
16	.09895	.10238	.10586	.10939	.11298	.11661	.12030	.12403	.12782
17	.09544	.09891	.10242	.10600	.10963	.11331	.11705	.12083	.12466
18	.09236	.09585	.09941	.10303	.10670	.11043	.11421	.11805	.12193
19	.08962	.09316	.09675	.10041	.10413	.10790	.11173	.11561	.11955
20	.08718	.09076	.09439	.09809	.10185	.10567	.10955	.11348	.11746
21	.08500	.08861	.09229	.09602	.09983	.10370	.10762	.11159	.11562
22	.08304	.08669	.09041	.09419	.09803	.10194	.10590	.10993	.11400
23	.08128	.08496	.08871	.09254	.09642	.10037	.10438	.10845	.11257
24	.07968	.08340	.08719	.09105	.09498	.09897	.10302	.10713	.11130
25	.07823	.08198	.08581	.08971	.09368	.09771	.10181	.10596	.11017
26	.07690	.08069	.08456	.08850	.09251	.09658	.10072	.10491	.10916
27	.07570	.07952	.08343	.08740	.09145	.09556	.09973	.10397	.10826
28	.07459	.07845	.08239	.08640	.09049	.09464	.09885	.10312	.10745
29	.07358	.07747	.08145	.08550	.08962	.09380	.09806	.10236	.10673
30	.07265	.07658	.08059	.08467	.08883	.09305	.09734	.10168	.10608
31	.07179	.07575	.07980	.08392	.08811	.09236	.09668	.10106	.10550
32	.07100	.07500	.07907	.08323	.08745	.09174	.09610	.10051	.10497
33	.07027	.07430	.07841	.08259	.08685	.09118	.09556	.10000	.10450
34	.06960	.07366	.07780	.08201	.08630	.09066	.09508	.09955	.10407
35	.06897	.07306	.07723	.08148	.08580	.09019	.09464	.09914	.10369
36	.06839	.07251	.07672	.08099	.08534	.08976	.09424	.09876	.10334
37	.06786	.07200	.07624	.08054	.08492	.08937	.09387	.09843	.10303
38	.06736	.07153	.07580	.08013	.08454	.08901	.09354	.09812	.10275
39	.06689	.07110	.07539	.07975	.08418	.08868	.09324	.09784	.10249
40	.06646	.07069	.07501	.07940	.08386	.08838	.09296	.09759	.10226
41	.06606	.07032	.07466	.07908	.08356	.08811	.09271	.09736	.10205
42	.06568	.06997	.07434	.07878	.08329	.08786	.09248	.09715	.10186
43	.06533	.06964	.07404	.07850	.08303	.08762	.09227	.09696	.10169
44	.06501	.06934	.07376	.07825	.08280	.08741	.09208	.09678	.10153
45	.06470	.06906	.07350	.07801	.08259	.08722	.09190	.09663	.10139
46	.06441	.06880	.07326	.07779	.08239	.08704	.09174	.09648	.10126
47	.06415	.06855	.07304	.07759	.08221	.08688	.09160	.09635	.10115
48	.06390	.06833	.07283	.07740	.08204	.08673	.09146	.09623	.10104
49	.06366	.06811	.07264	.07723	.08188	.08659	.09134	.09613	.10094
50	.06344	.06791	.07246	.07707	.08174	.08646	.09123	.09603	.10086

$$a_{\overline{n}|}^{-1} \text{ at } r = \frac{r}{1-(1+r)^{-n}}$$

* See Sec. 21. One (1) now will purchase an annuity $r/[1-(1+r)^{-n}]$ at the end of each year for n years, beginning one year hence (assuming that in successive years the amount not yet paid out earns interest at rate r, expressed as decimal, compounded annually). In this table the rate r is expressed in percent.

The total amount accumulated at the end of n years is 1 if an annuity $s_{\overline{n}|}^{-1}$ is deposited at the end of each successive year (beginning one year hence) and if interest at rate r, compounded annually, is paid on the accumulated deposit at the end of each year.

$$s_{\overline{n}|}^{-1} = a_{\overline{n}|}^{-1} - r$$

Table 26 LOGARITHMS FOR INTEREST COMPUTATIONS

r	1 + r	log (1 + r)	r	1 + r	log (1 + r)
½%	1.005	00216 60617 56508	5½%	1.055	02325 24596 33711
1%	1.010	00432 13737 82643	6%	1.060	02530 58652 64770
1½%	1.015	00646 60422 49232	6½%	1.065	02734 96077 74757
2%	1.020	00860 01717 61918	7%	1.070	02938 37776 85210
2½%	1.025	01072 38653 91773	7½%	1.075	03140 84642 51624
3%	1.030	01283 72247 05172	8%	1.080	03342 37554 86950
3½%	1.035	01494 03497 92937	8½%	1.085	03542 97381 84548
4%	1.040	01703 33392 98780	9%	1.090	03742 64979 40624
4½%	1.045	01911 62904 47073	9½%	1.095	03941 41191 76137
5%	1.050	02118 92990 69938	10%	1.100	04139 26851 58225

The amount A of principal P at compound interest after n years is $A = P(1 + r)^n$.

Table 27 AMERICAN EXPERIENCE MORTALITY TABLE

x	l_x	d_x	q_x	p_x
10	100 000	749	0.007 490	0.992 510
11	99 251	746	0.007 516	0.992 484
12	98 505	743	0.007 543	0.992 457
13	97 762	740	0.007 569	0.992 431
14	97 022	737	0.007 596	0.992 404
15	96 285	735	0.007 634	0.992 366
16	95 550	732	0.007 661	0.992 339
17	94 818	729	0.007 688	0.992 312
18	94 089	727	0.007 727	0.992 273
19	93 362	725	0.007 765	0.992 235
20	92 637	723	0.007 805	0.992 195
21	91 914	722	0.007 855	0.992 145
22	91 192	721	0.007 906	0.992 094
23	90 471	720	0.007 958	0.992 042
24	89 751	719	0.008 011	0.991 989
25	89 032	718	0.008 065	0.991 935
26	88 314	718	0.008 130	0.991 870
27	87 596	718	0.008 197	0.991 803
28	86 878	718	0.008 264	0.991 736
29	86 160	719	0.008 345	0.991 655
30	85 441	720	0.008 427	0.991 573
31	84 721	721	0.008 510	0.991 490
32	84 000	723	0.008 607	0.991 393
33	83 277	726	0.008 718	0.991 282
34	82 551	729	0.008 831	0.991 169
35	81 822	732	0.008 946	0.991 054
36	81 090	737	0.009 089	0.990 911
37	80 353	742	0.009 234	0.990 766
38	79 611	749	0.009 408	0.990 592
39	78 862	756	0.009 586	0.990 414
40	78 106	765	0.009 794	0.990 206
41	77 341	774	0.010 008	0.989 992
42	76 567	785	0.010 252	0.989 748
43	75 782	797	0.010 517	0.989 483
44	74 985	812	0.010 829	0.989 171
45	74 173	828	0.011 163	0.988 837
46	73 345	848	0.011 562	0.988 438
47	72 497	870	0.012 000	0.988 000
48	71 627	896	0.012 509	0.987 491
49	70 731	927	0.013 106	0.986 894
50	69 804	962	0.013 781	0.986 219
51	68 842	1 001	0.014 541	0.985 459
52	67 841	1 044	0.015 389	0.984 611
53	66 797	1 091	0.016 333	0.983 667
54	65 706	1 143	0.017 396	0.982 604
55	64 563	1 199	0.018 571	0.981 429
56	63 364	1 260	0.019 885	0.980 115
57	62 104	1 325	0.021 335	0.978 665
58	60 779	1 394	0.022 936	0.977 064
59	59 385	1 468	0.024 720	0.975 280
60	57 917	1 546	0.026 693	0.973 307
61	56 371	1 628	0.028 880	0.971 120
62	54 743	1 713	0.031 292	0.968 708
63	53 030	1 800	0.033 943	0.966 057
64	51 230	1 889	0.036 873	0.963 127
65	49 341	1 980	0.040 129	0.959 871
66	47 361	2 070	0.043 707	0.956 293
67	45 291	2 158	0.047 647	0.952 353
68	43 133	2 243	0.052 002	0.947 998
69	40 890	2 321	0.056 762	0.943 238
70	38 569	2 391	0.061 993	0.938 007
71	36 178	2 448	0.067 665	0.932 335
72	33 730	2 487	0.073 733	0.926 267
73	31 243	2 505	0.080 178	0.919 822
74	28 738	2 501	0.087 028	0.912 972
75	26 237	2 476	0.094 371	0.905 629
76	23 761	2 431	0.102 311	0.897 689
77	21 330	2 369	0.111 064	0.888 936
78	18 961	2 291	0.120 827	0.879 173
79	16 670	2 196	0.131 734	0.868 266
80	14 474	2 091	0.144 466	0.855 534
81	12 383	1 964	0.158 605	0.841 395
82	10 419	1 816	0.174 297	0.825 703
83	8 603	1 648	0.191 561	0.808 439
84	6 955	1 470	0.211 359	0.788 641
85	5 485	1 292	0.235 552	0.764 448
86	4 193	1 114	0.265 681	0.734 319
87	3 079	933	0.303 020	0.696 980
88	2 146	744	0.346 692	0.653 308
89	1 402	555	0.395 863	0.604 137
90	847	385	0.454 545	0.545 455
91	462	246	0.532 468	0.467 532
92	216	137	0.634 259	0.365 741
93	79	58	0.734 177	0.265 823
94	21	18	0.857 143	0.142 857
95	3	3	1.000 000	0.000 000

Based on 100,000 living at age 10 years (x = age in years):
l_x = number of original 100,000 who live to reach age x
d_x = number of original 100,000 who live to reach age x but die before age $(x + 1)$
q_x = probability of dying before age $(x + 1)$ if alive at age x. $q_x = d_x/l_x$
p_x = probability of living to age $(x + 1)$ if alive at age x. $p_x = 1 - q_x$

CONSTANTS, EQUIVALENTS, AND CONVERSION FACTORS

Table 28 DECIMAL EQUIVALENTS OF COMMON FRACTIONS

1/64 =	0.015625
1/32 = 2/64 =	.03125
3/64 =	.046875
1/16 = 2/32 = 4/64 =	.0625
5/64 =	.078125
3/32 = 6/64 =	.09375
7/64 =	.109375
1/8 = 4/32 = 8/64 =	0.125
9/64 =	.140625
5/32 = 10/64 =	.15625
11/64 =	.171875
3/16 = 6/32 = 12/64 =	.1875
13/64 =	.203125
7/32 = 14/64 =	.21875
15/64 =	.234375
1/4 = 8/32 = 16/64 =	0.25
17/64 =	.265625
9/32 = 18/64 =	.28125
19/64 =	.296875
5/16 = 10/32 = 20/64 =	.3125
21/64 =	.328125
11/32 = 22/64 =	.34375
23/64 =	.359375
3/8 = 12/32 = 24/64 =	0.375
25/64 =	.390625
13/32 = 26/64 =	.40625
27/64 =	.421875
7/16 = 14/32 = 28/64 =	.4375
29/64 =	.453125
15/32 = 30/64 =	.46875
31/64 =	.484375

1/2 = 16/32 = 32/64 =	0.50
33/64 =	.515625
17/32 = 34/64 =	.53125
35/64 =	.546875
9/16 = 18/32 = 36/64 =	.5625
37/64 =	.578125
19/32 = 38/64 =	.59375
39/64 =	.609375
5/8 = 20/32 = 40/64 =	0.625
41/64 =	.640625
21/32 = 42/64 =	.65625
43/64 =	.671875
11/16 = 22/32 = 44/64 =	.6875
45/64 =	.703125
23/32 = 46/64 =	.71875
47/64 =	.734375
3/4 = 24/32 = 48/64 =	0.75
49/64 =	.765625
25/32 = 50/64 =	.78125
51/64 =	.796875
13/16 = 26/32 = 52/64 =	.8125
53/64 =	.828125
27/32 = 54/64 =	.84375
55/64 =	.859375
7/8 = 28/32 = 56/64 =	0.875
57/64 =	.890625
29/32 = 58/64 =	.90625
59/64 =	.921875
15/16 = 30/32 = 60/64 =	.9375
61/64 =	.953125
31/32 = 62/64 =	.96875
63/64 =	.984375

SQUARE ROOTS OF CERTAIN COMMON FRACTIONS

N	$\sqrt{N}$	N	$\sqrt{N}$	N	$\sqrt{N}$
1/2	0.7071	1/8	0.3536	1/16	0.2500
1/3	0.5774	3/8	0.6124	3/16	0.4330
2/3	0.8165	5/8	0.7906	5/16	0.5590
1/4	0.5000	7/8	0.9354	7/16	0.6614
3/4	0.8660	1/9	0.3333	9/16	0.7500
1/5	0.4472	2/9	0.4714	11/16	0.8292
2/5	0.6325	4/9	0.6667	13/16	0.9014
3/5	0.7746	5/9	0.7454	15/16	0.9682
4/5	0.8944	7/9	0.8819	1/32	0.1768
1/6	0.4082	8/9	0.9428	1/64	0.1250
5/6	0.9129	1/12	0.2887	1/50	0.1414
1/7	0.3780	5/12	0.6455		
2/7	0.5345	7/12	0.7638		
3/7	0.6547	11/12	0.9574		
4/7	0.7559				
5/7	0.8452				
6/7	0.9258				

N	Log N	N	Log N
$\pi = 3.14159265$	0.4971499	$\pi^2 = 9.86960440$	0.9942997
$2\pi = 6.28318531$	0.7981799	$\frac{1}{\pi^2} = 0.10132118$	9.0057003–10
$4\pi = 12.56637061$	1.0992099	$\sqrt{\pi} = 1.77245385$	0.2485749
$\frac{\pi}{2} = 1.57079633$	0.1961199	$\frac{1}{\sqrt{\pi}} = 0.56418958$	9.7514251–10
$\frac{\pi}{3} = 1.04719755$	0.0200286	$\sqrt{\frac{3}{\pi}} = 0.97720502$	9.9899857–10
$\frac{4\pi}{3} = 4.18879020$	0.6220886	$\sqrt{\frac{4}{\pi}} = 1.12837917$	0.0524551
$\frac{\pi}{4} = 0.78539816$	9.8950899–10	$\sqrt[3]{\pi} = 1.46459189$	0.1657166
$\frac{\pi}{6} = 0.52359878$	9.7189986–10	$\frac{1}{\sqrt[3]{\pi}} = 0.68278406$	9.8342834–10
$\frac{1}{\pi} = 0.31830989$	9.5028501–10	$\sqrt[3]{\pi^2} = 2.14502940$	0.3314332
$\frac{1}{2\pi} = 0.15915494$	9.2018201–10	$\sqrt[3]{\frac{3}{4\pi}} = 0.62035049$	9.7926371–10
$\frac{3}{\pi} = 0.95492966$	9.9799714–10	$\sqrt[3]{\frac{\pi}{6}} = 0.80599598$	9.9063329–10
$\frac{4}{\pi} = 1.27323954$	0.1049101		

e = Napierian base	= 2.71828183	0.43429448
$M = \log_{10} e$	= 0.43429448	9.63778431—10
$1/M = \log_e 10$	= 2.30258509	0.36221569
$180/\pi$ = degrees in 1 radian	= 57.2957795	1.75812263
$\pi/180$ = radians in 1°	= 0.01745329	8.24187737—10
$\pi/10800$ = radians in 1′	= 0.0002908882	6.46372612—10
$\pi/648000$ = radians in 1″	= 0.000004848136811095	4.68557487—10
sin 1″	= 0.000004848136811076	4.68557487—10
tan 1″	= 0.000004848136811133	4.68557487—10
centimeters in 1 (U.S.) ft	= 30.4800	1.4840150
feet in 1 cm	= 0.0328084	8.5159850—10
inches in 1 m	= 39.3700	1.5951663
pounds in 1 kg	= 2.20462	0.3433343
kilograms in 1 lb	= 0.453592	9.6566657—10
cubic inches in 1 (U.S.) gal	= 231	2.3636120
g (average value)	= 32.16 ft/sec/sec	1.5073
g (legal)	= 980.665 cm/sec^2	2.9915207
weight of 1 cu ft of water	= 62.425 lb (max. density)	1.7953586
weight of 1 cu ft of air	= 0.0807 lb (at 32°F)	8.907—10
ft-lb per sec in 1 hp	= 550	2.7403627
kg-m per sec in 1 hp	= 76.0404	1.8810445
watts in 1 hp (legal)	= 745.70	2.8725649

π = 3.14159 26535 89793 23846
e = 2.71828 18284 59045 23536
M = 0.43429 44819 03251 82765
$1/M$ = 2.30258 50929 94045 68402
$\log_{10} \pi$ = 0.49714 98726 94133 85435
$\log_{10} M$ = 9.63778 43113 00536 78912

Table 30 COMPLETE ELLIPTIC INTEGRALS, K AND E, FOR DIFFERENT VALUES OF THE MODULUS k

$\sin^{-1}k$	K	E	$\sin^{-1}k$	K	E	$\sin^{-1}k$	K	E
0°	1.5708	1.5708	50°	1.9356	1.3055	81°.0	3.2553	1.0338
1	1.5709	1.5707	51	1.9539	1.2963	81.2	3.2771	1.0326
2	1.5713	1.5703	52	1.9729	1.2870	81.4	3.2995	1.0314
3	1.5719	1.5697	53	1.9927	1.2776	81.6	3.3223	1.0302
4	1.5727	1.5689	54	2.0133	1.2681	81.8	3.3458	1.0290
5	1.5738	1.5678	55	2.0347	1.2587	82.0	3.3699	1.0278
6	1.5751	1.5665	56	2.0571	1.2492	82.2	3.3946	1.0267
7	1.5767	1.5649	57	2.0804	1.2397	82.4	3.4199	1.0256
8	1.5785	1.5632	58	2.1047	1.2301	82.6	3.4460	1.0245
9	1.5805	1.5611	59	2.1300	1.2206	82.8	3.4728	1.0234
10	1.5828	1.5589	60	2.1565	1.2111	83.0	3.5004	1.0223
11	1.5854	1.5564	61	2.1842	1.2015	83.2	3.5288	1.0213
12	1.5882	1.5537	62	2.2132	1.1920	83.4	3.5581	1.0202
13	1.5913	1.5507	63	2.2435	1.1826	83.6	3.5884	1.0192
14	1.5946	1.5476	64	2.2754	1.1732	83.8	3.6196	1.0182
15	1.5981	1.5442	65	2.3088	1.1638	84.0	3.6519	1.0172
16	1.6020	1.5405	65.5	2.3261	1.1592	84.2	3.6852	1.0163
17	1.6061	1.5367	66.0	2.3439	1.1545	84.4	3.7198	1.0153
18	1.6105	1.5326	66.5	2.3622	1.1499	84.6	3.7557	1.0144
19	1.6151	1.5283	67.0	2.3809	1.1453	84.8	3.7930	1.0135
20	1.6200	1.5238	67.5	2.4001	1.1408	85.0	3.8317	1.0127
21	1.6252	1.5191	68.0	2.4198	1.1362	85.2	3.8721	1.0118
22	1.6307	1.5141	68.5	2.4401	1.1317	85.4	3.9142	1.0110
23	1.6365	1.5090	69.0	2.4610	1.1272	85.6	3.9583	1.0102
24	1.6426	1.5037	69.5	2.4825	1.1228	85.8	4.0044	1.0094
25	1.6490	1.4981	70.0	2.5046	1.1184	86.0	4.0528	1.0086
26	1.6557	1.4924	70.5	2.5273	1.1140	86.2	4.1037	1.0079
27	1.6627	1.4864	71.0	2.5507	1.1096	86.4	4.1574	1.0072
28	1.6701	1.4803	71.5	2.5749	1.1053	86.6	4.2142	1.0065
29	1.6777	1.4740	72.0	2.5998	1.1011	86.8	4.2744	1.0059
30	1.6858	1.4675	72.5	2.6256	1.0968	87.0	4.3387	1.0053
31	1.6941	1.4608	73.0	2.6521	1.0927	87.2	4.4073	1.0047
32	1.7028	1.4539	73.5	2.6796	1.0885	87.4	4.4811	1.0041
33	1.7119	1.4469	74.0	2.7081	1.0844	87.6	4.5609	1.0036
34	1.7214	1.4397	74.5	2.7375	1.0804	87.8	4.6477	1.0031
35	1.7312	1.4323	75.0	2.7681	1.0764	88.0	4.7427	1.0026
36	1.7415	1.4248	75.5	2.7998	1.0725	88.2	4.8478	1.0021
37	1.7522	1.4171	76.0	2.8327	1.0686	88.4	4.9654	1.0017
38	1.7633	1.4092	76.5	2.8669	1.0648	88.6	5.0988	1.0014
39	1.7748	1.4013	77.0	2.9026	1.0611	88.8	5.2527	1.0010
40	1.7868	1.3931	77.5	2.9397	1.0574	89.0	5.4349	1.0008
41	1.7992	1.3849	78.0	2.9786	1.0538	89.1	5.5402	1.0006
42	1.8122	1.3765	78.5	3.0192	1.0502	89.2	5.6579	1.0005
43	1.8256	1.3680	79.0	3.0617	1.0468	89.3	5.7914	1.0004
44	1.8396	1.3594	79.5	3.1064	1.0434	89.4	5.9455	1.0003
45	1.8541	1.3506	80.0	3.1534	1.0401	89.5	6.1278	1.0002
46	1.8691	1.3418	80.2	3.1729	1.0388	89.6	6.3509	1.0001
47	1.8848	1.3329	80.4	3.1928	1.0375	89.7	6.6385	1.0001
48	1.9011	1.3238	80.6	3.2132	1.0363	89.8	7.0440	1.0000
49	1.9180	1.3147	80.8	3.2340	1.0350	89.9	7.7371	1.0000

$$K = \int_0^{\pi/2} \frac{dx}{\sqrt{1 - k^2 \sin^2 x}} \qquad E = \int_0^{\pi/2} \sqrt{1 - k^2 \sin^2 x}\, dx$$

Table 31 A TABLE OF CONVERSION FACTORS (WEIGHTS AND MEASURES)

To convert from	To	Multiply by
Acres (British)	sq. meters	4046.849
Acres (U. S.)	sq. miles	0.0015625
Acres (U. S.)	sq. yards	4840
Ares	sq. meters	100
Ares	sq. yards	119.60
Barrels, oil	gallons (U. S.)	42
Barrels (U. S., dry)	cu. inches	7056
Barrels (U. S., dry)	quarts (dry)	105.0
Barrels (U. S., liquid)	gallons	31.5
Bars	dynes/sq. cm.	1.000×10^{6}
Bars	pounds/sq. inch	14.504
Board feet	cu. feet	1/12
B. T. U. (mean)	calories, gram (mean)	251.98
B. T. U. (mean)	foot pounds	777.98
B. T. U. (mean)	horse power hours	3.9292×10^{-4}
B. T. U. (mean)	joules (Abs.)	1054.8
B. T. U. (mean)	kilowatt hours	2.930×10^{-4}
B. T. U. (mean)	kg. meters	107.56
Bushels (British, dry)	liters	36.3677048
Bushels (U. S., dry), (level)	cu. inches	2150.42
Bushels (U. S., dry)	liters	35.238329
Bushels (U. S., dry)	pecks	4
Calories, gram (mean)	B. T. U. (mean)	3.9685×10^{-3}
Centimeters	inches (U. S.)	0.393700
Centimeters	meters	10^{-2}
Centimeters	yards (British)	0.01093614
Centimeters	yards (U. S.)	0.01093611
Chains (surveyors' or Gunter's)	yards	22
Circular inches	sq. inches	0.78540
Circular mils	sq. inches	7.854×10^{-7}
Cords	cord feet	8
Cord feet	cu. feet	16
Cubic cm.	cubic inches	0.061023
Cubic ft. (U. S.)	cubic inches	1728
Cubic ft. (U. S.)	cubic yards	0.037037
Cu. inches (British)	bushels (British)	4.5081×10^{-4}
Cu. inches (British)	cubic cm.	16.3870253
Cu. inches (U. S.)	cubic cm.	16.387162
Cu. inches (U. S.)	cu. ft. (U. S.)	5.78704×10^{-4}
Cu. inches (British)	gallons (British)	0.003606
Cu. inches (U. S.)	gallons (U. S.)	0.0043290
Cu. inches (U. S.)	liters	0.0163868
Cu. inches (U. S.)	quarts (U. S., dry)	0.0148808
Cubic yds. (British)	cubic ft.	27
Cubic yds. (U. S.)	cubic ft.	27
Cubic yds. (British)	cubic meters	0.76455285
Cubic yds. (U. S.)	cubic meters	0.76455945
Decameter	meters	10
Decigrams	grams	0.1
Decimeter	meter	10^{-1}
Degrees	minutes	60
Degrees (See Table 8)	radians	0.0174533
Dekagrams	grams	10
Drams (apothecaries' or troy)	ounces (avoirdupois)	0.1371429

To convert from	To	Multiply by
Drams (apothecaries' or troy)	ounces (troy)	0.125
Drams (avoirdupois)	ounces (avoirdupois)	0.0625
Drams (avoirdupois)	ounces (troy)	0.056966146
Drams (U. S., fluid or apoth.)	cubic cm.	3.6967
Dynes	pounds	2.2481×10^{-6}
Ergs	B. T. U. (mean)	9.4805×10^{-11}
Ergs	foot pounds	7.3756×10^{-8}
Ergs	joules	1×10^{-7}
Ergs	kg. meters	1.0197×10^{-8}
Ergs/sec.	horse power	1.3410×10^{-10}
Ergs/sec.	K. W.	1×10^{-10}
Fathoms	feet	6
Feet (U. S.)	inches (U. S.)	12
Feet (U. S.)	meters	0.3048006096
Furlongs	miles (U. S.)	0.125
Gallons (British)	cubic inches	277.41
Gallons (British)	liters	4.5459631
Gallons (British)	quarts (British, liquid)	4
Gallons (U. S.)	cubic inches	231
Gallons (U. S.)	liters	3.78533
Gallons (U. S.)	quarts (U. S., liquid)	4
Gills (British)	cubic cm.	142.07
Gills (British)	pints (British, liquid)	0.25
Gills (U. S.)	cubic cm.	118.294
Gills (U. S.)	pints (U. S., liquid)	0.25
Grains	drams (avoirdupois)	0.03657143
Grains	drams (troy)	0.016667
Grains	milligrams	64.798918
Grains	ounces (avoirdupois)	0.0022857
Grains	ounces (troy)	0.0020833
Grams	ounces (avoirdupois)	0.0352739
Grams	ounces (troy)	0.0321507
Hectometers	meters	100
Hogsheads (British)	cubic ft.	10.114
Hogsheads (U. S.)	cubic ft.	8.42184
Hogsheads (U. S.)	gallons (U. S.)	63
Horse power	B. T. U. (mean)/min.	42.418
Horse power	calories, kg. (mean)/min.	10.688
Horse power	foot pounds/min.	33000
Horse power	foot pounds/sec.	550
Horse power	horse power (metric)	1.0139
Horse power	K. W. (g = 980.665)	0.74570
Horse power, electrical (U.S. or British)	watts (Abs.)	746.00
Horse power (metric)	horse power (U. S.)	0.98632
Hours (mean solar)	minutes (mean solar)	60
Hundredweights (long)	pounds	112
Hundredweights (long)	tons (long)	0.05
Hundredweights (short)	ounces (avoirdupois)	1600
Hundredweights (short)	pounds	100
Hundredweights (short)	tons (metric)	0.0453592
Hundredweights (short)	tons (long)	0.0446429
Inches (British)	centimeters	2.539998
Inches (U. S.)	centimeters	2.540005
Inches (U. S.)	feet (U. S.)	1/12
Joules (Abs.)	B. T. U. (mean)	9.480×10^{-4}
Joules (Abs.)	ergs	1×10^{7}
Joules (Abs.)	foot pounds	0.73756
Kilograms	grams	1000

To convert from	To	Multiply by
Kilograms	tons (long)	9.84207×10^{-4}
Kilograms	tons (metric)	0.001
Kilograms	tons (short)	0.0011023112
Kilometers	meters	1000
Kilometers	miles (nautical)	0.539593
Kilometers	miles (U. S.)	0.6213699495
Kilowatt hours	B. T. U. (mean)	3413.0
Kilowatt hours	foot pounds	2.6552×10^{6}
Kilowatt hours	joules (Abs.)	3.6000×10^{6}
Kilowatts	B. T. U. (mean)/min.	56.884
Kilowatts	foot pounds/min.	44254
Kilowatts	horse power	1.3410
Kilowatts	watts	1000
Knots (per hour)	feet/hour	6080.20
Knots (per hour)	miles/hour	1.15155
Leagues (nautical)	nautical miles	3
Leagues (statute)	statute miles	3
Links (surveyors' or Gunter's)	inches	7.92
Liters	cubic inches	61.025
Liters	gallons (British)	0.219976
Liters	gallons (U. S.)	0.26417762
Liters	quarts (British, liquid)	0.87990
Liters	quarts (U. S., dry)	0.908096
Liters	quarts (U. S., liquid)	1.056681869
Meters	feet (U. S.)	3.280833333
Meters	inches (British)	39.370113
Meters	inches (U. S.)	39.3700
Meters	yards (U. S.)	1.093611
Microns	meters	1×10^{-6}
Miles (nautical)	feet	6080.20
Miles (nautical)	kilometers	1.85325
Miles (nautical)	miles (U. S., statute)	1.1516
Miles (U. S., statute)	feet	5280
Miles (U. S. statute)	kilometers	1.609347219
Miles/hour	feet/min.	88
Miles/hour	knots (per hour)	0.8684
Milligrams	grains	0.01543236
Milligrams	grams	0.001
Millimeters	meters	0.001
Millimicrons	meters	1×10^{-9}
Mils	inches	0.001
Minims (British)	cubic cm.	0.059192
Minims (U. S., fluid)	cubic cm.	0.061612
Minutes (angle) (See Table 8)	radians	2.90888×10^{-4}
Minutes (angle)	seconds	60
Minutes (time)	seconds	60
Myriagrams	grams	10000
Myriameters	meters	10000
Ounces (avoirdupois)	grams	28.349527
Ounces (avoirdupois)	pounds (avoirdupois)	1/16
Ounces (avoirdupois)	pounds (troy)	0.075954861
Ounces (British, fluid)	cubic cm.	28.4130
Ounces (U. S., fluid)	cubic cm.	29.5737
Ounces (U. S., fluid)	liters	0.0295729
Ounces (U. S., fluid)	pints (U. S., liquid)	1/16
Pecks (British)	cubic inches	554.6
Pecks (British)	liters	9.091901
Pecks (U. S.)	bushels	0.25
Pecks (U. S.)	cubic inches	537.605
Pecks (U. S.)	liters	8.809582
Pecks (U. S.)	quarts (dry)	8

To convert from	To	Multiply by
Pennyweights	grains	24
Pennyweights	grams	1.55517
Pennyweights	ounces (avoirdupois)	0.0548571
Pennyweights	ounces (troy)	0.05
Pints (British, liquid)	cubic cm.	568.26
Pints (British, liquid)	quarts (British)	0.5
Pints (U. S., dry)	cubic cm.	550.61
Pints (U. S., dry)	quarts (U. S., dry)	0.5
Pints (U. S., liquid)	cubic cm.	473.179
Pints (U. S., liquid)	gallons (U. S.)	0.125
Pints (U. S., liquid)	quarts (U. S., liquid)	0.5
Poundals	pounds	0.031081
Pounds (avoirdupois)	kilograms	0.4535924277
Pounds (avoirdupois)	ounces (avoirdupois)	16
Pounds (avoirdupois)	ounces (troy)	14.5833
Pounds	poundals	32.174
Pounds (troy)	kilograms	0.3732418
Pounds (troy)	ounces (avoirdupois)	13.165714
Pounds (troy)	ounces (troy)	12
Quarts (British, liquid)	cubic cm.	1136.521
Quarts (U. S., dry)	cubic cm.	1101.23
Quarts (U. S., dry)	cubic inches	67.2006
Quarts (U. S., dry)	pecks (U. S.)	0.125
Quarts (U. S., dry)	pints (dry)	2
Quarts (U. S., liquid)	cubic cm.	946.358
Quarts (U. S., liquid)	cubic inches	57.749
Quarts (U. S., liquid)	gallons (U. S.)	0.25
Quarts (U. S., liquid)	liters	0.946333
Radians (See Table 10)	degrees	57.29578
Rods (Surveyors' measure)	yards	5.5
Scruples	grains	20
Seconds	minutes	1/60
Sq. cm.	sq. inches	0.15500
Sq. feet (British)	sq. meters	0.09290289
Sq. feet (U. S.)	sq. inches	144
Sq. feet (U. S.)	sq. meters	0.09290341
Sq. inches (British)	sq. cm.	6.4515898
Sq. inches (U. S.)	sq. cm.	6.4516258
Sq. inches (U. S.)	sq. feet (U. S.)	1/144
Sq. meters	sq. inches	1550.0
Sq. meters	sq. yards (British)	1.195992
Sq. meters	sq. yards (U. S.)	1.195985
Sq. miles	acres	640
Sq. yards (British)	sq. meters	0.836126
Sq. yards (U. S.)	sq. feet	9
Sq. yards (U. S.)	sq. meters	0.83613
Tons (long)	hundredweights (short)	22.400
Tons (long)	pounds (avoirdupois)	2240
Tons (long)	pounds (troy)	2722.22
Tons (long)	tons (metric)	1.0160470
Tons (long)	tons (short)	1.12000
Tons (metric)	tons (long)	0.984207
Tons (metric)	tons (short)	1.10231
Tons (short)	hundredweights (short)	20
Tons (short)	kilograms	907.1846
Tons (short)	tons (long)	0.892857
Tons (short)	tons (metric)	0.907185
Watts (Abs.)	B. T. U. (mean)/min.	0.056884
Watts (Abs.)	ergs/sec.	1×10^7
Watts (Abs.)	joules/sec.	1
Yards (British)	meters	0.9143992
Yards (U. S.)	feet	3
Yards (U. S.)	meters	0.91440183

FOUR-PLACE TABLES

Table 32 COMMON LOGARITHMS OF TRIGONOMETRIC FUNCTIONS

Deg.	Log Rad	Log Sin	Log Cos	Log Tan	Log Ctn	Log Sec	Log Csc		
0	------	------	10.0000	------	-------	10.0000	-------	10.1961	90
1	8.2419	8.2419	9.9999	8.2419	11.7581	10.0001	11.7581	10.1913	89
2	8.5429	8.5428	9.9997	8.5431	11.4569	10.0003	11.4572	10.1864	88
3	8.7190	8.7188	9.9994	8.7194	11.2806	10.0006	11.2812	10.1814	87
4	8.8439	8.8436	9.9989	8.8446	11.1554	10.0011	11.1564	10.1764	86
5	8.9408	8.9403	9.9983	8.9420	11.0580	10.0017	11.0597	10.1713	85
6	9.0200	9.0192	9.9976	9.0216	10.9784	10.0024	10.9808	10.1662	84
7	9.0870	9.0859	9.9968	9.0891	10.9109	10.0032	10.9141	10.1610	83
8	9.1450	9.1436	9.9958	9.1478	10.8522	10.0042	10.8564	10.1557	82
9	9.1961	9.1943	9.9946	9.1997	10.8003	10.0054	10.8057	10.1504	81
10	9.2419	9.2397	9.9934	9.2463	10.7537	10.0066	10.7603	10.1450	80
11	9.2833	9.2806	9.9919	9.2887	10.7113	10.0081	10.7194	10.1395	79
12	9.3211	9.3179	9.9904	9.3275	10.6725	10.0096	10.6821	10.1340	78
13	9.3558	9.3521	9.9887	9.3634	10.6366	10.0113	10.6479	10.1284	77
14	9.3880	9.3837	9.9869	9.3968	10.6032	10.0131	10.6163	10.1227	76
15	9.4180	9.4130	9.9849	9.4281	10.5719	10.0151	10.5870	10.1169	75
16	9.4460	9.4403	9.9828	9.4575	10.5425	10.0172	10.5597	10.1111	74
17	9.4723	9.4659	9.9806	9.4853	10.5147	10.0194	10.5341	10.1052	73
18	9.4971	9.4900	9.9782	9.5118	10.4882	10.0218	10.5100	10.0992	72
19	9.5206	9.5126	9.9757	9.5370	10.4630	10.0243	10.4874	10.0931	71
20	9.5429	9.5341	9.9730	9.5611	10.4389	10.0270	10.4659	10.0870	70
21	9.5641	9.5543	9.9702	9.5842	10.4158	10.0298	10.4457	10.0807	69
22	9.5843	9.5736	9.9672	9.6064	10.3936	10.0328	10.4264	10.0744	68
23	9.6036	9.5919	9.9640	9.6279	10.3721	10.0360	10.4081	10.0680	67
24	9.6221	9.6093	9.9607	9.6486	10.3514	10.0393	10.3907	10.0614	66
25	9.6398	9.6259	9.9573	9.6687	10.3313	10.0427	10.3741	10.0548	65
26	9.6569	9.6418	9.9537	9.6882	10.3118	10.0463	10.3582	10.0481	64
27	9.6732	9.6570	9.9499	9.7072	10.2928	10.0501	10.3430	10.0412	63
28	9.6890	9.6716	9.9459	9.7257	10.2743	10.0541	10.3284	10.0343	62
29	9.7042	9.6856	9.9418	9.7438	10.2562	10.0582	10.3144	10.0272	61
30	9.7190	9.6990	9.9375	9.7614	10.2386	10.0625	10.3010	10.0200	60
31	9.7332	9.7118	9.9331	9.7788	10.2212	10.0669	10.2882	10.0127	59
32	9.7470	9.7242	9.9284	9.7958	10.2042	10.0716	10.2758	10.0053	58
33	9.7604	9.7361	9.9236	9.8125	10.1875	10.0764	10.2639	9.9978	57
34	9.7734	9.7476	9.9186	9.8290	10.1710	10.0814	10.2524	9.9901	56
35	9.7859	9.7586	9.9134	9.8452	10.1548	10.0866	10.2414	9.9822	55
36	9.7982	9.7692	9.9080	9.8613	10.1387	10.0920	10.2308	9.9743	54
37	9.8101	9.7795	9.9023	9.8771	10.1229	10.0977	10.2205	9.9662	53
38	9.8217	9.7893	9.8965	9.8928	10.1072	10.1035	10.2107	9.9579	52
39	9.8329	9.7989	9.8905	9.9084	10.0916	10.1095	10.2011	9.9494	51
40	9.8439	9.8081	9.8843	9.9238	10.0762	10.1157	10.1919	9.9408	50
41	9.8547	9.8169	9.8778	9.9392	10.0608	10.1222	10.1831	9.9321	49
42	9.8651	9.8255	9.8711	9.9544	10.0456	10.1289	10.1745	9.9231	48
43	9.8753	9.8338	9.8641	9.9697	10.0303	10.1359	10.1662	9.9140	47
44	9.8853	9.8418	9.8569	9.9848	10.0152	10.1431	10.1582	9.9046	46
45	9.8951	9.8495	9.8495	10.0000	10.0000	10.1505	10.1505	9.8951	45
		Log Cos	Log Sin	Log Ctn	Log Tan	Log Csc	Log Sec	Log Rad	Deg.

For degrees indicated in the left-hand column use the column headings at the top. For degrees indicated in the right-hand column use the column headings at the bottom.

The −10 portion of the characteristic of the logarithm is not printed but must be written down whenever such a logarithm is used.

Deg.	Rad	Sin	Cos	Tan	Ctn	Sec	Csc		
0	0.0000	0.0000	1.0000	0.0000	------	1.0000	------	1.5708	90
1	0.0175	0.0175	0.9998	0.0175	57.290	1.0002	57.299	1.5533	89
2	0.0349	0.0349	0.9994	0.0349	28.636	1.0006	28.654	1.5359	88
3	0.0524	0.0523	0.9986	0.0524	19.081	1.0014	19.107	1.5184	87
4	0.0698	0.0698	0.9976	0.0699	14.301	1.0024	14.336	1.5010	86
5	0.0873	0.0872	0.9962	0.0875	11.430	1.0038	11.474	1.4835	85
6	0.1047	0.1045	0.9945	0.1051	9.5144	1.0055	9.5668	1.4661	84
7	0.1222	0.1219	0.9925	0.1228	8.1443	1.0075	8.2055	1.4486	83
8	0.1396	0.1392	0.9903	0.1405	7.1154	1.0098	7.1853	1.4312	82
9	0.1571	0.1564	0.9877	0.1584	6.3138	1.0125	6.3925	1.4137	81
10	0.1745	0.1736	0.9848	0.1763	5.6713	1.0154	5.7588	1.3963	80
11	0.1920	0.1908	0.9816	0.1944	5.1446	1.0187	5.2408	1.3788	79
12	0.2094	0.2079	0.9781	0.2126	4.7046	1.0223	4.8097	1.3614	78
13	0.2269	0.2250	0.9744	0.2309	4.3315	1.0263	4.4454	1.3439	77
14	0.2443	0.2419	0.9703	0.2493	4.0108	1.0306	4.1336	1.3265	76
15	0.2618	0.2588	0.9659	0.2679	3.7321	1.0353	3.8637	1.3090	75
16	0.2793	0.2756	0.9613	0.2867	3.4874	1.0403	3.6280	1.2915	74
17	0.2967	0.2924	0.9563	0.3057	3.2709	1.0457	3.4203	1.2741	73
18	0.3142	0.3090	0.9511	0.3249	3.0777	1.0515	3.2361	1.2566	72
19	0.3316	0.3256	0.9455	0.3443	2.9042	1.0576	3.0716	1.2392	71
20	0.3491	0.3420	0.9397	0.3640	2.7475	1.0642	2.9238	1.2217	70
21	0.3665	0.3584	0.9336	0.3839	2.6051	1.0711	2.7904	1.2043	69
22	0.3840	0.3746	0.9272	0.4040	2.4751	1.0785	2.6695	1.1868	68
23	0.4014	0.3907	0.9205	0.4245	2.3559	1.0864	2.5593	1.1694	67
24	0.4189	0.4067	0.9135	0.4452	2.2460	1.0946	2.4586	1.1519	66
25	0.4363	0.4226	0.9063	0.4663	2.1445	1.1034	2.3662	1.1345	65
26	0.4538	0.4384	0.8988	0.4877	2.0503	1.1126	2.2812	1.1170	64
27	0.4712	0.4540	0.8910	0.5095	1.9626	1.1223	2.2027	1.0996	63
28	0.4887	0.4695	0.8829	0.5317	1.8807	1.1326	2.1301	1.0821	62
29	0.5061	0.4848	0.8746	0.5543	1.8040	1.1434	2.0627	1.0647	61
30	0.5236	0.5000	0.8660	0.5774	1.7321	1.1547	2.0000	1.0472	60
31	0.5411	0.5150	0.8572	0.6009	1.6643	1.1666	1.9416	1.0297	59
32	0.5585	0.5299	0.8480	0.6249	1.6003	1.1792	1.8871	1.0123	58
33	0.5760	0.5446	0.8387	0.6494	1.5399	1.1924	1.8361	0.9948	57
34	0.5934	0.5592	0.8290	0.6745	1.4826	1.2062	1.7883	0.9774	56
35	0.6109	0.5736	0.8192	0.7002	1.4281	1.2208	1.7434	0.9599	55
36	0.6283	0.5878	0.8090	0.7265	1.3764	1.2361	1.7013	0.9425	54
37	0.6458	0.6018	0.7986	0.7536	1.3270	1.2521	1.6616	0.9250	53
38	0.6632	0.6157	0.7880	0.7813	1.2799	1.2690	1.6243	0.9076	52
39	0.6807	0.6293	0.7771	0.8098	1.2349	1.2868	1.5890	0.8901	51
40	0.6981	0.6428	0.7660	0.8391	1.1918	1.3054	1.5557	0.8727	50
41	0.7156	0.6561	0.7547	0.8693	1.1504	1.3250	1.5243	0.8552	49
42	0.7330	0.6691	0.7431	0.9004	1.1106	1.3456	1.4945	0.8378	48
43	0.7505	0.6820	0.7314	0.9325	1.0724	1.3673	1.4663	0.8203	47
44	0.7679	0.6947	0.7193	0.9657	1.0355	1.3902	1.4396	0.8029	46
45	0.7854	0.7071	0.7071	1.0000	1.0000	1.4142	1.4142	0.7854	45
		Cos	Sin	Ctn	Tan	Csc	Sec	Rad	Deg.

For degrees indicated in the left-hand column use the column headings at the top. For degrees indicated in the right-hand column use the column headings at the bottom.

N	0	1	2	3	4	5	6	7	8	9	Proportional parts				
											1	2	3	4	5
10	0000	0043	0086	0128	0170	0212	0253	0294	0334	0374	4	8	12	17	21
11	0414	0453	0492	0531	0569	0607	0645	0682	0719	0755	4	8	11	15	19
12	0792	0828	0864	0899	0934	0969	1004	1038	1072	1106	3	7	10	14	17
13	1139	1173	1206	1239	1271	1303	1335	1367	1399	1430	3	6	10	13	16
14	1461	1492	1523	1553	1584	1614	1644	1673	1703	1732	3	6	9	12	15
15	1761	1790	1818	1847	1875	1903	1931	1959	1987	2014	3	6	8	11	14
16	2041	2068	2095	2122	2148	2175	2201	2227	2253	2279	3	5	8	11	13
17	2304	2330	2355	2380	2405	2430	2455	2480	2504	2529	2	5	7	10	12
18	2553	2577	2601	2625	2648	2672	2695	2718	2742	2765	2	5	7	9	12
19	2788	2810	2833	2856	2878	2900	2923	2945	2967	2989	2	4	7	9	11
20	3010	3032	3054	3075	3096	3118	3139	3160	3181	3201	2	4	6	8	11
21	3222	3243	3263	3284	3304	3324	3345	3365	3385	3404	2	4	6	8	10
22	3424	3444	3464	3483	3502	3522	3541	3560	3579	3598	2	4	6	8	10
23	3617	3636	3655	3674	3692	3711	3729	3747	3766	3784	2	4	6	7	9
24	3802	3820	3838	3856	3874	3892	3909	3927	3945	3962	2	4	5	7	9
25	3979	3997	4014	4031	4048	4065	4082	4099	4116	4133	2	4	5	7	9
26	4150	4166	4183	4200	4216	4232	4249	4265	4281	4298	2	3	5	7	8
27	4314	4330	4346	4362	4378	4393	4409	4425	4440	4456	2	3	5	6	8
28	4472	4487	4502	4518	4533	4548	4564	4579	4594	4609	2	3	5	6	8
29	4624	4639	4654	4669	4683	4698	4713	4728	4742	4757	1	3	4	6	7
30	4771	4786	4800	4814	4829	4843	4857	4871	4886	4900	1	3	4	6	7
31	4914	4928	4942	4955	4969	4983	4997	5011	5024	5038	1	3	4	5	7
32	5051	5065	5079	5092	5105	5119	5132	5145	5159	5172	1	3	4	5	7
33	5185	5198	5211	5224	5237	5250	5263	5276	5289	5302	1	3	4	5	7
34	5315	5328	5340	5353	5366	5378	5391	5403	5416	5428	1	2	4	5	6
35	5441	5453	5465	5478	5490	5502	5514	5527	5539	5551	1	2	4	5	6
36	5563	5575	5587	5599	5611	5623	5635	5647	5658	5670	1	2	4	5	6
37	5682	5694	5705	5717	5729	5740	5752	5763	5775	5786	1	2	4	5	6
38	5798	5809	5821	5832	5843	5855	5866	5877	5888	5899	1	2	3	5	6
39	5911	5922	5933	5944	5955	5966	5977	5988	5999	6010	1	2	3	4	5
40	6021	6031	6042	6053	6064	6075	6085	6096	6107	6117	1	2	3	4	5
41	6128	6138	6149	6160	6170	6180	6191	6201	6212	6222	1	2	3	4	5
42	6232	6243	6253	6263	6274	6284	6294	6304	6314	6325	1	2	3	4	5
43	6335	6345	6355	6365	6375	6385	6395	6405	6415	6425	1	2	3	4	5
44	6435	6444	6454	6464	6474	6484	6493	6503	6513	6522	1	2	3	4	5
45	6532	6542	6551	6561	6571	6580	6590	6599	6609	6618	1	2	3	4	5
46	6628	6637	6646	6656	6665	6675	6684	6693	6702	6712	1	2	3	4	5
47	6721	6730	6739	6749	6758	6767	6776	6785	6794	6803	1	2	3	4	5
48	6812	6821	6830	6839	6848	6857	6866	6875	6884	6893	1	2	3	4	5
49	6902	6911	6920	6928	6937	6946	6955	6964	6972	6981	1	2	3	4	4
50	6990	6998	7007	7016	7024	7033	7042	7050	7059	7067	1	2	3	3	4
51	7076	7084	7093	7101	7110	7118	7126	7135	7143	7152	1	2	3	3	4
52	7160	7168	7177	7185	7193	7202	7210	7218	7226	7235	1	2	3	3	4
53	7243	7251	7259	7267	7275	7284	7292	7300	7308	7316	1	2	2	3	4
54	7324	7332	7340	7348	7356	7364	7372	7380	7388	7396	1	2	2	3	4
N	0	1	2	3	4	5	6	7	8	9	1	2	3	4	5

N	0	1	2	3	4	5	6	7	8	9	Proportional parts 1	2	3	4	5
55	7404	7412	7419	7427	7435	7443	7451	7459	7466	7474	1	2	2	3	4
56	7482	7490	7497	7505	7513	7520	7528	7536	7543	7551	1	2	2	3	4
57	7559	7566	7574	7582	7589	7597	7604	7612	7619	7627	1	1	2	3	4
58	7634	7642	7649	7657	7664	7672	7679	7686	7694	7701	1	1	2	3	4
59	7709	7716	7723	7731	7738	7745	7752	7760	7767	7774	1	1	2	3	4
60	7782	7789	7796	7803	7810	7818	7825	7832	7839	7846	1	1	2	3	4
61	7853	7860	7868	7875	7882	7889	7896	7903	7910	7917	1	1	2	3	3
62	7924	7931	7938	7945	7952	7959	7966	7973	7980	7987	1	1	2	3	3
63	7993	8000	8007	8014	8021	8028	8035	8041	8048	8055	1	1	2	3	3
64	8062	8069	8075	8082	8089	8096	8102	8109	8116	8122	1	1	2	3	3
65	8129	8136	8142	8149	8156	8162	8169	8176	8182	8189	1	1	2	3	3
66	8195	8202	8209	8215	8222	8228	8235	8241	8248	8254	1	1	2	3	3
67	8261	8267	8274	8280	8287	8293	8299	8306	8312	8319	1	1	2	3	3
68	8325	8331	8338	8344	8351	8357	8363	8370	8376	8382	1	1	2	3	3
69	8388	8395	8401	8407	8414	8420	8426	8432	8439	8445	1	1	2	3	3
70	8451	8457	8463	8470	8476	8482	8488	8494	8500	8506	1	1	2	3	3
71	8513	8519	8525	8531	8537	8543	8549	8555	8561	8567	1	1	2	3	3
72	8573	8579	8585	8591	8597	8603	8609	8615	8621	8627	1	1	2	3	3
73	8633	8639	8645	8651	8657	8663	8669	8675	8681	8686	1	1	2	2	3
74	8692	8698	8704	8710	8716	8722	8727	8733	8739	8745	1	1	2	2	3
75	8751	8756	8762	8768	8774	8779	8785	8791	8797	8802	1	1	2	2	3
76	8808	8814	8820	8825	8831	8837	8842	8848	8854	8859	1	1	2	2	3
77	8865	8871	8876	8882	8887	8893	8899	8904	8910	8915	1	1	2	2	3
78	8921	8927	8932	8938	8943	8949	8954	8960	8965	8971	1	1	2	2	3
79	8976	8982	8987	8993	8998	9004	9009	9015	9020	9025	1	1	2	2	3
80	9031	9036	9042	9047	9053	9058	9063	9069	9074	9079	1	1	2	2	3
81	9085	9090	9096	9101	9106	9112	9117	9122	9128	9133	1	1	2	2	3
82	9138	9143	9149	9154	9159	9165	9170	9175	9180	9186	1	1	2	2	3
83	9191	9196	9201	9206	9212	9217	9222	9227	9232	9238	1	1	2	2	3
84	9243	9248	9253	9258	9263	9269	9274	9279	9284	9289	1	1	2	2	3
85	9294	9299	9304	9309	9315	9320	9325	9330	9335	9340	1	1	2	2	3
86	9345	9350	9355	9360	9365	9370	9375	9380	9385	9390	1	1	2	2	3
87	9395	9400	9405	9410	9415	9420	9425	9430	9435	9440	1	1	2	2	3
88	9445	9450	9455	9460	9465	9469	9474	9479	9484	9489	0	1	1	2	2
89	9494	9499	9504	9509	9513	9518	9523	9528	9533	9538	0	1	1	2	2
90	9542	9547	9552	9557	9562	9566	9571	9576	9581	9586	0	1	1	2	2
91	9590	9595	9600	9605	9609	9614	9619	9624	9628	9633	0	1	1	2	2
92	9638	9643	9647	9652	9657	9661	9666	9671	9675	9680	0	1	1	2	2
93	9685	9689	9694	9699	9703	9708	9713	9717	9722	9727	0	1	1	2	2
94	9731	9736	9741	9745	9750	9754	9759	9763	9768	9773	0	1	1	2	2
95	9777	9782	9786	9791	9795	9800	9805	9809	9814	9818	0	1	1	2	2
96	9823	9827	9832	9836	9841	9845	9850	9854	9859	9863	0	1	1	2	2
97	9868	9872	9877	9881	9886	9890	9894	9899	9903	9908	0	1	1	2	2
98	9912	9917	9921	9926	9930	9934	9939	9943	9948	9952	0	1	1	2	2
99	9956	9961	9965	9969	9974	9978	9983	9987	9991	9996	0	1	1	2	2
N	0	1	2	3	4	5	6	7	8	9	1	2	3	4	5

References

Treatises and Textbooks

Bellman, R., *Introduction to Matrix Analysis,* McGraw-Hill Book Company, New York, 1960.

Birkhoff, G., and MacLane, S., *A Survey of Modern Algebra,* The Macmillan Company, New York, 1941.

Burington, R. S., and Torrance, C. C., *Higher Mathematics,* McGraw-Hill Book Company, New York, 1939.

Carslaw, H. S., and Jaeger, J. C., *Operational Methods in Applied Mathematics,* 2d ed., Oxford University Press, London, 1948.

Churchill, R. V., *Fourier Series and Boundary Value Problems,* McGraw-Hill Book Company, New York, 1941.

Churchill, R. V., *Operational Mathematics,* 2d ed., McGraw-Hill Book Company, New York, 1958.

Conte, S. D., *Elementary Numerical Analysis,* McGraw-Hill Book Company, New York, 1965.

Gardner, M. F., and Barnes, J. L., *Transients in Linear Systems,* Vol. I (contains tables of transforms), John Wiley & Sons, Inc., New York, 1942.

Hart, W. L., *The Mathematics of Investment,* rev. ed., D. C. Heath and Company, New York, 1929.

Hildebrand, F. B., *Introduction to Numerical Analysis,* McGraw-Hill Book Company, New York, 1956.

Moore, J. T., *Elements of Linear Algebra and Matrix Theory,* McGraw-Hill Book Company, New York, 1968.

Morse, P. M., and Feshbach, H., *Methods of Theoretical Physics,* Vols. I, II, McGraw-Hill Book Company, New York, 1953.

Pease, Marshall G., *Methods of Matrix Algebra,* Academic Press, Inc., New York and London, 1965.

Scarborough, J. B., *Numerical Mathematical Analysis,* 6th ed., The Johns Hopkins Press, Baltimore, Md., 1966.

Shields, P. C., *Linear Algebra,* Addison-Wesley Publishing Company, Inc., Reading, Mass., 1964.

Tables

Abramowitz, M., and Stegan, I. A. (eds.), *Handbook of Mathematical Functions,* National Bureau of Standards, Applied Mathematics Series, 55, Government Printing Office, 1964.

Burington, R. S., and May, D. C., *Handbook of Probability and Statistics with Tables,* 2d ed., McGraw-Hill Book Company, New York, 1970.

Doetsch, G., *Handbuch der Laplace Transformation,* Band I, II, III, Birkhauser, Basel, 1950.

Doetsch, G., Kniess, H., and Voelker, D., *Tabellen zur Laplace Transformation, und Anleitung zum Gebrauch,* Springer-Verlag, Berlin and Gottingen, 1947.

Erdélyi, A. (ed.), *Higher Transcendental Functions,* Vols. I, II, III, McGraw-Hill Book Company, New York, 1953, 1955.

Erdélyi, A., Magnus, W., Oberhettinger, F., and Tricomi, F. G., *Tables of Integral Transforms,* Vols. I, II, McGraw-Hill Book Company, New York, 1954.

Glover, J. W., *Tables of Applied Mathematics in Finance, Insurance and Statistics,* George Wahr Publishing Company, Ann Arbor, Mich., 1923.

Kent, F. C., and Kent, M. E., *Compound Interest and Annuity Tables,* McGraw-Hill Book Company, New York, 1963.

Lipkin, L., Feinstein, I. K., and Derrick, L., *Accountants Handbook of Formulas and Tables,* Prentice-Hall, Inc., Englewood Cliffs, N.J., 1963.

Magnus, W., Oberhettinger, F., and Soni, R. P., *Formulas and Theorems for the Special Functions of Mathematical Physics,* Springer-Verlag, New York, 1966.

McCollum, P. A., and Brown, B. F., *Laplace Transform Tables and Theorems,* Holt, Rinehart and Winston, Inc., New York, 1965.

General Guide to Tables

Fletcher, A., Miller, J. C. P., Rosenfield, L., and Comrie, L. J., *An Index of Mathematical Tables,* 2d ed. (in 2 vols.), Addison-Wesley Publishing Company, Inc., Reading, Mass., 1962.

Greenwood, J. A., and Hartley, H. O., *Guide to Tables in Mathematical Statistics,* Princeton University Press, Princeton, N.J., 1962.

National Bureau of Standards, *Applied Mathematics Series,* Government Printing Office. (There are many comprehensive mathematical and statistical tables in this series.)

GLOSSARY OF SYMBOLS

Symbol	*Name or meaning*
Algebra	
$\|a\|$	absolute value of a
$n!$	n factorial
$0!$	zero factorial, $0! = 1$
$\binom{n}{r} = \frac{n!}{(n-r)!r!}$	binomial coefficient; $\binom{n}{r}$ is coefficient of $a^{n-r}b^r$ in expansion of $(a+b)^n$
$\sqrt[n]{a} = a^{1/n}$	the positive nth root of a, $a > 0$
$\sqrt[q]{a^p} = a^{p/q}$	the positive qth root of a^p, $a > 0$
$\sum_{i=1}^{n} a_i$	$\sum_{i=1}^{n} a_i = a_1 + a_2 + \cdots + a_n$
Sets	
$\in$	belongs to; is an element of; is a member of
$\notin$	does not belong to; is not an element of; is not a member of
$\supset$	contains
$\subseteq$	is included in; is a subset of
$\subset$	is a proper subset of; is properly included in; *inclusion*
$\not\subset$	is not a proper subset of; is not included in
$\cup$	union; cup; join; logical sum
$\cap$	intersection; cap; logical product
A' or C_UA	complement of A relative to universe U
$\emptyset$	the null set; the empty set
$=$	is equal to; equals; is identical with; is equivalent to
$\neq$	is not equal to; not equal
$\approx$ or $\leftrightarrow$	is equivalent to
$\not\approx$	is not equivalent to
$\{\ \}$	used to represent a set
$\{x\}$	the set of objects x
$\|$ or :	such that; for which
$\{x \mid x$ has property $p\}$	the set of objects x having the property p
$\{x \mid \quad \}$	a set builder, that is, a set consisting of all x such that
$[a,b]$	the set $a \leqq x \leqq b$
$[a,b)$	the set $a \leqq x < b$

$(a,b]$	the set $a < x \leqq b$
(a,b)	the set $a < x < b$
$\langle x, y \rangle$	ordered pair
$A \times B$	cartesian product of A and B, A cross B

$\cup A_r$ or $\bigcup_{r=1}^{n} A_r = A_1 \cup A_2 \cup \cdots \cup A_n$

$\cap A_r$ or $\bigcap_{r=1}^{n} A_r = A_1 \cap A_2 \cap \cdots \cap A_n$

$\aleph_0$	*cardinal number* of set of all natural numbers 1, 2, 3, . . . ; aleph null
$\aleph$	*cardinal number* of set of all real numbers; aleph

Logic

$\wedge$, &	and; *conjunction*
$\vee$	inclusive or; *inclusive disjunction*
$\underline{\vee}$	exclusive or; *exclusive disjunction*
$\rightarrow$ or $\Rightarrow$	implies; *implication*
$\leftrightarrow$ or $\Leftrightarrow$	is equivalent to; *equivalence*
$\sim$ or $'$	not; *negation*
$\forall$	for all, for every; *universal quantifier*
$\forall_x$	for all x
$\exists$	there exists at least one; for some; *existential quantifier*
$\exists_x$	there exists an x such that; for some x
$\exists \vert$	there exists uniquely
E!v	there exists exactly one v such that; *uniqueness quantifier*
$\vdash.$	it is asserted that

Relations

R	has the *relation* R to
$\not R$ or R'	does not have the relation R to
$\breve{R}$	the *converse* of relation R
$\approx$ or $=$	is identical (or equivalent) to; *identity* or *equivalence* relation
$\not\approx$ or $\neq$	is not identical (or not equivalent) to; *diversity relation*
$\subset$	is included in
R^{-1}	the *inverse* of relation R
$x\,R\,y$	x stands in the (binary) relation R to y

$x \not R y$	x does not stand in the (binary) relation R to y
$\leftrightarrow$	one-to-one reciprocal correspondence
$\rightarrow$ or $\curvearrowright$	corresponds to; gives; is transformed into; has for an image
$\xrightarrow{T}$ or $\overset{T}{\curvearrowright}$	passage by means of T; is transformed by T into
$F: S \rightarrow T$	F maps S into T

Functions

$f(x)$	*value* of function f at x
f^{-1}	*inverse* of function f
$f \circ g$	*composite* of function f and function g
$f^{-1} \circ f = f \circ f^{-1} = E$	
$f: (x, y)$	the function f whose ordered pairs are (x, y)

Probabilities

$P(E)$	probability of E
$P(E \cap F)$ or $P(E \wedge F)$	probability of both E and F
$P(E \mid F)$	probability of E, given F
$P(E \cup F)$ or $P(E \vee F)$	probability of E or F or both
$P(E \veebar F)$	probability of E or F, but not both E and F

Algebraic Structures

$\circ$, $\odot$, $\oplus$, $\otimes$, $\star$, $\cup$, $\cap$, $+$, $-$	symbols for *operations*
z or 0	zero
u or 1	unity
a^{-1}	*inverse* of a
$\circledless$	symbol used in ordering
$\circledgtr$	symbol used in ordering

Vectors

$\boldsymbol{V}_1 \cdot \boldsymbol{V}_2$	scalar product
$\boldsymbol{V}_1 \times \boldsymbol{V}_2$	vector product
∇S, grad S	gradient of S
div $\boldsymbol{V}$, $\nabla \cdot \boldsymbol{V}$	divergence of $\boldsymbol{V}$

curl $\boldsymbol{V}$, rot $\boldsymbol{V}$, $\nabla \times \boldsymbol{V}$	curl of $\boldsymbol{V}$
$\nabla^2 S$, div grad S	divergence of grad S

Matrices

$\begin{pmatrix} a_{11} \cdots a_{1n} \\ \cdots\cdots\cdots \\ a_{n1} \cdots a_{nn} \end{pmatrix}$ or $[a_{rs}]$ or (a_{ij}) or A	*matrix*
$(x_1, x_2, \ldots, x_n)$ or $\overrightarrow{x_p}$ or $\|x_1, x_2, \ldots, x_n\|$	row matrix, row vector
$\begin{Bmatrix} y_1 \\ \cdots \\ y_n \end{Bmatrix}$ or $y_p\uparrow$ or $\begin{Vmatrix} y_1 \\ \cdots \\ y_n \end{Vmatrix}$	column matrix, column vector
$\det A = \|A\| = \begin{vmatrix} a_{11} \cdots a_{1n} \\ \cdots\cdots\cdots \\ a_{n1} \cdots a_{nn} \end{vmatrix}$	*determinate* of matrix A

INDEX OF NUMERICAL TABLES

SUBJECT INDEX